Photochemical Reactors

# Photochemical Reactors

## Theory, Methods, and Applications of Ultraviolet Radiation

*Ernest R. Blatchley III*
*Purdue University*
*Indiana, USA*

WILEY

This edition first published 2023


*Registered Office*
John Wiley & Sons, Inc., 111 River Street, Hoboken, NJ 07030, USA

For details of our global editorial offices, customer services, and more information about Wiley products visit us at www.wiley.com.

Wiley also publishes its books in a variety of electronic formats and by print-on-demand. Some content that appears in standard print versions of this book may not be available in other formats.



*Library of Congress Cataloging-in-Publication Data Applied for:*
Hardback ISBN: 9781119871309

Cover Design: Wiley
Cover Images: Courtesy of Trojan Technologies and Ernest R. Blatchley III

Set in 9.5/12.5pt STIXTwoText by Straive, Pondicherry, India

# Contents

# Preface

As compared to other common reactor types, development and application of photochemical reactors have been relatively recent. The fundamental principles that guide our current understanding of photochemistry and photochemical reactors started to emerge only within roughly the last century. Despite this relatively short period of development, interest in photochemical reactors has expanded rapidly, especially over the last 20–30 years.

Applications of photochemical reactors, which are dominated by ultraviolet (UV) radiation sources, are expanding rapidly. However, common misunderstandings or simplifications of the behavior of these systems have negatively affected the implementation of these technologies. Implementation of UV-based reactor systems has also been hampered by a lack of any comprehensive reference or textbook to describe these systems.

A central goal of this work is to provide a comprehensive summary of the theory, methods, and common applications of UV radiation. Heavy emphasis has been placed on the refereed literature as the sources of information for this book; however, for parts of the material presented herein, it was necessary to cite documents from the gray literature.

The book is intended for use as a reference for practicing engineers and scientists, and as a textbook for a graduate-level class in engineering. Readers are assumed to have a formal educational background in calculus and differential equations, general chemistry, calculus-based physics (Newtonian mechanics), and fluid mechanics; previous educational or practical experience related to process engineering or physico/chemical processes of environmental engineering will be beneficial, though not required.

This book was prepared as a series of three logically ordered modules. When taught as a 1-semester class, each module can be taught effectively in roughly five weeks. Module 1 (Chapters 1–4) presents the background and history of the development of principles of photochemistry and photochemical reactors. The principles presented in Module 1 are foundational to the remainder of the book. Module 2 (Chapters 4–8) presents detailed descriptions of the numerical and experimental (empirical) methods that are used to characterize and quantitatively describe photochemical reactions and photochemical reactor dynamics. Module 3 (Chapters 9–11) presents detailed summaries of three common groups of applications, including disinfection of water and wastewater, photochemical reactors designed to impart chemical changes on water and wastewater, and the use of UV to disinfect air and surfaces. Below are brief summaries of the material included in each of the chapters of this book.

*Chapter 1. Background and History* – This chapter includes descriptions of early, foundational discoveries in photochemistry, including the "Laws of Photochemistry," the people and personalities involved in their development, and some intriguing philosophical points that were raised by some of the pioneers in this field. The chapter also includes descriptions of important milestones in the development of our current understanding and application of photochemistry and photochemical reactors.

*Chapter 2. Photochemical Reactions* – Most students will be familiar with the logic and methods that are used to describe the mechanisms and kinetics of (thermal) chemical reactions. These principles are summarized as a refresher, followed by parallel presentation of analogous principles that are used to define our understanding of the mechanisms and kinetics of photochemical reactions. Important similarities, differences, and connections between chemical and photochemical reactions are also presented.

*Chapter 3. Photochemical Reactor Theory* – Students with experience in reactor engineering or physico/chemical processes of engineering will be familiar with the logic and mathematical basis of basic reactor models that used to describe the dynamic behavior of thermal and biochemical reactors. These principles are presented briefly, followed by parallel presentation of the principles that are used to define our understanding of the dynamics of photochemical reactors. Important similarities, differences, and connections between chemical and photochemical reactors are also presented.

*Chapter 4. Ultraviolet Radiation Sources* – The fundamental principles of physics and chemistry that govern the generation and spatial distribution of radiant energy from natural and "artificial" sources of UV radiation are presented. Most artificial sources involve processes that convert input electrical power into output UV photon power; the basic principles that govern the behavior of these sources are presented.

*Chapter 5. Actinometry and Radiometry* – This chapter provides a detailed summary of the methods and instruments that are used to measure radiant energy from UV sources. Chemical actinometry represents a family of methods that are used to measure UV energy based on known photochemical behavior. Radiometric methods rely on detectors that convert input photon energy into an output electronic signal.

*Chapter 6. Numerical Models for Simulation of Photochemical Reactor Behavior* – The dynamic behavior of photochemical reactors is governed by the dose distribution that is delivered by that system. In turn, the dose distribution is governed by the spatial distribution of radiant energy and fluid mechanical behavior within the irradiated zone of the reactor. These attributes of photochemical reactors can be accurately simulated using contemporary computational tools, including fluence rate field models and computational fluid dynamics, both of which are summarized in this chapter. The chapter also includes a summary of integrated applications of these numerical methods for simulation of photochemical reactor dynamics.

*Chapter 7. Validation of Photochemical Reactors* – Applications of UV photoreactors often require that they meet standards of performance. Common methods that have been developed to define the performance of photochemical reactors are presented in this chapter, along with discussion of some of the benefits and limitations of these methods. Modern and emerging methods for characterization of photochemical reactor performance are also presented.

*Chapter 8. Methods for Quantification of Microbial Responses to UVC Irradiation* – Methods that are used to quantify and characterize the effects of UVC radiation on microbes need to account for the mechanisms that govern these effects. This chapter includes descriptions of the methods that have been developed to account for this mechanistic behavior for several groups of microbes, including bacteria, viruses, protozoa, and algae. The implications of the use of inappropriate methods are also described.

*Chapter 9. UV Disinfection of Drinking Water and Municipal Wastewater* – These represent the most common applications of UV-based photoreactors. The fundamental principles that govern the design and application of these systems are presented, including summaries of important behavioral characteristics of these systems. Differences and similarities among systems that are used for disinfection of drinking water, municipal wastewater, and intermediate applications are also presented.

*Chapter 10. Photolysis and Advanced Reaction Processes for Control of Trace Contaminants* – UV-based reactors are often used to alter the chemical characteristics of water and wastewater. The fundamental principles that are used to simulate and design reactors that are used for direct photolysis, advanced oxidation processes, and advanced reduction processes are presented.

*Chapter 11. UV Disinfection of Air and Surfaces* – The use of UVC-based systems for disinfection of air and surfaces was first demonstrated in the 1930s and has since been shown to be effective for reducing the risks of exposure to airborne and surface-associated pathogens. The emergence of hospital-acquired infections, antibiotic-resistant microbes, and the COVID-19 pandemic have heightened interest in these systems. This chapter provides a summary of the fundamental principles that are used to design and implement UVC-based systems for disinfection of air and surfaces.

It is likely that applications of photochemistry and photochemical reactors will continue to expand into other areas of engineering and the physical sciences. The principles and methods described in this book will provide a foundation for understanding their dynamic behavior, but adaptations may be required to account for differences between these new applications and those that are described herein.

Ernest R. Blatchley III
Purdue University – West Lafayette, IN

## Acknowledgements

I started writing this book during a sabbatical leave from Purdue University. Prior to leaving for this sabbatical, a graduate student in one of my classes asked a simple question about my plans for this book: "fiction or nonfiction?" As you might guess, this student was a bit of a wise-guy, but I appreciated his wry smile as he asked this question. Among my goals in writing this book was to be comprehensive and rigorously correct in presenting the current state-of-knowledge as related to photochemical reactors. So, to answer this question posed by my former student (who will remain un-named), nonfiction!

The process of writing this book was a tremendous learning experience for me and one that would not have been possible without the support of many other people. First and foremost are the contributions of many engineers and scientists at Trojan Technologies. Special thanks to Linda Gowman for championing this project and for her steadfast interest. I would also like to thank Ted Mao and Mark Kustermans who continued to support this project after Linda retired. I am indebted to the many current and former employees of Trojan Technologies who provided support for this project in the form of formal reviews of individual chapters and informal conversations. These include: Jim Robinson, Mike Sasges, Gord Knight, Gang Fang, and Domenico Santoro; for anyone who provided comments or other input that I have not named here, please accept my sincere apologies. I am also grateful to Professor Ronnie Gehr of McGill University whose thoughtful, detailed comments and suggestions led to notable improvements of several chapters.

For the spring 2022 semester, I developed an online graduate-level class based on the material presented in this book. Students in this class were gracious and helpful in identifying errors within the text, which were completely my fault. I would like to thank them for their patience and willingness to help: Sudharshan Anandan, Kevin Brown, Xiaosu Ding, Rachel Gehr, Chunxu Huang, Kartikeya Pandey, Satya Patra, and Jamaie Scott.

Barbara Blatchley, Professor of Psychology at Agnes Scott College and one of my sisters, has also been an important source of inspiration for me. She is a gifted author and has written both a textbook (*Statistics in Context*) and a great book on the subject of luck (*What are the Chances: Why We Believe in Luck*). I regularly bent her ear to get advice on this project. Her guidance was important to me, not only in moving me toward completion of this project, but also in doing it the right way. Thank you Barbie!

Finally, I want to thank my wife, Marcia Blatchley, for her interest and support of this project. The process of writing this book has been a huge consumer of my time, but she has always been patient, supportive, and understanding, even (or especially) in the times when I was pulling my hair out.

## About the Companion Website

This book is accompanied by a companion website.

**www.wiley.com/go/blatchley/photochemicalreactors**

This website includes the entire textbook.

# 1

# Background and History

## 1.1 Introduction

Photochemistry is defined as "a branch of chemistry that deals with the effect of radiant energy in producing chemical changes." [1] In a more formal sense, photochemistry involves any overall process:

$$R + h\nu \rightarrow R^* \rightarrow P \tag{1.1}$$

where $R$ represents a molecule that absorbs the energy of a photon ($h\nu$) to yield an electronically excited molecule ($R^*$), which in turn reacts to yield one or more products ($P$) [2]. As we will see in Chapter 2 and other parts of this book, there are many other possible fates for electronically excited intermediates, $R^*$. In some cases, the excess energy within $R^*$ is dissipated by physical processes that may return the reactant molecule to its ground state. These may be viewed as examples of photophysical processes: [2]

$$R + h\nu \rightarrow R^* \rightarrow R \tag{1.2}$$

Photochemical reactors include natural and engineered systems in which one or more photochemical processes take place to bring about changes to system composition. Common changes within these systems may involve chemical or microbial components of the system, but in principle any component or phase of the system may be susceptible to photochemical change. In most applications, we are concerned with changes to fluid (i.e. liquid or gas) composition; however, photochemical and photophysical processes may also be induced within solids or on solid surfaces.

As compared with other fields of science, our understanding of the fundamental principles of photochemistry and photochemical reactor theory is relatively young. Much of the foundational material that informs our current understanding of photochemistry and photochemical reactors has been formulated over the last century. As a point of reference, calculus was developed independently by Isaac Newton and Gottfried Leibniz in the mid-17th century. Newton also played critical roles in establishment of many of the basic principles of modern physics and mechanics; he defined the "laws of motion" in 1687 in the *Principia Mathematica Philosophiae Naturalis* (Latin for "Mathematical Principles of Natural Philosophy").

This first chapter is intended to provide brief introductions to photochemistry and photochemical reactors, as well as the people and personalities involved in the development of the fundamental principles that inform our understanding of these processes. In addition to providing

*Photochemical Reactors: Theory, Methods, and Applications of Ultraviolet Radiation*, First Edition. Ernest R. Blatchley III.

Companion website: www.wiley.com/go/blatchley/photochemicalreactors

historical context, we will see that the people who developed these principles represented a remarkable cross section of humanity. However, among many (not all) of these individuals, several admirable personal attributes were evident. In general, the people who informed our current understanding of photochemical processes had remarkably clear, profound perceptions of the world that surrounds us and were able to distill their understanding of nature and natural phenomena into clear, concise statements. Some were able to describe these phenomena in mathematical terms, and many had a passion to communicate their understanding to others. These pioneers of photochemistry and photochemical processes were often progressive, courageous individuals who were willing to challenge scientific dogma, sometimes at the cost of significant professional risk.

## 1.2 Early Applications, Discoveries

In the 5th century BC, Hippocrates prescribed heliotherapy (i.e. sunbathing) for medical and psychological purposes [3]. This practice has continued until today, although a formal understanding of the specific medical conditions that were improved by this practice did not emerge until the 1700s. It was not until the 1800s that adverse human health effects were linked to exposure to sunlight.

Isaac Newton's contributions to science were not limited to the discovery of calculus or the formulation of the laws of motion. In 1666, he showed that white sunlight could be divided into colors; he coined the term "spectrum" to describe this range of colors and developed an instrument that involved optical elements, including a prism, which may have been the first spectrometer [4].

Ancient civilizations understood that sunlight provides visibility, warmth, health, and vitality; however, their understanding of these phenomena was almost entirely empirical. Physical explanations evolved through mythology. The first report of an organic photochemical reaction was attributed to Trommsdorff (1834), who reported that crystals of α-Santonin turned yellow and "burst" when exposed to sunlight [5]. Between the 17$^{th}$ century and early 20$^{th}$ century, scientific principles related to electromagnetic radiation and photochemistry evolved considerably [3, 6].

In 1614, Sala noted that sunlight turned silver nitrate ($AgNO_3$) crystals black. More than a century and a half later (1777), Scheele observed that paper soaked in silver chloride ($AgCl$) darkened when exposed to sunlight. Though it was not evident at the time, this discovery was related to chemical (film) photography that would emerge in the late 19th century. Subsequent experiments by Ritter (1801) indicated that radiation just beyond the violet portion of the electromagnetic spectrum was more effective than other colors (wavelengths) at promoting this reaction. Radiation in this region of the spectrum was referred to as "deoxidizing" rays; today, we identify this as ultraviolet (UV) radiation.

A seminal contribution was provided by Kirchhoff and Bunsen in 1859 in the form of the spectroscope, an early form of a spectrometer (see Figure 1.1) [7]. This device was used to demonstrate that atoms absorb and emit electromagnetic radiation at characteristic wavelengths. Based on observations made with this device, they speculated that gaps in the spectrum of solar radiation were attributable to selective absorption by constituents of earth's atmosphere. This device and the observations made by Kirchhoff and Bunsen provided the first clear links between chemical

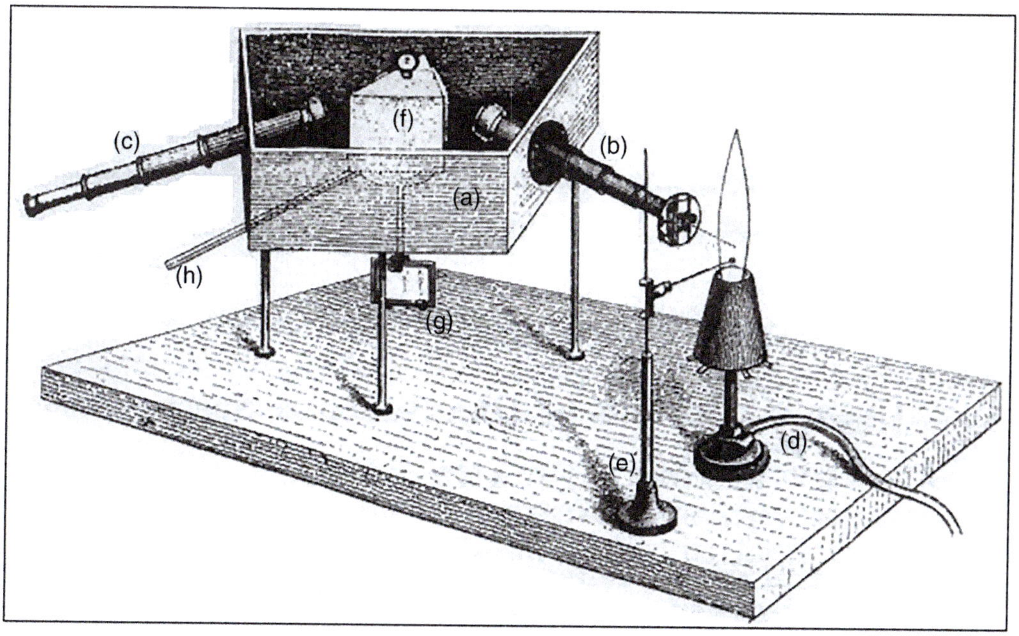


**Figure 1.1** Schematic illustration of spectroscope developed by Kirchhoff and Bunsen [7]. (a) was an internally blackened, trapezoidal box; two small telescopes (b and c) were directed at a prism (f). The eyepiece of (b) was removed and replaced by a plate in which a slit formed by two brass edges was adjusted at the focus of the objective lens. Light was provided by a device later identified as a Bunsen burner (d). A sample of the material of interest was held in the flame by a platinum wire mounted on a stand (e). The prism rested on a brass plate; the prism and a mirror (g) were rotated using handle (h). *Source:* Kirchhoff and Bunsen [7], figure 2 (p.91)/with permission of Taylor & Francis.

composition and spectroscopic behavior. As such, this device represents the foundation for contemporary analytical spectroscopy.

In 1865, Maxwell proposed that light and sound both belong to a larger energy spectrum that has wave-like properties. The term "electromagnetic" wave was assigned because Maxwell believed that both were generated by the interactions of electric and magnetic fields. Maxwell's theory was later supported by Hertz who demonstrated the existence of microwaves, which fell beyond (i.e. at wavelengths longer than) the UV, visible, and infrared (IR) portions of the spectrum. These findings established the basis for the wave theory of light, which still informs some aspects of our current understanding of the behavior of electromagnetic radiation.

## 1.3 Development of Modern Principles of Photochemistry

The observations and discoveries leading up to the work of Maxwell and Hertz formed the foundation of modern photochemistry. However, formalization of many of the fundamental principles of photochemistry required a number of other profound insights and discoveries. In many cases, the people who made these discoveries emerged as icons of the scientific world and are recognized among the giants of scientific history.

Max Planck was a German physicist who studied under Kirchhoff and Helmholz [8]. He succeeded Kirchhoff as Professor at the University of Berlin. Like many of the other scientists who participated in the development of principles that define photochemistry, he made numerous, wide-ranging scientific contributions, largely in the areas of thermodynamics and the behavior of electromagnetic radiation. Perhaps his greatest contribution came from his examination of radiation emitted from heated bodies, in which he pointed out that the principles that had been used to describe the behavior of radiation to that point were incapable of describing the emission spectrum from a "black body." This led to the development of quantum theory, and ultimately to Planck being awarded the Nobel Prize in Physics in 1919.

As part of his "miracle year" (1905), Albert Einstein published a paper that addressed the photoelectric effect; when subjected to intense irradiation, electrons are ejected from metal atoms [9]. This observation was inconsistent with the wave theory of light and supported the quantum theory that Planck had promoted. Ultimately, this work earned Einstein the Nobel Prize in Physics in 1922. Interestingly, the Nobel Committee was reluctant to recognize the importance of this work, the only work of Einstein that he himself would describe as "revolutionary." The Nobel Committee never recognized Einstein for the development of the theories of special and general relativity, which arguably have had greater impact on modern physics.

Although perhaps not as well known as Planck or Einstein, Giacomo Ciamician has been recognized as the founder of photochemistry. He was a progressive thinker and was not constrained by theories that had been established by his predecessors. As an example, he published a paper in *Science* in 1912[a] in which he suggested that solar energy should be advanced as an alternative to combustion of fossil fuels, including coal [10]. He described coal as the "most concentrated form of solar energy," but pointed out that there were limited reserves of coal and other fossil fuels. He noted that the Sahara Desert receives the solar equivalent of $6 \times 10^9$ tons of coal/day, and that photochemical transformations based on this resource could be used to address energy needs. Ciamician suggested many ways in which solar energy could be used to address energy needs, including photoelectric batteries, solar heating, increased agricultural production, solar fuels, and renewable energy.

He made these claims at a time when photography was essentially the only practical application of photochemistry. His assertions in this paper were of course valid, but even today would be viewed by many as progressive, despite our current knowledge of other problems associated with fossil fuel combustion, such as greenhouse gas emission. He concluded his paper with a remarkably insightful question: "*Would it not be advantageous to make better use of radiant energy?*" For this work and other aspects of his work, he has also been given the moniker of the Father of Green Chemistry. It is remarkable to consider how far ahead of his time he was.

Ciamician completed his PhD at Universität Gießen in 1880 [11]. He accepted a faculty position at Università di Bologna where he was joined by Paolo Silber, who worked as his assistant. Ciamician was a gifted theorist, whereas Silber's gifts were primarily as an experimentalist. They complemented each other well, and were remarkably productive in their research, especially as related to photochemistry. Ciamician and Silber conducted most of their experiments on the terrace of Ciamician's laboratory, using ambient solar radiation (see Figure 1.2). The focus of their work was on the photochemical behavior of plant tissues and plant extracts.

Ciamician was apparently an engaging lecturer and an excellent teacher. His research was viewed as a platform to promote industrial activity, especially in developing countries. He was

**Figure 1.2** Giacomo Ciamician on the terrace at his laboratory at the Università di Bologna with numerous flasks being exposed to ambient solar radiation. *Source:* Photo from Nebbia and Kauffman (2007) [11]. Courtesy of Department of Chemistry "Giacomo Ciamician" – Alma Mater Studiorum - Università di Bologna.

nominated for the Nobel Prize in Chemistry nine times, though he never won. He was appointed to the Italian Senate by the King of Italy based on his scientific authority.

The period from the late 1800s until the 1930s might be considered as a "golden age" of science in Western Europe, especially in Germany. At the time, scientists were revered by society for their creativity and contributions. The culture within common society, as well as within the scientific community, placed tremendous value on the thought leaders of the time. Scientific journals and conferences were established as fora to promote intellectual debate and discussion.

Prominent examples of this cultural aspect were the Solvay Conferences (*Conseils Solvay*), which were initiated by Ernest Solvay in 1911. These conferences, in which a clear emphasis was placed on *discussion* as opposed to *presentation*, continue today. Solvay conferences are organized around a single, unifying theme and involve thought leaders who are brought in by invitation [12]. Figure 1.3 is an image of the attendees of the Fifth Solvay Conference, which was held in Brussels, Belgium, in 1927. The men and women who participated in this conference represent the *Who's Who* of the scientific community that existed at the time and include many iconic figures.

Sadly, the remarkable accomplishments of this era were stunted by political upheaval in Europe in the 1930s, especially in Germany. At the time, roughly one-quarter of German scientists were non-Aryan. The *Civil Service Law* (1933) that was implemented by the National Socialist German

**Figure 1.3** Image of attendees at the Fifth Solvay International Conference on **Electrons and Photons**, held in Brussels, Belgium, October 1927. Of the 27 attendees, 19 eventually earned a Nobel Prize. Attendees included: **Back:** Auguste Piccard, Émile Henriot, Paul Ehrenfest, Édouard Herzen, Théophile de Donder, Erwin Schrödinger, JE Verschaffelt, Wolfgang Pauli, Werner Heisenberg, Ralph Fowler, and Léon Brillouin. **Middle:** Peter Debye, Martin Knudsen, William Lawrence Bragg, Hendrik Anthony Kramers, Paul Dirac, Arthur Compton, Louis de Broglie, Max Born, and Niels Bohr. **Front:** Irving Langmuir, Max Planck, Marie Curie, Hendrik Lorentz, Albert Einstein, Paul Langevin, Charles-Eugène Guye, CTR Wilson, and Owen Richardson [13]. *Source:* Benjamin Couprie/Wikimedia Commons/Public Domain.

Workers' (i.e. Nazi) Party resulted in expulsion of many prominent scientists from Germany. Among those affected by this law were Albert Einstein, Max Born, Eugene Wagner, James Franck, Hans Bethe, Felix Bloch, Otto Stern, Rudolph Peierls, Lise Meitner, Samuel Goldsmit, Hans Krebs, and Edward Teller [14].

Another prominent German scientist who was affected by the *Civil Service Law* was Alexander Schönburg. He was dismissed by the Nazis in 1934 and moved to Egypt by 1937 [15]. He was then appointed as Professor of Chemistry at Cairo University where he conducted research on phototherapy to cure skin diseases for the rest of his career. His work was among the first applications of photochemistry in medicine. Most of his experiments involved the use of ambient sunlight, an abundant natural resource in Cairo. The focus of his work was on preparative organic photochemistry, especially medicinal herbs of Islamic culture. Near the end of his career, he wrote a book entitled *Preparative Organic Photochemistry*, which is still available in print.

Political intrusions into the scientific community have been common occurrences throughout history and continue even today. As related specifically to photochemistry, there were many other examples of personalities that clashed during the development of foundational principles of photochemistry, including the "laws" of photochemistry.

## 1.4 Laws of Photochemistry: People and Personalities

The principles that are now identified as the "laws" of photochemistry were each developed by pairs of men. The relationships between these pairs involved a wide range of personal interactions, from two people who did not know each other at all, to two people who apparently hated each other, to a Professor/Student pair who complemented each other well and worked collaboratively for many years. The laws themselves are summarized briefly below, along with some anecdotal information about the personalities of the people who formulated them. Images of the men to whom these laws are attributed are presented in Figure 1.4. Note that among these images, only Grotthus appears as a sketch; this is because he died before photography became widely available. The image of Bunsen and Roscoe includes Gustav Kirchhoff; although he is not acknowledged as an active participant in the development of the law of reciprocity (often referred to as the Bunsen–Roscoe Law [BRL]), he collaborated extensively with Bunsen and Roscoe.

**Figure 1.4** Images of scientists who are credited with establishment of the laws of photochemistry. (a) Theodor von Grotthuss [16]. *Source:* Jazi/Wikimedia Commons/Public Domain; (b) John Draper [17]. *Source:* Science Source/Pexels; (c) Johannes Stark [18]. *Source:* A. B. Lagrelius & Westphal, Stockholm/ Wikimedia Commons/Public Domain; (d) Albert Einstein [19]. *Source:* ParentRap/Pixabay; (e) Gustav Kirchhoff, Robert Bunsen and Henry Roscoe [20]. *Source:* Henry Roscoe/Wikimedia Commons.

### 1.4.1 Grotthuss–Draper Law: First Law of Photochemistry

The first law of photochemistry was developed independently by Baron Freiherr Christian Johann Dietrich Theodor von Grotthuss and John W. Draper. The law states that "light" (photons) must be absorbed for a photochemical reaction to take place. This law, which may now appear to be obvious, was originally proposed by Grotthuss, then restated by Draper roughly 40 years later as a result of completely independent work.

Grotthuss was born into nobility in Courland (now part of Latvia) in 1785 [21, 22]. For most of his career, his research focus was on electrochemistry, specifically galvanic cells and electrolysis. His work with electrochemistry was foundational to the understanding of the topic, and was confirmed by Faraday roughly 40 years after it was originally presented by Grotthuss. Near the end of his short career, Grotthuss studied the behavior of salt solutions and their responses to light. In this work, he was able to demonstrate that these reactions only proceeded when light was absorbed. He extended this work to develop an early photometer based on photobleaching of an alcoholic iron (III) thiocyanate (rhodanide) solution. The principle of this system represents the foundation of experiments that we now refer to as chemical actinometry. His career was cut short by health problems which prevented him from conducting his work. In turn, this led to depression; he committed suicide at the age of 37.

John W. Draper was born in England in 1811. He moved to the United States in 1832 and studied medicine at the University of Pennsylvania, where he completed his MD in 1836. Draper was apparently unaware of the work of Grotthuss and used the results of experiments involving exposure of gas-phase $Cl_2$ and $H_2$ to sunlight to develop what turned out to be a restatement of the first law of photochemistry, as described above. Like Grotthuss, he developed an early form of a photometer; in Draper's case, the device employed the same gas-phase reaction of $Cl_2$ and $H_2$ he used to quantify incident radiation [21, 23].

Draper accepted a position at the Medical School of New York University (NYU) and became deeply interested in photography. He is credited with the first photograph of a human face, that of his sister Dorothy Catherine Draper, as well as the first photograph of the moon. Much of his work in the area of photography was aimed at improving the speed of the photographic process. His name appears prominently in writings of the history of photography.

The founder of the American Chemical Society (ACS), Charles F. Chandler, invited Draper to be the first President of ACS, a position that Draper accepted. In 2001, ACS designated its founding and the accomplishments of Draper as a National Historical Chemical Landmark at NYU, one of 40 such designations that existed at the time [24]. The John W. Draper Interdisciplinary Master's Program in Humanities and Social Thought was established within the Graduate School of Arts and Science at NYU to honor him.

### 1.4.2 Stark–Einstein Law: Second Law of Photochemistry

Johannes Stark has been credited with providing the first description of what is now known as the second law of photochemistry, also known as the Stark–Einstein law, the photochemical equivalence law or photoequivalence law. It states that a quantum (now referred to as a photon) of light energy can excite at most one molecule. Stark published papers in 1908 and 1911 that addressed this issue and made distinctions between direct and indirect (e.g. photosensitized) reactions [25]. However, Stark mentioned this behavior only in qualitative terms, without providing a rigorous proof.

Stark was a gifted experimentalist and made seminal scientific contributions. He won the Nobel Prize in Physics in 1919 for his discovery of the effect of electric fields on the emission of photons from molecules. This work represented an important contribution to quantum granularity [14].

Stark's accomplishments as a scientist were overshadowed by his staunch support of far-right nationalistic views and the Nazi party. When Hitler became Chancellor of the Third Reich, Stark was named as the President of the Physical and Technical Institute in Berlin [14]. He was later appointed as President of the German Research Foundation, and in this position he withdrew funding in the area of theoretical physics. He did this in part to make it difficult for some of Germany's most prominent physicists, many of whom were non-Aryan, to continue their work. Stark was so zealous in his desire to implement his nationalist agenda that he even attacked Werner Heisenberg, who was said to have been politically neutral.

Stark was particularly focused on attacking and discrediting Einstein's work. He and Phillip Lenard, who won the Nobel Prize in Physics in 1905 for his work on cathode rays, attacked Einstein regularly. They viewed relativity as a "Jewish fraud." A (translated) quote attributed to Stark suggested that: "Einstein's relativity theories were basically nothing more than an accumulation of artificial formulas based on arbitrary definitions and transformations of space and time coordinates." [26] Neither Stark nor Lenard was as scientifically agile as Einstein – few people have been. Professional jealousy may have also contributed to their animosity toward Einstein.

In many respects, Albert Einstein was the polar opposite of Stark and Lenard. He was a Jewish, pacifistic, internationalist, and a nonconformist. In 1905 (his "miracle year"), Einstein published several papers that established relativity and the photoelectric effect; he completed these papers while working as a patent clerk. The first of these papers presented the particle theory of light and launched the concept of the "wave–particle duality." This concept describes electromagnetic radiation as having attributes of waves and particles. Einstein published a paper in 1912 that "made definite" the postulate of Stark, now known as the second law of photochemistry. He did this by providing a rigorous thermodynamic proof of photon–molecule equivalence [23]. This law has been recognized as being central to photochemistry since Einstein's 1912 publication [25, 27].

The equivalence of photons and molecules also extends to nomenclature that is commonly used in photochemistry. The term "einstein," which is used to describe Avogadro's number of photons, was originally proposed by Bodenstein and Wagner [28], and was adopted for general use by the International Union of Pure and Applied Chemistry (IUPAC) [29]. Perhaps unfortunately, IUPAC now describes the term "einstein" as obsolete; they have suggested use of the phrase "one mole of photons" as a replacement for the term "einstein."

Einstein had a gift for being able to visualize and explain remarkably complex aspects of nature and the world that surrounds us. He was also able to provide rigorous proofs for these behaviors, many of which form the foundation of modern physics. He had the courage to challenge long-held scientific misconceptions and eventually earned global acclaim for his scientific discoveries. His expulsion from Germany along with many other non-Aryan scientists ended up backfiring on the Nazi party, in the sense that many of these scientists worked for enemies of Germany during World War II.

### 1.4.3 Bunsen–Roscoe Law: The Law of Reciprocity

The BRL indicates that the extent of a purely photochemical process depends on the dose (fluence) of radiation delivered to the target, independent of the time required to deliver that dose. The BRL is sometimes referred to as the Third Law of Photochemistry or a quantitative extension of the first two laws. Unlike the First and Second Laws of Photochemistry, the BRL was developed as a result of extensive collaboration between its developers (Bunsen and Roscoe), who were remarkable individuals in their own right.

Robert Wilhelm Eberhard Bunsen was born in 1811 in Göttingen, Germany [30]. Like many of the other pioneers of photochemistry, he was something of a Renaissance man. He entered the University of Göttingen at the age of 17 where he earned his PhD, at age 19. He took a lectureship at the University of Göttingen and spent much of his early career working with arsenic. One of his early discoveries was a cure for arsenic poisoning. Later in his career, he used this compound on himself after an explosion occurred in his lab that cost him sight in his right eye and nearly killed him. He later invented the zinc-carbon battery (*aka* Bunsen battery).

Bunsen began his collaboration with Gustav Kirchhoff in the 1850s after Bunsen had accepted the Chair in Chemistry and Kirchhoff was named to the Chair of Physics, both at Heidelberg University. They developed the flame photometer (see Figure 1.1) for evaluation of the emission spectra of various elements and used this device to discover Cesium (1860) and Rubidium (1862). Other researchers used the same method to subsequently discover Indium (1863), Helium (1868), Europium (1896), Gallium (1875), and Hafnium (1922).

Sir Henry Enfield Roscoe was born to nonconformist parents in London, England, in 1833. He studied at University College London, which at the time was the academic research institution in England that was open to men who were denied admission to the older English universities based on religious beliefs [31]. After completion of his degree in London, he traveled to Heidelberg to study under Bunsen [32]. After completing his doctorate, he remained at Heidelberg for several years where he collaborated with Bunsen and Kirchhoff. His work with Bunsen on the photochemical behavior of the gas mixture of $Cl_2$ and $H_2$ led to definition of the BRL. Since that time, the Law of Reciprocity has been applied to many other photochemical processes, including microbial inactivation and (chemical) photography.

Later in his career, Roscoe accepted a Professorship in Chemistry at Owens College in Manchester, now known as the University of Manchester [31]. At the time of his appointment to this position, the reputation of the chemistry department at Owens College was poor; today, the program enjoys a strong reputation, in large part because of the leadership of Roscoe.

Roscoe was apparently a passionate teacher, and his teaching interests extended well beyond the university classroom. One illustration of this attribute was his initiation of a program to provide "intellectual recreation" to the unemployed of the working class during the Cotton Famine in England. These lectures regularly drew audiences of several thousand and led Roscoe to create *Science Lectures for the People*. He was knighted in 1884 and served in Parliament from 1885 to 1895 as a member of the Liberal Party.

The collaboration among Kirchhoff, Bunsen, and Roscoe represents an illustration of the potential for benefit from sustained collaboration. It is perhaps interesting to speculate about how a similar partnership among Stark and Einstein might have yielded benefits. They were two brilliant men who had different, largely complementary gifts, who chose to focus their professional lives on closely related research topics. A partnership between them could have been remarkable.

The relationship between Stark and Einstein also provides a clear illustration of some of the problems that can emerge from situations in which politics and ideology clash with science. There are some circumstances in which science can and should inform policy. There are far fewer situations in which policy should inform science, and oftentimes when this is attempted, the results are disastrous. This is a lesson that appears to remain unlearned by some today.

## 1.5 Natural Photochemical Processes

Life on earth depends on photosynthesis, a process by which organisms use ambient (sun)light to promote reactions that convert carbon dioxide and water into carbohydrates and molecular oxygen (see equation 1.3).[b]

$$CO_2 + H_2O \xrightarrow{\text{light}} C(H_2O) + O_2 \tag{1.3}$$

The products of photosynthesis are crucial to essentially all forms of life on earth; many forms of life could not exist in the absence of these compounds, including humans.

It is worth noting that unlike many other photochemical processes we will consider in this book, photosynthesis is promoted by absorption of visible radiation. Chlorophyll within photosynthetic organisms reflects green light but absorbs light in the red and blue portions of the spectrum. This absorbed radiation energy is then used to promote various reactions that yield the outcomes described by equation 1.3.

Another common photochemical reaction that takes place in nature is dimerization within nucleic acids (DNA and RNA). These reactions are caused by absorption of radiation at wavelengths shorter than approximately 320 nm (i.e. UVC and UVB radiation). Because ambient, tropospheric solar radiation can include radiation with wavelengths as short as roughly 290 nm, photodimerization is a process that is common in nature; most "natural" photochemical damage to nucleic acids is attributable to exposure to UVB radiation. Not surprisingly, many organisms that are exposed to ambient solar UVB radiation have developed patterns of behavior or physiological mechanisms to counteract this process. Some organisms have adopted migratory patterns that minimize their exposure to UVB radiation, while others have developed processes to reverse UV-induced damage; still others have developed metabolic processes that yield chemicals that function as natural sunscreens [33]. Photoinduced damage to DNA and RNA plays important roles in damage to skin and eyes; these reactions contribute to conditions such as skin cancer [34].

As we will see below and in Chapters 7–9, photochemical damage to DNA and RNA is the foundation of UV disinfection processes. However, in most of these systems, one or more sources of UVC radiation are used to drive the process. These "artificial" sources of UVC radiation are used because photodimerization reactions tend to be much more efficient in the UVC range than in the UVB range.

Vision is a natural photochemical process that relies on visible light [35, 36]. The sense of vision depends on a photochemical reaction that causes *cis-trans* isomerization of retinal (*aka* retinaldehyde) in the protein rhodopsin. This is one of the fastest photochemical reactions known, and is also a remarkably efficient process with a quantum yield[c] ($\Phi$) of approximately 0.67 $\frac{\text{molecule}}{\text{photon}}$.

## 1.6 Atmospheric Chemistry

The chemical composition of earth's atmosphere is profoundly influenced by photochemistry. Most gas-phase reactions that influence the composition of our atmosphere are the direct or indirect result of photochemical reactions. In a very real sense, photochemistry drives atmospheric chemistry. Interestingly, the composition of our atmosphere also has profound effects on all life forms on earth, not only because of the effects of direct exposure to the atmosphere, but also because of the physical effects of various atmospheric constituents on our climate and the spectrum of solar radiation we are exposed to.

An important illustration of these facts is found in the behavior of ozone ($O_3$) in earth's atmosphere. Ozone is a reactive gas that is produced and consumed in our atmosphere by a wide range of processes, many of which behave as cycles. Sydney Chapman was among the first to characterize these reaction sequences. He discovered a series of reactions that are now known as the *Chapman Cycle* (see equations 1.4): [37]

$$\begin{aligned} O_2 + h\nu\,(\lambda \lesssim 242\,\text{nm}) &\rightarrow O + O \\ O + O_2 + M &\rightarrow O_3 + M \\ O + O_3 &\rightarrow O_2 + O_2 \\ O_3 + h\nu\,(\lambda \lesssim 320\,\text{nm}) &\rightarrow O + O_2 \\ O + O + M &\rightarrow O_2 + M \end{aligned} \tag{1.4}$$

The reactions illustrated in equations 1.4 result in production *and* consumption of ozone. Many other reactions have been identified that participate in the production and consumption of ozone in earth's atmosphere [38].

Ozone in the stratosphere is critical as it represents essentially the only atmospheric constituent that is able to absorb radiation in the wavelength range 200 nm $\lesssim \lambda \lesssim$ 300 nm. As described above, radiation in this wavelength range (UVC and UVB) is known to cause damage to nucleic acids and proteins, and is also known to promote skin cancer and other adverse health effects. Therefore, any process that depletes stratospheric ozone has the potential to increase terrestrial exposure to damaging UV radiation.

Among the other reactions that influence ozone concentration in earth's atmosphere are several processes that involve oxygenated species of hydrogen and nitrogen [37]. These reactions involve compounds that are introduced to earth's atmosphere by natural and anthropogenic processes. A groundbreaking investigation by Harold Johnston [39] demonstrated that direct injection of nitrogen oxides from the exhaust of supersonic transport (SST) aircraft, which were being contemplated for production in the early 1970s, would lead to reductions in stratospheric ozone content of roughly 50%. These changes in stratospheric ozone were identified to result from a catalytic cycle in which ozone was destroyed:

$$\begin{aligned} NO + O_3 &\rightarrow NO_2 + O_2 \\ NO_2 + O &\rightarrow NO + O_2 \\ \hline \text{net}: O + O_3 &\rightarrow O_2 + O_2 \end{aligned} \tag{1.5}$$

Johnston was a brilliant, articulate chemist, who was also remarkably humble. His 1971 paper was the first to define a clear link between human activity and the composition of the stratosphere, a part of the earth's atmosphere that very few humans ever visit. His paper was also intriguing in that the proof he provided was presented in a manner that was easily digested and interpreted by people who lacked an advanced degree in chemistry. A memoir written a few years after his death commented on the comprehensive presentation of the paper, as well as the manner in which he was able to communicate complex chemical phenomena to the lay public and policy makers [40]. Johnston felt that it was important that science be used to inform policy.

Near the end of his career, Johnston published an interesting review article on the subject of atmospheric ozone. In addition to providing a summary of some of the highlights of the chemical discoveries that had been made relative to ozone, he provided some insightful comments on his interactions with other scientists, and human nature in general.

Though he is widely credited with establishing the link between the SST program and the potential for stratospheric ozone depletion, he made it clear that he felt Paul Crutzen was a codiscoverer; Crutzen had written papers in 1971 and 1972 that had also addressed this problem.

A few years later, Mario Molina and Sherwood Rowland discovered another catalytic cycle that had the potential to deplete stratospheric ozone. This process was initiated by photolysis of chlorofluorocarbons (CFCs) to yield a chlorine atom:

$$CF_2Cl_2 + h\nu\ (\lambda \lesssim 240\,\text{nm}) \rightarrow CF_2Cl + Cl \tag{1.6}$$

A critical aspect of reaction 1.6 is the requirement for radiation with wavelengths less than about 240 nm, which is only available at altitudes above about 20 km (i.e. in the stratosphere), within earth's "ozone layer." The chlorine atom that results from this reaction can then participate in a catalytic cycle that destroys ozone:

$$\begin{array}{l} Cl + O_3 \rightarrow ClO + O_2 \\ ClO + O \rightarrow Cl + O_2 \\ \hline \text{net}: O + O_3 \rightarrow O_2 + O_2 \end{array} \tag{1.7}$$

The similarity of this sequence to that of sequence 1.5 is striking. Before publishing their results, Molina and Rowland met with Johnston in Berkeley. They discussed the similarities in the reaction sequences described above and Johnston suggested some improvements to their paper. Molina and Rowland published their work in 1974, in what is now viewed as another landmark paper [41]. In 1995, Crutzen, Molina, and Rowland shared the Nobel Prize in Chemistry "for their work in atmospheric chemistry, particularly concerning the formation and decomposition of ozone." Their work and work related to it led to the understanding of the roles of human activity on the ozone layer. More importantly, this body of work led to the creation of the *Montreal Protocol on Substances that Deplete the Ozone Layer* (aka, *Montreal Protocol*) and several subsequent modifications thereof. The Montreal Protocol was perhaps the first international treaty to address an environmental problem than spanned political boundaries. One consequence of the *Montreal Protocol* was a global ban on the production of CFCs and other ozone-depleting compounds, except for a few specialized applications.

Many of the scientists who described the effects of human activities on stratospheric ozone were persecuted by representatives of the industries that were adversely affected by their findings. Johnston had an opportunity to view and experience this for many years and developed some remarkable insights regarding human nature that he included as the conclusion to his review paper on ozone chemistry:

> "When new scientific developments impinge on society, business, and politics, mechanisms other than the scientific method come into play. Sometimes it works this way: Scientist A publishes an article. Interest group B, with or without distorting the article, uses it to advance its cause and makes demands that conflict with the interests of group C. Group C hastily attacks A's person and motives. Both C and A feel outraged. Typically, neither B nor C understands the science of the original article. On the other hand, when X files a lawsuit against Y and newspaper reporters ask Y for comment, the usual answer is, "No comment until I have studied their suit." Surely this is a better model for C to follow when A's science is used as a weapon by B. However, if one person assumes the role of both A and B, that person is in politics already."

Unfortunately, the sequence of events described by Johnston has played out regularly in circumstances where scientific discoveries have inconvenienced or threatened various individuals, industries, or groups. However, Johnston's commentary on this unfortunate aspect of human nature provides an example to lead people through future encounters of this nature.

## 1.7 Early Discoveries and Applications

As with most scientific discoveries, there is usually a lag period between a discovery and the development of practical applications. In the case of photochemistry, among the most prominent early applications to emerge were in the areas of photography and disinfection. Both applications remain prominent today, though important aspects of their applications have changed considerably since their origin.

### 1.7.1 Photography

The *camera obscura* was developed in the 13th–14th centuries as a device for capturing an image on a wall or piece of paper [42]. The device involved a dark, closed space (box) with a small opening in one wall to allow light to enter. The image from this opening could then be viewed or copied by hand. This method was apparently used as an aid by some artists at the time.

The first photographic image was captured by Joseph Nicéphore Niépce in the mid-1820s. His device involved a metal plate coated in "bitumen of Judea," onto which an image was projected. The idea was that the bitumen would harden as a result of exposure to light. The unhardened (unexposed) material could then be washed away. After polishing, the remaining material functioned as a negative that could be used for printing.

More than 150 processes were developed since Niépce's invention to allow image development in chemical photography [43]. Many of these methods involved the use of silver halide salts. Ultimately, methods were established to allow the silver halides to be deposited on thin plastic films.

Today, most images are collected using digital photography, wherein an image is digitized to allow storage and printing. Two solid-state devices are used to accomplish image digitization: the *Charge Coupled Device* (CCD) and the *Complementary Metal-Oxide Semiconductor* (CMOS); the devices behave similarly, but most cameras involve CCDs [44].

Willard S. Boyle and George E. Smith of Bell Labs invented the CCD in 1969; The Nobel Prize in Physics 2009 was divided, one half awarded to Charles Kuen Kao "for groundbreaking achievements concerning the transmission of light in fibers for optical communication," the other half jointly to Willard S. Boyle and George E. Smith "for the invention of an imaging semiconductor circuit – the CCD sensor". The CCD sensor includes a grid of photosites, each corresponding to an individual pixel. When a CCD-based digital camera is turned on, the photosites are charged to a uniform voltage. When the shutter of the camera is opened, light is imposed on the CCD causing electrons from each photosite to be ejected, as defined by the photoelectric effect (recall that this was the discovery that earned Albert Einstein his Nobel Prize in Physics). The voltage drop at each photosite is proportional to the number of photons imposed on that site, and this voltage drop can be recorded as a number (i.e. digitized) through an analog-to-digital converter. Color is simulated by interpolation of images that are collected through the use of multiple colored filters, often red, green, and blue (RGB).

### 1.7.2 Disinfection Science

Disinfection of water was among the first practical applications of photochemistry. Arthur Downes and Thomas Blunt published a series of papers in which they described the ability of sunlight exposure to suppress growth in microbial communities [45–47]. They observed that dissolved oxygen played a role in the disinfection process involving ambient sunlight. Through the use of colored glass as a set of optical filters, they also noted that blue/violet light was most effective in bringing about these observed effects.

A series of papers by Saturnin Arloing in 1885 addressed the behavior of the spores of *Bacillus anthracis* in response to exposure to sunlight [48–50]. These experiments were risky to perform, in that he was working with the microbe that is responsible for the disease anthrax; the discovery by Robert Koch of *B. anthracis* as the etiologic agent in anthrax had been made only a few years prior to Arloing's work.[d] Sunlight exposure was demonstrated to suppress the growth of *B. anthracis*. Of particular interest was the observation that inoculation of guinea pigs with *B. anthracis* spores that had previously been UV-irradiated generated immunity among these animals to subsequent exposure to virulent *B. anthracis*. This was perhaps the first demonstration of the use of UV irradiation for the development of a disease vaccine.

Barnard and Morgan [51] used the output from metal and carbon arc lamps to evaluate the responses of six species of bacteria to UV radiation. The output from these sources was conditioned by passing a beam through a quartz lens and prism, thereby allowing examination of the effects of various wavelengths on these bacteria. The results of their work were qualitative or perhaps semi-quantitative, in that they documented the growth of bacteria on portions of agar plates that were subjected to exposure to radiation from the sources described above. Nonetheless, their results indicated that bacterial inactivation was entirely attributable to radiation within the UV portion of the spectrum, and that radiation from wavelengths of 226 to 329 nm was effective for inactivation of bacteria.

Thiele and Wolf demonstrated that molecular oxygen is required for microbial inactivation from UVA exposure, but was not required with radiation in the UVB or UVC portions of the spectrum [3]. This added to the work of Downes and Blunt and provided evidence of different mechanisms being involved among these various parts of the electromagnetic spectrum.

So-called "artificial" light sources were developed across much of the 19th century [3]. Thomas Edison received his first patent for an incandescent light in 1879, and these devices dominated commercial lighting soon thereafter. Peter Cooper Hewitt was among those who sought to make improvements over incandescent sources. He developed a mercury discharge lamp in the late 1890s and formed the Cooper Hewitt Vapor Lamp Company, with support from George Westinghouse, to manufacture these lamps starting in 1902 [52].

Frederick Gates published a series of three papers in 1929 and 1930 that provided quantitative information regarding the responses of bacteria (primarily *S. aureus*) to UV radiation across the UVC and UVB portions of the spectrum [53–55]. By the use of a "large quartz monochromator of special design," he was able to isolate lines of radiation at 230.2, 234.5, 237.9, 248.2, 253.6, 267.5, 280.4, 289.4, 296.7, and 302.2 nm from a mercury arc lamp.[e] By the use of "thermopiles of special design," he was able to accurately quantify the dose of radiation applied to his samples, which involved cultures of *S. aureus* that were grown on agar plates. While some aspects of the design of the experiments of Gates have been improved on, his work may represent the first reported action spectrum for microbial responses to UV irradiation. By plotting the bacterial "survival" responses for experiments conducted at each wavelength on a single graph, he observed that each curve had the same general shape. This implied that the mechanism of inactivation was the

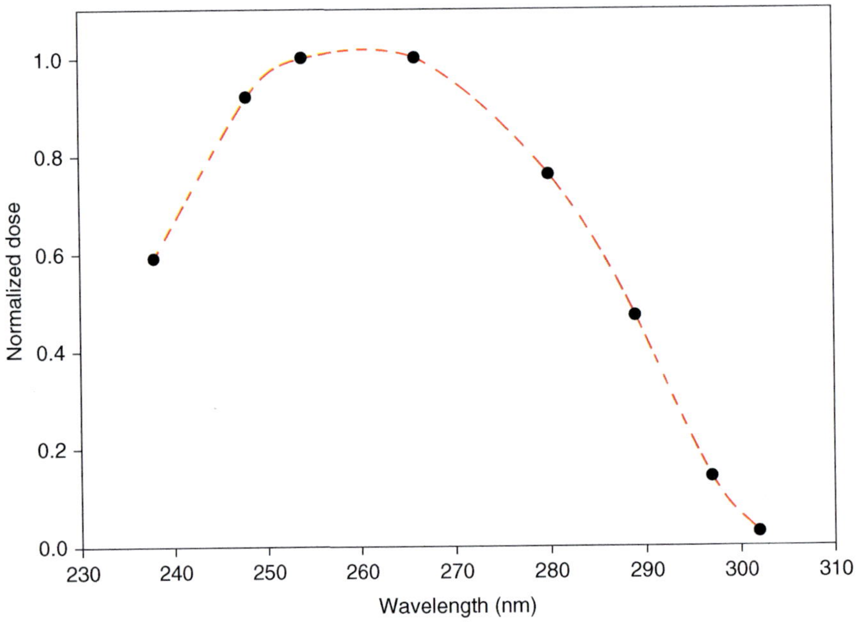


**Figure 1.5** Graphical representation of *S. aureus* inactivation action spectrum derived from the work of Gates [54].

same for each of these wavelengths. Similar behavior was reported for *E. coli*. Gates hypothesized that inactivation was attributable to photochemical damage to a specific biomolecule that was common to both bacteria.

The data from Gates' experiments allowed identification of the doses of UV radiation required to accomplish 1.0 $\log_{10}$ unit (i.e. 90%) inactivation of *S. aureus* in Gates' experiments. These data were used to develop an action spectrum from his data (see Figure 1.5). This action spectrum takes the form of the inverse of the dose required to accomplish 1.0 $\log_{10}$ unit of inactivation as a function of the corresponding wavelength. The dose at the wavelength corresponding to most effective inactivation (254 and 266 nm, in this case) was used as the basis of normalization. The normalized dose required to accomplish 1.0 $\log_{10}$ units of inactivation was calculated as described in equation 1.8:

$$\text{Normalized Dose} = \frac{\text{Dose required for 1 } \log_{10} \text{ inactivation at } \lambda}{\text{Dose required for 1 } \log_{10} \text{ inactivation at 254 nm}} \tag{1.8}$$

Although Gates did not present his data in the form used in Figure 1.5, the logic used in his experiments may have provided the first action spectrum. An action spectrum is a graphical representation of biological effect as a function of wavelength.

The work of Gates included a number of other observations that represented important contributions to our knowledge of UV disinfection behavior. He noted that the BRL was applicable for bacterial inactivation, except in the limit of small extents of inactivation, where some deviation was observed. He also noted that the rate and extent of inactivation was largely independent of temperature, pH, and polarization of light. Lastly, he noted that while UVC and UVB radiation were effective for inactivation of bacteria, radiation at wavelengths of 334 and 366 nm (UVA) was ineffective for bacterial inactivation. As it turns out, these observations supported previous research findings and motivated subsequent experiments that provided explanations for these behaviors.

One explanation for deviations from BRL behavior is the ability of some microbes to repair UV-induced damage. Kelner used spores of *Streptomyces griseus* and vegetative cells of *E. coli* to examine the reversal of UV-induced damage that resulted from subsequent exposure to visible light; he coined the term "photorepair" to describe this process [56, 57]. He reported that visible light ($\lambda < 510$ nm) was effective for recovery of bacteria that had been previously exposed to UVC radiation, but that radiation from the IR portion of the spectrum had no effect on the exposed bacteria. He also noted that the extent of reactivation was proportional to dose of visible light. In other words, photorepair was identified to be a photochemical process; as with inactivation responses, a "dose–response" relationship could be defined for photorepair.

Continued research allowed for the discovery of the enzyme photolyase in 1958 as a chemical that contributed to the photorepair process [58]. It was later discovered that this enzyme is responsible for repair of one of the two common forms of UV-induced damage – specifically, cyclobutane pyrimidine dimers (CPDs). A separate photolyase (enzyme) was later discovered to be responsible for repair of another common group of photoproducts, known as 6-4 photoproducts.

In 1960, Beukers and Berends [59] conducted experiments in which they subjected frozen solutions of thymine, one of the four pyrimidine nucleobases that comprise DNA, to UV irradiation. Analysis of the products of this reaction indicated the same elemental composition as the parent compound. Molecular weight determination suggested that the product was a dimer. IR spectroscopy indicated the presence of a cyclobutane structure in the product. Based on these results, Beukers and Berends concluded that the photoproduct was a thymine dimer (see Figure 1.6); a similar dimer was shown to be formed as a result of UV irradiation of thymine and uracil.

Research related to biochemical responses to UVC irradiation continued into the 1960s and included specific details of this behavior. Hanawalt and Setlow [60] conducted a study in which they examined the responses of *E. coli* to UVC irradiation. However, their work was different than previous studies in that they examined a range of physiological responses among these bacterial samples. In addition to the conventional measures of growth that are commonly applied, they also conducted experiments to examine the uptake of labeled ($^{35}$S) substrates to allow measurement of protein synthesis. Their results demonstrated that protein synthesis continued after $UV_{265}$ irradiation at doses that were able to effectively halt reproduction. This indicated that bacteria retained some physiological functions following exposure to UVC radiation. In other words, bacteria were still "alive" following UVC exposure, but they were incapable of reproduction. The doses of UVC radiation that were necessary to prevent replication of bacteria were much smaller than those that were required to "kill" bacteria. This observation is a key characteristic of UV-based disinfection processes, and one that is in some respects different from the responses of microbes to other disinfectants (Figure 1.7).

Until the late 1990s, the consensus opinion in the scientific community was that UV irradiation was ineffective for inactivation of the common protozoan parasites belonging to the families *Cryptosporidium* and *Giardia*. Various studies had reported that it was possible to inactivate these

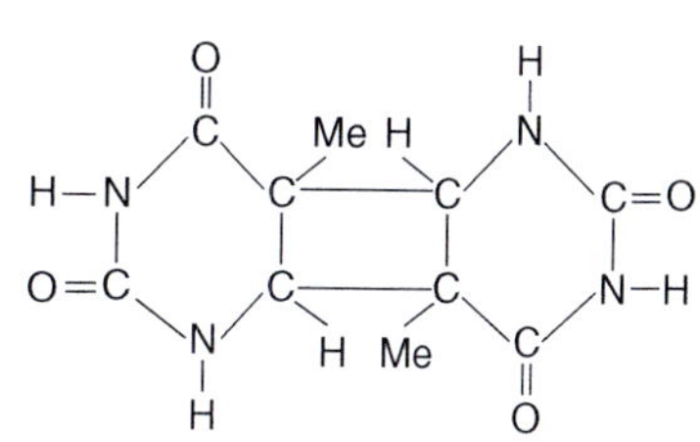


**Figure 1.6** Structure of thymine–thymine dimer proposed by Beukers and Berends [59] to result from $UV_{254}$ irradiation of thymine. *Source:* Beukers and Berends [59]/with permission of ELSEVIER.

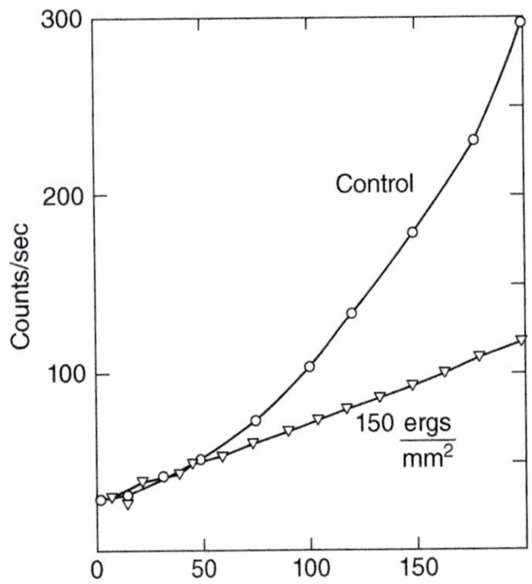


**Figure 1.7** Illustration of $^{35}S$ uptake by a control (unexposed) culture of *E. coli* and a sample of *E. coli* exposed to $UV_{265}$ irradiation (dose = $150\,\frac{ergs}{mm^2} = 1.5\,\frac{mJ}{cm^2}$) [60]. Uptake of $^{35}S$ was used as a surrogate for protein synthesis by the exposed samples. The rate of $^{35}S$ uptake remained essentially constant across the entire period of incubation for the $UV_{265}$ irradiated sample, thereby suggesting no growth in the culture. On the other hand, the rate of $^{35}S$ uptake by the control culture increased exponentially, matching the growth of the culture. *Source:* Hanawalt and Setlow [60], figure 4 (p.289)/with permission of ELSEVIER.

microbial pathogens; however, the UV doses required to accomplish effective control were perceived to be too high to be of practical significance. This conclusion was based on experiments that had been conducted using indirect measures of protozoan responses to UV irradiation, such as inclusion or exclusion of vital dyes, or *in-vitro* excystation [61, 62]. The application of these assays suggested that the $UV_{254}$ doses required to accomplish effective inactivation of these pathogens were 2–3 orders of magnitude greater than those required to inactivate most other waterborne microbes.

The pioneering work of Clancy et al. [63, 64], Craik et al. [65], Slifko et al. [66], and others demonstrated that when infectivity assays were used, based on mammalian hosts or human epithelial cells grown in tissue culture, UV irradiation was quite effective for inactivation of *Cryptosporidium*. Similar studies involving the use of mammalian hosts demonstrated that UV irradiation was also effective for inactivation of *Giardia* [67, 68]. As a result of these studies, UV irradiation is now viewed as the disinfectant of choice for inactivation of *Cryptosporidium* and *Giardia*.

### 1.7.3 Engineering Applications of UV Radiation in Drinking Water Disinfection

Although the scientific understanding of the effects of UV radiation on microbes was rapidly evolving in the early 1900s, the development of artificial sources of UV radiation paved the way for large-scale applications of UV disinfection. The first demonstration of UV disinfection at a municipal scale involved a system that was installed in Marseilles, France, in August 1910 [69]. Henri et al. fabricated a "zig-zag" reactor in the Laboratory of Physiology at the Sorbonne (see Figure 1.8) [70]. The reactor design was selected as a means of minimizing the distance through which radiation from the four 220 V, 3 amp Westinghouse Cooper Hewitt lamps in the system would need to travel, based on the recognition that disinfection efficacy improved as the target microbes and viruses were brought close to a lamp.

The first system at Marseille was designed to treat water at a flow rate of 125 $m^3$/hr, which was sufficient to supply the potable water needs of the community (population ~ 20,000 at the time; roughly 150 L/person·d). The system was reported to achieve roughly 3.5 $\log_{10}$ units of bacterial inactivation, corresponding to normalized energy consumption of 36 W·hr/$m^3$. For perspective, this

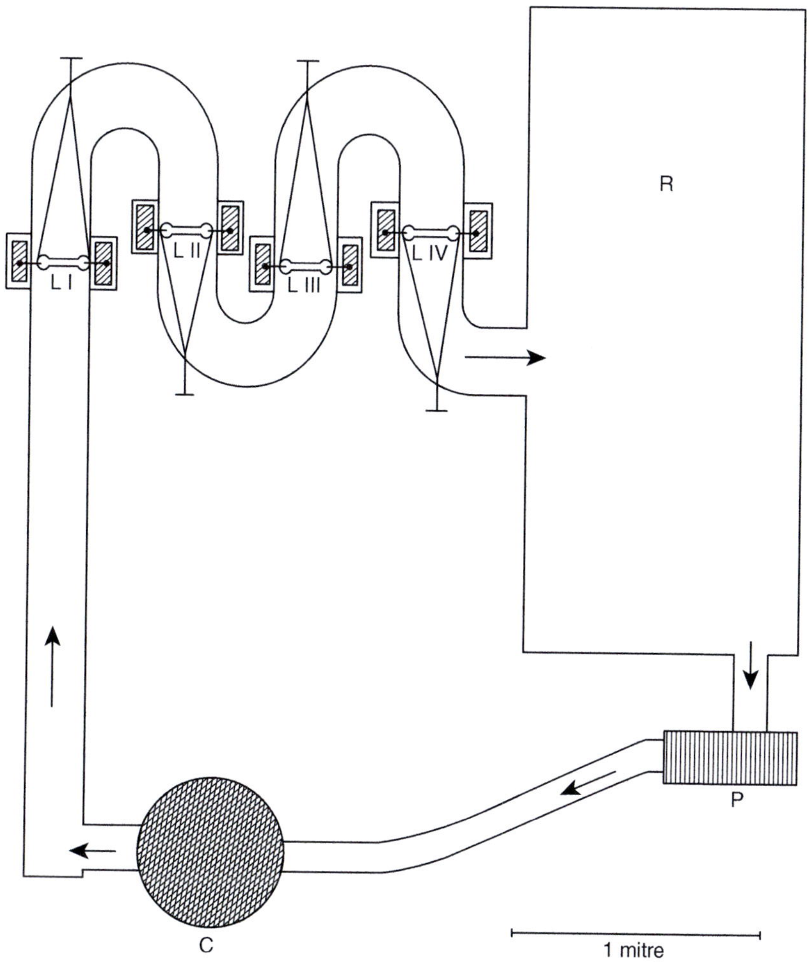


**Figure 1.8** Schematic illustration of first UV disinfection system installed in Marseilles, France [70]. Water was treated in this system by pumping water through an array of four UV lamps arranged in a "zig-zag" pattern. Lamps were installed across the pipe cross section at four locations, as indicated above. *Source:* Henri et al. [70]/with permission of ELSEVIER.

power consumption rate is similar to that of contemporary UV photoreactors used for water or wastewater disinfection; however, contemporary designs are based on more stringent treatment requirements than those that existed in the early 1900s.

Henri et al. continued their work by developing a second reactor, which was designed to yield improved efficiency [71]. Their second reactor system (see Figure 1.9) comprised a semicircular cross section with several internal "chicanes" (baffles). This second reactor was operated at a flow rate of approximately 25 $m^3$/hr, corresponding to roughly 20% of the flow rate that was applied to the reactor illustrated in Figure 1.8. In a limited set of measurements of the efficacy of this system, Henri et al. reported influent and effluent qualities of 30–300 "germs/$cm^3$" and 1 "germ/$cm^3$," respectively, while coliform bacteria were reduced from 50 to 1000/L to "no" (i.e. below detection

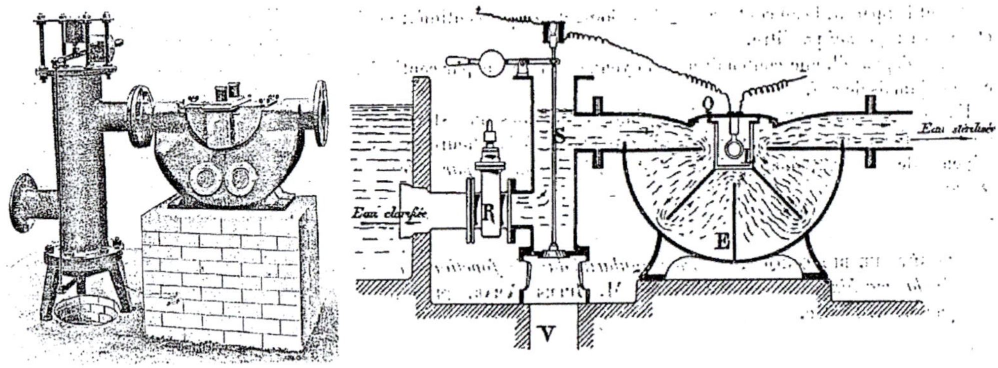


**Figure 1.9** Schematic illustrations of semicircular reactor prototype of Henri et al. [71] Sketch of reactor in actual treatment setting (left); cutaway sketch of reactor interior (right). *Source:* Henri et al. [71]/with permission of ELSEVIER.

limit) values in the effluent. For these conditions of operation, normalized energy consumption was reported to be 26 W·hr/$m^3$.

A similar effort to implement UV disinfection for drinking water was conducted in Henderson, Kentucky (USA) [72]. Until 1916, the City of Henderson had used water from the Ohio River directly, with no treatment. In 1916, the City issued a joint contract to the Pittsburgh Filter Manufacturing Company and the R.U.V. Company to install sedimentation, filtration, and UV disinfection [73]. The treatment system was designed to treat water at a nominal maximum flow rate of 470 $m^3$/hr. This flow rate was sufficient to meet the potable water needs of the entire population of Henderson, which was approximately 17,000 at the time (roughly 670 L/person·d). The UV disinfection system comprised three "legs," each including five UV units.[f] Profile and plan sketches of the system are included in Figure 1.10.

As with the reactors that were installed in Marseilles, the Henderson system was internally baffled to promote close approaches of the target microbes to the lamps. The system was reported

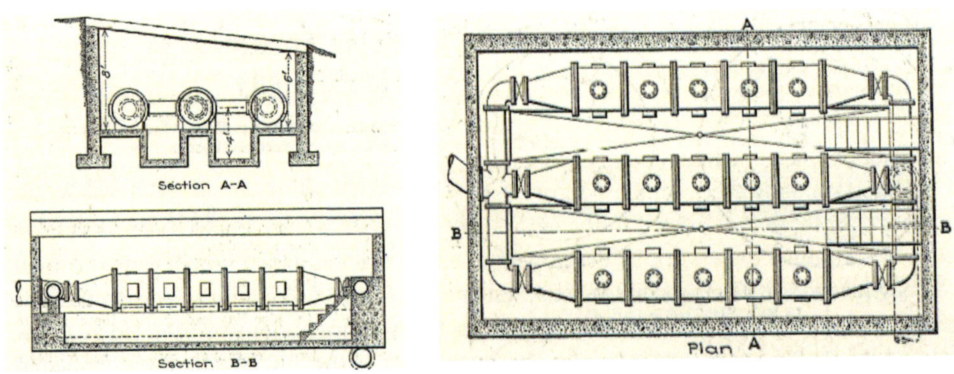


**Figure 1.10** Profile (left) and plan (right) sketches of UV disinfection system installed at Henderson, Kentucky. The footprint for the system was approximately 5 m × 9 m. *Source:* Smith [72]/with permission of SAGE Publications.

to accomplish roughly 2 $\log_{10}$ units of inactivation of total bacteria and *B. coli*.[g] For these operating conditions, normalized energy consumption was approximately 24 W·hr/$m^3$.

The systems that were installed at Marseilles, France and Henderson, Kentucky amounted to large-scale experiments with prototype reactors. Many of the features of these reactors appear in UV disinfection systems that exist today. However, the UV lamps that were available at the time were inefficient and unreliable.

Another important factor was that the availability of UV lamps in the 1910s was almost coincident with the availability of chlorine. Although molecular chlorine ($Cl_2$) was first produced by Carl Wilhelm Scheele in 1774, its first use in water treatment was probably in Middelkerke, Belgium, in 1902; applications of chlorine in water disinfection among other cities in Europe and the United States followed soon thereafter [75]. In the early years of the 20th century, chlorination was easier and less expensive to install and operate than UV systems, and because of this, UV disinfection was limited to niche applications until at least the 1980s [69].

Chlorination still remains as a dominant form of disinfection for drinking water, municipal wastewater, swimming pool water, and other media, largely because of its low cost, ease of application, and familiarity among users. However, beginning with the seminal work of Rook in which haloforms were discovered to result from chlorination of natural waters [76], interest in alternatives to chlorination has increased. One consequence of this change of perspective was a re-emphasis on the use of UV for drinking water disinfection.

In the early 1980s, Günther Schenck (Max Planck Institute for Radiation Chemistry) and Heinz Bernhardt (Association of Drinking Water Reservoirs, ATT) collaborated to conduct a project supported by ATT and the German government to examine UV disinfection of drinking water [69]. They concluded that a nominal dose of 40 mJ/$cm^2$ (400 J/$m^2$) would provide safe water, even when effects of photoreactivation were considered. They suggested biodosimetry with *E. coli* and *B. subtilis* spores to validate reactor systems and quantify their performance. Testing facilities were established in Vienna and Bonn for biodosimetry.

The Deutscher Verein des Gas- und Wasserfaches (DVGW) Standard W294 and its Austrian counterpart, Österreichisches Normungsinstitut (ÖNORM) M5873, were established to describe validation requirements in terms of support documentation, sensors, command and control, and the challenge test. *B. subtilis* spores were identified for use in reduction equivalent dose (RED)[h] measurement, while *E. coli* were selected to confirm 4 $\log_{10}$ inactivation, including repair. The US EPA published the *Ultraviolet Disinfection Guidance Manual for the Final Long Term 2 Enhanced Surface Water Treatment Rule* (*UVDGM*) which provides guidance for state regulatory agencies in the United States regarding validation of UV disinfection systems for drinking water [77]. Methods defined in the *UVDGM* are similar, though not identical to those of DVGW and ÖNORM.

### 1.7.4 Engineering Applications of UV Radiation in Disinfection of Municipal Wastewater

Wastewater management practices have evolved considerably through the history of human civilization [78]. Among early civilizations, there was a recognition of the importance of sanitation and systems such as pit latrines, canals, and sewers were built to address these concerns. Among the most impressive of these early systems were the sewers of Rome, including the famous *Cloaca Maxima*, which carried runoff and sewage away from central Rome to the River Tiber. However, from the time of the Fall of the Roman Empire until the late 1800s, sanitation practices deteriorated. As a result of industrialization, urbanization, and human population growth, epidemics became increasingly common in urban areas, especially in Europe. London, England experienced a cholera epidemic in 1854 that was made famous by Dr. John Snow and his identification of contaminated water as the cause. Just a few years later (1858), an especially hot, dry summer caused the River Thames to deteriorate into a particularly unpleasant condition; the London press labeled the event the "Great Stink." [79] Similar events took place in urban areas around the world.

These changes in human civilization, together with urbanization and dramatic increases in human population, motivated the development of systems to improve the management of wastewater from residential, commercial, and industrial activities. Beginning in the mid- to late-1800s, systems were developed to remove or transform particulate and chemical pollutants from municipal wastewaters. Disinfection processes were also included in wastewater treatment systems beginning as early as the 1890s [78], but their widespread application would not emerge until many decades later.

In the United States, *The Federal Water Pollution Control Act* (*FWPCA*) of 1948 was the first major law to address water pollution. However, the original version of this law had relatively little effect on wastewater treatment practices [80]. The 1972 amendments to the *FWPCA*, which became known as the *Clean Water Act* (*CWA*), brought about major changes to wastewater treatment practices. These included provisions to allow the United States Environmental Protection Agency (EPA) to establish discharge permit limitations to regulate pollutant concentrations in point source discharges and the establishment of a grants program to support construction of municipal wastewater treatment systems. Among the regulatory limits that were commonly applied were standards for microbial content, usually based on the concentration of viable indicator bacteria, such as *E. coli*, fecal coliforms, or *Enterococcus*.

The disinfectant of choice in most early municipal wastewater disinfection systems was chlorine. At the time (early 1970s), gaseous chlorine was relatively inexpensive and chlorination systems were relatively simple to design, build, and operate. However, concerns related to the negative attributes of chlorination motivated efforts to find alternatives to chlorine [81]. These concerns included safety issues associated with gaseous chlorine, recognition of the toxicity of residual chlorine and the need for dechlorination systems, similar recognition of toxicity associated with byproducts of chlorination, and the identification of microbial pathogens that were resistant to chlorination.

The use of UV systems for wastewater disinfection began in the late 1970s [81]. Much of the early work with these systems was conducted with the support of grants from the EPA that allowed implementation of full-scale UV hardware at municipal wastewater treatment facilities in the United States. Among these was the installation of an open-channel UV systems with lamps oriented horizontally, perpendicular to flow, at the Northwest Bergen Wastewater Treatment Plant in Waldwick, NJ [82]. The same system design was successfully demonstrated at the Port Richmond WWTP in New York City [83]. Soon thereafter, a modular open-channel system with lamps oriented horizontally and parallel to flow was successfully demonstrated at a WWTP in Tillsonburg, Ontario, Canada [84]. Profile sketches of this system are provided in Figure 1.11.

The systems involved in these early studies employed low-pressure Hg lamps, which are still commonly used today. Since that time, additional developments included automatic cleaning systems to remove (inorganic) foulants and systems that were designed around medium-pressure Hg lamps. More recent innovations in the designs of these systems involved the use of mercury amalgam lamps (*aka*, low-pressure, high-output or LPHO lamps) and diagonal lamp orientations. Figure 1.12 illustrates these features.

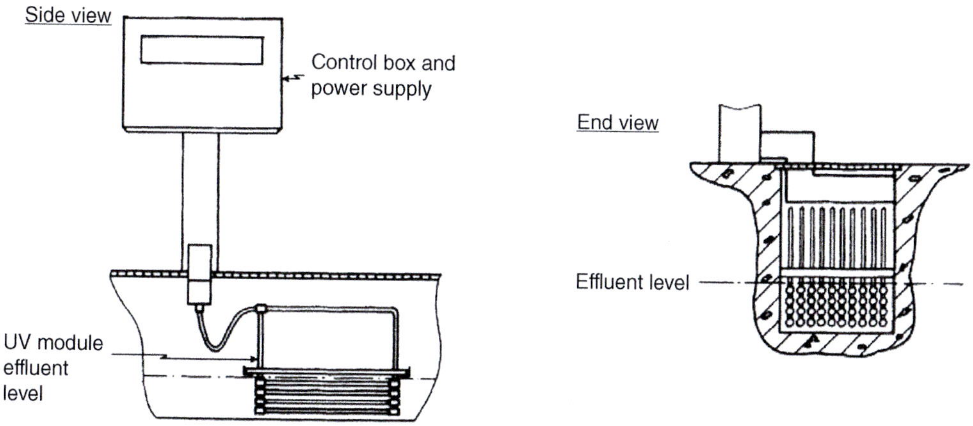


**Figure 1.11** Profile sketches of modular, horizontal, open-channel system that was tested at the Tillsonburg, Ontario WWTP. *Source:* Whitby et al. [84]/with permission of John Wiley & Sons.

**Figure 1.12** Photographic images of an automatic cleaning system (top left), an open-channel UV disinfection system with MP lamps (top right), a lamp module from a system with LPHO lamps (bottom left), and an artist's rendering of an open-channel system with a diagonal lamp orientation. *Source:* Images courtesy of Trojan Technologies.

## 1.8 Contemporary Applications

Photochemical processes are being used in many current engineering applications that include those described above, as well as a number of others. In general terms, most photochemical reactor systems involve the use of UV radiation. Photochemical reactor systems are most likely to emerge as viable options when the photochemical reactions of interest are rapid and lead to desirable changes, with little or no other change to the system. Brief descriptions of several contemporary engineering applications of photochemical reactors are presented below.

### 1.8.1 Disinfection of Water

Disinfection of aqueous media is the most common application of photochemical reactor systems. The vast majority of these systems are based on sources of UVC radiation.

Among drinking water systems, much of the interest in UV disinfection relates to the effectiveness of this process against the protozoan parasites belonging to the *Cryptosporidium* and *Giardia* genera. UV-based systems have also been demonstrated to be effective against most vegetative bacteria; however, the spore forms of some bacteria can be relatively resistant to UV inactivation.

The behavior of viruses represents a priority area for many disinfection processes, including those based on UV irradiation. Some viruses have demonstrated resistance to UV inactivation. Notable among these are the Adenoviruses. However, Adenoviruses are responsible for only a small fraction of the disease burden associated with viruses. On a global scale, roughly 20% of all cases of acute gastroenteritis have been attributed to the Human Norovirus, which has been identified as the etiologic agent for 685 million cases per year, 200 million of which occur among children less than five years old. This leads to an estimated 50,000 child deaths every year, nearly all of which occur in developing countries [85].

Another troubling aspect of the Human Norovirus was that it is difficult to grow in the laboratory. This has made it difficult to quantify the efficacy of disinfectants against the virus. However, this problem may have been solved in 2016 by Ettayebi et al. [86] With access to a method that allows propagation of the Human Norovirus, it may be possible to quantitatively define the dose–response behavior for UV and other disinfectants against Human Norovirus. However, efforts to replicate the method of Ettayebi et al. have been largely unsuccessful.

The characteristics of UV-based disinfection also favor its application as components of systems that are used to address the issue of water supply in developing countries. Many people in developing countries lack access to safe, affordable drinking water. This contributes to a chronic disease burden and a number of related public health and social problems, many of which are particularly prevalent among women and children in developing countries. UV-based disinfection has been implemented using solar UV radiation directly [87–90], as well as through the use of commercial UV reactor systems that are powered by electrical energy from solar panels.

Although municipal-scale applications of UV-based disinfection for potable water and wastewater are common today, many other UV disinfection applications involving water have emerged. For example, water shortages in populated areas have motivated the development of wastewater reuse facilities. Reuse endpoints include irrigation, water for non-potable applications (e.g. toilet flushing), all the way up to potable water production (i.e. direct potabilization). For these applications, UV-based treatment is often identified as a desirable component of the system because of its behavior as a broad-spectrum antimicrobial process, small footprint, low cost, and limited potential to generate disinfection byproducts (DBPs).

Disinfection of water in recreational water settings has also emerged as a common application. UV applications for water treatment in swimming pools are motivated by two treatment endpoints. First, UV is the disinfectant of choice against *Cryptosporidium parvum*, which has been identified as the etiologic agent for the majority of Recreational Water Illnesses of microbial origin in treated water settings (i.e. swimming pools and splash parks) [91]. Second, UV-based processes are able to degrade inorganic chloramines [92], which are common to chlorinated pools and have been associated with respiratory ailments and other adverse human health outcomes. However, UV-based processes also affect many other aspects of air and water chemistry in chlorinated swimming pools [93].

Another emerging application of UV for water treatment relates to the requirement to treat ballast water on ships. Ballast water represents a vector for transmission of nonindigenous invasive species. The International Maritime Organization has implemented a requirement for ballast water to be treated before discharge as a means of limiting these invasions [94]. Treatment systems that incorporate UV and one or more forms of physical separation (e.g. filtration) have been shown to be effective for meeting the IMO treatment requirements.

UV-based treatment processes have been developed for many other applications to disinfection of liquid aqueous media. The details of these applications will be governed by the same laws of physics, chemistry, and microbiology as those associated with the applications described above. However, the endpoints of treatment are likely to be different. As such, the UV systems used to meet these objectives are likely to involve different configurations than the systems that are designed for the applications described above.

### 1.8.2 Direct Photolysis in Water Treatment

At their most basic level, UV photoreactors are implemented to bring about changes to system composition as a result of one or more photochemical reactions. In many cases, these photochemical reactions initiate other processes that allow a treatment objective to be met. However, in some cases, the fundamental photochemical reaction is the process of interest.

The most common applications of direct photolysis in water treatment involve compounds that are photolabile. In general terms, this means that the target compound has attributes that allow it to absorb incident (usually UV) radiation efficiently (i.e. relatively large value of molar absorption coefficient, $\varepsilon$), and under conditions such that the energy of the absorbed radiation is used effectively to bring about chemical change (i.e. relatively large quantum yield, $\Phi$), either by breaking an existing bond or forming a new bond. Examples of aqueous compounds that are known to undergo direct photolysis include the inorganic chloramines [92], and N-nitrosodimethylamine (NDMA) [95], but many other compounds exist that can be effectively photolyzed.

### 1.8.3 Disinfection of Air (UVGI) and Surfaces

UV radiation is known to be effective as a broad-spectrum antimicrobial and antiviral agent because of its mechanism of action. In a general sense, if germicidally active photons can be delivered to the target microbe or virus, then inactivation can take place, regardless of the surrounding medium. And because UV-based damage to nucleic acids and proteins tends to be efficient, the amount (dose) of radiation needed to accomplish microbial and viral inactivation tends to be small, relative to the dose required to bring about other changes to the medium.

These attributes motivate the use of so-called ultraviolet germicidal irradiation (UVGI) systems to disinfect air, often in critical settings, such as hospitals or schools. UVGI was first suggested by Wells et al. in a series of papers that were published starting in the 1930s [96–98]. UVGI systems can be effective for improvement of indoor air quality; these systems are often incorporated as part of a treatment system that includes other methods of treatment, such as filtration [99, 100].

A related application involves the use of UV radiation to disinfect surfaces. In UV systems that are used to disinfect air and surfaces, the target microbes and viruses are not present in an aqueous matrix, but rather are exposed to air. The relative humidity (RH) of ambient air is known to affect the sensitivity of microbes and viruses that are suspended in air or on surfaces. In general, microbes and viruses tend to become more sensitive to UV irradiation as RH decreases.

## 1.9 Market Size and Growth

The market for UV-based reactors has grown considerably since the 1980s and the forecast suggests rapid growth for at least the next five years. The current market for UV disinfection equipment has a reported value of $2.9 billion (US), with a projected compound annual growth rate (CAGR) of 12.3% through 2025 [101]. This projection for aggressive growth is attributable to the roles played by UV-based disinfection systems in reducing the incidence of infectious diseases as well as improvements in UV hardware, including alternative UV sources. Most of this growth is associated with the municipal market, but opportunities have also emerged in other health-related applications, including disinfection applications related to the COVID-19 pandemic. Historically, North America and Europe have been the regions where the most extensive use of UV-based systems has been observed; however, rapid growth in the use of UV disinfection systems is expected for the Asia-Pacific region.

## 1.10 Objectives for Book

Assembly of this book was motivated by the absence of a comprehensive book on the subject of photochemical reactors and photochemical reactor theory, as well as common misunderstandings and misinterpretations of the photochemistry and the behavior of photochemical reactor systems. Therefore, the objectives in assembling this book are to define and describe the fundamental, governing behavior of photochemical reactor systems, and to demonstrate the application of these principles in the design of UV photoreactors. Particular attention will be given to aspects of these systems that govern their performance.

It is not possible or practical to address all possible applications of UV radiation in a single book. With this in mind, the intent is to define fundamental governing processes and their corresponding mathematical descriptions. Commonly used methods and applications will also be presented as illustrations of the use of these fundamental principles. Illustrative examples will be presented in completed form and in the form of "problems" for readers to solve; these will be developed and presented with the goal of providing readers with opportunities to apply the principles and methods that are described in this book.

## 1.11 Approaches Used in Book

The central goal for this book is to provide a resource on photochemical reactors that is widely accessible and interpretable by engineers and scientists who are likely to encounter these systems.

Although considerable advances have been made with these systems, especially since the 1980s, there remain many areas of confusion. The misunderstandings that are evident around photochemical reactors (mostly based on UV radiation) are attributable to many factors, including the complexity of these systems, nomenclature and terminology, and historical "short-cuts" to process descriptions that have been widely applied.

There is no way around the complex nature of photochemical reactor systems. Their behavior is influenced by fluid mechanics, the optical properties of the systems and their components, and fundamental chemical characteristics of photochemical targets, each of which brings complexity. As such, it is necessary for the user to be familiar with each of these areas of study. The approach that will be followed in this book will be to provide an ordered, systematic presentation of these phenomena and the basic principles that govern each. Examples will be presented to illustrate the links among these phenomena and the methods used to simulate them.

The IUPAC has provided a glossary of terms for use in descriptions of photochemistry and photochemical processes [29]. When possible and practical, the terms and symbols presented by the IUPAC will be used in this book. However, in recognition of the history of use of some nonstandard terminology, these conventional terms will also be presented for comparison, as appropriate.

The focus of this book is on the behavior of photochemical reactors, not on pure photochemistry. Because of this, there are terms and symbols that are required that will not be addressed by the IUPAC glossary. When possible, terms and symbols used to describe these other aspects of the system (e.g. flow rate, physical dimensions, etc.) will be chosen to be consistent with the technical literature on reactor behavior and process engineering.

Historically, many aspects of photochemical reactor behavior have been simplified or abridged as a means of simplifying process descriptions. These "short-cuts" have often involved well-intended efforts to provide digestible methods of describing reactor behavior. In some cases, these efforts have led to considerable confusion as well. Perhaps the clearest example of this relates to the word "dose," which is used to describe photon delivery to the targets in the system. Unfortunately, many of the simplified descriptions of photon delivery (dose) have led to fundamental misunderstandings about the behavior of photochemical reactor systems. As appropriate, these short-cuts will be presented and summarized, along with text to address the implications of these short-cuts and the magnitude of the errors associated with them. As we will see, in some situations these simplified approaches may be appropriate. However, it is important for those involved in the implementation of these systems to understand when such a short-cut can be applied, and to have a basic understanding of the error associated with these approaches.

Following this introduction to photochemical reactors, photochemistry, and some of the history of these subjects, the book will include three major sections. The first section will address basic theoretical concepts. It includes chapters to address the fundamental principles of photochemical reactions and those that relate to photochemical reactors. This section also includes a chapter to address various sources of UV radiation and the basic principles of physics that govern their behavior.

The next section addresses methods that are used to characterize and quantify the behavior of UV photoreactors. Chapters are included to address the numerical and experimental methods that are used to describe photochemical reactor systems. These include a chapter to address numerical methods (models) that are used to describe the fluence rate field in photochemical reactor systems and integration of fluence rate field simulations with simulations of fluid mechanical behavior, usually by application of the tools of computational fluid dynamics. Subsequent chapters address empirical methods that are used to quantify reactor performance.

The last section addresses contemporary applications of UV in photochemical reactor systems. Chapters are included to address common applications, such as disinfection of drinking water and municipal wastewater, including reuse applications. Chapters are also included to address the use of UV radiation in advanced oxidation and advanced reduction processes, as well as in applications aimed at disinfection of air and surfaces.

## Notes

**a** As a sign of the culture that existed at the time of this publication, the issue of *Science* in which this paper was published also included a summary of *First International Eugenics Congress.*

**b** Note that equation 1.3 is intentionally presented in simplified form to indicate the formation of general carbohydrate molecules and molecular oxygen as products of photosynthesis.

**c** The definition and physical interpretation of the quantum yield will be addressed later in this book.

**d** Koch was a physician and a gifted microbiologist. He studied and identified the etiologic agents that were responsible for many diseases of microbial origin. The Nobel Prize in Physiology or Medicine 1905 was awarded to Robert Koch "for his investigations and discoveries in relation to tuberculosis."

**e** Details of the monochromator were not provided, but the system appeared to involve a quartz prism and several quartz lenses. Similarly, no details of the thermopiles were provided.

**f** The "legs" in this system were operated in parallel, while the "units" operated in series.

**g** *B. coli* was originally known as *Bacillus coli*, but was subsequently named *E. coli* B. Delbrück and Luria [74].

**h** The physical interpretation of RED and methods used to measure it will be presented in Chapter 7, Validation of Photochemical Reactors.

## References

1 Merriam-Webster. Merriam-Webster dictionary by Merriam-Webster. https://www.merriam-webster.com/ (accessed 2021).

2 Turro, N.J., Ramamurthy, V., and Scaiano, J.C. (**2009**). *Principles of Molecular Photochemistry: An Introduction*. Sausalito, CA: University Science Books.

3 Hockberger, P.E. (**2002**). A history of ultraviolet photobiology for humans, animals and microorganisms. *Photochemistry and Photobiology 76* (6): 561–579.

4 MIT Spectroscopy Laboratory. The era of classical spectroscopy. http://web.mit.edu/spectroscopy/history/history-classical.html (accessed 2021).

5 Natarajan, A., Tsai, C.K., Khan, S.I. et al. (**2007**). The photoarrangement of alpha-santonin is a single-crystal-to-single-crystal reaction: A long kept secret in solid-state organic chemistry revealed. *Journal of the American Chemical Society 129* (32): 9846.

6 Hockberger, P.E. (**2000**). The discovery of the damaging effect of sunlight on bacteria. *Journal of Photochemistry and Photobiology B-Biology 58* (2–3): 185–191.

7 Kirchhoff, G. and Bunsen, R. (**1861**). On chemical analysis by spectrum observations. *Quarterly Journal of the Chemical Society of London 13*: 270–289.

8 Nobleprize.org. The nobel prize in physics. https://www.nobelprize.org/prizes/physics/1918/planck/biographical/ (accessed 2021).

9 National Public Radio (NPR) Albert Einstein's year of miracles: Light theory. https://www.npr.org/2005/03/17/4538324/albert-einsteins-year-of-miracles-light-theory (accessed 2021).

10 Ciamician, G. (**1912**). The photochemistry of the future. *Science 36* (926): 385–394.

11 Nebbia, G. and Kauffman, G.B. (**2007**). Prophet of solar energy: A retrospective view of Giacomo Luigi Ciamician (1857–1922), the founder of green chemistry, on the 150th anniversary of his birth. *The Chemical Educator 12* (5): 362–369.

12 International Solvay Institutes. Welcome to the solvay institutes. http://www.solvayinstitutes.be/index.html (accessed 2021).

13 Rare Historical Photos. The Solvay Conference, probably the most intelligent picture ever taken, 1927. http://rarehistoricalphotos.com/solvay-conference-probably-intelligent-picture-ever-taken-1927/ (accessed 2021).

14 Ball, P. *Serving the Reich: The Struggle for the Soul of Physics Under Hitler*. Chicago: The University of Chicago Press.

15 Nada, A.A. (**1983**). Professor A. Schönberg and the history of photochemistry in Egypt. *Journal of Chemical Education 60* (6): 451–452.

16 Chaplin, M. Grotthus mechanism. http://www1.lsbu.ac.uk/water/grotthuss.html (accessed 2021).

17 Librarium, H. C. H. John W. Draper. http://www.historiccamera.com/cgi-bin/librarium/pm.cgi?action=display&login=draper (accessed 2021).

18 Werzburg, U. Johannes Stark. http://www.uni-wuerzburg.de/en/uniarchiv/ausstellungen/roentgenring-science-mile/nobel-laureates/johannes-stark-1919/ (accessed 2021).

19 Biography Biography Albert Einstein. https://www.biography.com/people/albert-einstein-9285408 (accessed 2021).

20 Images, G. Gustav Kirchhoff, Robert Bunsen, and Henry Roscoe, scientists. http://www.gettyimages.com/detail/news-photo/gustav-kirchhoff-robert-bunsen-and-henry-roscoe-scientists-news-photo/463921887#gustav-kirchhoff-robert-bunsen-and-henry-roscoe-scientists-c1860-left-picture-id463921887.

21 Albini, A. (**2016**). *Photochemistry: Past, Present and Future*. Heidelberg: Springer-Verlag.

22 Jaselskis, B., Moore, C.E., and von Smolinsk, A. (**2007**). Theodor von Grotthus (1785-1822) - A trail blazer. *Bulletin for the History of Chemistry 32* (2): 119–128.

23 Taylor, H.S. (**1928**). Quantum processes in photochemistry. *Journal of Physical Chemistry 32* (4): 516–528.

24 Morrissey, S.R. (**2001**). Celebrating the founding of ACS. *Chemical & Engineering News 79*: 59–60.

25 Allmand, A.J. (**1926**). Einstein's law of photochemical equivalence. *Transactions of the Faraday Society 21*: 438–452.

26 Buzzfeed News. 7 Times scientists threw shade on now widely accepted ideas. https://www.buzzfeed.com/alexkasprak/7-times-scientists-threw-shade-on-now-widely-accepted-ideas?utm_term=.irddwQ3Z5#.vh2ymqvnM (accessed 2021).

27 Rubin, M.B. and Braslavsky, S.E. (**2010**). Quantum yield: the term and the symbol. A historical search. *Photochemical & Photobiological Sciences 9* (5): 670–674.

28 Bodenstein, M. and Wagner, C. (**1929**). Ein Vorschlag für die Bezeichnung der Lichtmenge in der Photochemie. *Zeitschrift für Physikalische Chemie 3B*: 456–458.

29 Braslavsky, S.E. (**2007**). Glossary of terms used in photochemistry 3(rd) edition (IUPAC recommendations 2006). *Pure and Applied Chemistry 79* (3): 293–465.

30 Famous Scientists Robert Bunsen. https://www.famousscientists.org/robert-bunsen/ (accessed 2021).

31 Wisniak, J. (**2016**). Henry Enfield Roscoe. *Educación Química 27*: 240–248.

32 Remane, H. and Bartel, H.G. (**2016**). Henry Enfield Roscoe – Wegbereiter der Chemie im Großbritannien des 19. Jahrhunderts. *Chemkon 23* (3): 125–130.

**33** Karentz, D., McEuen, F.S., Land, M.C., and Dunlap, W.C. (**1991**). Survey of mycosporine-like amino-acid compounds in antarctic marine organisms - potential protection from ultraviolet exposure. *Marine Biology 108* (1): 157–166.

**34** Saladi, R.N. and Persaud, A.N. (**2005**). The causes of skin cancer: A comprehensive review. *Drugs of Today 41* (1): 37–53.

**35** Wang, Q., Schoenlein, R.W., Peteanu, L.A. et al. (**1994**). Vibrationally coherent photochemistry in the femtosecond primary event of vision. *Science 266* (5184): 422–424.

**36** Johnson, P.J.M., Farag, M.H., Halpin, A. et al. (**2017**). The primary photochemistry of vision occurs at the molecular speed limit. *Journal of Physical Chemistry B 121* (16): 4040–4047.

**37** Finlayson-Pitts, B.J., Pitts, J., and James, N. (**2000**). *Chemistry of the Upper and Lower Atmosphere: Theory, Experiments, and Applications*, 969. San Diego, CA: Academic Press.

**38** Burkholder, J.B., Sander, S.P., Abbatt, J. et al. (**2015**). *Chemical Kinetics and Photochemical Data for Use in Atmospheric Studies, Evaluation No. 18*. Pasadena, CA: Jet Propulsion Laboratory.

**39** Johnston, H. (**1971**). Reduction of stratospheric ozone by nitrogen oxide catalysts from supersonic transport exhaust. *Science 173* (3996): 517–522.

**40** Somorjai, G.A. (**2016**). Harold S. Johnston *1920–2012: A Biographical Memoir*. Washington, DC: National Academy of Sciences.

**41** Molina, M.J. and Rowland, F.S. (**1974**). Stratospheric sink for chlorofluoromethanes - chlorine atomic-catalysed destruction of ozone. *Nature 249* (5460): 810–812.

**42** Tolmachev, I. (**2019**). A history of photography part 1: The beginning. https://photography.tutsplus.com/articles/a-history-of-photography-part-1-the-beginning--photo-1908 (accessed 2021).

**43** Grundberg, A., Newhall, B., Rosenblum, N., and Gernsheim, H. E. R. History of photography. https://www.britannica.com/technology/photography (accessed 2021).

**44** Long, B. (**2009**). *Complete Digital Photography*, 5e. Boston, MA: Course Technology/Cengage Learning.

**45** Downes, A. and Blunt, T.P. (**1878**). On the influence of light upon protoplasm. *Proceedings of the Royal Society of London 28*: 199–212.

**46** Downes, A. and Blunt, T.P. (**1877**). The influence of light upon the development of bacteria. *Nature 16*: 218.

**47** Downes, A. and Blunt, T.P. (**1877**). Researches upon the effect of light upon bacteria and other organisms. *Proceedings of the Royal Society of London 26*: 488–500.

**48** Arloing, S. (**1885**). Influence du soleil sur la végétation, la végétabilité et la virulence des cultures du *Bacillus anthracis. Comptes Rendus Hebdomadaires des Seances de l'Académie des Sciences 101*: 535–537.

**49** Arloing, S. (**1885**). Influence du soleil sur la végétabilité des spores du B*acillus anthracis. Comptes Rendus Hebdomadaires des Seances de l'Académie des Sciences 101*: 511–513.

**50** Arloing, S. (**1885**). Influence de la lumière sur la végétation et la propriétés pathogènes du *Bacillusanthracis. Comptes Rendus Hebdomadaires des Seances de l'Académie des Sciences 100*: 378–381.

**51** Barnard, J.E. and de Morgan, H.R. (**1903**). Upon the bactericidal action of some ultra-violet radiations as produced by the continuous-current arc. *Proceedings of the Royal Society of London. Series B, Containing Papers of a Biological Character 72*: 126–128.

**52** National Museum of American History at the Smithsonian Institution. Cooper Hewitt mercury vapor lamp. http://americanhistory.si.edu/collections/search/object/nmah_995012 (accessed 2021).

**53** Gates, F.L. (**1930**). A study of the bactericidal action of ultra violet light III. The absorption of ultra violet light by bacteria. *Journal of General Physiology 14* (1): 31–42.

**54** Gates, F.L. (**1929**). A study of the bactericidal action of ultra violet light I. The reaction to monochromatic radiations. *Journal of General Physiology 13* (2): 231–248.

**55** Gates, F.L. (**1929**). A study of the bactericidal action of ultra violet high II. The effect of various environmental factors and conditions. *Journal of General Physiology 13* (2): 249–260.

**56** Kelner, A. (**1949**). Photoreactivation of ultraviolet-irradiated escherichia-coli, with special reference to the dose-reduction principle and to ultraviolet-induced mutation. *Journal of Bacteriology 58* (4): 511–522.

**57** Kelner, A. (**1949**). Effect of visible light on the recovery of streptomyces griseus conidia from ultraviolet irradiation injury. *Proceedings of the National Academy of Sciences of the United States of America 35* (2): 73–79.

**58** Sancar, A. (**1996**). No "end of history" for photolyases. *Science 272* (5258): 48–49.

**59** Beukers, R. and Berends, W. (**1960**). Isolation and identification of the irradiation product of thymine. *Biochimica Et Biophysica Acta 41* (3): 550–551.

**60** Hanawalt, P. and Setlow, R. (**1960**). Effect of monochromatic ultraviolet light on macromolecular synthesis in *Escherichia-coli. Biochimica Et Biophysica Acta 41* (2): 283–294.

**61** Campbell, A.T., Robertson, L.J., and Smith, H.V. (**1992**). Viability of cryptospordium-parvum oocysts - correlation of invitro excystation with inclusion or exclusion of fluorogenic vital dyes. *Applied and Environmental Microbiology 58* (11): 3488–3493.

**62** Rice, E.W. and Hoff, J.C. (**1981**). Inactivation of giardia-lamblia cysts by ultraviolet-irradiation. *Applied and Environmental Microbiology 42* (3): 546–547.

**63** Clancy, J.L., Hargy, T.M., Marshall, M.M., and Dyksen, J.E. (**1998**). UV light inactivation of cryptosporidium oocysts. *Journal American Water Works Association 90* (9): 92–102.

**64** Clancy, J.L., Bukhari, Z., Hargy, T.M. et al. (**2000**). Using UV to inactivate cryptosporidium. *Journal American Water Works Association 92* (9): 97–104.

**65** Craik, S.A., Weldon, D., Finch, G.R. et al. (**2001**). Inactivation of Cryptosporidium parvum oocysts using medium- and low-pressure ultraviolet radiation. *Water Research 35* (6): 1387–1398.

**66** Slifko, T.R., Huffman, D.E., Dussert, B. et al. (**2002**). Comparison of tissue culture and animal models for assessment of Cryptospridium parvum infection. *Experimental Parasitology 101* (2–3): 97–106.

**67** Campbell, A.T. and Wallis, P. (**2002**). The effect of UV irradiation on human-derived Giardia lamblia cysts. *Water Research 36* (4): 963–969.

**68** Mofidi, A.A., Meyer, E.A., Wallis, P.M. et al. (**2002**). The effect of UV light on the inactivation of Giardia lamblia and Giardia muris cysts as determined by animal infectivity assay (P-2951-01). *Water Research 36* (8): 2098–2108.

**69** Hoyer, O. (**2000**). The status of UV technology in Europe. *IUVA News 2*: 22–27.

**70** Henri, V. and Helbronner, A. (**1910**). de Recklinghousen; Max, Stérilisation de grandes quantités d'eau par les rayons ultraviolets. *Comptes Rendus de l'Academie des Sciences 150*: 932–934.

**71** Henri, V.H. (**1910**). André de Recklinghousen, Max, Nouvelles recherches sur la stérilisation de grandes quantités d'eau par les rayons ultraviolets. *Comptes Rendus de l'Academie des Sciences 151*: 677–680.

**72** Smith, A.T. (**1917**). Ultra-violet rays finish treatment of Henderson water-supply. *Engineering News Record 79*: 1021–1022.

**73** Henderson (KY) Water Utility. North water plant history and operation. https://www.hkywater.org/departments/water/north-water-plant-history (accessed 2021).

**74** Delbrück, M. and Luria, S.E. (**1942**). Interference between bacterial viruses. I. Interference between two bacterial viruses acting upon the same host, and the mechanism of virus growth. *Archives of Biochemistry 1*: 111–141.

75 White, G.C. (**1992**). *The Handbook of Chlorination and Alternative Disinfectants*, 3rde, xiv. New York: Van Nostrand Reinhold, 1308 p.

76 Rook, J.J. (**1974**). Formation of haloforms during chlorination of natural waters. *Water Treatment and Examination 23* (Part 2): 234–243.

77 EPA, U (**2006**). *Ultraviolet Disinfection Guidance Manual for the Final Long Term 2 Enhanced Surface Water Treatment Rule*. Washington, DC: United States Environmental Protection Agency.

78 Lofrano, G. and Brown, J. (**2010**). Wastewater management through the ages: a history of mankind. *Science of the Total Environment 408* (22): 5254–5264.

79 Solomon, S. (**2010**). *Water: The Epic Struggle for Wealth, Power, and Civilization*. New York: HarperCollins.

80 United States Environmental Protection Agency. History of the clean water act. https://www.epa.gov/laws-regulations/history-clean-water-act (accessed 2021).

81 Whitby, G.E. and Scheible, O.K. (**2004**). The history of UV and wastewater. *IUVA News 6*: 15–26.

82 Scheible, O.K. and Bassell, C.D. (**1981**). *Ultraviolet Disinfection of a Secondary Wastewater Treatment Plant Effluent*. Cincinnati, OH: US EPA.

83 Scheible, O.K., Forndran, A., and Leo, W.M. (**1983**). *Pilot Investigation of Ultraviolet Wastewater Disinfection at the New York City Port Richmond Plant*, 202–225. Cincinnati, OH: Second National Symposium on Municipal Wastewater Disinfection.

84 Whitby, G.E., Palmateer, G., Cook, W.G. et al. (**1984**). Ultraviolet disinfection of secondary effluent. *Journal Water Pollution Control Federation 56* (7): 844–850.

85 Lopman, B. A.; Steele, D.; Kirkwood, C. D.; Parashar, U. D., The vast and varied global burden of norovirus: prospects for prevention and control. *Plos Medicine* **2016**, *13*, (4): e1001999. doi:https://doi.org/10.1371/journal.pmed.1001999.

86 Ettayebi, K., Crawford, S.E., Murakami, K. et al. (**2016**). Replication of human noroviruses in stem cell-derived human enteroids. *Science 353* (6306): 1387–1393.

87 Meierhofer, R. and Landolt, G. (**2009**). Factors supporting the sustained use of solar water disinfection - Experiences from a global promotion and dissemination programme. *Desalination 248* (1–3): 144–151.

88 Meierhofer, R. and Wegelin, M. (**2002**). *Solar Water Disinfection -A Guide for the Application of SODIS*, 88. Dubendorf, Switzerland: Swiss Federal Institute of Environmental Science and Technology.

89 McGuigan, K.G., Conroy, R.M., Mosler, H.J. et al. (**2012**). Solar water disinfection (SODIS): A review from bench-top to roof-top. *Journal of Hazardous Materials 235*: 29–46.

90 Mbonimpa, E.G., Vadheim, B., and Blatchley III, E.R. (**2012**). Continuous-flow solar UVB disinfection reactor for drinking water. *Water Research 46* (7): 2344–2354.

91 Hlavsa, M.C., Roberts, V.A., Kahler, A.M. et al. (**2015**). Outbreaks of illness associated with recreational water — United States, 2011–2012. *MMWR 64* (24): 668–672.

92 Li, J. and Blatchley III, E.R. (**2009**). UV photodegradation of inorganic chloramines. *Environmental Science & Technology 43* (1): 60–65.

93 Zare Afifi, M. and Blatchley III, E.R. (**2016**). Effects of UV-based treatment on volatile disinfection byproducts in a chlorinated, indoor swimming pool. *Water research 105*: 167–177.

94 International Maritime Organization. (**2017**). International convention for the control and management of ships' ballast water and sediments, 2004: Guidance on methodologies that may be used for enumerating viable organisms for type approval of ballast water management systems. In IMO: London.

95 Stefan, M.I. and Bolton, J.R. (**2002**). UV direct photolysis of N-nitrosodimethylamine (NDMA): Kinetic and product study. *Helvetica Chimica Acta 85* (5): 1416–1426.

**96** Wells, W.F., Wells, M.W., and Wilder, T.S. (**1942**). The environmental control of epidemic contagion I An epidemiologic study of radiant disinfection of air in day schools. *American Journal of Hygiene 35* (1): 97–121.

**97** Wells, W.F. and Stone, W.R. (**1934**). On air-borne infection - Study III Viability of droplet nuclei infection. *American Journal of Hygiene 20* (3): 619–627.

**98** Wells, W.F. (**1934**). On air-borne infection - Study II droplets and droplet nuclei. *American Journal of Hygiene 20* (3): 611–618.

**99** Memarzadeh, F., Olmsted, R.N., and Bartley, J.M. (**2010**). Applications of ultraviolet germicidal irradiation disinfection in health care facilities: Effective adjunct, but not stand-alone technology. *American Journal of Infection Control 38* (5): S13–S24.

**100** Zhang, Y.P., Mo, J.H., Li, Y.G. et al. (**2011**). Can commonly-used fan-driven air cleaning technologies improve indoor air quality? A literature review. *Atmospheric Environment 45* (26): 4329–4343.

**101** Markets and Markets (**2021**). UV Disinfection equipment market. https://www.marketsandmarkets.com/Market-Reports/uv-disinfection-market-217291665.html (accessed 2021).

# 2

# Photochemical Reactions

## 2.1 Introduction

Photochemical reactors are used to bring about changes within gas, liquid, and solid materials. In engineering applications, one or more photochemical or photophysical processes are applied to these media to promote a reaction sequence of interest. For some applications, the photochemical processes can be remarkably selective; in other situations, photochemical processes can initiate reaction sequences that are essentially nonselective.

For all photochemical reactors, the processes of interest involve one or more photochemical reactions. Therefore, it is beneficial to have a basic understanding of the physical and chemical phenomena that are responsible for photochemical reactions. The objective of this chapter is to establish this foundation as a means of improving our understanding of photochemical reactor systems, as will be addressed in subsequent chapters.

The laws of photochemistry will be presented as a foundation, followed by a basic presentation of the energy associated with photochemical processes. A summary of (thermal) chemical kinetics is then presented; this section is included because there are many important parallels and distinctions between (thermal) chemical and photochemical reactions. In addition, the products of many photochemical reactions are unstable intermediates (e.g. free radicals) which tend to react rapidly through thermal chemical reaction processes, many of which are bimolecular. This chapter concludes with a presentation of the basic principles of photochemical kinetics. Fundamental kinetic expressions are derived for monochromatic and polychromatic radiation sources. Photochemical kinetic expressions are also derived on an energy basis and a photon basis.

## 2.2 Laws of Photochemistry

Photochemical reactions are fundamentally different from thermal chemical reactions in terms of the physical phenomena and forms of energy that initiate them. The fundamental "laws" that govern photochemical processes were postulated by a group of remarkable scientists in the mid- to late-19th and early 20th centuries.

*Photochemical Reactors: Theory, Methods, and Applications of Ultraviolet Radiation*, First Edition. Ernest R. Blatchley III.

Companion website: www.wiley.com/go/blatchley/photochemicalreactors

The Grotthuss–Draper law[a] (also known as the first law of photochemistry or the *principle of photochemical activation*) states that light (or more generally, electromagnetic radiation) must be absorbed by a system to bring about a photochemical change. The law was originally presented by Theodor Grotthuss in 1817, then again by John William Draper in 1842 as a result of independently conducted experiments. The Stark–Einstein law[b] (also known as the second law of photochemistry, the *photochemical equivalence law*, or the *photoequivalence law*) states that for each photon absorbed by a chemical system, only one molecule is activated for a photochemical (or photophysical) process.

At first glance, the first and second laws of photochemistry may appear to be simple and obvious. Yet, there are important, sometimes subtle implications of these laws. One example is that the energy of an absorbed photon must meet or exceed the energy of a bond that will be formed or broken if a photochemical reaction is to proceed. Conversely, this implies that photons with energy less than the bond energy cannot cause cleavage or formation of the bond in question. The second law also raises an important point about the fundamental nature of molecules and photons. When considering a chemical substance, molecules (ions, atoms, radicals, or complexes)[c] are viewed as the smallest discrete unit of the substance. It is not possible to reduce a substance to a smaller unit than a molecule without changing the fundamental properties of the substance. Likewise, the photon is the smallest discrete unit of electromagnetic radiation and a quantum represents the amount of energy associated with a photon. It is not possible to divide a quantum into a smaller quantity of energy, nor is it possible to absorb part of a photon's energy. Photon absorption is an all-or-nothing process. The second law of photochemistry also implies that all photochemical events are independent; therefore, the effects of these events are additive.

The Stark–Einstein law formalizes the "equivalence" between molecules and photons. Some of the terminology used in describing photochemical processes can be attributed to this relationship and its implications for photochemistry. Specifically, the term "einstein" has been used to represent Avogadro's number of photons, just as the term "mole" is used to represent Avogadro's number of molecules.[d] One molecule may absorb a single photon; similarly, one mole of a molecule may absorb one einstein of photons.

The Grotthuss–Draper law is in a sense qualitative in that it does not define a specific, quantitative relationship between the amount of radiation absorbed and the extent of the corresponding reaction. The Bunsen–Roscoe law provides this quantitative structure by stating that "in two photochemical reactions, if the product of the intensity *[sic]* of illumination and the time of exposure are equal, the quantities of chemical material undergoing change will be equal." [1] In today's terminology, the Bunsen–Roscoe law indicates that the extent of a photochemical reaction will be governed by the dose (fluence) of radiation absorbed by the target. As such, dose is the master variable used in defining the extent of a photochemical process. We will see the emergence of dose as a master variable in later discussions of photochemical reaction kinetics and in the discussion of photochemical reactor theory (Chapter 3).

## 2.3 Energy in Photochemical Processes

Photon energy is a function of its frequency ($\nu$) or wavelength ($\lambda$) and is formally defined by equation 2.1:

$$U_p = h \cdot \nu = h \cdot \frac{c_0}{\lambda} \tag{2.1}$$

where[e,f,g]

$U_p$ = photon energy $[=] \frac{\text{J}}{\text{photon}}$

$h$ = Planck constant $= 6.626 \times 10^{-34} \frac{\text{J} \cdot \text{s}}{\text{photon}}$

$\nu$ = frequency of radiation $[=]$ Hz or $\text{s}^{-1}$

$c_0$ = speed of light in vacuum $= 2.998 \times 10^{8} \frac{\text{m}}{\text{s}}$

$\lambda$ = wavelength of radiation $[=]$ nm.

By extension, the energy associated with an einstein of photons is defined by equation 2.2:

$$U = U_p \cdot N_A \tag{2.2}$$

where

$U$ = energy per einstein at wavelength $\lambda$ $[=] \frac{\text{J}}{\text{einstein}}$

$N_A$ = Avogadro's number $= 6.023 \times 10^{23} \frac{\text{photon}}{\text{einstein}}$.

Because photochemical reactions are initiated by absorption of photon energy, a necessary condition for a photochemical reaction to take place is that photon energy must meet or exceed the energy of at least one bond in the target molecule. If this condition is not met, then it is not possible for an existing bond to be broken or a new bond to be formed – in other words, no reaction can occur as a result of photon absorption when photon energy is less than bond energy. On the other hand, if photon energy exceeds bond energy ($\Delta H$), then a reaction could occur.

Bond energies within molecules are influenced by the nature of the participating atoms, their oxidation states, as well as the characteristics of neighboring atoms and functional groups. Therefore, it is not possible to define bond energies in absolute terms that will apply universally. On the other hand, it is possible to summarize typical or mean values of bond energy.

Equations 2.1 and 2.2 can be rearranged by algebra to calculate the wavelength of radiation corresponding to a given bond energy. Equation 2.3 illustrates the use of this relationship for calculation of the wavelength of radiation that corresponds to the average bond energy of the H–H bond $\left(4.36 \times 10^{5} \frac{\text{J}}{\text{mole}}\right)$. Figure 2.1 illustrates photon energy as a function of wavelength as calculated by the combination of equations 2.1 and 2.2, along with indications of approximate bond energies and corresponding wavelengths, as calculated by equation 2.3.

$$\lambda = \frac{hc_0}{\Delta H} = \frac{\left(6.626 \times 10^{-34} \frac{\text{J} \cdot \text{s}}{\text{photon}}\right) \cdot \left(2.998 \times 10^{8} \frac{\text{m}}{\text{s}}\right) \cdot \left(\frac{10^{9}\ \text{nm}}{\text{m}}\right) \cdot \left(6.023 \times 10^{23} \frac{\text{photon}}{\text{einstein}}\right) \cdot \left(\frac{\text{einstein}}{\text{mole}}\right)}{4.36 \times 10^{5} \frac{\text{J}}{\text{mole}}}$$
$$= 274\ \text{nm} \tag{2.3}$$

Before proceeding further, a few comments about the information presented in Figure 2.1 may be helpful for interpretation.

- Most photochemical reactions are initiated by absorption of photons from the ultraviolet (UV) and visible portions of the electromagnetic spectrum (see ranges in Figures 2.1 and 2.2).

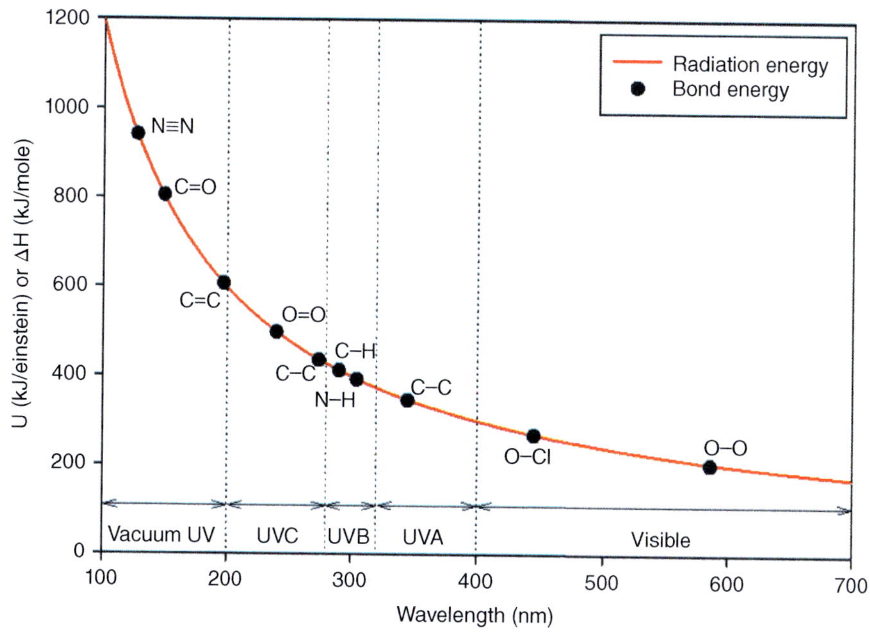


**Figure 2.1** Energy of electromagnetic radiation as a function of wavelength. Superimposed are typical values of bond energy for common bonds. Also indicated along the bottom of the graph are wavelength ranges corresponding to definitions of radiation types. *Source:* Based on Photolysis and the Hydroxyl Radical http://butane.chem.uiuc.edu/pshapley/environmental/l14/1.html.

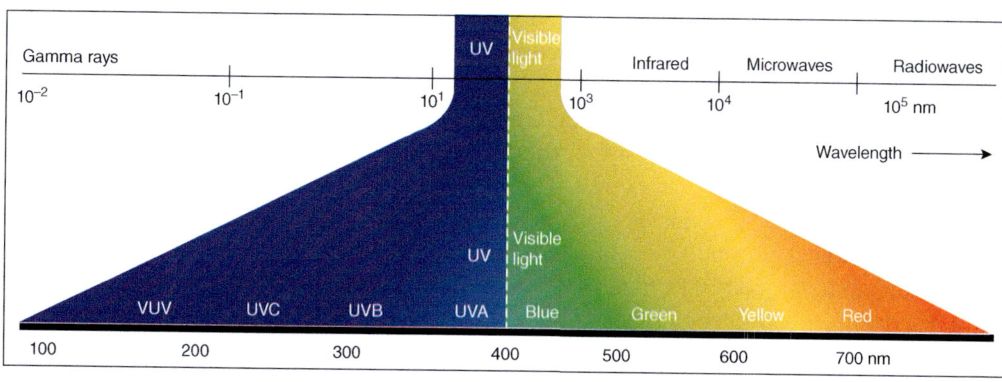


**Figure 2.2** Schematic illustration of electromagnetic (EM) spectrum, with particular focus on the UV and visible regions. *Source:* Albatross UV, Spectral Output, http://www.albatrossuv.com/UV-Spectral-Output.php. Visible radiation, corresponding nominally to 400 nm < $\lambda$ < 700 nm, defines the region of the EM spectrum to which the human sense of vision responds (see discussion in Chapter 1). The UV range is subdivided into several subranges which include: UVA (320 nm < $\lambda$ < 400 nm); UVB (280 nm < $\lambda$ < 320 nm); UVC (200 nm < $\lambda$ < 280 nm); and vacuum UV (100 nm < $\lambda$ < 200 nm). Note that various definitions of the wavelength boundaries for these regional definitions of UV radiation can be found in the literature.

- The spectrum of ambient solar radiation on earth will be governed by location, time of year, time of day, and atmospheric conditions, but in all cases, ambient solar radiation at near-sea-level altitudes will involve wavelengths of radiation above roughly 290 nm. Therefore, many of these bonds are non-photochemically active on earth without the application of radiation from an "artificial" radiation source (i.e. a source of radiation other than ambient sunlight). For example, the N≡N bond of molecular nitrogen requires radiation with wavelengths shorter than about 127 nm. Such radiation is only available in the upper reaches of earth's stratosphere or from specialized sources of radiation.
- Bond strength is strongly influenced by the number of bonds between atoms. For example, the double bond of molecular oxygen (O=O) has a bond strength such that only short wavelength radiation can cause bond cleavage. Cleavage of the O=O bond takes place in the earth's stratosphere and is largely associated with radiation in the wavelength range of 200 nm $\lesssim \lambda \lesssim$ 240 nm. Cleavage of this bond is critical to the formation of ozone ($O_3$) in this same region of the atmosphere. On the other hand, the O–O bond, which is found in the ozone molecule, has a substantially lower bond energy. As such, longer wavelengths of radiation are able to cleave the O–O bond. $O_3$ photolysis is also an important process in the earth's atmosphere [3].

The electromagnetic spectrum comprises all wavelengths of radiation from gamma to radio waves. Equation 2.1 defines the dependence of photon energy on the wavelength (or frequency) of radiation. Several common terms are used to describe radiation in various wavelength ranges of the electromagnetic spectrum. These are illustrated in Figure 2.2.

In examining the information presented in Figures 2.1 and 2.2, we see that virtually all photochemical reactions are initiated by absorption of radiation in the UV and visible portions of the EM spectrum. For radiation corresponding to longer wavelengths, photon energy is generally insufficient to promote a reaction. For shorter wavelengths, photon energy tends to be absorbed over short distances of the exposed medium. Moreover, photon energies in these ranges are often high enough to cause direct ionization of the absorbing molecule. As such, radiation in these regions is often referred to as "ionizing" radiation. Although important similarities exist between ionizing radiation and radiation from other parts of the EM spectrum (including UV and visible), important distinctions in the physical and chemical behavior among these radiation types exist as well. The focus of this book is on photochemical reactions. Consequently, the behavior of ionizing radiation (e.g. x-rays and $\gamma$ radiation) will not be addressed in detail herein.

Another important characteristic of photochemical reactions is that the rates of these processes tend to be (largely) independent of temperature. This is because the energy required to promote photochemical processes is attributable to photon absorption, rather than the thermal (kinetic) energy of molecules in the system. However, it is important to recognize that the concept of temperature-independence applies only to pure photochemical processes. As we will see, photochemical processes often lead to the formation of reactive intermediates, which in turn participate in thermal chemical reactions. These so-called "secondary" reactions display temperature-dependence that conforms to conventional behavior of other thermal chemical reactions.

## 2.4 Kinetics

The development of engineering applications of photochemistry requires a fundamental understanding of the rates of the basic processes involved. In this section, the kinetics of conventional (thermal) chemical reactions will be presented first, followed by a parallel presentation of the kinetics of photochemical reactions. This approach is motivated by several factors:

1) Most engineers and scientists have experience with the principles that define the kinetics of thermal chemical reactions. This background can serve as a foundation for understanding other reaction types, including photochemical reactions.
2) Important analogies exist between thermal and photochemical reactions.
3) Important distinctions between thermal and photochemical reactions are also evident.
4) In many cases, photophysical and/or photochemical processes will initiate one or more thermal chemical reactions.
5) Similar parallels and differences exist between thermal chemical reactors and photochemical reactors (principles to be developed in Chapter 3).

### 2.4.1 Kinetics of Thermal Chemical Reactions

Thermal chemical reactions depend on collisions among reactive entities. For purposes of illustration, we will consider the mechanics of an elementary, bimolecular reaction that takes place in aqueous solution.

Molecules in solution will move in response to the thermal energy of the medium in which they reside and the collisions they have with neighboring molecules. As the temperature of the solution increases, the thermal (kinetic) energy of the molecules will also increase. As two molecules approach each other, their electric fields begin to interact [4]. When the separation distance between the molecules decreases, the bonds within the molecules become increasingly strained, causing the chemical potential (energy) within the molecules to increase. In turn, these changes in chemical energy may cause changes in bond structure of the colliding molecules, leading to a chemical reaction. The rates of thermal chemical reactions depend on the frequency with which the reacting molecules collide, as well as "chemical" factors that affect the efficiency by which these collisions lead to a reaction.

Consider the following hypothetical bimolecular, elementary reaction:

$$B + C \longrightarrow \text{products} \tag{2.4}$$

In a strict sense, all reactions are reversible; however, for purposes of evaluating reaction kinetics, we may ignore the reverse reaction.[h] Principles of general chemistry teach us that the kinetics (rate) of reaction (2.4) can be described as follows:

$$r_B = \frac{\mathrm{d}[B]}{\mathrm{d}t} = -k \cdot [B] \cdot [C] \tag{2.5}$$

where

| | | | | |
|---|---|---|---|---|
| $r_B$ | = | rate of reaction with respect to reactant $B$ | [=] | $\frac{\text{mole}}{\text{L} \cdot \text{s}}$ |
| $[j]$ | = | activity of $j^{\text{th}}$ constituent in solution | [=] | $\frac{\text{mole}}{\text{L}}$ |
| $t$ | = | time | [=] | s |
| $k$ | = | second-order reaction rate constant | [=] | $\frac{\text{L}}{\text{mole} \cdot \text{s}}$. |

In dilute solutions, activity is closely approximated by molar concentration. Activity may be viewed as a measure of the "availability" of a reactant to participate in reaction. As activity increases, the rate of the reaction will increase because the frequency of bimolecular collisions will also increase.

As described above, a bimolecular reaction can be described as a series of events that lead to changes in the structure of reactant molecules. Arrhenius proposed a mathematical expression to define, in quantitative terms, the effects of temperature on the rate constant of a thermal chemical reaction:

$$k = P \cdot \exp\left(\frac{-E_a}{R \cdot T}\right) \tag{2.6}$$

where

| | | | | |
|---|---|---|---|---|
| $P$ | = | pre-exponential factor (*aka* frequency factor) | [=] | $\frac{\text{L}}{\text{mole} \cdot \text{s}}$ |
| $E_a$ | = | activation energy | [=] | $\frac{\text{kJ}}{\text{mole}}$ |
| $R$ | = | gas constant | = | $8.314\frac{\text{J}}{\text{K} \cdot \text{mole}}$ |
| $T$ | = | absolute temperature | [=] | K. |

Transition state theory suggests that bimolecular reactions will proceed via the development of an "activated" intermediate that is in equilibrium with the reactant molecules. In turn, the activated intermediate can participate in a reaction to lead to the formation of a product. These processes are described sequentially as follows:

$$B + C \rightleftharpoons BC^* \rightarrow \text{products} \tag{2.7}$$

where $BC^*$ represents the activated intermediate. $BC^*$ is an entity that has chemical properties that are different from the reactant molecules because of rearrangement of electron structure, relative to the parent molecules. The activation energy describes the energy associated with the fraction of collisions, $\exp\left(\frac{-E_a}{R \cdot T}\right)$, that have sufficient energy to yield $BC^*$.

Not all collisions between reactant molecules $B$ and $C$ will lead to production of $BC^*$ or ultimately to the reaction products. For some reactions, the orientation of the reactants at the time of collision may affect process efficiency. Other "chemical" factors that can affect the efficiency of reactant collisions include the viscosity and ionic strength of the medium in which they reside, the charges of the reactants, and pressure in the system [4, 5]. Collectively, these factors are accounted for in the pre-exponential factor.

Figure 2.3 is a classic schematic representation of the thermodynamics that govern the progress of a thermal chemical reaction. The activation energy represents a "barrier" to reaction progress. Bimolecular collisions that are able to meet or exceed this energy barrier may result in formation of the reaction product(s). Collisions that fail to meet or exceed this threshold will not yield product(s). Reaction progress may be facilitated by the addition of a catalyst, which may reduce the magnitude of $E_a$.

Reaction rate constants can vary by many orders of magnitude for elementary, bimolecular reactions, even when temperature is fixed. Table 2.1 provides a brief listing of reported rate constants for elementary, bimolecular reactions that are common in water treatment operations.

At the upper end of the values reported in Table 2.1, we observe rate constants of the order $10^{10}$ $\frac{\text{L}}{\text{mole} \cdot \text{s}}$, which are representative of so-called "diffusion limited" reactions. The rates of these reactions are limited only by the rate at which molecular collisions take place. For reactions with rate

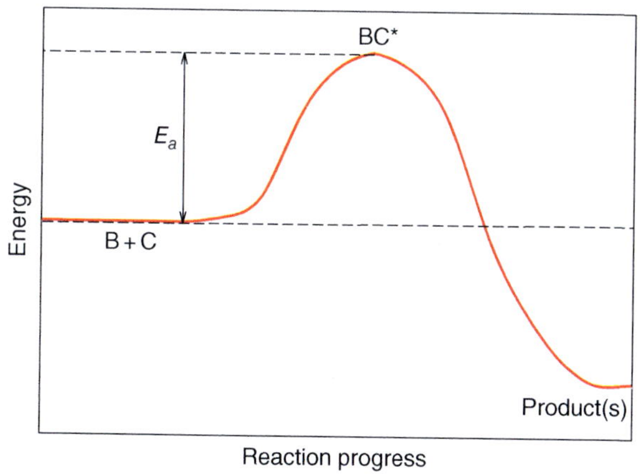


**Figure 2.3** Schematic illustration of thermal reaction progress and associated energy.

**Table 2.1** Summary of reported reaction rate constants [6–13] for elementary, bimolecular reactions among common or emerging water contaminants (Reactant 1) and common oxidants used (or formed) in water treatment (Reactant 2). The elementary reactions listed below are assumed to follow a sequence of: Reactant 1 + Reactant 2 → Product(s). Numerical values are reported rate constants at room temperature (L/mole·s). Note that "slow" indicates that the reaction rate was too slow to report a reliable value.

| | Reactant 2 | | |
|---|---|---|---|
| **Reactant 1 (attributes or applications)** | **$HO^{\cdot}$** | **$O_3$** | **HOCl** |
| 17α-Ethinylestradiol (synthetic steroid estrogen) | $9.8 \times 10^9$ | $1.8 \times 10^5$ | $3.5 \times 10^5$ |
| Atrazine (triazine herbicide) | $2.4–3.0 \times 10^9$ | 6–7.9 | Slow |
| $Br^-$ (bromide ion) | $1.1 \times 10^9$ | $2.6 \times 10^2$ | $1.6 \times 10^3$ |
| Caffeine (central nervous system stimulant) | $6.4 \times 10^9$ | 650 | Slow |
| Carbamazepine (antiepileptic) | $8.8 \times 10^9$ | $6.3 \times 10^5$ | $6 \times 10^6$ |
| Ibuprofen (non-steroidal anti-inflammatory drug) | $7.4 \times 10^9$ | 9.6 | <0.1 |
| $NH_4^+/NH_3$ (ammonium/ammonia) | $9 \times 10^7$ | 20 | $3 \times 10^6$ |
| N-N-diethyl-3-toluamide (DEET, insect repellent) | $4.6–7.5 \times 10^9$ | 0.123 | Slow |
| Sulfamethoxyzole (sulfonamide antibiotic) | $8.8 \times 10^9$ | $6.3 \times 10^5$ | $6 \times 10^6$ |

constants substantially below this value, one or more of the "chemical" factors listed above also constrains the observed reaction rate.

Another key feature of the rates of thermal chemical reactions is that they can all be logically described as functions of time. In fact, time may be thought of as the relevant *master variable* in evaluating progress of thermal chemical reactions. As we will see in Chapter 3, the analysis of thermal chemical reactor systems is then logically based on time as the independent (master) variable.

### 2.4.2 Kinetics of Photochemical Reactions: Monochromatic Radiation Source, Electromagnetic Energy Basis

The Stark–Einstein law teaches that molecules and photons can be viewed in an equivalent manner in photochemical reactions. In effect, photons may be viewed as "reactants," despite the fact that they do not contribute matter, only energy, to the process. The expressions (rate laws) used to describe photochemical kinetics build on this logic and the approach described above for elementary, bimolecular thermal chemical reactions. However, the physical phenomena that lead to a photochemical reaction are fundamentally different from those that promote thermal chemical reactions.

Photochemical reactions are initiated by absorption of a photon by a target molecule. For purposes of this discussion, we will consider a photochemical reaction in which a target molecule ($B$) absorbs a single photon and its corresponding quantum of energy. In schematic form, this initiates one or more photochemical and/or photophysical processes, as illustrated in Figure 2.4.

In a general sense, absorption of photon energy causes electronic excitation within the target molecule. If the energy of an absorbed photon exceeds the energy of at least one bond within the target molecule, it is possible, though not certain, that a subsequent reaction can take place. However, it is important to recognize that the energy needed to initiate a reaction is attributable to the absorbed photon, not to collisions with nearby molecules, as is the case in thermal chemical reactions.

The energy of the absorbed photon can be dissipated in many ways. The Jablonski diagram (Figure 2.5) was developed as a schematic illustration of the energy associated with this process, the various electronic energy states that can be involved, and the processes that may result from photon absorption.

A molecule in its ground state can absorb a photon to allow promotion of an electron to a higher energy orbital. This may result in formation of a singlet form of the molecule, in which electrons remain spin-paired but excited. The excess energy resulting from this absorption event may be released through fluorescence (photon emission), vibrational relaxation, quenching, or a reaction. It is also possible, though generally unlikely, that an electron within a singlet may undergo spin reversal to yield a triplet (non-spin-paired) excited state. The excess energy of a triplet can be released by processes that are similar to those that the singlet may experience, including phosphorescence (photon emission), vibrational relaxation, or reaction.

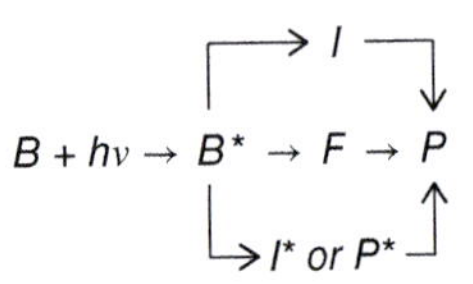


**Figure 2.4** Schematic illustration of the processes involved in a general photochemical reaction. B represents the target molecule; B* represents the target molecule in an electronically excited state. I is a reactive intermediate; F represents a "funnel"; I* and P* represent electronically excited intermediates and products, respectively. *Source:* Adapted from Turro et al. (2009).

Among the interesting features of the processes that lead to photochemical reactions is the fact that a molecule may be given access to all of its possible excited states, including multiple singlet and triplet states, as well as all of the various corresponding vibrational states. This is generally not the case with thermal chemical reactions.

The Jablonski diagram provides a schematic illustration of the various energy states that can result from photon absorption. In that regard, the vertical coordinate of the Jablonski diagram provides an illustration of the energy associated with each state of the target molecule. However, unlike the schematic illustration of the progression of a thermal reaction (see Figure 2.3), the horizontal axis in a Jablonski diagram has no physical meaning.

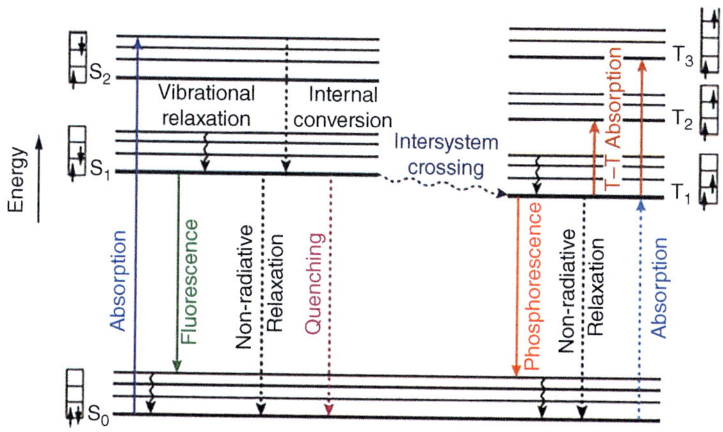


**Figure 2.5** Jablonski diagram used to describe an arbitrary process involving photon absorption Source: Xu et al. 2014 / Royal Society of Chemistry / CC BY-3.0. The target molecule starts from the ground state. Photon absorption leads to promotion of an electron to a higher energy orbital and formation of a singlet (e.g. $S_1$ or $S_2$). The energy of the singlet may be released by emission of a photon (fluorescence), vibrational relaxation (heat), quenching, or a photoreaction. It is also possible for intersystem crossing to occur, in which an excited electron "flips" (i.e. changes spin) to yield a triplet (e.g. $T_1$, $T_2$, or $T_3$). The energy of a triplet may be released through the processes of photon emission (phosphorescence), relaxation, or reaction.

In practical applications of photochemistry, it is unusual to work with individual molecules or photons. Rather, we generally work with solutions that contain large numbers of molecules, to which we may apply large numbers of photons. Calculations of photochemical kinetics are based on mathematical descriptions of photon energy absorption by target molecules and the corresponding transformation rate of the target molecules.

The Beer–Lambert law provides a quantitative description of this process (see equation 2.8) [16]. Specifically, for a solution containing a target molecule, the spectral radiant power of a collimated beam of radiation through the solution will decrease as follows:[i]

$$A(\lambda) = -\log_{10}\left(\frac{P(\lambda)}{P(\lambda)^0}\right) = \varepsilon(\lambda)\cdot[B]\cdot\ell \tag{2.8}$$

where

| | | | | |
|---|---|---|---|---|
| $A(\lambda)$ | = | absorbance at wavelength $\lambda$ | | |
| $P(\lambda)$ | = | transmitted spectral radiant power | [=] | W |
| $P(\lambda)^0$ | = | incident spectral radiant power | [=] | W |
| $\varepsilon(\lambda)$ | = | molar absorption coefficient at wavelength λ | [=] | $\frac{\text{L}}{\text{mole}\cdot\text{cm}}$ |
| $[B]$ | = | molar concentration of target molecule | [=] | $\frac{\text{mole}}{\text{L}}$ |
| $\ell$ | = | optical path length through absorbing medium | [=] | cm. |

For collimated radiation, the ratio $\frac{P(\lambda)}{P(\lambda)^0}$ can be replaced by the ratio of other measures of incident and transmitted radiant energy or power, including irradiance, fluence rate, photon irradiance, and photon fluence rate. These terms will be presented and defined later in this chapter.

A closely related term is defined as transmittance ($T(\lambda, \ell)$, see equation 2.9). Note that any description of transmittance should be accompanied by a specification of the corresponding optical path length ($\ell$). Note also that transmittance is a wavelength-dependent function.

$$T(\lambda,\ell) = \frac{P(\lambda)}{P(\lambda)^0} \tag{2.9}$$

The relationship between absorbance and transmittance is evident by comparison of equations 2.8 and 2.9:

$$A(\lambda) = -\log_{10}(T(\lambda,\ell)) \tag{2.10}$$

We will see several applications of the Beer–Lambert law, as well as several rearrangements of the basic mathematical expressions presented in equations 2.9 and 2.10.

Another point to consider is that the Beer–Lambert law is sometimes expressed in base $e$ (Napierian), rather than in base 10 (Decadic). When the Beer–Lambert law is applied, it should be done in a manner that explicitly indicates the basis of application (i.e. base 10 or base $e$). The Decadic forms of the terms in the Beer–Lambert law are most commonly applied in the literature, and will be used herein.

The Beer–Lambert law is applied when analyzing photochemical reactions and photoreactors in at least three basic ways:

1) To calculate the spatial distribution of radiant power (fluence rate or irradiance) within an irradiated solution.
2) To calculate the local or volumetric rate of photon absorption by the target molecule.
3) To calculate local concentration of an absorbing compound.

From the perspective of a photochemical process, the molar absorption coefficient, $\varepsilon(\lambda)$, describes the efficiency by which a target molecule absorbs incident radiation at wavelength $\lambda$. The first and second laws of photochemistry teach that photon absorption is required for a photochemical reaction to take place. However, the energy of an absorbed photon does not always lead to a photochemical event; there are many other possible fates of this energy.

Quantitatively, we must account for the fact that not all absorbed photons lead to a chemical change. The quantum yield, $\Phi(\lambda)$, is used to describe the efficiency by which absorbed photons promote a photochemical event. The quantum yield is formally applied to describe the ratio of *primary photochemical events* to the number of photons absorbed, where the term "primary photochemical event" refers to a reaction that results directly from the production of an electronically excited intermediate by the absorption of a photon. With this in mind, a primary quantum yield is defined as:

$$\Phi(\lambda) = \frac{\text{amount of reactant consumed or product formed by a photochemical event}}{\text{amount of photons}(\lambda)\text{ absorbed}} \tag{2.11}$$

The amounts referred to in the numerator and denominator of equation 2.11 must be expressed in equivalent units. Therefore, the ratio of the numerator to the denominator may be expressed as $\frac{\text{molecules}}{\text{photon}}$ or $\frac{\text{moles}}{\text{einstein}}$. The units on the quantum yield are often ignored or implied, but they should be interpreted as a ratio of these equivalent units.

For a primary photochemical event, there are well-defined limits on the quantum yield:[k]

$$0 \leq \Phi(\lambda) \leq 1 \frac{\text{molecules}}{\text{photon}} \tag{2.12}$$

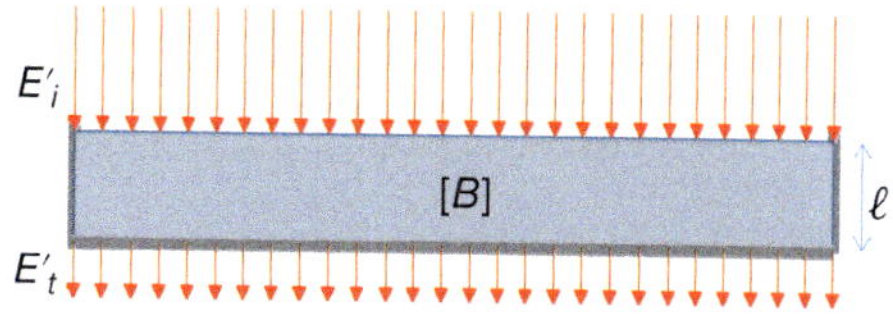


**Figure 2.6** Schematic illustration of experiment involving a single absorbing target compound, B, present at a molar concentration of [B]. Collimated, monochromatic radiation is imposed on a shallow, well-mixed reactor of depth = ℓ, at an incident fluence rate of $E'_i$; radiation is transmitted through the bottom of the solution at a fluence rate of $E'_t$. Fluence rate is one of several parameters that can be used to quantify the availability of photons. Other, similar parameters are listed in Table 2.2. **Note:** This schematic illustration does not address several optical and geometric factors that influence the behavior of real collimated beam systems and the interpretation of results of experiments based on these systems, including beam nonuniformity, deviations from true collimation, absorption of radiation through the depth of solution, as well as reflection and refraction of radiation at phase boundaries within the system. These attributes of system behavior will be addressed later in the book (see Chapter 5).

A quantum yield of zero implies that no photochemical reaction takes place as a result of photon absorption, whereas a quantum yield of $1\,\frac{\text{molecule}}{\text{photon}}$ for a primary photochemical event implies that all absorbed photons lead to conversion of the absorbing molecule from its original form to a product.

It should be noted here that quantum yields above $1\,\frac{\text{molecules}}{\text{photon}}$ are often reported in the literature. These should be interpreted as *effective quantum yields*, in that they provide evidence of one or more secondary reactions that lead to transformation of the parent compound and/or formation of a reaction product.[1] Effective quantum yields with values above $1\,\frac{\text{molecules}}{\text{photon}}$ often imply the existence of radical-mediated chain reactions.

To move toward more practical equations to describe the rates of photochemical processes, we will consider a solution that contains a single absorbing compound. For purposes of this example, we will assume that the target compound is the only constituent in solution that absorbs incident radiation, that a single photochemical event can result from absorption, and that the product(s) of this reaction do not absorb incident radiation. To further simplify, we will also assume that the incident radiation is collimated, uniform in incident fluence rate ($E'_i$), and monochromatic. Also, we will assume that the vessel (reactor) that contains the target compound is well-mixed[m], cylindrical, with a surface area of A, shallow with depth of $\ell$, and a volume of $\forall$. Note that by this definition, $\forall = A \cdot \ell$. Figure 2.6 provides a schematic illustration of this situation.

The schematic presented in Figure 2.6 is intended to illustrate one of the simplest experiments in which a photochemical reaction may take place. For any photochemical reactor system, equation 2.13 can be used to describe the rate of a primary photochemical process that takes place within it:

$$r_B = -\frac{d[B]}{dt} = I_a \cdot \Phi(\lambda) \tag{2.13}$$

where

| | | | | |
|---|---|---|---|---|
| $r_B$ | = | rate of reaction of the target molecule, $B$ | [=] | $\frac{\text{mole}}{\text{L} \cdot \text{s}}$ |
| $[B]$ | = | activity of $B$ in solution | [=] | $\frac{\text{mole}}{\text{L}}$ |

$t$ = time [=] s

$I_a$ = volumetric rate of photon absorption by $B$ [=] $\frac{\text{einstein}}{\text{L} \cdot \text{s}}$

$\Phi$ = primary quantum yield for reaction of interest [=] $\frac{\text{mole}}{\text{einstein}}$.

Equation 2.13 is applicable to essentially any system in which a photochemical reaction takes place. However, it is complicated by the term $I_a$, which requires that the volumetric rate of photon absorption *by the target compound* be accounted for. For the reactor system illustrated in Figure 2.6, $I_a$ can be defined as follows:

$$I_a = \frac{A \cdot (E_i' - E_t')}{\forall \cdot U} \tag{2.14}$$

where

$A$ = surface area of reactor [=] $\text{cm}^2$

$E_i'$ = incident fluence rate [=] $\text{W/m}^2$

$E_t'$ = transmitted fluence rate [=] $\text{W/m}^2$

$\forall$ = reactor volume [=] $\text{cm}^3$

$U$ = energy per einstein of radiation [=] $\frac{\text{J}}{\text{einstein}}$

= $N_A \cdot U_p$

$N_A$ = Avogadro's number = $6.023 \times 10^{23} \frac{\text{photons}}{\text{einstein}}$.

Fluence rate is formally defined as the total radiant power, $P$, incident from all directions onto a small sphere divided by the cross-sectional area of that sphere [2]. The fluence rate can be thought of as an analog to activity or molar concentration of a chemical reactant, in that it describes the availability of photons to participate in a reaction.

For the simple photoreactor indicated in Figure 2.6, equation 2.14 may be simplified slightly because its geometry dictates that $\forall = A \cdot \ell$:

$$I_a = \frac{(E_i' - E_t')}{\ell \cdot U} \tag{2.15}$$

Furthermore, the Beer–Lambert law allows the values of incident and transmitted fluence rate to be related to each other (equation 2.8):

$$E_t' = E_i' \cdot 10^{-\varepsilon \cdot [B] \cdot \ell} \tag{2.16}$$

Substitution of this expression into the numerator of equation 2.15 and then in turn into equation 2.13 yields:

$$r_B = -\frac{d[B]}{dt} = E_i' \cdot \frac{\left(1 - 10^{-\varepsilon \cdot [B] \cdot \ell}\right)}{\ell \cdot U} \cdot \Phi \tag{2.17}$$

Unfortunately, equation 2.17 is mathematically and geometrically difficult to interpret as a general governing equation. On the other hand, two important limiting conditions can be evaluated for which there are relatively simple, closed-form solutions to equation 2.17 that yield important insights into the behavior of photochemical reactions and photochemical reactor systems.

#### 2.4.2.1 Limiting Case 1: Opaque Solution

For the first limiting case, we will consider a situation in which the target compound is present in solution at a concentration (activity) that is sufficiently high to cause the solution to be opaque to incident radiation. Mathematically, this condition is met when:

$$\varepsilon \cdot [B] \cdot \ell \rightarrow \infty \tag{2.18}$$

In other words, a condition of an opaque solution will be met when the product $\varepsilon \cdot [B] \cdot \ell$ (Absorbance, previously referred to as optical density[n]) becomes very large. This can be accomplished with a compound that is a strong absorber (i.e. $\varepsilon \rightarrow \infty$), or for a high concentration of the target compound, or a long optical path length, or some combination of these factors.

For this situation, all incident radiation is absorbed by the target compound (i.e. $E'_t = 0$). Mathematically, this allows simplification of equation 2.17 to yield:

$$r_B = -\frac{\mathrm{d}[B]}{\mathrm{d}t} = \frac{E'_i \cdot \Phi}{\ell \cdot U} \tag{2.19}$$

In other words, the rate of conversion of the target compound is directly proportional to the incident fluence rate and is zero-order with respect to the reactant concentration. Equation 2.19 may be rearranged to yield another form of the governing equation for the kinetics of a photochemical reaction in an opaque solution:

$$r'_B = -\frac{\mathrm{d}[B]}{E'_i \cdot \mathrm{d}t} = \frac{\Phi}{\ell \cdot U} \tag{2.20}$$

This equation includes a term $E'_i \cdot dt$ that represents the incremental *dose* (or *fluence*) of radiation imposed on the target chemical over the time increment d$t$. When considered over the entire period of exposure, the dose can be formally defined as:

$$D = \int_0^{\tau} E'_i \cdot \mathrm{d}t \tag{2.21}$$

where

$D$ = dose of incident radiation [=] $\frac{\mathrm{W}}{\mathrm{m}^2}$

$\tau$ = period of exposure [=] s.

A plot of target concentration as a function of dose yields a straight line with a slope of $-\frac{\Phi}{\ell \cdot U}$(see Figure 2.7). Therefore, for these experimental conditions, target compound concentration can be monitored as a function of applied dose to allow estimation of the quantum yield for the reaction of interest by linear regression of the data as described above and in Figure 2.7.

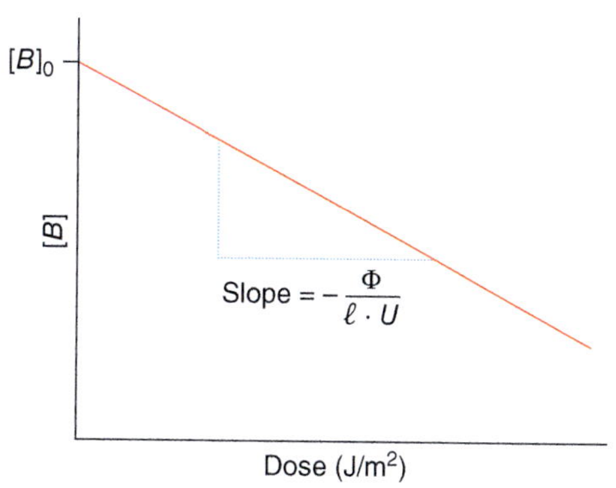


**Figure 2.7** Schematic illustration of the results of an experiment in which a single absorbing, reacting compound is present under circumstances that yield a solution that is opaque to incident radiation. The slope of the "best-fit" line, forced through the point defined by the original concentration, $[B]_0$ and a dose of 0 is equal to $-\frac{\Phi}{\ell \cdot U}$ When the slope of this regression line is known, the quantum yield can be estimated by algebra.

#### 2.4.2.2 Limiting Case 2: Transparent Solution

The second limiting case is the logical opposite of the first. A "transparent" solution is one for which the absorbance approaches zero:

$$\varepsilon \cdot [B] \cdot \ell \to 0 \tag{2.22}$$

As before, any combination of $\varepsilon$, $[B]$, or $\ell$ that satisfies equation 2.22 will conform to this condition. In a strict sense, if $\varepsilon \cdot [B] \cdot \ell \equiv 0$, then we know from the first law of photochemistry that no photochemical reaction can take place because no photons would be absorbed. So, application of this limiting condition requires that the absorbance *approaches* zero but is not equal to zero. In practical terms, the derivation presented below will yield acceptably low error when $\varepsilon \cdot [B] \cdot \ell \lesssim 0.02$.

Application of this condition to equation 2.17 can be accomplished through the use of a Taylor series expansion. The Taylor series expansion of the exponential function is:

$$e^z = \sum_{i=0}^{\infty} \frac{z^i}{i!} = 1 + \frac{z}{1!} + \frac{z^2}{2!} + \frac{z^3}{3!} + \cdots \tag{2.23}$$

When applied to base 10, this yields:

$$10^z = 1 + \frac{z \cdot \ln(10)}{1!} + \frac{[z \cdot \ln(10)]^2}{2!} + \frac{[z \cdot \ln(10)]^3}{3!} + \cdots \tag{2.24}$$

In the limit as $z \to 0$, the higher-order terms become negligible and equation 2.24 reduces to:

$$\lim_{z \to 0} 10^z \approx 1 + z \cdot \ln(10) = 1 + 2.303 \cdot z \tag{2.25}$$

For our application, the exponent of our function is $z = -\varepsilon \cdot [T] \cdot \ell$. Substitution of this identity back into equation 2.17 and rearrangement yields a governing equation for the case of a transparent solution:

$$r_B = -\frac{d[B]}{dt} = \frac{2.303 \cdot E_i' \cdot \varepsilon \cdot [B]}{U} \cdot \Phi \tag{2.26}$$

Notice that equation 2.26 indicates that for the condition $\varepsilon \cdot [B] \cdot \ell \to 0$, the rate of a pure photochemical reaction is first-order with respect to the reactant concentration. As with the first limiting condition, we can rearrange this equation by dividing both sides of the equation by the incident fluence rate:

$$r_B' = -\frac{d[B]}{E_i' \cdot dt} = \frac{2.303 \cdot \varepsilon \cdot [B]}{U} \cdot \Phi \tag{2.27}$$

Again we see that reaction progress can be defined as a function of the applied dose. Separation of variables and integration allows explicit definition of the dose-dependent concentration of the target compound:

$$\int \frac{d[B]}{[B]} = \frac{2.303 \cdot \varepsilon \cdot}{U} \cdot \Phi \cdot \int E_i' \cdot dt \tag{2.28}$$

Integrating both sides of this equation, we find

$$\ln \frac{[B]}{[B]_0} = -\frac{2.303 \cdot \varepsilon \cdot \Phi \cdot D}{U} \tag{2.29}$$

or

$$[B] = [B]_0 \cdot \exp\left(-\frac{2.303 \cdot \varepsilon \cdot \Phi \cdot D}{U}\right) \tag{2.30}$$

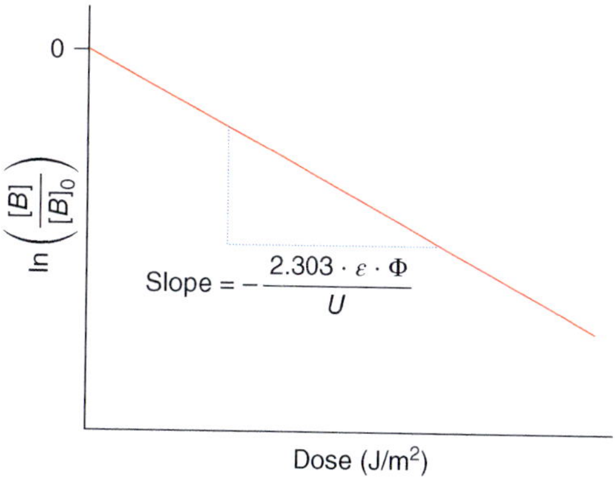


**Figure 2.8** Schematic illustration of the results of an experiment in which a single absorbing, reacting compound is present under circumstances that yield a solution that is nearly transparent (i.e. $\varepsilon \cdot [B] \cdot \ell \lesssim 0.02$) to incident radiation. The slope of the "best-fit" line, forced through the origin is $-\frac{2.303 \cdot \varepsilon \cdot \Phi}{U}$.

Equation 2.29 indicates that a plot of $\ln \frac{[B]}{[B]_0}$ as a function of applied dose ($D$) should yield a straight line through the origin with slope $= -\frac{2.303 \cdot \varepsilon \cdot \Phi}{U}$. Graphically, this is illustrated in Figure 2.8. Because $\varepsilon$ and $U$ can be determined independently, the quantum yield can then be estimated by algebraic manipulation of this slope expression.

$$\Phi = \frac{-\,\text{slope} \cdot U}{2.303 \cdot \varepsilon} \tag{2.31}$$

It is important to recognize that the quantum yield that is being estimated by this approach is the same as the quantum yield that would be estimated from an experiment involving an opaque solution. In other words, if two experiments were conducted to estimate $\Phi$ for a given reaction, with one experiment being conducted using an opaque solution and the second experiment being conducted with a nearly transparent solution, the estimate of $\Phi$ should be identical.

Another important outcome of this analysis is that despite the fact that the same reaction is being promoted under both limiting conditions, the overall reaction order is different for these cases. For the case of an opaque solution, we anticipate a photochemical reaction that is zero-order with respect to the concentration of the target compound (see equation 2.20), regardless of whether we consider time or dose to be the independent variable. On the other hand, for the case of a transparent solution, the reaction is first-order with respect to the concentration of the target compound (see equations 2.26 and 2.27). Again, this behavior is observed regardless of our choice of time or dose as the independent variable.

For the case of an opaque solution, equations 2.19 and 2.20 indicate that for pure photochemical processes, reaction progress can be represented as a function of time or dose, respectively. Time is perhaps the more intuitive (or familiar) independent variable to use for reaction kinetics; the fact that dose can also be used as the independent (master) variable to describe photochemical reaction progress is a consequence of the Stark–Einstein law and the Roscoe–Bunsen law. Similarly, for the case of transparent solutions, equations 2.26 and 2.27 also indicate that time and dose can be used as the independent variable. In fact, time and dose may be applied as the independent variable for any direct photochemical process, regardless of the optical condition of the system.

Another important implication of the case of a transparent solution emerges from an analysis of the geometric constraints of the system illustrated in Figure 2.6. Mathematically, this limiting condition will apply for any circumstance in which $\varepsilon \cdot [B] \cdot \ell \to 0$. An example of such a system is one in which the surface area of the reactor becomes infinitesimally small (i.e. $A \to 0$) and the depth of the reactor also becomes infinitesimally small (i.e. $\ell \to 0$). In geometric terms, this corresponds to a point in space. Therefore, equations 2.26 and 2.27 apply rigorously to a point in space. This implies that the *local* rate of a photochemical reaction is directly proportional to the *local* fluence rate. As we will see in Chapter 3, fluence rates within photochemical reactor systems often demonstrate steep spatial gradients. As such, these systems are characterized by equally steep gradients in local reaction rates.

The case of a nearly transparent solution also allows examination of the factors that affect (local) photochemical reaction rates. Equation 2.26 can be rearranged to yield a form that is consistent with the expression that is used to describe the kinetics of an elementary, bimolecular reaction (see equations 2.5 and 2.6).

$$r_B = -\frac{\mathrm{d}[B]}{\mathrm{d}t} = \frac{2.303 \cdot \varepsilon \cdot \Phi}{U} \cdot E_i' \cdot [B] \tag{2.32}$$

The quotient on the right side of this equation can be viewed as a rate constant ($k$):

$$r_B = -\frac{\mathrm{d}[B]}{\mathrm{d}t} = k \cdot E_i' \cdot [B] \tag{2.33}$$

$$k = \frac{2.303 \cdot \varepsilon \cdot \Phi}{U} \tag{2.34}$$

As with elementary, bimolecular reactions, the rate constant can be decomposed to identify the processes that affect reaction kinetics. In the case of these thermal reactions, we saw previously that the rate constant accounts for the fraction of collisions that have enough energy to overcome the thermodynamic barrier to reaction progress (activation energy), and chemical factors (e.g. orientation, steric factors, etc.) that could also limit reaction progress.

In the case of a photochemical reaction, the availability of the target molecule is represented conventionally as $[B]$, while the availability of photons is described by $E_i'$. The rate constant is decomposed to account for factors that affect reaction success. Specifically, the molar absorption coefficient ($\varepsilon$) describes the efficiency by which the target molecule absorbs incident photons, while the quantum yield ($\Phi$) describes the efficiency by which absorbed photons promote reaction.

#### 2.4.2.3 Kinetics of Photochemical Reactions: Photon Basis

The second law of photochemistry established an equivalence between molecules and photons. They both represent the smallest, discrete unit of their respective entities (chemical reactant and electromagnetic radiation, respectively); they cannot be divided into smaller entities than these units and retain their physical or chemical characteristics. In the context of a photochemical reaction, they are both modeled as "reactants" in that the intrinsic rate of a photochemical process is described by the activity (~molar concentration) of the molecule and the fluence rate of incident photons.

Given this equivalence, descriptions of photochemical kinetics based on direct accounting of photon availability may represent a more rational method than the methods described above, which are based on time as the master variable. As we will see in the material to follow, electromagnetic energy-based and photon-based descriptions of photochemical reactions are nearly identical for reactions that are initiated by monochromatic sources of radiation. The decision to apply energy-based or photon-based kinetic expressions is moot for monochromatic sources of radiation, as these two approaches provide equivalent information for monochromatic systems. On the other hand, for polychromatic sources, the photon-based approach provides a clearer, perhaps more intuitive description of photochemical reaction kinetics than one based on electromagnetic energy.

Many terms are available to describe the availability of photons to participate in a process. Before proceeding further, several of the most common terms will be presented and compared. Figure 2.9 illustrates the general concepts of irradiance and fluence rate. Irradiance is defined as the "radiant power, $P$, of all wavelengths incident from all upward directions on a small element of surface containing the point under consideration divided by the area of the element (dA)," while fluence rate is defined as the "total radiant power, P, incident from all directions onto a small sphere divided by the cross-sectional area of that sphere (dA)." [2] For a collimated beam of radiation, these two terms are identical; however, for other reactor systems, these two terms may have different physical interpretations. Fluence rate is the more general term and it will be applied in most applications in this text.

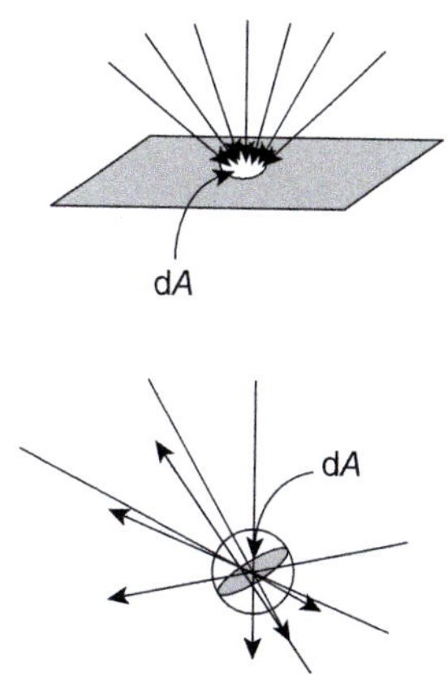


**Figure 2.9** Schematic illustration of the formal definitions of irradiance (top) and fluence rate (bottom). *Source:* [17], Bolton and Stefan 2002 / with permission of Springer Nature.

Table 2.2 provides a summary of common terms used to describe the availability of photons at a given location. These definitions are based on electromagnetic energy or specifically on photons.

We have seen previously that photochemical reactions are initiated by absorption of photon energy by a molecule. The efficiency by which absorbed photon energy is used to initiate change in the target molecule

**Table 2.2** Summary of common terms used to describe photon availability. The superscript prime (′) indicates fluence basis; the subscript $p$ indicates a photon basis, while the subscript $\lambda$ indicates a spectral property (i.e. wavelength-dependence). The upper limit of integration ($\tau$) on the fluence (dose) definitions represents the period of exposure.

| Term | Symbol | Units |
|---|---|---|
| Irradiance | $E$ | $\frac{\mathrm{W}}{\mathrm{m}^2}$ |
| Fluence rate | $E'$ | $\frac{\mathrm{W}}{\mathrm{m}^2}$ |
| Photon irradiance | $E_p$ | $\frac{\mathrm{einstein}}{\mathrm{m}^2 \cdot \mathrm{s}}$ |
| Photon fluence rate | $E'_p$ | $\frac{\mathrm{einstein}}{\mathrm{m}^2 \cdot \mathrm{s}}$ |
| (Dose) fluence | $(D) = H = \int_0^{\tau} E(t) \cdot \mathrm{d}t$ | $\frac{\mathrm{J}}{\mathrm{m}^2}$ |
| Photon (Dose) fluence | $(D_p) = H_p = \int_0^{\tau} E(t) \cdot \mathrm{d}t$ | $\frac{\mathrm{J}}{\mathrm{m}^2}$ |
| Spectral irradiance | $E_\lambda$ | $\frac{\mathrm{W}}{\mathrm{m}^2 \cdot \mathrm{nm}}$ |
| Spectral fluence rate | $E'_\lambda$ | $\frac{\mathrm{W}}{\mathrm{m}^2 \cdot \mathrm{nm}}$ |
| Spectral photon irradiance | $E_{p,\lambda}$ | $\frac{\mathrm{einstein}}{\mathrm{m}^2 \cdot \mathrm{nm} \cdot \mathrm{s}}$ |
| Spectral photon fluence Rate | $E'_{p,\lambda}$ | $\frac{\mathrm{einstein}}{\mathrm{m}^2 \cdot \mathrm{nm} \cdot \mathrm{s}}$ |

is described by the quantum yield. When we consider a primary photochemical process (i.e. an elementary photochemical reaction), the upper limit on the quantum yield is:

$$\Phi_{\max} \equiv 1\frac{\text{molecule}}{\text{photon}} = 1\frac{\text{mole}}{\text{einstein}} \tag{2.35}$$

However, for most reactions, the primary quantum yield is less than $1\frac{\text{mole}}{\text{einstein}}$.

With this in mind, it is then possible to develop a mathematical expression for the rate of a primary photochemical process on the basis of photons. To do this, we will again consider the system illustrated in Figure 2.6, where a solution containing a single absorbing molecule will undergo a photochemical reaction. However, in this case, we will describe photon availability using the photon fluence rate ($E'_p$). Equation 2.13 still applies for this approach; for clarity, it is repeated here:

$$r_B = -\frac{d[B]}{dt} = I_a \cdot \Phi \tag{2.13}$$

In the previous development, we quantified the volumetric rate of photon absorption ($I_a$) on the basis of incident and transmitted fluence rate $\left(\frac{\text{W}}{\text{m}^2}\right)$. For this development, we will use incident and transmitted photon fluence rate to quantity the volumetric rate of photon absorption:

$$I_a = \frac{A \cdot \left(E'_{p,i} - E'_{p,t}\right)}{\forall} = \frac{\left(E'_{p,i} - E'_{p,t}\right)}{\ell} \tag{2.36}$$

where

$E'_{p,i}$ = incident photon fluence rate $[=]\frac{\text{einstein}}{\text{m}^2 \cdot \text{s}}$

$E'_{p,t}$ = transmitted photon fluence rate $[=]\frac{\text{einstein}}{\text{m}^2 \cdot \text{s}}$.

Given these definitions, we can now proceed to develop kinetics expressions to address the same two special cases we considered previously.

#### 2.4.2.4 Limiting Case 1: Opaque Solution

For an opaque solution, all photons imposed on the reactor will be absorbed by the target compound, B. Therefore, $E'_{p,t} = 0$. Substitution of this information into equation 2.36, with subsequent substitution into equation 2.13 allows definition of a governing equation to describe reaction kinetics for this system:

$$r_B = -\frac{d[B]}{dt} = \frac{E'_{p,i}}{\ell} \cdot \Phi \tag{2.37}$$

Not surprisingly, this equation is very similar to equation 2.19, except that no conversion from energy to photons is included in the expression. Like equation 2.19, equation 2.37 indicates that the rate of decay of the photoreactive compound is zero-order for the situation of an opaque solution (when opacity is due to photon absorption by the target molecule *only*). Therefore, an experiment could be conducted wherein the concentration of $B$ is measured as a function of time (or the applied photon fluence). A plot of reactant concentration as a function of applied photon fluence should yield a straight line with slope $-\frac{\Phi}{\ell}$.

#### 2.4.2.5 Limiting Case 2: Transparent Solution

For this situation, we must again apply the Beer–Lambert law for $\varepsilon \cdot [B] \cdot \ell \rightarrow 0$. Analysis of the behavior of the Beer–Lambert law in this limit is identical to that presented previously (equations 2.22–2.26)) and will not be repeated here. When applied to equation 2.13, this yields:

$$r_B = -\frac{\mathrm{d}[B]}{\mathrm{d}t} = 2.303 \cdot \varepsilon \cdot \Phi \cdot E'_{p,i} \cdot [B] \tag{2.38}$$

Mathematically, this is very similar to equation 2.32 except that the photon energy is now accounted for directly. The use of this expression for purposes of estimating a quantum yield would be similar to that described in Figure 2.8, except that the graphical estimation of quantum yield would be conducted by finding a best-fit line through the origin of a plot of $ln\left(\frac{[B]}{[B]_0}\right)$ vs. photon fluence ($\frac{\text{einstein}}{\text{m}^2}$). Such a plot would be identical to Figure 2.8, except for a change of scale on the horizontal axis from fluence ($\frac{\text{J}}{\text{m}^2}$) to photon fluence ($\frac{\text{einstein}}{\text{m}^2}$), and the slope of the curve from this experiment would be $-2.303 \cdot \varepsilon \cdot \Phi$.

## 2.5 Summary of Expressions to Describe Photochemical Kinetics: Monochromatic Radiation Sources

We have now considered several methods for describing the kinetics of a photochemical reaction in a simple collimated beam experiment with a single absorbing compound and a monochromatic radiation source. This represents one of the most basic applications of photochemistry, but the governing equations provide a foundation for examination of photochemical reaction progress in other, more complex systems. Table 2.3 provides a summary of these expressions, along with the basic expression used to describe the kinetics of an elementary, bimolecular thermal chemical reaction.

### 2.5.1 Kinetics of Photochemical Reactions: Polychromatic Radiation Source, Photon Basis

Practical photoreactors often employ polychromatic radiation sources. A common example is a reactor based on a medium-pressure (MP) mercury (Hg) lamp. These devices emit radiation across a broad spectrum that includes UV, visible, and infrared (IR) radiation. Figure 2.10 illustrates a typical (normalized) output spectrum for a MP Hg lamp.

The spectrum presented in Figure 2.10 illustrates relative spectral irradiance (dimensionless) as a function of wavelength. Photochemical reactions often demonstrate wavelength-dependence. As such, simulations of the behavior of these systems requires that wavelength-dependent attributes of the system be accounted for, including the spectral characteristics of the source of radiation, as well as those of the physico/chemical properties of the reactants.

Table 2.2 includes several parameters that are commonly applied to define the spectral characteristics of polychromatic radiation sources, including spectral irradiance ($E_\lambda$), spectral fluence rate ($E'_\lambda$), spectral photon irradiance ($E_{p,\lambda}$), and spectral photon fluence rate ($E'_{p,\lambda}$). For each of these variables, the units are defined on a per-wavelength basis.

To quantify the effects of polychromatic radiation on photochemical reaction rates, we will again apply the principle of photon–molecule equivalence that is taught by the second law of

**Table 2.3** Summary of kinetic expressions for elementary thermal chemical and photochemical reactions.

| Reaction type | Simple reaction expression | Expanded reaction expression | Kinetic expression | Comments |
|---|---|---|---|---|
| Thermal chemical | $B + C \rightarrow$ products | $B + C \rightarrow BC^* \rightarrow$ products | Basic form: $\frac{\mathrm{d}[B]}{\mathrm{d}t} = -k \cdot [B] \cdot [C]$<br>Expanded form: $\frac{\mathrm{d}[B]}{\mathrm{d}t} = -A \cdot \exp\left(-\frac{E_a}{R \cdot T}\right) \cdot [B] \cdot [C]$ | Activity represents reactant availability to participate in reaction<br><br>Exponential term describes the fraction of collisions between $B$ and $C$ that can overcome activation energy ($E_a$)<br><br>$A \cdot [B] \cdot [C]$ describes the rate of collision between reactants $B$ and $C$ |
| Photochemical (electromagnetic energy basis) | $B + h\nu \rightarrow$ products | $B + h\nu \rightarrow B^* \rightarrow$ products | $\frac{\mathrm{d}[B]}{\mathrm{d}t} = -\frac{E_i \cdot \Phi}{\ell \cdot Q_e}$<br>$\frac{\mathrm{d}[B]}{\mathrm{d}t} = -\frac{2.303 \cdot E_i \cdot \varepsilon \cdot [B] \cdot \Phi}{Q_e}$ | Opaque solution containing a single compound (B) that absorbs incident, monochromatic radiation<br><br>Solution containing a single compound at a concentration such that $\varepsilon \cdot [B] \cdot \ell \rightarrow 0$ subjected to monochromatic radiation<br><br>$\varepsilon \cdot [B]$ describes the efficiency of photon absorption by the target molecule, B<br><br>$\Phi$ describes the efficiency by which absorbed photons result in transformation of the target molecule, B |
| Photochemical (photon basis) | $B + h\nu \rightarrow$ products | $B + h\nu \rightarrow B^* \rightarrow$ products | $\frac{\mathrm{d}[B]}{\mathrm{d}t} = -\frac{E_i \cdot \Phi}{\ell}$<br>$\frac{\mathrm{d}[B]}{\mathrm{d}t} = -2.303 \cdot E_i \cdot \varepsilon \cdot [B] \cdot \Phi$ | See comments in two boxes above |

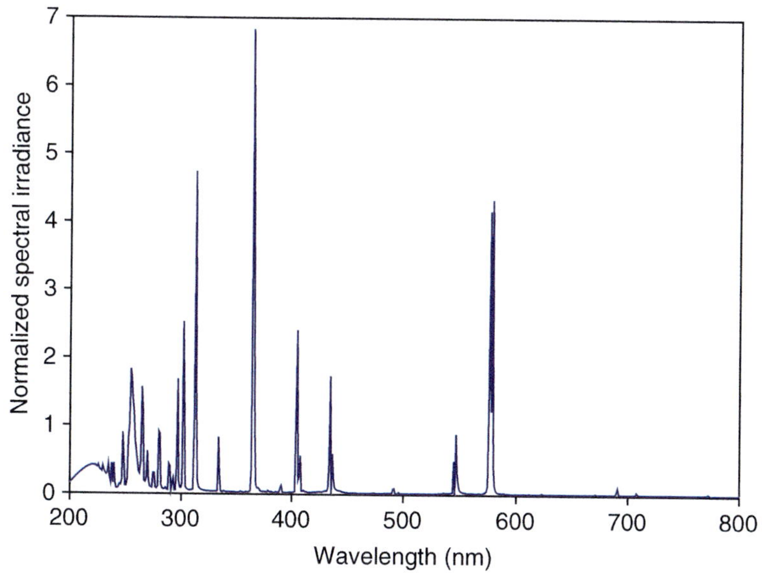


**Figure 2.10** Normalized output spectrum for a typical medium-pressure Hg lamp. Spectral irradiance at $\lambda$ = 254 nm was used as the basis for normalization. *Source:* Data from Trojan Technologies.

photochemistry. With that in mind, it is perhaps most intuitive to examine photochemical kinetics using the spectral photon fluence rate $\left(E'_{p,\lambda}\right)$ as the variable of choice to quantify photon availability (units of $\frac{\text{einstein}}{\text{m}^2 \cdot \text{nm} \cdot \text{s}}$).

An important corollary of the second law of photochemistry is that all photochemical events are independent and therefore their effects are additive [18]. Mathematically, this allows extension of the expressions presented above to describe the kinetics of a photochemical process based on polychromatic radiation.

We will again consider kinetics of a system for which a beam of collimated, uniform radiation is imposed on the flat surface of a shallow, well-mixed reactor containing a single absorbing/reacting compound, as illustrated in Figure 2.6. For the case of an opaque solution, equation 2.37 can be modified to account for spectral attributes of the system as follows:

$$r_B = -\frac{\mathrm{d}[B]}{\mathrm{d}t} = \int_{\lambda_1}^{\lambda_2} \frac{E'_{p,\lambda} \cdot \Phi(\lambda)}{\ell} \cdot \mathrm{d}\lambda \tag{2.39}$$

Equation 2.39 is a direct translation of equation 2.37, wherein the wavelength-dependent (spectral) characteristics of the radiation source and reactant have been accounted for by the terms $E'_{p,\lambda}$ and $\Phi(\lambda)$, respectively. The effects of exposure to a polychromatic source are accounted for by adding the contributions of all wavelengths that promote the reaction. In equation 2.39, this is defined by the wavelength range: $\lambda_1 \leq \lambda \leq \lambda_2$, which represents the limits of integration for the function. The quantum yield, $\Phi(\lambda)$, has been modified with a parenthetical term to indicate wavelength-dependence.

For the case of a transparent solution, equation 2.38 can be translated in a similar manner to account for polychromatic behavior:

$$r_B = -\frac{d[B]}{dt} = \int_{\lambda_1}^{\lambda_2} 2.303 \cdot \varepsilon(\lambda) \cdot \Phi(\lambda) \cdot E'_{p,\lambda} \cdot [B] \cdot d\lambda \tag{2.40}$$

Equations 2.39 and 2.40 require that information be available to describe the spectral characteristics of the imposed radiation ($E'_{p,\lambda}$), the spectral characteristics of photon absorption by the target molecule ($\varepsilon(\lambda)$), and the wavelength-dependence of the quantum yield ($\Phi(\lambda)$).

Implicit in the application of all equations presented herein to describe photochemical kinetics is an understanding that the variables used to describe these terms are dimensionally consistent, but may require unit conversions to develop meaningful descriptions of photochemical dynamics.

## 2.6 Summary

Photochemical reactions require that photons with sufficient energy be absorbed by a target molecule. If this condition is met, then it is possible (though not certain) that a photochemical reaction can take place.

The rate of a photochemical reaction is governed by three basic phenomena:

- Delivery of a photon to the target molecule: Our ability to deliver photons to target molecules is influenced by many factors in a system, including the power of the radiation source, the presence of other constituents in a system that could "compete" for the photons of interest by absorption, and the geometry of the system that is subjected to irradiation. In this chapter, we have considered only a simple system in which these attributes were either not present, or were accounted for in simple terms because of the system definition. For some photochemical reactors, these terms can complicate the analysis considerably.
- Absorption of a photon by a target molecule: This is a necessary condition for a photochemical reaction to take place. The ability of a molecule to absorb radiation at a given wavelength is quantified by its molar absorption coefficient, $\varepsilon(\lambda)$, which is governed by the structure of the molecule and the nature of the bonds from which electrons may be promoted through the act of absorbing photon energy.
- Use of the absorbed photon energy to promote a reaction: As indicated by the Jablonski diagram, there are many possible fates for the photon energy absorbed by a molecule; among these are chemical reactions. Because of this, and because of the discrete nature of photons (recall that photon absorbance is an "all-or-nothing" proposition), a fraction of the absorbed photons will lead to a photochemical event. We quantify this fraction using the quantum yield ($\Phi$), subject to the constraints that $0 \leq \Phi \leq 1 \frac{\text{molecule}}{\text{photon}}$.

These attributes of photochemistry are in some respects obvious conclusions of the fundamental laws of photochemistry. However, in many of the systems that we will consider in the remainder of this book, the implications of these steps are nonobvious and may require fairly complex mathematical descriptions.

## Notes

**a** Like many "laws," there are exceptions to the first law of photochemistry. An important example is the process known as photosensitization, in which a molecule in the immediate vicinity of the target molecule will absorb a photon, then transfer the absorbed energy to the target molecule by a subsequent collision.

**b** Here again, exceptions to this "law" have been observed. For example, some sources of electromagnetic radiation, including powerful lasers, can impose photons on a target at sufficiently fast rate to allow so-called "multi-photon" processes to occur. In these processes, more than one photon may be absorbed by a target molecule before the energy associated with the first absorbed photon(s) is dissipated through a photochemical or photophysical process. These intense sources of radiation are unusual; in most circumstances, the Stark–Einstein law will apply.

**c** From this point forward, the term "molecule" will be used to refer to a reacting chemical entity, unless a related term, such as ion, atom, radical, or complex is needed for clarity.

**d** The term "einstein" is widely used to represent Avogadro's number (i.e. one mole) of photons. However, the IUPAC has recently declared the term to be obsolete.

**e** The symbol [=] is to be interpreted as "has units of."

**f** The convention to be followed in this book is to use symbols and terms defined by the International Union of Pure and Applied Chemists (IUPAC): Braslavsky [2]. In some cases, this will involve use of the same symbol or similar symbols to represent different variables. When this happens, modifications to standard nomenclature or symbols may be used to promote clarity.

**g** Unit conversions will not always be included in equations, but will be required in calculations for dimensional consistency. Variables will be expressed in the units that are most commonly applied for their presentation. Therefore, in the application many equations in this book it will be necessary to apply unit conversions to achieve dimensional consistency. In the case of equation 2.1, unit conversions will be needed to account for expressions of length in units of m and nm.

**h** The kinetics of the reverse process may be examined by applying similar logic. For the reaction illustrated by equation 2.4, "products" would become the "reactant(s)" of the reverse reaction. A rate constant distinct from the forward rate constant would be used to describe the rate of the reverse reaction.

**i** The parenthetical term $(\lambda)$ is included here to indicate wavelength-dependence. $(\lambda)$ will be applied to variables for which wavelength-dependence is relevant; it may be omitted from some equations to reduce clutter and promote clarity.

**j** With the recent declaration of "einstein" as an obsolete term, the units $\frac{\text{molecules}}{\text{photon}}$ may be the most appropriate to associate with a quantum yield, but the units $\frac{\text{moles}}{\text{einstein}}$ are common in the literature.

**k** IUPAC convention calls for the quantum yield to be expressed as a dimensionless quantity. Units are included herein to reinforce the notion that the quantum yield is a ratio of molecules to photons (or moles to einsteins).

**l** Chain reactions are often involved in photochemical processes that are described by $\Phi(\lambda) > 1$. In such a process, a primary photochemical event can yield one or more reactive intermediates (often radicals) which can go on to participate in a chain reaction.

**m** The physical interpretation of the well-mixed assumption will be addressed in detail in Chapter 3, along with a discussion of the implications and limitations of this assumption.

**n** The term "optical density" is obsolete and its use is discouraged.

## References


1 Albini, A. (**2016**). *Photochemistry: Past, Present and Future*. Heidelberg: Springer-Verlag.

2 Braslavsky, S.E. (**2007**). Glossary of terms used in photochemistry 3(rd) edition (IUPAC Recommendations 2006). *Pure and Applied Chemistry 79* (3): 293–465.

3 Finlayson-Pitts, B.J., Pitts, J., and James, N. (**2000**). *Chemistry of the Upper and Lower Atmosphere: Theory, Experiments, and Applications*, 969. San Diego, CA: Academic Press.

4 Benjamin, M.M. (**2002**). *Water Chemistry*, 1e. Boston: McGraw-Hill.

5 Stumm, W. and Morgan, J.J. (**1995**). *Aquatic Chemistry: Chemical Equilibria and Rates in Natural Waters*. New York: Wiley.

6 Benitez, F.J., Acero, J.L., Garcia-Reyes, J.F. et al. (**2013**). Determination of the reaction rate constants and decomposition mechanisms of ozone with two model emerging contaminants: DEET and nortriptyline. *Industrial & Engineering Chemistry Research 52* (48): 17064–17073.

7 Broseus, R., Vincent, S., Aboulfadl, K. et al. (**2009**). Ozone oxidation of pharmaceuticals, endocrine disruptors and pesticides during drinking water treatment. *Water Research 43* (18): 4707–4717.

8 Kong, X.J., Jiang, J., Ma, J. et al. (**2016**). Degradation of atrazine by UV/chlorine: efficiency, influencing factors, and products. *Water Research 90*: 15–23.

9 Sun, P.Z., Lee, W.N., Zhang, R.C., and Huang, C.H. (**2016**). Degradation of DEET and caffeine under UV/chlorine and simulated sunlight/chlorine conditions. *Environmental Science & Technology 50* (24): 13265–13273.

10 Weng, S.C., Sun, P.Z., Ben, W.W. et al. (**2014**). The presence of pharmaceuticals and personal care products in swimming pools. *Environmental Science & Technology Letters 1* (12): 495–498.

11 Wols, B.A. and Hofman-Caris, C.H.M. (**2012**). Review of photochemical reaction constants of organic micropollutants required for UV advanced oxidation processes in water. *Water Research 46* (9): 2815–2827.

12 Lee, Y. and von Gunten, U. (**2010**). Oxidative transformation of micropollutants during municipal wastewater treatment: comparison of kinetic aspects of selective (chlorine, chlorine dioxide, ferrate(VI), and ozone) and non-selective oxidants (hydroxyl radical). *Water Research 44* (2): 555–566.

13 Deborde, M. and von Gunten, U. (**2008**). Reactions of chlorine with inorganic and organic compounds during water treatment - kinetics and mechanisms: A critical review. *Water Research 42* (1–2): 13–51.

14 Turro, N.J., Ramamurthy, V., and Scaiano, J.C. (**2009**). *Principles of Molecular Photochemistry: An Introduction*. Sausalito, CA: University Science Books.

15 Xu, H., Chen, R.F., Sun, Q. et al. (**2014**). Recent progress in metal-organic complexes for optoelectronic applications. *Chemical Society Reviews 43* (10): 3259–3302.

16 Swinehart, D.F. (**1962**). The Beer-Lambert law. *Journal of Chemical Education 39* (7): 333–335.

17 Bolton, J.R. and Stefan, M.I. (**2002**). Fundamental photochemical approach to the concepts of fluence (UV dose) and electrical energy efficiency in photochemical degradation reactions. *Research on Chemical Intermediates 28* (7–9): 857–870.

18 Bolton, J.R., Mayor-Smith, I., and Linden, K.G. (**2015**). Rethinking the concepts of fluence (UV Dose) and fluence rate: the importance of photon-based units - a systemic review. *Photochemistry and Photobiology 91* (6): 1252–1262.

# 3

# Photochemical Reactor Theory

## 3.1 Introduction

The most common contemporary applications of photochemical reactors involve processes to alter (improve) water and air composition. Like all reactor systems, the behavior of these reactors is governed by the fundamental principles of reactor theory. However, photochemical reactors differ from most conventional chemical reactors in several ways. First, the intrinsic rates of photochemical reactions tend to be fast, relative to many chemical reactions. Second, photochemical reactor systems are often characterized by remarkably strong gradients in fluence rate, which translate into equally strong gradients in local reaction rate. Third, photochemical reactors are characterized by short residence times and small volumes, as compared to their chemical (or biochemical) analogs. These differences among chemical reactors[a] and photochemical reactors can have profound implications in the development and application of mathematical models that are used to describe their behavior.

Models that are used to describe the behavior of reactor systems are based on the principles of mass (material), heat, and momentum balances. These balance principles have remarkably broad applications. As illustrated in the classic text by Bird, Stewart, and Lightfoot [1] and by many other authors, they also demonstrate profound parallelism in their logic and mathematical development.

Important similarities and differences exist between chemical and photochemical reactor models. Among practicing engineers, the models that are used to describe chemical reactors tend to be more familiar than those that are used to describe photochemical reactors. In addition, photochemical reactions are often used to initiate chemical reactions in reactor systems. The objective of this chapter is to present basic principles of photochemical reactor theory. This information will serve as a foundation for all subsequent chapters that address methods and applications of photochemical reactors.

This chapter will begin with brief coverage of fundamental principles and applications of material (molar or mass) balances. This will be followed by a presentation of some of the basic models that are used to describe the dynamic behavior of chemical reactor systems. The remainder of the chapter will provide a detailed presentation of photochemical reactor theory, with illustrations of applications to a few common reactor types. As appropriate, similarities and differences between chemical and photochemical reactor models will be pointed out. Lastly, an interactive model is presented for the reader to apply; this model includes animation that responds to user input and is intended to link a visual representation of reactor behavior with the mathematical expressions that are developed to describe the dynamics of photochemical reactor systems.

*Photochemical Reactors: Theory, Methods, and Applications of Ultraviolet Radiation*, First Edition. Ernest R. Blatchley III.

Companion website: www.wiley.com/go/blatchley/photochemicalreactors

## 3.2 Basic Principles of Material (Molar or Mass) Balance

Material balances represent a family of tools that may be thought of as analogs to those used for management of bank accounts. In the latter, an account holder may want to keep track of the money that is available in their account. This will be influenced by processes by which money enters the account (income), those responsible for the rate at which money leaves the account (expenses), and processes that take place within the account that also influence the balance of the account (e.g. interest or penalties). Any approach used to manage money in an account will need to include methods to simulate these processes.

The principles of material balances are illustrated in schematic form in Figure 3.1. In this illustration, we consider the fate of a single constituent in the region defined by the light blue color. We refer to this as the "control region" (CR) for our model. The CR may be thought of as an analog to the bank account, in that we are interested in simulating the dynamic behavior of a conservative entity (money in the case of a bank account; amount of material in a material balance) in this region. For purposes of this illustration, we will consider the boundaries of our CR to be fixed.[b]

The CR illustrated in Figure 3.1 is three-dimensional, as is the case with many material balances. However, it is possible to develop material balances based on two-dimensional (surfaces), one-dimensional (lines), or zero-dimensional (points) objects. In some circumstances, there may be reasons to develop a material balance based on a CR of *fractal* dimension. Regardless of the dimension of the CR, the principles of material balance are the same. Namely, the objective is to develop and apply mathematical descriptions of all phenomena that are responsible for input to the CR, output from the CR, and "reaction" processes that take place within the CR that affect the amount of the conservative entity of interest in the CR.

Material balances are based on the principle that matter (mass) is conservative. In other words, matter cannot be created or destroyed, but may be changed in form. When applied to a CR, this implies that we need to develop and apply mathematical expressions to account for material conservation.

A basic form of the material balance equation is presented in Equation 3.1 in word form, where the subscript $j$ is used to indicate that the expression applies for an individual constituent. If multiple constituents are to be examined in a system, then separate material balance equations will be needed for each constituent.

$$\begin{pmatrix}\text{time rate of change}\\\text{of constituent moles}\\\text{within CR}\end{pmatrix}_j = \begin{pmatrix}\text{molar input}\\\text{rate to CR}\end{pmatrix}_j - \begin{pmatrix}\text{molar output}\\\text{rate from CR}\end{pmatrix}_j \pm \begin{pmatrix}\text{effects of}\\\text{reactions}\\\text{within CR}\end{pmatrix}_j \tag{3.1}$$

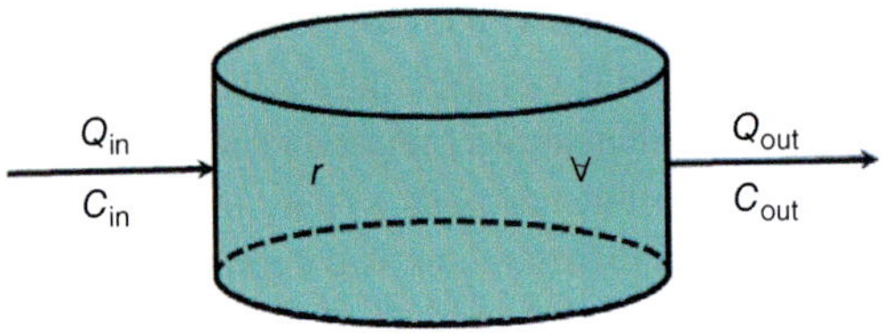


**Figure 3.1** Schematic illustration of material balance applied to an arbitrary region. The light blue area in this schematic illustrates the CR. Note that a CR can have essentially any shape and is not limited to three-dimensional objects.

The left side of equation 3.1 describes the dynamic behavior of the number of moles of the constituent of interest ($j$) within the CR.[c] The right side of the equation includes terms to describe the molar rate at which processes bring $j$ across the CR boundary into the CR (input rate), the molar rate at which processes take $j$ across the CR boundary to remove it from the CR (output rate), and the molar rates of processes (e.g. reactions) that take place within the CR that affect the amount of $j$ within the CR. Successful application of a material balance to describe the dynamic behavior of a system requires that appropriate mathematical expressions be defined for each of the terms in equation 3.1.

A general mathematical form of equation 3.1 can be presented as follows:

$$\frac{\mathrm{d}M_j}{\mathrm{d}t} = \Phi_{i,j} - \Phi_{o,j} \pm \langle r_j \rangle \cdot \forall \tag{3.2}$$

where

| | | | | |
|---|---|---|---|---|
| $M_j$ | = | amount[d] of constituent $j$ within CR | [=] | mole |
| $t$ | = | time | [=] | time |
| $\Phi_{i,j}$ | = | molar input rate of constituent $j$ to CR | [=] | $\frac{\text{mole}}{\text{time}}$ |
| $\Phi_{o,j}$ | = | molar output rate of constituent $j$ from CR | [=] | $\frac{\text{mole}}{\text{time}}$ |
| $\langle r_j \rangle$ | = | spatially averaged reaction rate(s)$_j$ within CR | [=] | $\frac{\text{mole}}{\text{volume} \cdot \text{time}}$ |
| $\forall$ | = | volume of CR | [=] | volume. |

The reaction term, $\langle r_j \rangle$, describes the kinetics of any process(es) that take place within the CR that produce (+) or consume (−) the constituent of interest within the CR, thereby affecting the number of moles of the constituent within the CR; these often take the form of reaction processes. Reaction terms are kinetic expressions that define how the concentration of the constituent of interest changes within the CR (see Chapter 2). When the spatial average of this behavior is properly accounted for, this term will describe the effects of reactions on the concentration of the constituent of interest within the entire CR (in units of $\frac{\text{moles}}{\text{volume}\cdot\text{time}}$ or $\frac{\text{concentration}}{\text{time}}$). We must therefore multiply these reaction rate terms by the volume of the CR to develop the molar-balance equation in consistent units.

A subtle, but important feature of equation 3.2 is the spatial-averaging function that is applied to describe the effects of reactions that take place within the CR, $\langle r_j \rangle$. For a general reactor system, it is common for the kinetics of a given reaction to display spatial variability. When this happens, it is necessary to account for the effects of this spatial variability in describing the behavior of the CR. As we will see later in this chapter, photochemical reactors are often characterized by strong gradients in local reaction rate. Because of this, it is necessary to develop mathematical approaches that accurately account for this spatial variability. We will return to this issue later in the chapter.

For the schematic illustration provided in Figure 3.1, equation 3.2 may be translated as follows:[e]

$$\frac{\mathrm{d}M}{\mathrm{d}t} = Q_{\text{in}} \cdot C_{\text{in}} - Q_{\text{out}} \cdot C_{\text{out}} \pm \langle r \rangle \cdot \forall \tag{3.3}$$

where

$$Q_{in} = \text{volumetric flow rate of fluid entering CR} \quad [=] \quad \frac{\text{volume}}{\text{time}}$$

$$C_{in} = \text{concentration of constituent in fluid entering CR} \quad [=] \quad \frac{\text{mole}}{\text{volume}}$$

$$Q_{out} = \text{volumetric flow rate of fluid leaving CR} \quad [=] \quad \frac{\text{volume}}{\text{time}}$$

$$C_{out} = \text{concentration of constituent in fluid leaving CR} \quad [=] \quad \frac{\text{mole}}{\text{volume}}.$$

It is important to note that all terms in this equation must be expressed in consistent units. In fact, most mistakes that are made in developing material balances are related to this issue. For this application, each term (product) in equation 3.3 will have units of $\frac{\text{moles}}{\text{time}}$.

The first term on the right side of this equation ($Q_{in} \cdot C_{in}$) describes the molar rate at which the constituent of interest enters the CR by *advection* (movement associated with bulk fluid motion, sometimes referred to as *convection*). Similarly, the second term ($Q_{out} \cdot C_{out}$) describes the molar rate at which the constituent of interest leaves the CR by advection. The last term on the right side describes the effects of reactions that take place within the CR on the number of moles of constituent within the CR. Like all processes in a material balance, these transport and reaction processes are assumed to act independently, and therefore their effects are additive.

Equations 3.2 and 3.3 are very general, though remarkably powerful tools for use in describing the behavior of natural and engineered systems. They form the basis of fundamental chemical reactor theory, several basic examples of which are presented in the next section.

## 3.3 Basic Chemical Reactor Models

Reactor models are developed to provide mathematical descriptions of the dynamic behavior of reactor systems. Because they are models, they are by definition simplifications of reality. Model predictions will always be accompanied by error; the magnitude of that error will depend on the degree to which the model accurately reflects the true behavior of the system it is designed to simulate. In many situations, the implicit and explicit assumptions that accompany reactor models will also influence their accuracy.

Chemical reactors demonstrate a wide range of characteristics. Not surprisingly, models of chemical reactor behavior demonstrate a similarly wide range of characteristics and complexity. The goal of this section is to present some of the fundamental reactor models and to discuss some of their implications.

### 3.3.1 Batch Reactor Model

Figure 3.2 provides a schematic illustration of a batch reactor. A comparison of Figures 3.1 and 3.2 reveals fundamental differences between these schematic illustrations. Specifically, the batch reactor does not include any input or output, which means that there is no material transport across the boundary of the CR. The batch reactor schematic also includes an impeller icon, which implies that the region in question (the CR, which in this case corresponds to the reactor volume) is well-mixed.

The well-mixed assumption is implicit in the application of the batch reactor model and carries a number of implications that have direct bearing on our analysis of the system as well as the ability of the model to accurately describe the dynamic behavior of the system. Specifically, the well-mixed assumption implies that composition within the region in question is spatially uniform; in other words, the concentration of any constituent in the system is the same at all locations within the system at a given time. If we were able to take a photograph to indicate the composition of the system at a given time, the image would indicate the same composition (concentration, temperature, etc.) at all locations. In turn, this implies that any reaction that takes place in the system proceeds at a spatially uniform rate. In applying the basic material balance model (equation 3.2 or 3.3), this allows us to disregard the spatial-averaging function (i.e. $\langle r \rangle = r$), which greatly simplifies our application of the model.

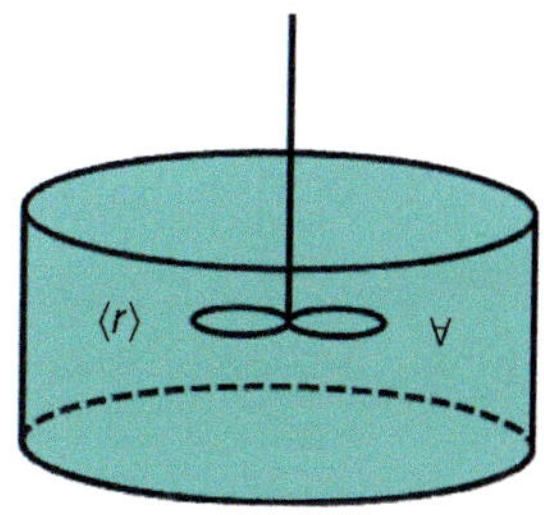


**Figure 3.2** Schematic illustration of (well-mixed) batch reactor. The light blue area in this schematic illustrates the CR.

Based on these characteristics, the general material balance equation can be simplified as follows to describe the dynamic behavior of a batch reactor:

$$\frac{dM}{dt} = r \cdot \forall \tag{3.4}$$

If we assume that the volume ($\forall$) of the reactor system remains constant, then we can divide both sides of equation 3.4 by $\forall$. Recognizing that $C = \frac{M}{\forall}$, which again is assumed to be spatially uniform in this system, we get:

$$\frac{dM}{\forall \cdot dt} = \frac{dC}{dt} = r \tag{3.5}$$

Equation 3.5 indicates that the time rate-of-change of the constituent concentration in a batch reactor is governed exclusively by the rate(s) of any reaction(s) that take place within the reactor. Equation 3.5 also indicates that if we are able to accurately measure the time-dependent concentration of the constituent of interest in a batch reactor, we can use those data to describe the kinetics of the reaction(s) that take place within the system, $r$. For this reason, batch reactors are often used in laboratories to quantify reaction kinetics.

In a strict sense, the well-mixed assumption will never apply in a system that has finite reactor dimensions. However, there are many systems that closely approach the physical interpretation of a well-mixed system. The well-mixed assumption will apply if the timescale for mixing is short relative to the timescale for the reaction. What does this mean? In a well-mixed system, a particle or fluid element will be able to sample all locations within the system in a characteristic time that is much shorter than a characteristic time that is used to describe reaction kinetics, such as the half-life of the reaction. This means that it is possible for a system to be considered well-mixed with respect to a slow reaction, but non-well-mixed with respect to a fast reaction.

The well-mixed assumption is important in the application of many reactor models, and in the interpretation of data from experiments. Anyone who uses a model that applies or infers a well-mixed condition should be aware of the implications of this assumption.

The reaction term ($r$) is included to account for the effects of any reaction(s) that take place in the system. In many chemical reactor systems, reaction kinetics can be described by first-order or pseudo-first-order kinetics.[f] If we assume a single first-order, consumptive reaction is taking place within the system, then:

$$r = -kC \tag{3.6}$$

where

$k$ = first-order reaction rate constant [=] $\frac{\text{L}}{\text{mole} \cdot \text{s}}$

$C$ = reactant concentration [=] $\frac{\text{mole}}{\text{L}}$.

Substituting this expression into our mass-balance equation (equation 3.5), we find:

$$\frac{dC}{dt} = -kC \tag{3.7}$$

Rearranging and integrating we find:

$$\ln\left(\frac{C}{C_0}\right) = -k \cdot t \tag{3.8}$$

This suggests that a plot of $\ln\left(\frac{C}{C_0}\right)$ vs $t$ should yield a straight line through the origin with a slope of $-k$ (see Figure 3.3).

### 3.3.2 Ideal Continuous-Flow Stirred Tank Reactor Model (CFSTR)

The continuous-flow stirred tank reactor (CFSTR) model is commonly applied for simulation of the behavior of systems for which fluid flow is allowed to take place into and out of the reactor. Figure 3.4 provides a schematic illustration of an ideal CFSTR.[g]

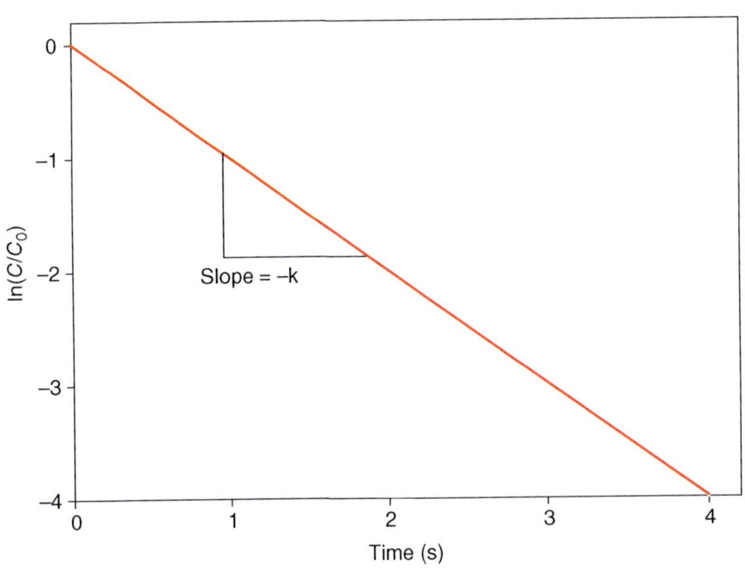


**Figure 3.3** Graphical interpretation of behavior of a batch reactor under conditions of a first-order, consumptive reaction. The slope of "best-fit" line through the origin will equal the negative of the reaction rate constant.

In contrast with the batch reactor, the CFSTR model allows for advective flow into and out of the CR. But because the system is assumed to be well-mixed, composition in the CR is spatially uniform and it is again possible to eliminate spatial-averaging of the reaction rate. Conveniently, this allows the entire reactor volume to be used as the CR.

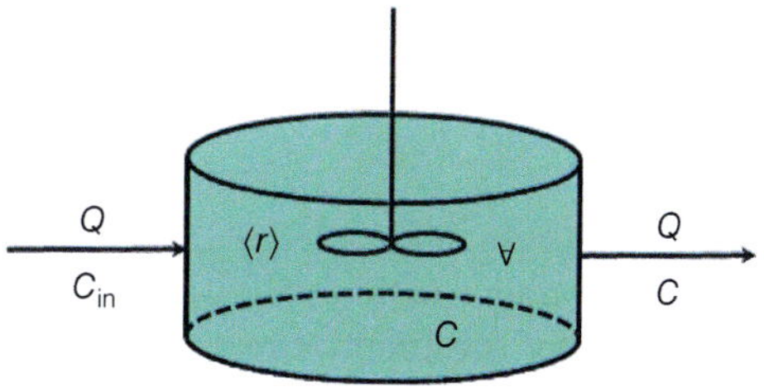


**Figure 3.4** Schematic illustration of an ideal CFSTR. The well-mixed assumption allows the entire internal volume of the reactor (light blue region in the schematic) to be used as the CR.

Another assumption that is inferred by Figure 3.4 is that the volumetric flow rates ($Q$) into and out of the CR are the same.[h] This allows us to remove the subscripts "in" and "out" on the volumetric flow rate terms that were included in Figure 3.1 and equation 3.3. In addition, the volume of fluid contained within the CR ($\forall$) will remain constant under this condition. The well-mixed assumption also requires that the concentration of the constituent of interest in the reactor must be equal to the concentration leaving the reactor ($C$). Collectively, these characteristics of the system allow us to rewrite equation 3.3 as follows:

$$\frac{\mathrm{d}M}{\mathrm{d}t} = Q \cdot C_{\mathrm{in}} - Q \cdot C \pm r \cdot \forall \tag{3.9}$$

As described previously, we will assume that reactor geometry is fixed in our simulations. Recognizing that $M$ can be decomposed as the product of volume and concentration, we may write:

$$M = \forall \cdot C \tag{3.10}$$

If we include this decomposition in the numerator of the left side of equation 3.9, apply the product rule from calculus, then recognize that because system volume does not change with time (i.e. $\frac{\mathrm{d}\forall}{\mathrm{d}t} = 0$), we find that the left side can be reduced to:

$$\frac{\mathrm{d}M}{\mathrm{d}t} = \frac{\mathrm{d}(\forall \cdot C)}{\mathrm{d}t} = \forall\frac{\mathrm{d}C}{\mathrm{d}t} + C\frac{\mathrm{d}\forall}{\mathrm{d}t} = \forall\frac{\mathrm{d}C}{\mathrm{d}t} \tag{3.11}$$

Substituting this result back into equation 3.9, we get:

$$\forall\frac{\mathrm{d}C}{\mathrm{d}t} = Q \cdot C_{\mathrm{in}} - Q \cdot C \pm r \cdot \forall \tag{3.12}$$

Dividing this result by system volume allows us to develop a general governing equation to describe dynamic behavior of an ideal CFSTR:

$$\frac{\mathrm{d}C}{\mathrm{d}t} = \frac{Q \cdot C_{\mathrm{in}} - Q \cdot C}{\forall} \pm r \tag{3.13}$$

It is often useful to combine terms in reactor models. An important example is the ratio of volume to volumetric flow rate, which is often referred to as the mean hydraulic detention time ($\theta$):

$$\theta = \frac{\forall}{Q} \tag{3.14}$$

Substituting this identity into equation 3.13 yields a fundamental form of the ideal CFSTR model:

$$\frac{dC}{dt} = \frac{C_{in} - C}{\theta} \pm r \tag{3.15}$$

As an illustration of the application of this governing equation, we will consider again an example of a first-order consumptive reaction (see equation 3.6). Substituting this kinetic expression into equation 3.15, we find:

$$\frac{dC}{dt} = \frac{C_{in} - C}{\theta} - k \cdot C \tag{3.16}$$

As will be the case with many material balance and reactor models, the governing equation is differential. Because equation 3.16 is first-order with respect to time, we need to define one time-dependent condition for concentration to solve this equation, such as an initial condition. For the example presented below, we will assume that the reactant concentration present in the system at $t = 0$ is identical to the influent concentration, $C_{in}$. A method to solve equation 3.16 based on this initial condition is presented in Appendix 3.A. This solution provides a time-dependent description of the "start-up" behavior of an ideal CFSTR with a first-order consumptive reaction taking place:

$$C = \frac{C_{in}}{1 + k \cdot \theta} + C_{in} \cdot \exp\left(\frac{-t}{1 + k \cdot \theta}\right) \cdot \left(\frac{k \cdot \theta}{1 + k \cdot \theta}\right) \tag{3.17}$$

The exponential term in the solution (equation 3.17) describes the time-dependence of the behavior of this reactor under start-up conditions. As time becomes large (mathematically, as $t \to \infty$), the exponential term becomes vanishingly small, and we approach a *steady-state* condition, which is described by the first term on the right side of equation 3.17.

In a formal sense, a steady-state condition is one in which the composition of a system becomes time-invariant (i.e. constant) as a result of the combined effects of transport processes and reactions. It is important to recognize that steady-state and *equilibrium* are two different conditions. Equilibrium describes a condition in which the rates of forward and reverse reactions are identical (because the chemical potentials of reactants and products are equal). Equilibrium and steady-state represent fundamentally different conditions in all but a few trivial cases.

For most people, a graphical solution to a complex equation, like equation 3.17, is helpful for understanding its implications. To examine this, we will consider a situation in which a reactor is operated under conditions such that the product of the reaction rate constant and the mean hydraulic detention time is one (note that this product is dimensionless):

$$k \cdot \theta = 1$$

Figure 3.5 illustrates the time-dependent behavior of this ideal CFSTR under this start-up condition. In other words, Figure 3.5 is a plot of the solution presented in equation 3.17 for the case of $k \cdot \theta = 1$.

Concentration and time have been non-dimensionalized in Figure 3.5 to allow generalization of the results, which suggests that the reactor system will need to be operated for the equivalent of at least 4–5 detention times ($\theta$) before a steady-state condition is approached.

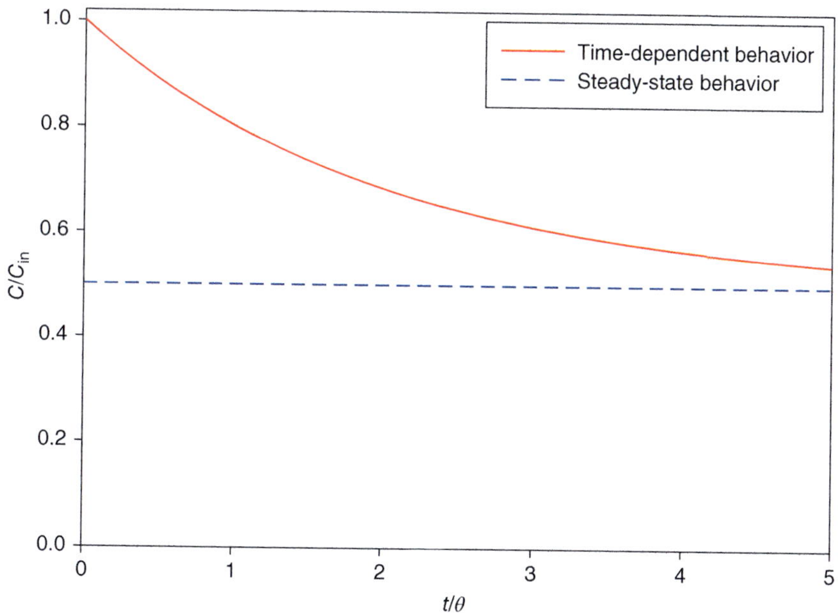


**Figure 3.5** Approach of ideal CFSTR to steady-state. Model simulation was conducted for operating conditions corresponding to $k{\cdot}\theta = 1$. The vertical axis in this simulation represents dimensionless concentration, where $C$ has been normalized by dividing by $C_{in}$. The horizontal axis represents dimensionless time, where $t$ has been normalized by dividing by $\theta$.

If interpreted literally, equation 3.17 suggests that an infinite amount of time will be required to reach a steady-state condition. Alternatively, steady-state will be reached when the time-dependent term in equation 3.17 becomes zero. For the ideal CFSTR, this implies that the steady-state condition for a reactor in which a single first-order consumptive reaction is taking place is defined by:

$$C_{\text{steady-state}} = \frac{C_{\text{in}}}{1 + k \cdot \theta} \tag{3.18}$$

*Conversion* is a term that can be used to describe reactor performance. Normally, reactor performance is evaluated based on steady-state conditions. The formal definition of conversion is:

$$\text{Conversion} = 1 - \frac{C_{\text{steady-state}}}{C_{\text{in}}} \tag{3.19}$$

So, for an ideal CFSTR, conversion is defined as:

$$\text{Conversion} = 1 - \frac{1}{1 + k \cdot \theta} = \frac{k \cdot \theta}{1 + k \cdot \theta} \tag{3.20}$$

As it turns out, the ideal CFSTR model provides an estimate of the theoretical lower limit on reactor performance. In systems that lack significant dead zones, the CFSTR model can be used to describe the minimum conversion to be expected for a reactor for a given operating condition (i.e. $k$ and $\theta$). Even if a reactor does not conform to the assumptions of the ideal CFSTR model, it is often useful to define the minimum conversion that can be expected.

### 3.3.3 Ideal Plug-Flow Reactor Model (PFR)

The CFSTR model describes the behavior of a system in which a condition of complete-mix is assumed to apply throughout the reactor volume (which in the case of a CFSTR coincides with the CR). This may be thought of as one end of the theoretical mixing spectrum. The plug-flow reactor (PFR) model represents the other end of this spectrum for a system that involves continuous flow, in that it implies no mixing in the direction of flow. On the other hand, the PFR model also requires that reaction rates be uniform in any cross section of the reactor; in effect, this requires that an assumption of complete-mix be applied perpendicular to the direction of flow.

Figure 3.6 provides a schematic illustration of a PFR. In practice, systems that approach a plug-flow condition often take the form of a long, thin reactor. For this particular model, we will assume that the volumetric flow rate into and out of the reactor system is identical ($Q$) and constant.

The assumptions of the PFR model imply that reaction rates are uniform in the cross section of the reactor. Because of this, we generally choose to model a PFR using a CR of differential volume. Specifically, we will use a CR of length $\Delta x$ and cross-sectional area of A, corresponding to the cross-sectional area of the reactor. Our differential CR is illustrated as the hatched area in Figure 3.6.

We will apply a material balance to the differential CR to develop an equation to describe the behavior of this ideal PFR. For purposes of this illustration, in which we will again consider an example involving a first-order, consumptive reaction, we will develop this in finite difference form:

$$\frac{\Delta M}{\Delta t} = V_x \cdot A \cdot C|_x - V_x \cdot A \cdot C|_{x+\Delta x} \pm r \cdot A \cdot \Delta x \tag{3.21}$$

As before, the left side of this equation describes the time rate-of-change of the amount (number of moles) of the material of interest in the CR. The first product on the right side describes the advective molar rate at which the constituent enters the CR, while the second product on the right side describes the advective molar rate at which the constituent leaves the CR. The last product describes the effect of reaction(s) on the amount of material in the CR. Note that because the CR has differential length, which we will evaluate in the limit as $\Delta x \rightarrow 0$, the reaction rate within the CR becomes spatially constant, so that we can eliminate the volume-averaging function.

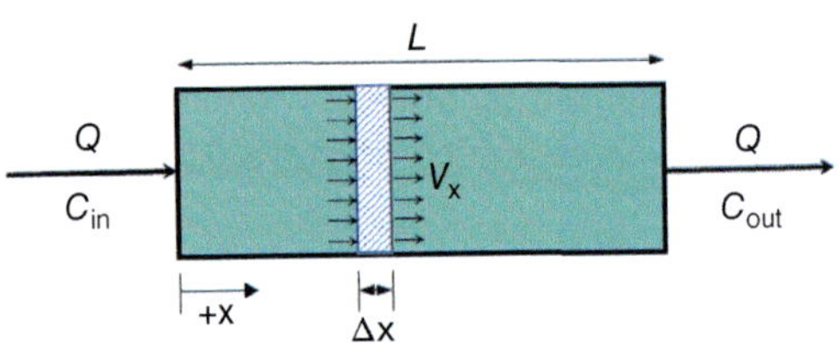


Figure 3.6 Schematic illustration of an ideal plug-flow reactor. The reactor is assumed to have a cross-sectional area of A, which is constant along the entire length of the reactor. All fluid elements and reactant molecules in them are assumed to move at the same velocity through the system, $V_x$ (a constant, in this example). The light blue area in this schematic illustrates the reactor volume. Because of the assumptions of the PFR model, we select our CR as the hatched area of differential length ($\Delta x$), bounded by the vertical lines and the walls of the reactor.

Dividing both sides of equation 3.35 by the product $A \cdot \Delta x$, which represents the volume of our CR, and recognizing that $\frac{M}{A \cdot \Delta x}$ represents concentration in the CR, we get:

$$\frac{\Delta C}{\Delta t} = -\left(\frac{V_x \cdot C}{\Delta x}\bigg|_{x+\Delta x} - \frac{V_x \cdot C}{\Delta x}\bigg|_{x}\right) \pm r \tag{3.22}$$

We then evaluate this equation in the limits as $\Delta t \rightarrow 0$ and $\Delta x \rightarrow 0$, and apply the definition of the derivative to yield a partial differential equation that describes the dynamic behavior of a reactant within our CR:[i]

$$\frac{\partial C}{\partial t} = -V_x \cdot \frac{\partial C}{\partial x} \pm r \tag{3.23}$$

To illustrate the application of equation 3.23, we will continue with an example in which a first-order consumptive reaction takes place within the system. Also, we will examine the steady-state condition of this reactor, such that $\frac{\partial C}{\partial t} = 0$. Under these conditions, equation 3.37 reduces to an ordinary differential equation:

$$0 = -V_x \cdot \frac{dC}{dx} - kC \tag{3.24}$$

Algebraic manipulation yields:

$$V_x \cdot \frac{dC}{dx} = -kC \tag{3.25}$$

which in turn can be rearranged to yield:

$$\frac{dC}{C} = -\frac{k}{V_x} \cdot dx \tag{3.26}$$

We can then integrate this equation across the length of the reactor, subject to the limits of integration illustrated in Figure 3.6:

$$\int_{C_{in}}^{C_{out}} \frac{dC}{C} = -\frac{k}{V_x} \cdot \int_0^L dx \tag{3.27}$$

Giving us a governing equation for the ideal PFR at steady-state with a first-order consumptive reaction:

$$\ln\left(\frac{C_{out}}{C_{in}}\right) = -\frac{k \cdot L}{V_x} \tag{3.28}$$

We may also recognize that because all fluid elements (and reactant molecules) move at the same velocity through the system ($V_x$), the ratio $\frac{L}{V_x}$ actually represents the mean hydraulic detention time in the system:

$$\frac{L}{V_x} = \theta \tag{3.29}$$

Substitution of this result back into equation 3.28 provides a tool to estimate the steady-state performance of a PFR for this condition:

$$\ln\left(\frac{C_{out}}{C_{in}}\right) = -k \cdot \theta \tag{3.30}$$

If we apply the definition of conversion from equation 3.29, we find that the conversion in an ideal PFR for these conditions is described by:

$$\text{Conversion} = 1 - \exp(-k \cdot \theta) \tag{3.31}$$

In some respects, the ideal PFR model describes an optimum reactor condition, in that the result of equation 3.31 represents a theoretical upper limit on conversion for the case of a reactor in which a first-order, consumptive reaction takes place; this is because all fluid elements reside in the reactor for the same time, $\theta$. When combined with the result of equation 3.20, which describes the conversion in an ideal CFSTR (theoretical lower limit on conversion for the same case), we have simple mathematical expressions that can be used to "bracket" reactor performance. To complete these calculations, we require accurate estimates of the reaction rate constant and the mean hydraulic detention time in the reactor. For some systems, the upper and lower limits on conversion (equations 3.31 and 3.20, respectively) will be quite close; under these circumstances, these two models may be sufficient to describe reactor performance. In other words, when these two estimates of reactor performance are close to each other, knowledge of $k$ and $\theta$ may be sufficient to describe the behavior of the system, even if mixing behavior in the actual system does not conform to either the ideal CFSTR or ideal PFR model. On the other hand, for some reactor systems, the estimates of conversion provided by equations 3.31 and 3.20 may be unacceptably far apart. When this happens, these simple, idealized models may not be sufficient to accurately describe reactor behavior; we may have to apply more detailed models to describe mixing behavior in the system to develop accurate predictions of reactor performance.

### 3.3.4 CFSTR Cascade Model

Some reactor systems display mixing behavior that is intermediate to the extremes described by the CFSTR and PFR models. A number of different modeling approaches have been developed to describe these systems. For purposes of this analysis, we will consider the CFSTR cascade model.

Specifically, we will consider perhaps the most basic form of the CFSTR cascade, in which a reactor system will comprise a series of $n$ identical well-mixed CFSTRs. Figure 3.7 provides a schematic illustration of such a system, along with relevant terms and symbols.

If we examine the general schematic provided in Figure 3.7, it is evident that the CFSTR cascade model reduces identically to the CFSTR when $n = 1$. Though it is a little more difficult to prove and perhaps to visualize, it turns out that the CFSTR cascade model is identical to the PFR model for the case of $n = \infty$. This behavior will be proven below for the steady-state condition. An important conclusion of this analysis is that by varying the parameter $n$ across the range of possible values (integer values in the range $1 \le n \le \infty$), we can simulate the behavior of a wide range of reactor types and behaviors.

Quantitatively, we can describe the behavior of the CSFTR cascade by applying material balances to each cell in the system. Following the approach described above for individual CFSTRs, and again using a first-order consumptive model in our simulation, we see for cell 1 that:

$$\frac{\forall}{n}\cdot\frac{dC_1}{dt} = Q(C_{in} - C_1) - k\cdot C_1\cdot\frac{\forall}{n} \text{ or } \frac{dC_1}{dt} = \frac{n\cdot(C_{in} - C_1)}{\theta} - k\cdot C_1 \tag{3.32}$$

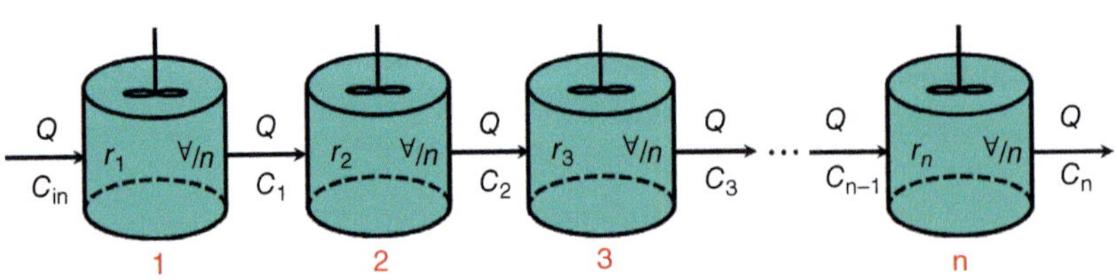


**Figure 3.7** Schematic illustration of CFSTR cascade. In this system, a series of $n$ identical reactors (cells), each with volume $\forall/_n$ is operated. The volumetric flow rate ($Q$) into and out of each cell is assumed to be identical and constant. The effluent from one cell becomes the influent for the next cell. Each cell is well-mixed. The terms in red font indicate the cell identifier used in model development.

where $\forall$ represents the total system volume and $\theta$ represents the mean detention time based on total system volume, i.e. $\theta = \frac{\forall}{Q}$. Applying similar logic to the other cells, we find:

$$\frac{\forall}{n} \cdot \frac{dC_2}{dt} = Q(C_1 - C_2) - k \cdot C_2 \cdot \frac{\forall}{n} \text{ or } \frac{dC_2}{dt} = \frac{n \cdot (C_1 - C_2)}{\theta} - k \cdot C_2 \tag{3.33}$$

$$\frac{\forall}{n} \cdot \frac{dC_3}{dt} = Q(C_2 - C_3) - k \cdot C_3 \cdot \frac{\forall}{n} \text{ or } \frac{dC_3}{dt} = \frac{n \cdot (C_2 - C_3)}{\theta} - k \cdot C_3 \tag{3.34}$$

Or more generally for the $n^{\text{th}}$ cell in the cascade:

$$\frac{\forall}{n} \cdot \frac{dC_n}{dt} = Q(C_{n-1} - C_n) - k \cdot C_n \cdot \frac{\forall}{n} \text{ or } \frac{dC_n}{dt} = \frac{n \cdot (C_{n-1} - C_n)}{\theta} - k \cdot C_n \tag{3.35}$$

The CFSTR cascade model yields a set of $n$ differential equations to describe overall reactor performance. To solve this system of equations, we require a time-dependent concentration condition (e.g. initial condition) for each cell in the cascade. More often, however, the CFSTR cascade model is used to describe steady-state behavior of a reactor; under a steady-state condition, the time-derivatives become zero and the model reduces to a set of $n$ algebraic equations.

### 3.3.5 Effects of Mixing on Performance in Chemical Reactors – Material Balance Approach

We will now consider the behavior of a reactor with arbitrary mixing behavior. We will assume that mixing can be accounted for by a series of equally sized, well-mixed cells, as illustrated in Figure 3.7. Mixing behavior within this system lies somewhere between those that would be predicted by the ideal CFSTR or PFR models. And because mixing affects reactor performance, we expect that conversion in this arbitrary reactor will be intermediate to the predictions of the ideal CFSTR and ideal PFR models, which as described above represent the theoretical lower and upper limits, respectively, on reactant conversion.

We will explore two methods for analysis of conversion in reactor systems that span the range of mixing behavior corresponding to $1 \leq n \leq \infty$ in the CFSTR cascade model. For both approaches, we will analyze system behavior based on the assumption of a first-order consumptive reaction taking place in the system.

For the first method of analysis, we will consider steady-state conversion in CFSTR cascade models by direct application of the material balance approach. To do this, we will focus on an arbitrary cell within the cascade illustrated in Figure 3.7. Under steady-state conditions, a mass-balance on this cell will yield the following:

$$0 = \frac{\forall}{n} \cdot \frac{dC_i}{dt} = Q(C_{i-1} - C_i) - k \cdot C_i \cdot \frac{\forall}{n} \tag{3.36}$$

The left-hand side of the equation is equal to zero because of the steady-state assumption. Notice that the volume terms in this equation are divided by the number of cells ($n$) in the system. This is because the system comprises $n$ equally sized cells of total volume $\forall$ with a total residence time of $\theta = \frac{\forall}{Q}$. Rearrangement of equation 3.36 yields:

$$k \cdot C_i = \frac{n \cdot (C_{i-1} - C_i)}{\theta} \tag{3.37}$$

Grouping terms we find:

$$C_i \cdot \left(k + \frac{n}{\theta}\right) = \frac{n \cdot C_{i-1}}{\theta} \tag{3.38}$$

Further rearrangement yields:

$$\frac{C_i}{C_{i-1}} = \frac{n}{\theta \cdot \left(k + \frac{n}{\theta}\right)} = \frac{n}{n + k \cdot \theta} = \frac{1}{1 + \frac{k \cdot \theta}{n}} \tag{3.39}$$

The behavior described by equation 3.39 is general for all cells in the system. Therefore, we can use it to describe the behavior of every cell in the system. Additionally, we can describe the overall behavior of the reactor as follows:

$$\frac{C_n}{C_{\text{in}}} = \frac{C_1}{C_{\text{in}}} \cdot \frac{C_2}{C_1} \cdot \frac{C_3}{C_2} \cdots \frac{C_n}{C_{n-1}} \tag{3.40}$$

Each of the quotients on the right side of equation 3.40 will be identical (see equation 3.39). Therefore, we can use this information to relate effluent concentration to influent concentration in this system:

$$\frac{C_n}{C_{\text{in}}} = \left(\frac{1}{1 + \frac{k \cdot \theta}{n}}\right)^n \tag{3.41}$$

Equation 3.41 describes the steady-state performance of a CFSTR cascade comprising $n$ cells under conditions of a first-order consumptive reaction. When we consider the situation of a single CFSTR (i.e. $n = 1$) and the general solution provided in equation 3.41, it is analogous to a comparison of simple interest and compound interest in a bank account. Compound interest will have greater yield because of the multiplicative effect of interest. In the case of a reactor, we observe greater conversion in the case of $n$ small CFSTRs than we will in a single CFSTR with the same volume because of an analogous multiplicative (compounding) effect.

We can apply equation 3.19 to develop a general expression to describe conversion in a CFSTR cascade:

$$\text{Conversion} = 1 - \frac{C_{\text{steady}-\text{state}}}{C_{\text{in}}} = 1 - \left(\frac{1}{1 + \frac{k \cdot \theta}{n}}\right)^n \tag{3.42}$$

As a check, we observe that equation 3.42 reduces exactly to equation 3.20 for the case of $n = 1$, which corresponds to the ideal CFSTR mode. Similarly, it can be shown that equation (3.42) predicts the steady-state conversion of an ideal PFR, which corresponds to the case of $n = \infty$ (not shown here).

Overall performance of a reactor system will be influenced by the rate of the reaction(s) that take place in the system, the size of the reactor relative to its flow rate, and mixing behavior in the system. For the CFSTR cascade model, these attributes of a system are captured by the parameters $k$, $\theta$, and $n$, respectively. Figure 3.8 provides a graphical illustration of the effects of these parameters on overall reactor performance.

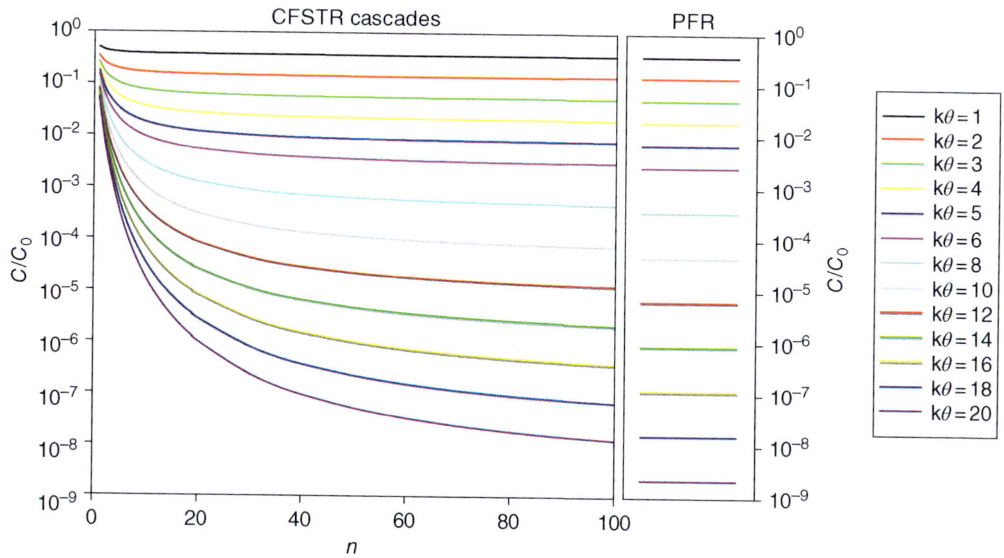


**Figure 3.8** (left panel) Fraction of reactant remaining $\left(\frac{C}{C_0}\right)$ as a function of number of cells ($n$) in CFSTR cascades for a range of values of $k \cdot \theta$ (see legend at right). Also included (center panel) are calculated fractions of reactant remaining for an ideal PFR with the same conditions. Note that the vertical axes on the left and center panels are presented on a $\log_{10}$ scale.

The product $k \cdot \theta$ is a dimensionless parameter that characterizes the extent of possible reaction progress. Reactions will proceed far toward completion if the reaction rate is fast (large $k$), if the reactor is large relative to the flow rate (large $\theta$), or both. The range of $k \cdot \theta$ (a dimensionless parameter) presented in Figure 3.8 is intentionally large, perhaps to the point of being unrealistic, to illustrate the effects of this product on reactor performance.

As described previously, the mixing behavior of a CFSTR cascade is also governed by the parameter $n$, such that systems with $n = 1$ correspond to the ideal CFSTR model, whereas systems with $n = \infty$ correspond to the ideal PFR. The range of $n$ values included in Figure 3.8 is intentionally broad, so as to illustrate the effects of this parameter on reactor performance.

The range of demands for reactor performance in real applications is quite broad. In general, some reactors are designed to accomplish only modest degrees of reactant conversion. When this is the case, the performance of the reactor is not strongly affected by mixing behavior. This is evident for the curves in Figure 3.8 that correspond to $k \cdot \theta = 1$, the smallest value of this dimensionless parameter considered here. For such a system, actual performance is largely unaffected by the value $n$. Also, the calculated performance of the system corresponding to $n = 1$ (ideal CFSTR) is not markedly different from the predicted performance for $n = \infty$ (ideal PFR).

On the other hand, in some systems the demands for treatment can be quite severe. For example, with disinfection systems, wherein the objective of performance is to inactivate microbial or viral pathogens, it is common for systems to be expected to reliably achieve reductions of reactant (microbial or viral) concentration on the order of 3–5 $\log_{10}$ units. For such a system, overall performance can be dramatically affected by mixing behavior; in the CFSTR cascade model, mixing behavior is described by the parameter $n$. Moreover, we see that predictions of reactor performance for finite values of $n$ diverge from the ideal PFR case increasingly as predicted performance increases.

The central messages from Figure 3.8 are as follows:

- Performance of chemical reactor systems is influenced by $k$ (intrinsic reaction rate[s]), $\theta$ (i.e. reactor size relative to flow rate), and $n$ (mixing behavior).
- The lower and upper limits on reactor performance are defined by the ideal CFSTR and PFR models, respectively. For modest extents of reaction, the predictions of these two models are likely to be close; but as the extent of reaction increases, mixing behavior becomes an increasingly important consideration in evaluating reactor behavior.

### 3.3.6 Time as a Master Variable – Residence Time Distribution Function

In all of the expressions presented above, the independent (master) variable has been time. The behavior of a chemical reactor system depends on the amount of time provided for reactant chemicals to interact with each other. In chemical reactor systems for which the ideal mixing models (CFSTR, PFR, and CFSTR cascade) are applied, time-dependence can be accounted for by the mean hydraulic detention time, $\theta$. However, there are circumstances wherein more detailed information about reaction time is required for reactor analysis.

For continuous-flow reactor systems, some fluid elements will reside in the reactor for a short period of time, while others will reside for longer periods of time. More generally, there will be a *residence time distribution* (RTD) that will characterize a reactor for a given operating condition.

Formally, we define the RTD for a reactor as a probability density function. Following this definition, which derives from general principles of probability theory, the required characteristics of an RTD include the following:

$$E(t) \cdot \mathrm{d}t = \text{fraction of fluid elements with residence time between } t \text{ and } t + \mathrm{d}t \tag{3.43}$$

And because $E(t)$ is a probability density function:

$$\int_0^{\tau} E(t) \cdot \mathrm{d}t = \text{fraction of fluid elements with residence time} \leq \tau \tag{3.44}$$

$$\int_0^{\infty} E(t) \cdot \mathrm{d}t \equiv 1 \tag{3.45}$$

The limits of integration in equations 3.44 and 3.45 represent the range of possible values. In both equations, the lower limit is 0 because negative residence time has no physical meaning. The upper limit in equation 3.45 represents the maximum time that a fluid element could reside in our reactor. Figure 3.9 provides a graphical illustration of the RTD and its relevant terms.

The shape of the RTD for a reactor depends on the volumetric flow rate imposed on the reactor ($Q$), its volume ($\forall$), and mixing behavior within the system. For the CFSTR cascade model, the RTD is governed by the parameter $n$. Equation 3.46 provides a general expression for the normalized RTD in a CFSTR cascade:[j]

$$E(t) \cdot \theta = \frac{n^n}{(n-1)!} \cdot \left(\frac{t}{\theta}\right)^{n-1} \cdot \exp\left(-\frac{n \cdot t}{\theta}\right) \tag{3.46}$$

Figure 3.10 illustrates dimensionless RTDs for a wide range of CFSTR cascades, ranging from $n = 1$ to $n = \infty$. For the case of $n = 1$, which corresponds with the ideal CFSTR, we see that the RTD is broad and follows simple exponential decay function. This implies that some fluid elements will reside in the reactor for a short time, allowing relatively little time for reaction progress. For this same system, a small fraction of fluid elements will reside for comparatively long periods of

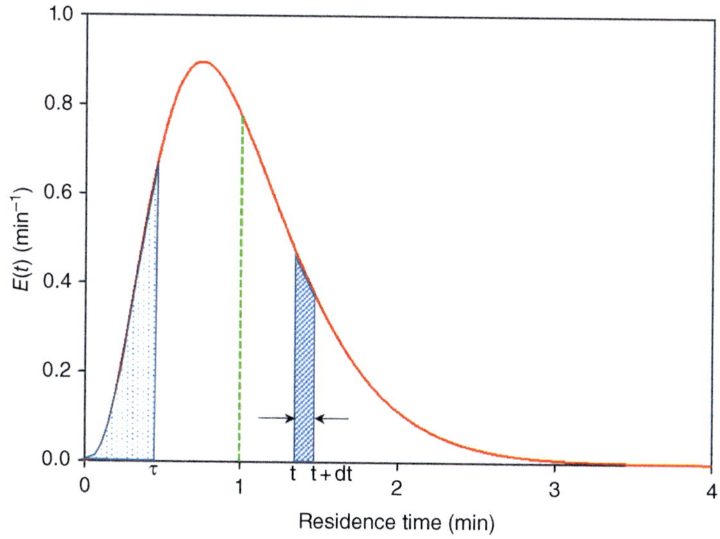


**Figure 3.9** Illustration of residence time distribution function for an arbitrary reactor with a mean hydraulic detention time ($\theta$) of one minute, indicated by the green, vertical dashed line. Because the RTD is a density distribution function, we know that the area under the curve must be 1. Formally, we describe this by the integral: $\int_0^\infty E(t) \cdot dt \equiv 1$ (equation 3.45). The stippled area represents the fraction of fluid elements with a residence time $\leq \tau$, formally defined by the integral: $\int_0^\tau E(t) \cdot dt$ (equation 3.44). The diagonally hatched area represents the fraction of fluid elements with residence time between times $t$ and $t + dt$, formally described by the product $E(t) \cdot dt$ (equation 3.43).

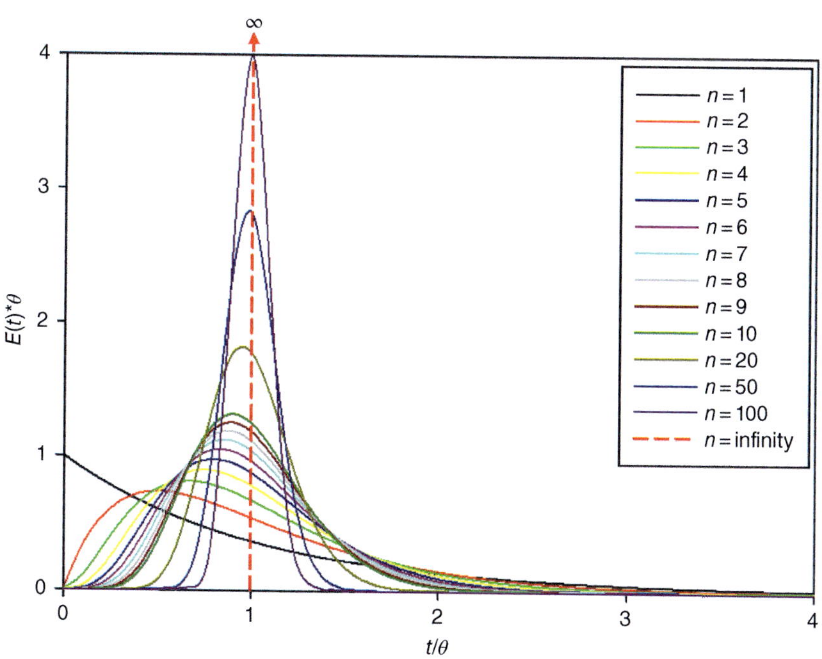


**Figure 3.10** Normalized RTDs for CFSTR cascades for values of n ranging from 1 to $\infty$. Note that the centroid of the RTD, $\bar{t}$, may not be equal to the mean hydraulic detention time, $\theta$.

time; for these fluid elements, the reaction of interest may approach completion. Overall reactor performance is determined by the weighted-average of reaction progress among all fluid elements.

For the case of $n = \infty$, which corresponds with the ideal PFR, the RTD is defined by a delta ("spike") function, such that all fluid elements reside in the reactor for the same time, $t = \theta$. This turns out to be the most efficient reactor possible, for a given value of $\theta$.

For values of $n$ between 1 and $\infty$, we observe mixing behavior that is intermediate to the CFSTR and PFR models. This variation in mixing behavior plays an important role in reactor performance. The relationships between reactor performance (Figure 3.8) and mixing behavior (Figure 3.10) are important to understand. Students will be given opportunities to explore these behaviors and their interrelationships in problems that will be given at the end of this chapter.

It should be noted that real reactors will fail to conform strictly to the RTDs that are illustrated in Figure 3.10. For example, the case of $n = 1$ requires that some fluid elements have zero residence time within the reactor. At the other extreme, the case of $n = \infty$ requires that all fluid elements have exactly the same residence time within the system. The intermediate cases ($1 < n < \infty$) require that each cell within the cascade be well-mixed. Despite the constraints of these literal interpretations, there are many reactor systems for which mixing behavior can be closely approximated by the CFSTR cascade model. As such, there is merit to the application of this model for descriptions of reactor RTD and reactor performance.

## tpb 6pt Effects of Mixing on Performance in Chemical Reactors – Segregated-Flow Model

A second method of analysis involves application of the *segregated-flow model*, which is illustrated in schematic form in Figure 3.11. In this model, fluid and reactant molecules are assumed to enter the reactor as small, well-mixed parcels of fluid (i.e. well-mixed batch reactors) that do not exchange material with other fluid parcels while in the reactor (i.e. segregated flow). However, immediately upon leaving the system, the parcels are assumed to mix with each other. This is an example of what is sometimes referred to as "latest possible mixing" within a reactor system [2].

The fluid parcels may each reside in the system for different amounts of time. It is assumed that the fluid parcels display a RTD that matches that of the reactor. Some fluid elements will follow a path through the reactor that allows a short residence time, and by extension relatively little time for reaction to proceed, while other fluid elements will experience a relatively long residence time and will have a relatively long time for reaction to proceed. The overall performance of the reactor is calculated as the weighted-average of the performance of the segregated fluid parcels. The weighting function is the RTD.

Mathematically, the segregated-flow model may be described as follows:

$$\left(\frac{C}{C_0}\right)_{\text{reactor}} = \int_0^{\infty} \left(\frac{C(t)}{C_0}\right)_{\text{batch}} \cdot E(t) \cdot \mathrm{d}t \tag{3.47}$$

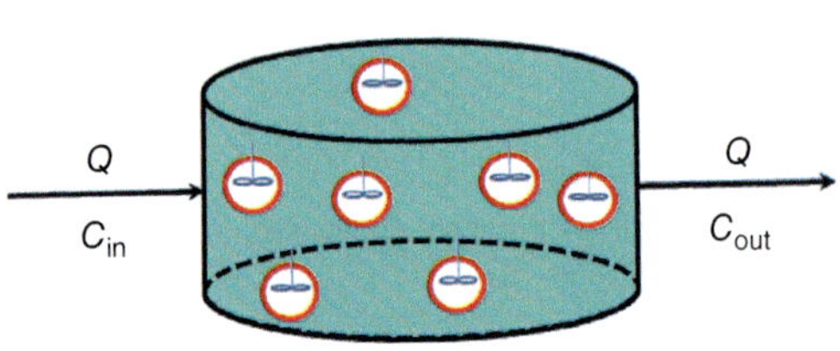


**Figure 3.11** Schematic illustration of segregated-flow model. The model can be applied to any system with arbitrary mixing behavior. Flow is assumed to arrive in small, well-mixed packets of fluid which do not interact in the reactor system. The packets are assumed to combine (mix) at the exit of the reactor.

where

$\left(\frac{C}{C_0}\right)_{\text{reactor}}$ = fraction of reactant remaining in the reactor effluent

$\left(\frac{C(t)}{C_0}\right)_{\text{batch}}$ = fraction of reactant remaining after a given exposure time in a batch reactor

= intrinsic kinetics of reaction of interest

$E(t)$ = residence time distribution function [=] $\text{min}^{-1}$

$t$ = time [=] min.

As an illustration, we will apply the segregated-flow model to predict the performance of an ideal CFSTR in which a first-order consumptive reaction is taking place. For a first-order consumptive reaction, the time-dependent degradation of a reactant in a batch reactor is described by:

$$\left(\frac{C(t)}{C_0}\right)_{\text{batch}} = \exp(-k \cdot t) \tag{3.48}$$

As demonstrated in Appendix 3.B, the RTD function for an ideal CFSTR is:

$$E(t) = \frac{1}{\theta} \cdot \exp\left(-\frac{t}{\theta}\right) \tag{3.49}$$

Substituting the information from equations 3.48 and 3.49 into equation 3.47, we find:

$$\left(\frac{C}{C_0}\right)_{\text{reactor}} = \int_0^\infty \exp(-k \cdot t) \cdot \frac{1}{\theta} \cdot \exp\left(-\frac{t}{\theta}\right) \cdot dt \tag{3.50}$$

The integrand can be simplified as:

$$\left(\frac{C}{C_0}\right)_{\text{reactor}} = \int_0^\infty \frac{1}{\theta} \cdot \exp\left[-t \cdot \left(k + \frac{1}{\theta}\right)\right] \cdot dt \tag{3.51}$$

Integrating, we find:

$$\left(\frac{C}{C_0}\right)_{\text{reactor}} = \frac{-1}{\theta} \cdot \left(\frac{1}{\left(k + \frac{1}{\theta}\right)}\right) \cdot \exp\left[-t \cdot \left(k + \frac{1}{\theta}\right)\right]\Bigg|_0^\infty \tag{3.52}$$

Evaluating this function at the limits integration and subtracting, we find:

$$\left(\frac{C}{C_0}\right)_{\text{reactor}} = \frac{1}{(1 + k \cdot \theta)} \tag{3.53}$$

The result presented in equation 3.53 is identical to the result we calculated by mass-balance in equation 3.18. If we were to apply the segregated-flow model to a CFSTR cascade, we would find a result that is identical to equation 3.41.

The segregated-flow model allows calculation of reactor performance based on knowledge of the RTD and intrinsic reaction kinetics. This calculation depends on the assumption of segregated flow, as described above. For first-order reactions, the segregated-flow model yields an exact prediction of reactor behavior; this is because the rate of reaction is linearly dependent on reactant concentration, and therefore the time of mixing is of no consequence. However, if we consider processes that involve reaction kinetics other than first-order, the segregated-flow model will yield results that are incorrect, but sometimes still useful. Specifically, if we define reaction order by the symbol β, then the following generalizations apply:

- For systems in which $\beta < 1$, the segregated-flow model yields a lower limit on reactor performance (i.e. actual conversion in the reactor will be greater than the prediction of the segregated-flow model).
- For systems in which $\beta = 1$, the segregated-flow model yields an accurate prediction of reactor performance (i.e. actual conversion in the reactor will be equal to the prediction of the segregated-flow model).
- For systems in which $\beta > 1$, the segregated-flow model yields an upper limit on reactor performance (i.e. actual conversion in the reactor will be less than the prediction of the segregated-flow model).

## 3.4 Models for Photochemical Reactors

Material balance principles can be used to describe and predict the behavior of any reactor system; however, photochemical reactors behave differently than conventional chemical reactors in many ways. First, photochemical reactions are initiated by fundamentally different mechanisms than chemical reactions. Second, the rates of photochemical reactions are often (but not always) faster than those of chemical reactions. Third, photochemical reactors are often characterized by steep gradients in local reaction rate.

The laws of photochemistry have taught us that:

- Radiation (photons) must be absorbed by a target for a reaction to proceed.
- Photon energy must meet or exceed bond energy to allow a reaction to proceed.
- Only one photon may be absorbed by a target molecule.
- The extent of a photoreaction depends on the dose delivered to the target(s), not the amount of time that is required to deliver that dose.

We have also seen a general expression that can be used to describe the rate of a photochemical process:

$$r_B = -\frac{d[B]}{dt} = I_a \cdot \Phi \tag{3.54}$$

where

$r_B$ = rate of reaction with respect to the target, B $[=] \frac{\text{mole}}{\text{L} \cdot \text{s}}$

$[B]$ = activity (concentration) of B in solution $[=] \frac{\text{mole}}{\text{L}}$

$t$ = time $[=]$ s

$I_a$ = volumetric rate of photon absorption by B $[=] \frac{\text{einstein}}{\text{L} \cdot \text{s}}$

$\Phi$ = primary quantum yield for reaction of interest $[=] \frac{\text{mole}}{\text{einstein}}$.

Equation 3.54 is relatively simple, and for a few photochemical reactor systems, it is fairly straightforward to apply. However, in many photochemical reactors, the term $I_a$ can be complicated. In general, $I_a$ depends on the strength (power) of the radiation source(s), the geometry of the system in which the target is exposed, and the optical characteristics of the media and materials that comprise the system.

The objective of this section is to explore these factors for some common reactor types, to demonstrate approaches that can be used to simulate or predict photochemical reactor performance, and to address limitations and sources of error in these calculations. As we will see, the logic we will apply in these calculations is an extension of the logic used to describe/predict the behavior of chemical reactors, but includes terms to account for the nature of photochemical reactions.

### 3.4.1 Dose as the Independent (Master) Variable

With chemical reactions, reaction progress is logically defined as a function of time. Similarly, with chemical reactors, process efficiency is described as a function of time, or more generally by the RTD. Time is the independent variable in both cases; time may be viewed as the *master variable* in chemical reaction processes.

On the other hand, the laws of photochemistry have taught us that progress in a photochemical reaction is governed by the dose of radiation delivered to the photochemical target. Similarly, process efficiency in photochemical reactors is governed by the distribution of doses delivered to the target; the dose distribution will govern photochemical reactor behavior. In that sense, dose (or the dose distribution) may be viewed as the master variable for photochemical reactions and photochemical reactors.

Our goal in describing photochemical processes is to apply methods that allow accurate estimation of dose delivery to targets in the system. In general, we will do this by describing the dose distribution delivered by the system, followed by integration of this distribution with an expression to describe the intrinsic kinetics of the photochemical process, using the logic of the segregated-flow model. However, there are some circumstances where a description of photochemical reaction process based on time (rather than dose) will make sense. One example is a situation in which photochemical and thermal chemical reactions take place simultaneously in a system; comparisons of reaction progress in these two reactor types should be done on a time basis, as a convenient "dose" analog for (thermal) chemical reaction progress does not exist. A second example is a system, such as a collimated beam, in which the fluence rate delivered to the target is essentially constant and uniform. Under this circumstance, it is equally appropriate to consider reaction progress as a function of time or dose.

### 3.4.2 Batch Reactor

We have seen in Chapter 2 that with an ideal, shallow, well-mixed batch reactor subjected to a uniform, collimated beam of radiation, the observed rate of a photochemical process will be influenced by the optical characteristics of the system.[k] Specifically, for the case of an opaque solution ($\varepsilon \cdot [B] \cdot \ell \to \infty$), reactant degradation will follow zero-order behavior, regardless of whether we choose to use time or dose as the independent variable (see equations 2.19 and 2.20, respectively). For a transparent solution ($\varepsilon \cdot [B] \cdot \ell \to 0$), reaction progress is first-order in reactant concentration, again regardless of whether we choose to use time or dose as the independent variable (see equations 2.26 and 2.27, respectively).

Recall also that the condition $\varepsilon \cdot [B] \cdot \ell \to 0$ has general application in photochemical reactor systems. In particular, if we allow the surface area of exposure (A) and the depth of the exposure ($\ell$) to both approach zero, then the system in question (the CR) becomes a point. Therefore, for a point in a system, the (local) photochemical reaction rate becomes:[l]

$$r_B = -\frac{d[B]}{dt} = \frac{2.303 \cdot E' \cdot \varepsilon \cdot [B]}{U} \cdot \Phi \tag{3.55}$$

Or equivalently:

$$\frac{d[B]}{E' \cdot dt} = -\frac{2.303 \cdot \varepsilon \cdot [B]}{U} \cdot \Phi \tag{3.56}$$

Equations 3.55 and 3.56 can be applied to a point in any photochemical reactor. However, their application to describe overall photoreactor behavior requires detailed information regarding the spatial distribution of the fluence rate, $E'$. For the time being, we will postpone descriptions of the models and measurement methods that are used to simulate or measure the fluence rate field in a reactor and the models of fluid flow that need to be coupled with them for purposes of predicting the behavior of continuous-flow UV photoreactors. Instead, we will consider the qualitative aspects of the fluence rate field and its implications for photoreactor performance.

### 3.4.3 Fluence Rate Fields in Photoreactors

UV radiation can be generated by a wide range of natural and manufactured sources (see Chapter 4). For most applications, the fluence rate within a system will demonstrate strong spatial variation, with sharp gradients in one or more coordinate directions.

Figure 3.12 provides several examples of fluence rate fields within photoreactor systems that have been measured or calculated. The details of the methods by which the fluence rate field can be characterized by measurement or simulation will be presented later in this book (see Chapters 5 and 6).

In contemporary photoreactors, the fluid to be treated will flow past one or more sources of UV radiation. For UV systems used to treat liquids, the radiation sources are often encased in a quartz envelope to provide protection of electrical components from the surrounding fluid, while allowing effective transmission of radiation generated by the source to the fluid as it flows past the radiation source.

From Chapter 2, we know that the local rate of a photochemical reaction is directly proportional to the local fluence rate. Therefore, the strong gradients in fluence rate that are evident in Figure 3.12 translate to equally strong spatial gradients in local photochemical reactions rates. For example, in Figure 3.12a–c, we see that the local fluence rate near the midpoint of a lamp is roughly 10x higher than the fluence rate near the end of the lamp, for locations at a fixed radial distance from the lamp. In Figure 3.12d and e, we see that local fluence rate varies by roughly an order of magnitude within a cross section from the surface of the quartz sleeve that surrounds a lamp to the location roughly half way between two lamps in an array. For perspective, the lamps in the array illustrated in Figures 3.11d and e are located on 7.5 cm centers. This information indicates that over distances of a few cm, we will also see gradients in local photochemical reaction rate of a factor 10; in some systems, we may see variation of 100x over a similar distance.

Most photoreactors are operated as flow-through systems, in which the fluid to be treated is pumped past one or more sources of radiation. The flow field in these systems, which is almost always in the turbulent regime, presents another level of complexity in the analysis of photoreactor behavior. Figure 3.13 illustrates these issues for a photoreactor in which the lamps are oriented vertically, with flow moving perpendicular to the lamp axis. This is one of many existing photoreactor configurations.

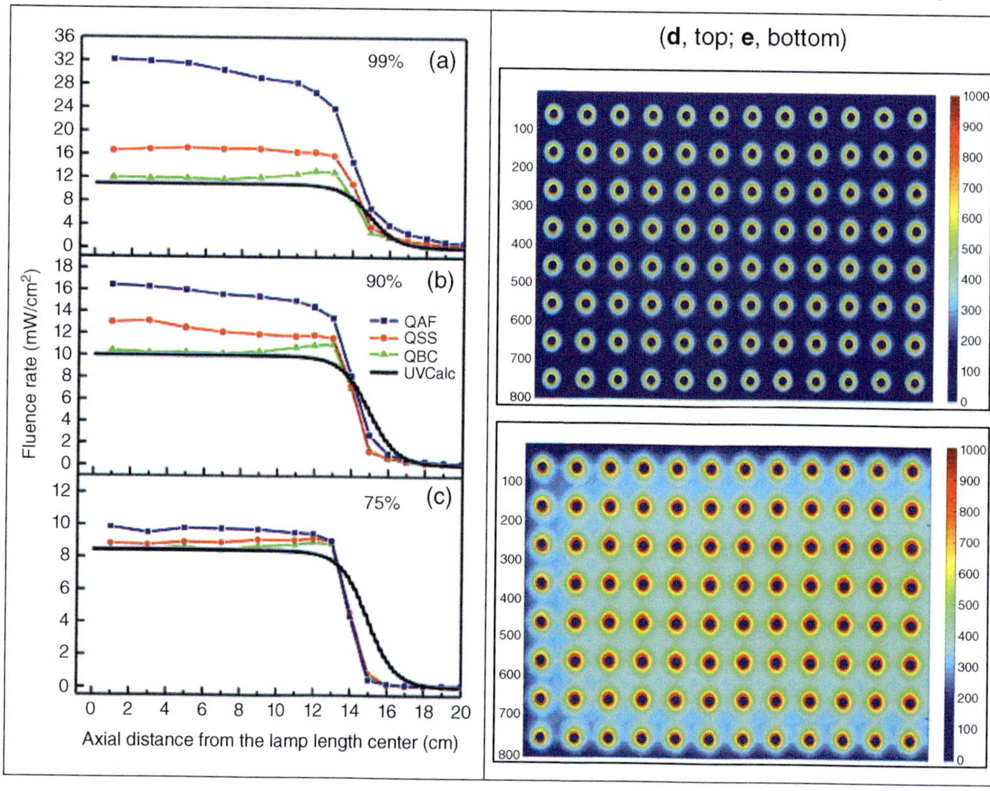


**Figure 3.12** Measured and simulated fluence rate fields in photoreactors. Panels a, b, and c illustrate measurements of fluence rate at a fixed radial distance (8 mm) from the quartz sleeve of a low-pressure Hg lamp using a micro-fluorescent silica detector (MFSD) and the numerical model UVCalc. For these distributions, the outer wall of the reactor was coated with materials of varying specular reflectance at 254 nm: quartz/aluminum foil (QAF, 80.5%); quartz/stainless steel (QSS, 26.2%); and quartz/black cloth (QBC; 11.1%) [3]. Panels d and e illustrate fluence rate fields calculated by the method of ray tracing in the cross section of a 12 × 8 array of low-pressure, high output Hg lamps under two different transmittance conditions: 70% (d) and 80% (e) through a 1.00 cm optical path length [4]. *Source (a-c):* Li et al. 2012/with permission of Elsevier.

Interaction of the flow field and the fluence rate field results in delivery of a wide range of UV doses. Figure 3.14 illustrates this for the reactor described in Figure 3.13.

Fluid flow through photoreactors is typically within the turbulent regime. In addition, there are often structures within the flow field that are oriented perpendicular to the nominal flow direction, which lead to the development of regions of vortex shedding (see Figure 3.13 bottom). Because of these features, the flow field in these systems includes randomness. One consequence of this is that no two particle trajectories through a photoreactor will be the same, even if their initial positions are identical.

Another aspect of particle trajectories through these systems is that they all yield a different UV dose. For the four trajectories illustrated in Figure 3.14, UV dose received by the particles varies by almost two orders of magnitude. Particle trajectory A is characterized by a received dose of 1.9 $mJ/cm^2$, whereas trajectory B yields a dose of 73.2 $mJ/cm^2$. By simulation of a large number of trajectories through such a system, it is possible to develop an estimate of the *dose distribution* delivered by a photoreactor (see Figure 3.15).

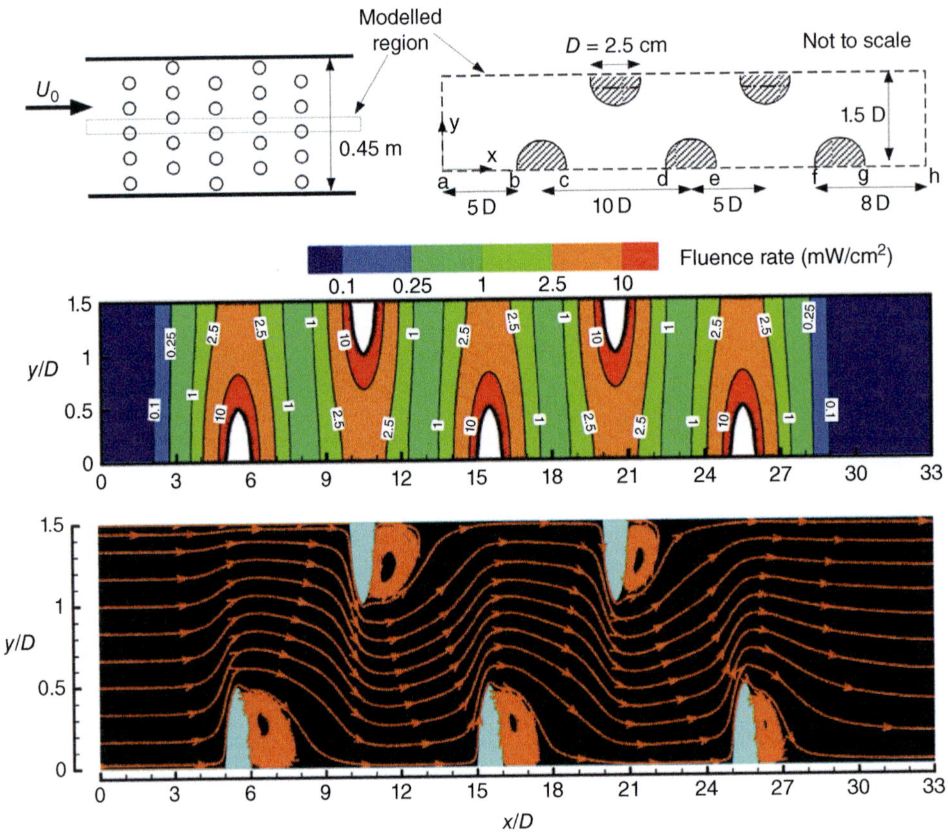


**Figure 3.13** Schematic illustration of UV photoreactor with vertical lamp orientation (images from Lyn et al. [5].). Top panel (left) illustrates a plan view of the system, as well as the central region of the reactor that was the focus of the modeling effort. The parameter *D* (top right) represents the diameter of the quartz sleeve that surrounded each lamp; in this system, *D* = 2.5 cm. Center panel represents a simulation of the fluence rate field as developed by the Line Source Integration (LSI) model [6]. Bottom panel illustrates the stream lines for the flow through the system, as simulated by computational fluid dynamics. The blue-shaded areas represent the lamp sleeves; the scales of the horizontal and vertical axes (both of which are normalized against *D*) cause apparent distortion from their conventional circular cross section. *Source:* Based on Lyn et al. 1999.

### 3.4.4 Effects of Mixing on Performance of UV Photoreactors

The dose distribution described by Figure 3.15 provides an illustration of the actual behavior of an operating UV photoreactor. This particular dose distribution has a bimodal character; as we will see, this distribution will lead to relatively inefficient process performance.

For applications such as disinfection, wherein the goal of treatment will involve 3–5 $\log_{10}$ units of conversion of the target, the low-dose portion of the dose distribution will govern overall reactor behavior because most of the organisms that "survive" (i.e. retain viability) in the effluent will have followed trajectories that yielded these low doses. As it turns out, most of these trajectories will be found along the wall of the reactor (see trajectory A in Figure 3.14).

This observation led to modifications of the reactor described above by inclusion of flow conditioning devices to alter these low-dose trajectories (see Figure 3.16). Modifications of the reactor wall forced particles that started at a near-wall location to move closer to one or more lamps as they moved through the irradiated zone, leading to improvements in reactor performance [8].

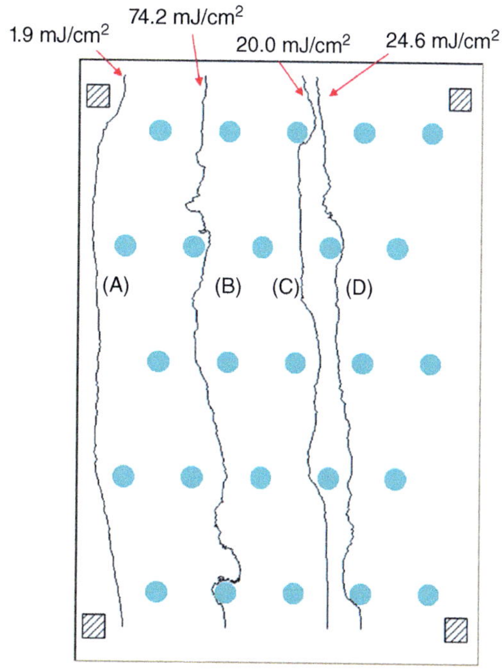


**Figure 3.14** Simulated particle trajectories through a photoreactor with a vertical lamp orientation. Flow in this system is nominally from bottom to top on this image (plan view) [7]. Solid lines illustrate simulated trajectories, as developed by a random walk model. Doses for each trajectory were calculated by mapping the fluence rate field onto each trajectory (see Figure 3.13) and integrating according to: dose = $\int_0^{\tau} E' \cdot dt$, where $\tau$= residence time of particle within irradiated zone of reactor. Capital letters (A, B, C, D) refer to each of four particle trajectories that are illustrated above. *Source:* Chiu et.al. 1999/The American Society of Civil Engineers.

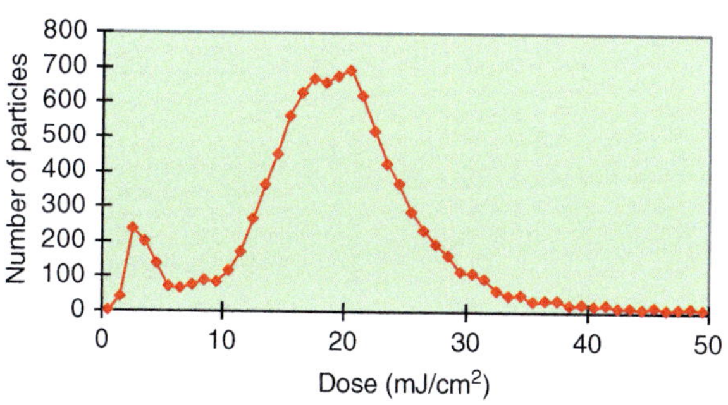


**Figure 3.15** Estimate of UV dose distribution delivered by a photoreactor illustrated in Figures 3.13 and 3.14 for a given operating condition [7]. The dose distribution estimate was developed by mapping the simulated trajectories of 10,000 particles through the irradiated zone of the photoreactor described in these preceding figures. For this simulation, the flow field was measured at a large number of points within the irradiated zone using laser Doppler velocimetry (LDV). These measurements were then applied as input to a random walk model to simulate individual trajectories, followed by mapping to the fluence rate field from an LSI simulation. *Source:* Chiu et al. 1999, figure 8 (p.14)/American Society of Civil Engineers.

Severin et al. [9, 10] conducted mathematical and experimental analysis of a simple annular reactor to examine the effects of mixing on reactor performance. The reactor that was the focus of their investigations is illustrated in Figure 3.17. The reactor housed a single low-pressure Hg lamp that was surrounded by a quartz sleeve. The reactor was built to include two impeller shafts that were rotated at high speed to allow an approach to a condition of complete-mix within the reactor volume. In the first phase of their study, one shaft that included five 3-bladed impellers was driven by a motor rated at 1550 rpm; as will be described below, this condition allowed a

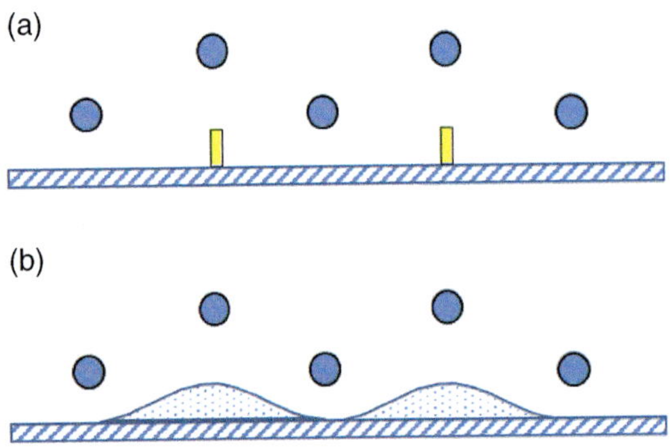


Figure 3.16 Schematic illustrations (plan view) of modifications made to walls of a vertical UV reactor to improve the dose distribution delivered by the reactor by diverting particles away from regions of the reactor characterized by low fluence rate [8]. (a) Perpendicular baffles; (b) sinusoidal waveforms. *Source:* Chiu et al. 1999, figure 1 (p.460)/American Society of Civil Engineers.

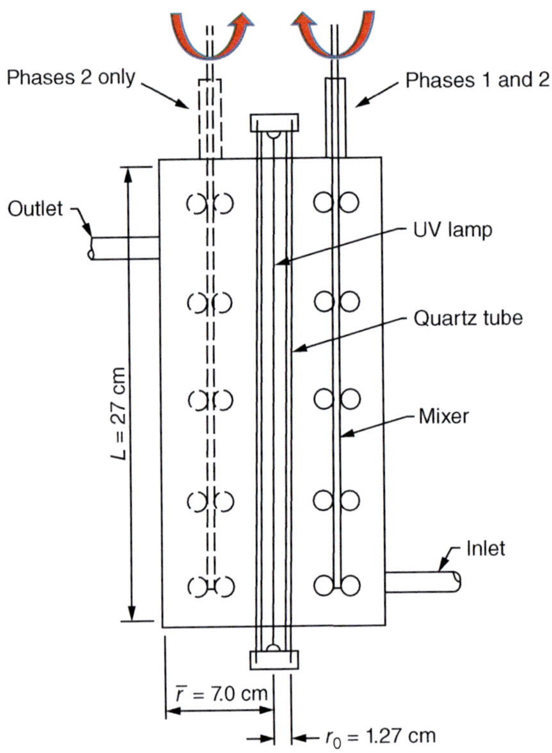


Figure 3.17 Schematic illustration of well-mixed annular photoreactor (modified from Severin et al. [9, 10]). The reactor had a total volume of 4.01 L. Impeller shafts were installed in the reactor to promote aggressive mechanical mixing. Impellers were rotated at high speed in opposing directions to promote a well-mixed condition.

condition of complete-mix to be approached based on a tracer test, but failed to meet a well-mixed condition with respect to UV disinfection. Because of this, a subsequent study phase involved modification of the reactor to include a second impeller shaft. For the second phase, the impeller shafts were both driven by motors that were rated at 6000 rpm; the impellers were rotated in opposite directions to promote mixing. This condition was observed to conform to a well-mixed condition with respect to disinfection. The reactor was operated over a range of flow rates, such that mean hydraulic detention time in their experiments ranged from $20\ s \leq \theta \leq 260\ s$. Absorbance at the characteristic wavelength of the Hg lamp was also adjusted over a wide range.

Several reactor models were developed in parallel with these experiments to describe overall reactor behavior. Included among these were the following:

- Completely mixed reactor – This model assumed that the reactor space was well-mixed in the radial and longitudinal directions. This model conforms to the case of the ideal CFSTR and as such represents a theoretical lower limit on performance.
- Ideal plug-flow – This model assumes complete mixing in the radial direction and no mixing in the longitudinal direction. This corresponds with the ideal PFR model and as such represents a theoretical upper limit on performance.

For modeling purposes, Severin et al. also presented a dimensionless group to describe performance of their UV disinfection system: $m \cdot k \cdot E_0' \cdot \theta^{m}$

Where

| | | | | |
|---|---|---|---|---|
| $E_0'$ | = | fluence rate at surface of quartz sleeve | [=] | $\frac{mW}{cm^2}$ |
| $m$ | = | ratio of average fluence rate within reactor to $E_0'$ | | |
| $k$ | = | rate constant for inactivation of microorganism | [=] | $\frac{cm^2}{mJ}$ |
| $\theta$ | = | mean hydraulic detention time within reactor | [=] | s. |

Severin et al. examined disinfection as the reaction of interest in their system. For their analysis, they invoked the series-event model, which is first-order with respect to the concentration of the target microbe.[n]

Figure 3.18 illustrates observed behavior and the predictions of the limiting models described above for inactivation of *E. coli* and f2 phage during phase 1 of the study of Severin et al. Despite the aggressive mechanical mixing that was applied during this portion of the experiment, the system deviated

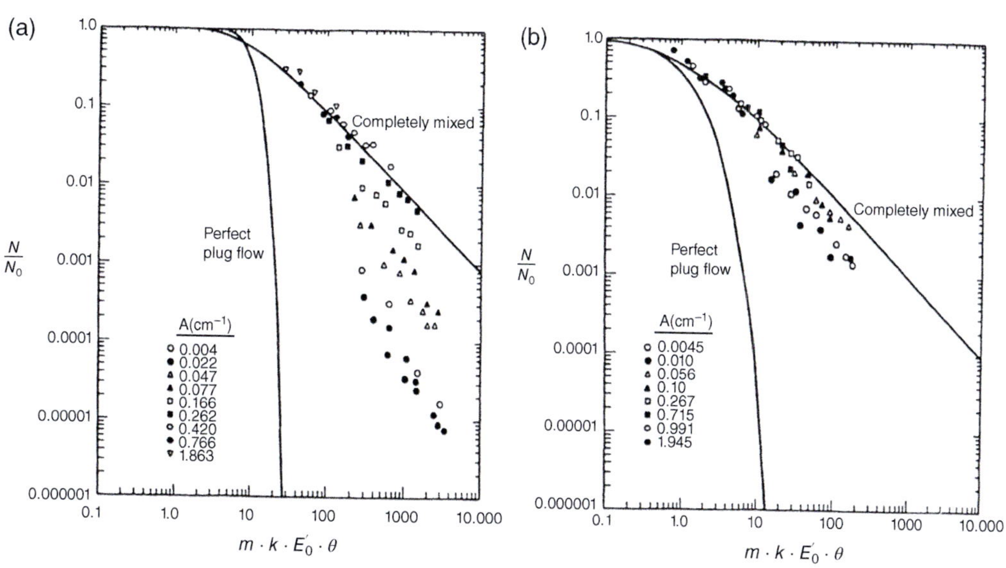


**Figure 3.18** Observed behavior for mechanically mixed reactor of Severin et al. 1984 during phase 1 of their study. Also included are predictions of reactor behavior based on the completely mixed and ideal PFR reactor models. (a) Measured and simulated behavior under original mixing condition; (b) measured and simulated behavior under modified (aggressive) mixing condition.

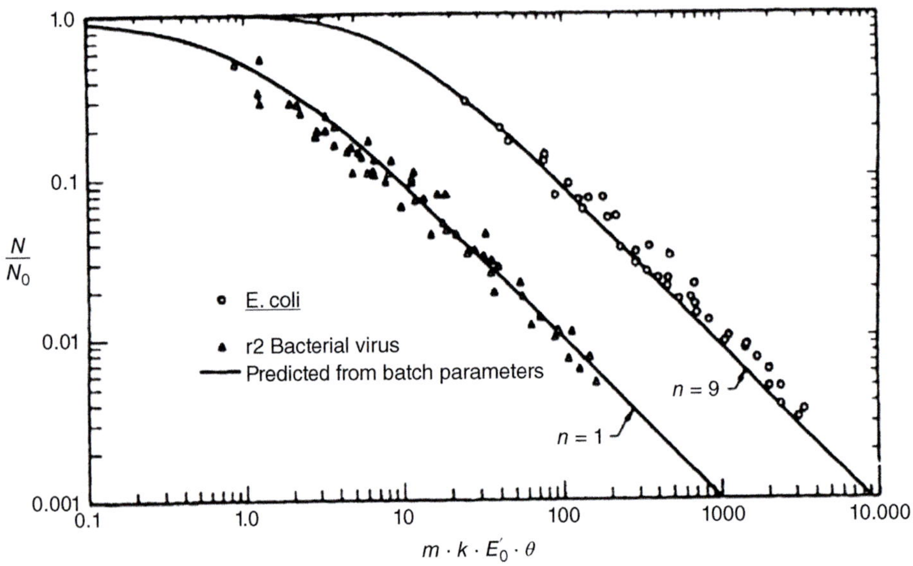


**Figure 3.19** Comparison of observed inactivation of *E. coli* and f2 phage with predictions of well-mixed reactor model. *Source:* Adapted from Severin et al. 1984.

substantially from the prediction of the well-mixed (CFSTR) model. Observed behavior was intermediate to the theoretical limits of performance predicted by these two limiting models.

As described above, Severin et al. added a second impeller device in phase 2 of their study to promote mixing and the approach to a well-mixed condition. The extreme measures taken in this second phase of their work allowed their reactor to closely approach the well-mixed condition, as illustrated by their comparisons of measured inactivation responses and predictions of the well-mixed reactor model (see Figure 3.19) and other measurements.

Several important lessons are evident from the work described above:

- Mixing behavior has profound effects on performance of UV photoreactors.
- Because most photochemical reactions of interest are rapid, the mixing conditions required to approach a well-mixed condition are severe. Moreover, since the well-mixed condition corresponds to a theoretical lower limit on performance, there is little motivation to promote such a condition in a practical photoreactor system, at least when high reactant conversion is a design objective.
- The ideal PFR corresponds with the theoretical upper limit on performance. This extreme is also difficult to reach; no commercially available reactor has reached this limit, but contemporary reactor designs have improved to allow this condition to be approached. Efficient UV photoreactors generally involve internal geometric features that promote controlled, strategic mixing in critical areas of the reactor, so as to reduce the likelihood of a fluid element receiving a low dose of radiation in traversing the irradiated zone of the reactor.

The work of Severin et al. [9, 10] demonstrates that it is possible to approach the limit of a well-mixed photoreactor; however, it appears to require aggressive mixing of the entire irradiated zone. The ability to approach a well-mixed condition is helpful for now in that it facilitates a direct comparison of the behaviors of chemical and photochemical reactor systems through the use of the CFSTR cascade model.

Borrowing from the geometry and internal mixing characteristics of the reactor used by Severin et al. Figure 3.20 provides a conceptual illustration of a photochemical CFSTR cascade. Each cell

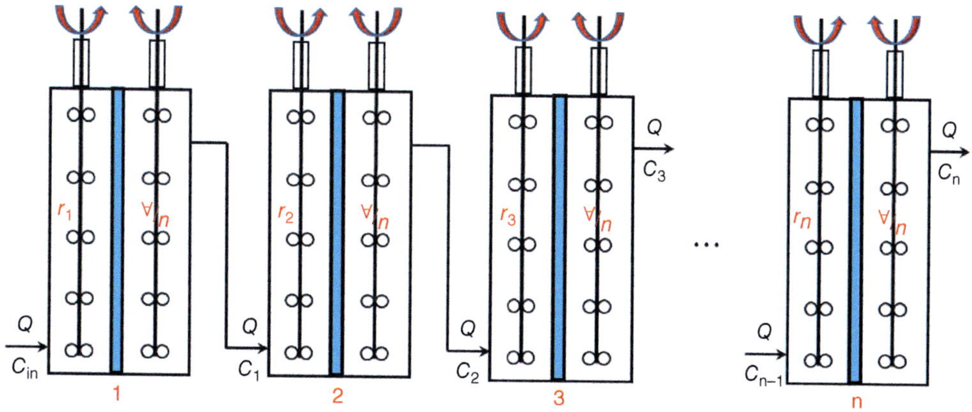


**Figure 3.20** Conceptual illustration of a photochemical CFSTR cascade. Each cell within the cascade is assumed to be well-mixed. As such, composition and the reaction rate within each cell are spatially uniform. Each cell in the cascade has a volume of $\frac{\forall}{n}$, where ∀=total system volume. The volumetric flow rate into and out of each cell is identical and constant, with a value of $Q$.

within this cascade would be aggressively mixed to allow spatial uniformity of composition and reaction rate within each cell.

As with the chemical CFSTR system, a material balance approach can be used to predict the performance of the photochemical CFSTR cascade. This will begin with application of the basic material balance equation to the $i^{\text{th}}$ cell in the cascade, assumed to be operated at steady-state:

$$0 = \frac{\forall}{n} \cdot \frac{dC_i}{dt} = Q(C_{i-1} - C_i) - k \cdot E'_{\text{avg}} \cdot C_i \frac{\forall}{n} \tag{3.57}$$

This equation is similar to equation 3.36, which described the behavior of the $i^{\text{th}}$ cell in a chemical CFSTR cascade system. However, two important mathematical differences are evident. First, the rate constant ($k$) in the chemical reactor has been replaced by the product $k \cdot E'_{\text{avg}}$, which is used to describe the rate of the photochemical reaction within the system. Strictly speaking, the rate of the photochemical reaction within each cell will demonstrate spatial gradients that are identical to the gradients in local fluence rate; however, for the systems included above, the aggressive mechanical mixing in the system will overcome this behavior, resulting in (effectively) spatially uniform composition. The second difference is that the "rate constant" in equation 3.57 is defined as:

$$k = \frac{2.303 \cdot \varepsilon \cdot \Phi}{U} \tag{3.58}$$

Rearrangement of equation 3.57 yields:

$$0 = \frac{n \cdot (C_{i-1} - C_i)}{\theta} - k \cdot E'_{\text{avg}} \cdot C_i \tag{3.59}$$

Separation of variables by algebra then yields:

$$k \cdot E'_{\text{avg}} \cdot \theta \cdot C_i = n \cdot (C_{i-1} - C_i) \tag{3.60}$$

Then:

$$C_i \cdot \left(n + k \cdot E'_{\text{avg}} \cdot \theta\right) = n \cdot C_{i-1} \tag{3.61}$$

Lastly:

$$\frac{C_i}{C_{i-1}} = \frac{n}{n + k \cdot E'_{\text{avg}} \cdot \theta} = \frac{1}{1 + \dfrac{k \cdot E'_{\text{avg}} \cdot \theta}{n}} \tag{3.62}$$

As before, if we recognize that overall reactor performance can be described by the following product:

$$\frac{C_n}{C_{\text{in}}} = \frac{C_1}{C_{\text{in}}} \cdot \frac{C_2}{C_1} \cdot \frac{C_3}{C_2} \cdots \frac{C_n}{C_{n-1}} \tag{3.63}$$

The overall system performance is described by:

$$\frac{C_n}{C_{\text{in}}} = \left( \frac{1}{1 + \dfrac{k \cdot E'_{\text{avg}} \cdot \theta}{n}} \right)^n \tag{3.64}$$

Equation 3.64, which describes the performance of a photochemical CFSTR cascade, is mathematically identical to equation 3.41, which describes analogous behavior in a chemical CFSTR cascade. As such, we should expect the solutions of these two equations to be similar, both in mathematical and graphical terms.

The CFSTR model can be used to illustrate a range of dose distributions that could be relevant in UV photoreactors. Figure 3.21 illustrates this behavior in the form of normalized UV dose distributions delivered by (hypothetical) CFSTR photoreactor cascades.

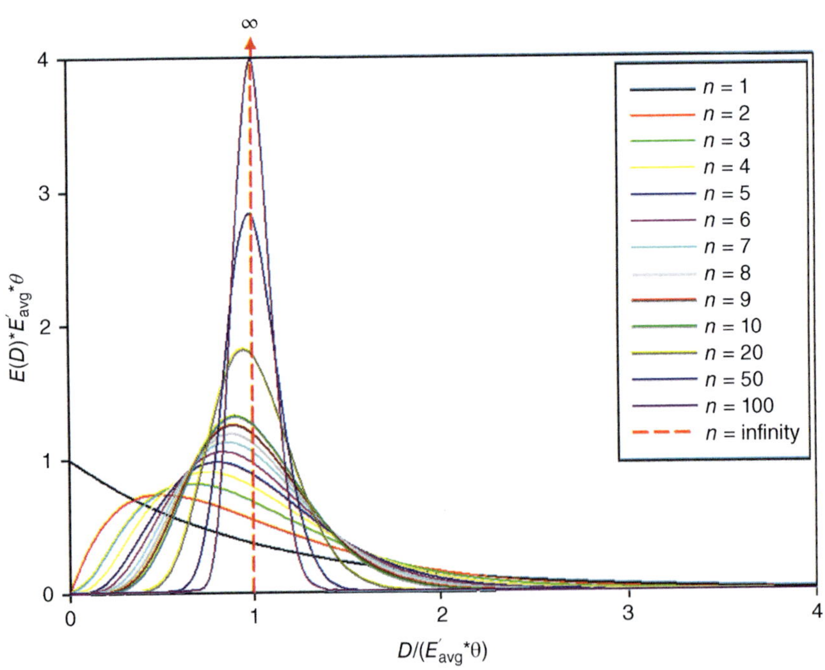


**Figure 3.21** Normalized UV dose distributions for CFSTR cascades for values of $n$ ranging from 1 to $\infty$. Normalization is based on the product $E'_{\text{avg}} \cdot \theta$. Note that $E'_{\text{avg}} \cdot \theta$ will be close, but probably not identical to the average of the dose distribution, $D_{\text{avg}}$, for reasons that are similar to those that result in $\bar{t} \neq \theta$ in RTDs (see Figure 3.10).

As with the RTDs that were presented in Figure 3.10, the extremes of the distributions illustrated in Figure 3.21 (specifically, $n = 1$ and $n = \infty$) are probably unrealistic and are difficult to approach in real reactors. They are included here largely for purposes of illustration. Realistic examples of dose distributions delivered by UV photoreactors will be presented in later chapters.

In examining the dose distributions from Figure 3.21, several important features emerge. Considering first the case of the well-mixed system ($n = 1$), we observe substantial "short-circuiting" of the reactor in that a large fraction of the fluid elements that traverse the system will receive a low dose of radiation; in fact, roughly half of the fluid elements will receive a dose less than $E'_{\mathrm{avg}} \cdot \theta$. Dose received by a fluid element can be calculated as:

$$\mathrm{dose} = \int_0^{\tau} E'(t) \cdot \mathrm{d}t \tag{3.65}$$

where

| | | | | |
|---|---|---|---|---|
| $E'(t)$ | = | fluence rate received by fluid element | [=] | $\frac{\mathrm{mW}}{\mathrm{cm}^2}$ |
| $t$ | = | time | [=] | s |
| $\tau$ | = | total exposure time (total time spent in irradiated zone) | [=] | s. |

A low dose can be the result of a short exposure time, exposure to low fluence rate, or both. The dose distribution corresponding to the well-mixed system also includes doses that are well above $E'_{\mathrm{avg}} \cdot \theta$. These will be the result of fluid trajectories through the irradiated zone that are characterized by a large value of $\tau$, high $E'(t)$, or both. Target reactants that are contained within these fluid elements are likely to react; however, the overall performance of the system will be poor because of the fluid elements that receive low doses. Even if there were a reason to design or build such a reactor, it would be difficult to do because of the mixing requirements.

At the other extreme is the behavior of the ideal PFR system ($n = \infty$). The dose distribution for this system is defined by a delta ($\delta$) function. Mixing conditions in an ideal PFR would be such that all fluid elements received the same dose, $E'_{\mathrm{avg}} \cdot \theta$.

### 3.4.5 Prediction of Performance in Photochemical Reactors – Segregated-Flow Model

Figure 3.22 provides a schematic illustration of the factors that affect performance of a UV photoreactor. The goal in describing these systems is to develop models that can be used to accurately predict reactor performance (i.e. concentration of the target compound in treated fluid, C). To develop such a prediction, we need information to describe the kinetics of the photochemical reactions that occur within the system, the concentration of the reactant in the fluid that enters the reactor, and the UV dose distribution delivered by the reactor. The dose distribution is governed by the combined effects of the flow field (i.e. fluid mechanics) and the fluence rate field. In turn, the flow field is governed by the applied flow rate and the geometry of the reactor (if we assume that fluid properties such as density and viscosity remain constant), while the fluence rate field will depend on the reactor geometry, lamp output power, and the optical characteristic of the materials and media in the system (e.g. UV transmittance of treated fluid).

The segregated-flow model can be used to describe the behavior of photochemical reactors. However, unlike its application to chemical reactors wherein the independent variable is time, for photochemical reactors the independent variable is dose. Equation 3.66 illustrates the mathematical form of the segregated-flow model for prediction of photochemical reactor performance:

$$\left(\frac{C}{C_0}\right)_{\mathrm{reactor}} = \int_0^{\infty} \left(\frac{C(D)}{C_0}\right)_{\mathrm{batch}} \cdot E(D) \cdot \mathrm{d}D \tag{3.66}$$

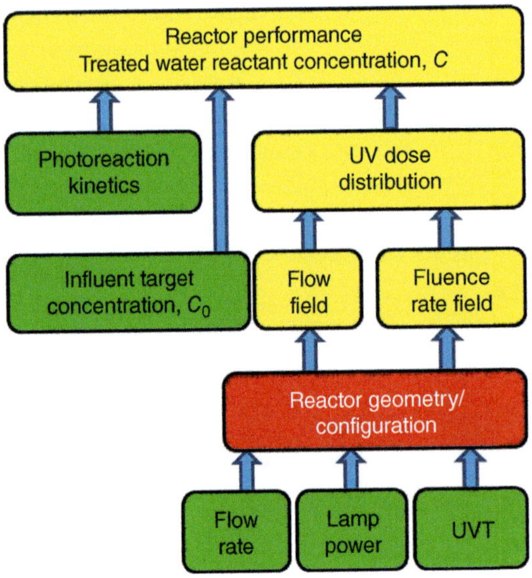


**Figure 3.22** Schematic illustration of factors that affect performance in UV photoreactors. Arrows represent the direction of dependence; yellow boxes indicate attributes of the system that depend on other features of the system; green boxes indicate attributes that are measurable, but that may demonstrate variability; red box illustrates attributes that are measurable but fixed.

where

$\left(\frac{C}{C_0}\right)_{\text{reactor}}$ = fraction of reactant remaining in fluid treated by reactor

$\left(\frac{C(D)}{C_0}\right)_{\text{batch}}$ = fraction of reactant remaining after exposure to a given dose ($D$)
= intrinsic kinetics of photoreaction of interest

$E(D)$ = dose distribution function $[=]\ \frac{\text{m}^2}{\text{J}}$

$D$ = dose $[=]\ \frac{\text{J}}{\text{m}^2}$.

Note that the segregated-flow model can also be derived from principles of probability theory. This derivation is presented in Appendix 3.C.

The dose distribution function is an analog to the RTD that is often used to describe the behavior of chemical reactor systems. Like the RTD, the dose distribution function is a probability density function. As such, the following conditions apply:

$$E(D) \cdot \text{d}D = \text{fraction of fluid elements receiving dose between } D \text{ and } D + \Delta D \tag{3.67}$$

$$\int_0^{\delta} E(D) \cdot \text{d}D = \text{fraction of fluid elements receiving dose} \leq \delta \tag{3.68}$$

$$\int_0^{\infty} E(D) \cdot \text{d}D \equiv 1 \tag{3.69}$$

Application of the segregated-flow model for prediction of photoreactor performance requires that the term $\left(\frac{C(D)}{C_0}\right)_{\text{batch}}$ be explicitly defined. As described above, the physical interpretation of this term is a mathematical description of the intrinsic kinetics of the reaction(s) of interest (i.e. a model of reaction kinetics). For purposes of the application of this model, our focus is on a small parcel of fluid, or perhaps even an individual particle or molecule in the system. With that perspective, it is evident that our perspective is equivalent to a point in space. As such, we can use equation 2.27 (energy-based) or equation 2.38 (photon-based) to describe the term $\left(\frac{C(D)}{C_0}\right)_{\text{batch}}$. We will use both equations at various points in this book to quantify photoreactor performance.

However, for the time being, we will examine a third approach that allows a direct parallel with the preceding discussion of performance of chemical reactor systems.

Equation 3.55 can be rearranged in a form that mimics the equation used to describe the kinetics of an elementary, bimolecular reaction:

$$\frac{d[B]}{dt} = -\left(\frac{2.303 \cdot \varepsilon \cdot \Phi}{U}\right) \cdot E' \cdot [B] \tag{3.70}$$

The quotient in parentheses $\left(\frac{2.303 \cdot \varepsilon \cdot \Phi}{U}\right)$ can be thought of as a rate constant. It includes terms to describe both the efficiency by which a target absorbs radiation ($\varepsilon$) and the efficiency by which absorbed radiation is used to promote a reaction ($\Phi$). The terms $E'$ and $[B]$ describe the availability of the two "reactants" (photons and the molecule B, respectively) to participate in this reaction. This is entirely analogous to the logic we used to describe the kinetics of an elementary, bimolecular chemical reaction (see equations 2.5 and 2.6).

With this logic in mind, equation 3.70 can be rewritten to conform to the conventional used in describing chemical kinetics:

$$\frac{d[B]}{dt} = -k \cdot E' \cdot [B] \tag{3.71}$$

where

$$k = \frac{2.303 \cdot \varepsilon \cdot \Phi}{U} \tag{3.72}$$

Equation 3.71 is first-order with respect to the reactant concentration. Therefore, it can be applied directly and rigorously to the segregated-flow model. Figure 3.23 illustrates the application of the segregated-flow model for prediction of steady-state reactor behavior in a several hypothetical

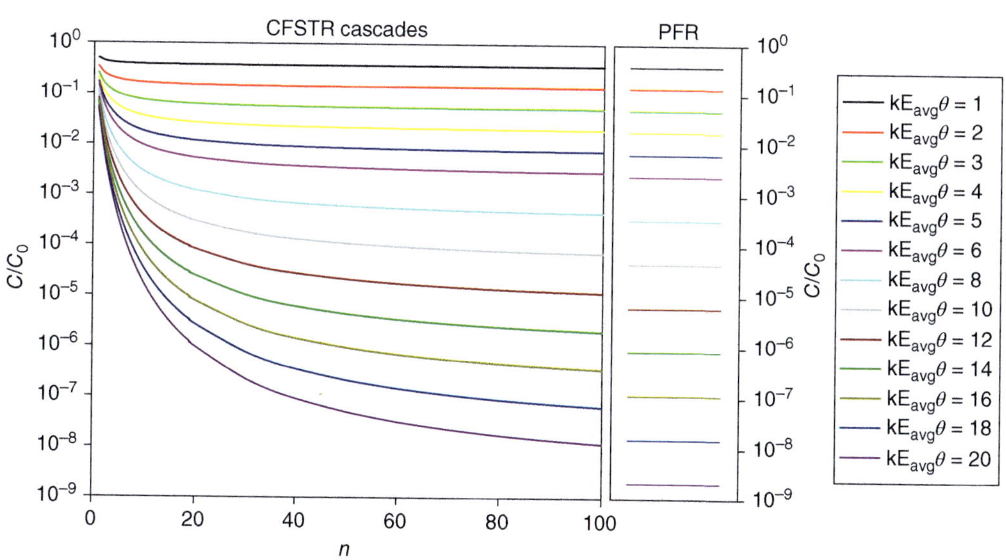


**Figure 3.23** (left panel) Fractional reactant conversion $\left(\frac{C}{C_0}\right)$ as a function of number of cells ($n$) in CFSTR photoreactor cascades for a range of values of $k \cdot E'_{avg} \cdot \theta$ (see legend at right). Also included (center panel) are calculated fractional reactant conversion values for an ideal PFR with the same conditions. Note that the vertical axes on the left and center panels are presented on a $\log_{10}$ scale.

CFSTR photoreactor cascades. Note that this figure has exactly the same shape and features as Figure 3.8, which illustrates predictions of steady-state performance of a chemical reactor system, as simulated by the CFSTR cascade model, with a first-order consumptive reaction. In fact, these two figures illustrate a number of important parallels that exist between chemical and photochemical reactors.

As before, in dealing with photoreactors for which the extent of conversion is fairly modest, say less than 1 $\log_{10}$ unit, the details of mixing behavior and the dose distribution become relatively unimportant. In such a case, estimates of reactor performance based on a detailed description of the dose distribution (and presumably the mixing behavior that describes it) will be similar to predictions based on an average dose delivered by the reactor. On the other hand, for systems where larger extents of conversion are required, the details of the dose distribution are important; accurate predictions of photoreactor performance under such conditions require accuracy in simulating or measuring the dose distribution as well as the fundamental chemical properties that govern the intrinsic kinetics of photoreactions, namely $\varepsilon$ and $\Phi$.

## 3.5 Executable Model

Simulations of photochemical reactor dynamics require tools that account for the dose distribution delivered by the reactor, the intrinsic kinetics of the reaction of interest, and the concentration of the photochemical target(s) (see Figure 3.22). In turn, accurate estimates of the dose distribution require detailed knowledge of fluid mechanics and the spatial distribution of radiant energy within the irradiated zone of the reactor. Contemporary analyses of photochemical reactors often involve the use of advanced numerical models that allow for detailed, accurate simulations of fluid mechanics and the spatial distribution of radiant energy, using tools such as computational fluid dynamics and ray tracing, respectively. Details of these and other numerical modeling methods will be presented in Chapter 6. When applied properly, these modeling techniques can provide detailed, accurate representations of the physics that govern photochemical reactor dynamics, and by extension the performance of these systems.

In broad terms, numerical simulations of fluid mechanical behavior in photochemical reactors need to account for advective transport within the irradiated zone, as well as the turbulence characteristics of the flow field; most continuous-flow UV photoreactors operate in the turbulent regime. Numerical models of the spatial distribution of radiant energy should account for the output power of the UV source, as well as the optical phenomena that govern the spatial distribution of radiant energy within the reactor system. Although models that incorporate thorough descriptions of these phenomena are very powerful for simulation of photoreactor dynamics and design, they also tend to be complicated.

As a pedagogical tool, a numerical model has been developed to capture the central features of these numerical simulations, without having to address some of the details that often accompany them.[o] This simplified model follows a Lagrangian (particle-centered) approach that allows visualization of the trajectories of individual particles as they traverse the irradiated zone of a simple UV photoreactor.

The model was developed to allow simulations of the behavior of a cylindrical reactor with a single UV lamp, located along its central axis. The model prompts the user to define lamp power, transmittance of fluid flowing through the reactor (presumed to be water), and volumetric flow rate through the reactor. The model also allows users to define reactor dimensions, characteristics of the computational grid, and some features of the graphical output. The fluence rate field for the system is simulated using a point-source summation (PSS) approach (see Chapter 6). The flow field is simulated as a parabolic function, onto which random fluctuations are superimposed. The flow field is interrogated to allow simulation of the trajectories of individual particles, based on an initial position and the characteristics of the flow field described above. Dose increments are calculated for each "step" a particle takes through the system.

Output of the model is presented on four figures that provide animation.[p] Figure 1 illustrates a cross section of the system with the fluence rate profile indicated by a colored spectrum; a two-dimensional representation of a particle trajectory is superimposed on this cross section. Figure 2 illustrates a longitudinal profile (*y*-*z*) of the particle trajectory; also included in Figure 2 is a panel that provides a running summary of the residence time ($t$), longitudinal position ($z$), dose gradient, and accumulated dose of the particle. When a particle reaches the end of the computational domain, the final values of these parameters are displayed. Figure 3 provides a radial profile (*r*-*z*) of a particle trajectory. Figure 4 provides a three-dimensional rendering of a particle trajectory.

This model was developed as a tool to illustrate the dynamic features of continuous-flow UV photoreactors. It is intended for use as a teaching tool and not for research and development. The simulations of particle trajectories and accumulated dose are qualitatively correct, but probably not representative of any real reactor. However, the model allows users to explore and visualize the effects of various input parameters on process dynamics in UV photoreactors. Students are encouraged to use this model to examine the responses of the simulated reactor to changes in input parameters. Problems will be provided at the end of this chapter to provide opportunities to apply the model and to interpret its implications.

## 3.6 Summary

The principles of material balance inform the development of mathematical models that are used to describe reactor performance. Classical reactor theory has been developed to describe the behavior of chemical reactors. These same principles inform the development of photochemical reactor theory. The primary differences between the models that are used to describe chemical reactor and photochemical reactor performance relate to the nature of the reactions that take place in these systems and the mathematical expressions that are used to describe them.

Application of these tools has revealed several generalizations about chemical and photochemical reactors and the factors that govern their performance:

- Overall reactor performance is governed by the intrinsic kinetics of the reactions of interest, the time/dose allowed for the reactions to progress, and internal mixing behavior within the reactor.
- The ideal CFSTR model provides a theoretical lower limit on reactor performance. This model implies a well-mixed condition throughout the reactor volume.
- The ideal PFR model provides a theoretical upper limit on reactor performance. This model implies a condition of no mixing in the direction of flow and spatially uniform reaction rate (i.e. spatially uniform reactant concentration or fluence rate field) in the directions perpendicular to the flow direction.
- Knowledge of the RTD (chemical reactor) or dose distribution (photochemical reactor) allows simulation of reactor performance by application of the segregated-flow model. These simulations should be accurate if the reactions of interest display kinetics that are first-order with respect to the reactant.

## Notes

**a** In the context of this chapter, the term "chemical reactor" will be used to refer to a system in which reactant conversion is brought about by one or more thermal chemical and/or biochemical reactions. The term "photochemical" reactor will be used to describe a system in which reactant conversion is initiated by a photochemical event that results in direct transformation of a reactant or formation of a reactive intermediate that causes a subsequent indirect transformation of a reactant.

**b** It is possible to develop a mass balance for a system with flexible (dynamic) boundaries, but systems with flexible boundaries are unusual in the UV applications we are working toward in this chapter and will bring additional complexity that may not be warranted in the presentation of this material.

**c** The subscript "$j$" refers to the $j^{th}$ constituent in this system. In other words, one mass-balance expression is required for each constituent of interest in the system.

**d** This form of the material balance has been established using moles as a measure of the amount of constituent. However, it may be equally valid to quantify the amount of a constituent with other terms, such as mass, number of organisms in a population, Curies of radioactive material, etc. The units we select to quantify the amount of material in a material balance will depend on the application and the material in question, but the basic form of material balance equations will remain.

**e** From this point forward, the subscript $j$ will be omitted for purposes of clarity.

**f** Pseudo-first-order conditions will apply when the concentration of one reactant varies with time. This can occur as a result of several circumstances, such as when one reactant that participates in a second-order reaction is present in large stoichiometric excess relative to the reactant of interest.

**g** The word "ideal" is included to emphasize that no real reactor system will ever reach this condition. In this case, the word ideal relates to the stirred (well-mixed) condition of the reactor. As described above, the well-mixed assumption implies that the timescale for mixing is short relative to the timescale for reaction. In practical terms, this assumption will hold when the system is aggressively mixed and the reaction(s) that take place within the system is (are) relatively slow. As either of these conditions is relaxed, the ability of the CFSTR model actual system performance will diminish.

**h** This is not a requirement of the CFSTR model, but is often assumed in reactor analysis.

**i** The velocity term, $V_x$, is brought outside of the derivative in this example because it is assumed to be constant.

**j** A formal derivation of this expression is provided in Appendix 3.B.

**k** As a reminder, the derivation of photochemical reaction kinetics presented in Chapter 2 for a batch reactor does not account for a number of factors that will affect the dynamic behavior of these systems. A more detailed, complete analysis of the behavior of these systems is presented in Chapter 5.

**l** Terms in equations 3.55 and 3.56 were defined in Chapter 2.

**m** Note: 1 mJ ≡ 1 mW·s.

**n** Details of the series-event model and other models of UV disinfection kinetics will be presented later in this book. For purposes of this analysis, the critical issue is that the kinetics of disinfection are modeled to display first-order kinetics with respect to the concentration of the target microbe or virus.

**o** The model is available as an executable program, written in Matlab. The model was developed by Mr. Nikhil Sangwan. The filename for this program is "Single_Lamp_Cylindrical_Reactor.m" and is available as an editable Matlab file.

**p** Figure numbers listed in this paragraph refer to Figure numbers that are assigned within the program, not Figures that appear within this chapter.

## References


1 Bird, R.B., Stewart, W.E., and Lightfoot, E.N. (**1960**). *Transport Phenomena*. New York: Wiley.
2 Levenspiel, O. (**1972**). *Chemical Reaction Engineering*, 2e. New York: Wiley.
3 Li, M.K., Qiang, Z.M., Bolton, J.R., and Ben, W.W. (**2012**). Impact of reflection on the fluence rate distribution in a UV reactor with various inner walls as measured using a micro-fluorescent silica detector. *Water Research 46* (11): 3595–3602.
4 Ahmed, Y.M.H.I.H. (**2017**). *Stochastic Evaluation of Disinfection Performance in Large-Scale, Open-Channel UV Photoreactors*. West Lafayette, IN: Purdue University.
5 Lyn, D.A., Chiu, K., and Blatchley III, E.R. (**1999**). Numerical modeling of flow and disinfection in UV disinfection channels. *Journal of Environmental Engineering-Asce 125* (1): 17–26.
6 Blatchley III, E.R. (**1997**). Numerical modelling of UV intensity: Application to collimated-beam reactors and continuous-flow systems. *Water Research 31* (9): 2205–2218.
7 Chiu, K., Lyn, D.A., Savoye, P., and Blatchley III, E.R. (**1999**). Integrated UV disinfection model based on particle tracking. *Journal of Environmental Engineering-Asce 125* (1): 7–16.
8 Chiu, K.P., Lyn, D.A., Savoye, P., and Blatchley III, E.R. (**1999**). Effect of UV system modifications on disinfection performance. *Journal of Environmental Engineering-Asce 125* (5): 459–469.
9 Severin, B.F., Suidan, M.T., and Engelbrecht, R.S. (**1984**). Mixing effects in UV disinfection. *Journal Water Pollution Control Federation 56* (7): 881–888.
10 Severin, B.F., Suidan, M.T., and Engelbrecht, R.S. (**1983**). Kinetic modeling of UV disinfection of water. *Water Research 17* (11): 1669–1678.
11 Walpole, R.E., Myers, R.H., Myers, S.L., and Ye, K. (**2012**). *Probability & Statistics for Engineers & Scientists*, 9e. Boston: Prentice Hall.


## Appendix 3.A Derivation of Relationship to Describe Transient (Start-up) Behavior in an Ideal CFSTR

The general governing equation to describe the dynamic behavior of an ideal CFSTR for the case of a single, consumptive, first-order reaction was presented previously as equation 3.16, copied below:

$$\frac{dC}{dt} = \frac{C_{in} - C}{\theta} - k \cdot C \tag{3.A.1}$$

We seek a closed-form (analytical) solution to this equation to describe the dynamic behavior of the reactor for a specific condition, namely reactor start-up. For this calculation, we will assume that the reactant concentration within the reactor is identical to the inlet concentration, and that the consumptive reaction is allowed to start at $t = 0$.

To develop this solution, we first apply an algebraic manipulation of equation 3.A.1 to isolate all concentration (C)-dependent terms on the left side of the equation:

$$\frac{dC}{dt} + C\left(k + \frac{1}{\theta}\right) = \frac{C_{in}}{\theta} \tag{3.A.2}$$

The right side of equation 3.A.2 is independent of concentration ($C$), but is a function of the independent variable (time); in this case, the right side is a constant.

Equation 3.A.2 is an example of a linear, first-order ordinary differential equation, which emerges in many engineering applications and can be solved by many approaches. A commonly applied method to solve such an equation involves the *integrating factor*. We first define the integrating factor ($\alpha$), which in this case will be the constant term by which concentration is multiplied on the left side of equation 3.A.2:

$$\alpha = k + \frac{1}{\theta} \tag{3.A.3}$$

Substitution of this new term back into equation 3.A.2 yields:

$$\frac{\mathrm{d}C}{\mathrm{d}t} + \alpha \cdot C = \frac{C_{\mathrm{in}}}{\theta} \tag{3.A.4}$$

We are looking for a solution (function) that allows the time-derivative and concentration terms to behave in a similar manner, so that the sum of these terms (left side of equation 3.A.4) can be equal to the right side of the equation, which in this case is a constant, $\frac{C_{\mathrm{in}}}{\theta}$. One function that displays this type of behavior is the exponential function. So, we will choose to multiply both sides of the equation by the term: $e^{\alpha \cdot t}$:

$$e^{\alpha \cdot t} \cdot \left(\frac{\mathrm{d}C}{\mathrm{d}t} + \alpha \cdot C\right) = \frac{C_{\mathrm{in}}}{\theta} \cdot e^{\alpha \cdot t} \tag{3.A.5}$$

However, we may also recognize that the following identity applies by the product rule:

$$\frac{\mathrm{d}(C \cdot e^{\alpha \cdot t})}{\mathrm{d}t} = \alpha \cdot C \cdot e^{\alpha \cdot t} + e^{\alpha \cdot t} \cdot \frac{\mathrm{d}C}{\mathrm{d}t} \tag{3.A.6}$$

But inspection of the right side (and left side) of equation 3.A.6 reveals that it is identical to the left side of equation 3.A.5:

$$\frac{\mathrm{d}(C \cdot e^{\alpha \cdot t})}{\mathrm{d}t} = e^{\alpha \cdot t} \cdot \left(\frac{\mathrm{d}C}{\mathrm{d}t} + \alpha \cdot C\right) \tag{3.A.7}$$

So, we can substitute the left side of equation 3.A.7 into equation 3.A.5:

$$\frac{\mathrm{d}(C \cdot e^{\alpha \cdot t})}{\mathrm{d}t} = \frac{C_{\mathrm{in}}}{\theta} \cdot e^{\alpha \cdot t} \tag{3.A.8}$$

Multiplying both sides of equation 3.A.8 by $\mathrm{d}t$ yields:

$$\mathrm{d}(C \cdot e^{\alpha \cdot t}) = \frac{C_{\mathrm{in}}}{\theta} \cdot e^{\alpha \cdot t} \cdot \mathrm{d}t \tag{3.A.9}$$

Integrating both sides of equation 3.A.9 gives us:

$$C \cdot e^{\alpha \cdot t} = \int \frac{C_{\mathrm{in}}}{\theta} \cdot e^{\alpha \cdot t} \cdot \mathrm{d}t \tag{3.A.10}$$

We then evaluate the indefinite integral in equation 3.A.10 to develop a general solution:

$$C \cdot e^{\alpha \cdot t} = \frac{C_{\mathrm{in}}}{\alpha \cdot \theta} \cdot e^{\alpha \cdot t} + \beta \tag{3.A.11}$$

where $\beta$ = constant of integration. We can evaluate $\beta$ by considering the condition of the reactor at our initial condition, $t = 0$. For this simulation, we have made the assumption that the reactant concentration in the reactor at $t = 0$ is the influent concentration, $C_{\mathrm{in}}$. Substitution of this value and its corresponding time ($t = 0$) into equation 3.A.11 yields:

$$C_{\mathrm{in}} = \frac{C_{\mathrm{in}}}{\alpha \cdot \theta} + \beta \tag{3.A.12}$$

With this information, we can solve for the constant of integration by algebra:

$$\beta = C_{\text{in}} \cdot \left(1 - \frac{1}{\alpha \cdot \theta}\right) \tag{3.A.13}$$

Substitution of the constant of integration back into the general solution (equation 3.A.11) gives us:

$$C \cdot e^{\alpha \cdot t} = \frac{C_{\text{in}}}{\alpha \cdot \theta} \cdot e^{\alpha \cdot t} + C_{\text{in}} \cdot \left(1 - \frac{1}{\alpha \cdot \theta}\right) \tag{3.A.14}$$

Now, we can divide both sides of this equation by $e^{\alpha \cdot t}$:

$$C = \frac{C_{\text{in}}}{\alpha \cdot \theta} + C_{\text{in}} \cdot e^{-\alpha \cdot t} \cdot \left(1 - \frac{1}{\alpha \cdot \theta}\right) \tag{3.A.15}$$

Lastly, we substitute the definition of the integrating factor (equation 3.A.3) into equation 3.A.15 and do a little bit of algebra to find an expression to define the time-dependent concentration of our reactant in the reactor for this condition, which corresponds with a "start-up" condition:

$$C = \frac{C_{\text{in}}}{1 + k \cdot \theta} + C_{\text{in}} \cdot \exp\left(\frac{-t}{1 + k \cdot \theta}\right) \cdot \left(\frac{k \cdot \theta}{1 + k \cdot \theta}\right) \tag{3.A.16}$$

## Appendix 3.B Derivation of Normalized Residence Time Distribution Functions for CFSTR Cascade Systems

The CFSTR cascade model can be used to describe serial reactor systems; it can also be used to simulate the behavior of arbitrary reactor systems by using the parameter "$n$" (number of cells in cascade) as a fitting parameter to describe mixing behavior that is "equivalent" to a CFSTR cascade.

A critical step in the use of the cascade model is description of the RTD function. The approach described below will allow definition of mathematical relationships to describe the RTD as a density distribution function for a single CFSTR, for two identical CFSTRs in series, and then for three identical CFSTRs in series. For each of these three systems, the volumetric flow rate ($Q$) and total system volume ($\forall$) will be constant. The behavior of the system will be examined in response to an ideal pulse input of an inert tracer of mass M at $t = 0$ to the first cell in the cascade. We will then apply material balance principles to simulate the time-dependent tracer concentration from the system. This information then leads to an RTD for each of these reactor systems, from which it will be possible to develop a general expression for the RTD of any cascade of n ideal CFSTRs in series by inspection.

So-called "tracer tests" are commonly used to evaluate mixing behavior in natural and engineered "reactor" systems. An inert tracer is added as a stimulus to the system, usually in a manner that is easy to describe mathematically. A common approach is to introduce a tracer mass M at $t = 0$ to a system (i.e. an ideal "pulse"), with monitoring of effluent tracer concentration as a function of time. Because the tracer is inert, the dynamic behavior of the tracer in the effluent is governed entirely by transport processes. As such, we can use the time-dependent effluent tracer concentration to provide a quantitative description of transport behavior within the system.

### 3.B.1 Single CFSTR

Figure 3.B.1 illustrates the "tracer test" experiment that will be used to evaluate the RTD for a single CFSTR. The experiment is conducted such that the concentration of tracer in the influent to the system ($C_{in}$) is always zero. The entire tracer mass (M) is introduced to the system at $t = 0$, such that the initial concentration in the reactor is $C_0 = \frac{M}{\forall}$.

A mass-balance applied to this system yields the following:

$$\forall \frac{dC}{dt} = Q(C_{in} - C) \tag{3.B.1}$$

But from above, we know that $C_{in} = 0$. Equation 3.B.1 can be simplified by eliminating this term. Rearranging and substituting the definition of mean hydraulic detention time ($\theta = \frac{\forall}{Q}$), we get:

$$\frac{dC}{dt} = -\frac{C}{\theta} \tag{3.B.2}$$

Using algebra to separate variables, we find:

$$\frac{dC}{C} = -\frac{dt}{\theta} \tag{3.B.3}$$

We then apply integration to equation 3.B.3 with the following limits of integration:

$$\int_{C_0}^{C} \frac{dC}{C} = -\int_0^t \frac{dt}{\theta} \tag{3.B.4}$$

The lower limits of integration in equation 3.B.4 represent the initial condition in the system, whereas the upper limits describe the condition of the system at an arbitrary time t after introduction of the tracer. Integration of equation 3.B.4 yields:

$$\ln\left(\frac{C(t)}{C_0}\right) = -\frac{t}{\theta} \text{ or } C(t) = \frac{M}{\forall} \cdot \exp\left(-\frac{t}{\theta}\right) \tag{3.B.5}$$

Our goal in conducting a tracer test experiment is to introduce a stimulus to the system that allows measurement of a system response that will inform the RTD delivered by the system under a given operating condition (i.e. flow rate). For the pulse addition of an inert tracer, all tracer molecules are introduced to the system at the same time. The time spent by each molecule can be quantified as follows:

$$\overline{C} \cdot dt = \frac{Q \cdot C \cdot dt}{M} = \frac{\text{mass of tracer with residence time between } t \text{ and } t + dt}{\text{total tracer mass}} \tag{3.B.6}$$

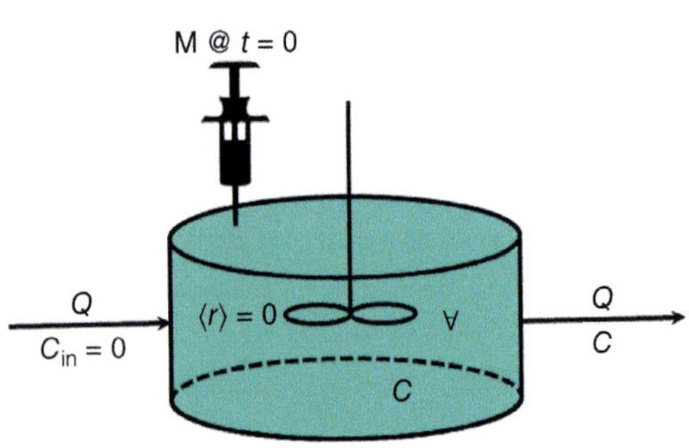


**Figure 3.B.1** Schematic illustration of tracer test applied to a single CFSTR of volume ∀. The entire tracer mass (*M*) is added to the reactor instantaneously at $t = 0$.

If we assume that tracer molecules and fluid elements within the reactor behave identically, then visual inspection of equations 3.43 and 3.B.6 indicates that:

$$\overline{C} = E \tag{3.B.7}$$

In other words, the time-course record of tracer concentration leaving the system can be used to calculate the RTD function:

$$E(t) = \frac{Q \cdot C(t)}{M} \tag{3.B.8}$$

Therefore, for the single CFSTR, we can calculate the RTD by substitution of equation 3.B.5 into equation 3.B.8:

$$E(t) = \frac{Q \cdot C(t)}{M} = \frac{Q}{M} \cdot \frac{M}{\forall} \cdot \exp\left(-\frac{t}{\theta}\right) = \frac{1}{\theta} \cdot \exp\left(-\frac{t}{\theta}\right) \tag{3.B.9}$$

As is the case with many engineering applications, there are benefits to providing a dimensionless representation of system behavior. In the case of RTD, it is possible to non-dimensionalize by multiplying both sides of equation 3.B.9 by $\theta$:

$$E(t) \cdot \theta = \exp\left(-\frac{t}{\theta}\right) \tag{3.B.10}$$

## 3.B.2 Cascade of Two Identical CFSTRs

Figure 3.B.2 illustrates the CFSTR cascade model for a system that comprises two identical CFSTRs in series, each with a volume of $\frac{\forall}{2}$ (i.e. total system volume = $\forall$).

In this example, we have a system that has the same overall volume (and $\theta$) as the first system, but now the system comprises two identical CFSTRs in series. In this tracer test, the entire tracer mass is introduced to cell 1 in the cascade. Therefore, the initial condition ($t = 0$) for this system is that the tracer concentration in cell 1 is $\frac{2 \cdot M}{\forall}$, while the initial tracer concentration in cell 2 is zero. Note that the initial concentration in cell 1 for this test is exactly two times greater than the initial condition for the tracer test with a single CFSTR. This is because in test described in Figure 3.B.2, the entire tracer mass is introduced and instantly dispersed in exactly ½ the volume described by the single CFSTR case.

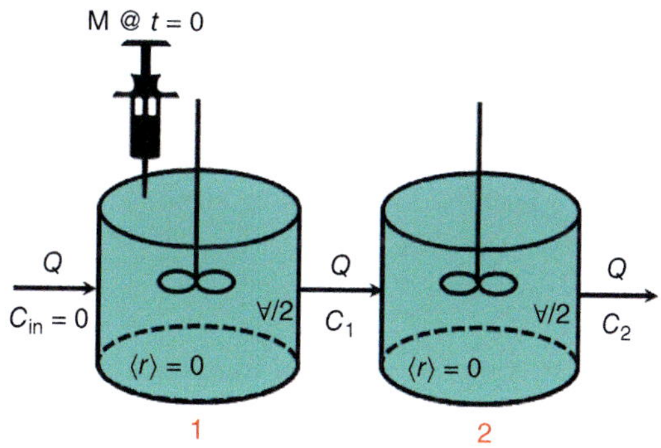


**Figure 3.B.2** Schematic illustration of CFSTR cascade with $n = 2$. Numbers indicated in red font in this schematic describe cell number. An inert tracer of mass $M$ is introduced instantaneously to cell 1 at $t = 0$.

We begin our analysis of the behavior of the tracer in this system by conducting a material balance around cell 1. This analysis will follow the logic applied above, except that the volume of the CR is now $\frac{\forall}{2}$, whereas in the previous example the CR volume was $\forall$:

$$\frac{\forall}{2} \cdot \frac{dC_1}{dt} = Q(C_{in} - C_1) \tag{3.B.11}$$

Again recognizing that the fluid coming into the first cell contains no tracer, we can rearrange equation 3.B.11:

$$\frac{dC_1}{C_1} = -2 \cdot \frac{dt}{\theta} \tag{3.B.12}$$

Integrating, we find:

$$\ln\left(\frac{C_1}{C_{1,0}}\right) = -2 \cdot \frac{t}{\theta} \tag{3.B.13}$$

Now, recognizing that the initial condition is defined as: $C_{1,0} = \frac{2 \cdot M}{\forall}$. Substituting this value into equation 3.B.13 and rearranging, we find:

$$C_1 = \frac{2 \cdot M}{\forall} \cdot \exp\left(-\frac{2 \cdot t}{\theta}\right) \tag{3.B.14}$$

Now, we move onto cell 2 in the cascade. For this system, our general mass balance yields:

$$\frac{\forall}{2} \cdot \frac{dC_2}{dt} = Q(C_1 - C_2) \tag{3.B.15}$$

Rearranging this equation, we find:

$$\frac{dC_2}{dt} = \frac{2 \cdot (C_1 - C_2)}{\theta} \tag{3.B.16}$$

where $\theta = \frac{\forall}{Q}$ (i.e. residence time is defined based on total system volume).

In this case, our influent concentration ($C_1$) is not zero, but rather is defined by equation 3.B.14. We will start toward a solution to this equation by algebraic manipulation to separate variables:

$$\frac{dC_2}{dt} + 2 \cdot \frac{C_2}{\theta} = \frac{2 \cdot C_1}{\theta} \tag{3.B.17}$$

Note that equation 3.B.17 has the same form as equation 3.A.2, namely that of a linear first-order ordinary differential equation. As before, we can solve this equation by applying the integrating factor approach. In this case, we will define the variable: $\alpha = \frac{2}{\theta}$. Substitution of this variable into equation 3.B.17 yields:

$$\frac{dC_2}{dt} + \alpha \cdot C_2 = \alpha \cdot C_1 \tag{3.B.18}$$

As in Appendix 3.A, the integrating factor is applied as follows:

$$C_2 \cdot e^{\alpha \cdot t} = \int e^{\alpha \cdot t} \cdot \alpha \cdot C_1 \cdot dt \tag{3.B.19}$$

Substitution of the solution from equation 3.B.14 into equation 3.B.19 yields:

$$C_2 \cdot e^{\alpha \cdot t} = \int e^{\alpha \cdot t} \cdot \alpha \cdot \frac{2 \cdot M}{\forall} \cdot e^{-\alpha \cdot t} \cdot dt \tag{3.B.20}$$

In this case, the exponential terms in the integrand cancel, yielding:

$$C_2 \cdot e^{\alpha \cdot t} = \int \alpha \cdot \frac{2 \cdot M}{\forall} \cdot \mathrm{d}t \tag{3.B.21}$$

Integration of equation 3.B.21 yields:

$$C_2 \cdot e^{\alpha \cdot t} = \alpha \cdot \frac{2 \cdot M}{\forall} \cdot t + \beta \tag{3.B.22}$$

where $\beta$ is a constant of integration. We can evaluate the constant of integration by noting that $C_2 = 0$ at $t = 0$ (see above). Substitution of these values into equation 3.B.22 indicates that $\beta = 0$. Therefore, equation 3.B.22 can be simplified to:

$$C_2 \cdot e^{\alpha \cdot t} = \alpha \cdot \frac{2 \cdot M}{\forall} \cdot t \tag{3.B.23}$$

Or by rearrangement:

$$C_2 = \alpha \cdot \frac{2 \cdot M}{\forall} \cdot t \cdot e^{-\alpha \cdot t} \tag{3.B.24}$$

Now, again using the identity from equation 3.B.8, we find the RTD for this system:

$$E(t) = \frac{Q \cdot C(t)}{M} = \frac{Q}{M} \cdot \alpha \cdot \frac{2 \cdot M}{\forall} \cdot t \cdot e^{-\alpha \cdot t} \tag{3.B.25}$$

Or

$$E(t) = \frac{2^2}{\theta} \cdot \frac{t}{\theta} \cdot \exp\left(-\frac{2 \cdot t}{\theta}\right) \tag{3.B.26}$$

Lastly, we can arrange this in dimensionless terms by multiplying both sides by $\theta$:

$$E(t) \cdot \theta = 2^2 \cdot \frac{t}{\theta} \cdot \exp\left(-\frac{2 \cdot t}{\theta}\right) \tag{3.B.27}$$

### 3.B.3 Cascade of Three Identical CFSTRs

Figure 3.B.3 illustrates the pulse test as applied to a cascade of three CFSTRs in series. The logic used in explaining this system's behavior is an extension of the discussion above. The difference between this test and the previous test is that now the entire system volume is equally divided among three cells. As a consequence, the initial condition is now defined as: $C_{1,0} = \frac{3 \cdot M}{\forall}$, while the initial concentrations in cells 2 and 3 will both be zero.

For this analysis, we will define $\alpha = \frac{3}{\theta}$ Using this definition and the logic presented above, we will apply separation of variables and the integrating factor method to yield the following expressions of tracer concentration leaving cells 1, 2, and 3 as functions of time:

$$C_1 = \frac{3 \cdot M}{\forall} \cdot \exp(-\alpha \cdot t) \tag{3.B.28}$$

$$C_2 = \frac{3 \cdot M}{\forall} \cdot \alpha \cdot t \cdot \exp(-\alpha \cdot t) \tag{3.B.29}$$

$$C_3 = \frac{3 \cdot M}{\forall} \cdot \alpha^2 \cdot \frac{t^2}{2} \cdot \exp(-\alpha \cdot t) \tag{3.B.30}$$

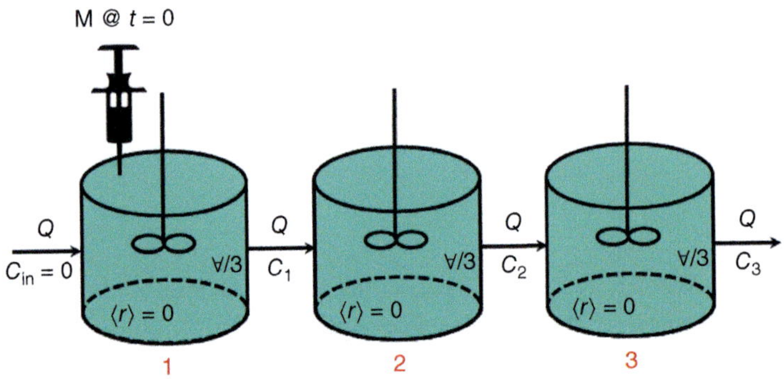


**Figure 3.B.3** Schematic illustration of CFSTR cascade with $n = 3$. Numbers indicated in red font in this schematic describe cell number. An inert tracer of mass $M$ is introduced instantaneously to cell 1 at $t = 0$.

As in the previous examples, we apply the identity from equation 3.B.8 to find the RTD for this system:

$$E(t) = \frac{Q \cdot C(t)}{M} = \frac{Q}{M} \cdot \frac{3 \cdot M}{\forall} \cdot \alpha^2 \cdot \frac{t^2}{2} \cdot \exp(-\alpha \cdot t) \tag{3.B.31}$$

By substitution of the definitions of $\theta$ and $\alpha$, we can use some rearrangement to define the dimensionless form of the RTD for this system:

$$E(t) \cdot \theta = \frac{3^3}{2} \cdot \left(\frac{t}{\theta}\right)^2 \cdot \exp\left(-\frac{3 \cdot t}{\theta}\right) \tag{3.B.32}$$

### 3.B.4 Generalization of Results to a Cascade of n CFSTRs in Series

In the preceding analysis, there are several patterns that emerge from the analysis. These patterns develop because of the physics that are assumed to apply in this system, and the corresponding mathematical relationships that are used to describe these behaviors. Table 3.B.1 provides a summary of the dimensionless RTD functions for cascades of 1, 2, and 3 CFSTRs in series.

Inspection of these expressions and examination of the mathematical processes that were used to generate them allows generalization of the normalized RTD function for CFSTR cascades. Specifically, for a CFSTR cascade containing $n$ equally sized cells in series, the normalized RTD is defined as:

$$E(t) \cdot \theta = \frac{n^n}{(n-1)!} \cdot \left(\frac{t}{\theta}\right)^{n-1} \cdot \exp\left(-\frac{n \cdot t}{\theta}\right) \tag{3.B.33}$$

Figure 3.B.4 illustrates the normalized RTD functions for a wide range of CFSTR cascade systems. The system defined by $n = 1$ represents a single CFSTR and is identical to the ideal CFSTR model. The number of cells in the cascade ($n$) increases, the RTD becomes narrower. The system corresponding to $n = \infty$ yields an RTD that behaves as a delta (δ) function, indicating that all elements of the system will have the same residence time in the system, defined by θ. This corresponds exactly with the definition of an ideal PFR.

In terms of process performance, the mixing behaviors described by the ideal CFSTR and PFR models represent the lower and upper limits of efficiency, respectively. RTDs that are intermediate to these extreme conditions are correspondingly intermediate in terms of mixing efficiency and

**Table 3.B.1** Summary of normalized RTD functions for CFSTR cascade systems.

| Number of cells in cascade | 1 | 2 | 3 |
|---|---|---|---|
| Normalized RTD function | $E(t)\cdot\theta = \exp\left(-\frac{t}{\theta}\right)$ | $E(t)\cdot\theta = 2^2\cdot\frac{t}{\theta}\cdot\exp\left(-\frac{2\cdot t}{\theta}\right)$ | $E(t)\cdot\theta = \frac{3^3}{2}\cdot\left(\frac{t}{\theta}\right)^2\cdot\exp\left(-\frac{3\cdot t}{\theta}\right)$ |
| Equation number | 3.B.10 | 3.B.27 | 3.B.32 |

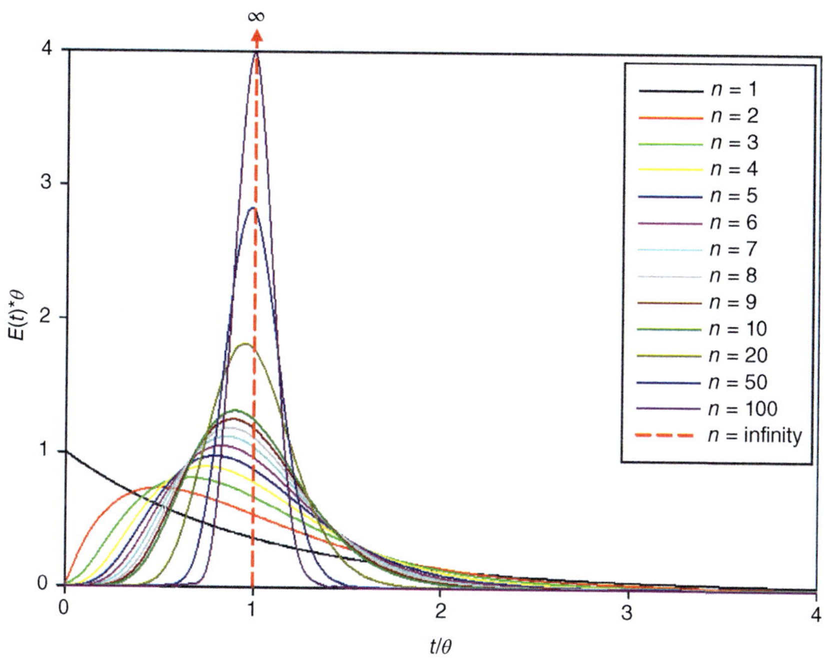


**Figure 3.B.4** Dimensionless RTD functions for CFSTR cascades containing $n$ cells. Note that the case corresponding to $n = \infty$ corresponds with the ideal PFR model, while the case for $n = 1$ is identical to the ideal CFSTR.

reactor performance. As such, the parameter $n$ can be used to describe mixing efficiency in an arbitrary reactor system.

## Appendix 3.C Proof of Segregated-flow Model Based on Probability Theory

We will define an event A as one in which a target molecule remains unchanged after passing through the irradiated zone of a photochemical reactor. In the terminology of probability theory:

$$P(A) = \text{probability of event } A \tag{3.C.1}$$

Similarly, we may define:

$$P(A|D_i) = \text{conditional probability of event } A, \text{given exposure to dose } D_i \tag{3.C.2}$$

From basic principles of probability theory:

$$P(A|D_i) = \frac{P(A \cap D_i)}{P(D_i)} \tag{3.C.3}$$

where $A \cap D_i$ describes the event where the target molecule remains unchanged and receives dose $D_i$. The denominator on the right side of equation 3.C.3 describes:

$$P(D_i) = \text{probability that target receives dose } D_i \tag{3.C.4}$$

We then apply the theorem of total probability: [11]

$$P(A) = \sum_i P(A \cap D_i) = \sum_i P(A|D_i) \cdot P(D_i) \tag{3.C.5}$$

We now explore the physical interpretation of the terms in equation 3.C.5.

- The left side of the equation, $P(A)$, describes the probability of a target molecule remaining unchanged after passing through the irradiated zone of a photochemical reactor. This translates directly to the fraction of target molecules that remain after passing through the reactor, or $\left(\frac{C}{C_0}\right)_{\text{reactor}}$.
- The conditional probability, $P(A|D_i)$, describes the probability of a target molecule remaining unchanged after exposure to a given dose, $D_i$. This translates directly to the intrinsic kinetics of the photoreaction of interest, or $\left(\frac{C(D)}{C_0}\right)_{\text{batch}}$.
- $P(D_i)$ describes the probability that a target molecule will receive a dose $D_i$. If we consider $E(D_i)$ to be a probability density function that we refer to as the dose distribution function (see definition in text), then $P(D_i) = E(D_i) \cdot \Delta D_i$.

Substituting these terms back into equation 3.C.5, we find:

$$P(A) = \sum_i P(A|D_i) \cdot E(D_i) \cdot \Delta D_i \tag{3.C.6}$$

where it is understood that $\Delta D_i$ are nonoverlapping dose intervals and all possible dose intervals are included in the summation. If we evaluate equation 3.C.6 in the limit as $\Delta D_i \rightarrow 0$, then:

$$\left(\frac{C}{C_0}\right)_{\text{reactor}} = \int_0^{\infty} \left(\frac{C(D)}{C_0}\right)_{\text{batch}} \cdot E(D) \cdot \mathrm{d}D \tag{3.C.7}$$

This is the segregated-flow model, equation 3.66.

# 4

# Ultraviolet Radiation Sources

## 4.1 Introduction

In principle, photochemical reactions and photophysical processes can be driven by radiation from essentially any segment of the electromagnetic spectrum. However, for most of the reactions that are relevant to photochemical reactors, ultraviolet (UV) radiation is required. Therefore, knowledge of the behavior and output characteristics of UV sources will be beneficial in understanding photochemical reactor systems. The focus of this chapter will be on natural and "artificial" sources of UV radiation. However, the fundamental physical and chemical processes that are responsible for generation of electromagnetic radiation apply to other regions of the spectrum, including infrared (IR) and visible radiation.

## 4.2 Incandescence

Any material that exists at a temperature above absolute zero will emit electromagnetic radiation by the process known as incandescence. As such, any "hot" material becomes a potential emitter. The term *blackbody* is used to describe an ideal thermal emitter, one that is in thermal equilibrium with its surroundings. A blackbody will absorb all radiation imposed on it; in other words, a blackbody expresses zero transmittance and zero reflectance. More importantly, from the perspective of this presentation, a blackbody is a perfectly diffuse radiator that will emit a spectrum of electromagnetic radiation that can be calculated by Planck's law [1], as described in equation 4.1:

$$M_\lambda(T) = \frac{2 \cdot \pi \cdot h \cdot c^2}{\lambda^5} \cdot \frac{1}{\exp\left(\frac{h \cdot c}{k \cdot \lambda \cdot T}\right) - 1} \tag{4.1}$$

where

$M_\lambda(T)$ = spectral radial emittance, per unit wavelength, into a hemisphere $[=] \ \frac{W}{m^2 \cdot nm}$

$h$ = Planck's constant ($6.62607 \times 10^{-34}$ W · s$^2$)

$c$ = speed of light in a vacuum ($2.998 \times 10^8 \ \frac{m}{s}$)

*Photochemical Reactors: Theory, Methods, and Applications of Ultraviolet Radiation*, First Edition. Ernest R. Blatchley III.

Companion website: www.wiley.com/go/blatchley/photochemicalreactors

| | | | | |
|---|---|---|---|---|
| λ | = | wavelength of radiation | [=] | m |
| $k$ | = | Boltzmann's constant $(1.3806 \times 10^{-23}\,\frac{W \cdot s}{K})$. | | |

By taking the first derivative of equation 4.1 with respect to λ and evaluating the resulting function at a value of zero, the wavelength corresponding to maximum spectral radial emittance can be defined (i.e. Wien's displacement law): [1]

$$\lambda_{\max} = \frac{b}{T} \tag{4.2}$$

where

| | | | | |
|---|---|---|---|---|
| $\lambda_{\max}$ | = | wavelength of maximum spectral emittance at a given temperature | [=] | nm |
| $b$ | = | Wien's displacement constant $(2.8978 \times 10^6\ \text{nm} \cdot \text{K})$. | | |

Figure 4.1 illustrates that the majority of radiation emitted by a blackbody is likely to be in the IR range. However, as material temperature increases, spectral contributions in the visible and UV portions of the spectrum also become evident. Personal experience also teaches that the color of a hot object is influenced by its temperature. Planck's law quantifies this behavior [2].

Incandescence forms the basis for many sources of radiation, including the common (incandescent) lamp or "light bulb," which produces visible light by electrical resistance heating of a metal filament. Lamp filaments are usually produced from tungsten, because it has the highest melting point (3655 K) of any metal. However, any "hot" material can function as an emitter.

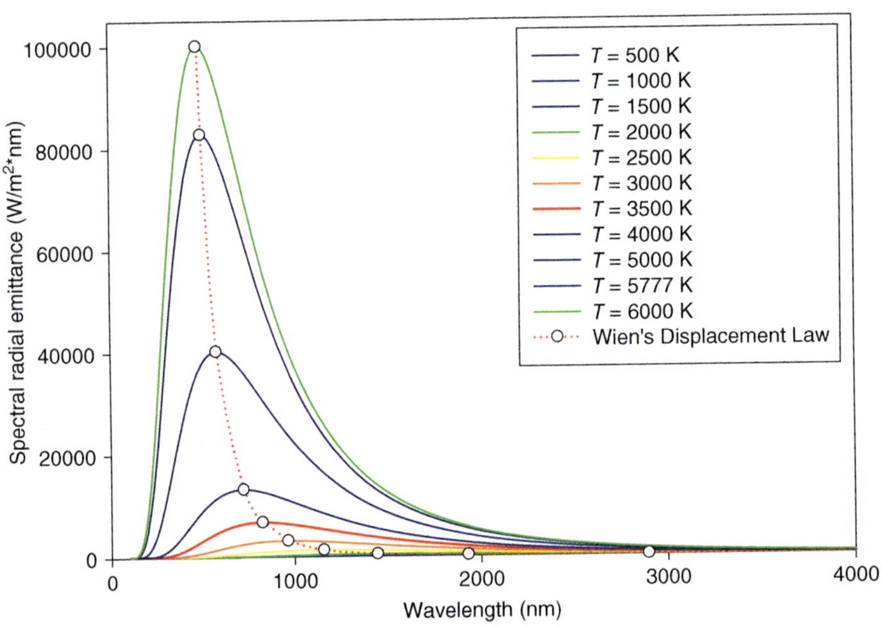


**Figure 4.1** Calculated emission spectra for a spherical blackbody emitter as a function of temperature, as predicted by Planck's law. Also included is an illustration of the wavelength of maximum emission, based on Wien's displacement law.

## 4.3 Solar Radiation

In many respects, the sun represents our most fundamental source of electromagnetic radiation in the IR, visible, and UV regions. The sun is a roughly spherical body primarily comprising hydrogen ($H_2$) gas [3]. Earth revolves around the sun on the ecliptic plane, a roughly elliptical orbit that has the sun at one of its foci. The amount of solar radiation that reaches the surface of the earth is inversely proportional to the square of the earth–sun distance ($r$). The mean earth–sun distance ($r_0$) is $1.496 \times 10^8$ km, which is defined formally as 1 astronomical unit (AU) [3]. In its annual orbit of the sun, the earth–sun distance varies from $0.983\ \text{AU} \leq r \leq 1.017\ \text{AU}$.

The outermost layer of the sun is referred to as the photosphere. If the photosphere is assumed to be a blackbody, it is possible to relate the measured spectrum of solar radiation received at the outer edge of earth's atmosphere to the temperature of the photosphere. In other words, it is possible to apply the blackbody assumption to simulate the spectrum of solar radiation imposed on earth's atmosphere by inverse application of Planck's law. Following this logic, the equivalent temperature of the sun's photosphere has been estimated to be somewhere between 5777 K and 6000 K.

The calculation described above needs to account for the surface area of the emitter (sun) and the earth–sun distance. Formally, the spectral irradiance imposed on the outer edge of earth's atmosphere can be calculated as follows:

$$I_{0,\lambda} = M_\lambda(T) \cdot \left(\frac{r_s}{r_0}\right)^2 \tag{4.3}$$

where

$I_{0,\lambda}$ = solar spectral irradiance on a surface normal to incident radiation at $r_0$ $[=]\ \frac{W}{m^2 \cdot nm}$

$r_s$ = mean radius of solar disk ($6.9598 \times 10^5$ km).

Equation 4.3 can be used to simulate the spectrum of radiation imposed on the outside of earth's atmosphere. The parenthetical term in equation 4.3 accounts for the surface area of the emitter and the distance between the emitter (sun) and the receptor site (earth). Figure 4.2 illustrates the results of this calculation for the sun corresponding to assumptions of spherical blackbody emission at 5777 K and 6000 K.

The spectrum of radiation received at the surface of the earth will also be influenced by the composition of the atmosphere and the thickness of the atmospheric layer through which incident solar radiation must travel. In considering this path length, it is useful to apply some common terminology that is used to describe the geometry of this system. First, the term *zenith angle* is used to describe the angle between the sun and the vertical; for a situation in which the sun is directly overhead, the zenith angle is zero. The zenith angle approaches 90° at sunrise and sunset. Second, the term *air mass* provides a relative indication of the optical path length through which solar radiation must pass, as compared to the thickness of the atmosphere. Therefore, for a zenith angle of zero, the air mass is defined as 1.0. As the zenith angle increases, so does the air mass. Figure 4.3 provides a schematic illustration of these concepts.

Using the terminology provided in Figure 4.3, the air mass ($m$) is defined as:

$$m = \frac{1}{\cos(\theta)} \tag{4.4}$$

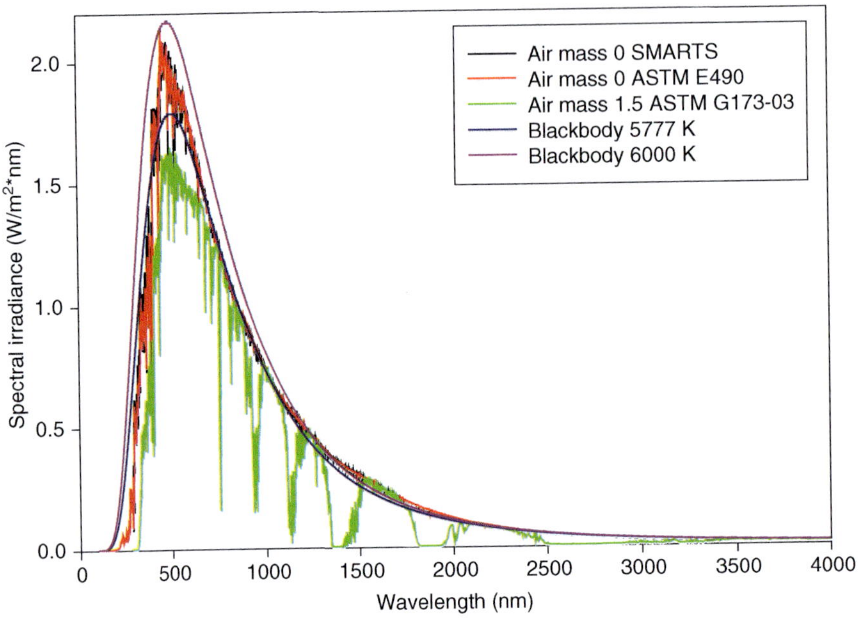


**Figure 4.2** Simulated spectra of spherical blackbody emitters at 5777 K and 6000 K, corresponding to the radius of the sun and the mean earth–sun distance, $r_0$. Also included for comparison are standard solar spectra for air mass zero (based on the SMARTS model and ASTM Standard E490), as well as the standard solar spectrum for air mass 1.5 (ASTM G173-03).

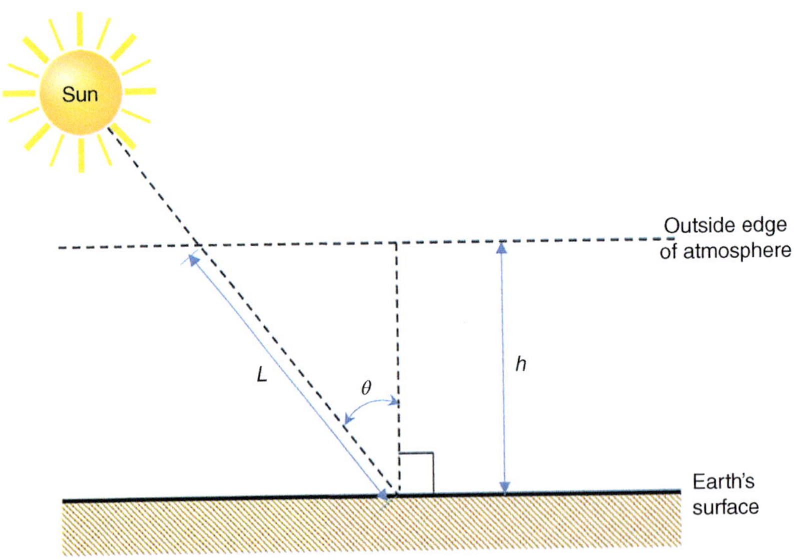


**Figure 4.3** Schematic illustration of terms used to describe the optical path length for solar radiation to penetrate earth's atmosphere. $L$ represents actual path length; $h$ represents the thickness of earth's atmosphere; $\theta$ represents zenith angle. Air mass ($m$) represents the ratio $\frac{L}{h}$.

The approximation provided by equation 4.4 works well for most conditions. However, as the zenith angle approaches 90°, the effects of earth's curvature start to become important. Demerjian et al. developed a formal correction for this term; a graphical illustration of equation 4.4 and the values of m corrected for curvature are provided in Figure 4.4 [2]. The estimate of air mass based on equation 4.4 is accurate for zenith angles corresponding to most daylight hours. Specifically, the error associated with equation 4.4 is negligible for zenith angles as high as 60°; as the zenith angle approaches conditions corresponding to sunrise and sunset, the error increases, with error of approximately 13% at a zenith angle of 86° [4].

The solar spectrum at air mass zero has been measured by satellites, space shuttle missions, high-altitude aircraft, rocket soundings, and ground-based solar telescopes [5]. Numerical models are also available to simulate this spectrum. A commonly used tool for this purpose is the Simple Model of the Atmospheric Radiative Transfer of Sunshine (SMARTS) [6], which is publicly available and can be downloaded. Measured and calculated standard solar spectra corresponding to air mass zero are included in Figure 4.2 for comparison with the predictions of the blackbody model.

In general, the spectra based on the assumption of a blackbody ($T = 5777$ K and 6000 K) have similar shapes and magnitudes, especially in the visible and IR ranges of the spectrum. However, the blackbody simulations substantially overestimate solar spectral irradiance in the UV region at air mass zero. Specifically, the blackbody simulations predict higher spectral irradiance than the measurements or simulated (SMARTS) spectra; moreover, the blackbody simulations suggest that the solar spectrum will include radiation at shorter wavelengths than indicated by measurements or the SMARTS model.

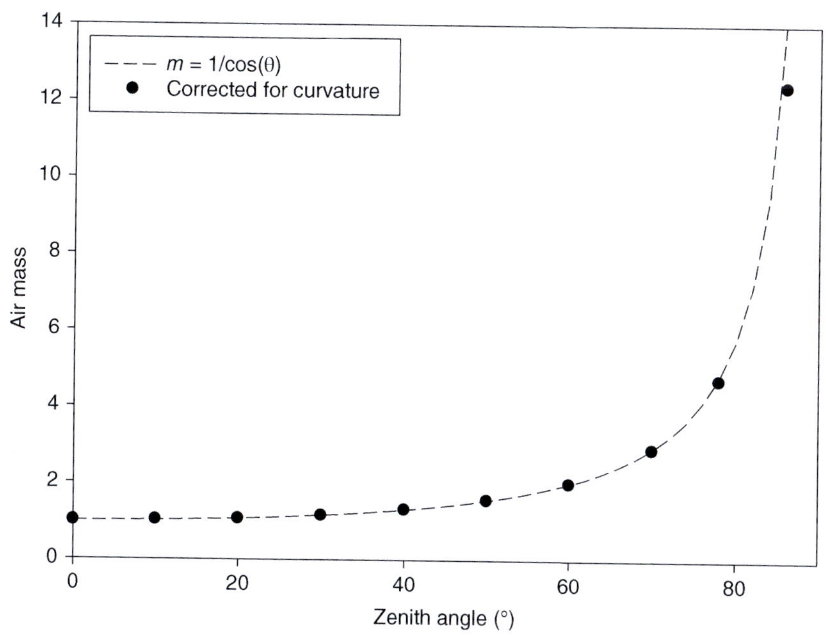


Figure 4.4 Air mass as a function of zenith angle based on the geometry presented in Figure 4.3 and corrected for earth's curvature.

A further comparison of solar spectra relates to the effect of air mass. Included in Figure 4.2 is a standard solar spectrum corresponding to air mass 1.5 [7]. This standard spectrum provides a representative description of solar spectral irradiance experienced at the surface of the earth for a zenith angle of 37°.

The air mass 1.5 solar spectrum is generally reduced relative to the air mass zero spectrum. In addition, there are several spectral bands in which spectral irradiance is close to zero; these are due to strong absorption by atmospheric constituents. These dips in the air mass 1.5 solar spectrum are particularly evident in the IR region, and are largely attributable to greenhouse gases, such as $CO_2$ and $H_2O$. More relevant to this presentation is the general suppression of UV radiation, which is caused by absorption and scattering by atmospheric constituents; ozone ($O_3$) plays a particularly important role in absorbing solar UV radiation.

The solar spectrum at altitudes close to mean sea level is characterized by steeply declining spectral irradiance down to the shortest wavelength for which measurable spectral irradiance is observed, roughly 290 nm. The nature of the solar spectrum is influenced by time of day, time of year (i.e. Julian day), location (latitude, longitude, and altitude), and atmospheric composition, including ozone, particulates, and cloud cover. Data to describe atmospheric ozone and particulate composition are available based on satellite measurements. Cloud cover is more difficult to quantify, in that it is characterized by fractional cover and cloud depth, both of which affect ambient solar spectral irradiance. In general, the effects of atmospheric composition on ambient solar spectral irradiance have been well defined, except for the effects of cloud cover.

Within the solar spectrum received at terrestrial locations are contributions in the UVA and UVB portions of the spectrum. For photochemical applications, radiation in the UVA and UVB portions of the spectrum can both be important. However, UVB radiation ($280\ \text{nm} \leq \lambda \leq 320\ \text{nm}$), which is characterized by shorter wavelengths than UVA ($320\ \text{nm} \leq \lambda \leq 400\ \text{nm}$), is more energetic and therefore capable of initiating reactions that cannot be initiated within the UVA range. As an example, UVB radiation is able to cause direct photochemical damage to nucleic acids (DNA and RNA), and as such can function as a disinfectant. This behavior also allows UVB radiation to cause damage to plant and animal tissues.

The U.S. Department of Agriculture has established a program and network of monitoring stations within the United States to measure ambient solar UVB radiation [8]. The network comprises a series of monitoring stations that have been established around the United States; monitoring stations within this network have also been established in Canada, China, Greece, Mexico, New Zealand, Norway, South Korea, and the United Arab Emirates.

The UV-B Monitoring and Research Program also includes tools to allow data from monitoring sites to be downloaded. In addition, a synthetic spectrum tool is available that allows the solar UV spectrum to be simulated based on curve-fitting of measured irradiance at several UV wavelengths to simulate a continuous transmittance curve for UV radiation. In turn, this information is convolved with a spectrum of extraterrestrial radiation to yield simulations of total, diffuse, and direct spectral irradiances [8].

## 4.4 Artificial Sources of UV Radiation

For most applications of UV radiation, the ambient solar spectrum does not include radiation at the wavelengths of interest or at values of spectral irradiance or fluence rate that are economically viable, without substantial amplification. Although there is tremendous, largely untapped potential to use solar radiation in engineering applications, among photochemical reactor systems, ambient solar UV radiation is used only in specialized applications. Because of this, most photochemical reactors employ "artificial" sources of UV radiation. In many cases, these sources accomplish conversion of electrical energy into photons in the UV region. Several mechanisms are available to accomplish this conversion.

The text below will include brief presentations of common or emerging sources of UV radiation, along with information pertaining to the mechanism(s) of photon generation. Characteristic output spectra for these sources are also provided as a means of understanding and comparing output from these devices.

### 4.4.1 Gas Discharge Lamps: Mercury Lamps

Mercury lamps fall into the general category of gas discharge lamps. A general schematic of their construction is provided in Figure 4.5. Establishment of a voltage between the electrodes in this system causes electrons ($e^-$) to migrate from the cathode to the anode. Interactions between electrons and the gases inside the lamp lead to emission of radiation.

The basic components of a gas discharge lamp include the lamp envelope, electrodes (cathode and anode), a noble gas or a mixture of noble gases, and a filling gas which could be doped in liquid form or solid mixture form at room temperature. An ideal lamp envelope will be impermeable to gases and chemically compatible with the filling gas. It must also transmit emitted radiation effectively, be an electrical insulator, and in some cases be able to withstand extended exposure to temperatures as high as 600–800°C [1]. Fused silica and quartz satisfy these requirements well, and because of this, these materials are commonly used to produce lamp envelopes for discharge lamps.

Tungsten (W) electrodes are usually used in gas discharge lamps. This is because of its high melting point and its low potential to emit gas-phase W that could lead to "end darkening" of the lamp envelope.

Before addressing the specific gases that are often used in gas discharge lamps, it is useful to consider the nature of the interactions between electrons and these gases. When an electron is caused to migrate in a gas-filled envelope, three general types of collisions can take place: [1]

1) an elastic collision in which minimal energy transfer takes place, largely because of the difference in mass between an electron and an atom,
2) a collision that results in electronic excitation of the atom, with corresponding loss of kinetic energy on the part of the electron, or
3) a collision that results in ionization of the atom, if electron energy is sufficient.

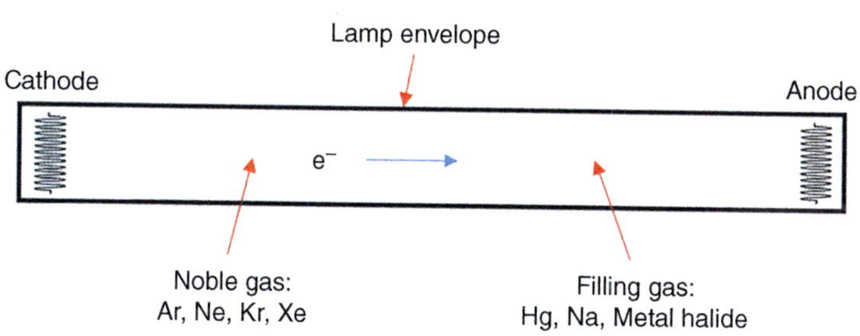


**Figure 4.5** General schematic of gas discharge lamp.

Among any period (row) in the periodic table of elements, the noble gases have the highest ionization energy because they have stable, filled valence electron shells. Many discharge lamps operate such that ionization of the noble gas does not take place. Rather, the objective may be to promote ionization of the filling gas.

With that in mind, an ideal filling gas will have the following characteristics: [1]

- low ionization energy: this implies a metallic element;
- expression of sufficiently high vapor pressure to facilitate collisions with electrons;
- chemically inert toward the other lamp components (i.e. electrodes and the lamp envelope);
- excited states (above ground) should correspond to useful wavelengths of radiation.

Mercury (Hg) satisfies these requirements well and because of this, many discharge lamps are based on Hg as a filling gas.

Figure 4.6 represents an energy-level diagram for Hg. This diagram illustrates the various energy levels (relative to the ground state of Hg) corresponding to possible excited electronic states for Hg. As such, it also provides information about the wavelengths of radiation that correspond to each of the electronic transitions.

As described above, Hg has physical and chemical characteristics that facilitate its use as a filling gas in gas discharge lamps. However the behavior of an Hg lamp is affected by the amount of Hg included in the system [1].

In low-pressure Hg lamps, the partial pressure of Hg will be on the order of $10^{-3}$–$10^{-2}$ Torr. Under these conditions, collisions between electrons and Hg atoms are relatively infrequent. This is especially true at ignition, when very little Hg will exist in the gas phase. As such, at start-up, most collisions with electrons will involve the noble gas, usually Ar, which can be electronically excited to a non-radiative state, with a corresponding energy of 11.55 eV. The energy of this excited Ar atom can be lost by collision with another atom. Collisions with Hg can result in ionization because its ionization energy is 10.43 eV. Most of the collisions between Ar and electrons result in heating; roughly 20% of the energy applied to a low-pressure Hg lamp results in heating, which in turn causes the temperature of the lamp to increase along with the vapor pressure of Hg. The saturation (equilibrium) vapor pressure of Hg in low-pressure lamps is governed by the temperature of the lamp (i.e. some liquid-phase Hg may exist within the envelope of an ignited low-pressure Hg lamp).

Collisions among electrons, Ar atoms, and Hg atoms lead to electronic excitation of Hg atoms. In a general sense, a set of reversible processes take place within the plasma[a] of a low-pressure Hg lamp; these processes can be summarized as follows:

$$\text{Hg}^*(\text{more excited}) \rightleftharpoons \text{Hg}^*(\text{less excited}) + h\nu \tag{4.5}$$

As written in equation 4.5, the forward process represents photon emission, whereas the reverse process represents photon absorption, sometimes referred to as "self-absorption" in a gas discharge lamp. Self-absorption becomes increasingly important as the concentration of mercury atoms at low excitation states, $\text{Hg}^*$(less excited), increases.

The output of a low-pressure Hg lamp is profoundly affected by temperature, largely because of the effect of temperature on the equilibrium vapor pressure of Hg (see Figure 4.7). At low temperature, the equilibrium vapor pressure of Hg is also low, resulting in a low probability of collisions between Hg and electrons. At high temperatures and corresponding equilibrium vapor pressures, self-absorption becomes increasingly important, which leads to inefficient generation of UV photons. The ideal partial pressure of Hg for a low-pressure lamp is approximately $7 \times 10^{-3}$ Torr, corresponding to a temperature of about 40°C [1].

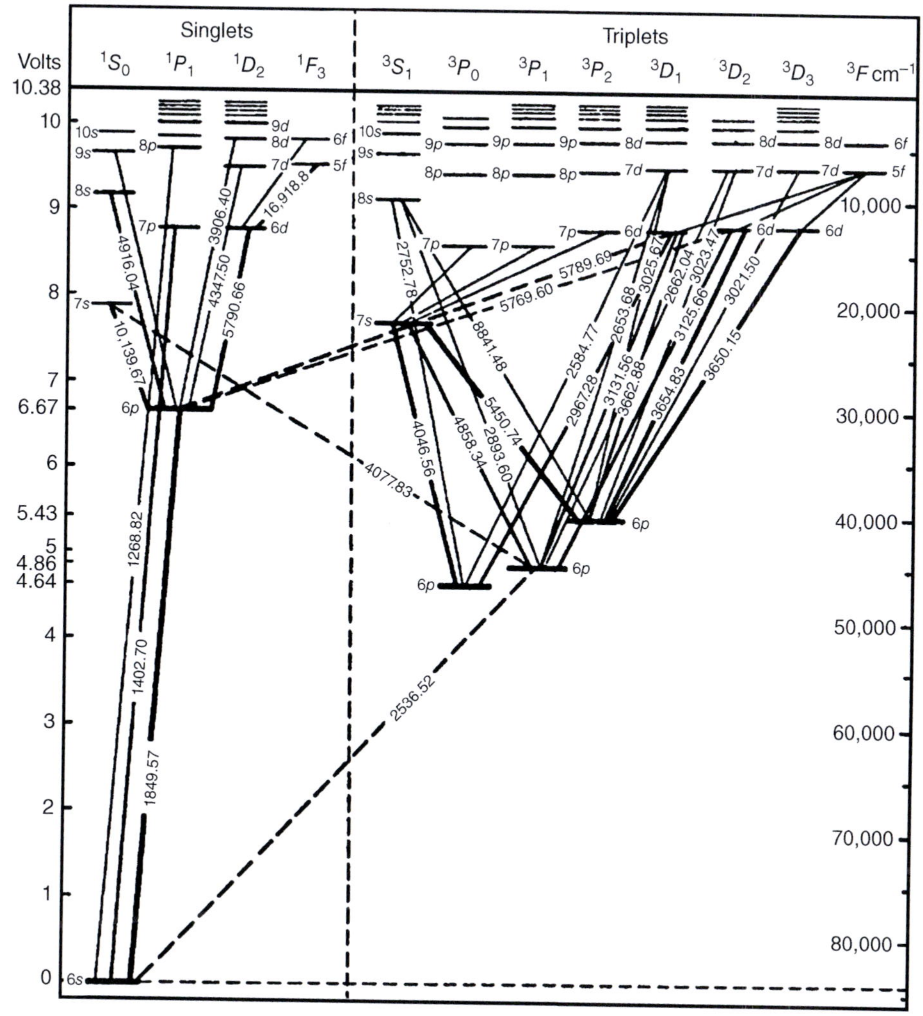


**Figure 4.6** Energy-level diagram (Grotrian diagram) for elemental mercury [9]. The left vertical axis describes the potential through which an electron must be accelerated to yield a given transition. The right vertical axis illustrates the energy scale in terms of wavenumber (cm$^{-1}$). Horizontal lines in the field of this graph illustrate energy levels (relative to ground) for various excited states. The lines connecting energy levels indicate the wavelength (in Angstroms) corresponding to each change in electronic state. *Source:* Herzberg 1944 / with permission of Dover Publications.

Output characteristics of low-pressure Hg lamps are affected by other attributes as well. Typical operating conditions for low-pressure Hg lamps are as follows: Ar partial pressure ≈ 3 T, current ≈ 400 mA, tube loading ≈ 0.5 W/cm, and wall temperature ≈ 40°C [1].

The output of low-pressure Hg lamps is dominated by two electronic transitions (see Figure 4.6). The dominant transition is $6^3P_1 \rightarrow 6^1S_0$, which corresponds to a wavelength of 253.7 nm. As will be discussed in subsequent chapters, radiation at this wavelength is well suited for a number of applications, including disinfection and some photolysis reactions. The transition from $6^1P_1 \rightarrow 6^1S_0$ corresponds to a wavelength of 185.0 nm. This radiation falls in the vacuum UV (VUV) range and is far more energetic than the radiation at 253.7 nm. VUV radiation can be used in some applications and

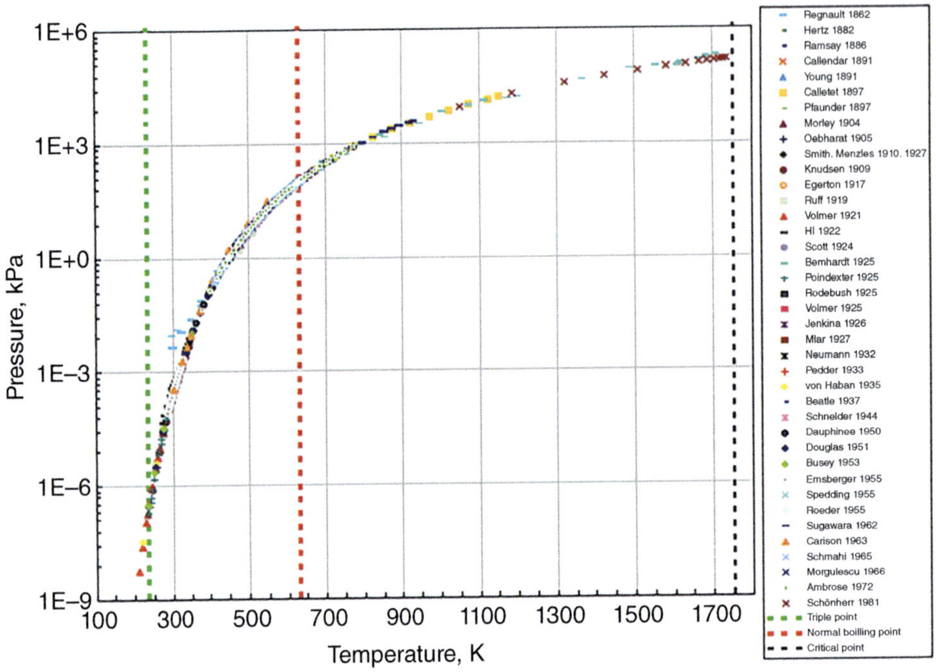


**Figure 4.7** Equilibrium vapor pressure of mercury as a function of temperature, as summarized by publications from 1862 to 1981 [10]. Also included in this graph is the triple point of Hg, which defines the condition in which the gas, liquid, and solid phases coexist in thermodynamic equilibrium.

is capable of promoting photolysis of a wide range of molecules, including water; photolysis of water leads to production of radical intermediates, which in turn can open reaction pathways that otherwise might not exist. Radiation at 185 nm can also cause photolysis of molecular oxygen ($O_2$), which in turn opens a range of other reaction pathways, including the reactions that lead to production of ozone ($O_3$). However, transmission of VUV radiation requires that the lamp envelope (and any other materials that surround the lamp) be made of materials that are characterized by high transmittance at the wavelength of interest.

Mercury amalgam lamps have been developed as alternatives to conventional low-pressure Hg lamps. In amalgam lamps, the mercury is associated with other metals to form a solid amalgam. The amalgam serves as a pressure buffer for Hg in the system. In addition, amalgam lamps can operate at higher temperatures (ca. 100°C) than conventional low-pressure Hg lamps. Collectively, these attributes allow amalgam lamps to generate more power than conventional low-pressure Hg lamps. However, the output spectrum of an amalgam lamp is essentially identical to that of a conventional low-pressure Hg lamp, in terms of the location and relative height of the peaks.

It is perhaps relevant to recognize that most conventional fluorescent lamps are low-pressure Hg lamps. However, there are two important differences between low-pressure Hg lamps that are intended for generation of UV radiation and those that are used as fluorescent lamps for general room lighting. First, the inside surface of a fluorescent lamp is coated with a material that will function as fluorophore. Specifically, the material should effectively absorb radiation that is emitted by excited Hg; the absorbed energy is re-emitted at longer (visible) wavelengths by fluorescence. Second, the envelope material for a fluorescent lamp can be an inexpensive form of glass; the envelope of a fluorescent lamp only needs to transmit visible radiation, not UV. In fact, transmission of UV radiation from these lamps is generally undesirable.

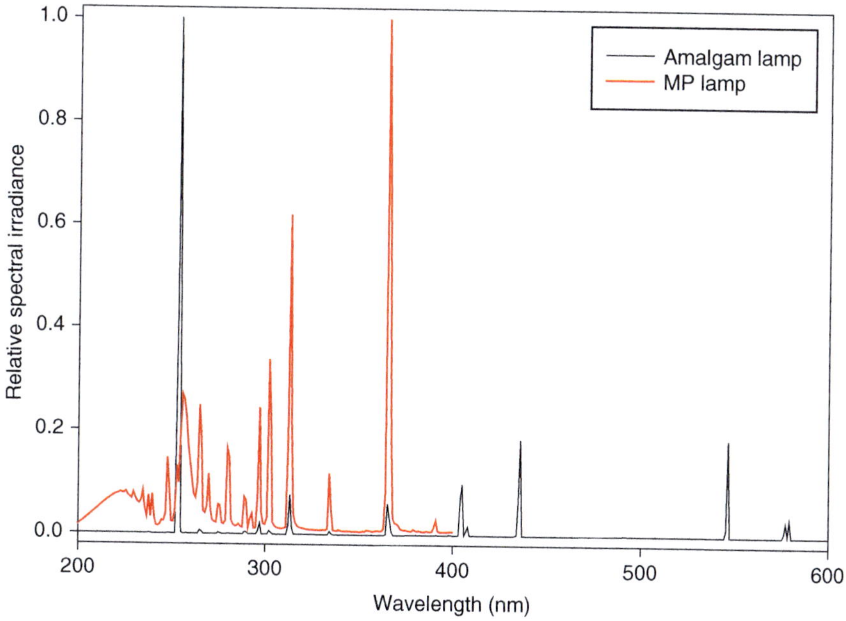


**Figure 4.8** Normalized output spectra of a mercury amalgam (low pressure) lamp and a medium-pressure (MP) Hg lamp. The spectra for each lamp are normalized against their respective spectral peaks. Data courtesy of Trojan Technologies.

Figure 4.8 illustrates a normalized output spectrum for a mercury amalgam lamp. A normalized spectrum is presented because it provides a general interpretation of lamp output; absolute lamp output will depend on the location of the measurement device relative to the lamp, the composition and optical characteristics of the media that exist between the lamp and the measurement device, and the conditions under which the lamp is operated. The normalized spectrum also facilitates comparisons with other radiation sources.

Medium-pressure Hg lamps operate at Hg partial pressures of roughly 75 Torr (approximately 0.1 atm). At this pressure, the physics of photon generation will include other processes that are not relevant in low-pressure Hg lamps. At start-up, the output of a medium-pressure lamp is similar to that of a low-pressure lamp, with dominant peaks at 254 and 185 nm. However, self-absorption is strong in the plasma of a medium-pressure lamp, causing substantial heating. Typical plasma temperatures inside medium-pressure Hg lamps are 5000–7000 K. As a result, there is a relevant incandescent component to the output of medium-pressure Hg lamps.

The output spectrum of a medium-pressure Hg lamp is characterized by a series of lines superimposed on a continuum (see Figure 4.8). The lines are attributable to electronic transitions that are defined by the energy-level diagram of Hg (see Figure 4.6). Thermal energy within this plasma is sufficient to assist in the promotion of Hg atoms to excited energy states. The process of "pressure broadening" is responsible for loss of some of the line characteristics of the output spectrum of these lamps. Pressure broadening involves collisions of Hg atoms with other particles in the system, including electrons, ions, and impurity atoms. These collisions cause changes in the energy distribution of Hg within the plasma, which in turn alters the output spectrum.

The output spectrum of a medium-pressure lamp has two interesting features in the UVC range. First, there is a broad hump from approximately 200–245 nm that is attributable to ion–electron recombination:

$$Hg^{+} + e^{-} \rightarrow Hg^{*} + h\nu \tag{4.6}$$

The ability of this reaction to yield photons across a range of wavelengths is believed to be attributable to variations in the kinetic energy of mercury ions and electrons in the plasma. A second broad peak is evident from wavelengths of 250–270 nm, with a notable dip at 254 nm. The dip is attributable to self-absorption, corresponding to the $6^1S_0 \rightarrow 6^3P_1$ transition [1].

Mercury lamps have long been the default option for generation of UV radiation. However, the inclusion of mercury in these devices presents risks that have motivated the development of alternative sources of UV radiation. Evidence of this pressure to develop new sources for room lighting and UVC radiation can be found in the *Minamata Convention on Mercury* [11]. This convention, which represents an international treaty, was developed with an objective of protecting human health and the environment from anthropogenic emissions of mercury. The convention was adopted in 2013 and has been signed by 128 countries (as of 2020), of which 105 have ratified. The convention calls for a number of measures to reduce mercury exposure, including a ban on the development of new mercury mines, reductions in the manufacture of mercury products, and the development of environmentally sound methods of storage for mercury. The convention allows production and use of some fluorescent lamps (Hg-based) to be ended, based on the amount of mercury they contain. The treaty does not explicitly address Hg-based lamps used for generation of UV radiation, and as such has no direct bearing on their production. However, other aspects of the convention will motivate reductions in the use of mercury, which by extension may promote the development of alternatives to Hg-based lamps for generation of UV radiation. A similar piece of legislation in Europe has also resulted in restrictions on the use of mercury [12]. As with the *Minamata Convention*, Hg-based lamps for general lighting are allowed, but there are restrictions on the mass of Hg that can be included in these lamps.

### 4.4.2 Light-Emitting Diodes (LEDs)

A light-emitting diode (LED) is a semiconducting device that emits electromagnetic radiation ("light") as a result of recombination of electrons and holes, following the principle of electroluminescence [13]. LEDs are small and can be efficient, robust sources of radiation. The first LEDs were developed in the early 1960s and involved emission of IR radiation. Visible LEDs soon followed, and in more recent years, LEDs have been produced that emit radiation in the UV portion of the electromagnetic spectrum.

All LEDs are diodes. Figure 4.9 provides a schematic illustration of the structure and behavior of a basic diode [13]. A diode is a device that allows electrical current to flow in one direction, if operated within its rated voltage range. To produce a diode, a semiconductor material (usually Si or Ge) is doped with materials to allow the formation of free electrons or holes. For n-type (negative) semiconductors, the dopant is usually a pentavalent impurity, while for p-type (positive) semiconductors, the dopant is usually trivalent. When n-type and p-type semiconductors are brought into contact, a so-called p-n junction is formed. Free electrons from the n-type material will diffuse toward the p-type material, while holes will diffuse from the p-type material toward the n-type material. The result is a depletion of charge carriers (electrons and holes) in the vicinity of the interface, and the formation of a "depletion zone" (Figure 4.9 a, b). This results in a barrier potential that prevents the flow of electrical current. The barrier potential is a material property; for Si, it has a value of roughly 0.7 V, while for Ge, it has a value of about 0.3 V. When operated as part of a DC circuit such that the positive terminal is connected to the p-type material and the negative terminal is connected to n-type (this is known as a *forward-biased* condition), electrostatic forces will cause electrons and holes to migrate toward the interface. If the applied voltage exceeds the barrier potential, then an electrical current will flow (Figure 4.9c). In a *reverse-biased* condition, the terminals of

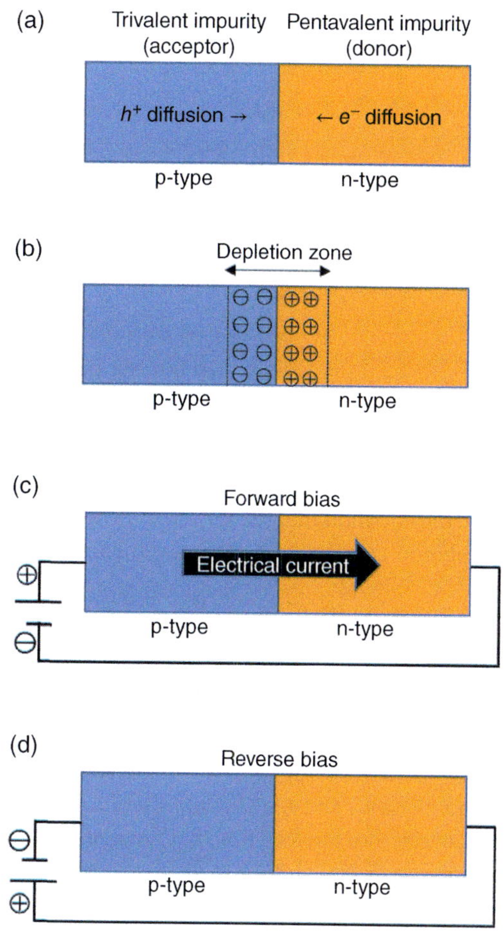


**Figure 4.9** Schematic illustration of the behavior of a basic diode. (a) Doped semiconductor materials (p-type and n-type) are brought into contact. Free electrons will diffuse toward p-type material, while holes will diffuse toward n-type material. The result is depletion of charge carriers in the area close to the interface between the materials (p-n junction, see b), with the formation of a "depletion zone," thereby preventing flow of electrical current. When the device is connected to an electrical circuit such that the positive terminal is connected to p-type material and the negative terminal connected to n-type material, electrostatic forces will promote the movement of electrons and holes toward the interface, causing a reduction in the width of the depletion zone (c). If a voltage above the barrier potential of the semiconductor is applied, then electrical current will flow under this forward-biased condition. However, if the electrical terminals are connected in the opposite sense, a reverse-biased condition (d), then no current will flow, unless a very high voltage is applied.

the source of electricity are connected in the opposite direction, causing migration of electrons and holes in the reverse direction (Figure 4.9d). This has the effect of preventing the flow of electrical current, except in cases where a high voltage is applied.

When operated with a forward bias, electrons and holes migrate toward the p-n junction. This causes recombination of electrons and holes; because conducting electrons are in the conduction band, they are at a higher energy state than they will be after recombination, where they will exist in the valence band. As such, recombination results in a release of energy. For some diodes, these transitions are radiative. The photon that is released by this type of recombination will have a wavelength that corresponds to the bandgap of the material. The bandgap will be influenced by the composition of the dopants.

UV LEDs are generally fabricated using group-III nitrides, such as GaN, InN, AlN, and alloys (mixtures) of these [14–17]. The bandgap for GaN (3.45 eV) corresponds to a photon at roughly 360 nm, while the bandgap for AlN (6.13 eV) corresponds to 202 nm [18]. Theoretically, these represent the upper and lower limits of emission wavelength for AlGaN semiconductors. It is possible to select the output wavelength by varying the composition of the $Al_xGa_{1-x}N$.

Angerer et al. developed AlGaN semiconductors that spanned a range of semiconductor alloy compositions. Also reported in their work were measured values of the bandgap for these materials

[19]. They fit a quadratic expression to their data to describe the relationship between bandgap and composition:

$$E_g^{Al_xGa_{1-x}N}(x) = x \cdot E_g^{AlN} + (1-x) \cdot E_g^{GaN} - b \cdot x(1-x) \tag{4.7}$$

where

| | | |
|---|---|---|
| $E_g^{Al_xGa_{1-x}N}(x)$ | = | bandgap of the AlGaN semiconductor [=] eV |
| $x$ | = | mole fraction of Al in $Al_xGa_{1-x}N$ |
| $E_g^{AlN}$ | = | bandgap of AlN (6.13 eV at room temperature) |
| $E_g^{GaN}$ | = | bandgap of GaN (3.45 eV at room temperature) |
| $b$ | = | direct gap bowing parameter (1.3 ± 0.2 eV at room temperature). |

By convention, bandgap energy values are reported in units of electron volts (eV). Recognizing that the movement of an electron between these bands results in conversion of electronic energy to photon energy, and converting energy in eV to Joules ($1\,J \equiv 6.242 \times 10^{18}$ eV), it is possible to rearrange the equation that relates photon energy to wavelength to allow calculation of the wavelength of a photon corresponding to a given bandgap:

$$\lambda = \frac{hc}{E} = \frac{(6.626 x\, 10^{-34} \mathrm{J} \cdot \mathrm{s}) \cdot \left(2.998 \times 10^8 \frac{\mathrm{m}}{\mathrm{s}}\right)}{E(\mathrm{eV}) \cdot \left(\frac{J}{6.242 \times 10^{18}\ \mathrm{eV}}\right)} \cdot \frac{10^9\ \mathrm{nm}}{\mathrm{m}} \tag{4.8}$$

Figure 4.10 provides an illustration of the bandgap and corresponding wavelength for $Al_XGa_{1-X}N$ semiconductors, as functions of mole fraction of aluminum ($x$) in the ternary AlGaN mixture, based on equations 4.7 and 4.8. The information presented in this figure suggests that, in principle, AlGaN semiconductors can be "tuned" to essentially any wavelength across the wavelength range from approximately 200 to 360 nm.

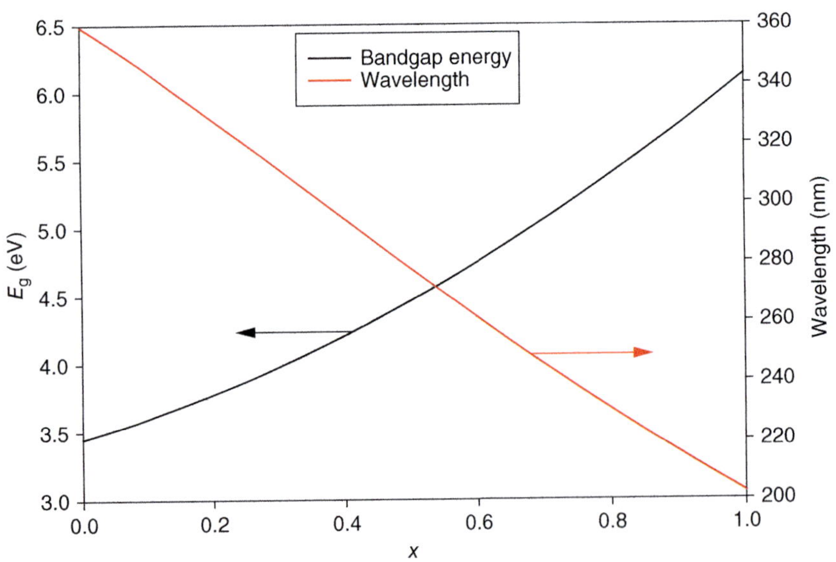


**Figure 4.10** Bandgap energy (left) and corresponding wavelength (right) as a function of mole fraction of aluminum ($x$) in $Al_xGa_{1-x}N$ films.

In practical terms, LEDs are not as well-behaved as might be inferred from Figure 4.10. Actual output from LEDs is usually characterized by a relatively narrow wavelength range, as opposed to a single wavelength. As with many optical devices, the spectral properties of UV LEDs are often characterized by bandwidth at half-maximum output. Figure 4.11 illustrates measured (relative) output spectra for several UV LEDs. The bandwidths at half-maximum relative output range from 10.9 to 14.3 nm.

UV LEDs have a number of advantages (or potential advantages) relative to conventional mercury lamps as sources of UVC radiation. These include the absence of mercury, compactness, robustness (durability), instant on/off capability, and the potential for higher efficiency and longer lifetimes than Hg lamps. In addition, UV LEDs offer directional capability that is lacking in most mercury lamp sources.

The primary disadvantages of UV LEDs, relative to Hg lamps, include low output power, low efficiency, and high cost. However, a common opinion in the LED industry is that given the interest in UV LEDs for applications in UV photoreactors, there is now incentive to improve the material properties of the semiconductors that form the foundation of UV LEDs, and to increase production, both of which should lead to improvements of these attributes.

Though there is some disagreement regarding terminology, the efficiency of LEDs is characterized by a range of parameters, including the following: [15, 20]

$$\text{Injection Efficiency} = \eta_{\text{Injection}} = \frac{\text{Number of electrons that recombine}}{\text{Number of electrons passing through device}} \tag{4.9}$$

$\eta_{\text{Injection}}$ describes the fraction of electrons injected into the active region of the LED that recombine with holes. Note that the numerator in this term accounts for all recombination events, including

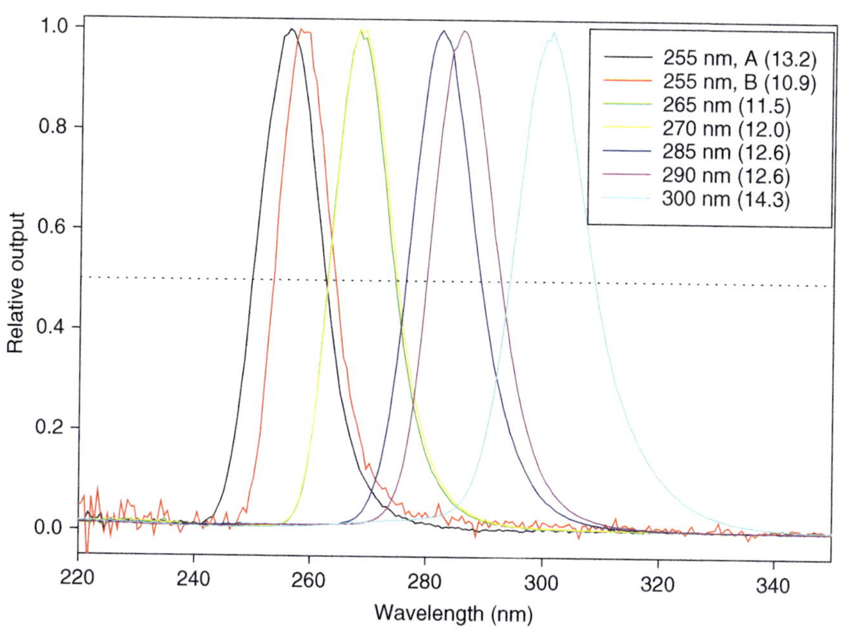


**Figure 4.11** Relative output spectra of UV LEDs. Values in the legend indicate the nominal peak wavelength of each LED, as indicated by the manufacturer. Horizontal reference line indicates 50% of maximum relative output. Parenthetical entries in legend indicate spectrum bandwidth at ½ maximum output. Data provided by AquiSense.

radiative and non-radiative events. A second measure of LED efficiency is Radiative Efficiency, defined as:

$$\text{Radiative Efficiency} = \eta_{\text{Radiative}} = \frac{R_{\text{sp}}}{R_{\text{sp}} + R_{\text{non-rad}}} \tag{4.10}$$

where

$R_{\text{sp}}$ = rate of recombination events that lead to spontaneous photon emission
$R_{\text{non-rad}}$ = non-radiative recombination rate.

The term $\eta_{\text{Radiative}}$ (sometimes referred to as Internal Quantum Efficiency, $\eta_{\text{IQE}}$) represents the fraction of electron–hole recombination events that result in photon emission.

$$\text{Extraction Efficiency} = \eta_{\text{Extraction}} = \frac{\text{Number of photons emitted from LED}}{\text{Number of photons generated in LED}} \tag{4.11}$$

$\eta_{\text{Extraction}}$ describes the fraction of photons generated within the active region of the LED that are emitted by the LED.

$$\text{External Quantum Efficiency} = \eta_{\text{EQE}} = \frac{\text{Number of photons emitted by LED}}{\text{Number of electons injected into LED}} \tag{4.12}$$

$\eta_{\text{EQE}}$ describes the efficiency of conversion of electrons into photons emitted by the LED. By analysis of the definitions provided above, it is possible to relate these terms:

$$\eta_{\text{EQE}} = \eta_{\text{Injection}} \cdot \eta_{\text{Radiative}} \cdot \eta_{\text{Extraction}} \tag{4.13}$$

From this derivation, it is evident that the efficiency by which an LED converts electrical energy into photon energy depends on several aspects of the device itself, which in turn are influenced by the material properties and geometry of the LED.

Beyond $\eta_{\text{EQE}}$, LEDs are also characterized in terms of the wall plug efficiency, $\eta_{\text{Wall}}$. Formally, $\eta_{\text{Wall}}$ is defined as follows:

$$\eta_{\text{Wall}} = \frac{\text{LED optical power}}{\text{electrical power supplied to device}} \tag{4.14}$$

However, this term may be decomposed to allow consideration of the factors that affect it. Specifically:

$$\eta_{\text{Wall}} = \eta_{\text{EQE}} \cdot \eta_{\text{Feeding}} \tag{4.15}$$

where $\eta_{\text{Feeding}}$ represents the "feeding efficiency":

$$\eta_{\text{Feeding}} = \frac{\text{mean energy of emitted photons}}{\text{total energy acquired by electron : hole pair from power source}} \tag{4.16}$$

The decompositions provided in equations 4.13 and 4.15 indicate the factors that affect the overall efficiency and performance of UV LEDs. The individual terms that comprise these relationships also inform research efforts that are aimed at improving $\eta_{\text{EQE}}$ and $\eta_{\text{Wall}}$. In general, both $\eta_{\text{EQE}}$ and $\eta_{\text{Wall}}$ decrease with the nominal peak wavelength of emission of UV LEDs. Figure 4.12 provides a clear illustration of this behavior. For the time being, the relatively low values of $\eta_{\text{EQE}}$ and $\eta_{\text{Wall}}$ for UVC LEDs represent barriers to their widespread implementation in UV photoreactors.

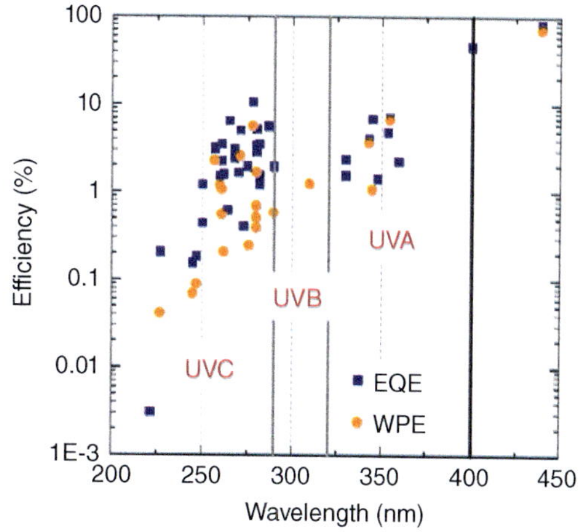


**Figure 4.12** Reported values of improving $\eta_{EQE}$ (EQE) and $\eta_{Wall}$ (WPE) for UV LEDs as a function of nominal peak wavelength of emission. *Source:* Zhang et al. [21], figure 1/with permission of Angel Business Communications publication.

### 4.4.3 Excimer Lamps

An *excimer* is a weakly bound, usually diatomic[b] electronically excited dimer that is nonbonding or weakly bonding in the ground state. A related definition is an *exciplex*, which represents an excited complex that has similar nonbonding characteristics in its ground state. Both terms appear in the literature, but often the term "excimer" is used to represent both groups; for purposes of simplicity, the term excimer will be used to represent formally defined excimers and exciplexes.

Photon emission from an excimer is attributable to spontaneous decomposition of an excimer into its constituent ground-state atoms. The transition from an excimer to the ground-state atoms is a radiative process and the photon that results from this process has a wavelength that corresponds to the energy gap between the excimer and the ground-state atoms. Figure 4.13 illustrates an energy diagram for xenon (Xe), which includes the excimer molecule ($Xe_2^*$), the ground-state atoms, and several other electronically excited forms of Xe.

A common configuration for an excimer lamp is the so-called dielectric barrier-discharge (DBD) lamp. DBDs were originally applied as ozone generators, but have more recently been applied as sources of UV radiation for other photochemical processes. Although many geometries for DBD systems are available, a common configuration involves an annular geometry (see Figure 4.14).

Using Figure 4.14 (left) as a guide and working outward radially, the components of an annular DBD excimer lamp are as follows: [22, 24, 25]

- An inner electrode is made of a conducting material (metal), often with an annular geometry.
- An inner quartz tube functions as a dielectric material between the two electrodes.
- The annular space between the two quartz tubes is filled with the gas(es) that is (are) used to generate the excimer. Excimer generation within this space occurs within the (pseudo) randomly distributed microdischarges, which can be thought of as small "lightning bolts" that develop between the electrodes. Electron densities within these microdischarges are on the order of $10^{14}$–$10^{15}$ per cm$^3$, with energies as high as 10 eV.
- An outer quartz tube functions as a dielectric barrier between the electrodes, but also allows effective transmission of UV photons that are generated within the annular space.
- A wire mesh electrode is placed at the outside radius of the outer quartz tube. A large electrical potential difference is established between the two electrodes to promote microdischarge formation. Electrical field strength in DBD lamps spans the range from 0.1 to 100 kV/cm.

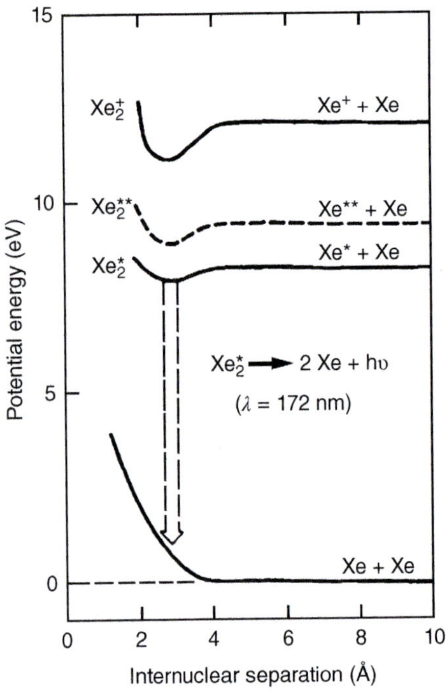


**Figure 4.13** Potential energy diagram for xenon. The transition between the excimer ($Xe_2^*$) and its constituent ground-state atoms (Xe + Xe) results in emission of a photon ($\lambda$ = 172 nm). *Source:* Based on Eliasson and Kogelschatz 1991.

The most common excimer lamps involve rare gas atoms (Rg), halogen atoms (X), or Rg/X combinations. The formation of excimers involves collisions (reactions) between Rg, X, or Rg/X with electrons. Excimer formation from rare gases usually involves argon (Ar), krypton (Kr), or xenon (Xe), while halogen atoms used in excimers include fluorine (F), chlorine (Cl), bromine (Br), and iodine (I).

The formation of an excimer generally involves a sequence of reactions that depend on the combination of gases used to form the excimer. In the case of rare gases, a simplified representation of the reactions that lead to excimer formation is as follows: [22].

$$Rg + e^- \rightarrow Rg^* + e^- \tag{4.17}$$

In reaction 4.17, a collision between a ground-state rare gas atom (Rg) and an electron ($e^-$) results in transfer of kinetic energy from the electron, with a corresponding gain of chemical (potential) energy by Rg to yield an excited rare gas atom (Rg∗). A three-body reaction then takes place to form the rare gas excimer ($Rg_2^*$):

$$Rg^* + Rg + Rg \rightarrow Rg_2^* + Rg \tag{4.18}$$

The stoichiometry indicated in reaction 4.18 indicates that the rate of $Rg_2^*$ formation depends on the square of the activity (pressure) of the rare gas; because of this, the behavior of excimer lamps is strongly influenced by pressure. It is relevant to note that there are many other reactions that take place in the microdischarges that result in the formation of other high-energy excited states (e.g. $Xe^{**}$, $Xe_2^{**}$, $Xe^+$, $Xe_2^+$, see Figure 4.13 for illustration).

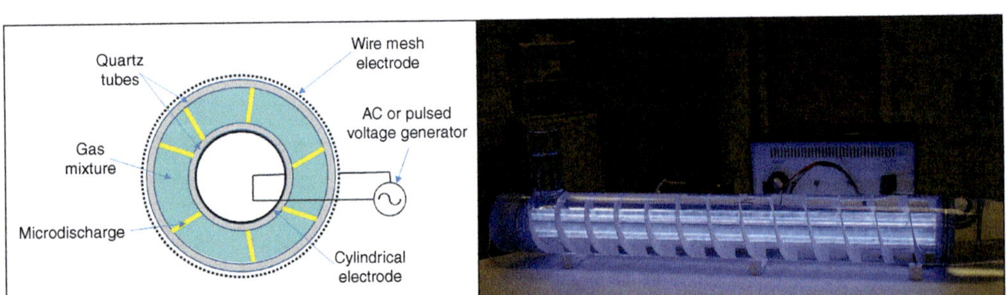


**Figure 4.14** (Left) Schematic illustration of annular DBD excimer lamp (end view). The constituent gases of the lamp are contained between two quartz cylinders. Pseudo-randomly distributed microdischarge emission into the annular region promotes generation of excimers. Spontaneous decomposition of the excimers results in release of photons. (Right) Photograph of an operating XeBr annular DBD excimer lamp inside an annular reactor that was designed for water disinfection. *Source:* Based on Naunovic et al. 2008.

Rare gas/halogen mixtures are commonly used to generate excimers; however, the chemistry of excimer formation in the case of these mixtures is slightly different than with rare gases alone. In particular, excimer formation will involve a so-called "harpoon" reaction, in which an excited rare gas atom will split a halogen molecule to form an excimer, RgX*: [26].

$$\mathrm{Rg}^* + X_2 \rightarrow \mathrm{Rg}X^* + X \tag{4.19}$$

Once an excimer has been formed, it will undergo spontaneous degradation to yield its constituent atoms and a photon:

$$\mathrm{Rg}_2^* \rightarrow \mathrm{Rg} + \mathrm{Rg} + h\nu \tag{4.20}$$

$$\mathrm{Rg}X^* \rightarrow \mathrm{Rg} + X + h\nu \tag{4.21}$$

The photons that are produced by reactions 4.20 and 4.21 have a wavelength that corresponds to the energy difference between their excimer and the constituent atoms in their ground state. Another important aspect of this reaction is that because there is no ground-state form of the excimer molecule, the products of reactions 4.20 and 4.21 do not absorb the photons that are emitted by these processes. As such, there is no self-absorption of produced photons in excimer lamps [22, 24, 25, 27]. This represents a potentially important advantage of excimer lamps relative to Hg lamps, for which self-absorption represents a source of energy inefficiency.

Output spectra of excimer lamps have been reported in the literature. Figure 4.15 provides several examples of these spectra. The output spectra of these lamps are characterized by a relatively

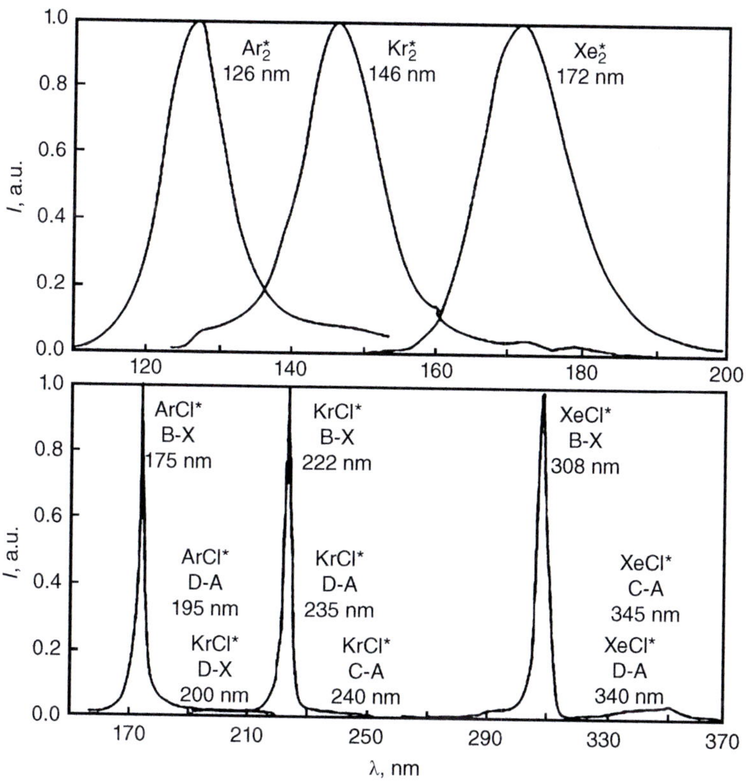


**Figure 4.15** (top) Relative output spectra of rare gas excimer lamps. (bottom) Relative output spectra of rare gas/halogen excimer lamps. *Source:* Sosnin et al. 2006, figure 3 (p.148)/with permission of ELSEVIER.

narrow bandwidth. In the case of rare gas excimers, the bandwidth at half-max ranges from 2 to 30 nm, whereas for rare gas/halogen excimers, the band width at half-max ranges from 2 to 15 nm [27, 28].

The location of the peak wavelength of emission of excimer lamps varies in a regular pattern, following trends that can be linked to their respective rows in the periodic table. Figure 4.16 illustrates these trends for diatomic excimers that include several examples of $Rg_2^*$, $X_2^*$, and $RgX^*$. The peak wavelength for each excimer is plotted as a function of the average periodic table row number for the constituent atoms. For example, for $Ar_2^*$, both atoms are in row 3, giving an average row number of 3, whereas for $ArF^*$, we find Ar in row 3 and fluorine in row 2, given an average row number of 2.5.

Within each group of excimers, consistent trends of peak wavelength vs. average periodic table row number are evident. For the rare gas excimers ($Rg_2^*$), the peak wavelength increases at a modest rate with row number; the slope of the line for halogen excimers ($X_2^*$) is similar to that of $Rg_2^*$, but starts at a higher value. Among the rare gas/halogen excimers ($RgX^*$), for a fixed halogen (X), the slopes of the relationships between peak wavelength and average row number are steeper than those of the homoatomic excimers ($Rg_2^*$ and $X_2^*$). On the other hand, when the rare gas (Rg) is fixed, a modest decreasing trend is observed.

The information presented in Figures 4.15 and 4.16 indicate one of the strengths of excimer lamps as UV sources. Specifically, these devices can be tailored to an application by selection of the excimer gas. Photochemical processes offer the opportunity to select a radiation source for which the photons will be used effectively to promote the reaction of interest, which usually happens in a wavelength range where relatively strong absorbance and quantum yield are displayed by the photochemical target. On the other hand, overall process efficiency benefits from selection of a wavelength that will be effectively transmitted by the medium that contains the photochemical targets.

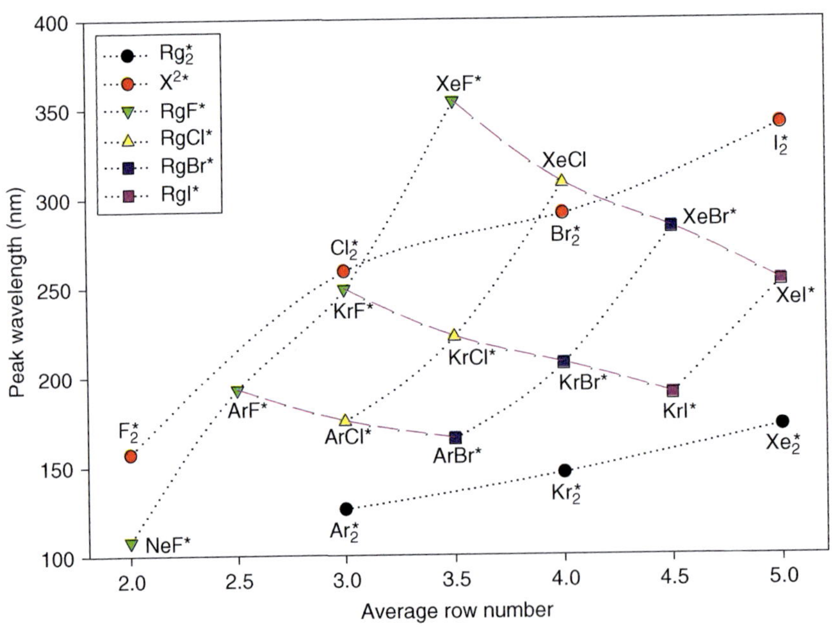


**Figure 4.16** Peak output wavelength as a function of average periodic table row number for constituent atoms. For the rare gas/halogen excimers ($RgX^*$), the black dotted lines illustrate changes of peak wavelength for excimers with a common halogen; the pink dashed lines illustrate these trends for excimers that contain a common rare gas. Data from IUPAC Gold Book 2019.

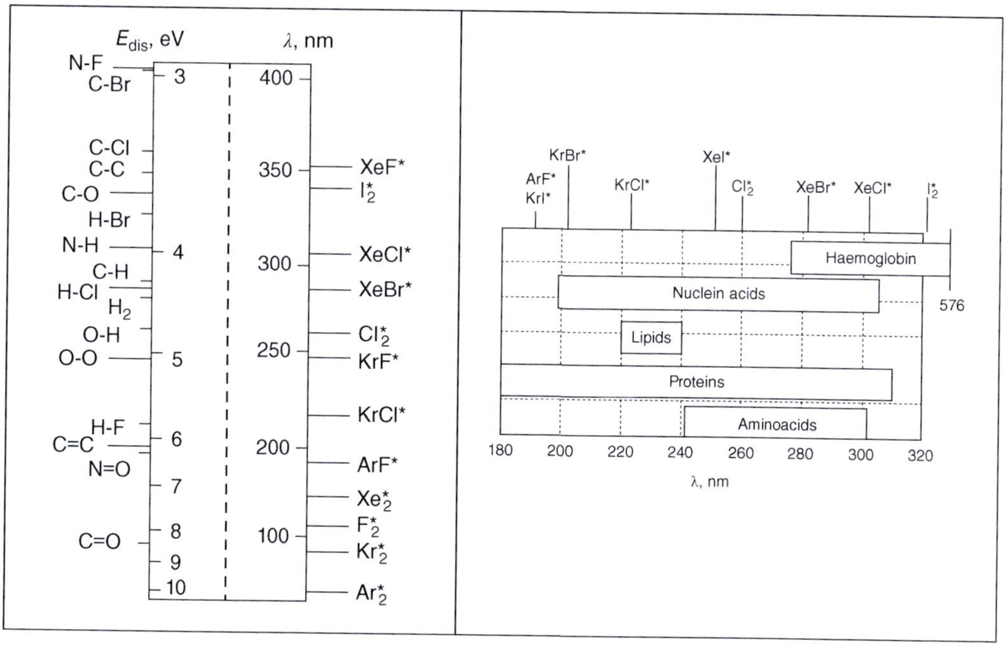


**Figure 4.17** (Left) Representative bond dissociation energies for common chemical groups with corresponding excimer peak wavelengths. (Right) Illustration of ranges of absorption by common biomolecules, with corresponding peak wavelengths of excimers. *Source:* Sosnin et al. 2006, figure 8,9 (p.151)/ with permission of ELSEVIER.

Therefore, it is desirable to have low absorbance (high transmittance) in the irradiated medium by the wavelengths of radiation emitted by the source. Figure 4.17 illustrates this behavior for two common groups of photochemical targets.

Excimer lamps have been reported to have overall efficiencies that range from roughly 5 to 20%, depending on gas mixture, operating conditions, and lamp geometry [25, 30]. Lamp lifetimes range from 4000 to 20,000 hours, depending on gas composition and patterns of use.

Excimer lamps are characterized by a relatively rapid start-up time. Lomaev et al. reported that excimer lamp output will reach a near-stationary output condition within 0.5–1 seconds after ignition [27].

Another important advantage of excimer lamps is the wide range of geometric configurations that can be created. The annular orientation illustrated in Figure 4.14 represents a common geometry; however, flat-screen sources have been developed that offer the potential for the development of a fundamentally different spatial distribution of radiation energy around the source than can be achieved by more conventional geometries.

Figure 4.18 illustrates a flat-panel $Xe_2^*$ lamp and its output characteristics. The device was developed by milling microcavities into a quartz plate that is then filled with excimer gases. Wire mesh electrodes are installed on either side of the flat plate to allow the formation of the excimer from constituent gases, with subsequent emission of photons. The output spectrum of this device is strongly influenced by gas pressure inside the microcavities.

The flat-panel device illustrated in Figure 4.18 provides an illustration of the geometric flexibility that is possible using excimer lamps. Such a flat-panel design allows for "tiling" of lamps to build a large flat-panel source. The emission pattern of a flat-panel device will be fundamentally different from that of a conventional cylindrical lamp geometry. This can provide advantages in terms of overall reactor efficiency, as will be discussed in subsequent chapters.

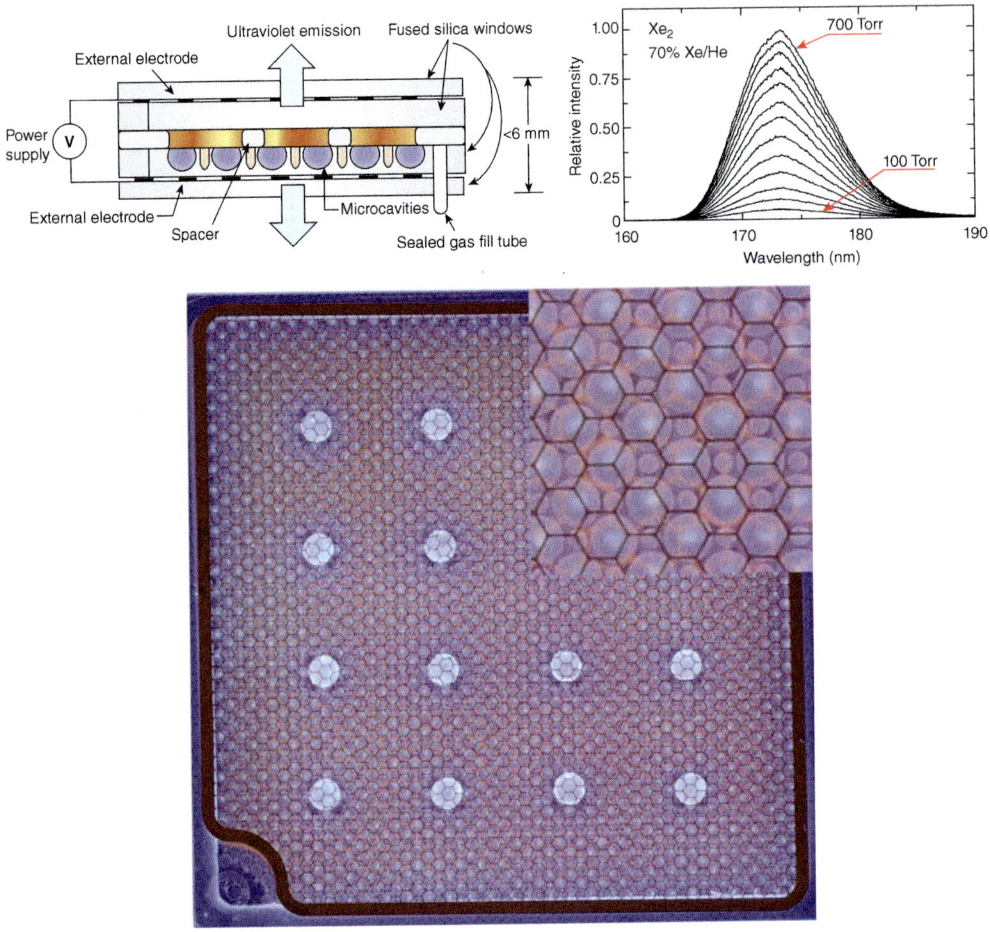


**Figure 4.18** (Top left) Schematic illustration of flat panel excimer lamp. Hemispherical or parabolic (cross-sectional) microcavities are milled into a quartz plate to allow filling with excimer gas(es). Electrodes are applied to both sides of the microcavities to allow excitation of excimer gases. (Top right) Relative output spectrum of flat-panel lamp for a range of gas pressures; output is normalized against peak output at $P = 700$ Torr. (Bottom) Photographic image of a 10 cm × 10 cm flat-panel xenon excimer lamp (approximately actual size). Inset image in the upper right of this photograph illustrates the hexagonal electrode structure that has been placed on top of the array of microcavities [31]. *Source:* Park et al. 2017 / AIP Publishing / CC BY-4.0. Source link: https://aip.scitation.org/doi/full/10.1063/1.4976530.

### 4.4.4 Lasers

The word "laser" is an acronym for the term **L**ight **A**mplification by the **S**timulated **E**mission of **R**adiation. Therefore, to understand lasers, it is necessary to understand the process of stimulated emission.

We will start by considering two possible electronic energy states for an atom, $E_1$ and $E_2$, such that $E_2 > E_1$ (see Figure 4.19). Three processes exist to allow transitions between $E_1$ and $E_2$. The first (process 1 in Figure 4.19) is absorption, where a photon at a frequency (wavelength) corresponding to the energy difference $E_2$–$E_1$ is absorbed by the atom, causing promotion of an electron from $E_1$ to $E_2$. The opposite process (process 2 in Figure 4.19) involves spontaneous emission of a photon at this

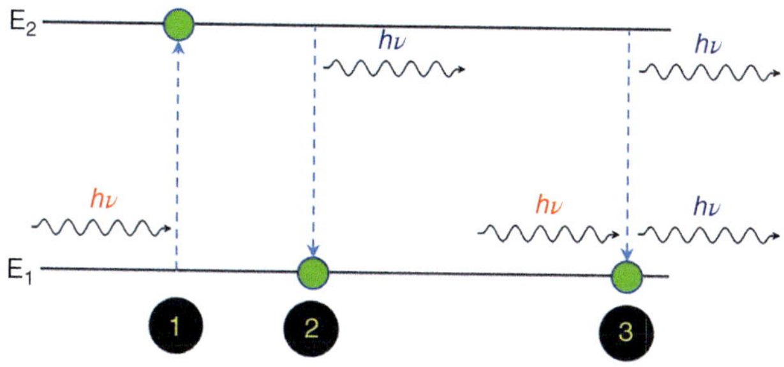


**Figure 4.19** Schematic illustration of two electronic energy levels of an atom, $E_1$ and $E_2$, such that $E_1 < E_2$. Three basic processes exist to allow electronic transitions between $E_1$ and $E_2$. Photons ($h\nu$) indicated by red font are absorbed, whereas those indicated by blue font are emitted. Process 1 involves absorption of a photon, resulting in promotion of an electron from $E_1$ to $E_2$. Process 2 represents spontaneous emission of a photon to cause the atom to relax from $E_2$ to $E_1$. Process 3 is one in which a photon is absorbed by the atom in the excited state to stimulate emission of a second photon at the same frequency (wavelength); process 3 is an illustration of stimulated emission.

same frequency (wavelength), causing the atom to relax from $E_2 \rightarrow E_1$. The third process is one in which a photon at this same frequency is absorbed by an atom in the excited state. Absorption of this photon causes emission of a photon at the same frequency (wavelength); the emitted photon will also be coherent with the initiating photon in this process, which means that the wave characteristics of the incident and emitted photons are in phase. Process 3 is an illustration of stimulated emission.

For purposes of this presentation, we will define $N_1$ as the population of atoms at $E_1$ and $N_2$ as the population of atoms at $E_2$. Under normal conditions, we expect $N_1 >> N_2$. For stimulated emission to be effective, it is necessary for $N_2 > N_1$, a condition known as population inversion. Under an inversion condition, it is possible for a single photon to cause co-emission of a coherent photon. Therefore, one of the requirements of lasers is that a mechanism to accomplish population inversion must exist in the system.

There are many mechanisms by which a population inversion condition can be accomplished. In general, it is fairly easy to achieve this condition when $E_2$–$E_1$ is small; however, when this energy difference is large, the energy losses of the system tend to be large. All lasers are characterized by low electrical efficiency, but because of this phenomenon, lasers that produce short wavelength radiation (e.g. UV) tend to be relatively inefficient.

Figure 4.20 provides a schematic illustration of basic laser components. Pumping represents a process whereby an external energy source is used to promote electrons to higher energy states, thereby allowing a condition of population inversion to be achieved. Pumping can be accomplished

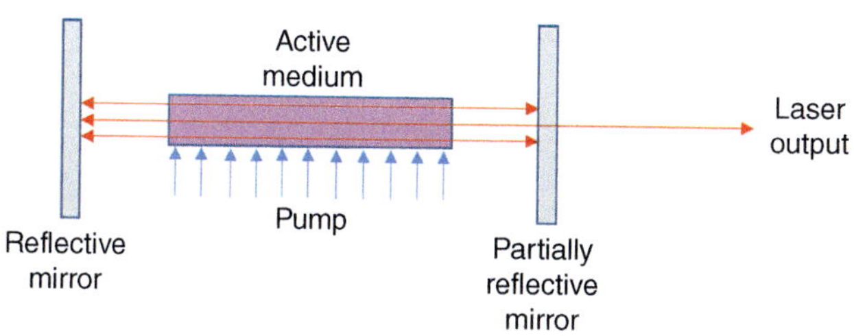


**Figure 4.20** Schematic illustration of laser components.

by several different methods, including other sources of intense radiation (e.g. flash lamps or other lasers), electrical current, or a chemical reaction. The pumping energy is applied to the active laser medium (*aka*, gain medium, lasing medium) wherein the population inversion condition is established. The active medium can be gas, liquid (often dye solutions), or (crystalline) solid.

In many situations where lasers are applied, there is a need to alter the output. A common device used for this purpose is a frequency multiplier, which is a nonlinear optical device that allows photons to be combined. The result of this combination is that output frequency is an integer multiple of input frequency, usually double or triple. For example, when the frequency of radiation is doubled, its wavelength is reduced by ½. So, a frequency multiplier can be used to convert radiation with a wavelength of 532 nm (green light) to radiation at 266 nm (UVC radiation) by doubling. This approach is common among lasers.

As compared to most other sources of UV radiation, lasers are expensive and inefficient. However, they generate radiation with unusual properties that are helpful in some specialized applications. Laser radiation is coherent, highly collimated, and (close to) monochromatic. Figure 4.21 illustrates an output spectrum of a 266 nm laser. The spectrum is characterized by a bandwidth of roughly 0.3 nm at half-maximum.

Laser radiation is also characterized by an output beam that has a small diameter and high photon flux. Because of this, some lasers allow for multiphoton processes to take place.

### 4.4.5 Upconversion

*Fluorescence* is a process wherein electromagnetic radiation is absorbed by a molecule, resulting in electronic excitation of the molecule, causing emission of radiation at a different wavelength. For most fluorescent molecules (*fluorophores*), the emitted radiation is characterized by a longer wavelength (lower energy) than the radiation that caused excitation. This is because some of the energy gained by the fluorophore as a result of photon absorption is lost by non-radiative mechanisms, such as heat. When the peak of the emission spectrum is observed at a longer wavelength than the peak of the excitation spectrum, it is referred to as a Stokes (red) shift, named after the Irish physicist and mathematician Sir George G. Stokes, who made seminal contributions in optics and fluid mechanics. Another contributing factor to this behavior is the fact that the typical lifetime

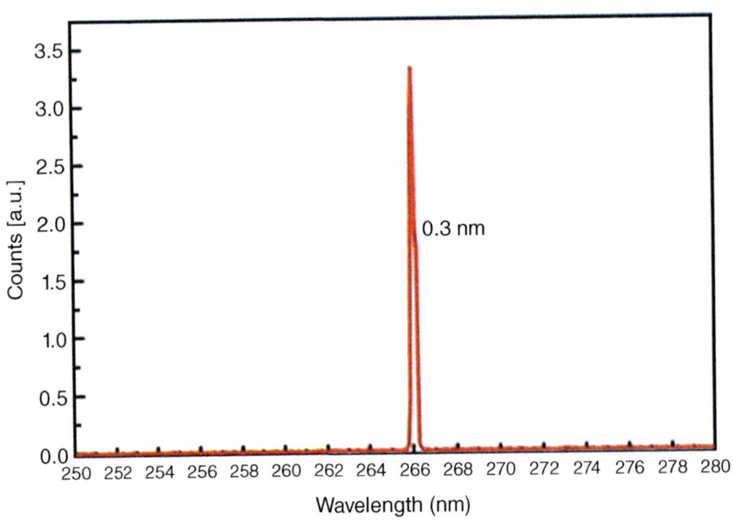


**Figure 4.21** Output spectrum of a UVC laser. *Source:* Zhai et al. 2013, figure 5 (p.3)/IOP Publishing.

of an electronically excited fluorophore is on the order of 0.5–20 ns [33]. As such, it is extremely unlikely that a second photon could be absorbed by an excited fluorophore before spontaneous emission occurs. By extension, this implies that it is not possible (or extremely unlikely) to accumulate photon energy within a fluorophore beyond that corresponding to a single photon.

Some materials are able to participate in luminescent processes that yield an anti-Stokes (blue) shift. An example of such a process is *upconversion* (UC), a process in which two or more low-energy photons are absorbed sequentially to reach an excited state, resulting in spontaneous emission of a high-energy photon [34, 35].

The basic principles that are responsible for UC have been examined since the 1960s and the process has now evolved to the point where practical applications are being developed. A related process is *downconversion* (DC), wherein a high-energy photon is absorbed leading to emission of two or more low-energy photons [36]. UC and DC are processes that offer the potential to use radiation from inexpensive, abundant sources of electromagnetic radiation, such as sunlight or the output from fluorescent lamps. UC and DC both allow for improved spectral matching between sources of radiation and photochemical targets.

UC was originally developed to accomplish conversion of IR radiation to visible radiation [34]. However, in recent years, various materials have been identified or developed that allow for visible→UV UC. Two groups of materials are able to accomplish this transition [37]. The first of these are organic sensitizers/acceptors, which are able to convert visible radiation into radiation in the UVA range. The second group of processes involve lanthanide (Ln)-doped inorganic crystalline solids, which are able to yield output radiation in the UVA, UVB, and UVC ranges. Because of their broader output spectral range, and because they are able to yield UVC radiation, these materials are probably more important to photoreactors than other UC materials.

The Ln-doped materials involve $Ln^{3+}$ ions as replacements within a crystalline matrix that function to activate the inorganic crystalline structure to allow UC [38]. Common crystalline materials used include fluorides ($LiLuF_4$, $LiYF_4$, $Lu_7O_6F_9$) and silicates ($Y_2SiO_5$, β-$Y_2Si_2O_7$). The lanthanides all express a trivalent oxidation state and involve partially filled 4f subshells, which are optically active in UC processes [36, 37].

Among the Ln-substituted materials, two mechanisms of excitation have been identified (see Figure 4.22) [35, 37]. The first is called *excited-state absorption* (ESA), wherein two or more

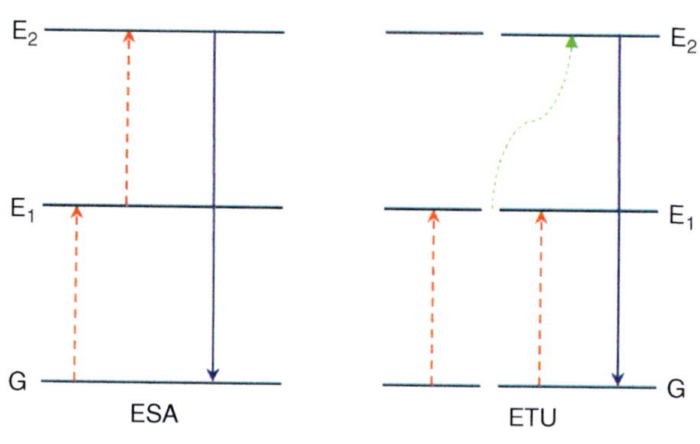


**Figure 4.22** Schematic illustration of upconversion mechanisms for $Ln^{3+}$-doped crystalline solids. G represents the ground state of the ion. In excited-state absorption (ESA), a low-energy photon is absorbed by the ground-state ion, resulting in excitation to an intermediate energy level, $E_1$ (red, dashed arrow). A subsequent photon is absorbed by the ion at $E_1$ allowing promotion to a higher energy state, $E_2$, which yields a high-energy photon by spontaneous emission (blue arrow). In energy-transfer upconversion (ETU), a low-energy photon causes promotion of ions at adjacent sites to $E_1$ in the solid matrix. Energy transfer between the ions at $E_1$ (green, dashed arrow) allows promotion of one of the sites to $E_2$, again yielding photon emission.

low-energy photons are absorbed sequentially to yield an intermediate excited state, then a radiative excited state. The second process is known as *energy-transfer upconversion* (ETU) and involves excitation of neighboring sites within a crystalline matrix, which then participate in energy transfer to yield the radiative excited state.

ESA and ETU both require that the ions in the intermediate excited state ($E_1$) have a relatively long lifetime. For $Ln^{3+}$ ions, the reported lifetimes are in the range of μs to ms. [36]

Among the lanthanides, the praseodymium ion ($Pr^{3+}$) is the most effective UC promoter because of its broad excitation range [38]. Figure 4.23 illustrates the upconversion emission spectra for $Pr^{3+}$-doped and co-doped materials, along with representative energy-level diagrams. These spectra illustrate that for all of the $Pr^{3+}$-doped materials, there was substantial emission in the UVC range. Interestingly, co-doping with $Li^+$ and $Gd^{3+}$ resulted in increases in overall UVC emission.

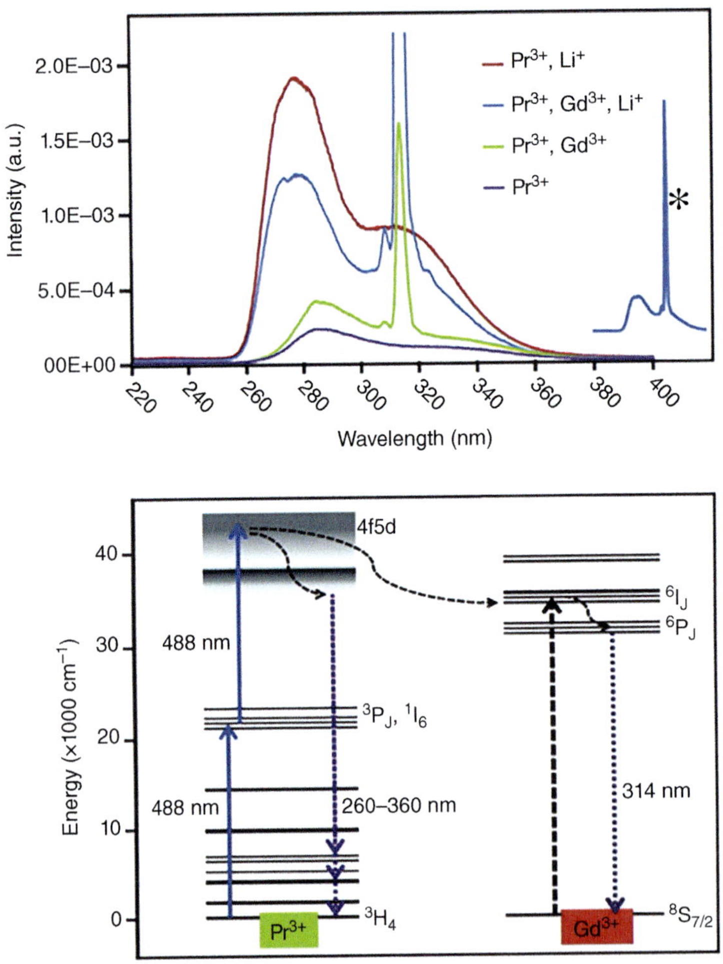


**Figure 4.23** (Top) Upconversion emission spectra of the $Pr^{3+}$-doped silicate powder material, $Y_2SiO5{:}Pr^{3+}$ as a result of excitation with 488 nm laser radiation; also included are upconversion emission spectra for these same materials co-doped with $Li^+$, $Gd^{3+}$, or both. Inset image illustrates the upconversion spectrum for $Y_2SiO_5{:}Pr^{3+},Gd^{3+},Li^+$. (Bottom) Energy-level diagram for $Pr^{3+}$ and $Gd^{3+}$ emissions. *Source:* Cates et al. 2011, figure 2 (p.3681)/American Chemical Society.

$Li^+$-co-doping resulted in a blue shift in the output spectra, with peaks of the output spectra shifting from 285 nm with the $Pr^{3+}$-doped material to 278 nm for the material that included $Li^+$-co-doping.

## 4.5 Summary

Photochemical processes depend on access to one or more sources of electromagnetic radiation, usually with output in the UV range. More specifically, most photochemical processes employ UVC radiation. Perhaps the most important source of electromagnetic radiation is the sun, which includes UVA and UVB radiation at near sea level altitudes. However, the lack of UVC radiation in ambient solar radiation motivates the use of artificial UV sources. Among these, the most commonly used are mercury lamps. These devices allow efficient conversion of input electrical power into output UV photons. However, a number of alternative sources of UV radiation exist and are in various stages of development. These include UV LEDs, excimer lamps, lasers, and UC. These alternative sources provide advantages and disadvantages, relative to conventional Hg lamps. For some of these alternatives, advancements in materials science are likely to lead to improvements in efficiency and output characteristics. In addition, some of these alternative sources provide advantages in terms of possible geometric configurations. Collectively, these attributes suggest that alternative UV sources will be used with increasing regularity in the future.

## Notes

**a** A plasma is a hot ionized gas comprising roughly equal concentrations of positively charged ions and electrons (negatively charged). Plasmas behave differently than neutral gases and are sometimes referred to as the "fourth state of matter." Because plasmas are made up of electrically charged particles, they are strongly influenced by electric and magnetic fields.
**b** Triatomic excimers also exist.

## References

1 Phillips, R. (**1983**). *Sources and Applications of Ultraviolet Radiation*. London, England: Academic Press.
2 Demerjian, K.L., Schere, K.L., and Peterson, J.T. (**1980**). Theoretical estimates of actinic (spherically integrated) flux and photolytic rate constants of atmospheric species in the lower troposphere. In: *Advances in Environmental Science and Technology*, vol. 10 (ed. J.N. Pitts Jr., R.L. Metcalf and D. Grosjean), 369–459. New York: Wiley.
3 Iqbal, M. (**1983**). *An Introduction to Solar Radiation*. Toronto, Ontario, Canada: Academic Press.
4 Finlayson-Pitts, B.J. and Pitts, J.N. (**2000**). *Chemistry of the Upper and Lower Atmosphere: Theory, Experiments, and Applications*, 969. San Diego, CA: Academic Press.
5 ASTM. (**2014**). Standard solar constant and zero air mass solar spectral irradiance tables. In E490−00a ed.
6 Gueymard, C. and National Renewable Energy Laboratory NREL SMARTS: Simple model of the atmospheric radiative transfer of sunshine. https://www.nrel.gov/rredc/smarts/ (accessed 2021).
7 ASTM. (**2012**). Standard tables for reference solar spectral irradiances: Direct normal and hemispherical on 37° tilted surface. In G173−03 ed.

8 USDA UV-B monitoring and research program. http://uvb.nrel.colostate.edu/UVB/index.jsf (accessed 2021).
9 Herzberg, G. (**1944**). *Atomic Spectra and Atomic Structure*. New York: Dover Publications.
10 Huber, M.L., Laesecke, A., and Friend, D.G. (**2006**). *The Vapor Pressure of Mercury*. Boulder, CO: NIST.
11 UNEP. (**2013**). Minamata convention on mercury: Text and annexes. https://www.mercuryconvention.org/en.
12 EU (**2011**). Directive 2011/65/EU of the European Parliament and of the Council of 8 June 2011 on the restriction of the use of certain hazardous substances in electrical and electronic equipment. *EUR-Lex: Official Journal of the European Union* L174/88–L174/110.
13 Held, G. (**2009**). *Introduction to Light Emitting Diode Technology and Applications*. Boca Raton, FL: CRC Press.
14 Song, K., Mohseni, M., and Taghipour, F. (**2016**). Application of ultraviolet light-emitting diodes (UV-LEDs) for water disinfection: A review. *Water Research 94*: 341–349.
15 Khan, A., Balakrishnan, K., and Katona, T. (**2008**). Ultraviolet light-emitting diodes based on group three nitrides. *Nature Photonics 2* (2): 77–84.
16 Ding, K., Avrutin, V., Ozgur, U., and Morkoc, H. (**2017**). Status of growth of group III-nitride heterostructures for deep ultraviolet light-emitting diodes. *Crystals 7* (10): 1–75.
17 Orton, J.W. and Foxon, C.T. (**1998**). Group III nitride semiconductors for short wavelength light-emitting devices. *Reports on Progress in Physics 61* (1): 1.
18 Yoshida, S., Misawa, S., and Gonda, S. (**1982**). Properties of $Al_XGa_{1-X}N$ films prepared by reactive molecular-beam epitaxy. *Journal of Applied Physics 53* (10): 6844–6848.
19 Angerer, H., Brunner, D., Freudenberg, F. et al. (**1997**). Determination of the Al mole fraction and the band gap bowing of epitaxial AlxGa1-xN films. *Applied Physics Letters 71* (11): 1504–1506.
20 Zhao, H.P., Liu, G.Y., Zhang, J. et al. (**2013**). Analysis of internal quantum efficiency and current injection efficiency in III-nitride light-emitting diodes. *Journal of Display Technology 9* (4): 212–225.
21 Zhang, Y., Krishnamoorthy, S., Akyol, F., and Rajan, S. (**2016**). Quantum tunneling boosts UV LED efficiency. https://compoundsemiconductor.net/article/99539/Quantum_Tunneling_Boosts_UV_LED_Efficiency%7BfeatureExtra%7D (accessed 2021).
22 Eliasson, B. and Kogelschatz, U. (**1991**). Modeling and applications of silent discharge plasmas. *Ieee Transactions on Plasma Science 19* (2): 309–323.
23 Naunovic, Z., Lim, S., and Blatchley III, E.R. (**2008**). Investigation of microbial inactivation efficiency of a UV disinfection system employing an excimer lamp. *Water Research 42* (19): 4838–4846.
24 Eliasson, B. and Kogelschatz, U. (**1988**). UV excimer radiation from dielectric-barrier discharges. *Applied Physics B-Photophysics and Laser Chemistry 46* (4): 299–303.
25 Sosnin, E.A., Oppenlander, T., and Tarasenko, V.F. (**2006**). Applications of capacitive and barrier discharge excilamps in photoscience. *Journal of Photochemistry and Photobiology C-Photochemistry Reviews 7* (4): 145–163.
26 Yu, Y.C., Setser, D.W., and Horlguchl, H. (**1983**). Thermochemical and kinetic-studies of the xenon halide B-state and C-state in 0.5-5 atm of buffer gas. *Journal of Physical Chemistry 87* (12): 2199–2209.
27 Lomaev, M.I., Sosnin, E.A., and Tarasenko, V.F. (**2012**). Excilamps and their applications. *Progress in Quantum Electronics 36* (1): 51–97.
28 Lomaev, M.I., Sosnin, E.A., and Tarasenko, V.F. (**2016**). Excilamps and their applications. *Chemical Engineering & Technology 39* (1): 39–50.
29 IUPAC (**2019**). IUPAC compendium of chemical terminology - the gold book. https://goldbook.iupac.org/ (accessed 2021).

30 Kogelschatz, U. (**2003**). Dielectric-barrier discharges: their history, discharge physics, and industrial applications. *Plasma Chemistry and Plasma Processing 23* (1): 1–46.

31 Park, S.J., Herring, C.M., Mironov, A.E. et al. (**2017**). 25 W of average power at 172 nm in the vacuum ultraviolet from flat, efficient lamps driven by interlaced arrays of microcavity plasmas. *Apl Photonics 2* (4): https://aip.scitation.org/doi/10.1063/1.4976530.

32 Zhai, S.Y., Wang, X.L., Wei, Y. et al. (**2013**). A compact efficient deep ultraviolet laser at 266 nm. *Laser Physics Letters 10* (4): 045402.

33 Terpetschnig, E. and Jameson, D. M. Fluorescence Lifetime (FLT). http://www.iss.com/resources/pdf/technotes/Fluorescence_Lifetime.pdf (accessed 2021).

34 Auzel, F. (**2004**). Upconversion and anti-stokes processes with f and d ions in solids. *Chemical Reviews 104* (1): 139–173.

35 Wang, F. and Liu, X.G. (**2009**). Recent advances in the chemistry of lanthanide-doped upconversion nanocrystals. *Chemical Society Reviews 38* (4): 976–989.

36 Cates, E.L., Chinnapongse, S.L., and Kim, J.H. (**2012**). Engineering light: advances in wavelength conversion materials for energy and environmental technologies. *Environmental Science & Technology 46* (22): 12316–12328.

37 Zhang, W.Y., Yang, S.L., Li, J. et al. (**2017**). Visible-to-ultraviolet Upconvertion: energy transfer, material matrix, and synthesis strategies. *Applied Catalysis B-Environmental 206*: 89–103.

38 Cates, E.L., Cho, M., and Kim, J.H. (**2011**). Converting visible light into UVC: Microbial inactivation by $Pr^{3+}$-activated upconversion materials. *Environmental Science & Technology 45* (8): 3680–3686.

# 5

# Actinometry and Radiometry

## 5.1 Introduction

Photochemical reactions depend on delivery of photons to the target. As described previously, the local rate of a photochemical reaction is directly proportional to the local fluence rate. More generally, the extent of a photochemical reaction is governed by the dose (fluence) of radiation delivered to the targets in the system. Therefore, the availability of tools to accurately measure the (local) fluence rate, or the dose delivered by a photochemical reactor, is important for interpretation of its behavior.

Kuhn et al. defined an *actinometer* as "a chemical system or a physical device by which the number of photons in a beam absorbed into the defined space of a chemical reactor can be determined integrally or per time." [1] Broadly speaking, this definition of actinometers could be interpreted to include any chemical for which the quantum yield and molar absorption coefficient are known at the wavelength(s) applied in the photoreactor. This definition could also include microorganisms or viruses for which UV dose–response behavior is known.

This chapter will include basic information regarding chemical actinometers and physical devices that are used to measure fluence rate, irradiance, or dose. Basic principles of chemical actinometry will be presented and summarized, along with information pertaining to several important chemical actinometers. Information will also be presented regarding the behavior and application of physical actinometer systems, with particular focus on radiometric instruments. Although the logic used in the application of microorganisms as actinometers is in many ways consistent with that of chemical actinometers, the presentation of microbial or viral actinometry (i.e., biodosimetry) will be given in Chapter 7, which addresses the use of biodosimetry for validation of reactor performance.

## 5.2 Chemical Actinometry

A chemical actinometer is a substance that is used to measure fluence rate or fluence (i.e. UV dose) by quantification of photochemical conversion. In principle, any photochemically active chemical can be used as a chemical actinometer, as long as the quantum yield and molar absorption coefficient are known at the wavelength(s) of interest.

For a substance that undergoes a pure photochemical reaction, the extent of conversion of the substance is directly proportional to the number of photons absorbed by the substance and the efficiency by which absorbed photons promote the reaction of interest. This principle is foundational to

*Photochemical Reactors: Theory, Methods, and Applications of Ultraviolet Radiation*, First Edition. Ernest R. Blatchley III.

Companion website: www.wiley.com/go/blatchley/photochemicalreactors

chemical actinometry and should form the basis for interpretation of results from the applications of a chemical actinometer in any system. As such, it is important to recognize that interpretation of data from the application of a chemical actinometer requires the use of a mathematical model that accurately accounts for photon absorption by the actinometer and efficiency of absorbed photon use to drive reactions. By extension, this implies that such a model will account for the physical and chemical properties of the actinometer, as well as the geometric and optical characteristics of the system in which it is applied.

Chemical actinometers have been developed for application in photochemical applications involving radiation that spans the entire UV and visible portions of the spectrum [1]; however, the focus in this chapter will be on actinometers that address UV radiation. In some respects, actinometers may be viewed as a reference standard for measurement of fluence rate or UV dose, in the sense that the photochemical behavior of a chemical will not change (i.e. degrade or evolve) with time. Once an actinometer has been developed, it should never require (re)calibration. In fact, chemical actinometry is often used as a primary standard for recalibration of physical devices, such as radiometers [2].

As with any photochemical process, the behavior of a chemical actinometer will depend on the geometry and optical characteristics of the system to which it is applied. For much of the discussion to follow, the focus will be on the application of a chemical actinometer in a shallow, well-mixed batch reactor under a (quasi) collimated beam (CB). CBs are commonly used in bench-scale laboratory experiments and may represent the simplest geometry for such applications.

Figure 5.1 illustrates a flat-plate CB [3]. This device utilizes the principle of a "baffled light track," in which a series of two or more circular apertures of the same size (diameter) are aligned along an axis that is aimed at a source of radiation [4]. For this system, circular apertures are milled to a size that is larger than the objects that are to be subjected to irradiation, so that the entire object is flooded with output from the CB. Ideally, the fluence rate from the CB will be essentially uniform over the surface of the irradiated object.

CBs allow for relatively straightforward interpretation of the results of photochemical experiments. However, as with all photochemical reactors, there are numerous factors that will influence the delivery of photons to the target, such as an actinometer. The nature and effects of these factors are summarized below, along with physical and mathematical interpretations of the relationships between these factors and photochemical kinetics.

### 5.2.1 Beam Nonuniformity

CBs are designed with a goal of promoting quantifiable, uniform irradiation of an object. However, the output from a CB will display nonuniformity, which in turn will affect local photochemical reaction rates and overall photochemical conversion.

The effects of CB output nonuniformity can be ameliorated by mixing of the contents of the reactor. A common configuration for a CB experiment involves exposure of a shallow, well-mixed liquid to the beam. The vessel used to contain the liquid is often a Petri dish. The exposure vessel should be smaller in diameter than the beam and it should be positioned at the center of the beam (see Figure 5.1).

A benefit of collimation is that it becomes relatively simple to measure incident irradiance delivered by the beam; often times these measurements are conducted by placing a radiometer detector at the center of the beam. Well-designed collimators will allow radiation to approach a condition of

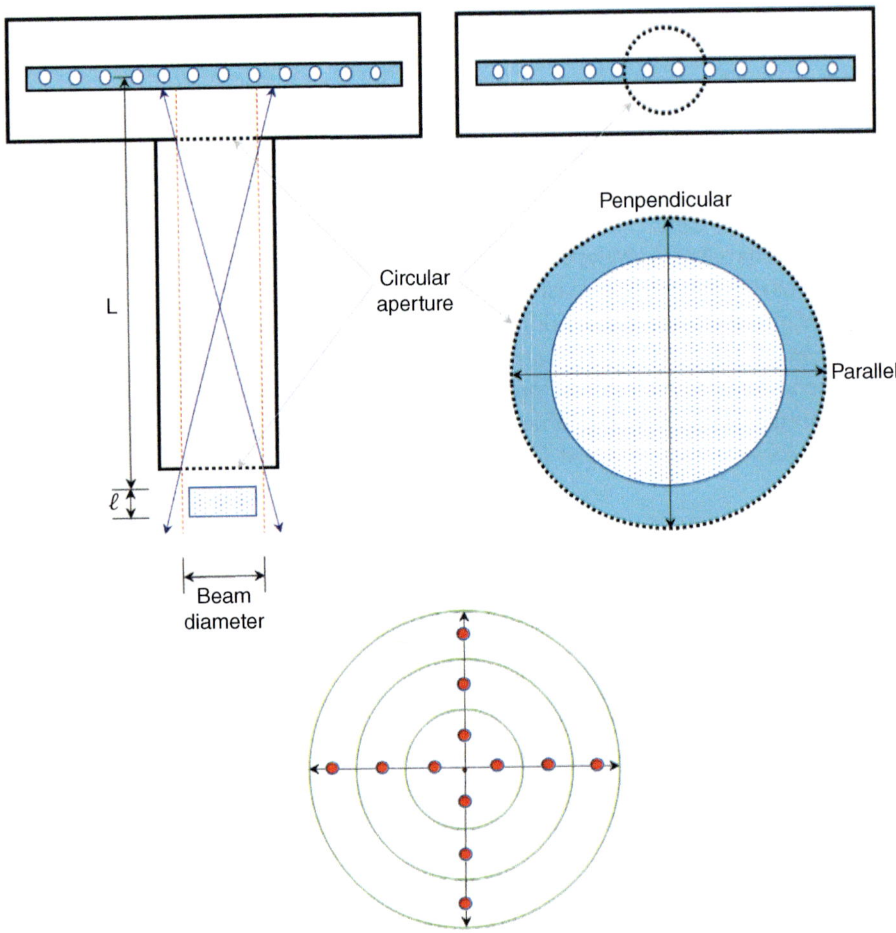


**Figure 5.1** Schematic illustration of flat-plate collimated beam [3]. Upper left drawing provides a profile view of the CB. The dashed, horizontal black lines illustrate vertically aligned circular apertures aimed at the central axis of the UV lamp. Vertical dashed red lines illustrate the beam diameter. Diagonal blue arrows indicate spatial extent of irradiation and the degree to which the output from the device deviates from perfect collimation. White circles indicate imaginary point sources used to simulate lamp output power. Upper right drawing illustrates plan view of the device. Lower right drawing illustrates the irradiated zone (pale blue) as well as an object that is being irradiated, in this case a shallow, well-mixed batch reactor placed within the beam (stippled area). Also included in this drawing are the parallel and perpendicular transects used for scanning of the beam. Bottom drawing also illustrates the parallel and perpendicular transects, along with locations of irradiance (or fluence rate) measurement using a detector that approaches point-wise measurements, such as a fiber optic detector. Each location will be associated with an imaginary annular ring that includes the measurement location.

collimation (i.e. parallel rays) with uniform fluence rate or irradiance at the outlet aperture. The approach to collimation is beneficial for measurements with radiometers because many of these devices often demonstrate a cosine response. As long as the incident angle is close to zero (i.e. perpendicular to irradiated surface), the cosine of the incident angle will be close to 1. For example, an incident angle of 10° will introduce roughly 1.5% error with a device that displays cosine dependence. Achieving an incident angle of less than 10° is easily accomplished by the use of a long collimator (see Figure 5.1, upper left image).

In addition to deviations from perfect collimation, there will be spatial variation in the irradiance delivered by a CB that may be caused by nonuniformity of the UV source, geometric attributes of the system, or variability in the optical characteristics of the materials used to produce the CB.

Figure 5.2 illustrates measured values of local irradiance within the output beam of a flat-plate CB. The values reported in this image were collected using a fiber optic radiometer detector with a circular sensor of diameter ~ 1 mm. The device was positioned at locations along the principal axes of the circular beam, with one axis being parallel to the lamp axis and the other being perpendicular to the lamp axis (see Figure 5.1, lower right and bottom images). Measurements of local irradiance were collected at an interval of 1.0 mm along both axes.

Beam nonuniformity is accounted for by the use of the "Petri factor," which is formally defined as the ratio of the average irradiance across the exposure surface to the irradiance at the center of the surface. When a Petri dish or other cylindrical container is used as the exposure vessel, the average irradiance across the exposure surface can be calculated by the area-weighted average of irradiance measurements across the surface of the irradiated object (see Example 5.1).

**Example 5.1**

***Problem Statement:*** Measured values of irradiance from Figure 5.2 are summarized below for the area corresponding to a 47 mm diameter Petri dish as an irradiation vessel (see numerical values highlighted in yellow). Based on these values, calculate the Petri factor for this vessel under the flat-plate CB. Assume that the vessel is centered in the beam.

| $r_i$ (mm) | $r_o$ (mm) | Annulus area, $A_i$ (mm$^2$) | Measured irradiance, $E$ (mW/cm$^2$) | | | | Average irradiance in annular area, $E_{avg,i}$ (mW/cm$^2$) | $A_i \times E_{avg,i}$ (mW·mm$^2$/cm$^2$) |
|---|---|---|---|---|---|---|---|---|
| 0 | 1 | 3.14 | 103 | 103 | 104 | 104 | 103.5 | 325 |
| 1 | 2 | 9.42 | 103 | 103 | 104 | 104 | 103.5 | 975 |
| 2 | 3 | 15.71 | 103 | 104 | 102 | 103 | 103 | 1618 |
| 3 | 4 | 21.99 | 103 | 104 | 102 | 103 | 103 | 2265 |
| 4 | 5 | 28.27 | 103 | 104 | 102 | 103 | 103 | 2912 |
| 5 | 6 | 34.56 | 103 | 104 | 102 | 103 | 103 | 3559 |
| 6 | 7 | 40.84 | 103 | 104 | 102 | 103 | 103 | 4207 |
| 7 | 8 | 47.12 | 103 | 104 | 102 | 102 | 102.75 | 4842 |
| 8 | 9 | 53.41 | 103 | 104 | 102 | 102 | 102.75 | 5488 |
| 9 | 10 | 59.69 | 103 | 103 | 103 | 102 | 102.75 | 6133 |
| 10 | 11 | 65.97 | 103 | 103 | 103 | 102 | 102.75 | 6779 |
| 11 | 12 | 72.26 | 103 | 104 | 102 | 101 | 102.5 | 7406 |
| 12 | 13 | 78.54 | 103 | 103 | 102 | 101 | 102.25 | 8031 |
| 13 | 14 | 84.82 | 102 | 104 | 101 | 100 | 101.75 | 8631 |
| 14 | 15 | 91.11 | 102 | 103 | 101 | 100 | 101.5 | 9247 |
| 15 | 16 | 97.39 | 102 | 104 | 101 | 100 | 101.75 | 9909 |
| 16 | 17 | 103.67 | 102 | 103 | 101 | 100 | 101.5 | 10,523 |
| 17 | 18 | 109.96 | 101 | 103 | 100 | 100 | 101 | 11,106 |
| 18 | 19 | 116.24 | 101 | 103 | 100 | 99 | 100.75 | 11,711 |
| 19 | 20 | 122.52 | 101 | 102 | 99 | 99 | 100.25 | 12,283 |

(Continued)

| $r_i$ (mm) | $r_o$ (mm) | Annulus area, $A_i$ (mm$^2$) | Measured irradiance, $E$ (mW/cm$^2$) | | | | Average irradiance in annular area, $E_{avg,i}$ (mW/cm$^2$) | $A_i \times E_{avg,i}$ (mW·mm$^2$/cm$^2$) |
|---|---|---|---|---|---|---|---|---|
| 20 | 21 | 128.81 | 101 | 102 | 98 | 99 | 100 | 12,881 |
| 21 | 22 | 135.09 | 101 | 102 | 98 | 98 | 99.75 | 13,475 |
| 22 | 23 | 141.37 | 100 | 101 | 98 | 97 | 99 | 13,996 |
| 23 | 24 | 147.65 | 99 | 100 | 96 | 97 | 98 | 14,470 |
| | Σ= | 1809.56 | | | | | Σ= | 182,772 |

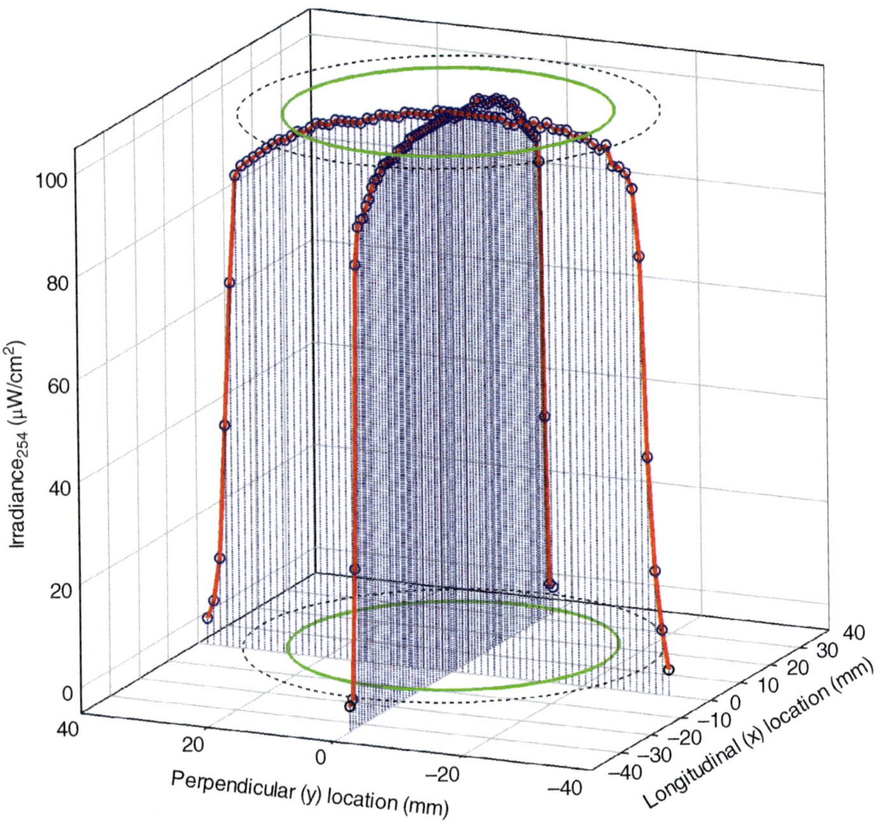


**Figure 5.2** Measured irradiance field under a 60 mm diameter flat-plate collimated beam. Measurements were collected at 1.0 mm increments along the two primary axes of the circular beam using a fiber optic radiometer detector and a micropositioning device. Dashed circles at the top and bottom of the figure illustrate the aperture of the device; solid green circles at the top and bottom of the figure indicate the location of a beam-centered, 47 mm diameter Petri dish, commonly used as an exposure vessel for samples under this beam. Note that the exposure vessel is smaller than the beam, so that the vessel is "flooded" with radiation from the beam of nearly uniform irradiance.

***Solution:*** The formal definition of the Petri factor is as follows:

$$P = \text{Petri Factor} = \frac{\text{Spatial average of irradiance across vessel surface}}{\text{Irradiance at beam center}} \tag{E.1}$$

The values in the numerator and denominator need to be calculated from the table of measurements.

The table of irradiance measurements represents values that were collected along the two primary axes of the beam. As illustrated in Figure 5.1 (bottom), the circular surface of the irradiated vessel can be subdivided into a set of annular areas, with each annulus being defined by an inside radius and an outside radius, $r_i$ and $r_o$, respectively. For each annular area, there are four irradiance measurements. The average irradiance for each annular ring, $E_{avg}$, can be calculated as the mean of the four values that were collected in the annulus. The spatial average of irradiance can be calculated as the area-weighted average of these measurements:

$$E_{\text{spatial avg}} = \frac{\sum_i A_i \cdot E_{\text{avg},i}}{\sum_i A_i} \tag{E.2}$$

where

$A_i$ = area of $i^{\text{th}}$ annular area

$E_{\text{avg},i}$ = mean of irradiance values for $i^{\text{th}}$ annular area.

The denominator in equation E.2 is given by the sum of the third column of the table above:

$$\sum_i A_i = 1809.56\ \text{mm}^2 \tag{E.3}$$

The numerator in equation E.2 is given by the sum of the ninth column in the table:

$$\sum_i A_i \cdot E_{\text{avg},i} = 182,772\ \frac{\text{mW} \cdot \text{mm}^2}{\text{cm}^2} \tag{E.4}$$

The ratio of the results from equations E.4 and E.3 yields the spatial average of irradiance across the surface of the irradiated vessel:

$$E_{\text{spatial avg}} = \frac{182,772\ \frac{\text{mW} \cdot \text{mm}^2}{\text{cm}^2}}{1809.56\ \text{mm}^2} = 101\ \frac{\text{mW}}{\text{cm}^2} \tag{E.5}$$

Finally, the Petri factor is calculated as the ratio of $E_{\text{spatial avg}}$ to irradiance measured at the center of the irradiated surface. The irradiance at the center of the irradiated surface is given by the mean of the four measurements that were collected closest to the beam centerline (see value highlighted in pink in table above).

$$P = \frac{101\ \frac{\text{mW}}{\text{cm}^2}}{103.5\ \frac{\text{mW}}{\text{cm}^2}} = 0.976 \tag{E.6}$$

The Petri factor is a characteristic of the CB (largely defined by the source of radiation and the geometry of the CB) and the geometry of the irradiation vessel. In most cases, the output of the UV source and the geometry of the CB will not change appreciably. If the same irradiation vessel is used in all experiments, then the Petri factor becomes a constant. Therefore, it can be calculated once and then applied in all subsequent experiments. However, it is good laboratory practice to conduct periodic checks of the output of the device as a means of confirming the value of the Petri factor.

### 5.2.2 Reflection and Refraction

Most photochemical reactors involve two or more media in contact, including the materials that comprise the source of radiation, the medium through which the radiation is transmitted to reach the target, and the medium in which the target is suspended or dissolved. Each of these media will be characterized by different optical properties, including the refractive index.

The refractive index represents the ratio of the speed of light in vacuum to the speed of light in the medium of interest:

$$n_i = \frac{c}{v_i} \tag{5.1}$$

where

$n_i$ = refractive index for $i^{\text{th}}$ medium
$c$ = speed of light in vacuum
$v_i$ = speed of light in $i^{\text{th}}$ medium.

A refractive index greater than 1 indicates a medium for which electromagnetic radiation ("light") travels more slowly than it does in a vacuum. More generally, a high refractive index describes a medium in which light travels relatively slowly; high refractive indexes are often seen with relatively dense materials.

The refractive index is a material property that can be measured. Refractive indexes are also functions of wavelength, although the variation across wavelengths in the UV and visible ranges is often modest. Figure 5.3 illustrates empirical relationships that have been used to describe the wavelength-dependence of refractive index among several media that are commonly encountered in photochemical reactor systems, including air [5], (liquid) water [6], quartz (fused silica) [7], and borosilicate glass [8].

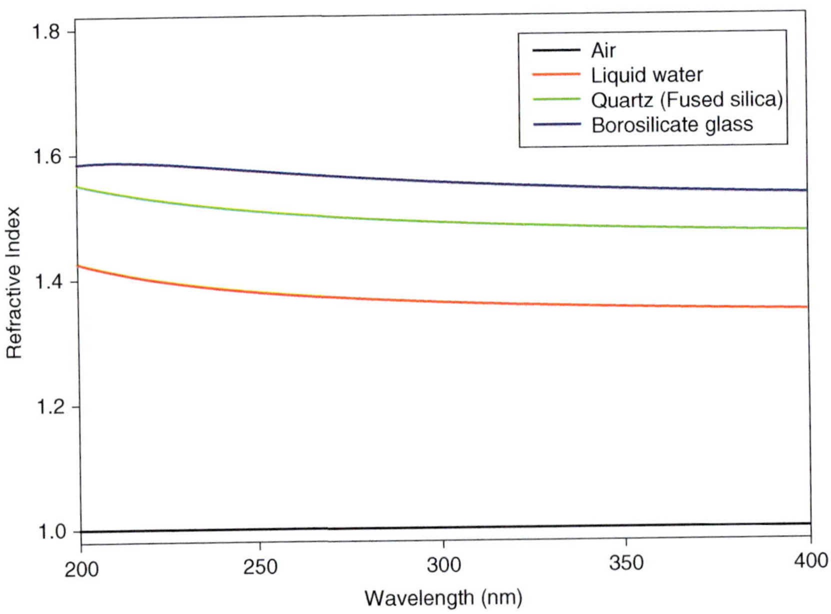


**Figure 5.3** Refractive index as a function of wavelength for media that are often used in photochemical reactors. Curves were developed from information provided by Edlén for air [5], Schiebener et al. for water [9], Malitson for quartz [7], and Polyanskiy for borosilicate glass [8].

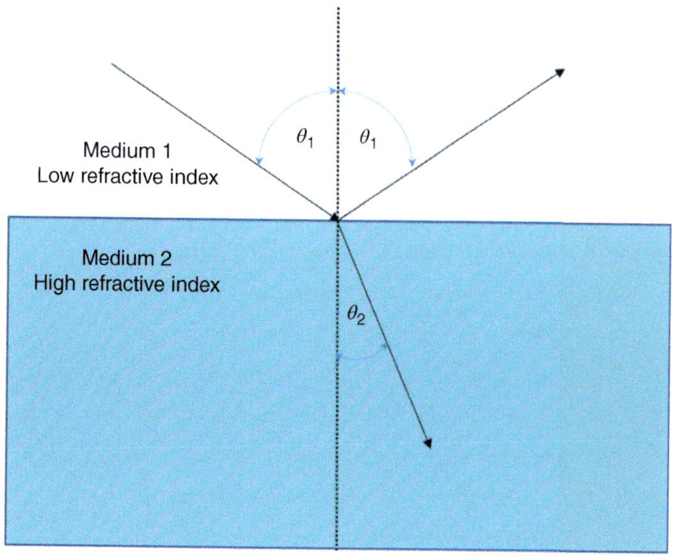


**Figure 5.4** Schematic illustration of the behavior of a beam of electromagnetic radiation imposed on an interface between two media of different refractive index.

Figure 5.4 illustrates the basic concepts of refraction and (specular) reflection at an interface between two media with different refractive indexes. A beam of light, or more generally electromagnetic radiation, is imposed on the interface at an incident angle, $\theta_1$. A portion of the incident radiation will be reflected by the surface, with an angle of reflection equal to the angle of incidence, $\theta_1$. The remaining radiation will cross the interface where it will encounter a medium with a different refractive index. The radiation will retain its frequency, but its direction will change. Therefore, when electromagnetic radiation passes through an interface between media with different refractive indexes, the angle of refraction ($\theta_2$) will be different than the incident angle.

Snell's law provides a quantitative description of the directional change that will result from a refractive event, where a beam of electromagnetic radiation is imposed on an interface between media with different refractive indexes:

$$n_1 \cdot \sin(\theta_1) = n_2 \cdot \sin(\theta_2) \tag{5.2}$$

For the situation illustrated in Figure 5.4 where a beam of radiation passes through a medium with low $n$ into a medium with high $n$, the angle of refraction will be less than the angle of incidence.

Although not often relevant in most bench-scale photoreactors, the process of refraction also allows the possibility of a situation known as *total internal reflection* in which refracted radiation can be trapped in a system. Specifically, when a ray is transmitted from a high-refractive-index medium to a medium with low refractive index, it is possible for the angle of refraction to meet or exceed 90°. For two media in which the refractive indexes of the first and second media are such that $n_1 > n_2$, it is possible to define an angle of incidence such that the angle of refraction is 90°. This is referred to as the critical angle, $\theta_c$. If we assume $\theta_2 = 90^o$, then by application of Snell's law and the knowledge that $sin(90^o) \equiv 1$, we can determine the critical angle:

$$\theta_c = \arcsin\left(\frac{n_2}{n_1}\right) \tag{5.3}$$

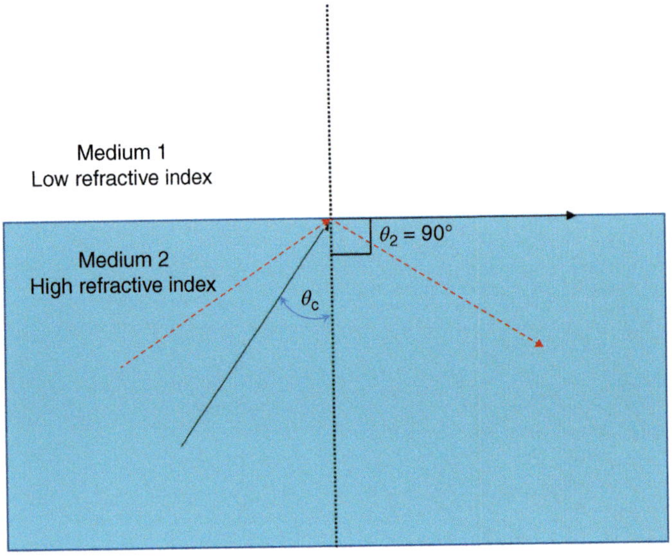


**Figure 5.5** Illustration of total internal reflection. When a ray of radiation is directed through a medium of high refractive index toward an interface with a material of lower refractive index, it is possible to achieve a condition where the angle of refraction reaches 90°. The critical angle, $\theta_c$, is the incident angle, $\theta_1$, for which the angle of refraction, $\theta_2 = 90°$. Therefore, when $\theta_1 > \theta_c$, all radiation will be reflected back into the originating medium. The red, dashed arrow illustrates a ray for which $\theta_1 > \theta_c$; the ray is reflected back into the high-refractive-index medium.

When the angle of incidence exceeds the critical angle, then all radiation imposed on the interface will be reflected back into the medium from which it was transmitted. *Total internal reflection* allows a beam of radiation to be retained by a medium and to be transmitted through a long optical path. In concept, this presents some interesting advantages in applications where the generation of photons represents a barrier to process effectiveness. Total internal reflection is the foundational principle behind fiber optic devices and has been used in some photoreactor designs. Figure 5.5 illustrates the concept of total internal reflection for a simple planar interface.

Because refractive indexes demonstrate wavelength-dependence, the critical angle between two media will also be a function of wavelength. Figure 5.6 illustrates the wavelength-dependence of critical angles for combinations of the media described previously in Figure 5.3.

Reflection and refraction are important optical phenomena that are relevant in many photochemical reactor systems. When a beam of light/electromagnetic radiation travelling through a particular medium is incident on the interface between that medium and a second medium, a fraction of the beam is reflected back into the first medium, and the remainder is refracted into the second medium. The fractions reflected and refracted depend on the angle of incidence and the refractive indexes of the two media. The fraction reflected can be calculated by application of the Fresnel equations. For the special case of normal incidence, as in a typical CB system, the Fresnel equations can be simplified as follows:

$$\rho = \left(\frac{n_1 - n_2}{n_1 + n_2}\right)^2 \tag{5.4}$$

where

$\rho$ = reflectance, or fraction of radiation reflected from an interface being subjected to normal (perpendicular) radiation.

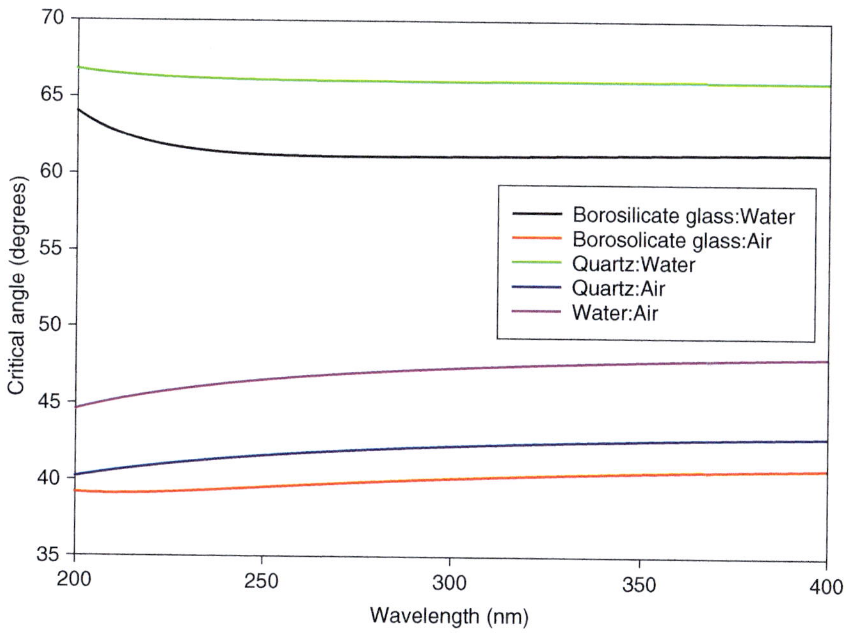


**Figure 5.6** Wavelength-dependence of critical angles for interfaces between fluids (air and water) and solids (quartz and borosilicate glass) that are often incorporated in UV photoreactors. Critical angles were calculated using equation 5.3, using data from Figure 5.3.

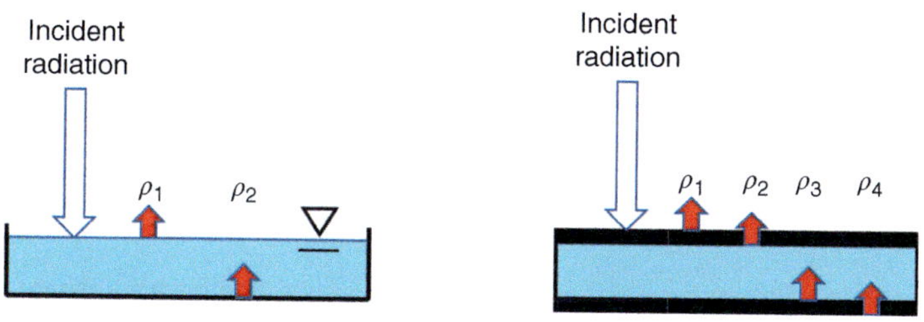


**Figure 5.7** Schematic illustrations of reflection from interfaces subjected to perpendicular radiation, as would be applied from a collimated beam. Left image illustrates a shallow reactor containing water with a free surface (e.g. Petri dish). In this situation, the incident radiation will encounter two media interfaces, each displaying reflectance ($\rho_1$ and $\rho_2$). The first reflection event will reduce the amount of radiation imposed on the liquid medium, while the second event will increase radiation exposure. Right image illustrates an enclosed reactor, such as a quartz cuvette, which is often used for exposure of volatile compounds to UV radiation under a collimated beam. For this situation, there are at least four primary reflection events. Note that reflected radiation may experience subsequent, secondary reflection events (not illustrated in these images). In most situations, it is appropriate to ignore secondary and higher order reflection events. Some advanced modeling methods (e.g. ray tracing) are able to efficiently account for these higher order reflection events.

Reflection of radiation from media interfaces can play an important role in photochemical processes. For example, even with simple reactor systems that are used in laboratory experiments, there can be multiple reflective reactions that take place within a multiphase system. Figure 5.7 illustrates some examples of reflective reactions for devices that are often subjected to irradiation from a CB in laboratory experiments. In these systems, it is common to account for the first reflective event and to ignore subsequent reactions. It should be noted that ignoring these subsequent

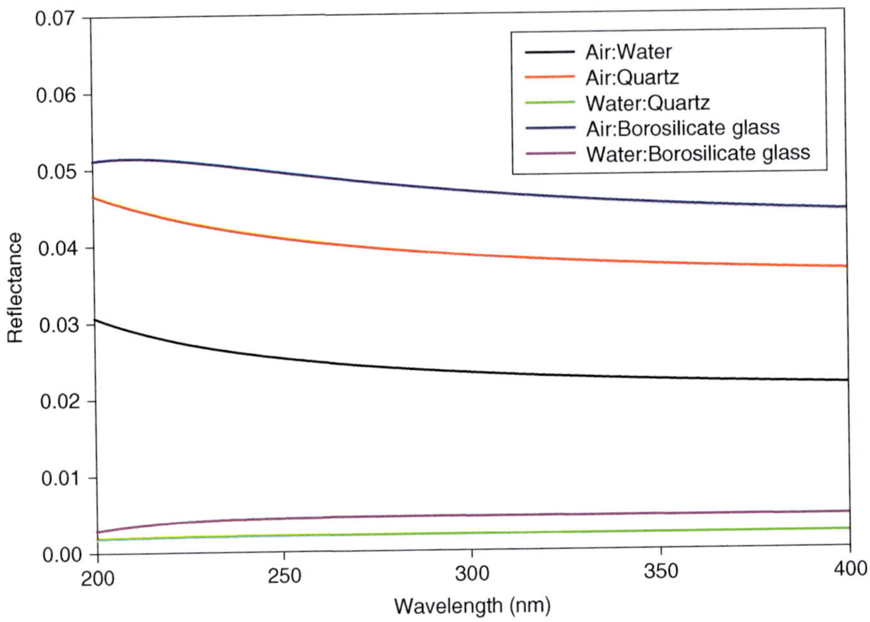


**Figure 5.8** Reflectance as a function of wavelength for interfaces that are common in experiments with collimated beams.

reflections represents a source of error in the analysis of the dynamics behavior of these systems, though this error is small in most situations.

Equation 5.4 dictates that wavelength-dependence among refractive indexes will translate to wavelength-dependence for reflectance. Figure 5.8 illustrates wavelength-dependence for reflectance at media interfaces that are common in CB systems.

### 5.2.3 Absorption

For a photochemical reaction to occur, photons must be absorbed by the target. However, absorption may also occur with nontarget constituents in the system, and regardless of the absorbing material, absorption leads to a monotonic decrease in fluence rate or irradiance in the direction of radiation propagation.

The Beer–Lambert law (see equation 2.8) provides a quantitative description of the effects of absorption. In a formal sense, the Beer–Lambert law is applicable when the molar concentration and molar absorption coefficient for all absorbing compounds are known. For a solution in which many absorbing compounds are present, the Beer–Lambert law can be modified, by recognition of the fact that absorbance is an additive property:

$$A(\lambda) = \sum_i \varepsilon(\lambda)_i \cdot [B]_i \cdot \ell_i \tag{5.5a}$$

where

| | | | |
|---|---|---|---|
| $A(\lambda)$ | = | absorbance (base 10) at wavelength $\lambda$ | |
| $\varepsilon(\lambda)_i$ | = | molar (decadic) absorption coefficient of $i^{\text{th}}$ constituent at wavelength $\lambda$ | $[=] \frac{\text{L}}{\text{mole} \cdot \text{cm}}$ |
| $[B]_i$ | = | molar concentration of $i^{\text{th}}$ constituent | $[=] \frac{\text{mole}}{\text{L}}$ |
| $\ell_i$ | = | optical path length through absorbing medium for of $i^{\text{th}}$ constituent | $[=]$ cm. |

Note that in most situations, the optical path length ($\ell_i$) will be the same for all constituents. Equation 5.5a is written to allow the most general application of the Beer–Lambert law. Also, note that the Beer–Lambert law is sometimes defined using base e, rather than base 10:

$$A_e(\lambda) = \sum_i \kappa(\lambda)_i \cdot [B]_i \cdot \ell_i \tag{5.5b}$$

where

$A_e(\lambda)$ = absorbance (base $e$) at wavelength $\lambda$

$\kappa(\lambda)_i$ = molar (Napierian) absorption coefficient of $i^{\text{th}}$ constituent at wavelength $\lambda$ [=] $\frac{\text{L}}{\text{mole} \cdot \text{cm}}$.

The base $e$ (Napierian) form is derived by calculus directly from the physical implications of Beer–Lambert behavior; namely that spatial gradient in irradiance (or fluence rate) is linearly related to the incident irradiance (or fluence rate) and the concentration of the absorbing compound(s)

$$\frac{dE}{dx} = -\kappa \cdot [B] \cdot E \tag{5.6}$$

However, the base 10 (decadic) form has become the default form of the Beer–Lambert law in many applications.

The composition of a solution will not always be defined. Under these circumstances, the Beer–Lambert law may be presented in a simplified form:

$$\frac{dE}{dx} = -\alpha(\lambda) \cdot E \tag{5.7}$$

where

$\alpha(\lambda)$ = Napierian absorption coefficient for absorbing medium [=] $\text{cm}^{-1}$.

The absorption coefficient is a collective parameter that is used to describe the absorption characteristics of a given medium, including contributions from all compounds that absorb at a given wavelength. As long as the composition of the absorbing medium remains constant, $\alpha(\lambda)$ will also remain constant at a given wavelength.

In a shallow, well-mixed batch reactor under a CB (see Figure 5.9), radiation will enter the liquid volume at a fluence rate value of $E_0$.[a] The presence of absorbing compounds in the sample will cause the fluence rate to decrease exponentially with distance ($x$) into the absorbing medium. At the bottom of the vessel, the fluence rate will be reduced to a value of $E_\ell$ (see Figure 5.9). Fluence rate will decrease in the direction of propagation according to equation 5.7.

If the vessel is well mixed (i.e. the exposure time is long relative to the characteristic mixing time in the vessel), then reactants in the system will have time to "sample" all parts of the reactor during their exposure period. Under this condition (a "well-mixed" condition), it is relevant to consider the average fluence rate within the vessel as a measure of the rate at which photons are imposed on the targets during the period of irradiation.

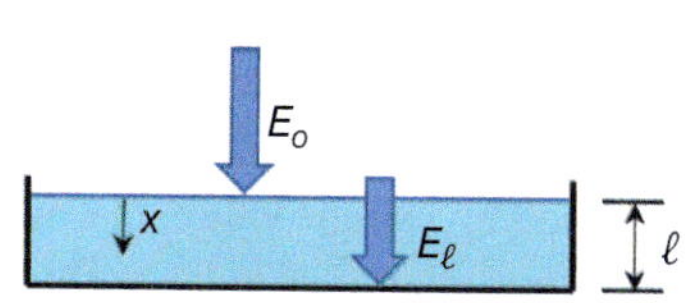


**Figure 5.9** Schematic illustration of shallow, well-mixed reaction vessel under a collimated beam. The solution depth in the vessel is ℓ. The coordinate system for this analysis is such that the x-direction corresponds to the direction of propagation of the incident beam of radiation.

The mean fluence rate (spatial average) within the vessel can be calculated by the depth-weighted values of fluence rate as follows:

$$E_{avg} = \frac{\int_0^{\ell} E_0 \cdot \exp(-\alpha x) \cdot dx}{\ell} \tag{5.8}$$

Integration and algebraic manipulation of equation 5.8 yields:

$$E_{avg} = \frac{E_0}{\alpha \cdot \ell}[1 - \exp(-\alpha \cdot \ell)] \tag{5.9}$$

The term "water factor" is used to correct for the effects of absorption by defining the ratio between incident and average fluence rate within the vessel:

$$W = \text{Water Factor} = \frac{E_{avg}}{E_0} = \frac{1 - \exp(-\alpha \cdot \ell)}{\alpha \cdot \ell} \tag{5.10}$$

However, as described above, the base 10 (decadic) form of Beer's law is commonly applied. Therefore, an equivalent term has been defined using the decadic form:

$$\text{Water Factor} = \frac{E_{avg}}{E_0} = \frac{1 - 10^{-a \cdot \ell}}{a \cdot \ell \cdot \ln 10} \tag{5.11}$$

where

$a$ = decadic absorption coefficient [=] $cm^{-1}$.

Equations 5.10 and 5.11 are equally valid methods of describing the effects of absorption on a photochemical reaction in a vessel under a CB. However, as is the case with all applications of the Beer–Lambert law, it is important to define whether calculations are to be performed using the Napierian (base $e$) or decadic (base 10) approach; both approaches yield the same results when properly applied.

### 5.2.4 Divergence

In a non-absorbing medium, irradiance decreases with the square of distance from an isotropic point source of radiation. In effect, the power from the point source is being diluted in space over a spherical surface, with the point source at its center. Under these circumstances, the output power from the source is evenly distributed across the surface of the sphere. The surface area of a sphere is given by $4\pi r^2$, where $r$ is the radius of the sphere. The fluence rate received at the surface of the sphere under these conditions is inversely proportional to the square of the (radial) distance from the point source (see Figure 5.10).

CBs are often constructed around cylindrical UV sources, such as Hg lamps. In a strict sense, distance from the lamp to a given receptor site is an ambiguous term because the output device is a cylindrical lamp of finite length. However, the output from a cylindrical lamp can be approximated as a series of colinear point sources (see Figure 5.1). For a relatively narrow aperture and a long length of collimation, i.e. the distance from the lamp axis to surface of the object being irradiated ($L$ in Figure 5.1), the distance from each hypothetical point source to the surface of the object being irradiated is close to $L$. And because radiation at a given point on this surface will be received from several point sources along the length of the lamp, it is convenient to assume that this distance is defined by a single value, $L$. This approximation introduces minimal error, as long as the ratio of $L$ to aperture diameter is large [3].

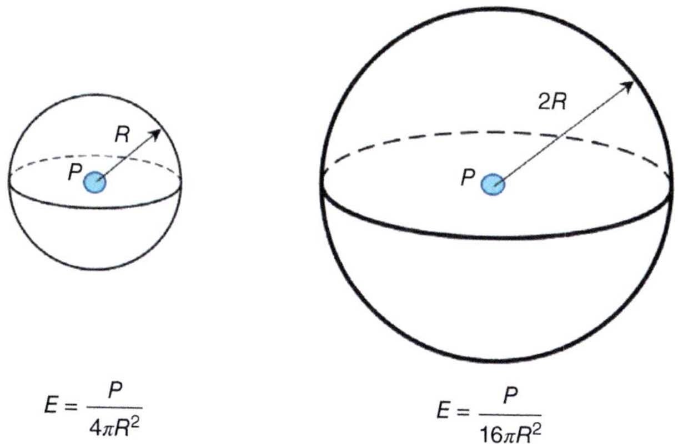


**Figure 5.10** Schematic illustration of fluence rate received at the surface of a sphere of a non-absorbing medium surrounding a point source with output power of P. Because the sphere on the right has twice the radius of the sphere on the left, the fluence rate received on the surface of the right sphere is ¼ of the fluence rate received on the surface of the sphere on the left.

The effects of distance from the radiation source will also affect the magnitude of irradiance delivered through the depth of the solution placed under the CB. For a solution of finite depth, $\ell$, irradiance of collimated radiation will decrease through the depth of the solution with the square of distance from the lamp axis. The average divergence factor can be defined by the ratio:

$$AD = \text{Average Divergence Factor} = \frac{\int_0^{\ell} \frac{L^2}{(L+x)^2} \cdot \mathrm{d}x}{\ell} = \frac{L}{L+\ell} \tag{5.12}$$

### 5.2.5 Absorption by Photoproducts

Ideally, the only compound that will absorb radiation in an application of chemical actinometry will be the actinometer itself. This simplifies the interpretation of data from a chemical actinometry experiment. However, this goal is not always met and needs to be accounted for.

When a reaction product absorbs radiation at the wavelength(s) of interest, it represents an internal optical filter in the system. In a general sense, this will decrease the rate at which photons are delivered to the chemical actinometer, and as such it will decrease the rate of photochemical conversion of the actinometer.

### 5.2.6 Extent of Absorption by Chemical Actinometer

As discussed in Chapter 2, the extent of absorption by the target molecule will affect the kinetics of observed phototransformation processes. The general expression to describe the rate of a photochemical process was presented in Chapter 2, and is repeated below:

$$r_B = -\frac{\mathrm{d}[B]}{\mathrm{d}t} = I_a \cdot \Phi(\lambda) \tag{5.13}$$

where

$r_B$ = rate of reaction with respect to the target, $B$ $[=] \ \frac{\text{mole}}{\text{L} \cdot \text{s}}$

$[B]$ = activity (concentration) of B in solution $[=] \ \frac{\text{mole}}{\text{L}}$

$t$ = time [=] s

$I_a$ = volumetric rate of photon absorption by B [=] $\frac{\text{einstein}}{\text{L} \cdot \text{s}}$

$\Phi$ = primary quantum yield for reaction of interest [=] $\frac{\text{mole}}{\text{einstein}}$.

Note that in this case, B represents the chemical actinometer itself.

In principle, this expression can be used to describe the kinetics of any photochemical process; the term $I_a$, which describes the volumetric rate of photon absorption by the target molecule, can be difficult to define in some systems. However, as indicated in Chapter 2, there are two limiting cases for which the interpretation of the volumetric rate of photon absorption is relatively easy to define, and for these two cases, the resulting kinetic expressions are also relatively simple. The first of these is an opaque solution (i.e., $\varepsilon \cdot [B] \cdot \ell \rightarrow \infty$), in which all incident photons are absorbed by the target molecule, leading to a condition of zero-order kinetics:

$$r_B = -\frac{d[B]}{dt} = P \cdot (1-\rho) \cdot AD \cdot \frac{E_i' \cdot \Phi}{\ell \cdot U} \tag{5.14}$$

The terms that precede the quotient on the right-hand-side of equation 5.14 are the relevant correction factors presented above. Note that for an opaque solution, the water factor ($W$) is not relevant because all incident photons are absorbed by the photochemical target.

The second limiting case is one of a (nearly) transparent solution, or one in which $\varepsilon \cdot [B] \cdot \ell \rightarrow 0$. Under these circumstances, the kinetics of a photochemical reaction in a well-mixed, shallow, batch reactor under a CB are described by:

$$r_B = -\frac{d[B]}{dt} = P \cdot (1-\rho) \cdot W \cdot AD \cdot \frac{2.303 \cdot E_i' \cdot \varepsilon \cdot [B]}{U} \cdot \Phi \tag{5.15}$$

As a reminder, the terms that precede the quotient on the right-hand-side of equation 5.15 are the correction factors discussed above.

Chemical actinometers are often applied for purposes of quantifying the behavior of photochemical reactor systems. In general terms, the observed behavior of a photochemical reactor will be governed by the integrated effects of the dose distribution and the kinetics of the photochemical reactions of interest.

Across the range of conditions that correspond to the limiting cases of opaque solutions and transparent solutions, there can be widely divergent changes in photoreactant conversion. One attribute of the system that will be profoundly affected by changes in optical conditions will be the fluence rate field. For solutions that are opaque to incident radiation or nearly so, photons emitted from the radiation source will penetrate only a short distance into solution, whereas for a transparent solution, the depth of photon penetration will be relatively large.

The segregated-flow model is often used to describe the behavior of purely photochemical reactor systems. As described in Chapter 3, the accuracy of predictions by this model depend on the extent of mixing within the system and the ability to describe mixing effects; however, the error introduced by such an analysis depends on the extent of reaction. For systems that are characterized by relatively slow kinetics and a small extent of conversion, mixing behavior has a relatively modest effect on overall photoreactor performance. On the other hand, for systems in which fast photoreactions are involved and the extent of conversion is large, mixing behavior will have a profound effect on overall reactor conversion.

Because of these attributes, applications of chemical actinometers for description of the behavior of photochemical reactor systems should involve optical conditions that mimic those expected in the intended application. Moreover, the chemical actinometer used in such experiments should mimic the actual photochemical kinetics displayed by the target molecule.

### 5.2.7 Polychromatic Behavior

For polychromatic systems, the effects of wavelength-dependent processes or attributes of the system must be accounted for. Again considering chemical actinometry in a conventional CB experiment with a shallow, well-mixed batch reactor, the dynamics of the photochemical process of interest are most easily described for the two limiting cases. It is important to note that the conditions that allow application of the two limiting cases are assumed to apply across the entire range of wavelengths imposed on the actinometer (see limits of integration below).

For the case of an opaque solution, kinetics are described by:

$$r_B = -\frac{\mathrm{d}[B]}{\mathrm{d}t} = \int_{\lambda_1}^{\lambda_2} P(\lambda) \cdot (1 - \rho(\lambda)) \cdot AD(\lambda) \cdot \frac{E'_{p,\lambda} \cdot \Phi(\lambda)}{\ell} \cdot \mathrm{d}\lambda \tag{5.16}$$

Note that the correction factors must also be accounted for across the incident wavelength range ($\lambda_1 \rightarrow \lambda_2$). Similarly, for a transparent solution, the kinetics are described as follows:

$$r_B = -\frac{\mathrm{d}[B]}{\mathrm{d}t} = \int_{\lambda_1}^{\lambda_2} P(\lambda) \cdot (1 - \rho(\lambda)) \cdot W(\lambda) \cdot AD(\lambda) \cdot 2.303 \cdot \varepsilon(\lambda) \cdot \Phi(\lambda) \cdot E'_{p,\lambda} \cdot [B] \cdot \mathrm{d}\lambda \tag{5.17}$$

### 5.2.8 Alternative Bench-Scale Reactors for Use With Chemical Actinometers

A CB system, coupled with a shallow, well-mixed batch reactor provides a simple method for bench-scale evaluation of photochemical kinetics, and for estimation of the quantum yield, a fundamental characteristic of any photochemical reaction. However, many other methods of irradiation and reactor (irradiated vessel) geometries can also be used for this purpose. Zepp presented two methods that can be applied to many of these systems based on a mathematical model that includes parameters that can be measured directly or calculated based on fundamental measurements [10].

Both methods are based on a general expression to describe the effects of a photoreaction on composition of a solution that contains a single photoreactive component. For our purposes, we will consider the behavior of the system when subjected to a single wavelength of radiation; the same approach can be extended to polychromatic behavior following the methods described above.

$$\frac{\mathrm{d}[T]}{\mathrm{d}t} = -\Phi_T \cdot E'_i \cdot \left(\frac{A}{\forall}\right) \cdot F_S \cdot F_T \tag{5.18}$$

where

| | | |
|---|---|---|
| $[T]$ | = | molar concentration of the target compound |
| $t$ | = | time |
| $\Phi_T$ | = | quantum yield for $T$ |
| $E'_i$ | = | fluence rate incident on the reaction vessel |
| $A$ | = | exposed surface area of reaction vessel |
| $\forall$ | = | volume of reacting fluid in vessel |
| $F_S$ | = | fraction of incident radiation absorbed by the system |
| $F_T$ | = | fraction of absorbed radiation absorbed by the target compound. |

The Beer–Lambert law can be applied to quantify $F_S$:

$$F_S = 1 - 10^{-(\alpha + \varepsilon_T[T])\ell} \tag{5.19}$$

where

$\alpha$ = decadic absorption coefficient for fluid in irradiated vessel, in the absence of the target ($cm^{-1}$)

$\varepsilon_T$ = molar (decadic) absorption coefficient for target molecule ($M^{-1} \cdot cm^{-1}$)

$\ell$ = effective optical path length for system.

The parameter $F_T$ can be estimated as follows:

$$F_T = \frac{\varepsilon_T[T]}{\alpha + \varepsilon_T[T]} \tag{5.20}$$

The approach defined by equations (5.18)–(5.20) is broadly applicable, but carries several implicit assumptions:

1) A single value of the incident fluence rate ($E'_i$) can be used to describe the entire system. If $E'_i$ is not constant either spatially or temporally, then this term would need to be modified to allow spatial or temporal integration.
2) A single value of the optical path length ($\ell$) can be used to describe the entire system. In a system that involves a CB and a shallow, well-mixed batch reactor, the value of $\ell$ is unambiguous – it represents the depth of solution in the dish. Equations (5.18)–(5.20) are intended for application to a reactor with an arbitrary shape, with irradiation being allowed from all directions. Under these conditions, some photons will encounter a long optical path length, while the path length available to other photons may be short. The parameter $\ell$ is used to represent an effective optical path length to account for this behavior. But given the nonlinear behavior of the Beer–Lambert law, this assumption will decrease in validity as the range of possible optical path lengths increases.
3) A single value of target compound concentration can be assumed to describe the entire reaction vessel. This condition will be met if the behavior of the system yields spatially uniform composition. An example of a condition that could lead to this outcome would be a well-mixed system.

These assumptions are likely to be valid in systems where $E'_i$ is uniform across the exposed surface area (A) and where the extent of photoreaction is modest, ideally 1 $\log_{10}$ unit of change or less. When these conditions are met, the effects of the dose distribution on overall reactor behavior and the error introduced by this approach will be modest.

If we assume that these assumptions are valid, then at least two approaches can be used to estimate the quantum yield for the target compound ($\Phi_T$) by such a reactor system. For the first approach, conditions will be applied such that the target compound is the only constituent of the system that absorbs radiation. In other words, $F_T \approx 1$. This condition will be met when $\alpha \ll \varepsilon_T[T]$ or when $\alpha \approx 0$. We will also assume that the system contains a dilute solution of the target compound, such that $F_S \rightarrow 0$. As described previously, a Taylor series expansion of equation (5.19) yields:

$$F_S = 2.303 \cdot \varepsilon_T[T]\ell \tag{5.21}$$

Substitution of these terms to describe $F_T$ and $F_S$ back into equation (5.18) yields an equation to describe the rate of change of the target compound in the reaction vessel:

$$\frac{d[T]}{dt} = -2.303 \cdot \Phi_T \cdot E'_i \cdot \left(\frac{A}{\forall}\right) \cdot \varepsilon_T[T]\ell \tag{5.22}$$

Separation of variables and integration of this first-order expression yields:

$$\ln\left(\frac{[T]}{[T]_0}\right) = -2.303 \cdot \Phi_T \cdot E'_i \cdot \left(\frac{A}{\forall}\right) \cdot \varepsilon_T \cdot \ell \cdot t \tag{5.23}$$

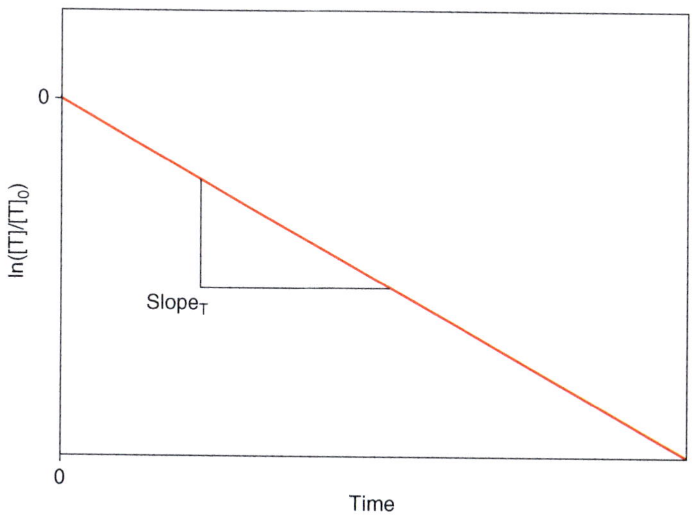


**Figure 5.11** Schematic illustration of the use of equation 5.24.

A plot of $\ln\left(\frac{[T]}{[T]_0}\right)$ vs. $t$ should yield a straight line that passes through the origin (see Figure 5.11) with a slope of:

$$\text{Slope}_T = -2.303 \cdot \Phi_T \cdot E_i' \cdot \left(\frac{A}{\forall}\right) \cdot \varepsilon_T \cdot \ell \tag{5.24}$$

Algebraic manipulation of equation (5.24) allows an estimate of the quantum yield to be developed:

$$\Phi_T = -\frac{\text{Slope}_T}{2.303 \cdot E_i' \cdot \left(\frac{A}{\forall}\right) \cdot \varepsilon_T \cdot \ell} \tag{5.25}$$

To complete the application of this equation, it is necessary to develop independent measurements or estimates of the terms in the denominator of equation (5.25). The molar absorption coefficient can be estimated based on a measured value of absorbance through a known optical path length at a known molar concentration of the target compound. The remaining terms in the denominator of equation (5.25) describe characteristics of the reactor system and can be estimated as follows:

#### 5.2.8.1 Rate of Photon Application per Unit Volume of Solution: $E_i' \cdot \left(\frac{A}{\forall}\right)$

We will retain the assumption of a system in which a single compound is the only constituent that absorbs incident radiation. In this case, we will expose a well-defined reference compound ($R$) or actinometer (i.e. one for which the basic photochemical reaction, quantum yield, and the molar absorption coefficient are all known) to radiation under conditions such that $F_S \equiv 1$ and $F_R \equiv 1$. In other words, the actinometer will be the only compound that absorbs incident radiation and the actinometer solution will be prepared to be opaque to incident radiation. Under these conditions, the rate of reference compound photodecay can be described as follows:

$$\text{rate}_{\max} = \left(\frac{\mathrm{d}[R]}{\mathrm{d}t}\right)_{\max} = -\Phi_R \cdot E_i' \cdot \left(\frac{A}{\forall}\right) \tag{5.26}$$

where

$[R]$ = molar concentration of reference compound

$\Phi_R$ = quantum yield for decay of reference compound.

Note also that this condition represents the maximum possible rate of reference compound photodecay, $\text{rate}_{\text{max}}$. If $[R]$ is measured as a function of time under conditions of exposure to constant incident irradiance, $E_i'$, then equation (5.26) can be rearranged to yield an estimate of the product $E_i' \cdot \left(\frac{A}{\forall}\right)$:

$$E_i' \cdot \left(\frac{A}{\forall}\right) = \frac{-\left(\frac{\mathrm{d}[R]}{\mathrm{d}t}\right)_{\max}}{\Phi_R} \tag{5.27}$$

#### 5.2.8.2 Effective Path Length: $\ell$

For this part of the procedure, we will retain the condition $F_S \equiv 1$ but operate the experiment under a range of concentrations of the reference compound. The general rate expression for decay of the reference compound under these conditions is:

$$\text{rate} = \frac{\mathrm{d}[R]}{\mathrm{d}t} = -\Phi_R \cdot E_i' \cdot \left(\frac{A}{\forall}\right) \cdot \left(1 - 10^{-\varepsilon_R[R]\ell}\right) \tag{5.28}$$

We next define the variable $X$ as follows:

$$X = \frac{\text{rate}}{\text{rate}_{\text{max}}} = 1 - 10^{-\varepsilon_R[R]\ell} \tag{5.29}$$

Algebraic manipulation of equation (5.29) yields:

$$-\log_{10}(1 - X) = \frac{\text{rate}}{\text{rate}_{\text{max}}} = \varepsilon_R[R]\ell \tag{5.30}$$

Therefore, a plot of $-\log_{10}(1 - X)$ as a function of $\varepsilon_R[R]$ should yield a straight line that passes through the origin, with slope = $\ell$ (see Figure 5.12).

The second approach suggested by Zepp involves experiments in which the rate of photodecay of a target compound ($T$) and a reference compound ($R$) are evaluated in the same photoreactor under otherwise identical conditions. Figure 5.13 provides a summary of this approach.

Both compounds are applied under conditions that approach zero absorbance of incident radiation. Because of this, expressions to describe the rate of photodecay of each compound can be defined easily (see row 1 of Figure 5.13). For dilute solutions, the rate of photodecay for the target and the reference compound can both be described by equation (5.23) (see row 2 of Figure 5.13). In turn, a plot of $\ln\left(\frac{[T]}{[T]_0}\right)$ vs. $t$ and $\ln\left(\frac{[R]}{[R]_0}\right)$ vs. $t$ should yield straight lines that pass through the origin (see row e of Figure 5.13), with slopes defined by the equations listed in row 4 of Figure 5.13.

The ratio of the slopes of these two graphs yields:

$$\frac{\text{Slope}_T}{\text{Slope}_R} = \frac{\Phi_T \cdot \varepsilon_T}{\Phi_R \cdot \varepsilon_R} \tag{5.31}$$

Rearrangement of equation (5.31) allows estimation of the quantum yield for the target compound:

$$\Phi_T = \frac{\text{Slope}_T}{\text{Slope}_R} \cdot \Phi_R \frac{\varepsilon_R}{\varepsilon_T} \tag{5.32}$$

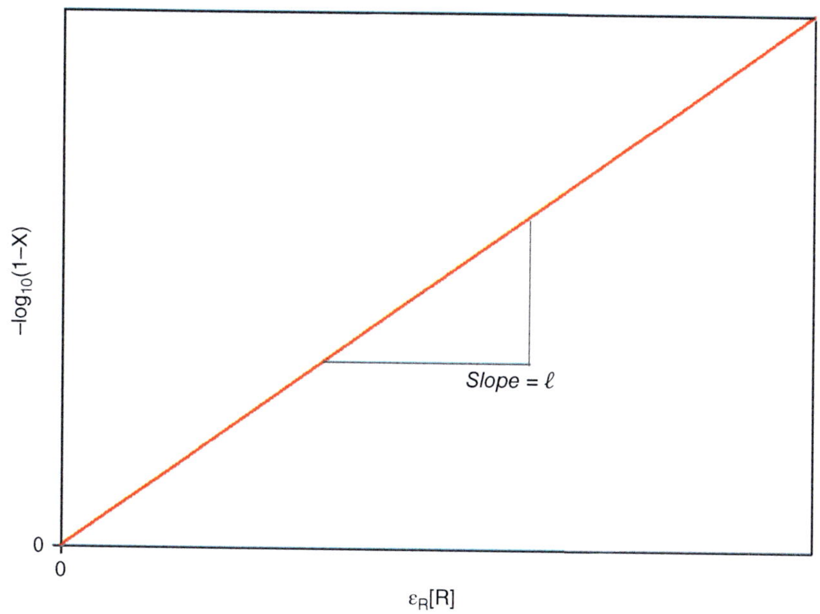


**Figure 5.12** Schematic illustration of the application of equation 5.30 for estimation of the effective path length, ℓ.

| Row | Target compound, T | Reference compound, R |
|---|---|---|
| 1 | $\frac{d[T]}{dt} = -2.303 \cdot \Phi_T \cdot E_i' \cdot \left(\frac{A}{\forall}\right) \cdot \varepsilon_T[T]\ell$ | $\frac{d[R]}{dt} = -2.303 \cdot \Phi_T \cdot E_i' \cdot \left(\frac{A}{\forall}\right) \cdot \varepsilon_R[R]\ell$ |
| 2 | $\ln\left(\frac{[T]}{[T]_0}\right) = -2.303 \cdot \Phi_T \cdot E_i' \cdot \left(\frac{A}{\forall}\right) \cdot \varepsilon_T \cdot \ell \cdot t$ | $\ln\left(\frac{[R]}{[R]_0}\right) = -2.303 \cdot \Phi_T \cdot E_i' \cdot \left(\frac{A}{\forall}\right) \cdot \varepsilon_R \cdot \ell \cdot t$ |
| 3 | 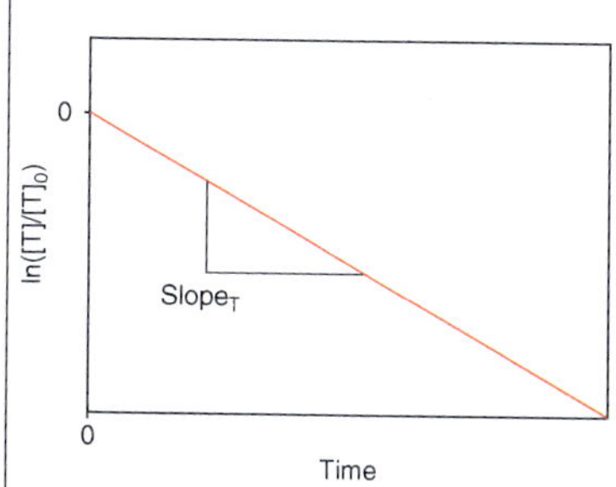  | 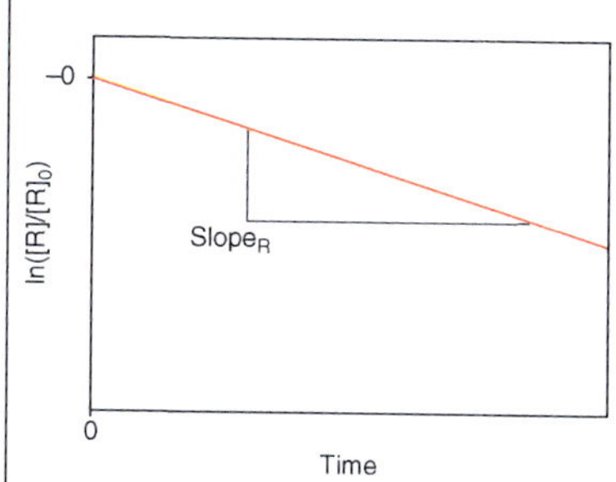  |
| 4 | $\text{Slope}_T = -2.303 \cdot \Phi_T \cdot E_i' \cdot \left(\frac{A}{\forall}\right) \cdot \varepsilon_T \cdot \ell$ | $\text{Slope}_R = -2.303 \cdot \Phi_R \cdot E_i' \cdot \left(\frac{A}{\forall}\right) \cdot \varepsilon_R \cdot \ell$ |

**Figure 5.13** Summary of logic used to estimate the quantum yield for a target compound using a merry-go-round reactor.

This second approach has been applied often to data from experiments that involve the use of a "merry-go-round" (MGR) reactor. The MGR reactor represents a common alternative to CBs. Under some circumstances, the MGR reactor may allow irradiation of larger volumes than is practical with conventional CB systems. A schematic illustration of a MGR reactor is presented in

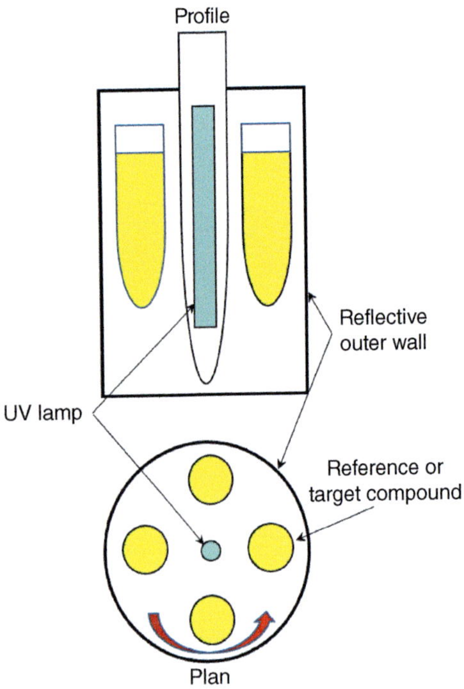


**Figure 5.14** Schematic illustration of merry-go-round (MGR) photoreactor; profile view (top) and plan view (bottom). In this system, a source of radiation (e.g. LP or MP Hg lamp, pale blue color) is placed on the central radial axis of the reactor. Multiple tubes containing the reference or target compound (yellow) are positioned at a uniform radial position from the lamp. The tubes should all have the same geometry and material properties, and should be filled to the same extent. The rack that positions the tubes is rotated around the lamp so as to promote uniform exposure to the UV radiation among all tubes during the period of exposure. The inside wall of the reactor is also made of a reflective material to further promote uniformity of applied fluence rate.

Figure 5.14. The MGR reactor has been applied for evaluation of photochemical kinetics of a wide range of compounds [11–14].

### 5.2.9 Chemical Actinometers Used With UV Photoreactors

The International Union of Pure and Applied Chemistry, IUPAC (Organic and Biomolecular Chemistry Division, Subcommittee on Photochemistry), has assembled a critical review of chemical actinometers [1]. This document provides an excellent summary of chemical actinometers that have been developed and documented for applications to gas, solid, liquid, and heterogeneous systems.

The IUPAC summary also includes a list of desirable characteristics for chemical actinometers, including:

- *Should be simple and well studied:* Ideally, the actinometer will undergo a pure photochemical reaction to yield a stable product that can be differentiated from the parent compound. If other, secondary reactions are involved, these should be well defined and easy to relate to the photochemical event that initiates the overall process. Quantum yields and molar absorption coefficients should be well defined.
- *Chemical components should be thermally stable and not susceptible to dark reactions:* Secondary reactions often complicate the interpretation of a chemical actinometry experiment. Ideally, the chemical actinometer will react only as a result of exposure to radiation.
- *Analytical methods should be simple, with direct spectrophotometry being preferred:* Chemical actinometry experiments will require the use of analytical methods to measure the dose-dependent concentration of the parent compound, the product(s) of reaction, or both. The availability of simple analytical methods for quantification of the chemical actinometer and/or its products can facilitate the use of the actinometer. Optical methods, including spectrophotometry and

measurements based on fluorescence, offer advantages of speed and simplicity, but are sometimes susceptible to interference by compounds that absorb radiation or fluoresce at the wavelength(s) used in the analytical method.

- *Actinometer should be sensitive:* Sensitive actinometers can be used to measure comparatively low doses of imposed radiation. Sensitivity will be governed by the product of the quantum yield and the molar absorption coefficient.
- *Evaluation of the number of photons absorbed should be simple and straightforward:* This objective will be met if the compound being measured is directly related to the primary photochemical event. Depending of the actinometer, this could be the chemical actinometer itself, a primary photoproduct, or a stable product of a secondary reaction.
- *Actinometer should be inexpensive, commercially available, and easy to synthesize and purify:* These attributes facilitate access to the actinometer.
- *Waste disposal should be simple:* Responsible use of any chemical, including a chemical actinometer, should consider disposal or reuse of waste products.

As is the case with most lists of ideal properties, there are very few chemical actinometers that completely satisfy all of these attributes. However, many actinometers have been identified that satisfy most of these characteristics. In general, these actinometers are the most widely used because of these attributes.

The text that follows provides brief summaries of some of the chemical actinometers that are commonly applied in UV photoreactors and UV photochemistry. The actinometers are presented roughly in their order of development by the scientific community. Additional details regarding these actinometers and their application can be found in the original papers that describe them.

#### 5.2.9.1 Uranyl Oxalate

The seminal paper of Leighton and Forbes [15] established uranyl oxalate as the default chemical actinometer for many years. Although others had used uranyl oxalate as an actinometer in experiments prior to their work, Leighton and Forbes were able to define the fundamental chemistry of the actinometer, including a set of detailed measurements of the quantum yield over a wide range of wavelengths.

The preparation and purification of uranyl oxalate involves a labor-intensive, multistep process that leads to the formation of the crystalline solid, which is then dissolved in water to function as an actinometer.

The basic stoichiometry of the photoreaction that represents the basis of this method was originally described as the photooxidation of oxalic acid to yield water, carbon dioxide, and carbon monoxide: [15]

$$H_2C_2O_4 \xrightarrow{h\nu} H_2O + CO_2 + CO \tag{5.33}$$

Subsequent research [16] indicated that two other reaction pathways were probably involved in degradation of oxalic acid or the oxalate ion, $C_2O_4^{2-}$:

$$H_2C_2O_4 \xrightarrow{h\nu} HCOOH + CO_2 \tag{5.34}$$

$$H_2C_2O_4 + U(VI) \xrightarrow{h\nu} H_2O + 2CO_2 + U(IV) \tag{5.35}$$

Mechanistic descriptions of the stoichiometric expressions described in equation (5.33)–(5.35) have also been developed [17].

Quantum yields reported for the uranyl oxalate actinometer are based on degradation of oxalic acid. The original analytical method used for this purpose was titration with potassium

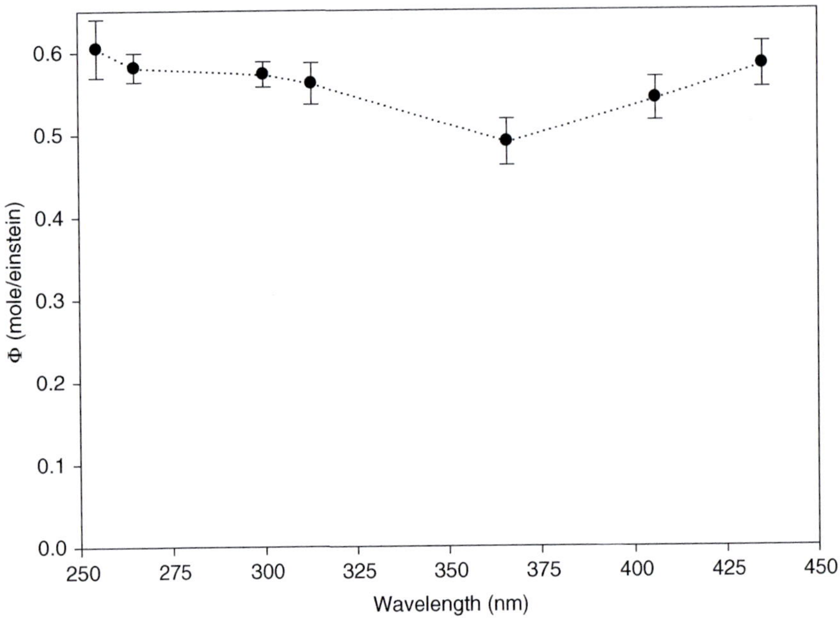


**Figure 5.15** Measured values of quantum yield as a function of wavelength for the uranyl oxalate actinometer [15]. Symbols represent the mean, while error bars represent the standard deviation of replicate experiments at each wavelength.

permanganate ($KMnO_4$) to determine the amount of oxalic acid remaining in solution [15]. However, later work indicated that an alternate method involving cerium (IV) sulfate as the oxidant, coupled with differential absorption spectroscopy yielded improved sensitivity [18].

The quantum yield for the uranyl oxalate actinometer has been quantified for wavelengths ranging from 435 to 254 nm [15]. These estimates were developed by using a monochromator to isolate narrow wavelength bands from a polychromatic radiation source, with each band being characterized by a narrow wavelength range. For each wavelength, replicate experiments were conducted. Figure 5.15 illustrates the results of these experiments. The quantum yield was observed to be fairly constant across this wavelength range, with mean values ranging from $0.49 - 0.61 \frac{\text{mole}}{\text{einstein}}$.

The uranyl oxalate actinometer is still used occasionally today, but because other options are available, many of which are much simpler to apply, it is no longer the default method. In addition, the use of this method requires disposal of wastes that contain uranium [1].

#### 5.2.9.2 Potassium Ferrioxalate

One of the important limitations of the uranyl oxalate actinometer is its low sensitivity. For low-power UV sources, this attribute dictates long exposure periods for accurate quantitation. This low sensitivity is actually desirable in applications involving high-power sources, but can be limiting otherwise.

The potassium ferrioxalate actinometer was developed as an alternative to uranyl oxalate; it has been demonstrated to be several hundred times more sensitive to UV radiation than uranyl oxalate, which leads to applications that involve much shorter exposure times than those required by the earlier method [19]. It has emerged as one of the most commonly applied chemical actinometers for UV radiation and may even be considered the "standard" for such measurements [2, 20].

The fundamental stoichiometry [2, 19] for the method involves photoreduction of $Fe^{3+}$ to $Fe^{2+}$:

$$2Fe(C_2O_4)_3^{3-} \xrightarrow{h\nu} 2Fe^{2+} + 5C_2O_4^{2-} + CO_2 \tag{5.36}$$

The $Fe^{2+}$ produced by this reaction is then reacted with 1,10 phenanthroline to yield a colored complex, which can be quantified by spectrophotometry ($\varepsilon_{510} = 11{,}100\ M^{-1} \cdot cm^{-1}$) [2].

Though not a requirement, the potassium ferrioxalate actinometer is often applied under conditions such that the actinometer solution is opaque to incident radiation. This condition allows complete absorption of incident photons by the actinometer. In effect, the actinometer becomes a "photon counter" under these conditions. A 6 mM solution of potassium ferrioxalate is opaque to radiation with wavelengths 254 nm $\leq \lambda \leq$ 390 nm through a 1 cm optical path, while a 0.15 M solution has been identified to achieve this condition for longer wavelengths [21]. The actinometer solution is stable in the dark, but will respond to conventional room lighting. Because of this, the preparation and application of potassium ferrioxalate needs to be conducted under red lighting or in a darkroom.

As with all actinometers, the quantum yield must be defined to allow translation of photoproduction or photodecay into applied radiant power. Experiments conducted to define a quantum yield must involve an alternative method of dose or power measurement to compare against. In addition, the experiment must involve a radiation source that is well characterized in terms of its spectral output.

Figure 5.16 provides a summary of quantum yield measurements that have been conducted for the potassium ferrioxalate actinometer. For wavelengths ranging from roughly 260 to 350 nm, the quantum yield for this actinometer is fairly constant, with a value of approximately 1.25 $\frac{\text{mole}}{\text{einstein}}$. At longer wavelengths, the quantum yield declines gradually to a value of roughly 0.85 $\frac{\text{mole}}{\text{einstein}}$ at 510 nm, after which there is a steep decline that approaches zero at wavelengths above about 580 nm.

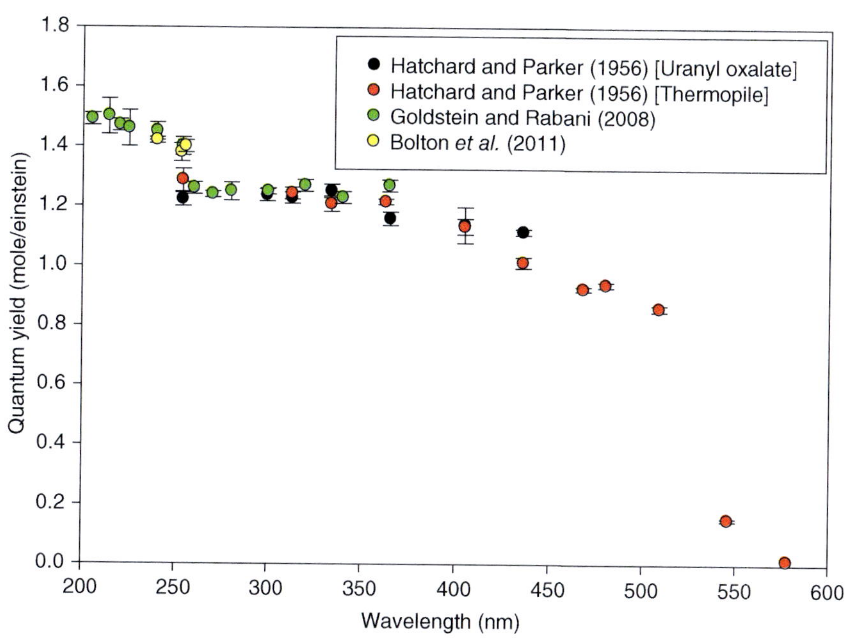


**Figure 5.16** Summary of quantum yield ($\Phi[Fe^{2+}]$) measurements for the potassium ferrioxalate actinometer. Hatchard and Parker developed estimates referenced to the uranyl oxalate actinometer and with reference to a thermopile [21]. The estimates by Goldstein and Rabani [20] and Bolton et al. [2] were referenced against calibrated Si photodiodes.

For the wavelength range 205–255 nm, the quantum yield is slightly elevated above the values seen for longer UVC wavelengths, with a nominal value of $1.4-1.5\,\frac{\text{mole}}{\text{einstein}}$ across this range. This behavior has been attributed to a second photochemical pathway [20], wherein an electron is abstracted from iron to yield $Fe^{2+}$:

$$Fe(C_2O_4)_3^{3-} \xrightarrow{h\nu} Fe(C_2O_4)_3^{2-} + e_{aq}^{-} \tag{5.37}$$

The accuracy of quantum yield determination will be influenced by the nature of the radiation source and the extent to which its spectral output has been conditioned and characterized. In addition, the accuracy of the instrument used to measure the power of the radiation source will influence the accuracy of a quantum yield estimate.

Hatchard and Parker used a polychromatic radiation source and a monochromator to deliver radiation to their actinometer solutions [21]. They applied two reference standards for measurement of incident radiation power. The first was the uranyl oxalate actinometer, which was selected (in part) because one of the motivations for their work was to develop an actinometer that improved on the characteristics of the uranyl oxalate actinometer. They also conducted experiments in which a calibrated thermopile was used to measure incident power.

Goldstein and Rabani used a low-pressure Hg lamp, coupled with optical filters to isolate the output line at 253.7 nm for some of their experiments [20]. For all other wavelengths, they used a Xe lamp coupled with optical filters or a monochromator to isolate wavelengths; they reported a bandwidth of approximately 4.7 nm for the monochromator. They used a Si photodiode that had been calibrated against a primary reference standard to measure applied radiant power.

In perhaps the most thorough experiments to date, Bolton et al. used a tunable laser to apply essentially monochromatic radiation in their experiments. Similar to the work of Goldstein and Rabani, they used a Si photodiode that had been calibrated against a primary reference standard for their measurements of radiant power. The reference standard that was used for this calibration was an absolute cryogenic radiometer (ACR, discussed below), which has been identified as the most accurate primary detector-based standard for these measurements [22, 23].

The measurements of quantum yield for the potassium ferrioxalate actinometer show good agreement among these various investigations for the wavelength ranges that overlap. Bolton et al. have suggested that a value of $\Phi_{253.7}(\text{Fe[II]}) = 1.39 \pm 0.02\frac{\text{mole}}{\text{einstein}}$ be used for this actinometer; this value appears to be independent of temperature for $5\,°\text{C} \leq T \leq 80\,°\text{C}$ [2].

Potassium ferrioxalate has a number of important attributes that motivate its use in actinometric applications for UV radiation, including a well-defined quantum yield across the UV spectrum and a relatively simple analytical method. However, important limitations of this method also exist, including a fairly labor-intensive synthesis process. Perhaps more importantly, potassium ferrioxalate is sensitive to normal room lighting, including radiation with wavelengths shorter than approximately 600 nm. As such, its application for measurement of UV radiation requires use in a space that does not impose radiation that confounds the interpretation of the test.

#### 5.2.9.3 Iodide/Iodate

Halide ions in aqueous solution have been reported to display similar photochemistry when subjected to UVC radiation: [24]

$$X^{-} \overset{h\nu}{\leftrightarrow} X^{-*} \rightarrow [X + e^{-}] \tag{5.38}$$

where $X^-$ represents a halide ion ($Cl^-$, $Br^-$, or $I^-$). The primary photochemical process leads to the formation of an electronically excited halide ion ($X^{-*}$), which reacts to yield a halogen atom (Cl, Br, or I) and a hydrated electron ($e^-$) in a solvent cage.[b]

Halide ions are inexpensive, widely available, and essentially nontoxic. Therefore, their use as a chemical actinometer provides some important advantages over many other actinometers. As we will see, the use of iodide ($I^-$) in a chemical actinometer provides some other advantages in applications involving UVC radiation.

Potassium iodide has been developed and applied as a chemical actinometer [25, 26]. The fundamental chemistry involved in the use of this actinometer is illustrated in Figure 5.17. Briefly, the process is initiated by a primary photochemical reaction in which iodide absorbs a photon to yield an electronically excited, hydrated iodide ion. Ultimately, this photochemical event leads to the formation of triiodide ($I_3^-$), which can be quantified easily by measurement of absorbance at 352 nm ($\varepsilon_{352} = 26{,}400\ M^{-1} \cdot cm^{-1}$). Additional details of this reaction mechanism are provided in Figure 5.17.

In degassed water and in the absence of any pH buffer, the quantum yield for $I_3^-$ formation at 253.7 nm has been reported to be $\Phi_{253.7}(I_3^-) \approx 0.05 \dfrac{\text{mole}}{\text{einstein}}$ [25]. The back reaction (Reaction 3 in Figure 5.17) dominates under these conditions and is largely responsible for this low quantum yield.

Electron scavengers have been added to solution as a means of improving the observed value of $\Phi(I_3^-)$. Initially, nitrous oxide ($N_2O$) was applied as an electron scavenger: [25, 27]

$$N_2O + e^- \rightarrow HO\cdot + OH^- + N_2 \tag{5.39}$$

From the perspective of $I_3^-$ formation, this reaction has two benefits. First, it competes for aqueous electrons, thereby reducing the importance of the back reaction. Second, it yields a hydroxyl radical ($HO\cdot$), which will react to yield $I_3^-$ as follows:

$$\begin{aligned} HO\cdot + I^- &\rightarrow I + OH^- \\ 2I + 2I^- &\rightarrow 2I_2^- \\ 2I_2^- &\rightarrow I_3^- + I^- \end{aligned} \tag{5.40}$$

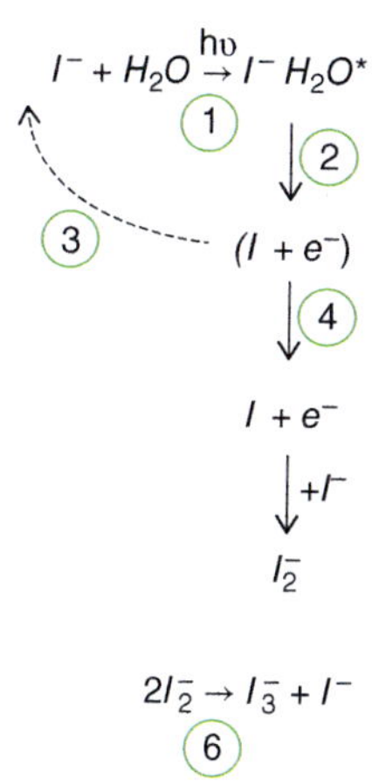


**Figure 5.17** Illustration of reaction mechanism for KI as an actinometer. Reaction 1 is the primary photochemical event, in which the iodide ion absorbs a photon to yield an electronically excited ion, which spontaneously reacts to yield an iodine atom and an electron in a solvent cage (Reaction 2). The caged complex can decompose via two pathways. The first pathway (Reaction 3) involves recombination of the iodine atom and the electron to yield iodide, while the second (Reaction 4) involves decomposition of the caged complex, resulting in migration of the electron out of the cage into the bulk solution. The iodine atom will react with iodide to yield $I_2^-$ (Reaction 5), which in turn will react with itself to yield triiodide ($I_3^-$), Reaction 6.

The overall stoichiometry for the $KI/N_2O$ actinometer is described as follows:

$$3I^- + N_2O + H_2O \xrightarrow{h\nu} I_3^- + N_2 + 2OH^- \tag{5.41}$$

Inclusion of $N_2O$ as an electron scavenger allowed the quantum yield to increase to $\Phi_{253.7}(I_3^-) \approx 0.235\ \frac{\text{mole}}{\text{einstein}}$ [25]. However, $N_2O$ is cumbersome to use because it must be dissolved by bubbling of $N_2O$ gas into an aqueous solution.

Potassium iodate ($KIO_3$) has also been applied as an electron scavenger and it has been shown to be superior to $N_2O$ in some respects [26]. The iodate ion, $IO_3^-$, scavenges electrons by the following process:

$$IO_3^- + e^- + 2H_2O \rightarrow IO^- + H_2O_2 + HO\cdot + OH^- \tag{5.42}$$

As with $N_2O$, there are multiple benefits to the inclusion of $IO_3^-$ as an electron scavenger. Like $N_2O$, it reduces the importance of the back reaction. In addition, it provides three mechanisms for the formation of $I_3^-$ in solution [26]. The first two of these are:

$$IO^- + H_2O + 2I^- \rightarrow I_3^- + 2OH^- \tag{5.43}$$

$$H_2O_2 + 3I^- \rightarrow I_3^- + 2OH^- \tag{5.44}$$

In addition, because $HO\cdot$ is formed in this reaction, process (5.40) (above) also contributes to $I_3^-$ formation. The overall stoichiometry for the $KI/KIO_3$ actinometer becomes:

$$8I^- + IO_3^- + 3H_2O \xrightarrow{h\nu} 3I_3^- + 6OH^- \tag{5.45}$$

The performance of this actinometer has been shown to depend on solution composition. By altering the composition of the actinometer solution, an optimum composition of 0.6 M KI, 0.1 M $KIO_3$, and 0.01 M $Na_2B_4O_7 \cdot 10H_2O$ (sodium tetraborate buffer for pH = 9.25) has been suggested and is commonly applied [26]. The pH buffer is included, in part, because $I^-$ undergoes a slow, spontaneous reaction in the dark to yield $I_3^-$. This reaction is suppressed at alkaline pH.

The quantum yield for triodide formation, $\Phi(I_3^-)$, has been shown to demonstrate temperature-dependence, which is attributed to secondary (thermal) reactions, especially the escape of hydrated electrons from the solvent cage (Reaction 4 in Figure 5.17). The temperature-dependence of $\Phi_{253.7}(I_3^-)$ is illustrated in Figure 5.18.

The $KI/KIO_3$ actinometer is particularly effective for evaluation of UVC radiation. The iodide ion ($I^-$) absorbs strongly at wavelengths shorter than approximately 290 nm, yet it is essentially non-absorbing for wavelengths longer than 330 nm. Therefore, it can be prepared and applied with conventional room lighting [26]. When subjected to the output from a low-pressure Hg lamp, the vast majority of $I_3^-$ formation is attributable to the line at 253.7 nm.

Goldstein and Rabani [20] and Bolton et al. [2] conducted experiments to define the wavelength-dependence of $\Phi(I_3^-)$. The methods used by these investigators were the same as those they applied in their work with potassium ferrioxalate, described above. Figure 5.19 illustrates this wavelength-dependence.

The reported quantum yield for this actinometer is steady at a value of approximately $0.92\frac{\text{mole}}{\text{einstein}}$ for wavelengths ranging from 200 to 250 nm. From 250 to 300 nm, there is a steady decline in the quantum yield to a value of approximately $0.25\frac{\text{mole}}{\text{einstein}}$ at about 300 nm. The data of Goldstein and Rabani indicate a modest rebound of the quantum yield at wavelengths ranging from 300 to 320 nm. Rahn et al. reported similar behavior in their work [28].

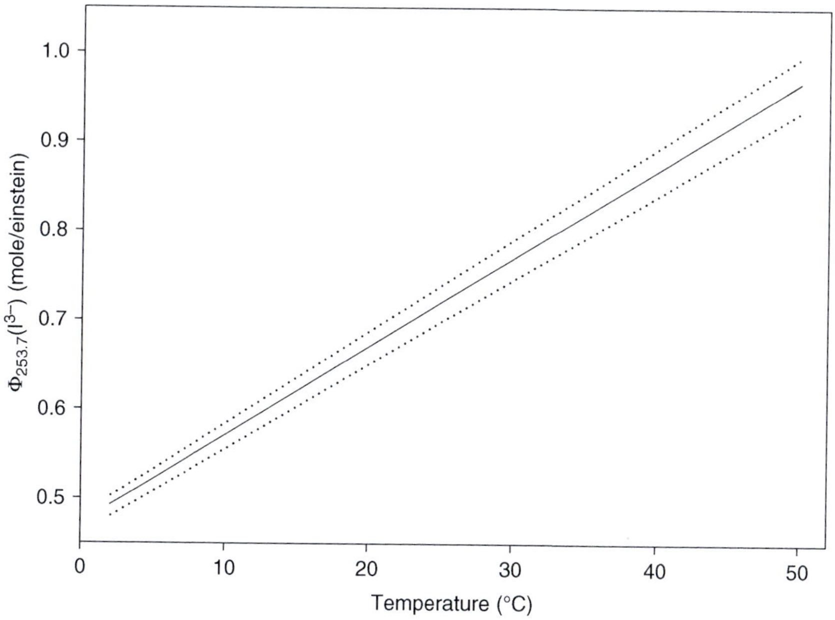


**Figure 5.18** Temperature-dependence of $\Phi_{253.7}(I_3^-)$ for the KI/$KIO_3$ actinometer, as recommended by Bolton et al. for 0 °C ≤ $T$ ≤ 50 °C [2]. The solid line represents the mean, while the dotted lines indicate recommended error bands.

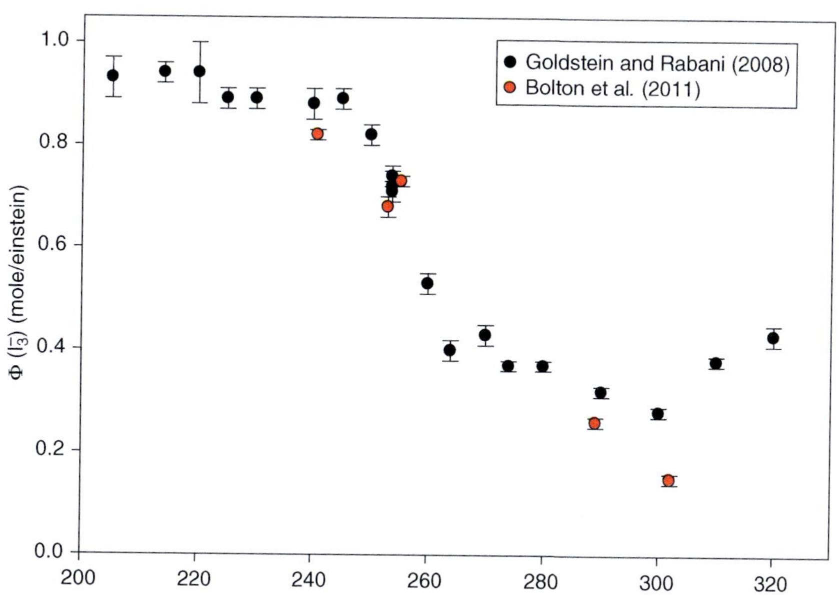


**Figure 5.19** Wavelength-dependence of $\Phi(I_3^-)$ [2, 20]. Actinometer composition was 0.6 M KI, 0.1 $KIO_3$, and 0.01 M $Na_2B_4O_7{\cdot}10H_2O$.

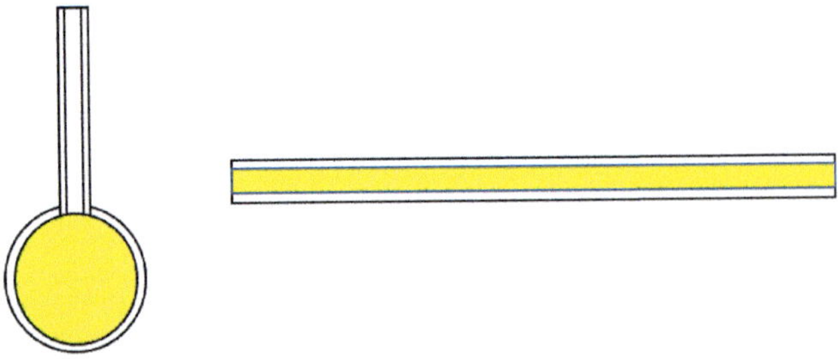

**Figure 5.20** Schematic illustration of vessels used for spherical (left) and tubular (right) actinometry. The yellow area represents the region that has been filled with the chemical actinometer. The walls of the vessels must be made from materials that are highly transmissive of the incident radiation.

One of the benefits of chemical actinometry, as compared with other methods of photon measurement, is that it can conform to essentially any vessel shape. As such, it can be applied in transparent vessels for evaluation of fluence delivered to a given location. Relevant applications of this approach include spherical actinometry and tubular actinometry, wherein an actinometer solution is placed in a small spherical or cylindrical container at a given location (see Figure 5.20). This allows fluence to be measured at that location. Deployment of the $KI/KIO_3$ actinometer in multiple spheres or tubes has been used to characterize the spatial distribution of fluence rate [29–31].

Local actinometry is a process that follows a similar strategy, but is performed on a continuous basis. Local actinometry is implemented to allow continuous measurement of the fluence rate delivered to a location within a photoreactor. It represents an alternative to conventional physical instruments that are used for this purpose.

Local actinometry is implemented by positioning a UV-transparent capillary tube at a well-described position in a reactor. Any position can be used, but a logical choice is a location within the reactor that limits reactor performance, such as a region that is characterized by relatively low fluence rate. An actinometer solution is pumped through the capillary, followed downstream by continuous measurement of a signal that can be related to the actinometer or a product of its photolysis. An important advantage of this approach is that it provides a continuous, real-time measurement of the effects of a photochemical reaction, and is not susceptible to electrical noise or interference, as is observed among physical actinometers.

The $KI/KIO_3$ actinometer is well suited for local actinometry because the reagents required for production of the actinometer are inexpensive and readily available. In addition, the photoproducts include forms of iodine that can be converted to molecular iodine ($I_2$), which has beneficial applications as a chemical disinfectant [32].

Figure 5.21 provides an illustration of an application of local actinometry that was developed for water treatment during long-term space travel. An example is the planned trip to Mars, with duration estimated to last 100–900 days. Given the payload restrictions that apply with long-term space travel and the life-support requirements for the astronauts, access to potable water will require extensive water reuse. And like most aspects of these missions, many aspects of human life support will involve closed-loop systems.

#### 5.2.9.4 2-Nitrobezaldehyde

The actinometers described above are useful for measurements of UVC radiation; potassium ferrioxalate also responds to longer wavelengths. There are many situations in which actinometry is

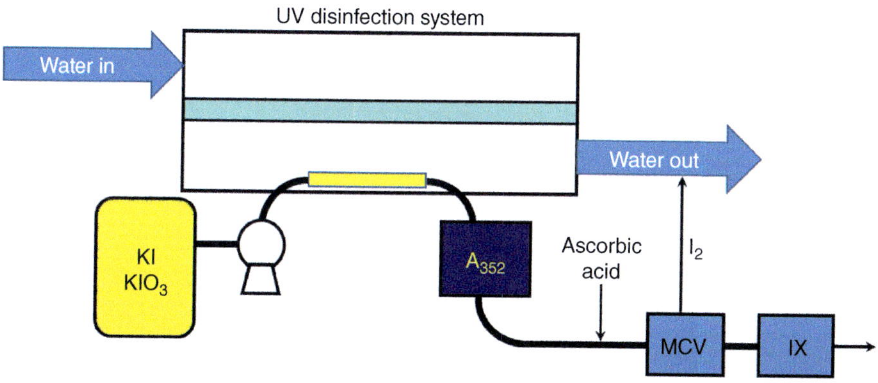


**Figure 5.21** Schematic illustration of an application of local actinometry, based on the $KI/KIO_3$ actinometer. *Source:* Adapted from Pennell 2005. [33] Water is pumped through a UV disinfection system. A solution of the $KI/KIO_3$ actinometer is pumped through a UV-transparent quartz capillary tube, located at a critical location in the reactor cross section. $I_3^-$ produced as a result of UV exposure in the capillary tube is monitored in real time by an online spectrophotometer, based on measurements of absorbance at 352 nm ($A_{352}$). Molecular iodine ($I_2$) formed by equilibrium with $I_3^-$ and by reaction with ascorbic acid is stored on a resin in a microbial check valve (MCV). Residual iodide is collected on an ion exchange (IX) column for subsequent reuse. After consumption, water will be treated for subsequent reuse (not shown), thereby allowing the system to function as a closed loop.

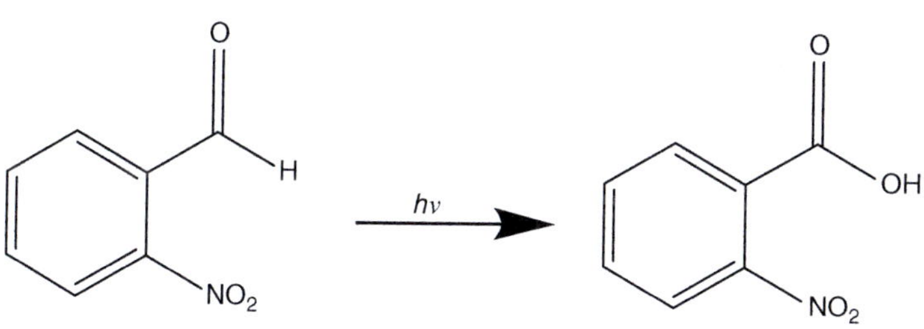


**Figure 5.22** UV-induced photoisomerization of 2-nitrobenzaldehyde (left) to 2-nitrosobenzoic acid (right).

relevant in applications of UVB and UVA radiation, including processes that take advantage of ambient solar UV radiation.

2-Nitrobenzaldehyde (2NB) has a long history of application, with the first studies of its photochemical behavior dating back to 1900 [34]. Absorption of radiation in the wavelength range 310 nm $\lesssim \lambda \lesssim$ 400 nm results in electronic excitation of the nitro ($-NO_2$) group on the 2NB molecule [35]. Absorption of photons in this wavelength range promotes an intramolecular rearrangement to form 2-nitrosobenzoic acid (see Figure 5.22). Because this reaction is intramolecular, it is largely unaffected by its surroundings. Consequently, 2NB has been used as an actinometer in solid, liquid, and gas-phase applications [36].

Common applications of this method involve quantification of 2NB by high-performance liquid chromatography (HPLC) [36]. However, other analytical methods have been applied, including infrared (IR) spectroscopy to follow loss of the $-NO_2$ group [37], and conductivity to follow the formation of the acid group in 2-nitrosobenzoic acid [38].

The quantum yield for the process described in Figure 5.22 has been measured by a number of researchers. Galbavy et al. conducted experiments to allow quantification of the quantum yield for wavelengths of radiation that include those observed in the ambient solar spectrum

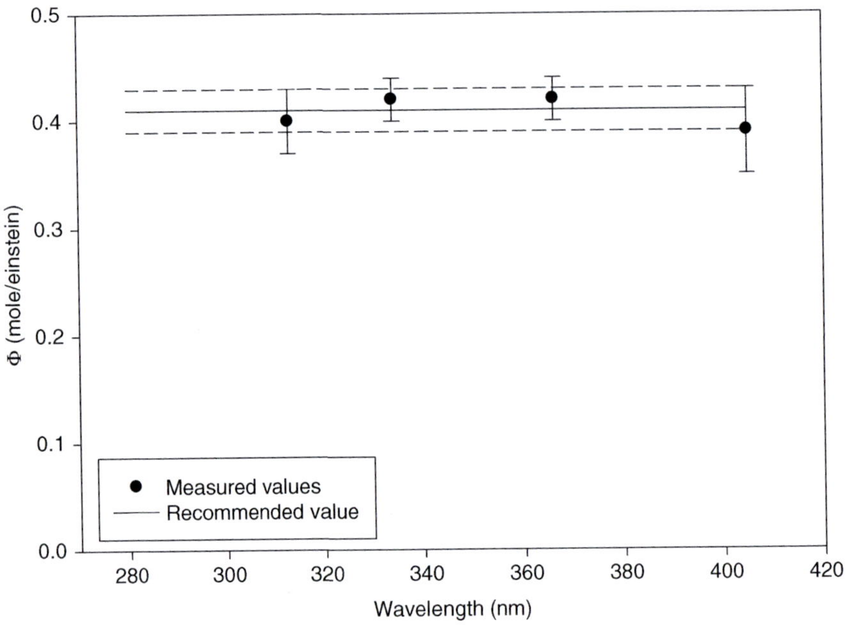


**Figure 5.23** Summary of measured values of the quantum yield for photoisomerization of 2NB [36]. Data points and error bars represent the mean and standard deviation of measured values. The solid and dashed horizontal lines represent values of the mean and standard deviation, respectively, of a value for Φ that has been recommended for the wavelength range of 280 nm $\lesssim \lambda \lesssim$ 405 nm. The recommended value is $\Phi = 0.41 \pm 0.02 \frac{\text{mole}}{\text{einstein}}$.

(see Figure 5.23). The quantum yield was observed to be essentially constant across the entire wavelength range illustrated in Figure 5.23. Galbavy et al. suggested a value of $\Phi = 0.41 \pm 0.02 \frac{\text{mole}}{\text{einstein}}$ for experiments involving 2NB in aqueous solution or dissolved in organic solvents [36]. A similar value was recommended for applications in ice.

The quantum yield for 2NB has also been observed to be essentially independent of temperature for $-25\,°\text{C} \leq T \leq 28\,°\text{C}$ [36, 39]. This has allowed the application of 2NB as an actinometer in a wide range of applications and media. For example, 2NB was incorporated into thin film samples of polymethylmethacrylate (PMMA) for use in quantifying penetration of solar UV radiation into snowpack [39].

#### 5.2.9.5 Nucleoside Actinometers

Nucleosides are subunits of nucleotides, which in turn are the building blocks of nucleic acids, DNA and RNA. Since many photobiological effects, including skin damage (e.g. skin cancer) and microbial inactivation (i.e. disinfection), involve photochemical damage to nucleic acids, it is logical to consider the development of chemical actinometers or "dosimeters" based on nucleosides. Moreover, because nucleosides have absorption spectra that tend to be similar to nucleic acids, it is possible to use them to develop meaningful estimates of dose delivery in monochromatic or polychromatic systems.

Wang examined the photochemistry of uracil and several uracil derivatives [40]. The general reaction was suggested to involve photo-rearrangement to yield a zwitterion, which then undergoes

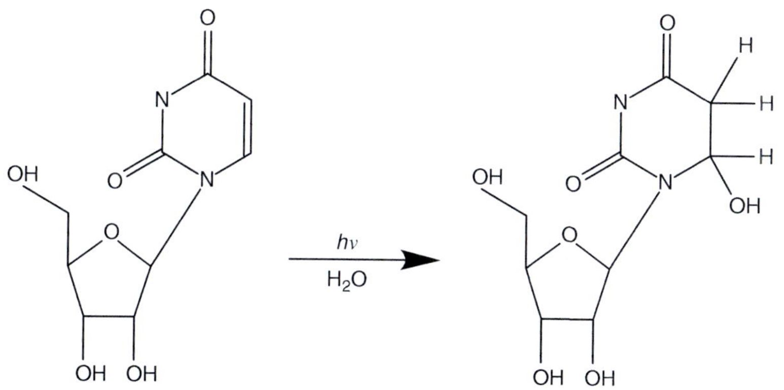


**Figure 5.24** Illustration of photohydration of uridine. *Source:* Adapted from [41].

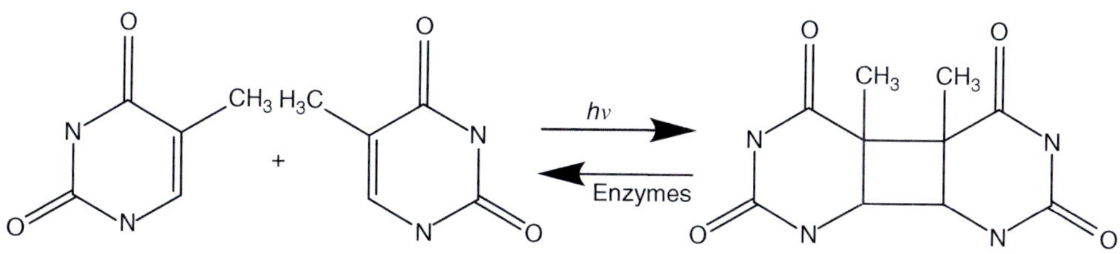


**Figure 5.25** Photodimerization of thymine. The forward reaction is photochemical, while the reverse reaction (photoreactivation) is promoted by enzymes.

spontaneous hydration. The overall process, referred to as photohydration, is illustrated for the case of uridine, which includes a uracil molecule, in Figure 5.24.

Uridine absorbs strongly in the UVC range, with peak absorption at 262 nm ($\varepsilon_{262}$ = 10,400 $M^{-1}\cdot cm^{-1}$) [42]. The photohydrate of uridine does not absorb appreciably at this wavelength, so it is convenient to monitor reaction progress based on absorbance at 262 nm. Uridine has been used as an actinometer for UVC radiation from monochromatic and polychromatic sources [41, 43, 44]. The quantum yield for this photohydration reaction is believed to be essentially constant across the germicidal UV spectrum, but has a relatively low reported value of $\Phi = 0.034 \frac{\text{mole}}{\text{einstein}}$ [41].

Thymine dimerization is known to play important roles in damage to skin and other external tissues, and also is critical to UV-based inactivation of microorganisms (i.e. disinfection). Che et al. developed an actinometer/dosimeter based on thymine dimerization that relies on this reaction, as illustrated in Figure 5.25 [45]. They observed four isomers of the dimer photoproduct and were able to detect them using chromatographic methods. This actinometer has the benefit of being based on a specific reaction that is known to be an important contributor to photobiological effects, but as with the uridine actinometer, the quantum yield for this reaction is quite low, with reported values of $\Phi = 0.001 - 0.01 \frac{\text{mole}}{\text{einstein}}$.

Bergstrom et al. examined methods for synthesis of pyrimidine nucleosides that involved UV-induced photoreactions [46]. One example of these reactions involved (E)-5-[2-methoxycarbonyl)ethenyl]cytidine (hereafter referred to as *S*), which undergoes *trans-cis* photoisomerization to yield an unstable intermediate, which then undergoes spontaneous ring closure to yield 3-β-D-ribofuranosyl-2,7-dioxopyrido[2,3-d]pyrimidine (hereafter referred to as *P*), as illustrated

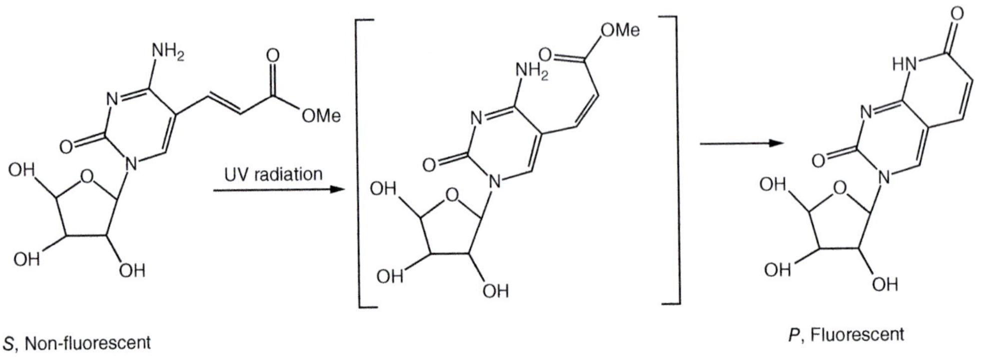


**Figure 5.26** UV photoreaction of S to yield P. *Source:* Adapted from [46]. Note that the starting material, S, is nonfluorescent, while the reaction product, P, is brightly fluorescent.

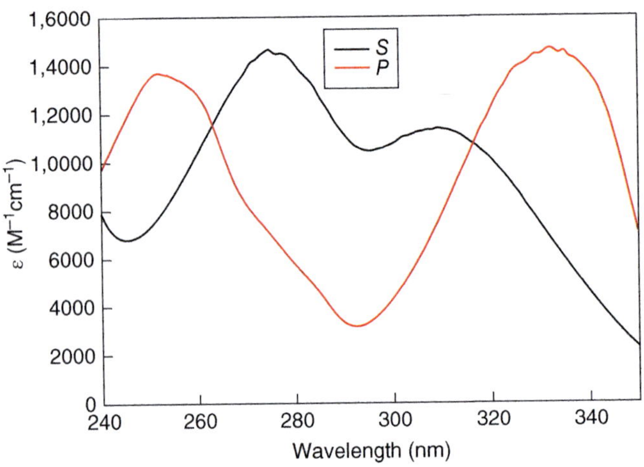


**Figure 5.27** Absorption spectra for S and P. *Source:* Adapted from [47].

in Figure 5.26. Ring closure prevents reversal of the photoisomerization reaction and yields a single product, *P*. *S* and *P* are extremely stable in the dark, with lifetimes of more than 10 years when stored in opaque containers.

*S* was developed as a chemical actinometer for solution-phase applications, and for Lagrangian Actinometry (LA), as will be described in Chapter 7. In the context of solution-phase applications, *S* offers two important advantages relative to other nucleoside actinometers. First, the $S \rightarrow P$ phototransformation yields a product that displays bright fluorescence (excitation maximum = 330 nm; emission maximum = 385 nm); this attribute is unusual among chemical actinometers [46, 47]. This allows online, continuous monitoring by fluorescence spectroscopy, which is a rapid, sensitive optical method. Second, the quantum yield for the reaction illustrated in Figure 5.26 is relatively high, with measured values of $\Phi = 0.693 - 0.850 \frac{\text{mole}}{\text{einstein}}$ across the UVC range [47]. However, a limitation of the *S* actinometer is that both the parent compound and its photoproduct absorb strongly across the UVC range (see Figure 5.27). Because of this, it is necessary to account for photon absorption by both *S* and *P* in applications of this actinometer.

## 5.3 Radiometry

Radiometry describes a family of methods by which electromagnetic radiation, usually in the IR, visible, and ultraviolet portions of the spectrum, is measured. These methods generally incorporate physical instruments that allow the development of a signal, usually electronic, that is influenced by and relatable to the quantity of radiation imposed on the detector. Of particular interest in the context of this book are radiometric measurements that are used to quantify ultraviolet radiation. Several different instrument types are available, not all of which will be covered herein.

Historically, the physical instruments used to quantify UV power have involved two methods. The first involves use of the photoconductive effect, wherein incident photons are used to promote electrons from the valance band to the conduction band of a material, with resulting change in an electrical signal (current). The second general method involves absorption of photons to promote heating of a material, with measurement of a change in temperature of the receiving material. These methods remain at the core of many radiometric measurements today. However, other methods have also emerged, including methods that are based on other optical signals, such as fluorescence. The focus of this chapter will be on methods that serve as reference standards, commonly applied devices, or emerging methods for quantification of incident UV radiation.

### 5.3.1 Absolute Cryogenic Radiometer (ACR)

The ACR was developed as an improvement over conventional silicon photodiodes, which had been used as a primary standard for measurements of optical power [48, 49]. Silicon photodiodes (described below) can be developed to have essentially 100% quantum efficiency, but are limited in terms of their accuracy and spectral range.

An ACR comprises three basic components [49]. The first is a cavity (usually copper) that is used to absorb incident radiation; the cavity will include a source of input electrical power to allow for electrical heating. The second component is a reference block that is maintained at a constant temperature. The third component is a weak thermal link between the absorbing cavity and the reference block.

The device operates based on the principle of electrical substitution. This involves an instrument wherein the absorbing cavity can be heated either by absorption of electromagnetic radiation or by electrical heating. The approach used for quantification of incident radiant power is to first allow the absorbing cavity to undergo an increase in temperature as a result of absorbing incident photons from the source of interest. Next, the device is heated to the same temperature by electrical heating. The electrical power required to reach this same equilibrium temperature condition is assumed to be equal to the power of the incident radiation.

A key characteristic of an ACR is the equivalence of the thermal response of the cavity to radiative and electrical heating. This equivalence is promoted by design to include the following attributes: [50]

- The absorbing cavity is designed to efficiently absorb incident photons.
- Applied electrical power is dissipated as heat.
- Temperature of reference sink remains constant during measurements.
- The detector (absorbing cavity) is shielded from other sources of radiation.

Another important attribute of the ACR is its operating temperature, which is typically in the vicinity of 4 K (liquid helium). This low temperature alters the physical properties of the material of the absorbing cell (copper) such that accuracy is improved by roughly 100X relative to previously developed primary standards.

The ACR is well suited for laser radiation, but has also been applied for other radiation sources. These instruments have been used in precision photochemistry experiments, wherein accurate determination of applied radiometric power is required. Examples include experiments wherein quantum yields are determined, such as with chemical actinometers [2, 36].

### 5.3.2 Thermopile

Thermopiles are solid-state devices that have been developed to allow measurement of energy input. They have been developed and are applied for measurement of many sources of energy, but in this context, the focus will be on their application for measurement of radiation energy, especially ultraviolet radiation.

Thermopiles comprise thermocouple junction pairs, usually arranged in series. To understand thermopiles, it is therefore beneficial to understand the basic behavior of thermocouples.

A typical thermocouple will include two dissimilar conductor materials (e.g. metals) in series. Seebeck observed that electrical current will flow between dissimilar metal junctions in contact when their junctions are held at different temperatures (see Figure 5.28). Electrons will tend to migrate from the hot junction to the cold junction in any conducting material. However, the ability of electrons to move under this thermal driving force will differ among conducting materials, and is a property of each material. By connecting two dissimilar conducting materials to the same two temperature conditions in a circuit, an electrical potential (voltage) is established. This is known as the Seebeck effect.

The magnitude of the voltage established by the Seebeck effect is linearly related to the difference in temperature between the hot and cold electrodes:

$$V = S \cdot \Delta T \tag{5.46}$$

where

$V$ = voltage established through circuit illustrated in Figure 5.28
$S$ = Seebeck coefficient
$\Delta T$ = difference in temperature between hot and cold electrodes.

The Seebeck coefficient is a function of temperature, as well as the material properties of the two conducting materials that comprise the thermocouple.

The voltage generated by a single thermocouple tends to be small. However, by connecting multiple ($n$) thermocouples in series, it is possible to increase the voltage as follows:

$$V = n \cdot S \cdot \Delta T \tag{5.47}$$

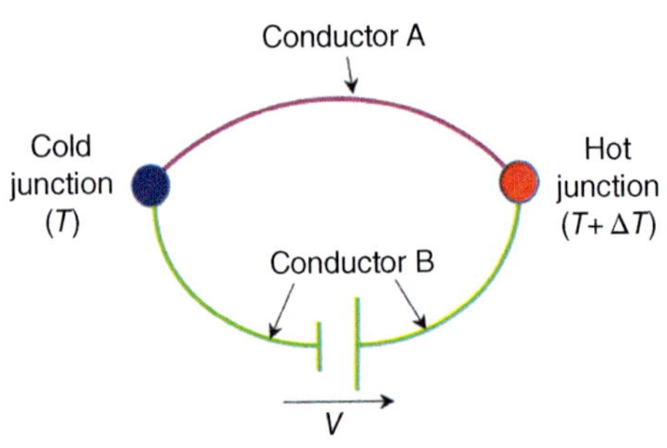


**Figure 5.28** Schematic illustration of Seebeck effect in a simple thermocouple. Green and purple lines represent dissimilar metals used to form junctions.

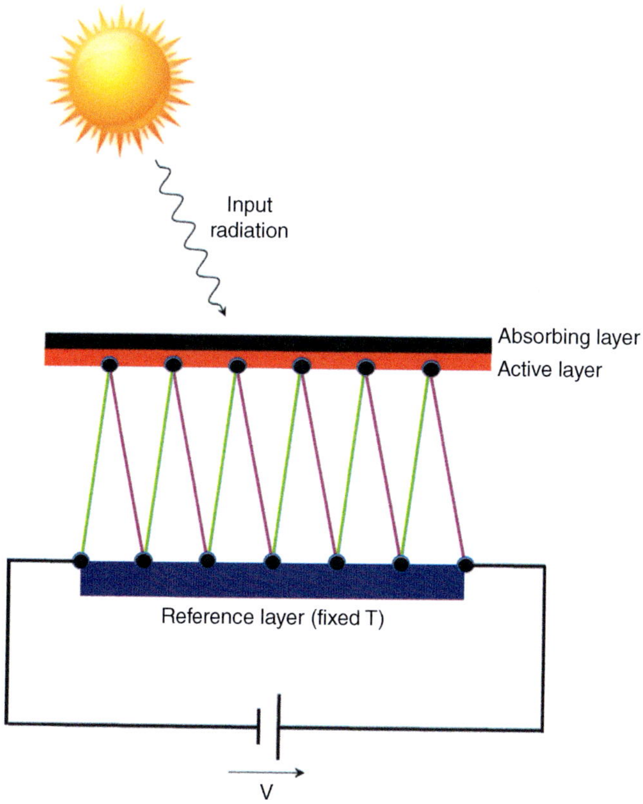


**Figure 5.29** Schematic illustration of a thermopile system as used to quantify a source of electromagnetic radiation. Green and purple lines represent dissimilar metals used to form junctions. Energy from the radiation source is imposed on the absorbing layer, which is in direct thermal contact with the active layer. Heating of the active layer causes its temperature to increase, relative to the reference layer, which is held at fixed (reference) temperature. The voltage generated by this system is proportional to the incident fluence rate.

Thermopiles can be used to quantify energy inputs from a wide range of sources. However, when used to quantify radiation, the radiation source itself becomes the source of energy input to the system, as illustrated in Figure 5.29. Typical thermopiles will involve many thermocouples in series, thereby facilitating measurement of photothermal heating via voltage measurements.

### 5.3.3 Photomultiplier Tube (PMT)

Photomultiplier tubes (PMTs) are used in a wide variety of applications where measurement of incident radiation is relevant. Their characteristics are well suited to applications that involve low fluence rate, but are applicable across a broad range of incident fluence rates.

The response of a PMT is initiated by photon absorption by a photocathode, resulting in primary electron emission (photoemission) of a photoelectron. The primary electron is then accelerated and focused onto a dynode array to allow the generation of secondary electrons, resulting in amplification of the original photoelectron. Primary and secondary electrons are collected at an anode, then they are sent to an external circuit for detection.

The basic components and common configurations of PMTs are illustrated in Figure 5.30, where two common modes of operation are illustrated [51]. The first is transmission mode (Figure 5.30, top),

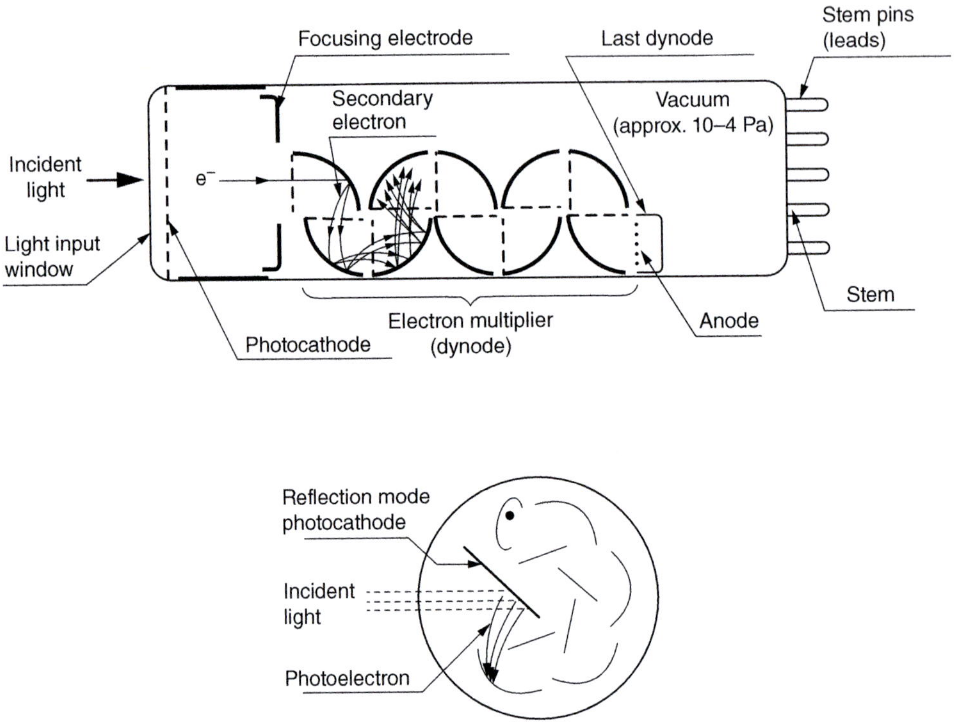


**Figure 5.30** Schematic illustrations of photomultiplier tubes. (Top) Transmission mode PMT; (bottom) reflection mode PMT. [52] *Source:* Reproduced with permission of Hamamatsu Photonics

wherein the photoelectrode is deposited as a thin film onto a window that is transparent to the radiation of interest. With this configuration, the photoelectron will be emitted in essentially the same direction as the incident beam of radiation. Reflection mode (Figure 5.30, bottom) is accomplished by the use of a photoelectrode as a reflective plate. In this mode, photoelectrons are reflected from the surface at an angle similar to that of the angle of incidence of photons onto the photoelectrode.

Regardless of the mode of operation, the basic system components and steps involved in the behavior of the instrument are similar [51]. The incident beam of radiation first encounters an entrance (input) window. The window should be made of a material that allows photons of interest to be transmitted, while potentially excluding radiation at wavelengths that will interfere with measurements or otherwise be non-beneficial; a monochromator can be used to isolate specific wavelengths of radiation. Incident photons will then strike the photocathode. If a photon has energy that exceeds the *work function*[c] of the photocathode material, then a photoelectron can be emitted (photoemission) to the vacuum space that surrounds the electrode. This primary photoelectron is key to detection and quantification of incident radiation. As such, the selection of the materials used to produce the window and the photoelectrode (cathode) will govern the wavelength range of photons that will cause the PMT to respond. Photocathodes are usually made of alkali metal semiconductors, as these materials are characterized by relatively low work functions.

Photoelectrons emitted from the photocathode are accelerated and focused onto a dynode array by a focusing electrode [51]. Each dynode in the array will yield multiple secondary electrons for each electron it receives, and each successive dynode is held at a more-positive electric potential than the previous dynode so that electrons travel "downstream" from dynode to dynode. Electron amplification is quantified in dynodes by the secondary emission ratio, $\delta$, which represents the number of secondary electrons per primary electron. For dynode materials used in PMTs, typical values of $\delta$ range from 2 to 100. Therefore, a single photoelectron imposed on a dynode array can be amplified considerably. This attribute allows PMTs to be used in low-light applications. The primary and secondary electrons that emerge from the dynode array are collected on the anode, then sent to an external circuit for quantification.

PMTs have response times on the order of 1–10 ns, and typically display a wide (linear) dynamic range [51]. However, deviations from linearity are observed when the instrument is subjected to high fluence rates.

### 5.3.4 Si Photodiode

Silicon (Si) photodiodes are part of a family of solid-state devices[d] that convert input radiation power (i.e. photons) into output electrical power (i.e. electrons). Irradiance can be calculated from measured power and the known surface area of the detector. Their behavior is in many ways the inverse of the behavior of light-emitting diodes (LEDs, see Chapter 4).

To create a Si photodiode, p-type and n-type semiconducting materials are brought into contact to form a p-n junction. When a photon with energy that exceeds the bandgap of the semiconducting material is absorbed by the diode, an electron ($e^-$) from the valence band is promoted to the conduction band of the material, leaving a positively charged hole ($h^+$). For every absorbed photon that has energy above the bandgap, an electron:hole ($e^- : h^+$) pair is formed. Electrostatic forces within the solid matrix will cause electrons to migrate away from the n-layer, while holes will migrate away from the p-layer. When this system is connected to an external circuit, electrical current will flow. The current will be directly related to the number of $e^- : h^+$ pairs formed.

The limits of spectral response of photodiodes are governed by the material properties of the photodiode itself, as well as the optical characteristics of any input optical elements (e.g. optical filters, monochromators, etc.) that are used with the device. At room temperature, silicon is characterized by a bandgap of 1.12 eV, corresponding to a wavelength of approximately 1100 nm [53]. For a Si photodiode that is applied without any input optical filters, the device will be responsive to radiation with wavelengths ranging from 200 nm $\lesssim \lambda \lesssim$ 1100 nm. Heating of these devices results in a shift of spectral response toward longer wavelengths ("red" shift), while cooling results in a shift toward shorter wavelengths ("blue" shift) [54].

Si photodiodes are characterized by linearity over as many as ten decades of input radiation power [4]. In addition, they tend to provide rapid response and be characterized by low noise and a long lifetime [53]. For these and other reasons, Si photodiodes are commonly used in laboratories and in field applications.

### 5.3.5 Spectroradiometer

A spectroradiometer is an instrument that is used to measure the spectral power characteristics of a radiation source. Common output from a spectroradiometric instrument is spectral irradiance ($E_\lambda[=]\frac{\mathrm{W}}{\mathrm{m}^2 \cdot \mathrm{nm}}$). As described previously, irradiance ($E$) is defined as: [55] "radiant power, P, of all wavelengths incident from all upward directions on a small element of surface containing the point under consideration divided by the area of the element." For comparison, spectral irradiance ($E_\lambda$) is defined as the derivative of irradiance with respect to wavelength:

$$E_\lambda = \frac{\mathrm{d}E(\lambda)}{\mathrm{d}\lambda} \tag{5.48}$$

An alternative explanation, and one that may be more intuitive, is that spectral irradiance is the irradiance associated with a narrow wavelength band. In fact, this explanation is consistent with the method by which many spectroradiometers produce a signal. Specifically, spectroradiometers often measure irradiance received within a large number of narrow wavelength bands.

Spectroradiometers are systems that comprise several common components. First, the system will include a set of input optics, which allow spectral and/or spatial conditioning of the imposed radiation before it enters the system. Input optics may include lenses or optical filters to accomplish this conditioning. The conditioned radiation is then imposed on a monochromator, which allows wavelength separation of the radiation. Most contemporary spectroradiometers use a diffraction grating to accomplish this task, but a prism could also be used. The monochromator will also include a device to allow application of a narrow wavelength band on subsequent optical elements.

After leaving the monochromator, radiation will then encounter a detector; common examples include thermopiles, PMTs, photodiodes, or CCD arrays. However, essentially any device that can detect photons could be used, including emerging devices (see discussion below regarding fluorescence-based devices) or even a chemical actinometer. Spectroradiometers will also include a computer interface, which is used to accomplish instrument control and for data acquisition.

As with other optical instruments, the components of a spectroradiometer should be chosen to be consistent with the intended application. Of particular importance is selection of components to allow effective detection of photons in the wavelength range of interest.

### 5.3.6 Micro Fluorescent Silica Detector (MFSD)

Historically, detectors used for measurement of UV radiation have been based on the principles of thermoelectric conversion (e.g. thermopiles), the photoelectric effect (e.g. PMTs, vacuum photodiodes), or photoconductivity (e.g. solid-state photodiodes). These devices allow rapid, effective measurement, but suffer from some important limitations, including size and geometric constraints. In principle, other aspects of the physics and chemistry of UV radiation could be employed for these measurements, but the development of instruments and methods based on these processes has proceeded at a modest rate.

A notable exception to this trend was initiated in 2011 by a group in China who presented the results of development of a new detector based on the principles of fluorescence (luminescence) [56]. The micro fluorescent silica detector (MFSD) is a small (length = 1.0 mm, outside diameter = 0.3 mm), germanium (Ge)-doped silica cylinder that provides a bright fluorescence response when exposed to UVC radiation. The Ge-doped silica material was prepared by chemical vapor deposition, yielding a product that contains roughly 10% Ge. The resulting material displays a strong fluorescence response for wavelengths in the range $210\ \text{nm} \lesssim \lambda \lesssim 280\ \text{nm}$. The fluorescence signal is transmitted by an optical fiber to a diode array for quantification.

The absence of a response for radiation with $\lambda \gtrsim 280$ nm indicates that the MFSD is nonresponsive to room lighting, which facilitates its application when room light or ambient sunlight are present or otherwise needed in the experiment. The device displays a linear relationship between excitation UV fluence rate and emission fluorescence, and is efficient at converting input UVC photons into fluorescence photons at longer wavelengths, with roughly 48% conversion at an input wavelength of 254 nm. The luminescent response of the MFSD is temperature-sensitive [56], with a 0.11% increase in signal strength per °C for $0^{\circ}\text{C} \leq T \leq 55^{\circ}\text{C}$.

The small size of the MFSD and its other characteristics have allowed it to be applied in a wide range of investigations of fluence rate fields in UV photoreactors. For example, it has been used to measure the fluence rate field in monochromatic [56] and polychromatic [57] UV photoreactors. It has been used to quantify effects of reflective surfaces on the fluence rate field in an operating UV photoreactor [58]. By coupling with numerical simulations of reactor behavior by computational fluid dynamics and fluence rate field modeling (CFD-I models, see Chapter 7), the online measurements of a MFSD were used to monitor the dose distribution delivered by a UV photoreactor [59]. It has also been used to quantify the effects of turbidity on the fluence rate field [60].

## 5.4 Summary

Photochemical processes are governed by the rate at which active photons are imposed on a system. As such, it is necessary to have methods that allow reliable measurements of the fluence rate of radiation imposed on a system. The most common methods available to conduct these measurements involve chemical actinometry and radiometry.

Chemical actinometers are chemical systems that have well-defined mechanisms of reactions, including accurate descriptions of their molar absorption coefficient and the quantum yield for their reaction. Applications of chemical actinometers must account for the physical, chemical, and optical phenomena that govern photon delivery to the actinometer. Chemical actinometers can be applied to entire photochemical reactor systems or to specific locations within them. A wide range of chemicals has been demonstrated for use as chemical actinometers. Selection of these chemicals or chemical systems should consider the characteristics of the radiation being measured, including its wavelength and the incident fluence rate. Chemical actinometers have an important advantage of being highly repeatable; this allows their use as primary standards for measurements of radiation. They are normally applied for quantification of incident fluence rate or fluence (dose) over a well-defined period of exposure, but can be adapted to online, near real-time measurements as well.

Radiometers are physical systems that are used to quantify incident fluence rate or fluence (dose). They are based on one or more physical measurements, such as a thermal response (i.e. temperature change), an electrical response (e.g. current or voltage), or an optical response (e.g. fluorescence or absorbance). Radiometric devices often provide rapid responses, which facilitate their use in real-time measurement and monitoring applications. These devices are often applied in a small space, sometimes approaching a point in a system.

## Notes

**a** Note that for a collimated beam, the values of fluence rate and irradiance will be identical.
**b** The square brackets [ ] are used to indicate a solvent cage.
**c** The work function is defined as the energy required to displace an electron from a solid medium to a surrounding vacuum.
**d** Examples of other common photodiodes include Germanium (Ge) and Silicon Carbide (SiC).

## References

1 Kuhn, H.J., Braslavsky, S.E., and Schmidt, R. (**2004**). Chemical actinometry. *Pure and Applied Chemistry 76* (12): 2105–2146.

2 Bolton, J.R., Stefan, M.I., Shaw, P.S., and Lykke, K.R. (**2011**). Determination of the quantum yields of the potassium ferrioxalate and potassium iodide-iodate actinometers and a method for the calibration of radiometer detectors. *Journal of Photochemistry and Photobiology a-Chemistry 222* (1): 166–169.

3 Blatchley III, E.R. (**1997**). Numerical modelling of UV intensity: Application to collimated-beam reactors and continuous-flow systems. *Water Research 31* (9): 2205–2218.

4 Ryer, A. (**1997**). *The Light Measurement Handbook*. Peabody, MA: International Light Technologies.

5 Edlén, B. (**1966**). The refractive index of air. *Metrologia 2* (2): 71–80.

6 Schiebener, P., Straub, J., Sengers, J., and Gallagher, J.S. (**1990**). Refractive-index of water and steam as a function of wavelength, temperature and density. *Journal of Physical and Chemical Reference Data 19* (3): 677–717.

7 Malitson, I.H. (**1965**). Interspecimen comparison of refractive index of fused silica. *Journal of the Optical Society of America 55* (10P1): 1205.

8 Polyanskiy, M. N. (**2008-2022**). Refractive index database. https://refractiveindex.info.

9 Schiebener, P., Straub, J., Sengers, J., and Gallagher, J.S. (**1990**). Refractive-index of water and steam as function of wavelength, temperature and density. *Journal of Physical and Chemical Reference Data 19* (3): 677–717.

10 Zepp, R.G. (**1978**). Quantum yields for reaction of pollutants in dilute aqueous solution. *Environmental Science & Technology 12* (3): 327–329.

11 Wegelin, M.; Canonica, S.; Mechsner, K.; Fleischmann; T.; Pesaro, F.; Metzler, A., Solar water disinfection: scope of the process and analysis of radiation experiments. *Journal of Water Supply: Research and Technology - AQUA* **1994,** *43*, (3), 154–169.

12 Canonica, S., Meunier, L., and Von Gunten, U. (**2008**). Phototransformation of selected pharmaceuticals during UV treatment of drinking water. *Water Research 42* (1–2): 121–128.

13 Canonica, S., Jans, U., Stemmler, K., and Hoigne, J. (**1995**). Transformation kinetics of phenols in water - photosensitization by dissolved natural organic material and aromatic ketones. *Environmental Science & Technology 29* (7): 1822–1831.

14 Chu, W. and Jafvert, C.T. (**1994**). Photodechlorination of polychlorobenzene congeners in surfactant micelle solutions. *Environmental Science & Technology 28* (13): 2415–2422.

15 Leighton, W.G. and Forbes, G.S. (**1930**). Precision actinometry with uranyl oxalate. *Journal of the American Chemical Society 52*: 3139–3152.

16 Carter, A.H. and Weiss, J. (**1940**). The transfer of excitation energy from uranium ions in solution. *Proceedings of the Royal Society of London Series a-Mathematical and Physical Sciences 171* (A958): 0351–0370.

17 Volman, D.H. and Seed, J.R. (**1964**). Photochemistry of uranyl oxalate. *Journal of the American Chemical Society 86* (23): 5095.

18 Pitts, J.N., Margerum, J.D., Taylor, R.P., and Brim, W. (**1955**). Determination of light intensities with a highly sensitive uranyl oxalate actinometer. *Journal of the American Chemical Society 77* (21): 5499–5501.

19 Parker, C.A. (**1953**). A new sensitive chemical actinometer .1. Some trials with potassium ferrioxalate. *Proceedings of the Royal Society of London Series a-Mathematical and Physical Sciences 220* (1140): 104–116.

20 Goldstein, S. and Rabani, J. (**2008**). The ferrioxalate and iodide-iodate actinometers in the UV region. *Journal of Photochemistry and Photobiology a-Chemistry 193* (1): 50–55.

21 Hatchard, C.G. and Parker, C.A. (**1956**). A new sensitive chemical actinometer. 2. Potassium ferrioxalate as a standard chemical actinometer. *Proceedings of the Royal Society of London Series a-Mathematical and Physical Sciences 235* (1203): 518–536.

22 Shaw, P.S., Lykke, K.R., Gupta, R. et al. (**1999**). Ultraviolet radiometry with synchrotron radiation and cryogenic radiometry. *Applied Optics 38* (1): 18–28.

23 Houston, J.M. and Rice, J.P. (**2006**). NIST reference cryogenic radiometer designed for versatile performance. *Metrologia 43* (2): S31–S35.

24 Jortner, J., Ottolenghi, M., and Stein, G. (**1964**). On photochemistry of aqueous solutions of chloride bromide + iodide ions. *Journal of Physical Chemistry 68* (2): 247.

25 Rahn, R.O. (**1993**). Use of potassium-iodide as a chemical actinometer. *Photochemistry and Photobiology 58* (6): 874–880.

26 Rahn, R.O. (**1997**). Potassium iodide as a chemical actinometer for 254 nm radiation: use of iodate as an electron scavenger. *Photochemistry and Photobiology 66* (4): 450–455.

27 Dainton, F.S. and Logan, S.R. (**1965**). Primary processes in photolysis of iodide ion in aqueous solution. *Proceedings of the Royal Society of London Series a-Mathematical and Physical Sciences 287* (1410): 281.

28 Rahn, R.O., Stefan, M.I., Bolton, J.R. et al. (**2003**). Quantum yield of the iodide-iodate chemical actinometer: dependence on wavelength and concentration. *Photochemistry and Photobiology 78* (2): 146–152.

29 Rahn, R.O. (**2013**). Fluence measurements employing iodide/iodate chemical actinometry as applied to upper-room germicidal radiation. *Photochemistry and Photobiology 89* (4): 816–818.

30 Rahn, R.O. (**2004**). Spatial distribution of upper-room germicidal UV radiation as measured with tubular actinometry as compared with spherical actinometry. *Photochemistry and Photobiology 80* (2): 346–350.

31 Rahn, R.O., Xu, P., and Miller, S.L. (**1999**). Dosimetry of room-air germicidal (254 nm) radiation using spherical actinometry. *Photochemistry and Photobiology 70* (3): 314–318.

32 Pennell, K.G., Naunovic, Z., and Blatchley III, E.R. (**2008**). Sequential inactivation of bacillus subtilis spores with ultraviolet radiation and iodine. *Journal of Environmental Engineering-Asce 134* (7): 513–520.

33 Pennell, K.G. (**2005**). *Sequential Water Disinfection Using UV Irradiation and Iodination for Long-Term Space Missions*. West Lafayette, IN: Purdue University.

34 Ciamician, G. and Silber, P. (**1900**). Chemische Lichtwirkungen. *Berichte der Deutschen Chemischen Gesellschaft 33*: 2911–2913.

35 Bouya, H., Al Rashidi, M., Roth, E. et al. (**2017**). Atmospheric degradation of 2-nitrobenzaldehyde: photolysis and reaction with OH radicals. *Atmospheric Environment 171*: 221–228.

36 Galbavy, E.S., Ram, K., and Anastasio, C. (**2010**). 2-Nitrobenzaldehyde as a chemical actinometer for solution and ice photochemistry. *Journal of Photochemistry and Photobiology a-Chemistry 209* (2–3): 186–192.

37 Pitts, J.N., Wan, J.K.S., and Schuck, E.A. (**1964**). Photochemical studies in alkali haliide matrix .I. O-Nitrobenzaldehyde actinometer + its application to kinetic study of photoreduction of benzophenone by benzhydrol in pressed potassium bormide disk. *Journal of the American Chemical Society 86* (18): 3606.

38 Leighton, P.A. and Lucy, F.A. (**1934**). The photoisomerization of the o-Nitrobenzaldehydes I. Photochemical results. *Journal of Chemical Physics 2* (11): 756–759.

39 Rowland, G.A. and Grannas, A.M. (**2011**). A solid-phase chemical actinometer film for measurement of solar UV penetration into snowpack. *Cold Regions Science and Technology 66* (2–3): 75–83.

40 Wang, S.Y. (**1962**). Analysis of the rate of the ultraviolet irradiation and reconstitution reactions of 1,3-dimethyluracil and uridine. *Photochemistry and Photobiology 1* (2): 135–145.

41 Zhang, J.Y., Boyd, I.W., and Esrom, H. (**1997**). UV intensity measurement for a novel 222 nm excimer lamp using chemical actinometer. *Applied Surface Science 109*: 482–486.

42 von Sonntag, C. and Schuchmann, H.P. (**1992**). UV disinfection of drinking water and by-product formation: some basic considerations. *Journal of Water Supply: Research and Technology - AQUA 41* (2): 67–74.

43 Linden, K.G. and Darby, J.L. (**1997**). Estimating effective germicidal dose from medium pressure UV lamps. *Journal of Environmental Engineering ASCE 123* (11): 1142–1149.

44 Jin, S., Mofidi, A.A., and Linden, K.G. (**2006**). Polychromatic UV fluence measurement using chemical actinometry, biodosimetry, and mathematical techniques. *Journal of Environmental Engineering ASCE 132* (8): 831–841.

45 Che, Y.S., Huang, J.S., Barnard, W., and Li, Y.H. (**1995**). Photoinduced dimerization of thymine in microemulsion for UVB radiation exposure measurements. *Analytica Chimica Acta 318* (1): 103–112.

46 Bergstrom, D.E., Inoue, H., and Reddy, P.A. (**1982**). Pyrido 2,3-D pyrimidine nucleosides - synthesis via cyclization of C-5-substituted cytidines. *Journal of Organic Chemistry 47* (11): 2174–2178.

47 Shen, C.Y., Fang, S.Y., Bergstrom, D.E., and Blatchley III, E.R. (**2005**). (E)-5- 2-(Methoxycarbonyl) ethenyl -cytidine as a chemical actinometer for germicidal UV radiation. *Environmental Science & Technology 39* (10): 3826–3832.

48 Gentile, T.R., Houston, J.M., and Cromer, C.L. (**1996**). Realization of a scale of absolute spectral response using the national institute of standards and technology high-accuracy cryogenic radiometer. *Applied Optics 35* (22): 4392–4403.

49 Gentile, T.R., Houston, J.M., Hardis, J.E. et al. (**1996**). National Institute of Standards and Technology high-accuracy cryogenic radiometer. *Applied Optics 35* (7): 1056–1068.

50 Martin, J.E., Fox, N.P., and Key, P.J. (**1985**). A cryogenic radiometer for absolute radiometric measurements. *Metrologia 21* (3): 147–155.

51 Hamamatsu Photonics (**2007**). *Photomultiplier Tubes: Basics and Applications*. Hamamatsu Photonics.

52 Hamamatsu Photonics. Photocathode technology. https://www.hamamatsu.com/eu/en/technology/innovation/photocathode/index.html (accessed 2021).

53 Corporation, H *Opto Semiconductor Handbook*. Hamamatsu.

54 AP Technologies (**2022**). *Photodiode Theory of Operation*. AP Technologies https://www.aptechnologies.co.uk/support/SiPDs/operation.

55 Braslavsky, S.E. (**2007**). Glossary of terms used in Photochemistry 3(rd) Edition (IUPAC Recommendations 2006). *Pure and Applied Chemistry 79* (3): 293–465.

56 Li, M.K., Qiang, Z., Li, T.G. et al. (**2011**). In situ measurement of uv fluence rate distribution by use of a micro fluorescent silica detector. *Environmental Science & Technology 45* (7): 3034–3039.

57 Li, M.K., Qiang, Z.M., Wang, C. et al. (**2017**). Experimental assessment of photon fluence rate distributions in a medium-pressure uv photoreactor. *Environmental Science & Technology 51* (6): 3453–3460.

58 Li, W.T., Li, M.K., Bolton, J.R. et al. (**2017**). Impact of inner-wall reflection on UV reactor performance as evaluated by using computational fluid dynamics: The role of diffuse reflection. *Water Research 109*: 382–388.

59 Li, M.K., Li, W.T., Qjang, Z.M., and Blatchley III, E.R. (**2017**). On-site determination and monitoring of real-time fluence delivery for an operating UV reactor based on a true fluence rate detector. *Environmental Science & Technology 51* (14): 8094–8100.

60 Li, M.K., Li, W.T., Wen, D. et al. (**2017**). Experimental evaluation of turbidity impact on the fluence rate distribution in a UV reactor using a microfluorescent silica detector. *Environmental Science & Technology 51* (22): 13241–13247.

# 6

# Numerical Models for Simulation of Photochemical Reactor Behavior

## 6.1 Introduction

UV photoreactors are used for a wide range of applications. In water treatment, they are used for disinfection, direct photolysis, advanced oxidation processes (AOPs), advanced reduction processes (ARPs), and combinations of these endpoints. For gas-phase systems, the most common applications involve disinfection, but as with systems used in liquid-phase applications, some gas-phase applications involve direct photolysis, promotion of free radical formation, or combinations.

In a general sense, the physical, chemical, and microbiological phenomena involved in these processes are reasonably well understood, but some are quite complicated. And given that the behavior of many of these systems depends on multiple (often-complicated) phenomena, the overall simulation of these systems can become quite challenging.

Despite these complicating factors, or perhaps because of them, mathematical (numerical) models have evolved for simulation of the performance of photochemical reactor systems; the accuracy of these models depends on the degree to which they capture the relevant physics, as well as other factors that relate to model implementation.

The logic behind these models has been presented in general terms previously (see Figure 3.22). Following this logic, the behavior of a photochemical reactor system is governed by:

- **Influent Composition:** In a process involving a purely photochemical endpoint, effluent (treated fluid) concentration of a photochemical target is directly related to influent (untreated fluid) concentration of the same target. When a photochemical process yields reactive intermediates, then the concentrations of other constituents in the system can also influence effluent quality. Nontarget constituents can also "compete" for photons from the radiation source(s), or for reactive intermediates that are formed in these processes, both of which will affect overall reaction progress.
- **Intrinsic Kinetics:** The intrinsic kinetics of a photochemical reaction define the rate of change of the concentration of the target as a function of the rate of photon absorption by the target. The fundamental response of a photoactive compound to exposure to UV radiation will be governed by its physico/chemical properties, including the molar absorption coefficient and the quantum yield. However, the apparent kinetics of photochemical processes will be influenced by the spatial distribution of radiant energy (i.e. fluence rate field). If reactive intermediates are generated by these photochemical processes, then rate expressions (with relevant rate constants) can be

*Photochemical Reactors: Theory, Methods, and Applications of Ultraviolet Radiation*, First Edition. Ernest R. Blatchley III.

Companion website: www.wiley.com/go/blatchley/photochemicalreactors

used to describe progress with these secondary reactions. It is also relevant that at the scale of a point in space, the kinetics of any pure photochemical reaction are first-order with respect to reactant concentration (or more formally activity) and the local fluence rate.

- **Dose Distribution:** The nature of contemporary photochemical reactor systems is such that there is a distribution of UV doses applied to photochemical targets as they traverse the irradiated zone of a photochemical reactor. This dose distribution is governed by the fluence rate distribution and fluid mechanical behavior within the reactor.

In a broader sense, influent composition is also influenced by upstream processes that define the fluid being treated as well as other processes that may affect composition. The intrinsic kinetics of photochemical reactions and secondary reactions can be defined using methods that have been presented previously (see Chapters 2 and 3). Therefore, the focus of this chapter will be on the mathematical (numerical) models that are used to simulate fluence rate fields and fluid mechanics in UV photoreactor systems.

With this in mind, the objectives for this chapter are as follows. First, the logic used to develop fluence rate ($E'$) field models is presented, along with examples of several $E'$ field models that are commonly applied. Second, the logic used in the development of models used to simulate fluid mechanical behavior is presented, along with derivations of the fundamental governing equations. The focus of this section will be on computational fluid dynamics (CFD), as applied in the simulation of UV photoreactors. Third, the use of integrated CFD-$E'$ field models will be presented for common applications. This chapter will conclude with general information to address the selection of model components for simulations of UV photoreactor performance.

## 6.2 Fluence Rate ($E'$) Field Models

The models that are used to describe the spatial distribution of radiant energy around one or more sources of UV radiation include two basic components:

- A sub-model to describe the physics of photon emission, including source geometry.
- A sub-model to describe optical behavior in the irradiated media that comprise the system.

These model components are described below, followed by summaries of fluence rate field modeling approaches that have employed these components.

### 6.2.1 Photon Emission Sub-Models

The concept of a point source is central to the logic and behavior of many fluence rate field models. In the limit of small size (i.e. as the size of a hypothetical radiation source approaches zero), a radiation source approaches the geometry of a point; the "point source" can be used as a building block for many sources of radiation. In other words, a real source of radiation can be simulated as a large number of point sources. With that in mind, the physics of radiation emission and transmission from a point source is foundational to many fluence rate field models.

Figure 6.1 provides schematic illustrations of four photon emission sub-models that have been applied in fluence rate field models. The schematic illustrations presented in Figure 6.1 are based on a cylindrical, low-pressure Hg lamp as the source of radiation. Analogous point-source models could be developed for other UV source types.

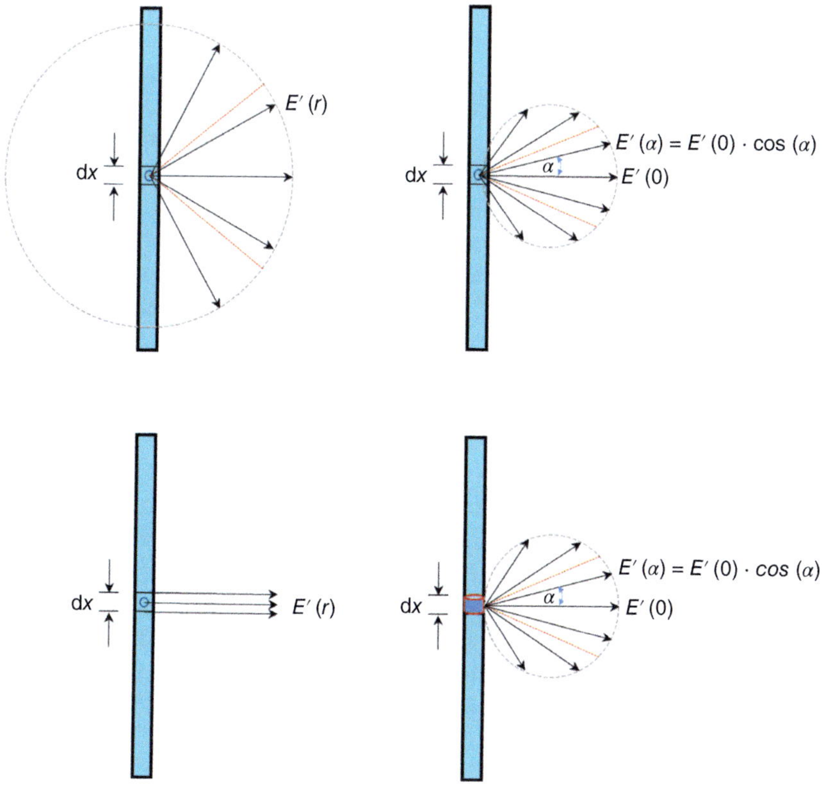


**Figure 6.1** Schematic illustrations of sub-models used to account for photon emission from sources of UV radiation. All four models are presented in a form that would be used to simulate output from a cylindrical mercury lamp, but variations on these models could be used for other source types and source geometries. **Upper left:** isotropic emission model, wherein radiation is emitted uniformly in all directions. Overall, the radiation source (lamp) can be simulated as a series of collinear point sources aligned along the central axis of the lamp; fluence rate contributions from each point source are governed by distance from each point source to the receptor site. **Upper right:** Lambertian emission model; fluence rate distributions from each point source demonstrate cosine-dependence. **Lower left:** radial emission model; photons from each point source are assumed to be emitted perpendicular to the axis of the radiation source. **Lower right:** segment-source model; linear array of differential cylindrical sources that emit from the surface of the cylindrical element; the emission pattern indicated in this diagram is Lambertian, but other emission patterns could be incorporated into the segment-source model. Directional restrictions can be incorporated into the isotropic and Lambertian models (see dashed red lines) to mimic emission patterns of actual sources; directional behavior is implied in the radial model.

The isotropic emission model (Figure 6.1, upper left) is based on the assumption that photons are emitted uniformly in all directions from all point sources that comprise the radiation source. For this example, point sources are located collinearly along the central axis of the lamp, but other spatial distribution patterns of point sources are possible. Fluence rate contributions from each point source vary as a function of distance (r) from the point source to the receptor site; the nature of the functional relationship between local fluence rate and r will depend on the physical (optical) phenomena that are included in the model (see "Sub-Models to Account for Optical Behavior" below). As with the isotropic emission model, the Lambertian emission model (Figure 6.1, upper right) is often applied with an assumption of point sources being located collinearly along the central axis of the lamp (again, other spatial distribution patterns of point sources are possible), but in addition to the distance-dependence used in the specular model, fluence rate contributions from each point source display cosine-dependence (i.e. fluence rate is proportional to the cosine of the angle of emission, relative to perpendicular). The radial emission model (Figure 6.1, lower left) works under the assumption that photon emission is always perpendicular to the axis of the source. The segment-source model assumes that a lamp can be simulated as a series of collinear, differential cylindrical sources; the emission pattern indicated in Figure 6.1, lower right, is Lambertian, but other emission patterns can be applied.

As described above, the radial model is based on an assumption that all radiation is emitted perpendicular to the source axis. This has the effect of placing a directional restriction on the output from the source. The spatial distribution of radiant energy from such a source will depend on the length of the source and its geometry.

In their most basic form, the isotropic and Lambertian models show no directional limitations (barriers). However, some UV sources have optical elements or packaging features that restrict or modify the direction of radiation propagation, usually by reflection, refraction, inclusion of an opaque surface, or some combination of these features. As such, it is sometimes appropriate to superimpose directional restrictions on specular and diffuse sources (see dashed red lines in Figure 6.1). In a numerical simulation of a fluence rate field, this can be accomplished by inclusion of terms to constrain output numerically, or by inclusion of terms to explicitly account for reflection or refraction.

### 6.2.2 Sub-Models to Account for Optical Behavior

The physical laws that govern the behavior of radiation at media interfaces and in absorbing media have been presented previously; however, those presentations were developed for specific applications, sometimes involving limiting circumstances or special cases. The following presentation is intended to be more general, and to expand on the application of these basic principles.

#### 6.2.2.1 Reflection and Refraction

Snell's law is used to describe the directional behavior of a beam of radiation imposed on an interface between two media with different refractive indexes. Figure 6.2 illustrates this basic behavior.

To reiterate (see equations 5.2 and 5.3), for a ray incident on an interface separating two media, Snell's law gives the relationship between the angles of incidence and refraction in terms of the refractive indexes of the two media.

$$n_1 \cdot \sin(\theta_1) = n_2 \cdot \sin(\theta_2) \tag{6.1}$$

where

$n_1$ = index of refraction for medium 1
$\theta_1$ = angle of incidence
$n_2$ = index of refraction for medium 2
$\theta_2$ = angle of refraction.

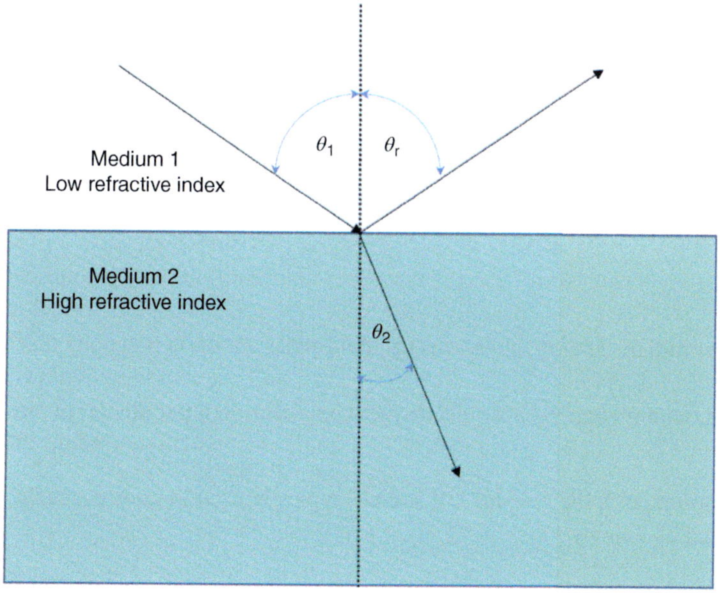


**Figure 6.2** Schematic illustration of the behavior of a ray of monochromatic radiation as it encounters an interface between two media with different refractive indexes. The direction of the refracted ray can be calculated by application of Snell's law. The fraction of radiation reflected and transmitted through the interface can be calculated by application of the Fresnel equations.

Equation 6.1 can be rearranged to allow definition of the angle of refraction as a function of the incident angle and the refractive indexes of the two media:

$$\theta_2 = \arcsin\left[\frac{n_1}{n_2} \cdot \sin(\theta_1)\right] \tag{6.2}$$

The law of reflection indicates that the angles of incidence and reflection are identical:

$$\theta_1 = \theta_r \tag{6.3}$$

The Fresnel equations enable calculation of the fraction of radiation energy that is reflected or transmitted through an interface. Two definitions are needed to address this behavior: [1]

$\rho$ = fraction of incident radiation reflected at an interface or discontinuity of refractive index, $n$ (reflectance)
$T$ = fraction of incident radiation transmitted through an interface or discontinuity (transmittance).

By definition:

$$\rho + T \equiv 1 \tag{6.4}$$

The magnitude of $\rho$ (and by extension, $T$) depends on polarization of radiant energy. The specific dependence is described as follows:

$$r_{\perp} = \frac{n_1 \cdot \cos\theta_1 - n_2 \cdot \cos\theta_2}{n_1 \cdot \cos\theta_1 + n_2 \cdot \cos\theta_2} \tag{6.5}$$

$$r_{\parallel} = \frac{n_2 \cdot \cos\theta_1 - n_1 \cdot \cos\theta_2}{n_1 \cdot \cos\theta_2 + n_2 \cdot \cos\theta_1} \tag{6.6}$$

where

$r_{\perp}$ = reflection coefficient of radiant energy whose electric field polarization is perpendicular ($\perp$) to the plane of incidence
$r_{\parallel}$ = reflection coefficient of radiant energy whose electric field polarization is parallel ($\parallel$) to the plane of incidence.

For randomly polarized radiation, as is the case for UV sources typically used in photoreactors, the reflectance can be calculated as follows:

$$\rho = \frac{1}{2}\left(r_{\perp}^2 + r_{\parallel}^2\right) \tag{6.7}$$

An important special case for this relationship applies for radiation imposed normal (perpendicular) to an interface or discontinuity. This special case, which was presented in the previous chapter (see equation 5.4), is derived here from the Fresnel equations. Specifically, for radiation imposed normal to a media interface, the angles of incidence and refraction are both zero (i.e. $\theta_1 = \theta_2 = 0$; $\cos\theta_1 = \cos\theta_2 = 1$). By substitution of this information into (equations 6.5–6.7), we find:

$$\rho = \frac{1}{2}\left[\left(\frac{n_1 - n_2}{n_1 + n_2}\right)^2 + \left(\frac{n_2 - n_1}{n_1 + n_2}\right)^2\right] = \frac{(n_1 - n_2)^2}{(n_1 + n_2)^2} \tag{6.7a}$$

More generally, the magnitude of $\rho$ depends on the angle of incidence and the indexes of refraction for the media that comprise the interface. For $E'$ calculations in most photochemical reactors, the most common refractive materials that will be involved in a reactor include air, quartz, and water.[a]

As a general approach, reflectance as a function of incident angle at media interfaces can be calculated as follows:

- At each wavelength, define values of refractive index ($n$) for the media that comprise the interface
- Use equation 6.2 to calculate the angle of refraction ($\theta_2$) as a function of the incident angle ($\theta_1$).
- Calculate the reflection coefficients ($r_{\perp}$ and $r_{\parallel}$) using equations 6.5 and 6.6.
- Calculate $\rho$ using equation 6.7.

Figure 6.3 illustrates the results of these calculations for air:quartz and quartz:water interfaces at three characteristic UVC wavelengths (220, 254, and 280 nm). As indicated in Chapter 5 (see Figure 5.3), refractive index is a weak function of wavelength for these media. The wavelength-

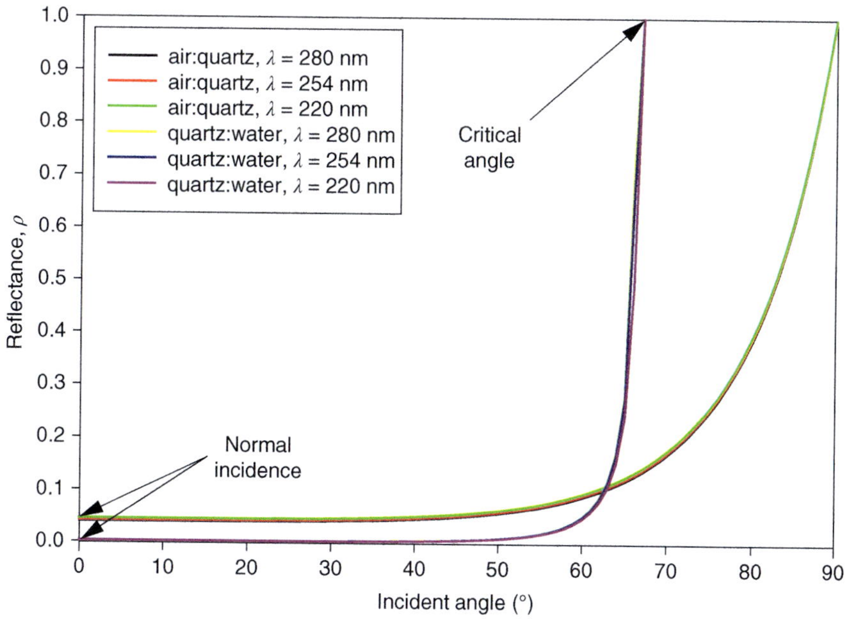


**Figure 6.3** Reflectance ($\rho$) as a function of incident angle for air:quartz and quartz:water interfaces. For each interface type, the weak dependence on wavelength is demonstrated by calculating $\rho$ at three representative wavelengths (220, 254, and 280 nm). Reflectance for radiation imposed normal to the interface is indicated by values corresponding to normal incidence (incident angle = 0°); the phenomenon of total internal reflection is demonstrated for the quartz:water interface by the existence of a finite critical angle.

dependence among these media translates to minimal changes in reflectance for these interfaces; to a first approximation, reflectance at these media interfaces does not demonstrate wavelength-dependence.

Reflectance for the special case of radiation imposed normal (perpendicular) to a media interface is given by equation 6.7a (also equation 5.4). In Figure 6.3, this corresponds to the situation of normal incidence. The values of reflectance calculated for these media interfaces are identical to those presented in Figure 5.11.

The concept of total internal reflection was also described in Chapter 5; it applies for circumstances where radiation is transmitted from a medium with relatively high refractive index to a medium with a relatively low refractive index (see Figure 5.5). In the context of the interfaces presented in Figure 6.3, this would apply for the quartz:water interface, but not for the air:quartz interface.[b] The critical angle for the quartz:water interface is approximately 66° at all three wavelengths evaluated in the UVC range. No critical angle is evident for the air:quartz interface because total internal reflection is not observed with this interface type.

Note that for a given ray path that encounters an interface, the fractions of reflected and transmitted radiation at the interface are the same regardless of whether the radiation is traveling from medium 1 to medium 2, or vice versa. Also, if $n_2 > n_1$ (i.e. no total internal reflection), an incident angle approaching 90° results in reflectance approaching unity (as seen in Figure 6.3 for air:quartz).

#### 6.2.2.2 Divergence/Dissipation

The effects of divergence (dissipation) (spreading with distance) were presented in Chapter 5 for the case of an isotropic emitter. The text in this section is intended to allow a broader, more general view of the effects of divergence on the fluence rate field that surrounds a point source, especially in the context of a radiation source in a photochemical reactor.

Figure 6.4 illustrates the effects of divergence on the fluence rate field surrounding a point source for isotropic and Lambertian emission models in non-absorbing media. For the isotropic point source, the fluence rate decreases simply with the square of distance from the source. On the other hand, for the Lambertian emitter, cosine-dependence must also be included in the calculation. Note that for the illustration presented in Figure 6.4, when this modeling approach is applied to a linear source, such as a mercury lamp, the coordinate system presented in Figure 6.4 implies that $\alpha = 0$ corresponds to directions perpendicular to the lamp axis, whereas $\alpha = \pm 90°$ corresponds to a direction parallel to the lamp axis. In general, the integral of fluence rate over the (real or hypothetical) surface of surrounding a point source in a non-absorbing medium must equal the power of the point source.

#### 6.2.2.3 Combination of Absorption (Beer–Lambert Law) and Reflection

Equations 5.5–5.7 illustrate the application of the Beer–Lambert law for solutions that contain one or more absorbing constituents. However, in photochemical reactor systems, radiation will often traverse several absorbing media, which usually exist as distinct phases. Therefore, the application of the Beer–Lambert law requires that these effects be accounted for.

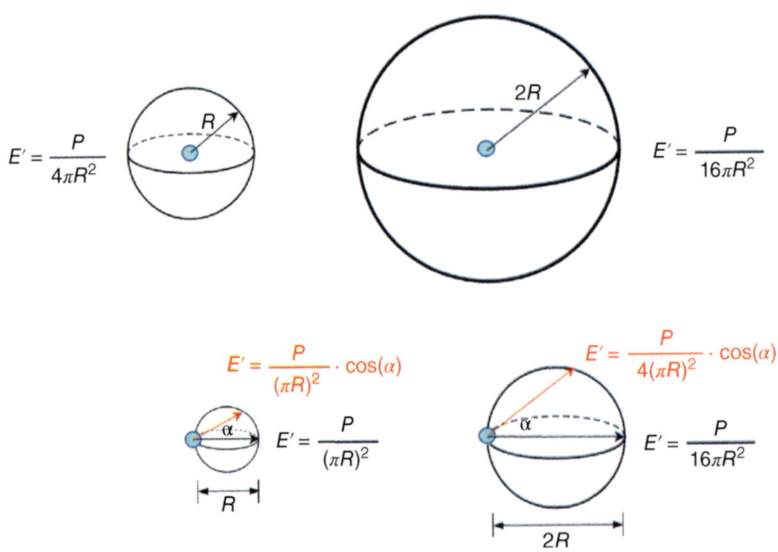


**Figure 6.4** Illustration of fluence rate field calculations in non-absorbing media around point sources for isotropic (top) and Lambertian (bottom) emitters. In the case of a isotropic point source, the fluence rate decreases with the square of distance ($R$) from the source. For a Lambertian emitter, cosine-dependence is imposed together with inverse-square dependence. The red arrow and red text are included to highlight the effects of cosine-dependence. Note that for the case of a Lambertian emitter, emission is illustrated only in one direction (i.e. to the right of the point source). When included in a model, emission from the "back side" of the point source could also need to be included.

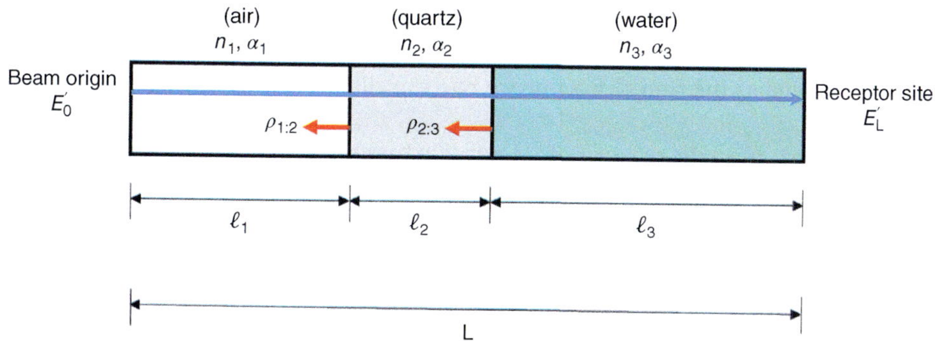


**Figure 6.5** Schematic illustration of the behavior of a collimated, monochromatic beam as it passes through three absorbing media.

For simplicity, we will consider the behavior of a single ray of collimated, monochromatic radiation; for such a source, no divergence will take place, thereby allowing us to focus entirely on the effects of absorption and reflection. Figure 6.5 illustrates the behavior of such a beam in a system where it is projected through three absorbing media: air, then quartz, then water. This sequence of media is common in UV photoreactors, though other media sequences can also be relevant and could be developed following similar logic.

The fluence rate along the beam will decrease exponentially according to the Beer–Lambert law within each absorbing medium. In addition, there will be reflection at each of the media interfaces. The scenario indicated in Figure 6.5 involves the radiation being imposed normal to each interface. The Fresnel equations are used to calculate reflectance at each media interface in the system. If radiation is imposed at a non-perpendicular angle, then the general forms of the Fresnel equations will need to be applied to calculate the reflectance (or transmittance), the angle of reflection, and the angle of refraction.

For the basic situation involved in Figure 6.5, the beam is assumed to start with a fluence rate of $E_0'$. After projection through the three media, the fluence rate at the receptor site $\left(E_L'\right)$ is:

$$E_L' = E_0' \cdot (1-\rho_{1:2}) \cdot (1-\rho_{2:3}) \cdot \exp(-\alpha_1\ell_1) \cdot \exp(-\alpha_2\ell_2) \cdot \exp(-\alpha_3\ell_3) \tag{6.8}$$

In equation 6.8, the effects of reflection at both media interfaces are included; also included are the effects of absorption through all three media. If additional interfaces or absorbing media are involved, then terms would need to be included to account for them.

### 6.2.3 Fluence Rate Field Models

Fluence rate field models are central to numerical simulations of photochemical reactor performance. They represent tools to accomplish spatial mapping of fluence rate within the irradiated zone of photochemical reactors. As with many model types, a wide range of fluence rate field models exist. The text below presents several fluence rate field models that are commonly applied, presented roughly in order of increasing complexity.

#### 6.2.3.1 Point-Source Summation/Line-Source Integration (PSS/LSI)

The point-source summation (PSS) model is based on the assumption that a tubular (cylindrical) UV radiation source can be simulated as an array of collinear point sources, which collectively emit the same power as the lamp being simulated [2]. The logic behind this model requires the development of an algorithm to describe the spatial distribution of radiant energy around a single point source of known output power and location. The fluence rate received at any location is estimated as the sum of the fluence rate contributions from all point sources in the system.

Among the factors that affect the accuracy of a simulation based on the PSS approach is the number of point sources used to simulate a UV source. The line-source integration (LSI) model was developed based on an assumption that a UV source could be simulated as a collinear array of point sources of infinitesimally small size [3]. Mathematically, the LSI model is equivalent to the PSS model in the limit as the number of point sources ($n$) approaches infinity.

The logic used in the development of the PSS and LSI models is illustrated in Figure 6.6. In the common form of the model illustrated in this figure, a set of collinear point sources is used to simulate the output from a lamp that includes a cylindrical plasma, as is found in most mercury lamps.

The PSS model is based on the assumption of isotropic emission by each point source. The model accounts for divergence and absorption, but does not include terms to account for reflection and refraction. The foundation of the model is an expression to describe the change in fluence rate with distance from a point source, based on the combined effects of divergence and absorption. At any location within the system, an estimate of the fluence rate is developed as the sum of the fluence rate contributions from all point sources in the system. The LSI model is based on identical logic, but the summation term is evaluated in the limit as $n \to \infty$ by evaluating the model in the limit as $dZ \to 0$ and applying integration, rather than summation.

The mathematical descriptions of the effects of divergence and absorption are essentially identical in the PSS and LSI models. Divergence is accounted for in the first term in the summation (PSS) or integrand (LSI) as the ratio of point-source power ($P/n$ or $P \cdot dZ/L$, respectively) to the surface area of a sphere of radius $\rho_i$ or $\rho$. Absorption is accounted for by inclusion of a term that provides an application of the Beer–Lambert law to describe absorption by quartz and water. For both media,

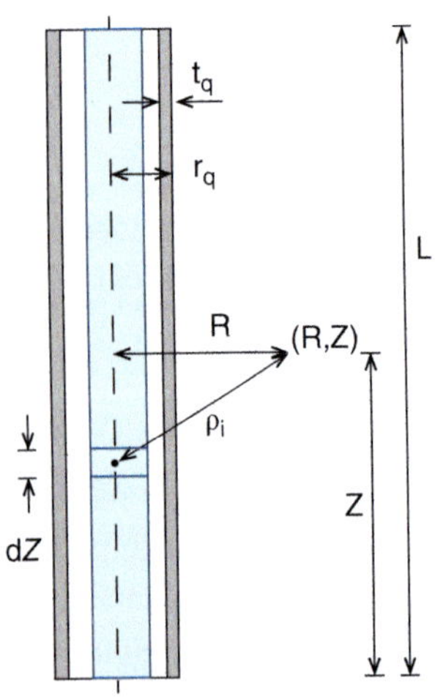


PSS

$$I(R, Z) = \sum_{i=1}^{n} \frac{P/n}{4\pi\rho_i^2} \cdot \exp\left[-\left(\alpha_q t_q + \alpha_w (R - r_q)\right) \frac{\rho_i}{R}\right]$$

LSI

$$I(R, Z) = \int_0^L \frac{P/L}{4\pi\rho^2} \cdot \exp\left[-\left(\alpha_q t_q + \alpha_w (R - r_q)\right) \frac{\rho}{R}\right] dz$$

**Figure 6.6** Schematic illustration of logic and terms used to describe the PSS and LSI models. This application of the model is presented to simulate the output from a lamp with a cylindrical plasma (light blue shaded region) housed within a quartz sleeve. The geometry of the lamp is simplified to that of a set of collinear point sources arranged along the axis of the lamp. In the PSS model, n point sources are used to simulate lamp output. In the LSI model, the "length" of each point source becomes differential, thereby allowing $n \to \infty$. The mathematical forms of the PSS and LSI models are indicated at right.

the optical path length is calculated as the product of the radial length of the media, multiplied by a term $\frac{\rho_i}{R}$ or $\frac{\rho}{R}$, using the logic of similar triangles.

A key to the application of the PSS and LSI models is definition of the term $\rho_i$ or $\rho$. The use of the model requires that these terms be written in a general manner based on other geometric variables that are used to describe the system. For the geometric variables described in Figure 6.6, this implies that $\rho_i$ or $\rho$ be defined in terms of R, Z, and L.

The PSS and LSI models are relatively simple and can be programmed into common spreadsheet software or mathematics packages. The PSS approach has been commonly applied to radiation sources with a cylindrical geometry, but the same logic can be applied to other source geometries. Examples of spreadsheet formulations of the PSS model are presented in Appendix 6.A.

The predictions of the PSS and LSI models have been demonstrated to be accurate in the far-field (i.e. relatively distant from the radiation source), but deviate from measured values in the near-field (i.e. relatively close to the radiation source) [3]. In addition, predictions from these models were found to deviate from goniometric measurements (in which irradiance is measured in various directions from the lamp) in the limit as the irradiance angle, $\theta$, approached 90° (see Figure 6.7) [4].

In the case of the near-field measurements, sources of error can be identified for both the PSS and LSI models, as well as the methods used to measure fluence rate. Specifically, neither of the models (PSS or LSI) accounts for reflection or refraction; in addition, both models involve simplification of the source geometry from a cylindrical plasma to a collinear collection of point sources. In the case of mercury lamps, most are produced with opaque end caps, thereby preventing emission along their axis, which is not accounted for in common applications of these models. These attributes will lead to errors that are likely to be greatest in the near-field. In addition, most radiometer detectors are characterized by a signal that shows angular dependence, often being close to cosine-dependence. This will represent a source of error that will also increase in magnitude in the near-field.

For goniometric measurements, the assumptions of the PSS and LSI models are also likely to contribute to error in fluence rate predictions for locations off the ends of the lamps (i.e. $\theta \rightarrow 90°$). Again, this is related to the simplification of the source geometry to that of a collinear set of point sources.

Within the cylindrical plasma of Hg lamps, radiation will be emitted from excited atoms as they relax back to their ground state. However, mercury lamps are also characterized by strong reabsorption of these photons, leading to a condition in which the majority of photons that escape the lamp envelope are probably generated near the lamp surface. The specific reactions involved in this process include (see also Figure 4.6):

$$\text{Excitation} \quad \text{Hg}\,6(^1S_0) + h\nu \rightarrow \text{Hg}\,6(^3P_1) \tag{6.9}$$

$$\text{Emission} \quad \text{Hg}\,6(^3P_1) \rightarrow \text{Hg}\,6(^1S_0) + h\nu \tag{6.10}$$

These processes contribute to an apparent directional character for Hg lamps. For the locations characterized by $\theta \rightarrow 90°$, as defined in Figure 6.7, the fluence rate contributions from a lamp are likely to be quite small, even if the end of the lamp was produced from a UV-transparent material.

The PSS and LSI models will demonstrate singularities in the limit as the distance from the lamp axis approaches zero (i.e. close to the lamp axis) [5]. Although the specific locations at which these singularities emerge as a numerical problem are generally not relevant to reactor performance (i.e. irradiated fluids do not have access to these areas), the existence of this mathematical problem represents a limitation for the PSS and LSI models.

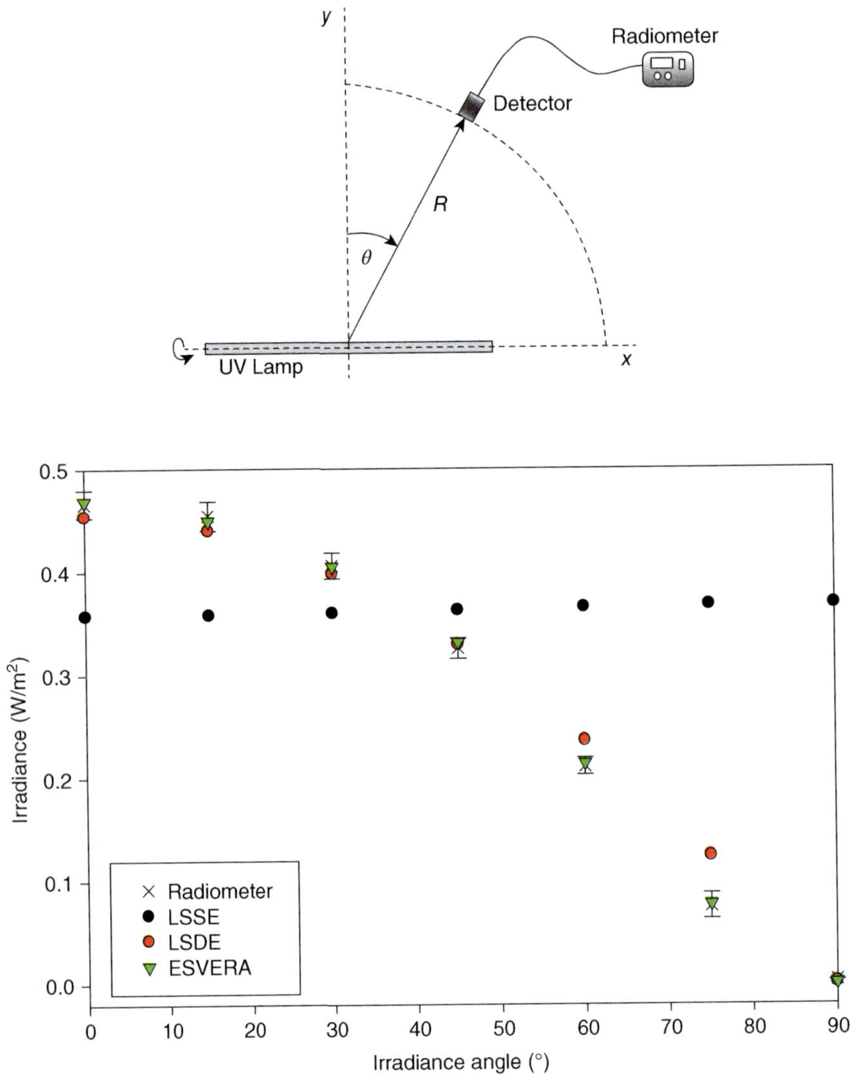


**Figure 6.7** (Top) Schematic illustration of setup used to conduct goniometric measurements; measurements from this setup were compared with simulations of fluence rate field models [4]. *Source:* Duran et. al 2010 / with permission of Elsevier. (Bottom) Comparison of measurements with predictions of numerical models. LSSE = line-source model with spherical (isotropic) emission; LSDE = line-source model with diffuse (Lambertian) emission; ESVERA = extensive source model with volumetric emission, accounting also for reflection, refraction, and absorption.

#### 6.2.3.2 Multiple Segment Source Summation (MSSS)

The multiple segment source summation (MSSS) model is based on an approach in which a cylindrical plasma is simulated as a collinear set of cylindrical elements of differential length (see Figure 6.1, lower right). It is commonly applied through the commercial software platform UVCalc® (Bolton Photosciences). Lamp output is simulated as originating from the surface of each differential cylindrical element. Emission from these surfaces is Lambertian (i.e. cosine-dependent). The commercial form of this model includes terms to account for reflection and refraction at the air:quartz and quartz:water interfaces; reflection at other material interfaces is not accounted for. The MSSS model, as implemented through UVCalc®, has been demonstrated to provide estimates of local fluence rate that are in better agreement with measurements by spherical actinometry and the micro-fluorescent silica detector (MFSD) than PSS or LSI models [4, 6–8].

In evaluating the agreement between model predictions and physical measurements, it is important to recognize the limitations and assumptions of the model and the measurements. Measurements of fluence rate by spherical actinometry are complicated by reflection and refraction that take place at the surfaces of the quartz spheres that are used to contain the actinometer solution. The MFSD is a small, cylindrical detector; its small size allows it to accept photons from essentially all incident angles (i.e. from across an entire solid angle of $4\pi$ steradians), but its cylindrical shape will yield error based on the formal definition of fluence rate, especially in the near-field.

As described above, the true behavior of a mercury plasma as a UV photon emitter is probably one in which isotropic emission occurs throughout the plasma as electronically excited mercury atoms relax to their ground state (equation 6.10). However, this process is counteracted by reabsorption of photons to allow promotion of ground-state mercury to an excited state (equation 6.9). The result of these two processes is that the majority of emitted photons probably originated by emission reactions that take place near the outside edge of the plasma. In that regard, the MSSS model appears to capture the physical behavior of these UV sources well.

The inclusion of a cosine-dependence term in the governing equation is probably not a precise description of the actual emission process; however, this term provides improvement in model predictions, relative to physical measurements, in comparison with model predictions based on isotropic emission. This is probably an empirical artifact of the details of the physics that govern lamp output, rather than a physically accurate description of these physics. Nonetheless, the inclusion of cosine-dependence, along with terms to account for reflection and refraction, yields model predictions that are in good agreement with reliable physical measurements. It has been noted that models that are based on isotropic emission throughout the plasma volume, together with terms to account for absorption, re-emission, reflection, and refraction yield model predictions that are closer to physical measurements than either PSS/LSI or MSSS [4, 9].

#### 6.2.3.3 Radiative Transfer Equation (RTE)

Remarkable parallels exist among mass, heat (energy), and momentum transport in terms of the actual processes that govern these phenomena, the logic used to solve problems in these areas, the mathematical expressions that are used to describe these phenomena, and the methods used to solve or develop approximate solutions to transport phenomena problems. However, the process of energy transport by the use of electromagnetic radiation is somewhat unique among transport phenomena, in that this mechanism of energy transport has no parallel in the fields of mass or momentum transport [10].

The radiative transfer equation (RTE) is used to describe energy transfer associated with electromagnetic radiation. It is often included as a part of an energy balance in the analysis of engineering systems. In the case of photochemical reactor systems, the RTE has been demonstrated to be effective for simulation of the spatial distribution of radiant energy.

The governing equation is presented below for the case of monochromatic radiation.

$$\underbrace{\frac{\mathrm{d}L_p\left(\vec{r},\vec{s}\right)}{\mathrm{d}s}}_{1} + \underbrace{(k+\sigma)L_p\left(\vec{r},\vec{s}\right)}_{2} = \underbrace{j^e\left(\vec{r}\right)}_{3} + \underbrace{\frac{\sigma}{4\pi}\int_{4\pi} L_p\left(\vec{r},\vec{s}\right)p\left(\vec{s}'\rightarrow\vec{s}\right)\mathrm{d}\Omega'}_{4} \tag{6.11}$$

where

$L_p$ = photon radiance[c]
$\vec{r}$ = position vector
$\vec{s}$ = propagation direction vector
$s$ = path length
$k$ = absorption coefficient
$\sigma$ = scattering coefficient
$j^e$ = emission (source) term
$p$ = phase function for in-scattering of photons
$\Omega'$ = solid angle about scattering direction vector ($\vec{s}'$).

The terms in equation 6.11 have the following physical interpretations:

1) Describes the change in photon radiance along the optical path $s$.
2) Describes the decrease of photon radiance that results from absorption and scattering away from the path.
3) Represents a radiation source. This term is usually zero, unless a source exists along a given path.
4) Describes the gain of photon radiance due to in-scattering.

Several numerical methods have been developed to solve the RTE, including Monte Carlo, Discrete Ordinate, and Finite Volume (FV) approaches. These methods are available as subroutines within many CFD packages and are convenient for use with coupled simulations of fluid mechanics and fluence rate fields.

Duran et al. implemented the RTE using a FV approach for simulation of fluence rate fields in several reactors with physical measurements [4]. These simulations were conducted using several photon emission models, including PSS, LSI, and MSSS. Their results indicated that cosine-dependence coupled with surface emission provides simulation results that are in good agreement with physical measurements, but that the agreement between these results was circumstantial (i.e. empirical), rather than an accurate description of the physics of photon emission from a mercury lamp.

In addition, they examined photon emission models that accounted for volumetric photon emission within a lamp plasma, together with absorption and re-emission. These models were found to provide incrementally better agreement with physical measurements than the other modeling approaches, probably because they simulate the actual physics of photon emission more closely than the other modeling approaches. In particular, they implemented a form of the RTE that accounted for the behavior of an extensive source with volumetric emission, in which the effects of reflection, refraction, and absorption within the plasma were accounted for (see ESVERA model predictions in Figure 6.7).

#### 6.2.3.4 Surface Power Apportionment for Cylindrical Excimer Lamps (SPACE)

As described above, the concept of a PSS can be applied broadly to simulate the output characteristics of essentially any source of radiation. In general, any logical grouping of point sources (i.e. consistent with geometry and output characteristics) can be used to describe the behavior of radiation sources with essentially any shape.

The Surface Power Apportionment for Cylindrical Excimer Lamps (SPACE) model was developed to allow simulation of the output characteristics of annular dielectric barrier discharge (DBD) excimer lamps (see Figure 4.14). The geometry of these sources is defined by two concentric cylindrical tubes made of a highly UV-transmissive dielectric material (e.g. quartz). A tubular

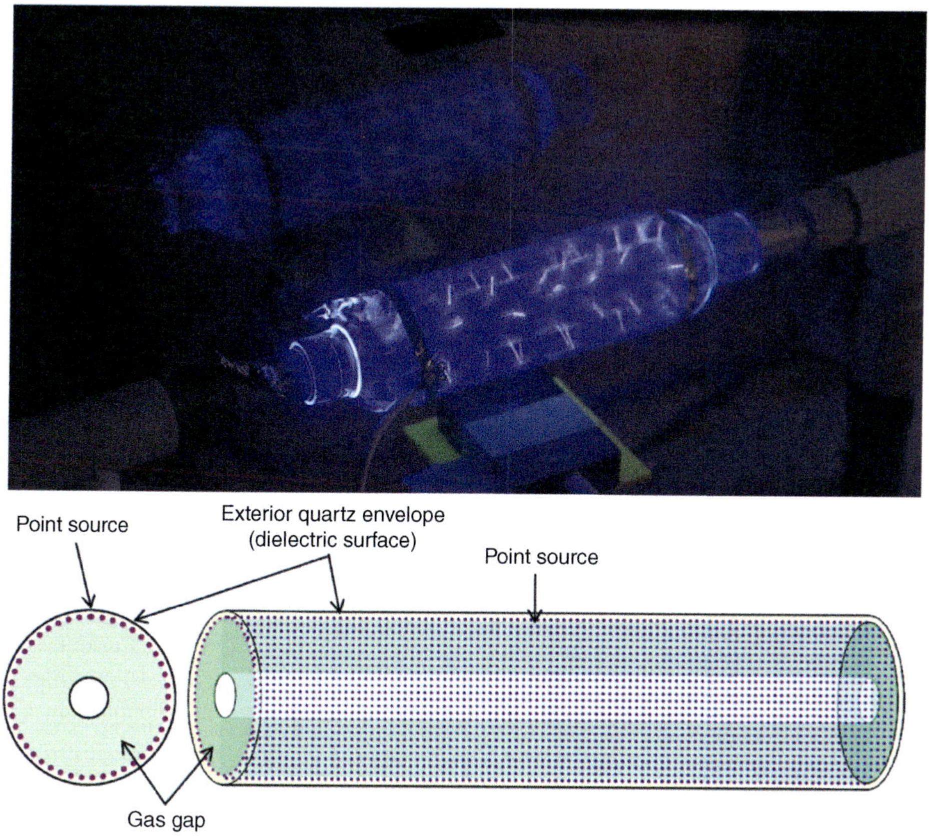


**Figure 6.8** (Top) Photographic image of DBD excimer lamp [11]. *Source:* Vanesa Rueda, Feb 2016/Master's Final Work: Optimum Transformer Turns Ratio of Power Supply for Dielectric Barrier Discharge Lamp. (Bottom) Schematic illustration of point-source approximation of DBD excimer lamp [12]. *Source:* Naunovic et al. 2008 / with permission of American Chemical Society

metallic electrode is installed inside the inner quartz tube, with an external wire mesh that serves as the second electrode. Filaments of electrical current (microdischarges) form between these electrodes, with termination near the wall of the annular space that is defined by the quartz tubes. Most of the electromagnetic radiation generated by these devices is released at or near the surface, at the terminus of each microdischarge. As a result of this geometry and physical behavior, the annular DBD excimer lamp closely resembles a collection of point-source discharges at the inside surface of the annular space (see Figure 6.8).

The SPACE model follows the point-source approach by including a set of point sources located on the inner surface of the outside quartz cylinder; point sources are uniformly distributed on this surface (see Figure 6.8). Each point source was assumed to behave as a Lambertian emitter.

The model followed a ray tracing approach (see below). Along each ray trajectory, the effects of absorption were accounted for; at each material interface, the effects of refraction and (specular) reflection were accounted for. For each point source in the system, as many as three pathways existed that could allow radiation to reach a given receptor site. These included a direct pathway that involved no reflection events, a pathway that involved specular reflection at the internal electrode surface, and a third pathway that involved specular reflection at the outer water of the reactor (see Figure 6.9).

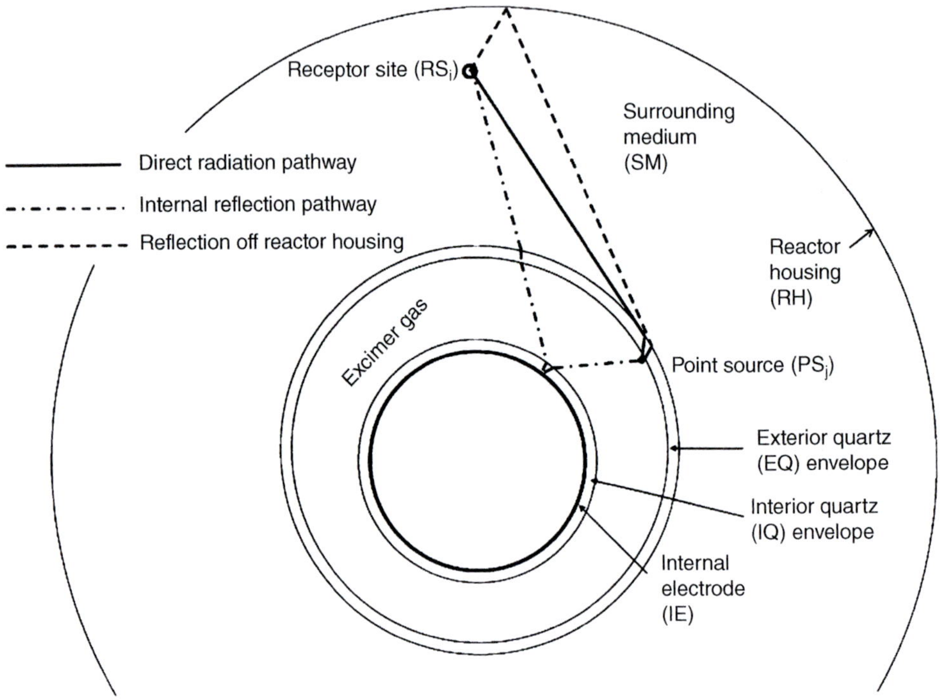


**Figure 6.9** Schematic illustration of possible paths from a point source to a receptor site in the surrounding medium (i.e. fluid being subjected to UV irradiation) in the SPACE model, including the direct radiation pathway, a pathway that involved reflection from the internal electrode surface, and a pathway that involved reflection off the outer housing (wall) of the reactor [12]. *Source:* Naunovic et al. 2008 / with permission of American Chemical Society. Note that this is a two-dimensional representation of the possible optical pathways between point sources and receptor sites. The SPACE model was developed in a fully three-dimensional form. Pathways involving additional (secondary or higher order) refraction or reflection events were assumed to provide negligible contributions, and were therefore ignored.

The SPACE model was developed based on the following logic:

- For each point source/receptor site (PS/RS) pair, pathways were identified that allowed radiation from the PS to reach the RS. Nonviable pathways (e.g. pathways that were blocked by the internal electrode or other opaque objects in the system) were excluded from subsequent calculations. Pathways were determined based on Snell's law and the Fresnel equations.
- Along each pathway, the effects of absorption in each absorbing medium were accounted for by the use of the Beer–Lambert law.
- Fluence rate contributions at a given RS from all contributing PSs in the system were summed to yield an estimate of the fluence rate at the RS.

Fluence rate calculations based on the SPACE model were compared with measurements of fluence rate by *local actinometry*, as illustrated in Figure 6.10. Local actinometry is a method wherein an actinometer solution is pumped through a UV-transparent capillary tube that is positioned at a known location relative to the radiation source. Photochemical conversion of the actinometer is used as a real-time measurement of fluence rate at the location of the capillary tube. This approach is valid if the capillary tube is short relative to the length of the radiation source, and if the capillary tube is aligned parallel to the axis of the radiation source.

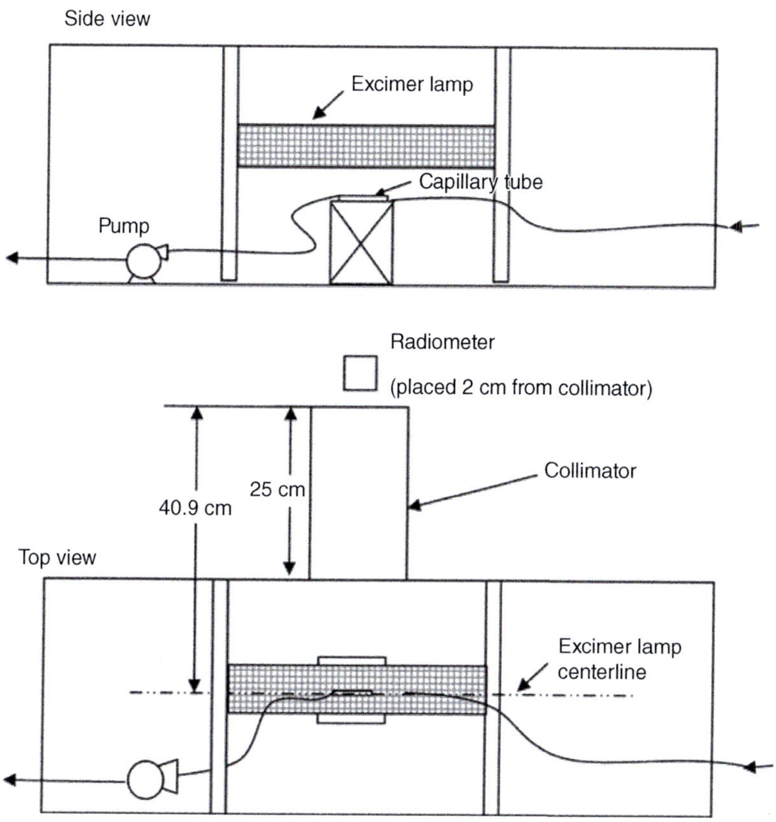


**Figure 6.10** Schematic illustrations of local actinometry method used to quantify fluence rate delivered by a DBD excimer lamp [12]. *Source:* Naunovic et al. 2008 / with permission of American Chemical Society. (Top) DBD XeBr excimer lamp was placed in a box such that the surrounding medium was air. An actinometer solution was pumped through a quartz capillary tube at a known radial position, parallel to the lamp axis. (Bottom) A collimating device was placed on the box to allow radiometric measurement of the fluence rate received at a fixed location. The SPACE model was applied in inverse form to calculate total output power from the lamp based on these radiometric measurements of fluence rate for the collimated radiation.

Figure 6.11 provides a comparison of the results of fluence rate calculations by the SPACE model with measurements by local actinometry. In all cases, the measurements and numerical predictions were in good agreement.

As with all comparisons of this type, it is relevant to consider sources of error in both the model and the measurement. In the model, the lamp was simulated as a set of uniformly distributed point sources. The measurements were based on photochemical conversion of a chemical actinometer in a quartz capillary tube of finite length.

#### 6.2.3.5 Ray Tracing

Ray tracing is a term used to describe a family of numerical methods wherein the output from a radiation source is simulated as a large number of rays emanating the source. Ray tracing software was originally developed for commercial lighting applications; however, the physical principles that govern the fate of visible radiation are identical to those that govern the behavior of UV radiation. As such, these commercial ray tracing packages can be adapted for use in simulating fluence rate fields in the vicinity of UV sources.

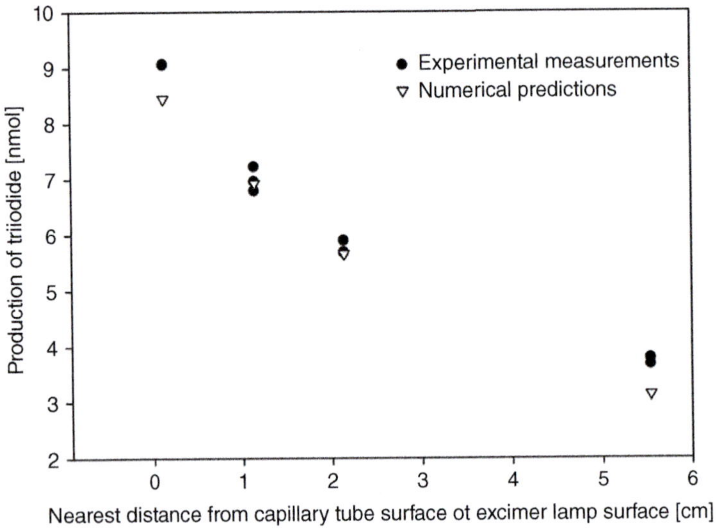


**Figure 6.11** Comparison of fluence rate measurements by local actinometry with corresponding predictions of fluence rate by the SPACE model for the reactor configuration illustrated in Figure 6.10.

Figure 6.12 illustrates the application of ray tracing for simulation of the fluence rate field around a low-pressure Hg lamp. Ray tracing calculations are conducted based on the following logic:

- The geometry of the radiation source, the surrounding media, and the remainder of the reactor (computational domain) are defined numerically (i.e. using CAD software).
- Physical and optical properties of each material in the system are assigned.
- For each ray, a starting location within the domain of the radiation source is selected. For sources that emit uniformly from within their source domain, the selection of the starting location is

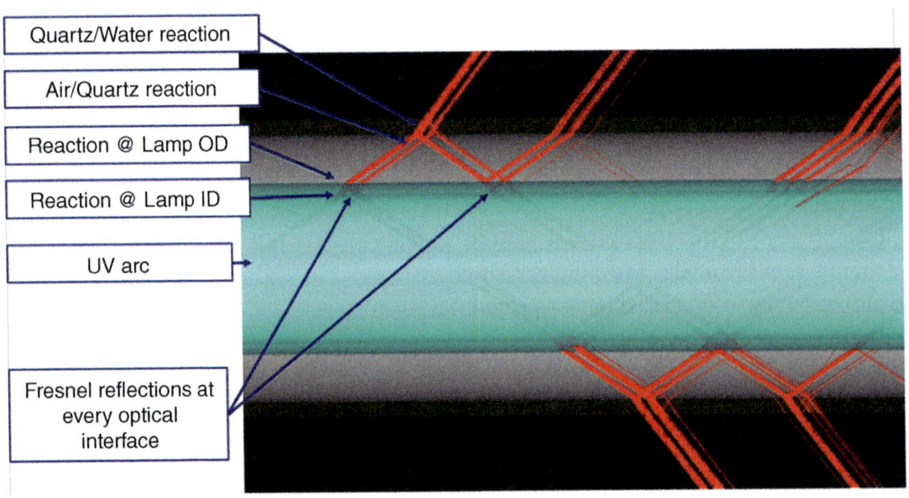


**Figure 6.12** Schematic illustration of the behavior of a single emission event from a low-pressure Hg lamp. Emission from a randomly selected location is initiated (left-center of image) at the outer edge of the plasma (UV arc). At each material interface, reflection and refraction events allow the ray to split. Along each ray, the effects of absorption are accounted for via the Beer–Lambert law [13]. *Source:* Ahmed et al. 2018 / with permission of American Chemical Society.

random. For sources where a known, nonuniform emission pattern exists, the nonuniform pattern is assigned numerically, following an approach that might be thought of as "guided" random selection. Specifically, the origin of each ray is selected according to an algorithm that includes randomness, but also a function to increase the likelihood of ray origin to locations associated with high emission. As the number of simulated rays becomes large, this method of guiding selection will yield an output pattern that accurately mimics the emission pattern of the radiation source.

- A direction of initial ray propagation from the point source is randomly selected. If the source is known to have a directional emission pattern, then constraints on the range of randomly selected initial ray directions are applied.
- Along each ray path, reflection and refraction are accounted for using Snell's law and the Fresnel equations. Absorption is accounted for in all transmitting media through the use of the Beer–Lambert law.
- An array of receptor sites or surfaces is defined. Radiant energy received at each receptor site or surface is tracked. In the case of ray tracing simulations for UV photoreactors wherein the objective is to simulate the fluence rate field, the receptor surfaces are defined as small-diameter, non-absorbing spheres, so as to mimic the formal definition of fluence rate. Following this definition, fluence rate at a receptor surface is defined as the total radiant energy received at that site, divided by the cross-sectional area of the spherical receptor.
- By simulating the fate of a large number of rays, the fluence rate field at each receptor site/surface can be simulated. Fluence rate estimates for locations between receptor spheres are developed based on multi-linear regression of fluence rate estimates for a set of the nearest spherical receptors.

Ray tracing is a stochastic process, in that it relies on the simulation of the fate of a large number of rays to converge on a fluence rate field estimate that accurately represents the true fluence rate field in the reactor. Ray tracing results have been compared with physical measurements based on spherical actinometer and the MFSD, which represent the best available measurement methods for local fluence rate. Examples of these comparisons are presented in Figure 6.13. In all cases, the results of ray tracing simulations were close to the corresponding physical measurements.

For large-scale reactors, the computational time required for ray tracing simulations can be problematic. As such, it may be desirable to incorporate strategies to simplify the ray tracing simulation and reduce the demands on simulation time. The following steps have been suggested to simplify ray tracing simulations for large-scale (i.e. multi-lamp) UV photoreactors [14].

***6.2.3.5.1 Distance to Negligible Contribution*** The Beer–Lambert law teaches that the irradiance (fluence rate) along a collimated ray of radiation will decrease exponentially because of the presence of absorbing compounds. If the absorption coefficient for a given medium is known, then a radial distance from a point source can be identified at which the fluence rate received from a point source has diminished to a negligible value. The criterion used to determine this distance is arbitrary, but could be a distance at which the fluence rate falls to less than 1% of the value adjacent to the point source. Once this distance has been identified, then the bounds of an array of receiving spheres can be defined around the radiation source to allow simulation of the fluence rate field around a radiation source.

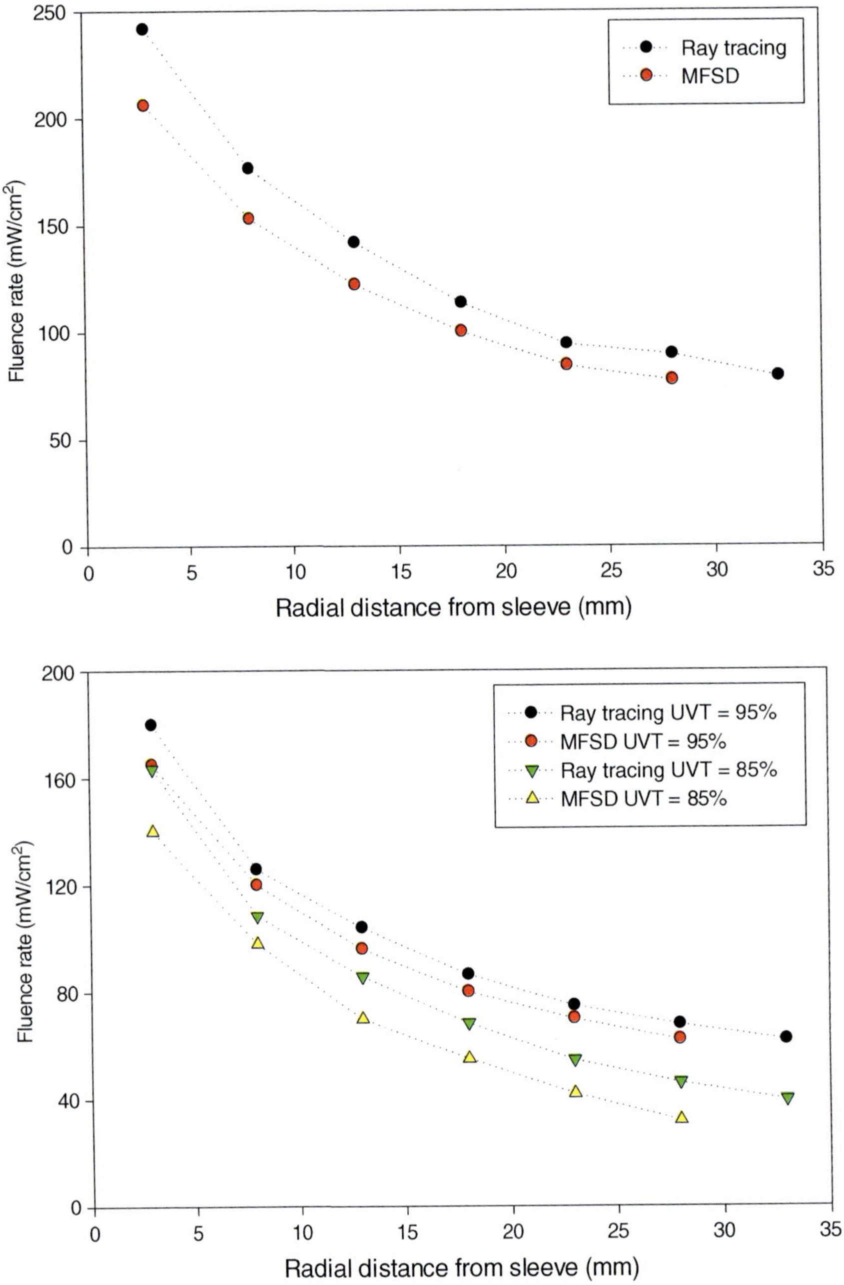


**Figure 6.13** Comparison of ray tracing simulation results with measurements by micro-fluorescent silica detector (MFSD) for a single-lamp reactor involving air as the surrounding medium (top) and water at two different values of transmittance as the surrounding medium (bottom) [13]. *Source:* Ahmed et al. 2018 / with permission of American Chemical Society.

**6.2.3.5.2 *Additivity*** As in other modeling approaches, the fluence rate contributions from individual rays can be assumed to be additive. Therefore, it is possible to simulate the fluence rate field around a single lamp and then superimpose that fluence rate field onto other, analogous fields from other lamps to generate a fluence rate field simulation that accounts for all lamps in the reactor. However, in doing so, it is important to recognize that fluence rate fields around lamps are influenced by the surrounding media. Therefore, it may be possible to identify lamps that have physical surroundings that mimic those of other lamps in a system.

Figure 6.14 illustrates this logic for a large-scale UV disinfection system that is used in municipal wastewater treatment. The reactor includes 14 banks of UV lamps, with each bank comprising 24 rows of 8 lamps each (192 lamps per bank). Within this array, there are six lamp positions that are unique. Position 1 represents the upper left and right corners; position 2 represents lamp positions along the left and right walls; position 3 represents lamp positions in the lower left and right corners, position 4 represents the non-corner lamps within the top row; position 5 represents the non-corner lamps within the bottom row; position 6 represents lamps in the central portion of the lamp array.

**6.2.3.5.3 *Symmetry*** Many UV systems display geometric features that allow symmetry arguments to be exploited to simplify ray tracing simulations. A symmetry plane is indicated in Figure 6.14 to illustrate this. Numerically, the effects of symmetry are accounted for in a ray tracing simulation by calculating the fluence rate field on one side of the symmetry plane, followed by application of a mirror image representation of the fluence rate field across the plane of symmetry. If the contributions of lamps on one side of the plane of symmetry are likely to be relevant to the fluence rate field on the other side, then a perfectly reflective (specular) surface can be assigned to the symmetry plane to account for these contributions.

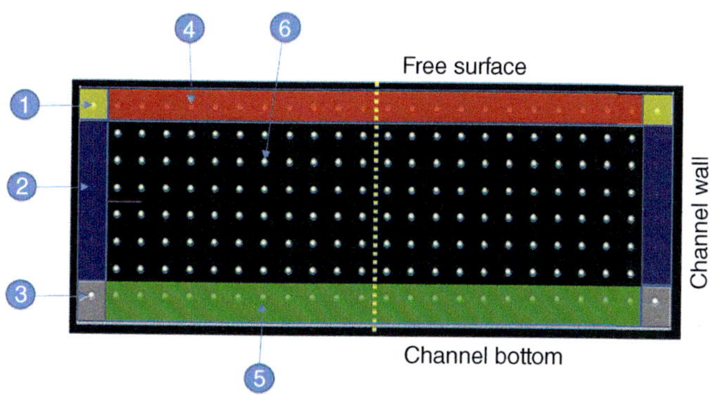


**Figure 6.14** Unique lamp positions within a 24 × 8 lamp array of an open-channel UV disinfection system. Fluence rate distributions around lamp positions adjacent to the free surface or reactor walls will be affected by reflection and refraction. To a first approximation, the six positions indicated in this figure are collectively representative of all 192 lamps in the array. The fluence rate field for the entire array is developed using ray tracing by assembling detailed simulations of the fluence rate fields surrounding these six lamp positions, followed by superposition of identical fluence rate field contributions around all other corresponding lamps. The dashed, yellow line represents a symmetry plane. *Source:* Modified from Ahmed et al. (2018) [14].

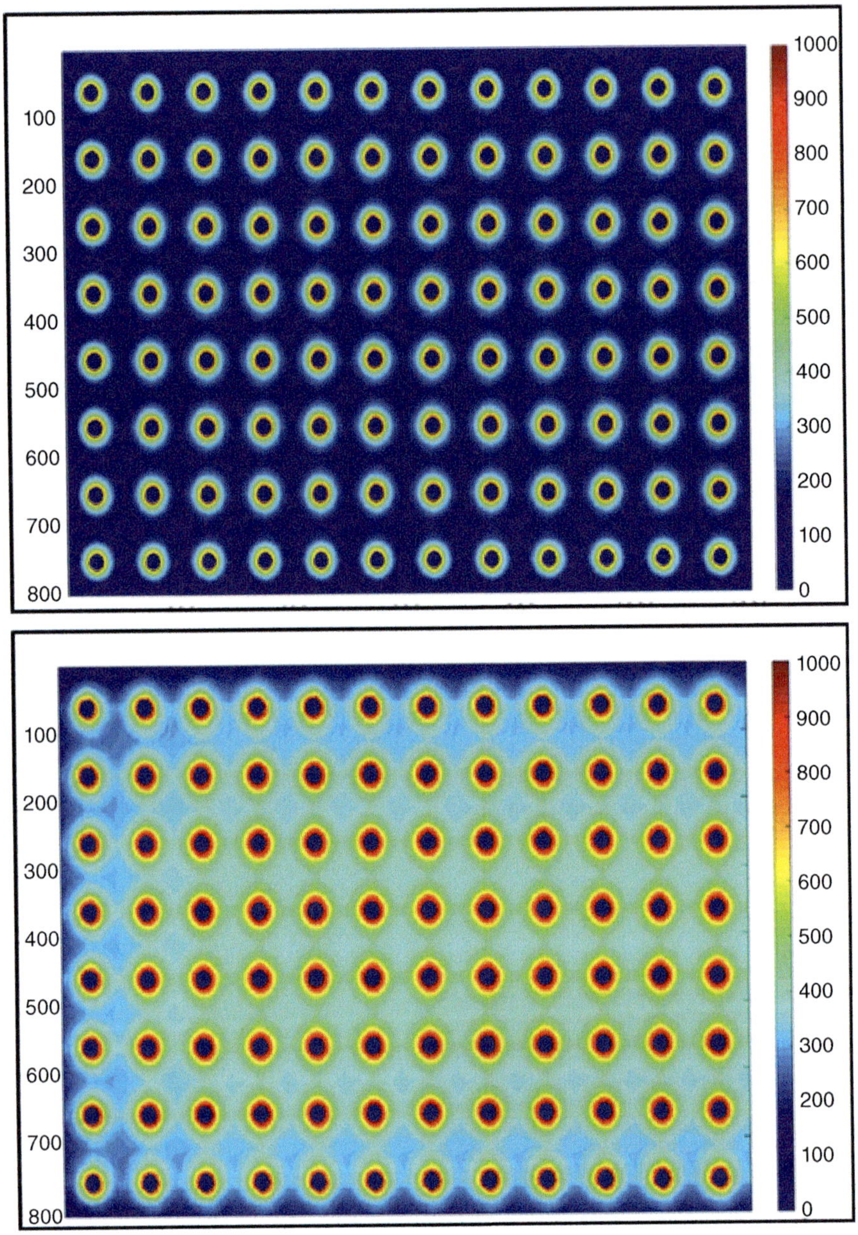


**Figure 6.15** Simulated fluence rate field for left-half of 24 × 8 lamp array corresponding to two UV transmittance conditions: (left) 70% UVT, (right) 80% UVT. Units on legend scale are mW/cm$^2$. *Source:* Images from [14].

Figure 6.15 provides two examples of fluence rate fields that were calculated for the 24 × 8 lamp array using ray tracing, with simplifications described above. Note that because of the symmetry argument, only half of the fluence rate field is illustrated in each case. The effects of UV transmittance on the fluence rate field are clearly evident in these images.

## 6.3 Computational Fluid Dynamics (CFD)

For most practical applications of UV radiation in photoreactors, the photochemical reactions that take place within these systems are characterized by rapid kinetics. In addition, most practical photochemical reactors are characterized by strong spatial gradients in local fluence rate; because photochemical reaction rates are directly proportional to the corresponding fluence rate, there are equally strong spatial gradients in local reaction rate within most practical photoreactors. This combination of attributes results in systems in which overall reactor performance, as characterized by reactant conversion, is profoundly affected by transport behavior (i.e. fluid mechanics, mixing). Therefore, process models for UV photoreactors generally require detailed information regarding fluid mechanical behavior within their irradiated zone.

Transport behavior in photochemical reactors is inherently multidimensional. The first effort to incorporate a detailed, multidimensional description of momentum transport into an analysis of a photochemical reactor was based on physical measurements of local velocity using laser Doppler velocimetry (LDV) [15]. LDV is an optical method that allows noninvasive measurement of local velocity at a rapid rate. By repeating these measurements a large number of times at a given location in the flow field within a physical model of an actual UV photoreactor, it was possible to quantify the mean and turbulence characteristics of the flow field at each location of measurement. These measurements were repeated at a large number of locations within the lamp array. In turn, these data were incorporated into a "random walk" model that allowed simulation of particle trajectories through the irradiated zone of the reactor. By mapping the fluence rate field (calculated separately) onto each simulated particle trajectory, it was possible to estimate trajectory-specific doses (see Figure 6.16). For each particle, the dose received was calculated by integrating[d] the time-history of fluence rate over its trajectory through the system as follows:

$$\text{Dose}_{\text{trajectory}} = \int_0^{\tau} E'(t) \cdot \mathrm{d}t \tag{6.12}$$

where

$\text{Dose}_{\text{trajectory}}$ = trajectory-specific dose [=] $\frac{\text{mJ}}{\text{cm}^2}$

$E'(t)$ = time-dependent fluence rate (governed by particle position within the fluence rate field) [=] $\frac{\text{mW}}{\text{cm}^2}$

$\tau$ = period of exposure [=] s.

The four trajectories illustrated in Figure 6.16 indicate substantial variations in the dose received by these particles. Trajectory (a) follows a path that remains relatively close to the reactor wall, where the fluence rate would be substantially lower than in the near-lamp areas. In contrast, trajectory (b) includes two segments where the particle is trapped on the downstream side of the lamp; this behavior is attributable to the periodic vortex (eddy) shedding that characterizes the region immediately downstream of a bluff body, such as a cylindrical lamp jacket. Because of this, the UV dose received by the particle that follows trajectory (b) was roughly 40 times greater than the dose corresponding to trajectory (a). Trajectories (c) and (d) indicate behavior that is intermediate to those of trajectories (a) and (b), with doses that are intermediate to those of trajectories (a) and (b).

The behavior illustrated in Figure 6.16 indicates that the combined effects of fluid mechanics and the fluence rate field will dictate that a *dose distribution* will be delivered by this reactor. By

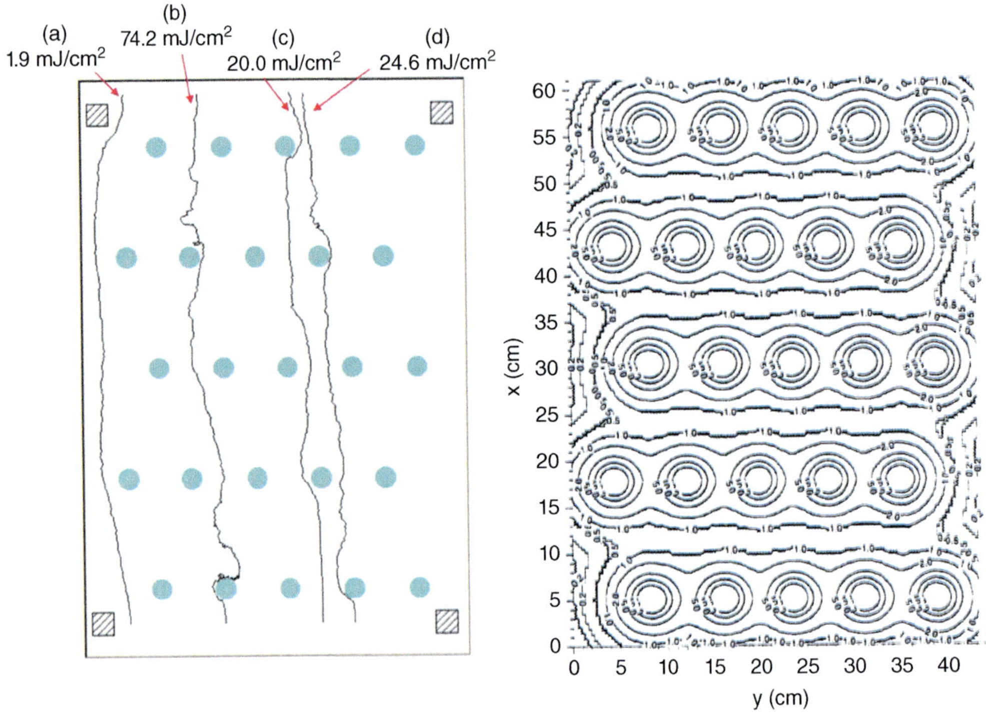


**Figure 6.16** (Left) Simulated particle trajectories based on random-walk model using LDV data. For this simulation, bulk fluid motion was from the bottom to the top of the reactor. Trajectory-specific UV doses for each particle are indicated (see numerical values across the top of the left panel); these values were based on mapping of the fluence rate field (right), calculated using the LSI model, onto each trajectory. UV doses received by each particle were calculated by integrating the time-history of the fluence rate over the period of exposure along their respective particle paths. *Source:* Chiu et al. 1999/American Society of Civil Engineers.

simulating the trajectories of a large number of particles, it is possible to estimate the dose distribution delivered by the reactor for a given operating condition. In general terms, the dose distribution estimate will converge on the "true" dose distribution as the number of trajectories included in the simulation increases. Figure 6.17 illustrates a dose distribution for the reactor

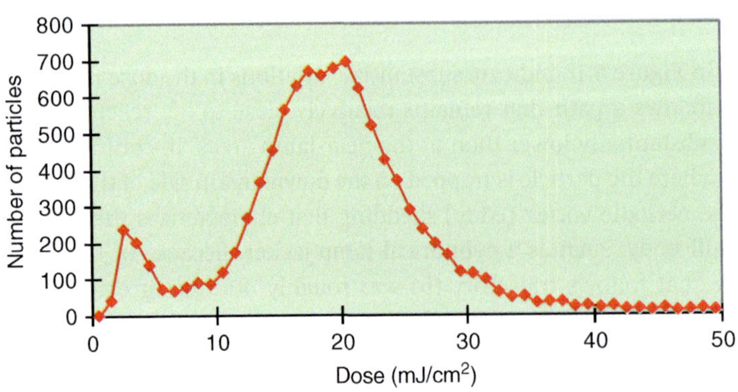


**Figure 6.17** UV dose distribution delivered by photoreactor illustrated in Figure 6.16, based on 10,000 simulated particle trajectories [15]. *Source:* Chiu et. al 1999/The American Society of Civil Engineers.

illustrated in Figure 6.16, based on 10,000 simulated particle trajectories. Integration of this dose distribution with the intrinsic kinetics of disinfection, based on the segregated-flow model, yielded estimates of reactor performance that were in good agreement with measured behavior [15]. This modeling approach, which followed a Lagrangian frame of reference (see section "Process Simulations by CFD-E' Field Modeling" below), also allowed identification of trajectories that yielded low doses within this distribution. By extension, this also allowed identification of reactor attributes that led to these low doses. In turn, this information was used to modify the reactor geometry, thereby leading to an improved dose distribution and better reactor performance [16].

The approach described above represents the foundation for analysis of the performance of photochemical reactor systems. It provides a detailed description of the physical behavior that governs the performance of these systems, including the (turbulent) flow field and the fluence rate field. However, this method is not likely to be applied broadly because it requires extensive, time-consuming LDV measurements using a physical model of an actual reactor.

Subsequently, most simulations of this nature have involved numerical simulations of fluid flow based on CFD. CFD represents a family of numerical methods that allow complex flow fields to be simulated in engineered and natural systems. Improvements in CFD software and computer hardware have allowed CFD to be used in many engineering applications.

The literature pertaining to CFD is extensive and a number of excellent reference texts have been developed to describe these methods and their application in detail. Readers who are interested in applying CFD in the analysis of photochemical reactor systems are encouraged to read books and other publications that address this topic [17–20]. The derivations of governing equations and related topics presented below are based on material presented in these references. In addition, many companies that produce CFD software also provide short-courses that address the details of the use of their products.

The goal for this section is to provide brief derivations of the basic equations that define CFD and to indicate how the results of CFD simulations can be linked to fluence rate field model results and kinetic models to predict photochemical reactor behavior.

### 6.3.1 Governing Equations: Fluid Mechanics

The governing equations for CFD simulations are based on the principles of mass, heat, and momentum conservation. If we consider a three-dimensional control region in a system that includes a moving fluid, a mass balance on the fluid itself can be developed as follows (see Figure 6.18).

In words, the mass balance on fluid in this control region can be defined as follows:

$$\begin{pmatrix} \text{time rate of change of} \\ \text{fluid mass in CR} \end{pmatrix} = \begin{pmatrix} \text{net rate of fluid mass flow} \\ \text{across CR boundaries} \end{pmatrix} \tag{6.13}$$

Using this logic, a formal mathematical description of the fluid mass balance in difference form can be defined as:

$$\forall \frac{\Delta \rho}{\Delta t} = \left(\rho u \Delta y \Delta z|_x - \rho u \Delta y \Delta z|_{x+\Delta x}\right) + \left(\rho v \Delta x \Delta z|_y - \rho v \Delta x \Delta z|_{y+\Delta y}\right) + \left(\rho w \Delta x \Delta y|_z - \rho w \Delta x \Delta y_{z+\Delta z}\right) \tag{6.14}$$

where $u$, $v$, and $w$ represent the components of velocity in the $x$, $y$, and $z$ directions in Cartesian space.[e]

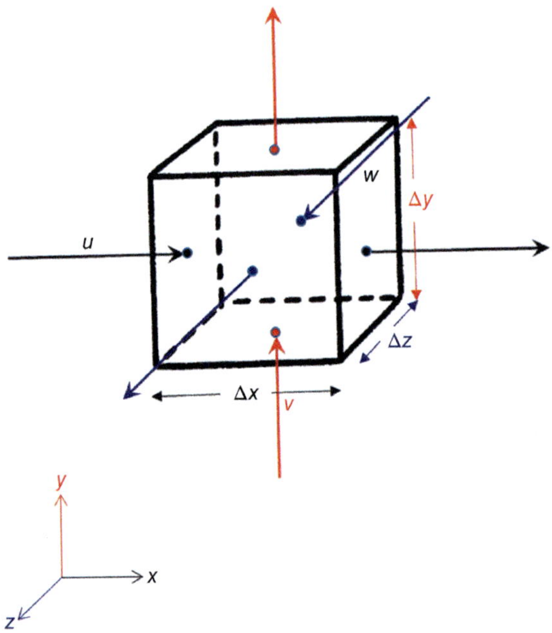


**Figure 6.18** Schematic illustration of mass balance applied to fluid flowing through a differential control region with dimensions Δx, Δy, Δz. Black, red, and blue arrows are used to indicate attributes that apply in the x, y, and z directions in Cartesian space. Similarly colored dots are included to indicate the nominal center of each plane through which fluid enters or leaves the control region.

Recognizing that the volume of the control region can be defined as:

$$\forall = \Delta x \Delta y \Delta z \tag{6.15}$$

We can divide both sides of equation 6.14 by $\Delta x \Delta y \Delta z$, then evaluate the resulting equation in the limits as $\Delta t \to 0$, $\Delta x \to 0$, $\Delta y \to 0, \Delta z \to 0$; subsequent application of the definition of the derivative allows derivation of the so-called "continuity equation."

$$\frac{\partial \rho}{\partial t} + \frac{\partial(\rho u)}{\partial x} + \frac{\partial(\rho v)}{\partial y} + \frac{\partial(\rho w)}{\partial z} = 0 \tag{6.16}$$

A second family of equations can be derived by applying the principles of conservation of momentum to a similar control region in a system with a moving fluid. We start by recognizing that momentum is defined as the product of mass and velocity. Newton's second law teaches that an object of mass $m$ will respond to applied forces as follows:

$$m\vec{a} = \sum \vec{F} \tag{6.17}$$

A wide range of forces can be applied to a fluid element of mass $m$; however, common forces to consider in systems that involve moving fluids include the force due to gravitational acceleration, the force due to differential pressure, and the shear forces applied to the control region. The force due to gravity is sometimes referred to as a body force, in that it applies to the entire body of the object. The forces due to differential pressure and fluid shear are referred to as "surface forces" because they are applied to the external surfaces of the object. We will consider these forces one at a time. These derivations will be presented for the x-direction only; similar principles can be applied to derive the analogous equations for the other coordinate directions. In all cases, we will consider the effects of forces that act on a differential fluid element with dimensions $\Delta x \Delta y \Delta z$ that is moving at the velocity of the surrounding fluid.

#### 6.3.1.1 Gravity

The gravitational force represents the weight of fluid contained in the control region. The gravitational force in the x-direction ($F_{g,\, x}$) is defined as:

$$F_{g,x} = \rho g_x \Delta x \Delta y \Delta z \tag{6.18}$$

where $g_x$ represents the x-component of gravitational acceleration.

#### 6.3.1.2 Differential Pressure

Figure 6.19 illustrates the control region being subjected to a differential pressure in the x-direction.

The net x-directed force resulting from the application of this differential pressure can be defined as:

$$F_{P,x} = \Delta y \Delta z \cdot \left(P|_x - P|_{x+\Delta x}\right) = -\frac{\Delta P}{\Delta x} \cdot \Delta x \Delta y \Delta z \tag{6.19}$$

The negative sign on the right-hand side of equation 6.19 indicates that the direction of net force is opposite of the direction of the gradient in pressure. In other words, if the pressure at location $x$ is less than the pressure at location $x + \Delta x$ (i.e. $\frac{\Delta P}{\Delta x} > 0$), the resultant force acts in the opposite (negative) direction.

#### 6.3.1.3 Shear Stress

The effects of x-directed shear stress are more complicated to consider than the effects of gravity and pressure, in the sense that these forces will act on all six faces of the cubic control region. Figure 6.20 illustrates a free-body diagram of the control region and the shear stresses that act on all six of its faces.

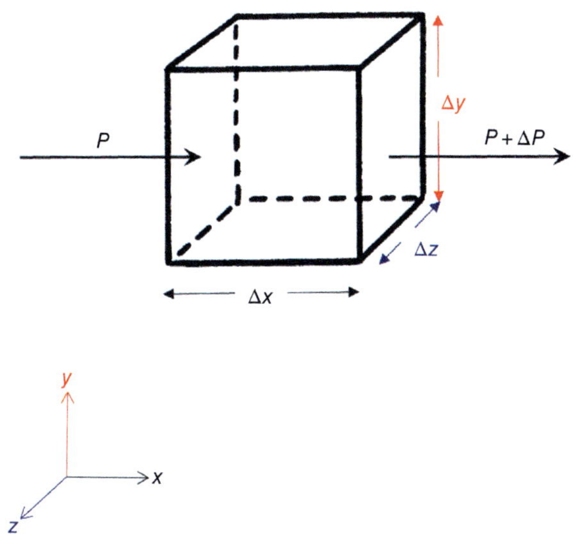


**Figure 6.19** Schematic illustration of differential control region being subjected to x-directed differential pressure.

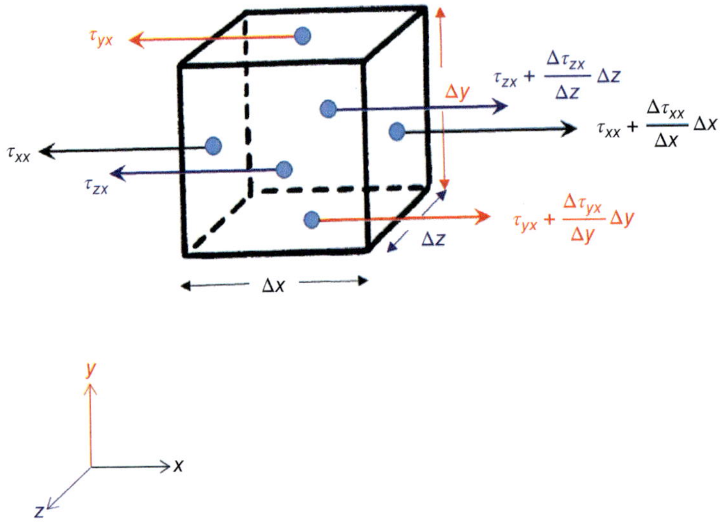


**Figure 6.20** Free-body diagram of differential fluid element being acted upon by x-directed shear forces. Colors of the arrows and corresponding text to indicate shear stress are coded to indicate the direction of momentum transport (see coordinate system sketch, lower left). Black arrows and text indicate x-directed stresses that act on planes that are perpendicular to the x-axis (i.e. y–z planes); red arrows and text indicate x-directed stresses that act on planes that are perpendicular to the y-axis (i.e. x–z planes); blue arrows and text indicate x-directed stresses that act on planes perpendicular to the z-axis (i.e. x–y planes). Dots are included to indicate nominal center of plane on which shear stress acts.

The net x-directed shear force ($F_{S,\,x}$) can be calculated as follows:

$$F_{S,x} = \left[\left(\tau_{xx} + \frac{\Delta\tau_{xx}}{\Delta x}\Delta x\right) - \tau_{xx}\right]\Delta y\Delta z + \left[\left(\tau_{yx} + \frac{\Delta\tau_{yx}}{\Delta y}\Delta y\right) - \tau_{yx}\right]\Delta x\Delta z + \left[\left(\tau_{zx} + \frac{\Delta\tau_{zx}}{\Delta z}\Delta z\right) - \tau_{zx}\right]\Delta x\Delta y \tag{6.20}$$

The subscripts on the shear stress variable indicate the direction of x-directed momentum transport. Therefore, the symbol $\tau_{yx}$ describes the flux of x-directed momentum in the y-direction.

The terms on the right side of equation 6.20 represent the net shear stress acting on the y–z plane, the x–z plane, and the x–y plane, respectively. Eliminating terms by subtraction, followed by rearrangement yields:

$$F_{S,x} = -\left[\frac{\Delta\tau_{xx}}{\Delta x} + \frac{\Delta\tau_{yx}}{\Delta y} + \frac{\Delta\tau_{zx}}{\Delta z}\right]\Delta x\Delta y\Delta z \tag{6.21}$$

As with the pressure term, the negative sign on the right side of this equation indicates that the force resulting from the gradient in shear operates in the opposite direction as this gradient. Referring back to equation 6.17, we will consider the net result of these forces on the acceleration of the fluid element (i.e. the change of its momentum). The mass of the fluid element may be decomposed into the following product:

$$m = \rho\Delta x\Delta y\Delta z \tag{6.22}$$

Because we have chosen to examine the behavior of this control region using a frame of reference that moves at the local velocity of the surrounding fluid, the x-directed acceleration experienced by the fluid element ($a$) can be defined as:

$$\vec{a} = \frac{Du}{Dt} = \frac{\partial u}{\partial t} + u\frac{\partial u}{\partial x} + v\frac{\partial u}{\partial y} + w\frac{\partial u}{\partial z} \tag{6.23}$$

Note that the term $\frac{Du}{Dt}$ is often referred to as the *substantial derivative* or the *total derivative* and results from the moving frame of reference.

Substituting the equations that describe the contributing forces, together with the equations that describe mass and acceleration into Newton's second law (equation 6.17), then dividing by $\rho\Delta x\Delta y\Delta z$, we find:

$$\frac{Du}{Dt} = -\frac{1}{\rho}\frac{\Delta P}{\Delta x} - \left(\frac{\Delta\tau_{xx}}{\Delta x} + \frac{\Delta\tau_{yx}}{\Delta y} + \frac{\Delta\tau_{zx}}{\Delta z}\right) \tag{6.24}$$

When this equation is evaluated in the limit as $\Delta x \to 0$, $\Delta y \to 0, \Delta z \to 0$, with application of the definition of the derivative, we find:

$$\frac{Du}{Dt} = -\frac{1}{\rho}\frac{\partial P}{\partial x} - \frac{1}{\rho}\left(\frac{\partial\tau_{xx}}{\partial x} + \frac{\partial\tau_{yx}}{\partial y} + \frac{\partial\tau_{zx}}{\partial z}\right) \tag{6.25}$$

Equation 6.25 is an *equation of motion* (for the x-direction).

Most UV photoreactors are used to treat liquid water or air, which both behave as incompressible, *Newtonian fluids* under the conditions that are relevant for most photochemical reactor systems. Therefore, the density of these fluids can be considered to be a constant. Shear stress in a Newtonian fluid is described by Newton's law of viscosity:

$$\tau_{yx} = -\mu\frac{\partial u}{\partial y} \tag{6.26}$$

where $\mu$ represents the absolute or dynamic viscosity of the fluid. Kinematic viscosity ($\nu$) is defined as the ratio of dynamic viscosity to density:

$$\nu = \frac{\mu}{\rho} \tag{6.27}$$

Substituting equations 6.26 and 6.27 into equation 6.25, with the understanding that we are dealing with a fluid of constant density, and expanding the definition of the substantial derivative, we find the x-directed equation of motion for a Newtonian fluid:

$$\frac{\partial u}{\partial t} + u\frac{\partial u}{\partial x} + v\frac{\partial u}{\partial y} + w\frac{\partial u}{\partial z} = -\frac{1}{\rho}\frac{\partial P}{\partial x} + \nu\left(\frac{\partial^2 u}{\partial x^2} + \frac{\partial^2 u}{\partial y^2} + \frac{\partial^2 u}{\partial z^2}\right) + g_x \tag{6.28}$$

Similarly, the equations of motion can be defined for the y-direction and the z-direction:

$$\frac{\partial v}{\partial t} + u\frac{\partial v}{\partial x} + v\frac{\partial v}{\partial y} + w\frac{\partial v}{\partial z} = -\frac{1}{\rho}\frac{\partial P}{\partial y} + \nu\left(\frac{\partial^2 v}{\partial x^2} + \frac{\partial^2 v}{\partial y^2} + \frac{\partial^2 v}{\partial z^2}\right) + g_y \tag{6.29}$$

$$\frac{\partial w}{\partial t} + u\frac{\partial w}{\partial x} + v\frac{\partial w}{\partial y} + w\frac{\partial w}{\partial z} = -\frac{1}{\rho}\frac{\partial P}{\partial z} + \nu\left(\frac{\partial^2 w}{\partial x^2} + \frac{\partial^2 w}{\partial y^2} + \frac{\partial^2 w}{\partial z^2}\right) + g_z \tag{6.30}$$

For completeness, the continuity equation can also be simplified for the case of an incompressible fluid:

$$\frac{\partial u}{\partial x} + \frac{\partial v}{\partial y} + \frac{\partial w}{\partial z} = 0 \tag{6.31}$$

Collectively, equations 6.28–6.30 represent the equations of motion for all three Cartesian coordinate directions for an incompressible, Newtonian fluid; they are also commonly referred to the *Navier–Stokes equations*. When combined with the continuity equation 6.31, these equations can be used to describe the flow of any incompressible, Newtonian fluid.

For laminar flow conditions, which are relevant in some photochemical reactor systems, it is sometimes possible to develop a closed-form, analytical solution to describe fluid motion. For example, for the case of laminar flow in a long, straight, circular tube, which is observed in some UV photoreactors that involve fluid flow through a capillary, velocity in the tube is entirely along the longitudinal axis of the tube. If this is identified as the x-direction, then the velocity profile in the tube is completely defined by the following simple expression (see definition sketch in Figure 6.21):

$$V_x(r) = -\frac{\mathrm{d}P}{\mathrm{d}x}\left(\frac{R^2 - r^2}{4\mu}\right) \tag{6.32}$$

where

$V_x(r)$ = x-directed fluid velocity at radial position r

$\frac{\mathrm{d}P}{\mathrm{d}x}$ = x-directed pressure gradient

$R$ = inside diameter of circular tube.

The pressure gradient ($\frac{\mathrm{d}P}{\mathrm{d}x}$) represents the "driving force" for fluid motion. The parabolic velocity profile for this physical situation is derived by applying a momentum balance to a differential annular ring of fluid in the circular tube, subject to a so-called "no-slip" boundary condition at the tube wall, which constrains velocity at that location to zero. The average velocity ($V_x$(avg)) for this condition is defined by:

$$V_x(\text{avg}) = -\frac{\mathrm{d}P}{\mathrm{d}x}\frac{R^2}{8\mu} \tag{6.33}$$

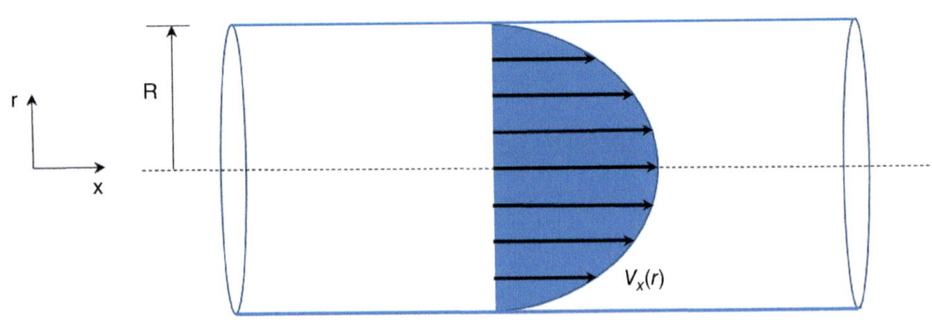


**Figure 6.21** Schematic illustration of parabolic velocity profile for laminar flow in a circular tube.

Before proceeding to consider other flow conditions, it is helpful to consider one of the methods by which the equation of continuity and the Navier–Stokes equation can be presented in more compact form. Specifically, the concept of index (tensor) notation will be applied as a method for condensing these equations, and other equations that will be developed later.

### 6.3.2 Index (Tensor) Notation

The governing equations for fluid mechanics for incompressible, Newtonian fluids (continuity: equation 6.31; Navier–Stokes: equations 6.28–6.30) and other equations to be presented below that address other aspects of transport behavior can be rather cumbersome to work with. Because of this, several conventions have been developed to allow these equations to be presented in a more compact form. Among these, the use of index (tensor) notation is perhaps the most common.

The rules for index notation are summarized in Table 6.1.

Within any expression, when the index is repeated, it implies that the sum is to be taken over all three coordinate directions [17]. Examples of the application of this convention are provided in Table 6.2.

**Table 6.1** Rules for application of tensor notation [17].

| Term | Definition | Comments |
|---|---|---|
| Space vector | $x_i \equiv \{x_1 \ x_2 \ x_3\}$ | This is used as a substitute for the three coordinate directions. For analysis in Cartesian coordinates, this translates to $\{x \quad y \quad z\}$. |
| Velocity vector | $u_i \equiv \{u_1 \quad u_2 \quad u_3\}$ | This is used as a substitute for velocity in the three coordinate directions. For analysis in Cartesian coordinates, this translates to $\{u \quad v \quad w\}$. |
| Stress tensor | $\tau_{ij} \equiv \begin{Bmatrix} \tau_{11} & \tau_{12} & \tau_{13} \\ \tau_{21} & \tau_{22} & \tau_{23} \\ \tau_{31} & \tau_{32} & \tau_{33} \end{Bmatrix}$ | The first index indicates a surface perpendicular to the coordinate direction, $x_i$, on which the stress acts.<br>The second index indicates the direction of stress.<br>The stress tensor is symmetric (i.e. $\tau_{ij} = \tau_{ji}$). |

**Table 6.2** Examples of the application of the repeated index convention in index notation [17].

| Example | Application in Cartesian coordinates |
|---|---|
| $u_i u_i \equiv \sum_{i=1}^{3} u_i u_i = u_1 u_1 + u_2 u_2 + u_3 u_3$ | This is used as shorthand notation for: $u \cdot u + v \cdot v + w \cdot w$. |
| $\frac{\partial u_i}{\partial x_i} \equiv \sum_{i=1}^{3} \frac{\partial u_i}{\partial x_i} = \frac{\partial u_1}{\partial x_1} + \frac{\partial u_2}{\partial x_2} + \frac{\partial u_3}{\partial x_3}$ | This is used as shorthand notation for: $\frac{\partial u}{\partial x} + \frac{\partial v}{\partial y} + \frac{\partial w}{\partial z}$ as applied in the continuity equation for incompressible fluids (equation 6.31). |
| $u_j \frac{\partial u_i}{\partial x_j} \equiv \sum_{i=1}^{3} u_j \frac{\partial u_i}{\partial x_j} = u_1 \frac{\partial u_i}{\partial x_1} + u_2 \frac{\partial u_i}{\partial x_2} + u_3 \frac{\partial u_i}{\partial x_3}$ | This is used as shorthand notation for: $u\frac{\partial u}{\partial x} + v\frac{\partial u}{\partial y} + w\frac{\partial u}{\partial z}$ as applied in the Navier–Stokes equations. |

When these conventions are applied to the continuity equation, it reduces to:

$$\frac{\partial u_i}{\partial x_i} = 0 \tag{6.34}$$

Similarly, the Navier–Stokes equations can be rewritten in index notation as:

$$\frac{\partial u_i}{\partial t} + u_j \frac{\partial u_i}{\partial x_j} = -\frac{1}{\rho}\frac{\partial P}{\partial x_i} + \nu \frac{\partial^2 u_i}{\partial x_j \partial x_j} + g_i \tag{6.35}$$

### 6.3.3 Governing Equations: Transport of Reactants

The primary motivation for application of CFD tools for simulation of the behavior of photochemical reactor behavior is to develop detailed, accurate numerical approximations of transport and reaction behavior within the irradiated zone of a reactor. The continuity equation and the Navier–Stokes equations are used for simulation of fluid mechanical behavior. In turn, this information is critical for simulations of transport of fluid-associated constituents.

The formal development of mathematical expressions to define the fate and transport of fluid-borne constituents is based on application of material balance principles. Figure 6.22 illustrates the application of this approach to a differential fluid element that is identical to the one used to develop the continuity equation (recall, this involved a mass balance on the fluid itself, rather than a dissolved or particle-associated constituent being carried by the surrounding fluid).

Recall that the general form of the mass-balance equation (in difference form) is:

$$\frac{\Delta M}{\Delta t} = \Phi_{\text{in}} - \Phi_{\text{out}} \pm \langle r \rangle \forall \tag{6.36}$$

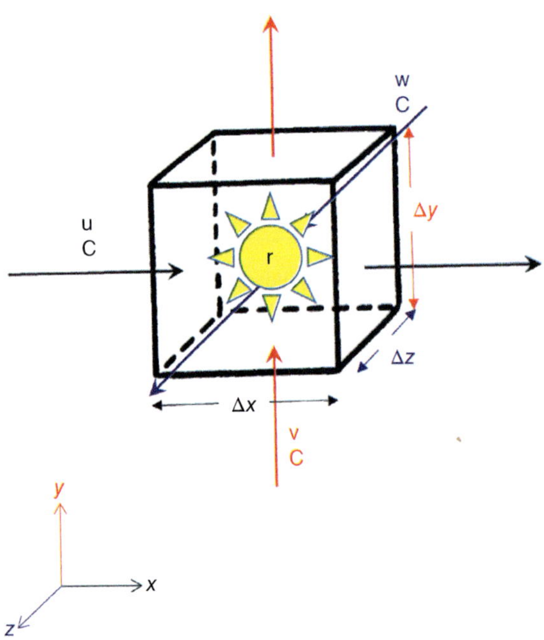


**Figure 6.22** Schematic illustration of the application of a mass balance for a fluid-borne constituent within a differential control volume. The term "r" is included to account for one or more (photochemical) reactions that produce or consume the constituent of interest within the control region. Terms associated with the principal axes are color-coded (see axis diagram, lower left).

where

$M$ = mass (amount) of constituent of interest within control region
$t$ = time
$\Phi_{in}$ = mass rate at which constituent of interest enters the control region
$\Phi_{out}$ = mass rate at which constituent of interest leaves the control region
$\langle r \rangle$ = spatial average of reaction rate(s) that produce or consume the constituent of interest within control region
$\forall$ = volume of control region.

Note that the terms $\Phi_{in}$ and $\Phi_{out}$ describe the net result of all transport processes that move the constituent of interest across the boundaries of the control region.

If we consider the system to be operating in the laminar regime, then the mass-balance (in difference form) becomes:

$$\begin{aligned}\Delta x \Delta y \Delta z \frac{\Delta C}{\Delta t} = {} & \left(uC|_x - uC|_{x+\Delta x}\right)\Delta y \Delta z + \left(vC|_y - vC|_{y+\Delta y}\right)\Delta x \Delta z + \left(wC|_z - wC|_{z+\Delta z}\right)\Delta x \Delta y \\ & - \mathcal{D}\left(\left.\frac{\partial C}{\partial x}\right|_x - \left.\frac{\partial C}{\partial x}\right|_{x+\Delta x}\right)\Delta y \Delta z - \mathcal{D}\left(\left.\frac{\partial C}{\partial y}\right|_y - \left.\frac{\partial C}{\partial y}\right|_{y+\Delta y}\right)\Delta x \Delta z \\ & - \mathcal{D}\left(\left.\frac{\partial C}{\partial z}\right|_z - \left.\frac{\partial C}{\partial z}\right|_{z+\Delta z}\right)\Delta x \Delta y \pm \langle r \rangle \Delta x \Delta y \Delta z \end{aligned} \tag{6.37}$$

where

$\mathcal{D}$ = molecular diffusivity of the constituent of interest.

$\mathcal{D}$ is a property of the constituent of interest and the fluid that surrounds it. In most circumstances, $\mathcal{D}$ can be considered to be a physical constant and an isotropic property (i.e. no directional dependence).

The first three terms on the right side of equation 6.37 describe the effects of advective transport of the constituent across the system boundary. The next three terms describe the effects of diffusive transport of the constituent across the system boundary. The last term describes the effects of reactions that take place within the control region that produce or consume the constituent of interest within the control region. If we divide both sides of equation 6.37 by $\Delta x \Delta y \Delta z$, evaluate the equation in the limits as $\Delta t \to 0$, $\Delta x \to 0$, $\Delta y \to 0$, and $\Delta z \to 0$, then apply the definition of the derivative, the result is an equation to describe the behavior of the constituent of interest within the control region:

$$\frac{\partial C}{\partial t} = \underbrace{-\left(\frac{\partial (uC)}{\partial x} + \frac{\partial (vC)}{\partial y} + \frac{\partial (wC)}{\partial z}\right)}_{\text{advection}} + \underbrace{\mathcal{D}\left(\frac{\partial^2 C}{\partial x^2} + \frac{\partial^2 C}{\partial y^2} + \frac{\partial^2 C}{\partial z^2}\right)}_{\text{diffusion}} \pm \underbrace{r}_{\text{reaction}} \tag{6.38}$$

The first term in the right side of equation 6.38 describes the net effect of advective transport on the time rate-of-change of constituent concentration within the control region; the second term describes the next effect of (molecular) diffusive transport on the time rate-of-change of constituent concentration within the control region; the third term describes the net effect of local reactions on the time rate-of-change of constituent concentration within the control region.

Note that the reaction term is now presented without the spatial averaging function. This is possible because we have applied this mass balance to a differential control region (i.e. evaluated in the limit as its dimensions all approach zero). In other words, this equation applies to a point in the reactor. By extension, this means that we need to be able to define the rate of reaction at each point in the system. In a photochemical process, the local reaction rate at a point is first-order with respect to the target reactant and proportional to the local fluence rate (see Chapter 2). But it is important to recognize that any number of reactions could take place within the control region. Valid kinetic expressions must be provided for all reactions (photochemical, thermochemical, biochemical, etc.) that affect the concentration of the constituent of interest within the control region.

A similar approach can be used to describe transport of any scalar quantity. Another common example is the application of an energy balance to allow simulation of the temperature profile within a system.

Following the conventions for index notation, the transport equation for a scalar quantity can be defined more generally as follows for a situation that involves laminar flow: [17]

$$\frac{\partial \phi}{\partial t} = \underbrace{-u_i \frac{\partial \phi}{\partial x_i}}_{\text{advection}} + \underbrace{\lambda \frac{\partial^2 \phi}{\partial x_i \partial x_i}}_{\text{diffusion}} \pm \underbrace{S_\phi}_{\text{source/sink}} \tag{6.39}$$

where

$\phi$ = scalar quantity (C or T)
$\lambda$ = diffusivity of scalar quantity (molecular diffusivity or thermal conductivity)
$S_\phi$ = source and/or sink terms that describe production and/or consumption of the scalar quantity within the control region.

### 6.3.4 Simulations for Systems Operating in the Turbulent Regime

The vast majority of photochemical reactors operate in the turbulent regime. Turbulence features of fluid motion can have profound effects on the rates of momentum, energy, and mass transport. As a result, numerical models that are used to simulate the behavior of these reactor systems need to account for the effects of turbulence.

Turbulence involves (pseudo) random motions that span a wide range of spatial and temporal scales. Energy transport in turbulent flows is described as a "cascade" of energy from large eddies to small eddies [17, 21]. Among the small eddies, energy dissipation is thought be largely associated with viscous forces.

In theory, it is possible to solve the governing equations (continuity, Navier–Stokes, scalar transport) for any flow field. In fact, this is accomplished through a method known as *direct numerical simulation* (DNS); however, DNS simulations require that the model be applied at temporal and spatial scales that are small enough to capture all of the relevant motions that characterize turbulence. For most applications, this level of detail implies a very detailed mesh, large amounts of computer memory, an extremely fast computer, and considerable CPU time.

A commonly applied alternative to DNS is the statistical approach suggested by Osborne Reynolds known as the *Reynolds decomposition*, wherein the expected value (mean) of a given parameter is separated from its fluctuations. Application of the Reynolds decomposition to the variables that are relevant in the governing equations described above, we find (in index notation):

$$u_i = \overline{u_i} + u_i' \tag{6.40}$$

$$P = \overline{P} + P' \tag{6.41}$$

$$\phi = \overline{\phi} + \phi' \tag{6.42}$$

Using the velocity terms as an example, the overbar $(\overline{u_i})$ represents the expected value (mean) of velocity, while the prime term $(u_i')$ represents deviations from the mean. The mean values of each parameter are defined as follows:

$$\overline{u_i} = \frac{1}{t_2 - t_1} \int_{t_1}^{t_2} u_i \cdot dt \tag{6.43}$$

$$\overline{P} = \frac{1}{t_2 - t_1} \int_{t_1}^{t_2} P \cdot dt \tag{6.44}$$

$$\overline{\phi} = \frac{1}{t_2 - t_1} \int_{t_1}^{t_2} \phi \cdot dt \tag{6.45}$$

where the averaging time $(t_2 - t_1)$ is chosen to be long relative to the timescale of turbulent motion in the system. In addition, the mean values of the deviation terms are all zero:

$$\overline{u_i'} = \frac{1}{t_2 - t_1} \int_{t_1}^{t_2} u_i' \cdot dt = 0 \tag{6.46}$$

$$\overline{P'} = \frac{1}{t_2 - t_1} \int_{t_1}^{t_2} P' \cdot dt = 0 \tag{6.47}$$

$$\overline{\phi'} = \frac{1}{t_2 - t_1} \int_{t_1}^{t_2} \phi' \cdot dt = 0 \tag{6.48}$$

When these definitions are applied to the governing equations described above, they are collectively referred to as the Reynolds Averaged Navier Stokes (RANS) equations (in index notation):

$$\frac{\partial \overline{u_i'}}{\partial t} = 0 \text{ (continuity)} \tag{6.49}$$

$$\frac{\partial \overline{u_i}}{\partial t} + \overline{u_j}\frac{\partial \overline{u_i}}{\partial x_j} = -\frac{1}{\rho}\frac{\partial \overline{P}}{\partial x_i} + \frac{\partial}{\partial x_j}\left[\nu \frac{\partial \overline{u_i}}{\partial x_j} - \overline{u_i' u_j'}\right] + g_i \text{ (Navier} - \text{Stokes)} \tag{6.50}$$

$$\frac{\partial \overline{\phi}}{\partial t} = -u_i \frac{\partial \overline{\phi}}{\partial x_j} + \frac{\partial}{\partial x_j}\left[\lambda \frac{\partial \overline{\phi}}{\partial x_j} - \overline{u_i' \phi'}\right] + S_\phi \text{ (Transport of a Scalar)} \tag{6.51}$$

The system of equations described by equations 6.49–6.51 is exact, but not closed because of two newly introduced terms. The physical interpretations of these terms are as follows:

- The product $\left(-\rho \overline{u_i' u_j'}\right)$ represents turbulent transport of $x_i$ momentum in the $x_j$ direction. This term functions as a stress on the fluid and is often referred to as a turbulent or Reynolds stress.
- The product $\left(-\rho \overline{u_i' \phi'}\right)$ describes the rate of turbulent transport of the scalar $(\phi)$ in the $x_i$ direction.

As it turns out, the turbulent transport rates for momentum, energy, and mass are generally much greater than their laminar analogs. As a result, the laminar terms are often ignored or otherwise excluded from simulation. In fact, the laminar (molecular) diffusivity terms for momentum, energy, and mass transport are often absent from the governing equations used in the analysis of systems that operate in the turbulent regime. However, the appearance of these two turbulent transport terms requires the inclusion of a system of equations to accomplish turbulence closure.

A wide range of turbulence closure models has been developed and applied. The focus of this review will be on the so-called k-ε model, which is among the most commonly applied of these models. It also shares characteristics with many of the other turbulence closure models.

The Boussinesq eddy-viscosity concept asserts that the turbulent stresses are proportional to mean velocity gradients. Mathematically, this concept is described as: [17]

$$-\overline{u_i'u_j'} = \nu_t\left(\frac{\partial u_i}{\partial x_j} + \frac{\partial u_j}{\partial x_i}\right) - \frac{2}{3}k\delta_{i,j} \tag{6.52}$$

where

$\nu_t$ = turbulent (eddy) viscosity
$k$ = kinetic energy
$= \frac{1}{2}\left(\overline{u_1^2} + \overline{u_2^2} + \overline{u_3^2}\right)$
$\delta_{i,j}$ = 1 for $i = j$
= 0 for $i \neq j$.

The physical interpretation of $\nu_t$ is a parameter that describes momentum transport in a turbulent flow field; the local rate of momentum transport is proportional to the local mean velocity gradient. The logic for this analogy follows from the random (or pseudo-random) nature of molecular diffusion and turbulent flows. A stochastic approach based on randomness yields Newton's law of viscosity in laminar flows; (pseudo) randomness in turbulent motion is assumed to result in an analogous dependence on velocity gradients. However, it is also important to recognize differences between Newton's law of viscosity and the Boussinesq eddy-viscosity concept; or more precisely, it is important to recognize that kinematic viscosity ($\nu$) is a property of a fluid, whereas the turbulent (eddy) viscosity ($\nu_t$) is a property of the fluid flow field, not the fluid exclusively.

By analogy, the eddy-diffusivity concept asserts that the local rate of turbulent mass transport can be described in a manner that is analogous to Fick's law or Fourier's law, which describe the rates of diffusive transport for mass and heat. Their turbulent analogs are described as follows:

$$-\overline{u_i'\phi'} = \Gamma\frac{\partial \phi}{\partial x_i} \tag{6.53}$$

where

$\Gamma$ = turbulent diffusivity for mass or heat.

The Reynolds analogy between transport of a scalar and momentum transport (i.e. mass transport vs. momentum transport, or heat transport and momentum transport) suggests that $\Gamma$ is linearly related to $\nu_t$:

$$\Gamma = \frac{\nu_t}{\sigma_t} \tag{6.54}$$

where

$\sigma_t$ = turbulent Schmidt number ($Sc_t$, mass transport) or turbulent Prandtl number $Pr_t$ (heat transport).

The turbulent Schmidt number ($Sc_t$) is a dimensionless ratio of the turbulent (eddy) viscosity to the turbulent diffusivity ($\mathcal{D}_t$):

$$Sc_t = \frac{\nu_t}{\mathcal{D}_t} \tag{6.55}$$

Turbulent diffusivity is a hypothetical construct used to relate the rate of local, turbulent mass transport to the local concentration gradient (equation 6.53). Unlike molecular diffusivity, which is a physical characteristic of a compound and the surrounding medium, turbulent diffusivity (like $\nu_t$) is a characteristic of the fluid flow field. Because of this, the turbulent diffusivity is also likely to demonstrate directional dependence. Therefore, the rate of turbulent mass transport can be described as: [22]

$$\overline{u_i'C'} = -\mathcal{D}_{t,i}\frac{\partial \overline{C}}{\partial x_i} \tag{6.56}$$

The Reynolds analogy suggests that the turbulent eddies that are responsible for momentum transport are also responsible for mass (or heat) transport. If this assumption is valid, then a value of $Sc_t \equiv 1$ should apply. In fact, this assumed value of $Sc_t$ has been used in simulations of mass transport in photochemical reactor systems where the focus is on the behavior of particulate constituents, such as microorganisms [23]. However, in simulations of the behavior of dissolved constituents in water, a value of $Sc_t = 0.7$ has been applied [24].

More broadly, the logic described above has been used to simulate mass transport behavior for aqueous liquid systems and gas-phase systems. Reported values of $Sc_t$ for these simulations have ranged from 0.1 to 2.11 for aqueous liquids, and from 0.3to 2.5 for simulations of constituent mass transfer in the atmosphere (i.e. gas phase) [22]. In the absence of quantitative information to guide the selection of a value of the turbulent Schmidt number, a value of $Sc_t \equiv 1$ may represent a logical starting point for these simulations. However, it should also be recognized that the concept of the turbulent Schmidt number relies on the validity of the Reynolds analogy of turbulent transport behavior. The fact that $Sc_t$ is generally reported to one, sometimes two, or perhaps even three significant digits may also be taken as an indication of the uncertainty that accompanies the use of this parameter.

The addition of the terms to describe the Reynolds stresses and turbulent scalar transport in the governing equations requires that additional equations be included in the model to accomplish closure. Many "turbulence closure models" are two-parameter models, and among these the k-ε model is one of the most commonly applied. Like other turbulence closure models, it is important to recognize that the k-ε model does not describe the actual mechanics of turbulence, but rather it describes the effects of turbulence [17].

The k-ε model has been described as a semiempirical model [17]. It is based the inclusion of two new parameters to describe the effects of turbulence: [20]

$$k = \text{turbulence kinetic energy} \tag{6.57}$$

$$\varepsilon = \text{turbulence kinetic energy dissipation rate} \tag{6.58}$$

The model itself comprises equations to describe transport of these two parameters within a turbulent flow field. The complete k-ε model comprises the following set of equations: [17]

$$\nu_t = c_\mu\frac{k^2}{\varepsilon} \tag{6.59}$$

$$\frac{\partial k}{\partial t} = \underbrace{-\overline{u_i}\frac{\partial k}{\partial x_i}}_{\text{advection}} + \underbrace{\frac{\partial}{\partial x_i}\left(\frac{\nu_t}{\sigma_k}\cdot\frac{\partial k}{\partial x_i}\right)}_{\text{diffusion}} + \underbrace{\nu_t\left(\frac{\partial \overline{u_i}}{\partial x_j} + \frac{\partial \overline{u_j}}{\partial x_i}\right)\cdot\frac{\partial \overline{u_i}}{\partial x_j}}_{P} + \underbrace{\beta g_i\frac{\nu_t}{\sigma_t}\cdot\frac{\partial \phi}{\partial x_i}}_{G} - \varepsilon \tag{6.60}$$

$$\frac{\partial \varepsilon}{\partial t} = \underbrace{-\overline{u_i}\frac{\partial \varepsilon}{\partial x_i}}_{\text{advection}} + \underbrace{\frac{\partial}{\partial x_i}\left(\frac{\nu_t}{\sigma_\varepsilon}\cdot\frac{\partial \varepsilon}{\partial x_i}\right)}_{\text{diffusion}} + \underbrace{c_{1\varepsilon}\frac{\varepsilon}{k}(P+G)(1 + c_{3\varepsilon}R_f) - c_{2\varepsilon}\frac{\varepsilon^2}{k}}_{\text{net generation}} \tag{6.61}$$

This system of equations introduces five empirical constants that need to be selected to complete a simulation. A set of these parameters was recommended by Launder and Spalding [20] and has remained in common use since that time. These recommended values are summarized in Table 6.3.

**Table 6.3** Recommended values for empirical constants in the k-ε model [20]. *Source:* Launder and Spalding 1983 / with permission of Elsevier.

| $C_\mu$ | $C_{1\varepsilon}$ | $C_{2\varepsilon}$ | $\sigma_k$ | $\sigma_\varepsilon$ |
|---|---|---|---|---|
| 0.09 | 1.44 | 1.92 | 1.0 | 1.3 |

The k-ε model has been used in a wide range of simulations for engineering systems and provides detailed information about the behavior of turbulent flow fields. However, many other turbulence models are available. The details of these models are beyond the scope of this text, but the reader should be aware that there are alternatives to the k-ε model.

### 6.3.5 Accuracy of Turbulence Models for Flow Field Simulations

CFD simulations are based on numerical approximations of the equations of motion and continuity, and as appropriate, a turbulence closure model. It is important to reiterate that turbulence closure models were developed to provide a description of the effects of turbulence, rather than a detailed accounting of the true physics that govern turbulence.

As with most numerical models, judgments of the accuracy of a model simulation are generally based on comparisons of model predictions with the results of analogous physical measurements. In comparing numerical results with physical measurements, it should be kept in mind that physical measurements also include error. In that regard, physical measurements are not "true" representations of a given phenomenon, but like numerical models they include error.

The text that follows provides illustrations of CFD simulations of fluid mechanical behavior in UV photoreactors along with corresponding physical measurements. The examples illustrated below are not exhaustive, but indicate a range of model complexity and UV photoreactor types.

#### 6.3.5.1 Open-Channel UV Photoreactor; Vertical Lamp Orientation

One of the earliest examples of a CFD simulation for a UV photoreactor was developed by Lyn *et al.* [25] for an open-channel reactor with a vertical lamp orientation; the original study was followed by a more detailed examination of fluid mechanical behavior within the same reactor [26]. Figure 6.23 is a definition sketch of the reactor system and the computational domain. Also included in this figure are definitions of the characteristic velocity ($U_0$) and characteristic length ($D$) that were used for normalization of the numerical model predictions. Note the orientation of the Cartesian coordinate system that was used to describe the computational domain, as well as the location of the origin.

A CFD model based on the RANS approach was applied, using a standard form of the k-ε model for turbulence closure. Model predictions were compared with measurements that were conducted on a physical model of the reactor using LDV [15].

In general, the model predictions were in good agreement with LDV measurements for most locations in the system (see Figures 6.24 and 6.25). However, model predictions deviated considerably from LDV measurements in the regions of the flow field that were characterized by the presence of unsteady flow features, including the regions located immediately downstream of the lamp/sleeve structures and the square support structures. It is well known that a pattern of nearly periodic vortex shedding will develop in the "wake" region of these structures, which are sometimes referred to as "bluff bodies" in the literature.

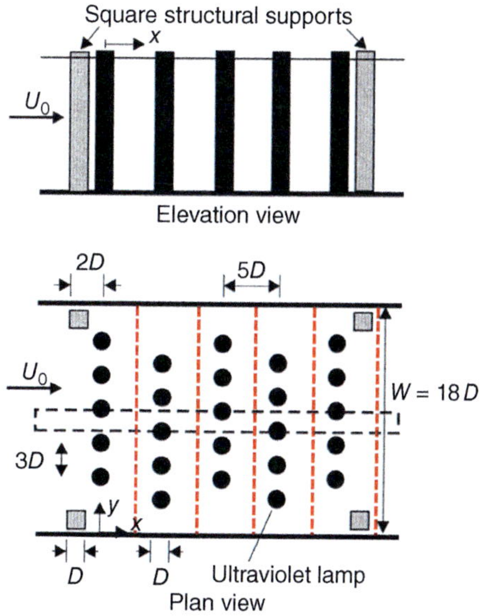


**Figure 6.23** Definition sketch of computational domain for vertical, open-channel UV disinfection reactor. The dashed (black) region in the bottom panel illustrates the central region of the reactor, which was the focus of an earlier investigation [25]. *Source:* Lyn 2004 / The American Society of Civil Engineers. *D* represents the diameter of a lamp sleeve, while $U_0$ represents the approach velocity; these two parameters were used for normalization of numerical model predictions. Dashed red lines represent approximate locations of cross sections for which velocity profile measurements were compared with model predictions (see Figures 6.24 and 6.25).

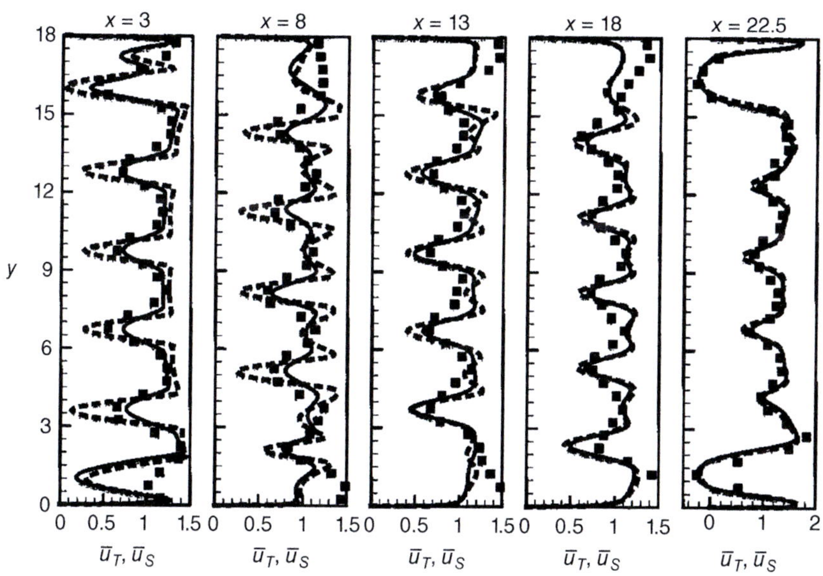


**Figure 6.24** Comparisons of average downstream (x-directed) velocity for five cross sections within an open-channel UV photoreactor with vertical lamp orientation [26]. *Source:* Lyn 2004/The American Society of Civil Engineers. Locations of cross sections are indicated by dimensionless streamwise coordinates (x) across the top of each panel. Dimensionless cross-stream locations (y) are indicated on the vertical axis. Dimensionless mean velocities based on LDV measurements (■), the time-average of the unsteady-flow simulation (–), and the steady RANS model (– – –) are presented.

Several general observations emerge from Figures 6.24 and 6.25:

- Agreement between measurements and simulations improves with distance into the lamp array (i.e. as the x-coordinate increases).
- Agreement between model predictions and measurement is better for the cross-stream (y-directed) component of velocity than the streamwise (x-directed) velocity component.

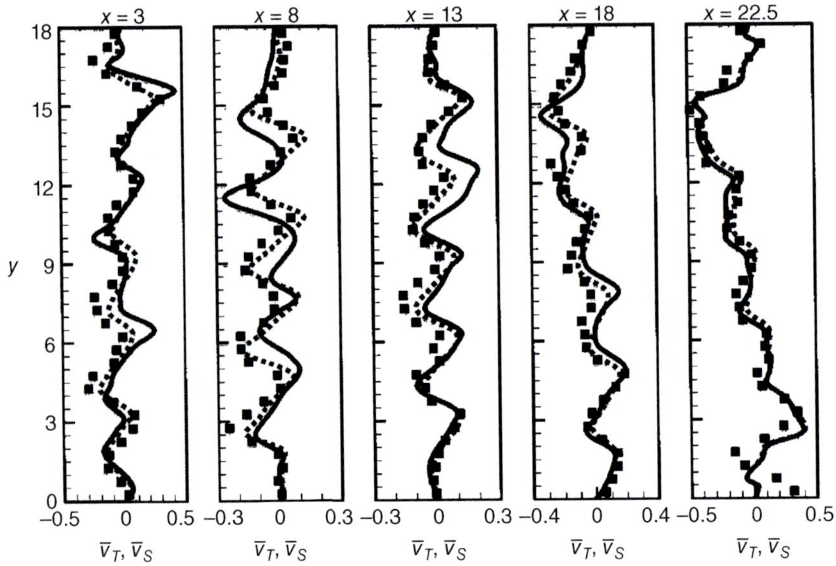


**Figure 6.25** Comparisons of average cross-stream (y-directed) velocity for five cross sections within an open-channel UV photoreactor with vertical lamp orientation [26]. Lyn 2004/The American Society of Civil Engineers. Locations of cross sections are indicated by dimensionless streamwise coordinates (x) across the top of each panel. Dimensionless cross-stream locations (y) are indicated on the vertical axis. Dimensionless mean velocities based on LDV measurements (■), the time-average of the unsteady-flow simulation (–), and the steady RANS model (- - -) are presented.

- Divergence between the model predictions and measurements is greatest in the "wake" regions of the array, where the flow field is characterized by unsteady flow structures.

Collectively, these observations suggest that the steady form of the RANS k-ε model provides an accurate description of the flow field for most parts of the computational domain. Simulation results are improved through the use of an unsteady simulation, but this is at a cost of considerably more computational effort and CPU time.

#### 6.3.5.2 Closed, Single-Lamp, Annular Reactors

As described previously, many other turbulence closure models can be applied in simulations of this type. Several comparisons of the abilities of alternative turbulence closure models to simulate hydraulic behavior in UV photoreactors have been published. In one such study, CFD simulations were applied with three common turbulence closure models, including the standard k-ε model, the realizable k-ε model, and a Reynolds stress model (RSM) [27]. These modeling approaches were used to simulate hydraulic behavior in two common single-lamp, annular reactors (see Figure 6.26). Single-lamp, annular reactors are among the most common UV photoreactor types and are generally applied at small scale, such as point-of-use applications.

For the L-shaped reactor, flow enters the reactor and experiences an immediate, though gentle change of direction as a result of encountering the upstream end of the lamp jacket. The flow will also encounter the lamp support structure, and then as the fluid approaches the outlet, it is forced to make a right-angle turn to exit the reactor. Away from these regions, the stationary components of the reactor system (lamp jacket and reactor wall) are oriented parallel to each other, and to the direction of incident flow. As a result, the flow field is characterized by relatively gentle changes of direction and magnitude, which simplifies the analysis by CFD. For this reactor type, CFD

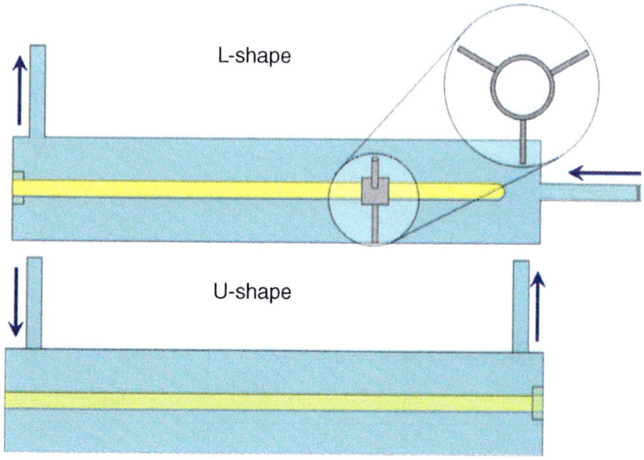


**Figure 6.26** Schematic illustrations of annular reactor types simulated by CFD models to compare the effects of turbulence closure model selection on hydraulic prediction [27]. Note direction of flow. Note also the lamp holder device (top) that was included in the simulations. *Source:* Sozzi and Taghipour 2006/with permission of Elsevier.

predictions based on all three turbulence-closure models yielded predictions that were in good agreement with particle image velocimetry (PIV) measurements in most locations. However, mesh quality was shown to influence model predictions, especially in regions where changes of velocity direction or magnitude were evident.

The U-shaped reactor presented greater challenges for simulation by CFD. In particular, the flow field at the inlet and the outlet of the system was forced to undergo right-angle turns. These rather sudden changes of direction tend to be more difficult to represent in numerical models than is the flow field in areas where the geometry of the system tends to straighten the flow (see previous paragraph). Visual inspection of the velocity field characterizations by CFD with the Realizable k-ε model and measurements by PIV indicate clear differences, especially in areas that were characterized by relatively rapid velocity gradients (changes in direction, magnitude, or both) (see Figure 6.27).

#### 6.3.5.3 Closed-Vessel, Cross-Flow, Four-Lamp Reactor

A subsequent study was conducted in which CFD simulations following a RANS approach were applied to allow comparisons for six turbulence closure models and their ability to predict hydraulic behavior in a closed-vessel reactor that housed four lamps oriented perpendicular to the incident direction of flow (see Figure 6.28) [28]. Turbulence models included in these simulations included two forms of the k-ε model, two forms of the k-ω model, the RSM, and a two-film model. Predictions from these models were compared with measurements by PIV.

In general, the models were variable in terms of their abilities to accurately simulate the flow field, based on comparisons with PIV measurements. All models provided reasonably accurate descriptions of the flow field away from the regions of the computational domain that were characterized by unsteady flow features. The k-ω and RSM closure models performed better than the k-ε and two-film models in the "wake" regions of the flow field; however, even these models yielded significant deviations from measured behavior in these regions. This is to be expected, given that these models all follow a RANS approach (i.e. averaging of the flow field characteristics is inherent to the model).

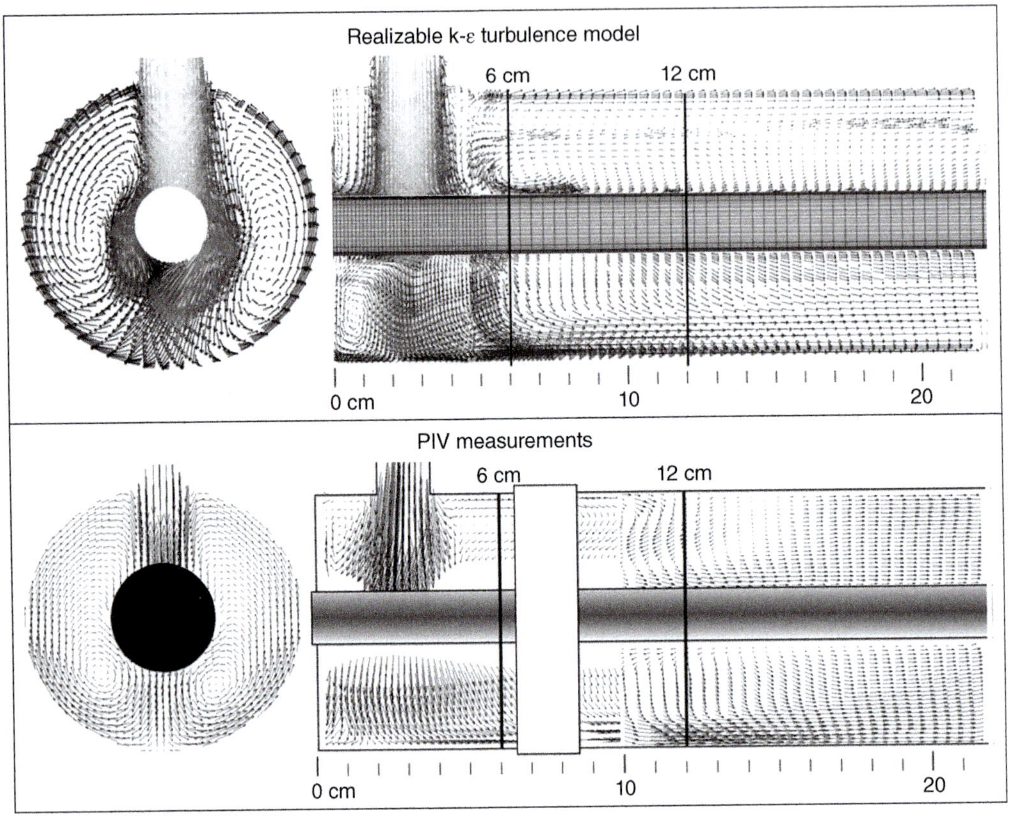


**Figure 6.27** Graphical images of CFD-predicted (top) and PIV measured (bottom) velocity fields in a U-shaped annular reactor [27]. For both data sets, the left image corresponds to a slice through the reactor inlet, while the right image corresponds to a slice along the central axis of the reactor in the longitudinal direction. *Source:* Sozzi and Taghipour 2006/with permission of Elsevier.

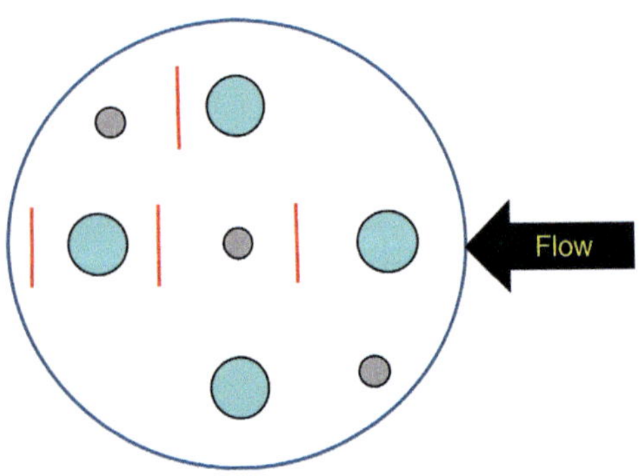


**Figure 6.28** Schematic illustration of reactor cross section for four-lamp, cross-flow reactor. *Source:* Redrawn from Liu et al. 2007. Light blue circles indicate locations of lamp/jacket structures; gray circles indicate horizontal rods; vertical red lines indicate locations where velocity field measurements by PIV were compared with CFD model predictions based on six turbulence closure models.

A broad range of alternatives to conventional RANS simulations exist. At one extreme are models based on DNS, in which the equations of motion and continuity are solved directly, but using an extremely fine computational mesh, so as to allow for simulation of even the smallest, relevant turbulent motions. DNS simulations can yield detailed, accurate descriptions of turbulent flow fields, but they can be very demanding in terms of computational time. The so-called "Large-Eddy Simulation" (LES) approach may be viewed as a compromise between conventional RANS simulations and DNS. LES approaches involve detailed simulations of the largest eddies, where the majority of the turbulent energy exists, while smaller scale turbulent motions are approximated by a model. LES is better able to capture the unsteady features of a flow field than RANS simulations, but at a lower computational cost that DNS [29].

A comparison of a RANS (k-ε) simulation with a simulation based on LES was conducted for several closed-vessel reactor types; for one of these reactors, predictions of hydraulic behavior based on these CFD approaches were compared with LDV measurements for an analogous physical model (see Figure 6.29) [30]. The LES model predictions were generally closer to the LDV measurements than the predictions of the k-ε model, especially in regions that were characterized by unsteady flow features: vortex shedding and recirculation (i.e. downstream of lamp/jacket assemblies).

#### 6.3.5.4 Reactors with Internal Baffles

The ability to accurately simulate unsteady flow field characteristics is not always important in modeling of UV photoreactors. However, it can be important in some situations, such as when internal baffles or static mixing devices are included in reactor design to improve overall reactor

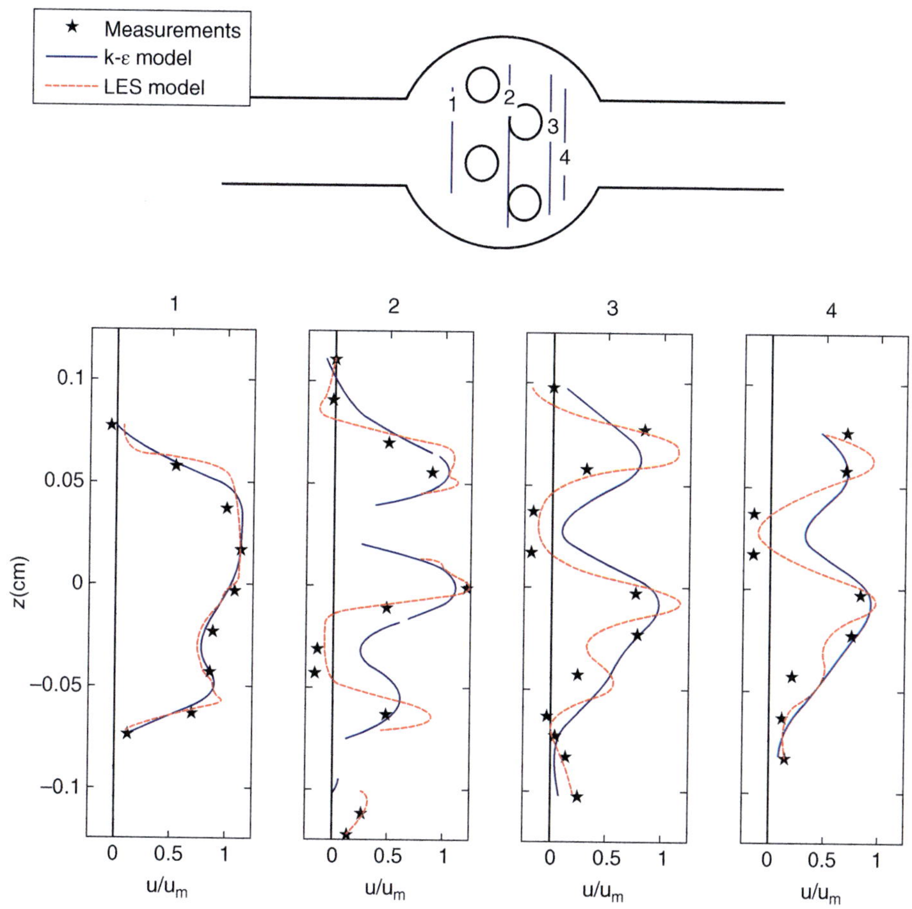


**Figure 6.29** Schematic illustration of cross section of 4-lamp, cross-flow reactor (top), with comparisons of streamwise velocity profiles at four locations (bottom) as measured by LDV and as simulated by a standard k-ε model and a large-eddy simulation (LES) model [30]. *Source:* Wols et al. 2010 / with permission of Elsevier.

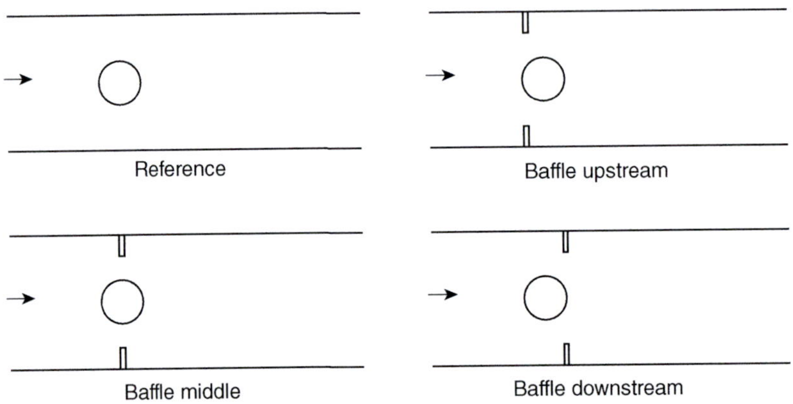


**Figure 6.30** Schematic illustrations of baffle locations assumed in study of four single-lamp hypothetical reactors by application of CFD using a RANS (k-ε) model and large-eddy simulation [30]. *Source:* Wols et al. 2010 / with permission of Elsevier.

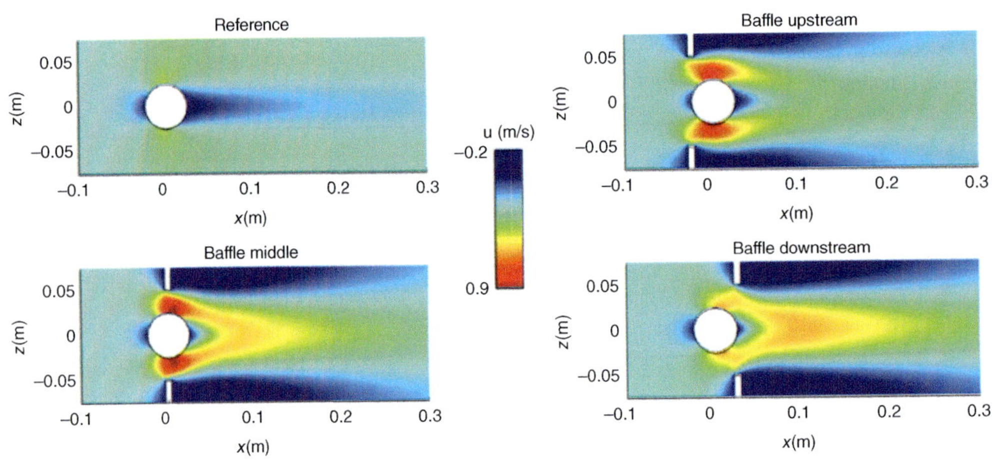


**Figure 6.31** Simulated (mean) streamwise velocity fields within four single-lamp, baffled reactors based on the RANS (k-ε) approach [30]. *Source:* Wols et al. 2010 / with permission of Elsevier.

behavior through induced mixing. Figure 6.30 illustrates a sequence of simple reactors that include internal baffles at various locations. CFD simulations of these hypothetical reactor types were conducted by RANS (k-ε) and LES [30]. Figures 6.31 and 6.32 illustrate the results of simulations for those four reactor types at a fixed approach velocity based on these two CFD approaches.

The basic features of the flow fields in these reactors are similar, but details differed among these CFD simulations: [30]

- Velocity increased around lamp in the presence or absence of baffles because of constriction of the flow path by the lamp/sleeve and the baffles (when present).
- Recirculation zones developed behind the lamp and baffles; stationary objects oriented nominally perpendicular to the flow function as bluff bodies.
- Baffles forced flow closer to lamp, but also increased fluid velocity near the lamp.
- For the reference case, simulation results produced by the k-ε and LES models were similar, except in the wake zone.

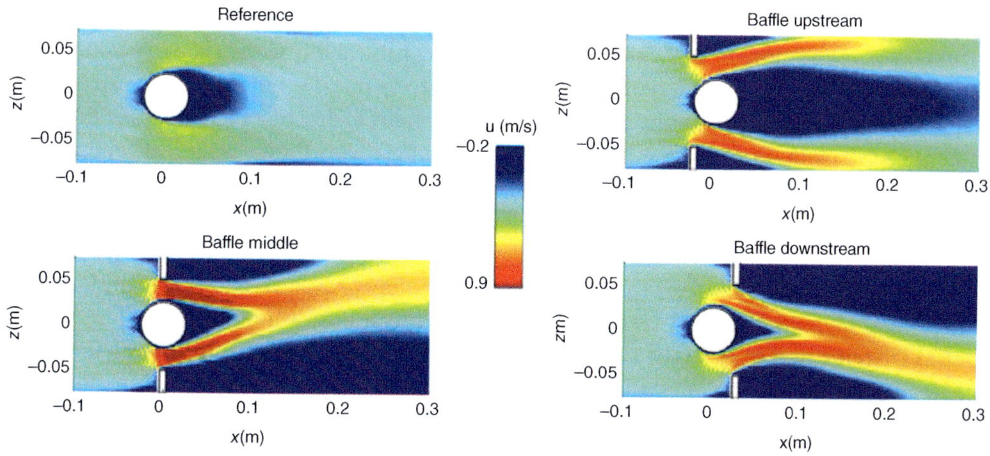


**Figure 6.32** Simulated (mean) streamwise velocity fields within four single-lamp, baffled reactors based on a large-eddy simulation [30]. Mean velocity was based on an ensemble average of the time-varying velocity field predicted by LES. *Source:* Wols et al. 2010 / with permission of Elsevier.

- For the baffled reactors, notable differences in the streamwise velocity field were evident in all recirculation zones. The recirculation zones predicted by LES were larger than those predicted by the k-ε model; LES model results also indicated negative velocity in the wake zone, which was not evident or less evident in the k-ε model predictions.
- Jets (regions of locally high, organized streamwise velocity) predicted by the LES model were larger and more pronounced than from the k-ε model. Stated another way, the k-ε model indicated more rapid diffusion of momentum than the LES simulation.
- Interestingly, the LES model predictions for the cases of reactors with "middle" and "downstream" baffles indicated a tendency of the jet to migrate to the reactor wall. Symmetry characteristics of the reactors would not suggest this tendency, but this behavior is consistent with the so-called Coandă effect, wherein a fluid jet tends to stay attached to a curved surface. A common illustration of this effect is observable by moving the curved surface of the back of a spoon into the downward jet of water from a faucet. The jet will "attach" to the spoon's surface and conform to its curvature, resulting in a change in the direction of the jet.

It is important to recognize that all CFD simulations are models of actual physical behavior, and that at some level, the model will fail to accurately describe this behavior. However, there are clear differences in the abilities of these various models to describe the fluid mechanics, especially in turbulent flow fields.

### 6.3.6 Process Simulations by CFD-E′ Field Modeling

The logic used to simulate process performance in photochemical reactors is illustrated in Figure 3.22. When adapted as a numerical model, this approach generally involves the inclusion of at least three basic sub-models:

1) **Intrinsic Kinetics:** Ideally, this will involve the use of mathematical expressions to describe the kinetics of the reactions of interest. The nature of these models is variable among the common photoreaction types. For processes based on pure photolysis, accurate descriptions of (local) reaction kinetics depend on accuracy in describing $\varepsilon$ and $\Phi$ (including wavelength-dependence,

as appropriate). For processes in which a primary photochemical reaction is used to promote the formation of one or more reactive intermediates (e.g. $HO^{\cdot}$), the ability to describe reaction kinetics depends on accuracy in describing the primary photochemical event (based on $\varepsilon$ and $\Phi$ for that reaction) as well as availability of accurate rate constants to describe the kinetics of the subsequent reactions that involve the intermediate. In contrast with these "chemical" endpoints, simulations of UV disinfection depend on accuracy of models that are used to describe the intrinsic kinetics of disinfection, often referred to as "dose–response" behavior. These models tend to be semiempirical, in that they generally describe changes in a measureable physiological endpoint among a population of target microbes as a function of dose, as opposed to the kinetics of the actual photobiochemical events that lead to microbial inactivation.

2) **Spatial Distribution of Radiant Energy:** As described in Chapter 3 (and elsewhere), the local kinetics of a photochemical process are directly proportional to the local fluence rate. Fluence rate field models are used to simulate the spatial distribution of radiant energy by accounting for the output power of the UV source and the physical (optical) behavior that governs the fate of this energy in its surroundings.
3) **Fluid Mechanics:** Mixing and transport behavior are critical to the behavior of all reactor systems. However, because photochemical reactions tend to be fast, and because most photochemical reactors are characterized by steep spatial gradients in local fluence rate (and equally steep gradients in local photochemical reaction rates), the effects of fluid mechanics on process performance in UV photoreactors can be more important than in corresponding chemical or biochemical reactor systems. Also, because most continuous-flow photochemical reactors operate in the turbulent regime, it is important to capture the effects of turbulence on mixing and transport behavior within the reactor system.

Numerical simulations of transport behavior, or of systems that involve fluid motion, can be developed from (at least) two common perspectives: Eulerian and Lagrangian. The Eulerian approach is one in which the system and its behavior are viewed from a fixed location, whereas the Lagrangian approach involves a frame of reference that moves with the velocity of the object being simulated.

Familiar examples of these perspectives may be found in video images of sporting events. As an example, consider the video that is found online at https://www.youtube.com/watch?v=paRQ7I-gYLhk. This video has essentially nothing to do with photochemical reactors, but provides clear illustrations of Eulerian and Lagrangian perspectives.[f] It is a video recording of final of the women's 8+ rowing event from the Olympics in Rio de Janeiro, Brazil. Parts of this video were shot using cameras that were placed at fixed locations, such as the 1000 m mark or the finish line; these are examples of an Eulerian perspective. On the other hand, other segments of this video were shot using cameras that were moving at essentially the same speed as the lead boat, either from a motorboat that was following the race, or from a camera that was mounted on an overhead system of cables that allowed the camera to move at the same speed as a boat; these are examples of a Lagrangian perspective. These two perspectives are useful for illustrating different aspects of the event. The Eulerian segments illustrate the relative behavior of all the boats in the race and their progress down the course. The Lagrangian segments illustrate the coordination of the athletes in a boat. Yet, despite these important differences, it is also important to recognize that the information gained from these two perspectives describes the same process. And perhaps more importantly, the overall result of this process is the same, regardless of perspective.

The Eulerian approach involves a fixed frame of reference; when generalized to an entire system, this approach yields "field" views of the system and the characteristics that are being modeled.

When applied to engineering systems, such as photochemical reactors, Eulerian simulations are often used to illustrate field views of fluid velocity, fluid kinetic energy, and the spatial distributions of scalar quantities such as concentration, temperature, or pressure. When applied for simulations of the fate and transport behavior of reactive constituents, Eulerian models include one additional mass-balance equation per reactant or product.

In contrast, Lagrangian simulations involve a moving frame of reference, usually based on the velocity of the object or control region being simulated. This is sometimes referred to as a "particle-centered" view. Lagrangian simulations of fate and transport behavior in engineering systems (e.g. photochemical reactors) generally involve the reaction simulations as post-processing operations. Specifically, it is common to solve the flow field first, then use the flow field simulation result together with a "particle-tracking" algorithm to simulate individual particle (reactant) trajectories. The rules of the reaction kinetics model(s) are then superimposed on the simulated trajectories to allow simulation of overall process behavior. Lagrangian simulations tend to require less CPU time than their analogous Eulerian simulations. In addition, Lagrangian simulations may be easier to understand and conceptualize.

When applied for simulations of photochemical reactors, the Lagrangian approach usually involves the development of a solution as a stepwise process [15, 16, 25, 31]. As described above, a CFD model is used to converge on a flow field simulation. Particle trajectories are then simulated by applying a force balance to "numerical particles" that are imposed on the system. A common strategy in these models is to simulate the (local) mean advective component of particle motion based on the (local) mean velocity of the flow field; the effects of local turbulence characteristics on particle motion are often simulated via a stochastic approach in which a form of a random-walk model is used to assign random components of local velocity based on the simulated characteristics of the flow field. The simulated fluence rate field is mapped onto each simulated particle trajectory, thereby allowing a trajectory-specific dose ($D_{\text{trajectory}}$) to be calculated as follows:[g]

$$D_{\text{trajectory}} = \int_0^{\tau} E'(t) \cdot \mathrm{d}t \tag{6.62}$$

where

$E'(t)$ = time-dependent fluence rate
$t$ = time
$\tau$ = period of exposure to UV radiation.

The process described above is repeated for a large population of particles that have starting locations that are representative of the influent cross section of the reactor system. As the number of particle simulations applied to the model increases, the dose distribution converges on a stable answer that is representative of the actual dose distribution delivered by the reactor for operating conditions that correspond to the model (e.g. Figure 6.17). The dose distribution is then integrated with an expression to describe intrinsic reaction kinetics through an analog of the segregated-flow model:[h]

$$\left(\frac{C}{C_0}\right)_{\text{reactor}} = \int_0^{\infty} \left(\frac{C}{C_0}\right)_{\text{batch}} E(D) \cdot \mathrm{d}D \tag{6.63}$$

where

$\left(\frac{C}{C_0}\right)_{\text{reactor}}$ = fraction of initial reactant concentration that remains in reactor effluent

$\left(\frac{C}{C_0}\right)_{batch}$ = intrinsic kinetics of reaction(s) of interest

$E(D)$ = dose distribution function (density form).

An Eulerian simulation will involve the continuity equation, the equations of motion, and a mass-balance equation for each constituent in the system (see equations 6.47–6.51) [25, 31, 32]. As appropriate, the flow field model may require the inclusion of a turbulence closure model.

Both approaches can be used to simulate the behavior of a system and in a general sense, they both provide integrated simulations of the same physical phenomena. The results of simulations based on these approaches can be useful in illustrating different aspects of system behavior; ideally, the overall results of the simulation should be the same for a given system and set of operating conditions. In practice, there can be subtle differences in the predictions of models based on Eulerian and Lagrangian simulations; however, these differences are usually associated with details of model implementation or assumptions that are applied (implicitly or explicitly).

For example, the segregated-flow model can be shown to hold rigorously for systems in which the reaction(s) of interest obey first-order kinetics. Deviations from first-order behavior can lead to incorrect solutions [33]. Another potential source of error relates to the assumptions used in a particle tracking simulation. The simulations of particle motion in a turbulent flow field require assumptions to be made regarding the statistical attributes of turbulence within the flow field. Any such assumption can lead to error in simulations of particle trajectories [23]. Eulerian approaches depend on an assumed link between turbulent momentum transport and turbulent transport of scalar quantities, as described by the Reynolds analogy (see equations 6.54 and 6.55, as well as accompanying text). The turbulent Schmidt number is used to describe this analogy in quantitative terms. If the Reynolds analogy were applied rigorously, then the value of the turbulent Schmidt number would always be 1. But reported values of this parameter deviate from unity. The link between momentum and mass transport in a turbulent flow field can represent an important source of error in these simulations [34]. And of course, any error in assignment of a kinetic expression or an input variable will also lead to error in the process simulation.

Models based on Eulerian and Lagrangian frames of reference can be useful for illustrating different aspects of the behavior of photochemical reactors. A Lagrangian model logically yields an estimate of the dose distribution as an intermediate result. Knowledge of the dose distribution allows for simulation of the extent of any photochemical reaction that may take place within a system; as such, modeling approaches that yield accurate estimates of the dose distribution can be very powerful for simulation of photoreactor behavior.

In contrast, Eulerian models are often used to illustrate the "field" behavior of a photoreactor. Figure 6.33 illustrates the received dose field within two single-lamp, annular reactors at fixed operating conditions. This method of visualization illustrates the accumulation of dose as fluid traverses the irradiated zone of the reactor. It should be noted that the dose distribution delivered by a photoreactor can be approximated in an Eulerian simulation by viewing a "slice" of the dose field across the outlet of the reactor. So, while the dose distribution is a natural outcome of a Lagrangian simulation, it is also possible to develop an estimate of the dose distribution by following an Eulerian approach.

### 6.3.7 Selection of Sub-Models

Numerical models are powerful tools that can be used to accurately simulate the combined effects of many complex physical phenomena. CFD-E′ field models are commonly applied in "numerical prototyping" of photochemical reactor systems as a method of reactor optimization [11, 36].

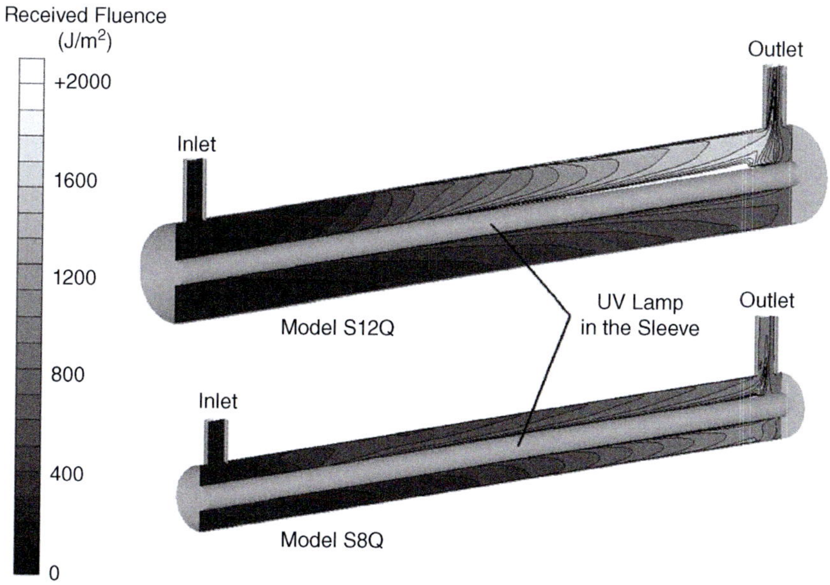


**Figure 6.33** Simulated dose fields within two single-lamp, annular photoreactors developed from an Eulerian perspective [35]. *Source:* Elyasi and Taghipour 2006/with permission of Elsevier.

Numerical models and computing hardware have advanced to the point where it is possible to conduct these simulations on numerical representations of reactors, use the results of these simulations to identify weaknesses in reactor behavior, and to conduct numerical tests of reactor modifications. Although ultimate validation of reactor performance relies on testing of physical prototypes, the approach of numerical prototyping is far less expensive and less time-consuming than the physical (empirical) testing methods. This approach has been further advanced by the integration of numerical prototyping with a numerical optimization platform for purposes of optimizing the design of UV disinfection reactors based on UV LEDs [37]. This particular approach was based on optimization criteria related to disinfection efficacy and input power requirements, but other optimization criteria could be applied and the approach could be expanded to other treatment endpoints.

Modelers are faced with many choices regarding the selection of sub-models and modeling strategies to be used in these simulations. There is no single combination of these sub-models that can be identified as the "best" choice for all applications. Decisions made about one sub-model can influence decisions about other sub-models. Modelers would be well advised to invest time in making these choices, so as to use tools that are well suited to a given application. The text that follows is intended to briefly address some of the factors that will affect these decisions.

#### 6.3.7.1 Reaction Kinetics Sub-Model

For processes that involve pure photolysis, the kinetics of the relevant photochemical reaction are governed by the molar absorption coefficient ($\varepsilon$) and the quantum yield ($\Phi$); when polychromatic sources of radiation are employed, the wavelength-dependent nature of these attributes must also be defined. When a photolytic process leads to the formation of one or more reactive intermediates, then rate expressions should be included to address all contributing elementary reactions. Lastly, in the case of UV disinfection processes, it is important to include dose–response models to accurately describe the responses of the target microbes or viruses to UV exposure over the range of doses to be expected within the dose distribution delivered by the reactor. These models are almost always semiempirical in nature, in that they do not describe the kinetics of the actual photobiochemical processes that lead to microbial inactivation, but rather they describe the effects of UV exposure on a measurable physiological endpoint that is related to UV exposure.

#### 6.3.7.2 Fluence Rate Field Sub-Model

A wide range of fluence rate field models is available in terms of their fidelity to the physical (optical) phenomena that govern the spatial distribution of radiant energy. Even simple fluence rate field models like PSS or LSI provide important information that is lost on methods that are based on spatial averaging of the fluence rate field. In general, as model components are added to address the governing physical phenomena in greater detail, and with greater fidelity, they become more complex and time-consuming to apply. In addition, increased model complexity is usually accompanied by an increase in the amount and quality of input data that are needed to support the model.

#### 6.3.7.3 CFD Sub-Model

In most applications, CFD models provide approximate solutions to the equations of motion and continuity. For the unusual situations in which UV photoreactors operate in the laminar regime, the flow field may become deterministic, and in some cases it is possible to describe fluid mechanics using a simple, closed-form, analytical solution. However, most UV photoreactors operate in the turbulent regime, and as such, it becomes necessary to incorporate components that allow for simulation of turbulent transport behavior.

#### 6.3.7.4 General Factors to Consider in Sub-Model Selection

Although the details of sub-model selection are likely to involve some problem-specific considerations, there are some factors that are quite general to these applications. These are discussed briefly below:

- Accuracy required in the simulation – When the expectations are for a simulation to provide a rough approximation of process performance, it may be possible to apply relatively simple models to describe system performance. On the other hand, when the demands or expectations of a simulation are such that high accuracy is required, then the use of a relatively complex model may be justified.
- Anticipated extent of reactant conversion – As described in Chapter 3, reactors that achieve relatively low conversion tend to be relatively insensitive to the details of reactor attributes that are simulated by these sub-models. Under those circumstances, simple models are often justified. However, as the extent of reaction progress increases, the overall performance of a reactor becomes increasingly sensitive to details of reaction kinetics, the fluence rate field, the flow field, or perhaps a combination of these factors. Because of this, simulations of reactors that are designed for high reactant conversion often require relatively complex sub-models to be used in their simulation.
- Accuracy and availability of input data – The quality of predictions made with numerical models is likely to be no better than the quality of the input data. If input data are unavailable or questionable in their accuracy, then the use of a highly detailed, complex model may be difficult to justify.
- Model cost – In general terms, the costs associated with model application increase with their complexity. Relevant costs include time of the modeler, CPU time, computational hardware and software, and the cost of supporting data.

- Availability of tools and methods to validate model predictions – A somewhat cynical (but correct) view of numerical models is that they are all incorrect, but some are useful. In more positive terms, it is correct to suggest that all models are simplifications of reality. The extent to which those simplifications introduce error, and the magnitude of those introduced errors, is highly variable. Ideally, model predictions should be subjected to validation by comparison with physical measurements. When possible, intermediate model outcomes should also be validated by physical measurements. For example, the dose distribution is a critical, intermediate output of CFD-E′ field models; in an ideal situation, dose distribution estimates developed by CFD-E′ field models should be validated using relevant physical measurements. Similarly, predicted reactant conversion should also be validated by physical measurements. However, it should also be noted that physical measurements are also characterized by errors, both in terms of their physical interpretation and the methods by which they are measured.

## 6.4 Summary

Numerical models used for simulation of the performance of photochemical reactors have evolved considerably over the last 20 years. These models generally incorporate sub-models to account for the fluence rate field, fluid mechanics, and the kinetics of relevant reactions. When integrated properly, often using an analog of the segregated-flow model, these numerical approaches can provide accurate predictions of process performance. As such, these models represent powerful tools for use in design of photochemical reactors and analysis of their performance. However, it is important to recognize that these methods involve the use of mathematical models, which are by definition simplifications of reality. So, despite the fact that these models can provide detailed information about reactor performance that is difficult or impossible to quantify using empirical methods, some skepticism of model results is warranted.

## Notes

**a** The refractive index for water may be affected by constituents in the water. However, very little information exists in the literature to address the dependence of the refractive index of aqueous solutions on solute composition. Therefore, the refractive index of an aqueous solution is generally assumed to be equal to that of pure water.

**b** Note that if the direction of radiation propagation was reversed, the opposite would be true. In other words, total internal reflection will be observed at a quartz:air interface, but not at a water:quartz interface.

**c** Photon radiance is defined as the number of photons per unit time that pass through a small transparent element of surface in a given direction from the source about a solid angle, divided by the solid angle and by the orthogonally projected area of the element in a plane normal to the given beam direction. Braslavsky [1]. Note also that the photon fluence rate ($E_{p,o}$) is calculated as: $\int_{4\pi} L_p \cdot d\Omega'$. Analogous definitions exist to relate *radiance* to *fluence rate*.

**d** Summation was used as an approximation to integration.

**e** Similar principles can be used to derive the governing equations in other coordinate systems.

**f** As a former rower from the United States, the author accepts that he has bias toward this sport and the outcome of this race. As an aside, the performance of the team from the United States in this event, both in terms of their preparation and execution, may have been the *perfect* rowing race.

**g** Note that the calculation of trajectory-specific doses often involves summation of dose increments associated with individual steps taken by the particle; when the steps are sufficiently small, this summation approach closely approximates integration.

**h** Applications of the segregated-flow model are usually based on a summation approximation to the integral described by equation (6.63).

**i** Note that the values of lamp length and lamp power used in this problem are roughly 10% of the values used in most contemporary reactors. These low values are used to simplify the calculations, while allowing the logic required for fluence rate models to be applied.

## References


1 Braslavsky, S.E. (**2007**). Glossary of terms used in photochemistry 3(rd) Edition (IUPAC Recommendations 2006). *Pure and Applied Chemistry 79* (3): 293–465.

2 Jacob, S.M. and Dranoff, J.S. (**1970**). Light intensity profiles in a perfectly mixed photoreactor. *AIChE Journal 16* (3): 359.

3 Blatchley, E.R. III (**1997**). Numerical modelling of UV intensity: Application to collimated-beam reactors and continuous-flow systems. *Water Research 31* (9): 2205–2218.

4 Duran, J.E., Taghipour, F., and Mohseni, M. (**2010**). Irradiance modeling in annular photoreactors using the finite-volume method. *Journal of Photochemistry and Photobiology a-Chemistry 215* (1): 81–89.

5 Irazoqui, A.H., Cerda, J., and Cassano, A.E. (**1973**). Radiation profiles in an empty annular photoreactor with a source of finite spatial dimensions. *AIChE Journal 19* (3): 460–467.

6 Li, M.K., Qiang, Z., Li, T.G. et al. (**2011**). In situ measurement of UV fluence rate distribution by use of a micro fluorescent silica detector. *Environmental Science & Technology 45* (7): 3034–3039.

7 Liu, D., Ducoste, J., Jin, S., and Linden, K. (**2004**). Evaluation of alternative fluence rate distribution models. *Journal of Water Supply Research and Technology-Aqua 53* (6): 391–408.

8 Quan, Y., Pehkonen, S.O., and Ray, M.B. (**2004**). Evaluation of three different lamp emission models using novel application of potassium ferrioxalate actinometry. *Industrial & Engineering Chemistry Research 43* (4): 948–955.

9 Imoberdorf, G.E., Taghipour, F., and Mohseni, M. (**2008**). Radiation field modeling of multi-lamp, homogeneous photoreactors. *Journal of Photochemistry and Photobiology a-Chemistry 198* (2–3): 169–178.

10 Bird, R.B., Stewart, W.E., and Lightfoot, E.N. (**1960**). *Transport Phenomena*. New York: Wiley.

11 Naunovic, Z. (**2006**). *Modeling and Design of an Ultraviolet Water Disinfection System for Long-term Space Missions*. West Lafayette, IN: Purdue University.

12 Naunovic, Z., Pennell, K.G., and Blatchley III, E.R. (**2008**). Development and performance of a fluence rate distribution model for a cylindrical excimer lamp. *Environmental Science & Technology 42* (5): 1605–1614.

13 Ahmed, Y.M., Jongewaard, M., Li, M., and Blatchley III, E.R. (**2018**). Raytracing for fluence rate simulations in ultraviolet photoreactors. *Environmental Science & Technology 52* (8): 4738–4745.

14 Ahmed, Y.M.H.I.H. (**2017**). *Stochastic Evaluation of Disinfection Performance in Large-Scale, Open-Channel UV Photoreactors*. West Lafayette, IN: Purdue University.

15 Chiu, K., Lyn, D.A., Savoye, P., and Blatchley III, E.R. (**1999**). Integrated UV disinfection model based on particle tracking. *Journal of Environmental Engineering ASCE 125* (1): 7–16.

16 Chiu, K.P., Lyn, D.A., Savoye, P., and Blatchley III, E.R. (**1999**). Effect of UV system modifications on disinfection performance. *Journal of Environmental Engineering ASCE 125* (5): 459–469.

17 Rodi, W. (**1993**). *Turbulence Models and Their Application in Hydraulics: A State-of-the-Art Review*. Rotterdam: International Association for Hydraulic Research.

18 Wendt, J.F. (**2009**). *Computational Fluid Dynamics: An Introduction*, 3e. Berlin: Springer-Verlag.

19 Versteeg, H.K. and Malalasekera, W. (**1995**). *An Introduction to Computational Fluid Dynamics: The Finite Volume Method*. Essex, England: Longman Scientific & Technical.

20 Launder, B.E. and Spalding, D.B. (**1974**). The numerical computation of turbulent flows. *Computer Methods in Applied Mechanics and Engineering 3*: 269–289.

21 Fischer, H.B., List, E.J., Koh, R.C.Y. et al. (**1979**). *Mixing in Inland and Coastal Waters*. New York: Academic Press.

22 Gualtieri, C.; Angeloudis, A.; Bombardelli, F.; Jha, S.; Stoesser, T., On the values for the turbulent schmidt number in environmental flows. *Fluids* **2017**, *2*, (2), 17; doi:https://doi.org/10.3390/fluids2020017.

23 Lyn, D.A. and Blatchley III, E.R. (**2005**). Numerical computational fluid dynamics-based models of ultraviolet disinfection channels. *Journal of Environmental Engineering ASCE 131* (6): 838–849.

24 Santoro, D., Raisee, M., Moghaddami, M. et al. (**2010**). Modeling hydroxyl radical distribution and trialkyl phosphates oxidation in UV-$H_2O_2$ photoreactors using computational fluid dynamics. *Environmental Science & Technology 44* (16): 6233–6241.

25 Lyn, D.A., Chiu, K., and Blatchley III, E.R. (**1999**). Numerical modeling of flow and disinfection in UV disinfection channels. *Journal of Environmental Engineering ASCE 125* (1): 17–26.

26 Lyn, D.A. (**2004**). Steady and unsteady simulations of turbulent flow and transport in ultraviolet disinfection channels. *Journal of Hydraulic Engineering ASCE 130* (8): 762–770.

27 Sozzi, D.A. and Taghipour, F. (**2006**). Computational and experimental study of annular photo-reactor hydrodynamics. *International Journal of Heat and Fluid Flow 27* (6): 1043–1053.

28 Liu, D., Wu, C., Linden, K., and Ducoste, J. (**2007**). Numerical simulation of UV disinfection reactors: Evaluation of alternative turbulence models. *Applied Mathematical Modelling 31* (9): 1753–1769.

29 Yang, Z.Y. (**2015**). Large-eddy simulation: past, present and the future. *Chinese Journal of Aeronautics 28* (1): 11–24.

30 Wols, B.A., Uijttewaal, W.S.J., Hofman, J. et al. (**2010**). The weaknesses of a k-epsilon model compared to a large-eddy simulation for the prediction of UV dose distributions and disinfection. *Chemical Engineering Journal 162* (2): 528–536.

31 Sozzi, D.A. and Taghipour, F. (**2006**). UV reactor performance modeling by Eulerian and Lagrangian methods. *Environmental Science & Technology 40* (5): 1609–1615.

32 Wols, B.A. and Hofman-Caris, C.H.M. (**2012**). Modelling micropollutant degradation in UV/$H_2O_2$ systems: Lagrangian versus Eulerian method. *Chemical Engineering Journal 210*: 289–297.

33 Levenspiel, O. (**1972**). *Chemical Reaction Engineering*, 2e. New York: Wiley.

34 Zhang, J., Tejada-Martinez, A.E., and Zhang, Q. (**2014**). Developments in computational fluid dynamics-based modeling for disinfection technologies over the last two decades: a review. *Environmental Modelling & Software 58*: 71–85.

35 Elyasi, S. and Taghipour, F. (**2006**). Simulation of UV photoreactor for water disinfection in Eulerian framework. *Chemical Engineering Science 61* (14): 4741–4749.

36 Naunovic, Z., Lim, S., and Blatchley III, E.R. (**2008**). Investigation of microbial inactivation efficiency of a UV disinfection system employing an excimer lamp. *Water Research 42* (19): 4838–4846.

37 Jenny, R.M., Jasper, M.N., Simmons, O.D. et al. (**2015**). Heuristic optimization of a continuous flow point-of-use UV-LED disinfection reactor using computational fluid dynamics. *Water Research 83*: 310–318.

## Appendix 6.A PSS Model Implementation in Spreadsheet Format

In many conventional UV systems, cylindrical lamps are positioned within a channel cross section with a goal of providing a uniformly high fluence rate within cross sections of the system that are parallel to the direction of flow. These lamps are encased in quartz "sleeves" to prevent direct contact of lamp electrodes with water, while allowing for effective transmission of output radiation to the surrounding liquid. Quartz is chosen for these applications because it tends to have high transmittance in the UVC range, but tends to be inert with respect to most chemical and photochemical reactions.

Obviously, the objective of producing a spatially uniform fluence rate field within a reactor cross section will favor short distances between lamps. Lamp placement will also affect head loss, which in many situations represents a critical design constraint; treatment scenarios often have very limited available head – overcoming these limitations with pumps is expensive and is generally difficult to justify from an economic standpoint. Head losses in UV systems tend to be small but will increase as lamp spacing decreases. Therefore, lamp geometry and spacing represent optimization issues in the design of photochemical reactors.

As an alternative to conventional low-pressure (LP) mercury (Hg) lamps, excimer lamps and light emitting diode (LED) sources have been developed; both of these lamp types can be developed as mercury-free UV sources, and both can be implemented in alternative geometries. For example, "flat-screen" UV sources based on LED and excimer lamp technologies have been developed. Fluence rate field models are important for evaluation of UV source geometry and spacing. A crude comparison of reactor designs based on these two UV source types is presented in the spreadsheet (Excel file) that accompanies this Appendix.

The schematic presented in Figure 6.A.1, left, represents a cross-sectional drawing of a simple 1-lamp system, based on a conventional LP Hg source; the schematic presented in Figure 6.A.1, right, represents a possible flat-screen alternative. For purposes of these fluence rate field

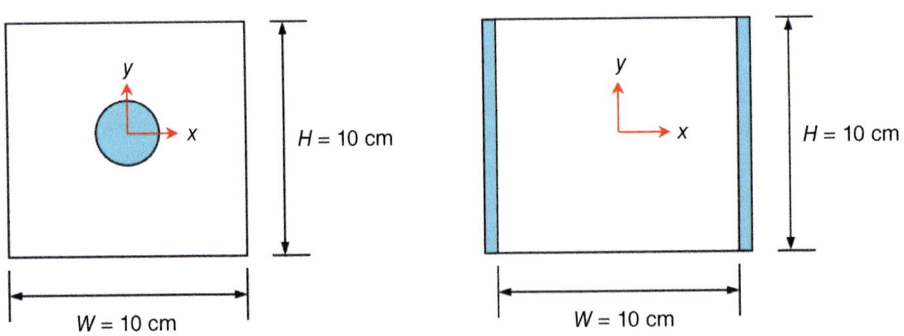


Figure 6.A.1 Schematic illustration of reactor cross sections. Left image represents system with a single cylindrical lamp located along the central, longitudinal axis of the reactor with a square cross section. Right image represents a system with flat-panel "screens" along the left and right walls of the reactor. Note coordinate system for each reactor simulation. Z-direction is into the plane of the page. Both UV source types are assumed to have a length ($L$) of 10 cm (into the page).

simulations, we will assume that the cylindrical lamp (left) has an output power of 10 W ($\lambda = 254$ nm) that is uniformly distributed along the 10 cm arc length. The lamp is encased in a quartz sleeve of diameter 2.5 cm, with a wall thickness of 1.5 mm.

For the flat-screen system, we will assume that identical flat-screen sources are placed on the vertical walls of system. Together, the flat-screens provide 10 W of output power (5.0 W each) at a characteristic wavelength, $\lambda = 254$ nm. Output power from these flat-screen sources will be assumed to be released in a uniform pattern along the entire length (10 cm) and width of each screen. We will also assume that the flat screens are each separated from the water by a quartz plate of thickness 1.5 mm.

For both lamp types, we will apply a point-source summation (PSS) model to simulate the fluence rate field in a single cross section, located at half the length of the reactor. For both simulations, we will assume the z-direction is into the page. In both cases, the UV sources have a length ($L$) in the direction of flow (i.e. into the page) of 10 cm. Therefore, the fluence rate cross sections we will simulate will be for a dimensionless longitudinal location of $\frac{z}{L} = 0.5$.

For both systems, we will use the following definitions of optical characteristics of the water in the system and the quartz that is used for the "sleeves" that are used as a barrier between the lamps and water:

$$\text{Water transmittance } (\lambda = 254\,\text{nm}) = 95\%(1.0\,\text{cm path})$$

$$\text{Quartz transmittance } (\lambda = 254\,\text{nm}) = 95\%(1.5\,\text{mm path})$$

For the LP Hg lamp simulation, the point sources used for the PSS simulation will be spaced at 1.00 cm intervals along the central axis of the lamp (which coincides with the central axis of the channel). These point sources will be located at $z = -4.5, -3.5, \ldots, 3.5, 4.5$ cm. This means that we will use ten (10) evenly spaced point sources to simulate the fluence rate field cross section for the LP Hg lamp system.

For the flat-panel reactor, we will use one hundred (100) evenly spaced point sources for each panel. Note that there are symmetry arguments that will be applied in these simulations that can simplify our calculations. Also, as suggested in Figure 6.A.1, right, we will choose the origin of our Cartesian coordinate system to be at the center of the plane at which these simulations will be conducted. Following this logic, the point sources you use for your simulations of the flat-panel reactor should be at ($y$,$z$) locations of (−4.5 cm, −4.5 cm), (−3.5 cm, −4.5 cm), ... (3.5 cm, 4.5 cm), (4.5 cm, 4.5 cm).

## Simulation of Fluence Rate Field in Cylindrical Lamp Reactor

The PSS method was applied to estimate the fluence rate at the grid locations in the upper-right quadrant of the fluence rate plane (see Figure 6.A.2). By symmetry arguments, these values were then translated to the other three quadrants of the fluence rate plane. The pink circles in Figure 6.A.2 represent the locations of the receptor sites in the fluence rate plane. The blue "stars" indicate the locations of the point sources that were used to simulate the lamp.

The origin of the coordinate system was selected as the center of the fluence rate plane, with the "$x$" and "$y$" directions being horizontal and vertical in the fluence rate plane, and the z-direction being parallel to the axis of the lamp. The lamp itself was represented as a series of colinear point sources. The point sources were located at ($x$,$y$) = (0,0) in this system, with $z$ values ranging from −4.5 cm to 4.5 cm, in 1.0 cm increments. The receptor sites were located at ($x$,$y$) values of 0.5–4.5

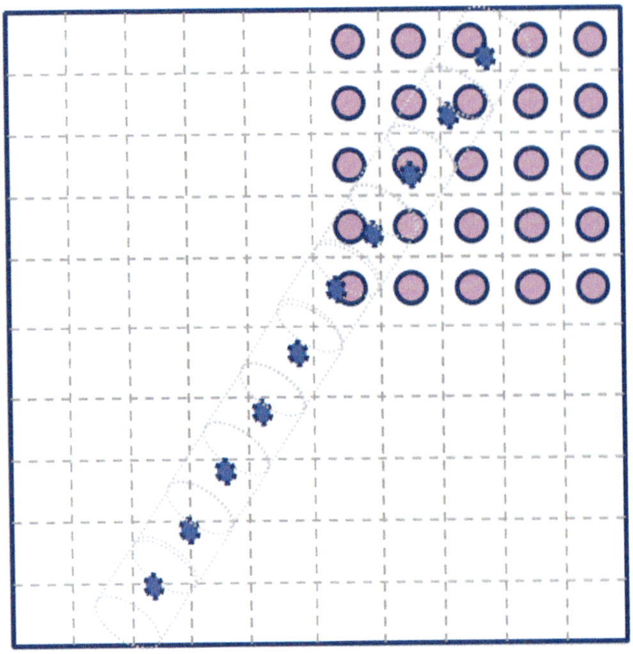

**Figure 6.A.2** Schematic representation of fluence rate plane located at $z = 0$ in reactor with cylindrical lamp. Also indicated in this schematic are the $x$ and $y$ directions in the Cartesian coordinate system, the locations of the receptor sites in the upper-right quadrant (pink circles), and the locations of the point sources along the central axis of the lamp (along the $z$-axis, indicated by blue stars).

cm, in 1.0 cm increments; the $z$ values of all receptor site locations were 0 in this coordinate system because the fluence rate plane was located at $z = 0$.

The physical characteristics of the lamp and reactor are included as constants in the spreadsheet labeled "Fluence Rate Calcs." For 95% quartz transmittance through a 0.15 cm optical path and 95% water transmittance through a 1.00 cm optical path, the absorbance coefficients for quartz and water as follows:

$$\alpha_q = \frac{-\ln\left(T_q\right)}{t_q} = \frac{-\ln(0.95)}{0.15\ \text{cm}} = 0.342\ \text{cm}^{-1}$$

$$\alpha_w = \frac{-\ln\left(T_w\right)}{\ell_w} = \frac{-\ln(0.95)}{1.00\ \text{cm}} = 0.0513\ \text{cm}^{-1}$$

From the coordinate locations of the point sources and the receptor sites, an array of calculations was then developed to define the distance from each point source to a given receptor site:

$$\rho_i = \sqrt{x_i^2 + y_i^2 + z_i^2}$$

where

$\rho_i$ = distance from $i^{\text{th}}$ point source to receptor site
$x_i$ = horizontal distance from $i^{\text{th}}$ point source to receptor site
= $x$ location of receptor site
$y_i$ = vertical distance from $i^{\text{th}}$ point source to receptor site

= $y$ location of receptor site
$z_i$ = longitudinal distance from $i^{\text{th}}$ point source to receptor site
= $z$ location of $i^{\text{th}}$ point source.

A second array of values was then developed to describe the fluence rate contributions from each point source to the given receptor site. Note that only half of the point sources (i.e. those with positive $z$-locations) were included in this set of calculations. Another symmetry argument was applied to perform the point source summation; specifically, because the fluence rate plane is a plane of symmetry, the fluence rate contributions from each of the five point sources with positive $z$-locations was calculated. The contributions of the point sources located at negative $z$-locations were included by symmetry. This was done by including a row at the end of each column that represented two times the sum of the contributions from the five point sources located at positive $z$-locations. These values are highlighted in yellow in the spreadsheet. Note that the fluence rate value corresponding to the location at $x = 0.5$ cm, $y = 0.5$ cm in the fluence rate plane was assigned a fluence rate value of 0 because it is inside the lamp, and therefore not part of the fluid domain. In turn, these values were used to define the fluence rate values within the upper-right quadrant of the plane. As described above, symmetry arguments were then used to define the fluence rate at other locations within the plane at $z = 0$. This information was then used to develop a contour graph of the fluence rate in this plane. This graph is included in the attached spreadsheet.

## Simulation of Fluence Rate Field in Flat-Screen Reactor

The simulation of the flat-screen reactor followed similar logic, but with allowance for the different geometry and configuration of the lamps. Figure 6.A.3 provides a schematic illustration of the approach used for these calculations.

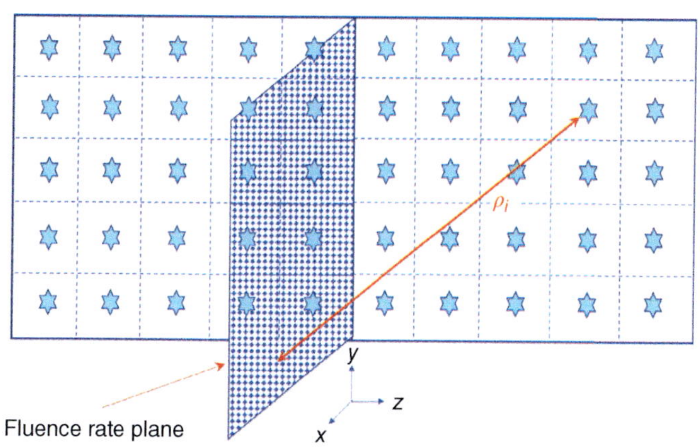


**Figure 6.A.3** Schematic representation of fluence rate plane located at $z = 0$ in reactor with flat-screen UV sources. Also indicated in this schematic are the $x$, $y$, and $z$ directions in the Cartesian coordinate system, and the locations of the point sources along one wall of the reactor (located in a plane parallel to the $z$-axis, indicated by blue stars). Receptor sites are in the fluence rate plane in a 10 × 10 grid (not indicated in schematic). An indication of the distance from an arbitrary point source to an arbitrary receptor site in the fluence rate plane is indicated by $\rho_i$.

At the top of the spreadsheet tab labeled "PSS, Flat Screen Reactor," numerical values and units for physical constants and reactor characteristics are assigned. The first field of numbers (spanning rows 24–73) defines the distance from each point source for the left-side flat screen in the positive $z$-domain to each receptor site on the fluence rate plane. The next field of numbers (spanning rows 80–129) represents calculations of the contribution of each point source for the left-side flat screen to the fluence rate received at each receptor site in the fluence rate plane. The next field of numbers (spanning rows 139–188) defines the distance from each point source for the right-side flat screen in the positive $z$-domain to each receptor site on the fluence rate plane. The next field of numbers (spanning rows 195–244) represents calculations of the contribution of each point source for the right-side flat screen to the fluence rate received at each receptor site in the fluence rate plane.

The last field of values (rows 251–301) provides the sum of contributions from all point sources (from both flat screens) to fluence rate received at each location within the fluence rate plane. Note that at the bottom of this field, a row of numbers highlighted in yellow is presented. This set of values represents the sum of fluence rate field contributions from all point sources from both flat screens. Note that these values represent 2X the sum of the fluence rate field calculations presented from the fields above. This is an application of a symmetry argument, recognizing that we have only accounted for point sources in the positive $z$-domain. The contributions of point sources in the negative $z$-domain will be mirror images of those from the positive $z$-domain. A graph was prepared to illustrate the fluence rate distribution in the fluence rate plane. This graph is included in the spreadsheet.

## Interpretation of Simulation Results

The flat-screen reactor yields a fluence rate distribution that is more uniform than that developed from the single-lamp reactor. This behavior is attributable to the broader, more-uniform distribution of UVC power from the flat-screen lamps than from the single, cylindrical lamp. In addition, the fluence rate field for the flat-screen reactor shows the highest values of fluence rate along the channel walls, with local minima at the centerline, at both the far upstream and downstream ends.

Numerical simulation of fluence rate field by PSS, Single-lamp reactor; $T = 65\%$.

**A1**

| | |
|---|---|
| $P = 10\,\text{W}$ | Lamp output power |
| $n = 10$ | Number of point sources |
| $L = 10\,\text{cm}$ | Lamp arc length |
| $D = 2.5\,\text{cm}$ | Lamp jacket diameter |
| $r = 1.25\,\text{cm}$ | Lamp jacket radius |
| $t_q = 0.15\,\text{cm}$ | Quartz jacket wall thickness |
| $T_q = 0.95$ | Quartz jacket transmittance |
| $\sigma_q = 0.342\,\text{cm}^{-1}$ | Quartz absorbance coefficient |
| $T_w = 0.65$ | Water transmittance (1.0 cm path length) |
| $\sigma_w = 0.431\,\text{cm}^{-1}$ | Water absorbance coefficient |

## B1

We will simulate the fluence rate (E′) field in the upper-right quadrant of the reactor only (other quadrants can be inferred by symmetry).

The origin for our coordinate system will be located in the center of the plane where we are calculating the fluence rate field ($z/L = 0$).

The 10 point sources are evenly spaced (1 cm increments) and located collinearly along the $z =$ axis at locations ±0.5 cm, ±1.5 cm, ±2.5 cm, ±3.5 cm, and ±4.5 cm.

The x-direction is horizontal across the fluence rate plane; the y-direction is vertical across the fluence rate plane; the z-direction is along the length of the lamp axis.

Distances from point sources to receptor sites are calculated using the Pythagorean theorem, as described below.

## C1

$\rho_i =$ distance from $i$th point source to receptor site (cm); $x$ and $y$ represent receptor site position in plane at $z/L = 0$; $z$ represents longitudinal distance from point source to plane at $z/L = 0$.

| **x (cm) →** | **0.5** | **0.5** | **0.5** | **0.5** | **0.5** | **1.5** | **1.5** | **1.5** | **1.5** | **1.5** | **2.5** | **2.5** | **2.5** | **2.5** | **2.5** | **3.5** | **3.5** | **3.5** | **3.5** | **3.5** | **4.5** | **4.5** | **4.5** | **4.5** | **4.5** |
|---|---|---|---|---|---|---|---|---|---|---|---|---|---|---|---|---|---|---|---|---|---|---|---|---|---|
| **y (cm) →** | **0.5** | **1.5** | **2.5** | **3.5** | **4.5** | **0.5** | **1.5** | **2.5** | **3.5** | **4.5** | **0.5** | **1.5** | **2.5** | **3.5** | **4.5** | **0.5** | **1.5** | **2.5** | **3.5** | **4.5** | **0.5** | **1.5** | **2.5** | **3.5** | **4.5** |
| **z (cm) ↓** | | | | | | | | | | | | | | | | | | | | | | | | | |
| 0.5 | 0.87 | 1.66 | 2.60 | 3.57 | 4.56 | 1.66 | 2.18 | 2.96 | 3.84 | 4.77 | 2.60 | 2.96 | 3.57 | 4.33 | 5.17 | 3.57 | 3.84 | 4.33 | 4.97 | 5.72 | 4.56 | 4.77 | 5.17 | 5.72 | 6.38 |
| 1.5 | 1.66 | 2.18 | 2.96 | 3.84 | 4.77 | 2.18 | 2.60 | 3.28 | 4.09 | 4.97 | 2.96 | 3.28 | 3.84 | 4.56 | 5.36 | 3.84 | 4.09 | 4.56 | 5.17 | 5.89 | 4.77 | 4.97 | 5.36 | 5.89 | 6.54 |
| 2.5 | 2.60 | 2.96 | 3.57 | 4.33 | 5.17 | 2.96 | 3.28 | 3.84 | 4.56 | 5.36 | 3.57 | 3.84 | 4.33 | 4.97 | 5.72 | 4.33 | 4.56 | 4.97 | 5.55 | 6.22 | 5.17 | 5.36 | 5.72 | 6.22 | 6.84 |
| 3.5 | 3.57 | 3.84 | 4.33 | 4.97 | 5.72 | 3.84 | 4.09 | 4.56 | 5.17 | 5.89 | 4.33 | 4.56 | 4.97 | 5.55 | 6.22 | 4.97 | 5.17 | 5.55 | 6.06 | 6.69 | 5.72 | 5.89 | 6.22 | 6.69 | 7.26 |
| 4.5 | 4.56 | 4.77 | 5.17 | 5.72 | 6.38 | 4.77 | 4.97 | 5.36 | 5.89 | 6.54 | 5.17 | 5.36 | 5.72 | 6.22 | 6.84 | 5.72 | 5.89 | 6.22 | 6.69 | 7.26 | 6.38 | 6.54 | 6.84 | 7.26 | 7.79 |

**D1**

Contributions of each point source to irradiance received at an individual receptor site (mW/cm$^2$).

| x (cm) → | 0.5 | 0.5 | 0.5 | 0.5 | 0.5 | 1.5 | 1.5 | 1.5 | 1.5 | 1.5 | 2.5 | 2.5 | 2.5 | 2.5 | 2.5 | 3.5 | 3.5 | 3.5 | 3.5 | 3.5 | 4.5 | 4.5 | 4.5 | 4.5 | 4.5 |
|---|---|---|---|---|---|---|---|---|---|---|---|---|---|---|---|---|---|---|---|---|---|---|---|---|---|
| y (cm) → | 0.5 | 1.5 | 2.5 | 3.5 | 4.5 | 0.5 | 1.5 | 2.5 | 3.5 | 4.5 | 0.5 | 1.5 | 2.5 | 3.5 | 4.5 | 0.5 | 1.5 | 2.5 | 3.5 | 4.5 | 0.5 | 1.5 | 2.5 | 3.5 | 4.5 |
| z (cm) ↓ | | | | | | | | | | | | | | | | | | | | | | | | | |
| 0.5 | 0.00 | 23.61 | 6.32 | 2.19 | 0.88 | 23.61 | 10.81 | 4.17 | 1.69 | 0.73 | 6.32 | 4.17 | 2.19 | 1.07 | 0.52 | 2.19 | 1.69 | 1.07 | 0.62 | 0.34 | 0.88 | 0.73 | 0.52 | 0.34 | 0.20 |
| 1.5 | 0.00 | 12.82 | 4.48 | 1.75 | 0.75 | 12.82 | 6.99 | 3.12 | 1.38 | 0.63 | 4.48 | 3.12 | 1.75 | 0.90 | 0.46 | 1.75 | 1.38 | 0.90 | 0.53 | 0.30 | 0.75 | 0.63 | 0.46 | 0.30 | 0.18 |
| 2.5 | 0.00 | 6.33 | 2.65 | 1.19 | 0.56 | 6.33 | 3.83 | 1.96 | 0.97 | 0.48 | 2.65 | 1.96 | 1.19 | 0.66 | 0.35 | 1.19 | 0.97 | 0.66 | 0.41 | 0.24 | 0.56 | 0.48 | 0.35 | 0.24 | 0.15 |
| 3.5 | 0.00 | 3.37 | 1.50 | 0.75 | 0.38 | 3.37 | 2.09 | 1.15 | 0.62 | 0.33 | 1.50 | 1.15 | 0.75 | 0.44 | 0.25 | 0.75 | 0.62 | 0.44 | 0.29 | 0.18 | 0.38 | 0.33 | 0.25 | 0.18 | 0.12 |
| 4.5 | 0.00 | 1.95 | 0.86 | 0.45 | 0.25 | 1.95 | 1.18 | 0.67 | 0.38 | 0.22 | 0.86 | 0.67 | 0.45 | 0.28 | 0.17 | 0.45 | 0.38 | 0.28 | 0.19 | 0.12 | 0.25 | 0.22 | 0.17 | 0.12 | 0.08 |
| sum | 0 | 96.15684 | 31.63339 | 12.67986 | 5.636416 | 96.15684 | 49.79037 | 22.1483 | 10.06468 | 4.771429 | 31.63339 | 22.1483 | 12.67986 | 6.736128 | 3.517206 | 12.67986 | 10.06468 | 6.736128 | 4.079238 | 2.348921 | 5.636416 | 4.771429 | 3.517206 | 2.348921 | 1.472415 |

**E1**

Simulated irradiance field (mW/cm$^2$) (1 cm increments, upper-right quadrant).

| *x* (cm) → <br> *y* (cm) ↓ | 0.5 | 1.5 | 2.5 | 3.5 | 4.5 |
|---|---|---|---|---|---|
| 4.5 | 5.64 | 4.77 | 3.52 | 2.35 | 1.47 |
| 3.5 | 12.68 | 10.06 | 6.74 | 4.08 | 2.35 |
| 2.5 | 31.63 | 22.15 | 12.68 | 6.74 | 3.52 |
| 1.5 | 96.16 | 49.79 | 22.15 | 10.06 | 4.77 |
| 0.5 | | 96.16 | 31.63 | 12.68 | 5.64 |

**F1**

Entire irradiance field (mW/cm$^2$) (1 cm increments), $T = 65\%$ (representative of municipal wastewater effluent).

| *x* (cm) → <br> *y* (cm) ↓ | −4.5 | −3.5 | −2.5 | −1.5 | −0.5 | 0.5 | 1.5 | 2.5 | 3.5 | 4.5 |
|---|---|---|---|---|---|---|---|---|---|---|
| 4.5 | 1.47 | 2.35 | 3.52 | 4.77 | 5.64 | 5.64 | 4.77 | 3.52 | 2.35 | 1.47 |
| 3.5 | 2.35 | 4.08 | 6.74 | 10.06 | 12.68 | 12.68 | 10.06 | 6.74 | 4.08 | 2.35 |
| 2.5 | 3.52 | 6.74 | 12.68 | 22.15 | 31.63 | 31.63 | 22.15 | 12.68 | 6.74 | 3.52 |
| 1.5 | 4.77 | 10.06 | 22.15 | 49.79 | 96.16 | 96.16 | 49.79 | 22.15 | 10.06 | 4.77 |
| 0.5 | 5.64 | 12.68 | 31.63 | 96.16 | | | 96.16 | 31.63 | 12.68 | 5.64 |
| −0.5 | 5.64 | 12.68 | 31.63 | 96.16 | | | 96.16 | 31.63 | 12.68 | 5.64 |
| −1.5 | 4.77 | 10.06 | 22.15 | 49.79 | 96.16 | 96.16 | 49.79 | 22.15 | 10.06 | 4.77 |
| −2.5 | 3.52 | 6.74 | 12.68 | 22.15 | 31.63 | 31.63 | 22.15 | 12.68 | 6.74 | 3.52 |
| −3.5 | 2.35 | 4.08 | 6.74 | 10.06 | 12.68 | 12.68 | 10.06 | 6.74 | 4.08 | 2.35 |
| −4.5 | 1.47 | 2.35 | 3.52 | 4.77 | 5.64 | 5.64 | 4.77 | 3.52 | 2.35 | 1.47 |

Cells with no entry are located inside lamp or lamp jacket

$I_{avg} = 19.14$ mW/cm$^2$
$I_{max} = 96.16$ mW/cm$^2$
$I_{min} = 1.47$ mW/cm$^2$
Count = 96

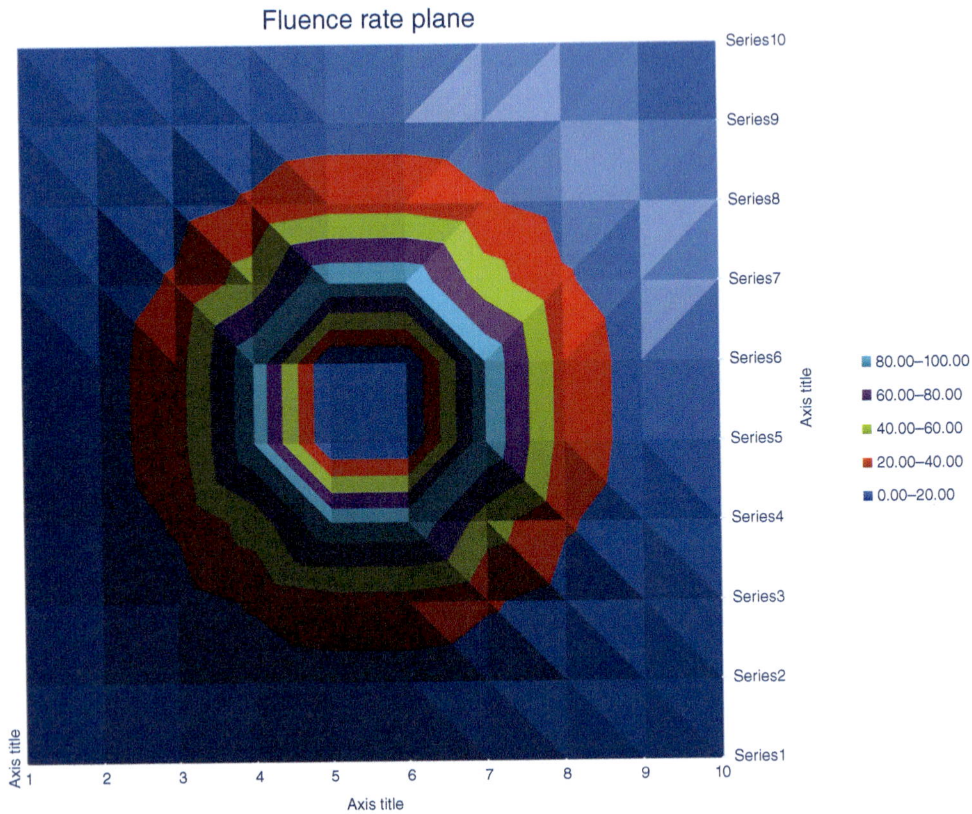


Numerical simulation of fluence rate field by PSS, Flat-screen reactor; $T = 65\%$.

## A2

| | |
|---|---|
| $P = 5\,\mathrm{W}$ | Lamp output power |
| $n = 50$ | Number of point sources |
| $L = 10\,\mathrm{cm}$ | Lamp arc length |
| $D = 2.5\,\mathrm{cm}$ | Lamp jacket diameter |
| $r = 1.25\,\mathrm{cm}$ | Lamp jacket radius |
| $t_q = 0.15\,\mathrm{cm}$ | Quartz jacket wall thickness |
| $T_q = 0.95$ | Quartz jacket transmittance |
| $\sigma_q = 0.341\,955\,296\,\mathrm{cm}^{-1}$ | Quartz absorbance coefficient |
| $T_w = 0.65$ | Water transmittance (1.0 cm path length) |
| $\sigma_w = 0.430\,782\,916\,\mathrm{cm}^{-1}$ | Water absorbance coefficient |

**B2**

We will simulate the fluence rate (E′) field in the upper-right quadrant of the reactor only (other quadrants can be inferred by symmetry).

The origin for our coordinate system will be located in the center of the plane where we are calculating the fluence rate field ($z/L = 0$).

The 100 point sources that represent each flat screen are evenly spaced at 1 cm increments within the plane of the left-side flat screen

The x-direction is horizontal across the fluence rate plane; the y-direction is vertical across the fluence rate plane; the z-direction is parallel to the plane of the flat screen.

Distances from point sources to receptor sites are calculated using the Pythagorean theorem, as described below.

Note that there will be a plane of symmetry at the fluence rate plane ($z/L = 0$), so calculations will be done for $0 < z/L < 0.5$, and then multiplied by 2, based on a symmetry argument.

Note also that calculations will be performed first for the left-side flat screen; a similar set of calculations will be conducted for the right-side flat screen; these results will then be added.

The origin for our coordinate system will be located in the center of the plane where we are calculating the fluence rate field ($z/L = 0$).

**C2**

Left-side flat screen (located at $x = -5.15$ cm); therefore, calculations of distance from point source to receptor site must account for this distance.

$\rho_i$ = distance from $i$th point source to receptor site (cm); $x_r$ and $y_r$ represent receptor site position in plane at $z/L = 0$; $z$ represents longitudinal distance from point source to plane at $z/L = 0$.

$y$ and $z$ represent point source locations for point sources in the left-side flat screen.

|  | $x_r$ (cm) → | 0.5 | 0.5 | 0.5 | 0.5 | 0.5 | 1.5 | 1.5 | 1.5 | 1.5 | 1.5 | 2.5 | 2.5 | 2.5 | 2.5 | 2.5 | 3.5 | 3.5 | 3.5 | 3.5 | 3.5 | 4.5 | 4.5 | 4.5 | 4.5 | 4.5 |
|---|---|---|---|---|---|---|---|---|---|---|---|---|---|---|---|---|---|---|---|---|---|---|---|---|---|---|
|  | $y_r$ (cm) → | 0.5 | 1.5 | 2.5 | 3.5 | 4.5 | 0.5 | 1.5 | 2.5 | 3.5 | 4.5 | 0.5 | 1.5 | 2.5 | 3.5 | 4.5 | 0.5 | 1.5 | 2.5 | 3.5 | 4.5 | 0.5 | 1.5 | 2.5 | 3.5 | 4.5 |
| $y$ (cm) ↓ | $z$ (cm) ↓ |  |  |  |  |  |  |  |  |  |  |  |  |  |  |  |  |  |  |  |  |  |  |  |  |  |
| −4.5 | 0.5 | 7.56 | 8.26 | 9.01 | 9.81 | 10.64 | 8.34 | 8.97 | 9.67 | 10.42 | 11.20 | 9.15 | 9.74 | 10.38 | 11.08 | 11.82 | 10.00 | 10.54 | 11.14 | 11.79 | 12.49 | 10.88 | 11.37 | 11.93 | 12.54 | 13.21 |
| −3.5 | 0.5 | 6.94 | 7.56 | 8.26 | 9.01 | 9.81 | 7.78 | 8.34 | 8.97 | 9.67 | 10.42 | 8.65 | 9.15 | 9.74 | 10.38 | 11.08 | 9.54 | 10.00 | 10.54 | 11.14 | 11.79 | 10.46 | 10.88 | 11.37 | 11.93 | 12.54 |
| −2.5 | 0.5 | 6.42 | 6.94 | 7.56 | 8.26 | 9.01 | 7.31 | 7.78 | 8.34 | 8.97 | 9.67 | 8.23 | 8.65 | 9.15 | 9.74 | 10.38 | 9.17 | 9.54 | 10.00 | 10.54 | 11.14 | 10.12 | 10.46 | 10.88 | 11.37 | 11.93 |
| −1.5 | 0.5 | 6.01 | 6.42 | 6.94 | 7.56 | 8.26 | 6.96 | 7.31 | 7.78 | 8.34 | 8.97 | 7.92 | 8.23 | 8.65 | 9.15 | 9.74 | 8.89 | 9.17 | 9.54 | 10.00 | 10.54 | 9.87 | 10.12 | 10.46 | 10.88 | 11.37 |
| −0.5 | 0.5 | 5.76 | 6.01 | 6.42 | 6.94 | 7.56 | 6.74 | 6.96 | 7.31 | 7.78 | 8.34 | 7.73 | 7.92 | 8.23 | 8.65 | 9.15 | 8.72 | 8.89 | 9.17 | 9.54 | 10.00 | 9.71 | 9.87 | 10.12 | 10.46 | 10.88 |
| 0.5 | 0.5 | 5.67 | 5.76 | 6.01 | 6.42 | 6.94 | 6.67 | 6.74 | 6.96 | 7.31 | 7.78 | 7.67 | 7.73 | 7.92 | 8.23 | 8.65 | 8.66 | 8.72 | 8.89 | 9.17 | 9.54 | 9.66 | 9.71 | 9.87 | 10.12 | 10.46 |
| 1.5 | 0.5 | 5.76 | 5.67 | 5.76 | 6.01 | 6.42 | 6.74 | 6.67 | 6.74 | 6.96 | 7.31 | 7.73 | 7.67 | 7.73 | 7.92 | 8.23 | 8.72 | 8.66 | 8.72 | 8.89 | 9.17 | 9.71 | 9.66 | 9.71 | 9.87 | 10.12 |
| 2.5 | 0.5 | 6.01 | 5.76 | 5.67 | 5.76 | 6.01 | 6.96 | 6.74 | 6.67 | 6.74 | 6.96 | 7.92 | 7.73 | 7.67 | 7.73 | 7.92 | 8.89 | 8.72 | 8.66 | 8.72 | 8.89 | 9.87 | 9.71 | 9.66 | 9.71 | 9.87 |
| 3.5 | 0.5 | 6.42 | 6.01 | 5.76 | 5.67 | 5.76 | 7.31 | 6.96 | 6.74 | 6.67 | 6.74 | 8.23 | 7.92 | 7.73 | 7.67 | 7.73 | 9.17 | 8.89 | 8.72 | 8.66 | 8.72 | 10.12 | 9.87 | 9.71 | 9.66 | 9.71 |
| 4.5 | 0.5 | 6.94 | 6.42 | 6.01 | 5.76 | 5.67 | 7.78 | 7.31 | 6.96 | 6.74 | 6.67 | 8.65 | 8.23 | 7.92 | 7.73 | 7.67 | 9.54 | 9.17 | 8.89 | 8.72 | 8.66 | 10.46 | 10.12 | 9.87 | 9.71 | 9.66 |
| −4.5 | 1.5 | 7.69 | 8.38 | 9.12 | 9.91 | 10.73 | 8.45 | 9.08 | 9.77 | 10.51 | 11.29 | 9.26 | 9.84 | 10.48 | 11.17 | 11.91 | 10.10 | 10.63 | 11.23 | 11.88 | 12.57 | 10.97 | 11.46 | 12.02 | 12.62 | 13.28 |
| −3.5 | 1.5 | 7.08 | 7.69 | 8.38 | 9.12 | 9.91 | 7.90 | 8.45 | 9.08 | 9.77 | 10.51 | 8.76 | 9.26 | 9.84 | 10.48 | 11.17 | 9.65 | 10.10 | 10.63 | 11.23 | 11.88 | 10.55 | 10.97 | 11.46 | 12.02 | 12.62 |
| −2.5 | 1.5 | 6.57 | 7.08 | 7.69 | 8.38 | 9.12 | 7.45 | 7.90 | 8.45 | 9.08 | 9.77 | 8.35 | 8.76 | 9.26 | 9.84 | 10.48 | 9.28 | 9.65 | 10.10 | 10.63 | 11.23 | 10.22 | 10.55 | 10.97 | 11.46 | 12.02 |
| −1.5 | 1.5 | 6.18 | 6.57 | 7.08 | 7.69 | 8.38 | 7.10 | 7.45 | 7.90 | 8.45 | 9.08 | 8.05 | 8.35 | 8.76 | 9.26 | 9.84 | 9.00 | 9.28 | 9.65 | 10.10 | 10.63 | 9.97 | 10.22 | 10.55 | 10.97 | 11.46 |
| −0.5 | 1.5 | 5.93 | 6.18 | 6.57 | 7.08 | 7.69 | 6.89 | 7.10 | 7.45 | 7.90 | 8.45 | 7.86 | 8.05 | 8.35 | 8.76 | 9.26 | 8.84 | 9.00 | 9.28 | 9.65 | 10.10 | 9.82 | 9.97 | 10.22 | 10.55 | 10.97 |
| 0.5 | 1.5 | 5.85 | 5.93 | 6.18 | 6.57 | 7.08 | 6.82 | 6.89 | 7.10 | 7.45 | 7.90 | 7.80 | 7.86 | 8.05 | 8.35 | 8.76 | 8.78 | 8.84 | 9.00 | 9.28 | 9.65 | 9.77 | 9.82 | 9.97 | 10.22 | 10.55 |
| 1.5 | 1.5 | 5.93 | 5.85 | 5.93 | 6.18 | 6.57 | 6.89 | 6.82 | 6.89 | 7.10 | 7.45 | 7.86 | 7.80 | 7.86 | 8.05 | 8.35 | 8.84 | 8.78 | 8.84 | 9.00 | 9.28 | 9.82 | 9.77 | 9.82 | 9.97 | 10.22 |
| 2.5 | 1.5 | 6.18 | 5.93 | 5.85 | 5.93 | 6.18 | 7.10 | 6.89 | 6.82 | 6.89 | 7.10 | 8.05 | 7.86 | 7.80 | 7.86 | 8.05 | 9.00 | 8.84 | 8.78 | 8.84 | 9.00 | 9.97 | 9.82 | 9.77 | 9.82 | 9.97 |
| 3.5 | 1.5 | 6.57 | 6.18 | 5.93 | 5.85 | 5.93 | 7.45 | 7.10 | 6.89 | 6.82 | 6.89 | 8.35 | 8.05 | 7.86 | 7.80 | 7.86 | 9.28 | 9.00 | 8.84 | 8.78 | 8.84 | 10.22 | 9.97 | 9.82 | 9.77 | 9.82 |
| 4.5 | 1.5 | 7.08 | 6.57 | 6.18 | 5.93 | 5.85 | 7.90 | 7.45 | 7.10 | 6.89 | 6.82 | 8.76 | 8.35 | 8.05 | 7.86 | 7.80 | 9.65 | 9.28 | 9.00 | 8.84 | 8.78 | 10.55 | 10.22 | 9.97 | 9.82 | 9.77 |
| −4.5 | 2.5 | 7.95 | 8.61 | 9.34 | 10.11 | 10.92 | 8.69 | 9.30 | 9.97 | 10.70 | 11.47 | 9.47 | 10.04 | 10.67 | 11.35 | 12.07 | 10.30 | 10.82 | 11.40 | 12.04 | 12.73 | 11.15 | 11.63 | 12.18 | 12.78 | 13.43 |
| −3.5 | 2.5 | 7.36 | 7.95 | 8.61 | 9.34 | 10.11 | 8.15 | 8.69 | 9.30 | 9.97 | 10.70 | 8.99 | 9.47 | 10.04 | 10.67 | 11.35 | 9.85 | 10.30 | 10.82 | 11.40 | 12.04 | 10.74 | 11.15 | 11.63 | 12.18 | 12.78 |
| −2.5 | 2.5 | 6.87 | 7.36 | 7.95 | 8.61 | 9.34 | 7.71 | 8.15 | 8.69 | 9.30 | 9.97 | 8.59 | 8.99 | 9.47 | 10.04 | 10.67 | 9.49 | 9.85 | 10.30 | 10.82 | 11.40 | 10.41 | 10.74 | 11.15 | 11.63 | 12.18 |
| −1.5 | 2.5 | 6.49 | 6.87 | 7.36 | 7.95 | 8.61 | 7.38 | 7.71 | 8.15 | 8.69 | 9.30 | 8.29 | 8.59 | 8.99 | 9.47 | 10.04 | 9.22 | 9.49 | 9.85 | 10.30 | 10.82 | 10.17 | 10.41 | 10.74 | 11.15 | 11.63 |
| −0.5 | 2.5 | 6.26 | 6.49 | 6.87 | 7.36 | 7.95 | 7.17 | 7.38 | 7.71 | 8.15 | 8.69 | 8.11 | 8.29 | 8.59 | 8.99 | 9.47 | 9.06 | 9.22 | 9.49 | 9.85 | 10.30 | 10.02 | 10.17 | 10.41 | 10.74 | 11.15 |
| 0.5 | 2.5 | 6.18 | 6.26 | 6.49 | 6.87 | 7.36 | 7.10 | 7.17 | 7.38 | 7.71 | 8.15 | 8.05 | 8.11 | 8.29 | 8.59 | 8.99 | 9.00 | 9.06 | 9.22 | 9.49 | 9.85 | 9.97 | 10.02 | 10.17 | 10.41 | 10.74 |
| 1.5 | 2.5 | 6.26 | 6.18 | 6.26 | 6.49 | 6.87 | 7.17 | 7.10 | 7.17 | 7.38 | 7.71 | 8.11 | 8.05 | 8.11 | 8.29 | 8.59 | 9.06 | 9.00 | 9.06 | 9.22 | 9.49 | 10.02 | 9.97 | 10.02 | 10.17 | 10.41 |
| 2.5 | 2.5 | 6.49 | 6.26 | 6.18 | 6.26 | 6.49 | 7.38 | 7.17 | 7.10 | 7.17 | 7.38 | 8.29 | 8.11 | 8.05 | 8.11 | 8.29 | 9.22 | 9.06 | 9.00 | 9.06 | 9.22 | 10.17 | 10.02 | 9.97 | 10.02 | 10.17 |
| 3.5 | 2.5 | 6.87 | 6.49 | 6.26 | 6.18 | 6.26 | 7.71 | 7.38 | 7.17 | 7.10 | 7.17 | 8.59 | 8.29 | 8.11 | 8.05 | 8.11 | 9.49 | 9.22 | 9.06 | 9.00 | 9.06 | 10.41 | 10.17 | 10.02 | 9.97 | 10.02 |
| 4.5 | 2.5 | 7.36 | 6.87 | 6.49 | 6.26 | 6.18 | 8.15 | 7.71 | 7.38 | 7.17 | 7.10 | 8.99 | 8.59 | 8.29 | 8.11 | 8.05 | 9.85 | 9.49 | 9.22 | 9.06 | 9.00 | 10.74 | 10.41 | 10.17 | 10.02 | 9.97 |

| | | | | | | | | | | | | | | | | | | | | | | | | | | |
|---|---|---|---|---|---|---|---|---|---|---|---|---|---|---|---|---|---|---|---|---|---|---|---|---|---|---|
| −4.5 | 3.5 | 8.32 | 8.95 | 9.65 | 10.40 | 11.19 | 9.03 | 9.62 | 10.27 | 10.98 | 11.72 | 9.79 | 10.33 | 10.94 | 11.61 | 12.32 | 10.59 | 11.09 | 11.67 | 12.29 | 12.96 | 11.42 | 11.89 | 12.42 | 13.01 | 13.65 |
| −3.5 | 3.5 | 7.76 | 8.32 | 8.95 | 9.65 | 10.40 | 8.51 | 9.03 | 9.62 | 10.27 | 10.98 | 9.32 | 9.79 | 10.33 | 10.94 | 11.61 | 10.15 | 10.59 | 11.09 | 11.67 | 12.29 | 11.02 | 11.42 | 11.89 | 12.42 | 13.01 |
| −2.5 | 3.5 | 7.29 | 7.76 | 8.32 | 8.95 | 9.65 | 8.09 | 8.51 | 9.03 | 9.62 | 10.27 | 8.93 | 9.32 | 9.79 | 10.33 | 10.94 | 9.80 | 10.15 | 10.59 | 11.09 | 11.67 | 10.69 | 11.02 | 11.42 | 11.89 | 12.42 |
| −1.5 | 3.5 | 6.94 | 7.29 | 7.76 | 8.32 | 8.95 | 7.78 | 8.09 | 8.51 | 9.03 | 9.62 | 8.65 | 8.93 | 9.32 | 9.79 | 10.33 | 9.54 | 9.80 | 10.15 | 10.59 | 11.09 | 10.46 | 10.69 | 11.02 | 11.42 | 11.89 |
| −0.5 | 3.5 | 6.72 | 6.94 | 7.29 | 7.76 | 8.32 | 7.58 | 7.78 | 8.09 | 8.51 | 9.03 | 8.47 | 8.65 | 8.93 | 9.32 | 9.79 | 9.38 | 9.54 | 9.80 | 10.15 | 10.59 | 10.31 | 10.46 | 10.69 | 11.02 | 11.42 |
| 0.5 | 3.5 | 6.65 | 6.72 | 6.94 | 7.29 | 7.76 | 7.51 | 7.58 | 7.78 | 8.09 | 8.51 | 8.41 | 8.47 | 8.65 | 8.93 | 9.32 | 9.33 | 9.38 | 9.54 | 9.80 | 10.15 | 10.27 | 10.31 | 10.46 | 10.69 | 11.02 |
| 1.5 | 3.5 | 6.72 | 6.65 | 6.72 | 6.94 | 7.29 | 7.58 | 7.51 | 7.58 | 7.78 | 8.09 | 8.47 | 8.41 | 8.47 | 8.65 | 8.93 | 9.38 | 9.33 | 9.38 | 9.54 | 9.80 | 10.31 | 10.27 | 10.31 | 10.46 | 10.69 |
| 2.5 | 3.5 | 6.94 | 6.72 | 6.65 | 6.72 | 6.94 | 7.78 | 7.58 | 7.51 | 7.58 | 7.78 | 8.65 | 8.47 | 8.41 | 8.47 | 8.65 | 9.54 | 9.38 | 9.33 | 9.38 | 9.54 | 10.46 | 10.31 | 10.27 | 10.31 | 10.46 |
| 3.5 | 3.5 | 7.29 | 6.94 | 6.72 | 6.65 | 6.72 | 8.09 | 7.78 | 7.58 | 7.51 | 7.58 | 8.93 | 8.65 | 8.47 | 8.41 | 8.47 | 9.80 | 9.54 | 9.38 | 9.33 | 9.38 | 10.69 | 10.46 | 10.31 | 10.27 | 10.31 |
| 4.5 | 3.5 | 7.76 | 7.29 | 6.94 | 6.72 | 6.65 | 8.51 | 8.09 | 7.78 | 7.58 | 7.51 | 9.32 | 8.93 | 8.65 | 8.47 | 8.41 | 10.15 | 9.80 | 9.54 | 9.38 | 9.33 | 11.02 | 10.69 | 10.46 | 10.31 | 10.27 |
| −4.5 | 4.5 | 8.78 | 9.39 | 10.06 | 10.78 | 11.54 | 9.46 | 10.02 | 10.65 | 11.33 | 12.06 | 10.19 | 10.71 | 11.30 | 11.95 | 12.64 | 10.96 | 11.45 | 12.00 | 12.61 | 13.27 | 11.76 | 12.22 | 12.74 | 13.32 | 13.94 |
| −3.5 | 4.5 | 8.26 | 8.78 | 9.39 | 10.06 | 10.78 | 8.97 | 9.46 | 10.02 | 10.65 | 11.33 | 9.74 | 10.19 | 10.71 | 11.30 | 11.95 | 10.54 | 10.96 | 11.45 | 12.00 | 12.61 | 11.37 | 11.76 | 12.22 | 12.74 | 13.32 |
| −2.5 | 4.5 | 7.82 | 8.26 | 8.78 | 9.39 | 10.06 | 8.57 | 8.97 | 9.46 | 10.02 | 10.65 | 9.37 | 9.74 | 10.19 | 10.71 | 11.30 | 10.20 | 10.54 | 10.96 | 11.45 | 12.00 | 11.06 | 11.37 | 11.76 | 12.22 | 12.74 |
| −1.5 | 4.5 | 7.49 | 7.82 | 8.26 | 8.78 | 9.39 | 8.27 | 8.57 | 8.97 | 9.46 | 10.02 | 9.10 | 9.37 | 9.74 | 10.19 | 10.71 | 9.95 | 10.20 | 10.54 | 10.96 | 11.45 | 10.83 | 11.06 | 11.37 | 11.76 | 12.22 |
| −0.5 | 4.5 | 7.29 | 7.49 | 7.82 | 8.26 | 8.78 | 8.09 | 8.27 | 8.57 | 8.97 | 9.46 | 8.93 | 9.10 | 9.37 | 9.74 | 10.19 | 9.80 | 9.95 | 10.20 | 10.54 | 10.96 | 10.69 | 10.83 | 11.06 | 11.37 | 11.76 |
| 0.5 | 4.5 | 7.22 | 7.29 | 7.49 | 7.82 | 8.26 | 8.03 | 8.09 | 8.27 | 8.57 | 8.97 | 8.88 | 8.93 | 9.10 | 9.37 | 9.74 | 9.75 | 9.80 | 9.95 | 10.20 | 10.54 | 10.65 | 10.69 | 10.83 | 11.06 | 11.37 |
| 1.5 | 4.5 | 7.29 | 7.22 | 7.29 | 7.49 | 7.82 | 8.09 | 8.03 | 8.09 | 8.27 | 8.57 | 8.93 | 8.88 | 8.93 | 9.10 | 9.37 | 9.80 | 9.75 | 9.80 | 9.95 | 10.20 | 10.69 | 10.65 | 10.69 | 10.83 | 11.06 |
| 2.5 | 4.5 | 7.49 | 7.29 | 7.22 | 7.29 | 7.49 | 8.27 | 8.09 | 8.03 | 8.09 | 8.27 | 9.10 | 8.93 | 8.88 | 8.93 | 9.10 | 9.95 | 9.80 | 9.75 | 9.80 | 9.95 | 10.83 | 10.69 | 10.65 | 10.69 | 10.83 |
| 3.5 | 4.5 | 7.82 | 7.49 | 7.29 | 7.22 | 7.29 | 8.57 | 8.27 | 8.09 | 8.03 | 8.09 | 9.37 | 9.10 | 8.93 | 8.88 | 8.93 | 10.20 | 9.95 | 9.80 | 9.75 | 9.80 | 11.06 | 10.83 | 10.69 | 10.65 | 10.69 |
| 4.5 | 4.5 | 8.26 | 7.82 | 7.49 | 7.29 | 7.22 | 8.97 | 8.57 | 8.27 | 8.09 | 8.03 | 9.74 | 9.37 | 9.10 | 8.93 | 8.88 | 10.54 | 10.20 | 9.95 | 9.80 | 9.75 | 11.37 | 11.06 | 10.83 | 10.69 | 10.65 |

D2

Contributions of each point source in left-side flat panel to irradiance received at an individual receptor site (mW/cm$^2$).

| $y$ (cm) ↓ | $x_r$ (cm) → | 0.5 | 0.5 | 0.5 | 0.5 | 0.5 | 1.5 | 1.5 | 1.5 | 1.5 | 1.5 | 2.5 | 2.5 | 2.5 | 2.5 | 2.5 | 3.5 | 3.5 | 3.5 | 3.5 | 3.5 | 4.5 | 4.5 | 4.5 | 4.5 | 4.5 |
|---|---|---|---|---|---|---|---|---|---|---|---|---|---|---|---|---|---|---|---|---|---|---|---|---|---|---|
| | $y_r$ (cm) → | 0.5 | 1.5 | 2.5 | 3.5 | 4.5 | 0.5 | 1.5 | 2.5 | 3.5 | 4.5 | 0.5 | 1.5 | 2.5 | 3.5 | 4.5 | 0.5 | 1.5 | 2.5 | 3.5 | 4.5 | 0.5 | 1.5 | 2.5 | 3.5 | 4.5 |
| | $z$ (cm) ↓ | | | | | | | | | | | | | | | | | | | | | | | | | |
| −4.5 | 0.5 | 0.01 | 0.00 | 0.00 | 0.00 | 0.00 | 0.00 | 0.00 | 0.00 | 0.00 | 0.00 | 0.00 | 0.00 | 0.00 | 0.00 | 0.00 | 0.00 | 0.00 | 0.00 | 0.00 | 0.00 | 0.00 | 0.00 | 0.00 | 0.00 | 0.00 |
| −3.5 | 0.5 | 0.01 | 0.01 | 0.00 | 0.00 | 0.00 | 0.00 | 0.00 | 0.00 | 0.00 | 0.00 | 0.00 | 0.00 | 0.00 | 0.00 | 0.00 | 0.00 | 0.00 | 0.00 | 0.00 | 0.00 | 0.00 | 0.00 | 0.00 | 0.00 | 0.00 |
| −2.5 | 0.5 | 0.01 | 0.01 | 0.01 | 0.00 | 0.00 | 0.01 | 0.00 | 0.00 | 0.00 | 0.00 | 0.00 | 0.00 | 0.00 | 0.00 | 0.00 | 0.00 | 0.00 | 0.00 | 0.00 | 0.00 | 0.00 | 0.00 | 0.00 | 0.00 | 0.00 |
| −1.5 | 0.5 | 0.02 | 0.01 | 0.01 | 0.01 | 0.00 | 0.01 | 0.01 | 0.00 | 0.00 | 0.00 | 0.00 | 0.00 | 0.00 | 0.00 | 0.00 | 0.00 | 0.00 | 0.00 | 0.00 | 0.00 | 0.00 | 0.00 | 0.00 | 0.00 | 0.00 |
| −0.5 | 0.5 | 0.02 | 0.02 | 0.01 | 0.01 | 0.01 | 0.01 | 0.01 | 0.01 | 0.00 | 0.00 | 0.00 | 0.00 | 0.00 | 0.00 | 0.00 | 0.00 | 0.00 | 0.00 | 0.00 | 0.00 | 0.00 | 0.00 | 0.00 | 0.00 | 0.00 |
| 0.5 | 0.5 | 0.02 | 0.02 | 0.02 | 0.01 | 0.01 | 0.01 | 0.01 | 0.01 | 0.01 | 0.00 | 0.01 | 0.00 | 0.00 | 0.00 | 0.00 | 0.00 | 0.00 | 0.00 | 0.00 | 0.00 | 0.00 | 0.00 | 0.00 | 0.00 | 0.00 |
| 1.5 | 0.5 | 0.02 | 0.02 | 0.02 | 0.02 | 0.01 | 0.01 | 0.01 | 0.01 | 0.01 | 0.01 | 0.00 | 0.01 | 0.00 | 0.00 | 0.00 | 0.00 | 0.00 | 0.00 | 0.00 | 0.00 | 0.00 | 0.00 | 0.00 | 0.00 | 0.00 |
| 2.5 | 0.5 | 0.02 | 0.02 | 0.02 | 0.02 | 0.02 | 0.01 | 0.01 | 0.01 | 0.01 | 0.01 | 0.00 | 0.00 | 0.01 | 0.00 | 0.00 | 0.00 | 0.00 | 0.00 | 0.00 | 0.00 | 0.00 | 0.00 | 0.00 | 0.00 | 0.00 |
| 3.5 | 0.5 | 0.01 | 0.02 | 0.02 | 0.02 | 0.02 | 0.01 | 0.01 | 0.01 | 0.01 | 0.01 | 0.00 | 0.00 | 0.00 | 0.01 | 0.00 | 0.00 | 0.00 | 0.00 | 0.00 | 0.00 | 0.00 | 0.00 | 0.00 | 0.00 | 0.00 |
| 4.5 | 0.5 | 0.01 | 0.01 | 0.02 | 0.02 | 0.02 | 0.00 | 0.01 | 0.01 | 0.01 | 0.01 | 0.00 | 0.00 | 0.00 | 0.00 | 0.01 | 0.00 | 0.00 | 0.00 | 0.00 | 0.00 | 0.00 | 0.00 | 0.00 | 0.00 | 0.00 |
| −4.5 | 1.5 | 0.00 | 0.00 | 0.00 | 0.00 | 0.00 | 0.00 | 0.00 | 0.00 | 0.00 | 0.00 | 0.00 | 0.00 | 0.00 | 0.00 | 0.00 | 0.00 | 0.00 | 0.00 | 0.00 | 0.00 | 0.00 | 0.00 | 0.00 | 0.00 | 0.00 |
| −3.5 | 1.5 | 0.01 | 0.00 | 0.00 | 0.00 | 0.00 | 0.00 | 0.00 | 0.00 | 0.00 | 0.00 | 0.00 | 0.00 | 0.00 | 0.00 | 0.00 | 0.00 | 0.00 | 0.00 | 0.00 | 0.00 | 0.00 | 0.00 | 0.00 | 0.00 | 0.00 |
| −2.5 | 1.5 | 0.01 | 0.01 | 0.00 | 0.00 | 0.00 | 0.01 | 0.00 | 0.00 | 0.00 | 0.00 | 0.00 | 0.00 | 0.00 | 0.00 | 0.00 | 0.00 | 0.00 | 0.00 | 0.00 | 0.00 | 0.00 | 0.00 | 0.00 | 0.00 | 0.00 |
| −1.5 | 1.5 | 0.01 | 0.01 | 0.01 | 0.00 | 0.00 | 0.01 | 0.01 | 0.00 | 0.00 | 0.00 | 0.00 | 0.00 | 0.00 | 0.00 | 0.00 | 0.00 | 0.00 | 0.00 | 0.00 | 0.00 | 0.00 | 0.00 | 0.00 | 0.00 | 0.00 |
| −0.5 | 1.5 | 0.02 | 0.01 | 0.01 | 0.01 | 0.00 | 0.01 | 0.01 | 0.01 | 0.00 | 0.00 | 0.00 | 0.00 | 0.00 | 0.00 | 0.00 | 0.00 | 0.00 | 0.00 | 0.00 | 0.00 | 0.00 | 0.00 | 0.00 | 0.00 | 0.00 |
| 0.5 | 1.5 | 0.02 | 0.02 | 0.01 | 0.01 | 0.01 | 0.01 | 0.01 | 0.01 | 0.01 | 0.00 | 0.00 | 0.00 | 0.00 | 0.00 | 0.00 | 0.00 | 0.00 | 0.00 | 0.00 | 0.00 | 0.00 | 0.00 | 0.00 | 0.00 | 0.00 |
| 1.5 | 1.5 | 0.02 | 0.02 | 0.02 | 0.01 | 0.01 | 0.01 | 0.01 | 0.01 | 0.01 | 0.01 | 0.00 | 0.00 | 0.00 | 0.00 | 0.00 | 0.00 | 0.00 | 0.00 | 0.00 | 0.00 | 0.00 | 0.00 | 0.00 | 0.00 | 0.00 |
| 2.5 | 1.5 | 0.01 | 0.02 | 0.02 | 0.02 | 0.01 | 0.01 | 0.01 | 0.01 | 0.01 | 0.01 | 0.00 | 0.00 | 0.00 | 0.00 | 0.00 | 0.00 | 0.00 | 0.00 | 0.00 | 0.00 | 0.00 | 0.00 | 0.00 | 0.00 | 0.00 |
| 3.5 | 1.5 | 0.01 | 0.01 | 0.02 | 0.02 | 0.02 | 0.01 | 0.01 | 0.01 | 0.01 | 0.01 | 0.00 | 0.00 | 0.00 | 0.00 | 0.00 | 0.00 | 0.00 | 0.00 | 0.00 | 0.00 | 0.00 | 0.00 | 0.00 | 0.00 | 0.00 |
| 4.5 | 1.5 | 0.01 | 0.01 | 0.01 | 0.02 | 0.02 | 0.00 | 0.01 | 0.01 | 0.01 | 0.01 | 0.00 | 0.00 | 0.00 | 0.00 | 0.00 | 0.00 | 0.00 | 0.00 | 0.00 | 0.00 | 0.00 | 0.00 | 0.00 | 0.00 | 0.00 |
| −4.5 | 2.5 | 0.00 | 0.00 | 0.00 | 0.00 | 0.00 | 0.00 | 0.00 | 0.00 | 0.00 | 0.00 | 0.00 | 0.00 | 0.00 | 0.00 | 0.00 | 0.00 | 0.00 | 0.00 | 0.00 | 0.00 | 0.00 | 0.00 | 0.00 | 0.00 | 0.00 |
| −3.5 | 2.5 | 0.01 | 0.00 | 0.00 | 0.00 | 0.00 | 0.00 | 0.00 | 0.00 | 0.00 | 0.00 | 0.00 | 0.00 | 0.00 | 0.00 | 0.00 | 0.00 | 0.00 | 0.00 | 0.00 | 0.00 | 0.00 | 0.00 | 0.00 | 0.00 | 0.00 |
| −2.5 | 2.5 | 0.01 | 0.01 | 0.00 | 0.00 | 0.00 | 0.00 | 0.00 | 0.00 | 0.00 | 0.00 | 0.00 | 0.00 | 0.00 | 0.00 | 0.00 | 0.00 | 0.00 | 0.00 | 0.00 | 0.00 | 0.00 | 0.00 | 0.00 | 0.00 | 0.00 |
| −1.5 | 2.5 | 0.01 | 0.01 | 0.01 | 0.00 | 0.00 | 0.01 | 0.00 | 0.00 | 0.00 | 0.00 | 0.00 | 0.00 | 0.00 | 0.00 | 0.00 | 0.00 | 0.00 | 0.00 | 0.00 | 0.00 | 0.00 | 0.00 | 0.00 | 0.00 | 0.00 |
| −0.5 | 2.5 | 0.01 | 0.01 | 0.01 | 0.01 | 0.00 | 0.01 | 0.01 | 0.00 | 0.00 | 0.00 | 0.00 | 0.00 | 0.00 | 0.00 | 0.00 | 0.00 | 0.00 | 0.00 | 0.00 | 0.00 | 0.00 | 0.00 | 0.00 | 0.00 | 0.00 |

| | | | | | | | | | | | | | | | | | | | | | | | | | | |
|---|---|---|---|---|---|---|---|---|---|---|---|---|---|---|---|---|---|---|---|---|---|---|---|---|---|---|
| 0.5 | 2.5 | 0.01 | 0.01 | 0.01 | 0.01 | 0.01 | 0.01 | 0.01 | 0.01 | 0.00 | 0.00 | 0.00 | 0.00 | 0.00 | 0.00 | 0.00 | 0.00 | 0.00 | 0.00 | 0.00 | 0.00 | 0.00 | 0.00 | 0.00 | 0.00 | 0.00 |
| 1.5 | 2.5 | 0.01 | 0.01 | 0.01 | 0.01 | 0.01 | 0.01 | 0.01 | 0.01 | 0.01 | 0.00 | 0.00 | 0.00 | 0.00 | 0.00 | 0.00 | 0.00 | 0.00 | 0.00 | 0.00 | 0.00 | 0.00 | 0.00 | 0.00 | 0.00 | 0.00 |
| 2.5 | 2.5 | 0.01 | 0.01 | 0.01 | 0.01 | 0.01 | 0.01 | 0.01 | 0.01 | 0.01 | 0.01 | 0.00 | 0.00 | 0.00 | 0.00 | 0.00 | 0.00 | 0.00 | 0.00 | 0.00 | 0.00 | 0.00 | 0.00 | 0.00 | 0.00 | 0.00 |
| 3.5 | 2.5 | 0.01 | 0.01 | 0.01 | 0.01 | 0.01 | 0.00 | 0.01 | 0.01 | 0.01 | 0.01 | 0.00 | 0.00 | 0.00 | 0.00 | 0.00 | 0.00 | 0.00 | 0.00 | 0.00 | 0.00 | 0.00 | 0.00 | 0.00 | 0.00 | 0.00 |
| 4.5 | 2.5 | 0.01 | 0.01 | 0.01 | 0.01 | 0.01 | 0.00 | 0.00 | 0.01 | 0.01 | 0.01 | 0.00 | 0.00 | 0.00 | 0.00 | 0.00 | 0.00 | 0.00 | 0.00 | 0.00 | 0.00 | 0.00 | 0.00 | 0.00 | 0.00 | 0.00 |
| −4.5 | 3.5 | 0.00 | 0.00 | 0.00 | 0.00 | 0.00 | 0.00 | 0.00 | 0.00 | 0.00 | 0.00 | 0.00 | 0.00 | 0.00 | 0.00 | 0.00 | 0.00 | 0.00 | 0.00 | 0.00 | 0.00 | 0.00 | 0.00 | 0.00 | 0.00 | 0.00 |
| −3.5 | 3.5 | 0.00 | 0.00 | 0.00 | 0.00 | 0.00 | 0.00 | 0.00 | 0.00 | 0.00 | 0.00 | 0.00 | 0.00 | 0.00 | 0.00 | 0.00 | 0.00 | 0.00 | 0.00 | 0.00 | 0.00 | 0.00 | 0.00 | 0.00 | 0.00 | 0.00 |
| −2.5 | 3.5 | 0.01 | 0.00 | 0.00 | 0.00 | 0.00 | 0.00 | 0.00 | 0.00 | 0.00 | 0.00 | 0.00 | 0.00 | 0.00 | 0.00 | 0.00 | 0.00 | 0.00 | 0.00 | 0.00 | 0.00 | 0.00 | 0.00 | 0.00 | 0.00 | 0.00 |
| −1.5 | 3.5 | 0.01 | 0.01 | 0.00 | 0.00 | 0.00 | 0.00 | 0.00 | 0.00 | 0.00 | 0.00 | 0.00 | 0.00 | 0.00 | 0.00 | 0.00 | 0.00 | 0.00 | 0.00 | 0.00 | 0.00 | 0.00 | 0.00 | 0.00 | 0.00 | 0.00 |
| −0.5 | 3.5 | 0.01 | 0.01 | 0.01 | 0.00 | 0.00 | 0.01 | 0.00 | 0.00 | 0.00 | 0.00 | 0.00 | 0.00 | 0.00 | 0.00 | 0.00 | 0.00 | 0.00 | 0.00 | 0.00 | 0.00 | 0.00 | 0.00 | 0.00 | 0.00 | 0.00 |
| 0.5 | 3.5 | 0.01 | 0.01 | 0.01 | 0.01 | 0.00 | 0.01 | 0.01 | 0.00 | 0.00 | 0.00 | 0.00 | 0.00 | 0.00 | 0.00 | 0.00 | 0.00 | 0.00 | 0.00 | 0.00 | 0.00 | 0.00 | 0.00 | 0.00 | 0.00 | 0.00 |
| 1.5 | 3.5 | 0.01 | 0.01 | 0.01 | 0.01 | 0.01 | 0.01 | 0.01 | 0.01 | 0.00 | 0.00 | 0.00 | 0.00 | 0.00 | 0.00 | 0.00 | 0.00 | 0.00 | 0.00 | 0.00 | 0.00 | 0.00 | 0.00 | 0.00 | 0.00 | 0.00 |
| 2.5 | 3.5 | 0.01 | 0.01 | 0.01 | 0.01 | 0.01 | 0.00 | 0.01 | 0.01 | 0.01 | 0.00 | 0.00 | 0.00 | 0.00 | 0.00 | 0.00 | 0.00 | 0.00 | 0.00 | 0.00 | 0.00 | 0.00 | 0.00 | 0.00 | 0.00 | 0.00 |
| 3.5 | 3.5 | 0.01 | 0.01 | 0.01 | 0.01 | 0.01 | 0.00 | 0.00 | 0.01 | 0.01 | 0.01 | 0.00 | 0.00 | 0.00 | 0.00 | 0.00 | 0.00 | 0.00 | 0.00 | 0.00 | 0.00 | 0.00 | 0.00 | 0.00 | 0.00 | 0.00 |
| 4.5 | 3.5 | 0.00 | 0.01 | 0.01 | 0.01 | 0.01 | 0.00 | 0.00 | 0.00 | 0.01 | 0.01 | 0.00 | 0.00 | 0.00 | 0.00 | 0.00 | 0.00 | 0.00 | 0.00 | 0.00 | 0.00 | 0.00 | 0.00 | 0.00 | 0.00 | 0.00 |
| −4.5 | 4.5 | 0.00 | 0.00 | 0.00 | 0.00 | 0.00 | 0.00 | 0.00 | 0.00 | 0.00 | 0.00 | 0.00 | 0.00 | 0.00 | 0.00 | 0.00 | 0.00 | 0.00 | 0.00 | 0.00 | 0.00 | 0.00 | 0.00 | 0.00 | 0.00 | 0.00 |
| −3.5 | 4.5 | 0.00 | 0.00 | 0.00 | 0.00 | 0.00 | 0.00 | 0.00 | 0.00 | 0.00 | 0.00 | 0.00 | 0.00 | 0.00 | 0.00 | 0.00 | 0.00 | 0.00 | 0.00 | 0.00 | 0.00 | 0.00 | 0.00 | 0.00 | 0.00 | 0.00 |
| −2.5 | 4.5 | 0.00 | 0.00 | 0.00 | 0.00 | 0.00 | 0.00 | 0.00 | 0.00 | 0.00 | 0.00 | 0.00 | 0.00 | 0.00 | 0.00 | 0.00 | 0.00 | 0.00 | 0.00 | 0.00 | 0.00 | 0.00 | 0.00 | 0.00 | 0.00 | 0.00 |
| −1.5 | 4.5 | 0.01 | 0.00 | 0.00 | 0.00 | 0.00 | 0.00 | 0.00 | 0.00 | 0.00 | 0.00 | 0.00 | 0.00 | 0.00 | 0.00 | 0.00 | 0.00 | 0.00 | 0.00 | 0.00 | 0.00 | 0.00 | 0.00 | 0.00 | 0.00 | 0.00 |
| −0.5 | 4.5 | 0.01 | 0.01 | 0.00 | 0.00 | 0.00 | 0.00 | 0.00 | 0.00 | 0.00 | 0.00 | 0.00 | 0.00 | 0.00 | 0.00 | 0.00 | 0.00 | 0.00 | 0.00 | 0.00 | 0.00 | 0.00 | 0.00 | 0.00 | 0.00 | 0.00 |
| 0.5 | 4.5 | 0.01 | 0.01 | 0.01 | 0.00 | 0.00 | 0.00 | 0.00 | 0.00 | 0.00 | 0.00 | 0.00 | 0.00 | 0.00 | 0.00 | 0.00 | 0.00 | 0.00 | 0.00 | 0.00 | 0.00 | 0.00 | 0.00 | 0.00 | 0.00 | 0.00 |
| 1.5 | 4.5 | 0.01 | 0.01 | 0.01 | 0.01 | 0.00 | 0.00 | 0.00 | 0.00 | 0.00 | 0.00 | 0.00 | 0.00 | 0.00 | 0.00 | 0.00 | 0.00 | 0.00 | 0.00 | 0.00 | 0.00 | 0.00 | 0.00 | 0.00 | 0.00 | 0.00 |
| 2.5 | 4.5 | 0.01 | 0.01 | 0.01 | 0.01 | 0.01 | 0.00 | 0.00 | 0.00 | 0.00 | 0.00 | 0.00 | 0.00 | 0.00 | 0.00 | 0.00 | 0.00 | 0.00 | 0.00 | 0.00 | 0.00 | 0.00 | 0.00 | 0.00 | 0.00 | 0.00 |
| 3.5 | 4.5 | 0.00 | 0.01 | 0.01 | 0.01 | 0.01 | 0.00 | 0.00 | 0.00 | 0.00 | 0.00 | 0.00 | 0.00 | 0.00 | 0.00 | 0.00 | 0.00 | 0.00 | 0.00 | 0.00 | 0.00 | 0.00 | 0.00 | 0.00 | 0.00 | 0.00 |
| 4.5 | 4.5 | 0.00 | 0.00 | 0.01 | 0.01 | 0.01 | 0.00 | 0.00 | 0.00 | 0.00 | 0.00 | 0.00 | 0.00 | 0.00 | 0.00 | 0.00 | 0.00 | 0.00 | 0.00 | 0.00 | 0.00 | 0.00 | 0.00 | 0.00 | 0.00 | 0.00 |
| sum | | 0.985 793 | 0.950 71 | 0.879 982 | 0.775 225 | 0.644 131 | 0.522014 | 0.503 357 | 0.466 53 | 0.413 388 | 0.348 222 | 0.281 216 | 0.271 349 | 0.252 119 | 0.224 788 | 0.191 635 | 0.153 704 | 0.148 483 | 0.138 38 | 0.124 142 | 0.106 961 | 0.085 065 | 0.082 289 | 0.076 941 | 0.069 436 | 0.060 394 |

E2

Right-side flat screen (located at $x = +5.15$ cm); therefore, calculations of distance from point source to receptor site must account for this distance.

$\rho_i$ = distance from $i$th point source to receptor site (cm); $x_r$ and $y_r$ represent receptor site position in plane at $z/L = 0.5$; $z$ represents longitudinal distance from point source to plane at $z/L = 0.5$.

$y$ and $z$ represent point source locations for point sources in the left-side flat screen.

| | $x_r$ (cm) → | 0.5 | 0.5 | 0.5 | 0.5 | 0.5 | 1.5 | 1.5 | 1.5 | 1.5 | 1.5 | 2.5 | 2.5 | 2.5 | 2.5 | 2.5 | 3.5 | 3.5 | 3.5 | 3.5 | 3.5 | 4.5 | 4.5 | 4.5 | 4.5 | 4.5 |
|---|---|---|---|---|---|---|---|---|---|---|---|---|---|---|---|---|---|---|---|---|---|---|---|---|---|---|
| | $y_r$ (cm) → | 0.5 | 1.5 | 2.5 | 3.5 | 4.5 | 0.5 | 1.5 | 2.5 | 3.5 | 4.5 | 0.5 | 1.5 | 2.5 | 3.5 | 4.5 | 0.5 | 1.5 | 2.5 | 3.5 | 4.5 | 0.5 | 1.5 | 2.5 | 3.5 | 4.5 |
| $y$ (cm) ↓ | $z$ (cm) ↓ | | | | | | | | | | | | | | | | | | | | | | | | | |
| −4.5 | 0.5 | 6.85 | 7.61 | 8.42 | 9.27 | 10.14 | 6.21 | 7.04 | 7.91 | 8.81 | 9.72 | 5.68 | 6.58 | 7.50 | 8.44 | 9.40 | 5.29 | 6.24 | 7.21 | 8.18 | 9.16 | 5.07 | 6.06 | 7.05 | 8.04 | 9.04 |
| −3.5 | 0.5 | 6.15 | 6.85 | 7.61 | 8.42 | 9.27 | 5.44 | 6.21 | 7.04 | 7.91 | 8.81 | 4.82 | 5.68 | 6.58 | 7.50 | 8.44 | 4.36 | 5.29 | 6.24 | 7.21 | 8.18 | 4.08 | 5.07 | 6.06 | 7.05 | 8.04 |
| −2.5 | 0.5 | 5.56 | 6.15 | 6.85 | 7.61 | 8.42 | 4.75 | 5.44 | 6.21 | 7.04 | 7.91 | 4.03 | 4.82 | 5.68 | 6.58 | 7.50 | 3.46 | 4.36 | 5.29 | 6.24 | 7.21 | 3.11 | 4.08 | 5.07 | 6.06 | 7.05 |
| −1.5 | 0.5 | 5.09 | 5.56 | 6.15 | 6.85 | 7.61 | 4.19 | 4.75 | 5.44 | 6.21 | 7.04 | 3.36 | 4.03 | 4.82 | 5.68 | 6.58 | 2.64 | 3.46 | 4.36 | 5.29 | 6.24 | 2.16 | 3.11 | 4.08 | 5.07 | 6.06 |
| −0.5 | 0.5 | 4.78 | 5.09 | 5.56 | 6.15 | 6.85 | 3.82 | 4.19 | 4.75 | 5.44 | 6.21 | 2.88 | 3.36 | 4.03 | 4.82 | 5.68 | 1.99 | 2.64 | 3.46 | 4.36 | 5.29 | 1.29 | 2.16 | 3.11 | 4.08 | 5.07 |
| 0.5 | 0.5 | 4.68 | 4.78 | 5.09 | 5.56 | 6.15 | 3.68 | 3.82 | 4.19 | 4.75 | 5.44 | 2.70 | 2.88 | 3.36 | 4.03 | 4.82 | 1.72 | 1.99 | 2.64 | 3.46 | 4.36 | 0.82 | 1.29 | 2.16 | 3.11 | 4.08 |
| 1.5 | 0.5 | 4.78 | 4.68 | 4.78 | 5.09 | 5.56 | 3.82 | 3.68 | 3.82 | 4.19 | 4.75 | 2.88 | 2.70 | 2.88 | 3.36 | 4.03 | 1.99 | 1.72 | 1.99 | 2.64 | 3.46 | 1.29 | 0.82 | 1.29 | 2.16 | 3.11 |
| 2.5 | 0.5 | 5.09 | 4.78 | 4.68 | 4.78 | 5.09 | 4.19 | 3.82 | 3.68 | 3.82 | 4.19 | 3.36 | 2.88 | 2.70 | 2.88 | 3.36 | 2.64 | 1.99 | 1.72 | 1.99 | 2.64 | 2.16 | 1.29 | 0.82 | 1.29 | 2.16 |
| 3.5 | 0.5 | 5.56 | 5.09 | 4.78 | 4.68 | 4.78 | 4.75 | 4.19 | 3.82 | 3.68 | 3.82 | 4.03 | 3.36 | 2.88 | 2.70 | 2.88 | 3.46 | 2.64 | 1.99 | 1.72 | 1.99 | 3.11 | 2.16 | 1.29 | 0.82 | 1.29 |
| 4.5 | 0.5 | 6.15 | 5.56 | 5.09 | 4.78 | 4.68 | 5.44 | 4.75 | 4.19 | 3.82 | 3.68 | 4.82 | 4.03 | 3.36 | 2.88 | 2.70 | 4.36 | 3.46 | 2.64 | 1.99 | 1.72 | 4.08 | 3.11 | 2.16 | 1.29 | 0.82 |
| −4.5 | 1.5 | 6.99 | 7.74 | 8.54 | 9.37 | 10.24 | 6.37 | 7.18 | 8.04 | 8.92 | 9.83 | 5.85 | 6.73 | 7.63 | 8.56 | 9.50 | 5.47 | 6.40 | 7.35 | 8.30 | 9.27 | 5.26 | 6.22 | 7.19 | 8.17 | 9.15 |
| −3.5 | 1.5 | 6.31 | 6.99 | 7.74 | 8.54 | 9.37 | 5.62 | 6.37 | 7.18 | 8.04 | 8.92 | 5.03 | 5.85 | 6.73 | 7.63 | 8.56 | 4.58 | 5.47 | 6.40 | 7.35 | 8.30 | 4.32 | 5.26 | 6.22 | 7.19 | 8.17 |
| −2.5 | 1.5 | 5.73 | 6.31 | 6.99 | 7.74 | 8.54 | 4.96 | 5.62 | 6.37 | 7.18 | 8.04 | 4.27 | 5.03 | 5.85 | 6.73 | 7.63 | 3.74 | 4.58 | 5.47 | 6.40 | 7.35 | 3.42 | 4.32 | 5.26 | 6.22 | 7.19 |
| −1.5 | 1.5 | 5.28 | 5.73 | 6.31 | 6.99 | 7.74 | 4.42 | 4.96 | 5.62 | 6.37 | 7.18 | 3.64 | 4.27 | 5.03 | 5.85 | 6.73 | 3.00 | 3.74 | 4.58 | 5.47 | 6.40 | 2.58 | 3.42 | 4.32 | 5.26 | 6.22 |
| −0.5 | 1.5 | 4.99 | 5.28 | 5.73 | 6.31 | 6.99 | 4.07 | 4.42 | 4.96 | 5.62 | 6.37 | 3.21 | 3.64 | 4.27 | 5.03 | 5.85 | 2.44 | 3.00 | 3.74 | 4.58 | 5.47 | 1.92 | 2.58 | 3.42 | 4.32 | 5.26 |
| 0.5 | 1.5 | 4.89 | 4.99 | 5.28 | 5.73 | 6.31 | 3.95 | 4.07 | 4.42 | 4.96 | 5.62 | 3.05 | 3.21 | 3.64 | 4.27 | 5.03 | 2.23 | 2.44 | 3.00 | 3.74 | 4.58 | 1.63 | 1.92 | 2.58 | 3.42 | 4.32 |
| 1.5 | 1.5 | 4.99 | 4.89 | 4.99 | 5.28 | 5.73 | 4.07 | 3.95 | 4.07 | 4.42 | 4.96 | 3.21 | 3.05 | 3.21 | 3.64 | 4.27 | 2.44 | 2.23 | 2.44 | 3.00 | 3.74 | 1.92 | 1.63 | 1.92 | 2.58 | 3.42 |
| 2.5 | 1.5 | 5.28 | 4.99 | 4.89 | 4.99 | 5.28 | 4.42 | 4.07 | 3.95 | 4.07 | 4.42 | 3.64 | 3.21 | 3.05 | 3.21 | 3.64 | 3.00 | 2.44 | 2.23 | 2.44 | 3.00 | 2.58 | 1.92 | 1.63 | 1.92 | 2.58 |
| 3.5 | 1.5 | 5.73 | 5.28 | 4.99 | 4.89 | 4.99 | 4.96 | 4.42 | 4.07 | 3.95 | 4.07 | 4.27 | 3.64 | 3.21 | 3.05 | 3.21 | 3.74 | 3.00 | 2.44 | 2.23 | 2.44 | 3.42 | 2.58 | 1.92 | 1.63 | 1.92 |
| 4.5 | 1.5 | 6.31 | 5.73 | 5.28 | 4.99 | 4.89 | 5.62 | 4.96 | 4.42 | 4.07 | 3.95 | 5.03 | 4.27 | 3.64 | 3.21 | 3.05 | 4.58 | 3.74 | 3.00 | 2.44 | 2.23 | 4.32 | 3.42 | 2.58 | 1.92 | 1.63 |
| −4.5 | 2.5 | 7.27 | 7.99 | 8.77 | 9.59 | 10.43 | 6.68 | 7.45 | 8.28 | 9.14 | 10.03 | 6.19 | 7.02 | 7.89 | 8.79 | 9.71 | 5.83 | 6.71 | 7.61 | 8.54 | 9.49 | 5.63 | 6.53 | 7.46 | 8.41 | 9.36 |
| −3.5 | 2.5 | 6.62 | 7.27 | 7.99 | 8.77 | 9.59 | 5.96 | 6.68 | 7.45 | 8.28 | 9.14 | 5.41 | 6.19 | 7.02 | 7.89 | 8.79 | 5.00 | 5.83 | 6.71 | 7.61 | 8.54 | 4.76 | 5.63 | 6.53 | 7.46 | 8.41 |
| −2.5 | 2.5 | 6.07 | 6.62 | 7.27 | 7.99 | 8.77 | 5.35 | 5.96 | 6.68 | 7.45 | 8.28 | 4.72 | 5.41 | 6.19 | 7.02 | 7.89 | 4.24 | 5.00 | 5.83 | 6.71 | 7.61 | 3.96 | 4.76 | 5.63 | 6.53 | 7.46 |
| −1.5 | 2.5 | 5.65 | 6.07 | 6.62 | 7.27 | 7.99 | 4.86 | 5.35 | 5.96 | 6.68 | 7.45 | 4.16 | 4.72 | 5.41 | 6.19 | 7.02 | 3.60 | 4.24 | 5.00 | 5.83 | 6.71 | 3.27 | 3.96 | 4.76 | 5.63 | 6.53 |
| −0.5 | 2.5 | 5.37 | 5.65 | 6.07 | 6.62 | 7.27 | 4.54 | 4.86 | 5.35 | 5.96 | 6.68 | 3.78 | 4.16 | 4.72 | 5.41 | 6.19 | 3.16 | 3.60 | 4.24 | 5.00 | 5.83 | 2.77 | 3.27 | 3.96 | 4.76 | 5.63 |
| 0.5 | 2.5 | 5.28 | 5.37 | 5.65 | 6.07 | 6.62 | 4.42 | 4.54 | 4.86 | 5.35 | 5.96 | 3.64 | 3.78 | 4.16 | 4.72 | 5.41 | 3.00 | 3.16 | 3.60 | 4.24 | 5.00 | 2.58 | 2.77 | 3.27 | 3.96 | 4.76 |
| 1.5 | 2.5 | 5.37 | 5.28 | 5.37 | 5.65 | 6.07 | 4.54 | 4.42 | 4.54 | 4.86 | 5.35 | 3.78 | 3.64 | 3.78 | 4.16 | 4.72 | 3.16 | 3.00 | 3.16 | 3.60 | 4.24 | 2.77 | 2.58 | 2.77 | 3.27 | 3.96 |
| 2.5 | 2.5 | 5.65 | 5.37 | 5.28 | 5.37 | 5.65 | 4.86 | 4.54 | 4.42 | 4.54 | 4.86 | 4.16 | 3.78 | 3.64 | 3.78 | 4.16 | 3.60 | 3.16 | 3.00 | 3.16 | 3.60 | 3.27 | 2.77 | 2.58 | 2.77 | 3.27 |

| | | | | | | | | | | | | | | | | | | | | | | | | | | |
|---|---|---|---|---|---|---|---|---|---|---|---|---|---|---|---|---|---|---|---|---|---|---|---|---|---|---|
| 3.5 | 2.5 | 6.07 | 5.65 | 5.37 | 5.28 | 5.37 | 5.35 | 4.86 | 4.54 | 4.42 | 4.54 | 4.72 | 4.16 | 3.78 | 3.64 | 3.78 | 4.24 | 3.60 | 3.16 | 3.00 | 3.16 | 3.96 | 3.27 | 2.77 | 2.58 | 2.77 |
| 4.5 | 2.5 | 6.62 | 6.07 | 5.65 | 5.37 | 5.28 | 5.96 | 5.35 | 4.86 | 4.54 | 4.42 | 5.41 | 4.72 | 4.16 | 3.78 | 3.64 | 5.00 | 4.24 | 3.60 | 3.16 | 3.00 | 4.76 | 3.96 | 3.27 | 2.77 | 2.58 |
| −4.5 | 3.5 | 7.67 | 8.36 | 9.10 | 9.89 | 10.72 | 7.11 | 7.85 | 8.64 | 9.46 | 10.32 | 6.65 | 7.43 | 8.26 | 9.13 | 10.01 | 6.32 | 7.14 | 8.00 | 8.89 | 9.80 | 6.14 | 6.98 | 7.85 | 8.76 | 9.68 |
| −3.5 | 3.5 | 7.06 | 7.67 | 8.36 | 9.10 | 9.89 | 6.45 | 7.11 | 7.85 | 8.64 | 9.46 | 5.94 | 6.65 | 7.43 | 8.26 | 9.13 | 5.57 | 6.32 | 7.14 | 8.00 | 8.89 | 5.35 | 6.14 | 6.98 | 7.85 | 8.76 |
| −2.5 | 3.5 | 6.55 | 7.06 | 7.67 | 8.36 | 9.10 | 5.88 | 6.45 | 7.11 | 7.85 | 8.64 | 5.32 | 5.94 | 6.65 | 7.43 | 8.26 | 4.90 | 5.57 | 6.32 | 7.14 | 8.00 | 4.66 | 5.35 | 6.14 | 6.98 | 7.85 |
| −1.5 | 3.5 | 6.15 | 6.55 | 7.06 | 7.67 | 8.36 | 5.44 | 5.88 | 6.45 | 7.11 | 7.85 | 4.82 | 5.32 | 5.94 | 6.65 | 7.43 | 4.36 | 4.90 | 5.57 | 6.32 | 7.14 | 4.08 | 4.66 | 5.35 | 6.14 | 6.98 |
| −0.5 | 3.5 | 5.91 | 6.15 | 6.55 | 7.06 | 7.67 | 5.15 | 5.44 | 5.88 | 6.45 | 7.11 | 4.50 | 4.82 | 5.32 | 5.94 | 6.65 | 4.00 | 4.36 | 4.90 | 5.57 | 6.32 | 3.70 | 4.08 | 4.66 | 5.35 | 6.14 |
| 0.5 | 3.5 | 5.82 | 5.91 | 6.15 | 6.55 | 7.06 | 5.06 | 5.15 | 5.44 | 5.88 | 6.45 | 4.39 | 4.50 | 4.82 | 5.32 | 5.94 | 3.87 | 4.00 | 4.36 | 4.90 | 5.57 | 3.56 | 3.70 | 4.08 | 4.66 | 5.35 |
| 1.5 | 3.5 | 5.91 | 5.82 | 5.91 | 6.15 | 6.55 | 5.15 | 5.06 | 5.15 | 5.44 | 5.88 | 4.50 | 4.39 | 4.50 | 4.82 | 5.32 | 4.00 | 3.87 | 4.00 | 4.36 | 4.90 | 3.70 | 3.56 | 3.70 | 4.08 | 4.66 |
| 2.5 | 3.5 | 6.15 | 5.91 | 5.82 | 5.91 | 6.15 | 5.44 | 5.15 | 5.06 | 5.15 | 5.44 | 4.82 | 4.50 | 4.39 | 4.50 | 4.82 | 4.36 | 4.00 | 3.87 | 4.00 | 4.36 | 4.08 | 3.70 | 3.56 | 3.70 | 4.08 |
| 3.5 | 3.5 | 6.55 | 6.15 | 5.91 | 5.82 | 5.91 | 5.88 | 5.44 | 5.15 | 5.06 | 5.15 | 5.32 | 4.82 | 4.50 | 4.39 | 4.50 | 4.90 | 4.36 | 4.00 | 3.87 | 4.00 | 4.66 | 4.08 | 3.70 | 3.56 | 3.70 |
| 4.5 | 3.5 | 7.06 | 6.55 | 6.15 | 5.91 | 5.82 | 6.45 | 5.88 | 5.44 | 5.15 | 5.06 | 5.94 | 5.32 | 4.82 | 4.50 | 4.39 | 5.57 | 4.90 | 4.36 | 4.00 | 3.87 | 5.35 | 4.66 | 4.08 | 3.70 | 3.56 |
| −4.5 | 4.5 | 8.18 | 8.82 | 9.53 | 10.29 | 11.08 | 7.65 | 8.34 | 9.09 | 9.88 | 10.70 | 7.23 | 7.95 | 8.73 | 9.55 | 10.41 | 6.93 | 7.68 | 8.48 | 9.33 | 10.20 | 6.76 | 7.53 | 8.35 | 9.20 | 10.08 |
| −3.5 | 4.5 | 7.61 | 8.18 | 8.82 | 9.53 | 10.29 | 7.04 | 7.65 | 8.34 | 9.09 | 9.88 | 6.58 | 7.23 | 7.95 | 8.73 | 9.55 | 6.24 | 6.93 | 7.68 | 8.48 | 9.33 | 6.06 | 6.76 | 7.53 | 8.35 | 9.20 |
| −2.5 | 4.5 | 7.13 | 7.61 | 8.18 | 8.82 | 9.53 | 6.52 | 7.04 | 7.65 | 8.34 | 9.09 | 6.02 | 6.58 | 7.23 | 7.95 | 8.73 | 5.65 | 6.24 | 6.93 | 7.68 | 8.48 | 5.45 | 6.06 | 6.76 | 7.53 | 8.35 |
| −1.5 | 4.5 | 6.77 | 7.13 | 7.61 | 8.18 | 8.82 | 6.13 | 6.52 | 7.04 | 7.65 | 8.34 | 5.59 | 6.02 | 6.58 | 7.23 | 7.95 | 5.19 | 5.65 | 6.24 | 6.93 | 7.68 | 4.97 | 5.45 | 6.06 | 6.76 | 7.53 |
| −0.5 | 4.5 | 6.55 | 6.77 | 7.13 | 7.61 | 8.18 | 5.88 | 6.13 | 6.52 | 7.04 | 7.65 | 5.32 | 5.59 | 6.02 | 6.58 | 7.23 | 4.90 | 5.19 | 5.65 | 6.24 | 6.93 | 4.66 | 4.97 | 5.45 | 6.06 | 6.76 |
| 0.5 | 4.5 | 6.47 | 6.55 | 6.77 | 7.13 | 7.61 | 5.79 | 5.88 | 6.13 | 6.52 | 7.04 | 5.22 | 5.32 | 5.59 | 6.02 | 6.58 | 4.79 | 4.90 | 5.19 | 5.65 | 6.24 | 4.55 | 4.66 | 4.97 | 5.45 | 6.06 |
| 1.5 | 4.5 | 6.55 | 6.47 | 6.55 | 6.77 | 7.13 | 5.88 | 5.79 | 5.88 | 6.13 | 6.52 | 5.32 | 5.22 | 5.32 | 5.59 | 6.02 | 4.90 | 4.79 | 4.90 | 5.19 | 5.65 | 4.66 | 4.55 | 4.66 | 4.97 | 5.45 |
| 2.5 | 4.5 | 6.77 | 6.55 | 6.47 | 6.55 | 6.77 | 6.13 | 5.88 | 5.79 | 5.88 | 6.13 | 5.59 | 5.32 | 5.22 | 5.32 | 5.59 | 5.19 | 4.90 | 4.79 | 4.90 | 5.19 | 4.97 | 4.66 | 4.55 | 4.66 | 4.97 |
| 3.5 | 4.5 | 7.13 | 6.77 | 6.55 | 6.47 | 6.55 | 6.52 | 6.13 | 5.88 | 5.79 | 5.88 | 6.02 | 5.59 | 5.32 | 5.22 | 5.32 | 5.65 | 5.19 | 4.90 | 4.79 | 4.90 | 5.45 | 4.97 | 4.66 | 4.55 | 4.66 |
| 4.5 | 4.5 | 7.61 | 7.13 | 6.77 | 6.55 | 6.47 | 7.04 | 6.52 | 6.13 | 5.88 | 5.79 | 6.58 | 6.02 | 5.59 | 5.32 | 5.22 | 6.24 | 5.65 | 5.19 | 4.90 | 4.79 | 6.06 | 5.45 | 4.97 | 4.66 | 4.55 |

F2

Contributions of each point source in right-side flat panel to irradiance received at an individual receptor site (mW/cm$^2$).

| $y$ (cm) ↓ | $x_r$ (cm) → / $z$ (cm) ↓ | 0.5 | 0.5 | 0.5 | 0.5 | 0.5 | 1.5 | 1.5 | 1.5 | 1.5 | 1.5 | 2.5 | 2.5 | 2.5 | 2.5 | 2.5 | 3.5 | 3.5 | 3.5 | 3.5 | 3.5 | 4.5 | 4.5 | 4.5 | 4.5 | 4.5 |
|---|---|---|---|---|---|---|---|---|---|---|---|---|---|---|---|---|---|---|---|---|---|---|---|---|---|---|
| | $y_r$ (cm) → | 0.5 | 1.5 | 2.5 | 3.5 | 4.5 | 0.5 | 1.5 | 2.5 | 3.5 | 4.5 | 0.5 | 1.5 | 2.5 | 3.5 | 4.5 | 0.5 | 1.5 | 2.5 | 3.5 | 4.5 | 0.5 | 1.5 | 2.5 | 3.5 | 4.5 |
| −4.5 | 0.5 | 0.01 | 0.01 | 0.00 | 0.00 | 0.00 | 0.01 | 0.01 | 0.00 | 0.00 | 0.00 | 0.02 | 0.01 | 0.01 | 0.00 | 0.00 | 0.03 | 0.01 | 0.01 | 0.00 | 0.00 | 0.04 | 0.02 | 0.01 | 0.00 | 0.00 |
| −3.5 | 0.5 | 0.02 | 0.01 | 0.01 | 0.00 | 0.00 | 0.03 | 0.01 | 0.01 | 0.00 | 0.00 | 0.04 | 0.02 | 0.01 | 0.01 | 0.00 | 0.07 | 0.03 | 0.01 | 0.01 | 0.00 | 0.09 | 0.04 | 0.02 | 0.01 | 0.00 |
| −2.5 | 0.5 | 0.02 | 0.02 | 0.01 | 0.01 | 0.00 | 0.05 | 0.03 | 0.01 | 0.01 | 0.00 | 0.09 | 0.04 | 0.02 | 0.01 | 0.01 | 0.15 | 0.07 | 0.03 | 0.01 | 0.01 | 0.23 | 0.09 | 0.04 | 0.02 | 0.01 |
| −1.5 | 0.5 | 0.03 | 0.02 | 0.02 | 0.01 | 0.01 | 0.08 | 0.05 | 0.03 | 0.01 | 0.01 | 0.17 | 0.09 | 0.04 | 0.02 | 0.01 | 0.37 | 0.15 | 0.07 | 0.03 | 0.01 | 0.70 | 0.23 | 0.09 | 0.04 | 0.02 |
| −0.5 | 0.5 | 0.04 | 0.03 | 0.02 | 0.02 | 0.01 | 0.11 | 0.08 | 0.05 | 0.03 | 0.01 | 0.28 | 0.17 | 0.09 | 0.04 | 0.02 | 0.86 | 0.37 | 0.15 | 0.07 | 0.03 | 2.80 | 0.70 | 0.23 | 0.09 | 0.04 |
| 0.5 | 0.5 | 0.05 | 0.04 | 0.03 | 0.02 | 0.02 | 0.12 | 0.11 | 0.08 | 0.05 | 0.03 | 0.35 | 0.28 | 0.17 | 0.09 | 0.04 | 1.29 | 0.86 | 0.37 | 0.15 | 0.07 | 8.45 | 2.80 | 0.70 | 0.23 | 0.09 |
| 1.5 | 0.5 | 0.04 | 0.05 | 0.04 | 0.03 | 0.02 | 0.11 | 0.12 | 0.11 | 0.08 | 0.05 | 0.28 | 0.35 | 0.28 | 0.17 | 0.09 | 0.86 | 1.29 | 0.86 | 0.37 | 0.15 | 2.80 | 8.45 | 2.80 | 0.70 | 0.23 |
| 2.5 | 0.5 | 0.03 | 0.04 | 0.05 | 0.04 | 0.03 | 0.08 | 0.11 | 0.12 | 0.11 | 0.08 | 0.17 | 0.28 | 0.35 | 0.28 | 0.17 | 0.37 | 0.86 | 1.29 | 0.86 | 0.37 | 0.70 | 2.80 | 8.45 | 2.80 | 0.70 |
| 3.5 | 0.5 | 0.02 | 0.03 | 0.04 | 0.05 | 0.04 | 0.05 | 0.08 | 0.11 | 0.12 | 0.11 | 0.09 | 0.17 | 0.28 | 0.35 | 0.28 | 0.15 | 0.37 | 0.86 | 1.29 | 0.86 | 0.23 | 0.70 | 2.80 | 8.45 | 2.80 |
| 4.5 | 0.5 | 0.02 | 0.02 | 0.03 | 0.04 | 0.05 | 0.03 | 0.05 | 0.08 | 0.11 | 0.12 | 0.04 | 0.09 | 0.17 | 0.28 | 0.35 | 0.07 | 0.15 | 0.37 | 0.86 | 1.29 | 0.09 | 0.23 | 0.70 | 2.80 | 8.45 |
| −4.5 | 1.5 | 0.01 | 0.00 | 0.00 | 0.00 | 0.00 | 0.01 | 0.01 | 0.00 | 0.00 | 0.00 | 0.02 | 0.01 | 0.01 | 0.00 | 0.00 | 0.03 | 0.01 | 0.01 | 0.00 | 0.00 | 0.03 | 0.02 | 0.01 | 0.00 | 0.00 |
| −3.5 | 1.5 | 0.01 | 0.01 | 0.00 | 0.00 | 0.00 | 0.02 | 0.01 | 0.01 | 0.00 | 0.00 | 0.04 | 0.02 | 0.01 | 0.01 | 0.00 | 0.05 | 0.03 | 0.01 | 0.01 | 0.00 | 0.07 | 0.03 | 0.02 | 0.01 | 0.00 |
| −2.5 | 1.5 | 0.02 | 0.01 | 0.01 | 0.00 | 0.00 | 0.04 | 0.02 | 0.01 | 0.01 | 0.00 | 0.07 | 0.04 | 0.02 | 0.01 | 0.01 | 0.12 | 0.05 | 0.03 | 0.01 | 0.01 | 0.17 | 0.07 | 0.03 | 0.02 | 0.01 |
| −1.5 | 1.5 | 0.03 | 0.02 | 0.01 | 0.01 | 0.00 | 0.06 | 0.04 | 0.02 | 0.01 | 0.01 | 0.13 | 0.07 | 0.04 | 0.02 | 0.01 | 0.25 | 0.12 | 0.05 | 0.03 | 0.01 | 0.41 | 0.17 | 0.07 | 0.03 | 0.02 |
| −0.5 | 1.5 | 0.04 | 0.03 | 0.02 | 0.01 | 0.01 | 0.08 | 0.06 | 0.04 | 0.02 | 0.01 | 0.20 | 0.13 | 0.07 | 0.04 | 0.02 | 0.47 | 0.25 | 0.12 | 0.05 | 0.03 | 0.99 | 0.41 | 0.17 | 0.07 | 0.03 |
| 0.5 | 1.5 | 0.04 | 0.04 | 0.03 | 0.02 | 0.01 | 0.09 | 0.08 | 0.06 | 0.04 | 0.02 | 0.23 | 0.20 | 0.13 | 0.07 | 0.04 | 0.62 | 0.47 | 0.25 | 0.12 | 0.05 | 1.52 | 0.99 | 0.41 | 0.17 | 0.07 |
| 1.5 | 1.5 | 0.04 | 0.04 | 0.04 | 0.03 | 0.02 | 0.08 | 0.09 | 0.08 | 0.06 | 0.04 | 0.20 | 0.23 | 0.20 | 0.13 | 0.07 | 0.47 | 0.62 | 0.47 | 0.25 | 0.12 | 0.99 | 1.52 | 0.99 | 0.41 | 0.17 |
| 2.5 | 1.5 | 0.03 | 0.04 | 0.04 | 0.04 | 0.03 | 0.06 | 0.08 | 0.09 | 0.08 | 0.06 | 0.13 | 0.20 | 0.23 | 0.20 | 0.13 | 0.25 | 0.47 | 0.62 | 0.47 | 0.25 | 0.41 | 0.99 | 1.52 | 0.99 | 0.41 |
| 3.5 | 1.5 | 0.02 | 0.03 | 0.04 | 0.04 | 0.04 | 0.04 | 0.06 | 0.08 | 0.09 | 0.08 | 0.07 | 0.13 | 0.20 | 0.23 | 0.20 | 0.12 | 0.25 | 0.47 | 0.62 | 0.47 | 0.17 | 0.41 | 0.99 | 1.52 | 0.99 |
| 4.5 | 1.5 | 0.01 | 0.02 | 0.03 | 0.04 | 0.04 | 0.02 | 0.04 | 0.06 | 0.08 | 0.09 | 0.04 | 0.07 | 0.13 | 0.20 | 0.23 | 0.05 | 0.12 | 0.25 | 0.47 | 0.62 | 0.07 | 0.17 | 0.41 | 0.99 | 1.52 |
| −4.5 | 2.5 | 0.01 | 0.00 | 0.00 | 0.00 | 0.00 | 0.01 | 0.01 | 0.00 | 0.00 | 0.00 | 0.01 | 0.01 | 0.00 | 0.00 | 0.00 | 0.02 | 0.01 | 0.01 | 0.00 | 0.00 | 0.02 | 0.01 | 0.01 | 0.00 | 0.00 |
| −3.5 | 2.5 | 0.01 | 0.01 | 0.00 | 0.00 | 0.00 | 0.02 | 0.01 | 0.01 | 0.00 | 0.00 | 0.03 | 0.01 | 0.01 | 0.00 | 0.00 | 0.04 | 0.02 | 0.01 | 0.01 | 0.00 | 0.05 | 0.02 | 0.01 | 0.01 | 0.00 |
| −2.5 | 2.5 | 0.02 | 0.01 | 0.01 | 0.00 | 0.00 | 0.03 | 0.02 | 0.01 | 0.01 | 0.00 | 0.05 | 0.03 | 0.01 | 0.01 | 0.00 | 0.07 | 0.04 | 0.02 | 0.01 | 0.01 | 0.10 | 0.05 | 0.02 | 0.01 | 0.01 |
| −1.5 | 2.5 | 0.02 | 0.02 | 0.01 | 0.01 | 0.00 | 0.04 | 0.03 | 0.02 | 0.01 | 0.01 | 0.08 | 0.05 | 0.03 | 0.01 | 0.01 | 0.13 | 0.07 | 0.04 | 0.02 | 0.01 | 0.20 | 0.10 | 0.05 | 0.02 | 0.01 |
| −0.5 | 2.5 | 0.03 | 0.02 | 0.02 | 0.01 | 0.01 | 0.06 | 0.04 | 0.03 | 0.02 | 0.01 | 0.11 | 0.08 | 0.05 | 0.03 | 0.01 | 0.21 | 0.13 | 0.07 | 0.04 | 0.02 | 0.33 | 0.20 | 0.10 | 0.05 | 0.02 |
| 0.5 | 2.5 | 0.03 | 0.03 | 0.02 | 0.02 | 0.01 | 0.06 | 0.06 | 0.04 | 0.03 | 0.02 | 0.13 | 0.11 | 0.08 | 0.05 | 0.03 | 0.25 | 0.21 | 0.13 | 0.07 | 0.04 | 0.41 | 0.33 | 0.20 | 0.10 | 0.05 |

| | | | | | | | | | | | | | | | | | | | | | | | | | | |
|---|---|---|---|---|---|---|---|---|---|---|---|---|---|---|---|---|---|---|---|---|---|---|---|---|---|---|
| 1.5 | 2.5 | 0.03 | 0.03 | 0.03 | 0.02 | 0.02 | 0.06 | 0.06 | 0.06 | 0.04 | 0.03 | 0.11 | 0.13 | 0.11 | 0.08 | 0.05 | 0.21 | 0.25 | 0.21 | 0.13 | 0.07 | 0.33 | 0.41 | 0.33 | 0.20 | 0.10 |
| 2.5 | 2.5 | 0.02 | 0.03 | 0.03 | 0.03 | 0.02 | 0.04 | 0.06 | 0.06 | 0.06 | 0.04 | 0.08 | 0.11 | 0.13 | 0.11 | 0.08 | 0.13 | 0.21 | 0.25 | 0.21 | 0.13 | 0.20 | 0.33 | 0.41 | 0.33 | 0.20 |
| 3.5 | 2.5 | 0.02 | 0.02 | 0.03 | 0.03 | 0.03 | 0.03 | 0.04 | 0.06 | 0.06 | 0.06 | 0.05 | 0.08 | 0.11 | 0.13 | 0.11 | 0.07 | 0.13 | 0.21 | 0.25 | 0.21 | 0.10 | 0.20 | 0.33 | 0.41 | 0.33 |
| 4.5 | 2.5 | 0.01 | 0.02 | 0.02 | 0.03 | 0.03 | 0.02 | 0.03 | 0.04 | 0.06 | 0.06 | 0.03 | 0.05 | 0.08 | 0.11 | 0.13 | 0.04 | 0.07 | 0.13 | 0.21 | 0.25 | 0.05 | 0.10 | 0.20 | 0.33 | 0.41 |
| −4.5 | 3.5 | 0.01 | 0.00 | 0.00 | 0.00 | 0.00 | 0.01 | 0.00 | 0.00 | 0.00 | 0.00 | 0.01 | 0.01 | 0.00 | 0.00 | 0.00 | 0.01 | 0.01 | 0.00 | 0.00 | 0.00 | 0.02 | 0.01 | 0.01 | 0.00 | 0.00 |
| −3.5 | 3.5 | 0.01 | 0.01 | 0.00 | 0.00 | 0.00 | 0.01 | 0.01 | 0.00 | 0.00 | 0.00 | 0.02 | 0.01 | 0.01 | 0.00 | 0.00 | 0.02 | 0.01 | 0.01 | 0.00 | 0.00 | 0.03 | 0.02 | 0.01 | 0.01 | 0.00 |
| −2.5 | 3.5 | 0.01 | 0.01 | 0.01 | 0.00 | 0.00 | 0.02 | 0.01 | 0.01 | 0.00 | 0.00 | 0.03 | 0.02 | 0.01 | 0.01 | 0.00 | 0.04 | 0.02 | 0.01 | 0.01 | 0.00 | 0.05 | 0.03 | 0.02 | 0.01 | 0.01 |
| −1.5 | 3.5 | 0.02 | 0.01 | 0.01 | 0.01 | 0.00 | 0.03 | 0.02 | 0.01 | 0.01 | 0.00 | 0.04 | 0.03 | 0.02 | 0.01 | 0.01 | 0.07 | 0.04 | 0.02 | 0.01 | 0.01 | 0.09 | 0.05 | 0.03 | 0.02 | 0.01 |
| −0.5 | 3.5 | 0.02 | 0.02 | 0.01 | 0.01 | 0.01 | 0.03 | 0.03 | 0.02 | 0.01 | 0.01 | 0.06 | 0.04 | 0.03 | 0.02 | 0.01 | 0.09 | 0.07 | 0.04 | 0.02 | 0.01 | 0.13 | 0.09 | 0.05 | 0.03 | 0.02 |
| 0.5 | 3.5 | 0.02 | 0.02 | 0.02 | 0.01 | 0.01 | 0.04 | 0.03 | 0.03 | 0.02 | 0.01 | 0.06 | 0.06 | 0.04 | 0.03 | 0.02 | 0.10 | 0.09 | 0.07 | 0.04 | 0.02 | 0.15 | 0.13 | 0.09 | 0.05 | 0.03 |
| 1.5 | 3.5 | 0.02 | 0.02 | 0.02 | 0.02 | 0.01 | 0.03 | 0.04 | 0.03 | 0.03 | 0.02 | 0.06 | 0.06 | 0.06 | 0.04 | 0.03 | 0.09 | 0.10 | 0.09 | 0.07 | 0.04 | 0.13 | 0.15 | 0.13 | 0.09 | 0.05 |
| 2.5 | 3.5 | 0.02 | 0.02 | 0.02 | 0.02 | 0.02 | 0.03 | 0.03 | 0.04 | 0.03 | 0.03 | 0.04 | 0.06 | 0.06 | 0.06 | 0.04 | 0.07 | 0.09 | 0.10 | 0.09 | 0.07 | 0.09 | 0.13 | 0.15 | 0.13 | 0.09 |
| 3.5 | 3.5 | 0.01 | 0.02 | 0.02 | 0.02 | 0.02 | 0.02 | 0.03 | 0.03 | 0.04 | 0.03 | 0.03 | 0.04 | 0.06 | 0.06 | 0.06 | 0.04 | 0.07 | 0.09 | 0.10 | 0.09 | 0.05 | 0.09 | 0.13 | 0.15 | 0.13 |
| 4.5 | 3.5 | 0.01 | 0.01 | 0.02 | 0.02 | 0.02 | 0.01 | 0.02 | 0.03 | 0.03 | 0.04 | 0.02 | 0.03 | 0.04 | 0.06 | 0.06 | 0.02 | 0.04 | 0.07 | 0.09 | 0.10 | 0.03 | 0.05 | 0.09 | 0.13 | 0.15 |
| −4.5 | 4.5 | 0.00 | 0.00 | 0.00 | 0.00 | 0.00 | 0.01 | 0.00 | 0.00 | 0.00 | 0.00 | 0.01 | 0.00 | 0.00 | 0.00 | 0.00 | 0.01 | 0.01 | 0.00 | 0.00 | 0.00 | 0.01 | 0.01 | 0.00 | 0.00 | 0.00 |
| −3.5 | 4.5 | 0.01 | 0.00 | 0.00 | 0.00 | 0.00 | 0.01 | 0.01 | 0.00 | 0.00 | 0.00 | 0.01 | 0.01 | 0.00 | 0.00 | 0.00 | 0.01 | 0.01 | 0.01 | 0.00 | 0.00 | 0.02 | 0.01 | 0.01 | 0.00 | 0.00 |
| −2.5 | 4.5 | 0.01 | 0.01 | 0.00 | 0.00 | 0.00 | 0.01 | 0.01 | 0.01 | 0.00 | 0.00 | 0.02 | 0.01 | 0.01 | 0.00 | 0.00 | 0.02 | 0.01 | 0.01 | 0.01 | 0.00 | 0.03 | 0.02 | 0.01 | 0.01 | 0.00 |
| −1.5 | 4.5 | 0.01 | 0.01 | 0.01 | 0.00 | 0.00 | 0.02 | 0.01 | 0.01 | 0.01 | 0.00 | 0.02 | 0.02 | 0.01 | 0.01 | 0.00 | 0.03 | 0.02 | 0.01 | 0.01 | 0.01 | 0.04 | 0.03 | 0.02 | 0.01 | 0.01 |
| −0.5 | 4.5 | 0.01 | 0.01 | 0.01 | 0.01 | 0.00 | 0.02 | 0.02 | 0.01 | 0.01 | 0.01 | 0.03 | 0.02 | 0.02 | 0.01 | 0.01 | 0.04 | 0.03 | 0.02 | 0.01 | 0.01 | 0.05 | 0.04 | 0.03 | 0.02 | 0.01 |
| 0.5 | 4.5 | 0.01 | 0.01 | 0.01 | 0.01 | 0.01 | 0.02 | 0.02 | 0.02 | 0.01 | 0.01 | 0.03 | 0.03 | 0.02 | 0.02 | 0.01 | 0.05 | 0.04 | 0.03 | 0.02 | 0.01 | 0.06 | 0.05 | 0.04 | 0.03 | 0.02 |
| 1.5 | 4.5 | 0.01 | 0.01 | 0.01 | 0.01 | 0.01 | 0.02 | 0.02 | 0.02 | 0.02 | 0.01 | 0.03 | 0.03 | 0.03 | 0.02 | 0.02 | 0.04 | 0.05 | 0.04 | 0.03 | 0.02 | 0.05 | 0.06 | 0.05 | 0.04 | 0.03 |
| 2.5 | 4.5 | 0.01 | 0.01 | 0.01 | 0.01 | 0.01 | 0.02 | 0.02 | 0.02 | 0.02 | 0.02 | 0.02 | 0.03 | 0.03 | 0.03 | 0.02 | 0.03 | 0.04 | 0.05 | 0.04 | 0.03 | 0.04 | 0.05 | 0.06 | 0.05 | 0.04 |
| 3.5 | 4.5 | 0.01 | 0.01 | 0.01 | 0.01 | 0.01 | 0.01 | 0.02 | 0.02 | 0.02 | 0.02 | 0.02 | 0.02 | 0.03 | 0.03 | 0.03 | 0.02 | 0.03 | 0.04 | 0.05 | 0.04 | 0.03 | 0.04 | 0.05 | 0.06 | 0.05 |
| 4.5 | 4.5 | 0.01 | 0.01 | 0.01 | 0.01 | 0.01 | 0.01 | 0.01 | 0.02 | 0.02 | 0.02 | 0.01 | 0.02 | 0.02 | 0.03 | 0.03 | 0.01 | 0.02 | 0.03 | 0.04 | 0.05 | 0.02 | 0.03 | 0.04 | 0.05 | 0.06 |
| sum | | 1.892<br>384 | 1.836<br>538 | 1.701<br>16 | 1.491<br>711 | 1.219<br>927 | 3.771<br>573 | 3.655<br>47 | 3.400<br>371 | 2.976<br>386 | 2.389<br>238 | 7.799<br>409 | 7.604<br>296 | 7.142<br>52 | 6.282<br>033 | 4.936<br>653 | 17.413<br>06 | 17.117<br>03 | 16.350<br>88 | 14.665<br>32 | 11.319<br>28 | 47.770<br>41 | 47.374<br>82 | 46.278<br>62 | 43.430<br>41 | 34.847<br>41 |

G2

Sum of contributions from both flat screens received at an individual receptor site (mW/cm$^2$).

| $y$ (cm) ↓ | $x_r$ (cm) →<br>$y_r$ (cm) →<br>$z$ (cm) ↓ | 0.5<br>0.5 | 0.5<br>1.5 | 0.5<br>2.5 | 0.5<br>3.5 | 0.5<br>4.5 | 1.5<br>0.5 | 1.5<br>1.5 | 1.5<br>2.5 | 1.5<br>3.5 | 1.5<br>4.5 | 2.5<br>0.5 | 2.5<br>1.5 | 2.5<br>2.5 | 2.5<br>3.5 | 2.5<br>4.5 | 3.5<br>0.5 | 3.5<br>1.5 | 3.5<br>2.5 | 3.5<br>3.5 | 3.5<br>4.5 | 4.5<br>0.5 | 4.5<br>1.5 | 4.5<br>2.5 | 4.5<br>3.5 | 4.5<br>4.5 |
|---|---|---|---|---|---|---|---|---|---|---|---|---|---|---|---|---|---|---|---|---|---|---|---|---|---|---|
| −4.5 | 0.5 | 0.01 | 0.01 | 0.01 | 0.00 | 0.00 | 0.02 | 0.01 | 0.01 | 0.00 | 0.00 | 0.02 | 0.01 | 0.01 | 0.00 | 0.00 | 0.03 | 0.02 | 0.01 | 0.00 | 0.00 | 0.04 | 0.02 | 0.01 | 0.00 | 0.00 |
| −3.5 | 0.5 | 0.02 | 0.01 | 0.01 | 0.01 | 0.00 | 0.03 | 0.02 | 0.01 | 0.01 | 0.00 | 0.05 | 0.02 | 0.01 | 0.01 | 0.00 | 0.07 | 0.03 | 0.02 | 0.01 | 0.00 | 0.09 | 0.04 | 0.02 | 0.01 | 0.00 |
| −2.5 | 0.5 | 0.04 | 0.02 | 0.01 | 0.01 | 0.01 | 0.05 | 0.03 | 0.02 | 0.01 | 0.01 | 0.09 | 0.05 | 0.02 | 0.01 | 0.01 | 0.16 | 0.07 | 0.03 | 0.02 | 0.01 | 0.23 | 0.09 | 0.04 | 0.02 | 0.01 |
| −1.5 | 0.5 | 0.05 | 0.04 | 0.02 | 0.01 | 0.01 | 0.08 | 0.05 | 0.03 | 0.02 | 0.01 | 0.17 | 0.09 | 0.05 | 0.02 | 0.01 | 0.38 | 0.16 | 0.07 | 0.03 | 0.02 | 0.70 | 0.23 | 0.09 | 0.04 | 0.02 |
| −0.5 | 0.5 | 0.07 | 0.05 | 0.04 | 0.02 | 0.01 | 0.12 | 0.08 | 0.05 | 0.03 | 0.02 | 0.29 | 0.17 | 0.09 | 0.05 | 0.02 | 0.87 | 0.38 | 0.16 | 0.07 | 0.03 | 2.80 | 0.70 | 0.23 | 0.09 | 0.04 |
| 0.5 | 0.5 | 0.07 | 0.07 | 0.05 | 0.04 | 0.02 | 0.13 | 0.12 | 0.08 | 0.05 | 0.03 | 0.35 | 0.29 | 0.17 | 0.09 | 0.05 | 1.29 | 0.87 | 0.38 | 0.16 | 0.07 | 8.45 | 2.80 | 0.70 | 0.23 | 0.09 |
| 1.5 | 0.5 | 0.07 | 0.07 | 0.07 | 0.05 | 0.04 | 0.12 | 0.13 | 0.12 | 0.08 | 0.05 | 0.29 | 0.35 | 0.29 | 0.17 | 0.09 | 0.87 | 1.29 | 0.87 | 0.38 | 0.16 | 2.80 | 8.45 | 2.80 | 0.70 | 0.23 |
| 2.5 | 0.5 | 0.05 | 0.07 | 0.07 | 0.07 | 0.05 | 0.08 | 0.12 | 0.13 | 0.12 | 0.08 | 0.17 | 0.29 | 0.35 | 0.29 | 0.17 | 0.38 | 0.87 | 1.29 | 0.87 | 0.38 | 0.70 | 2.80 | 8.45 | 2.80 | 0.70 |
| 3.5 | 0.5 | 0.04 | 0.05 | 0.07 | 0.07 | 0.07 | 0.05 | 0.08 | 0.12 | 0.13 | 0.12 | 0.09 | 0.17 | 0.29 | 0.35 | 0.29 | 0.16 | 0.38 | 0.87 | 1.29 | 0.87 | 0.23 | 0.70 | 2.80 | 8.45 | 2.80 |
| 4.5 | 0.5 | 0.02 | 0.04 | 0.05 | 0.07 | 0.07 | 0.03 | 0.05 | 0.08 | 0.12 | 0.13 | 0.05 | 0.09 | 0.17 | 0.29 | 0.35 | 0.07 | 0.16 | 0.38 | 0.87 | 1.29 | 0.09 | 0.23 | 0.70 | 2.80 | 8.45 |
| −4.5 | 1.5 | 0.01 | 0.01 | 0.00 | 0.00 | 0.00 | 0.02 | 0.01 | 0.01 | 0.00 | 0.00 | 0.02 | 0.01 | 0.01 | 0.00 | 0.00 | 0.03 | 0.01 | 0.01 | 0.00 | 0.00 | 0.03 | 0.02 | 0.01 | 0.00 | 0.00 |
| −3.5 | 1.5 | 0.02 | 0.01 | 0.01 | 0.00 | 0.00 | 0.03 | 0.02 | 0.01 | 0.01 | 0.00 | 0.04 | 0.02 | 0.01 | 0.01 | 0.00 | 0.06 | 0.03 | 0.01 | 0.01 | 0.00 | 0.07 | 0.03 | 0.02 | 0.01 | 0.00 |
| −2.5 | 1.5 | 0.03 | 0.02 | 0.01 | 0.01 | 0.00 | 0.04 | 0.03 | 0.02 | 0.01 | 0.01 | 0.07 | 0.04 | 0.02 | 0.01 | 0.01 | 0.12 | 0.06 | 0.03 | 0.01 | 0.01 | 0.17 | 0.07 | 0.03 | 0.02 | 0.01 |
| −1.5 | 1.5 | 0.04 | 0.03 | 0.02 | 0.01 | 0.01 | 0.07 | 0.04 | 0.03 | 0.02 | 0.01 | 0.13 | 0.07 | 0.04 | 0.02 | 0.01 | 0.25 | 0.12 | 0.06 | 0.03 | 0.01 | 0.41 | 0.17 | 0.07 | 0.03 | 0.02 |
| −0.5 | 1.5 | 0.06 | 0.04 | 0.03 | 0.02 | 0.01 | 0.09 | 0.07 | 0.04 | 0.03 | 0.02 | 0.20 | 0.13 | 0.07 | 0.04 | 0.02 | 0.48 | 0.25 | 0.12 | 0.06 | 0.03 | 0.99 | 0.41 | 0.17 | 0.07 | 0.03 |
| 0.5 | 1.5 | 0.06 | 0.06 | 0.04 | 0.03 | 0.02 | 0.10 | 0.09 | 0.07 | 0.04 | 0.03 | 0.24 | 0.20 | 0.13 | 0.07 | 0.04 | 0.63 | 0.48 | 0.25 | 0.12 | 0.06 | 1.52 | 0.99 | 0.41 | 0.17 | 0.07 |
| 1.5 | 1.5 | 0.06 | 0.06 | 0.06 | 0.04 | 0.03 | 0.09 | 0.10 | 0.09 | 0.07 | 0.04 | 0.20 | 0.24 | 0.20 | 0.13 | 0.07 | 0.48 | 0.63 | 0.48 | 0.25 | 0.12 | 0.99 | 1.52 | 0.99 | 0.41 | 0.17 |
| 2.5 | 1.5 | 0.04 | 0.06 | 0.06 | 0.06 | 0.04 | 0.07 | 0.09 | 0.10 | 0.09 | 0.07 | 0.13 | 0.20 | 0.24 | 0.20 | 0.13 | 0.25 | 0.48 | 0.63 | 0.48 | 0.25 | 0.41 | 0.99 | 1.52 | 0.99 | 0.41 |
| 3.5 | 1.5 | 0.03 | 0.04 | 0.06 | 0.06 | 0.06 | 0.04 | 0.07 | 0.09 | 0.10 | 0.09 | 0.07 | 0.13 | 0.20 | 0.24 | 0.20 | 0.12 | 0.25 | 0.48 | 0.63 | 0.48 | 0.17 | 0.41 | 0.99 | 1.52 | 0.99 |
| 4.5 | 1.5 | 0.02 | 0.03 | 0.04 | 0.06 | 0.06 | 0.03 | 0.04 | 0.07 | 0.09 | 0.10 | 0.04 | 0.07 | 0.13 | 0.20 | 0.24 | 0.06 | 0.12 | 0.25 | 0.48 | 0.63 | 0.07 | 0.17 | 0.41 | 0.99 | 1.52 |
| −4.5 | 2.5 | 0.01 | 0.01 | 0.00 | 0.00 | 0.00 | 0.01 | 0.01 | 0.00 | 0.00 | 0.00 | 0.02 | 0.01 | 0.01 | 0.00 | 0.00 | 0.02 | 0.01 | 0.01 | 0.00 | 0.00 | 0.03 | 0.01 | 0.01 | 0.00 | 0.00 |
| −3.5 | 2.5 | 0.02 | 0.01 | 0.01 | 0.00 | 0.00 | 0.02 | 0.01 | 0.01 | 0.00 | 0.00 | 0.03 | 0.02 | 0.01 | 0.01 | 0.00 | 0.04 | 0.02 | 0.01 | 0.01 | 0.00 | 0.05 | 0.03 | 0.01 | 0.01 | 0.00 |
| −2.5 | 2.5 | 0.02 | 0.02 | 0.01 | 0.01 | 0.00 | 0.03 | 0.02 | 0.01 | 0.01 | 0.00 | 0.05 | 0.03 | 0.02 | 0.01 | 0.01 | 0.08 | 0.04 | 0.02 | 0.01 | 0.01 | 0.10 | 0.05 | 0.03 | 0.01 | 0.01 |
| −1.5 | 2.5 | 0.03 | 0.02 | 0.02 | 0.01 | 0.01 | 0.05 | 0.03 | 0.02 | 0.01 | 0.01 | 0.08 | 0.05 | 0.03 | 0.02 | 0.01 | 0.14 | 0.08 | 0.04 | 0.02 | 0.01 | 0.20 | 0.10 | 0.05 | 0.03 | 0.01 |
| −0.5 | 2.5 | 0.04 | 0.03 | 0.02 | 0.02 | 0.01 | 0.06 | 0.05 | 0.03 | 0.02 | 0.01 | 0.12 | 0.08 | 0.05 | 0.03 | 0.02 | 0.21 | 0.14 | 0.08 | 0.04 | 0.02 | 0.33 | 0.20 | 0.10 | 0.05 | 0.03 |
| 0.5 | 2.5 | 0.04 | 0.04 | 0.03 | 0.02 | 0.02 | 0.07 | 0.06 | 0.05 | 0.03 | 0.02 | 0.13 | 0.12 | 0.08 | 0.05 | 0.03 | 0.25 | 0.21 | 0.14 | 0.08 | 0.04 | 0.41 | 0.33 | 0.20 | 0.10 | 0.05 |
| 1.5 | 2.5 | 0.04 | 0.04 | 0.04 | 0.03 | 0.02 | 0.06 | 0.07 | 0.06 | 0.05 | 0.03 | 0.12 | 0.13 | 0.12 | 0.08 | 0.05 | 0.21 | 0.25 | 0.21 | 0.14 | 0.08 | 0.33 | 0.41 | 0.33 | 0.20 | 0.10 |
| 2.5 | 2.5 | 0.03 | 0.04 | 0.04 | 0.04 | 0.03 | 0.05 | 0.06 | 0.07 | 0.06 | 0.05 | 0.08 | 0.12 | 0.13 | 0.12 | 0.08 | 0.14 | 0.21 | 0.25 | 0.21 | 0.14 | 0.20 | 0.33 | 0.41 | 0.33 | 0.20 |
| 3.5 | 2.5 | 0.02 | 0.03 | 0.04 | 0.04 | 0.04 | 0.03 | 0.05 | 0.06 | 0.07 | 0.06 | 0.05 | 0.08 | 0.12 | 0.13 | 0.12 | 0.08 | 0.14 | 0.21 | 0.25 | 0.21 | 0.10 | 0.20 | 0.33 | 0.41 | 0.33 |

| | | | | | | | | | | | | | | | | | | | | | | | | | | |
|---|---|---|---|---|---|---|---|---|---|---|---|---|---|---|---|---|---|---|---|---|---|---|---|---|---|---|
| 4.5 | 2.5 | 0.02 | 0.02 | 0.03 | 0.04 | 0.04 | 0.02 | 0.03 | 0.05 | 0.06 | 0.07 | 0.03 | 0.05 | 0.08 | 0.12 | 0.13 | 0.04 | 0.08 | 0.14 | 0.21 | 0.25 | 0.05 | 0.10 | 0.20 | 0.33 | 0.41 |
| −4.5 | 3.5 | 0.01 | 0.01 | 0.00 | 0.00 | 0.00 | 0.01 | 0.01 | 0.00 | 0.00 | 0.00 | 0.01 | 0.01 | 0.00 | 0.00 | 0.00 | 0.01 | 0.01 | 0.00 | 0.00 | 0.00 | 0.02 | 0.01 | 0.01 | 0.00 | 0.00 |
| −3.5 | 3.5 | 0.01 | 0.01 | 0.01 | 0.00 | 0.00 | 0.02 | 0.01 | 0.01 | 0.00 | 0.00 | 0.02 | 0.01 | 0.01 | 0.00 | 0.00 | 0.03 | 0.01 | 0.01 | 0.00 | 0.00 | 0.03 | 0.02 | 0.01 | 0.01 | 0.00 |
| −2.5 | 3.5 | 0.02 | 0.01 | 0.01 | 0.01 | 0.00 | 0.02 | 0.02 | 0.01 | 0.01 | 0.00 | 0.03 | 0.02 | 0.01 | 0.01 | 0.00 | 0.04 | 0.03 | 0.01 | 0.01 | 0.00 | 0.06 | 0.03 | 0.02 | 0.01 | 0.01 |
| −1.5 | 3.5 | 0.02 | 0.02 | 0.01 | 0.01 | 0.01 | 0.03 | 0.02 | 0.02 | 0.01 | 0.01 | 0.05 | 0.03 | 0.02 | 0.01 | 0.01 | 0.07 | 0.04 | 0.03 | 0.01 | 0.01 | 0.09 | 0.06 | 0.03 | 0.02 | 0.01 |
| −0.5 | 3.5 | 0.03 | 0.02 | 0.02 | 0.01 | 0.01 | 0.04 | 0.03 | 0.02 | 0.02 | 0.01 | 0.06 | 0.05 | 0.03 | 0.02 | 0.01 | 0.09 | 0.07 | 0.04 | 0.03 | 0.01 | 0.13 | 0.09 | 0.06 | 0.03 | 0.02 |
| 0.5 | 3.5 | 0.03 | 0.03 | 0.02 | 0.02 | 0.01 | 0.04 | 0.04 | 0.03 | 0.02 | 0.02 | 0.07 | 0.06 | 0.05 | 0.03 | 0.02 | 0.11 | 0.09 | 0.07 | 0.04 | 0.03 | 0.15 | 0.13 | 0.09 | 0.06 | 0.03 |
| 1.5 | 3.5 | 0.03 | 0.03 | 0.03 | 0.02 | 0.02 | 0.04 | 0.04 | 0.04 | 0.03 | 0.02 | 0.06 | 0.07 | 0.06 | 0.05 | 0.03 | 0.09 | 0.11 | 0.09 | 0.07 | 0.04 | 0.13 | 0.15 | 0.13 | 0.09 | 0.06 |
| 2.5 | 3.5 | 0.02 | 0.03 | 0.03 | 0.03 | 0.02 | 0.03 | 0.04 | 0.04 | 0.04 | 0.03 | 0.05 | 0.06 | 0.07 | 0.06 | 0.05 | 0.07 | 0.09 | 0.11 | 0.09 | 0.07 | 0.09 | 0.13 | 0.15 | 0.13 | 0.09 |
| 3.5 | 3.5 | 0.02 | 0.02 | 0.03 | 0.03 | 0.03 | 0.02 | 0.03 | 0.04 | 0.04 | 0.04 | 0.03 | 0.05 | 0.06 | 0.07 | 0.06 | 0.04 | 0.07 | 0.09 | 0.11 | 0.09 | 0.06 | 0.09 | 0.13 | 0.15 | 0.13 |
| 4.5 | 3.5 | 0.01 | 0.02 | 0.02 | 0.03 | 0.03 | 0.02 | 0.02 | 0.03 | 0.04 | 0.04 | 0.02 | 0.03 | 0.05 | 0.06 | 0.07 | 0.03 | 0.04 | 0.07 | 0.09 | 0.11 | 0.03 | 0.06 | 0.09 | 0.13 | 0.15 |
| −4.5 | 4.5 | 0.01 | 0.00 | 0.00 | 0.00 | 0.00 | 0.01 | 0.00 | 0.00 | 0.00 | 0.00 | 0.01 | 0.00 | 0.00 | 0.00 | 0.00 | 0.01 | 0.01 | 0.00 | 0.00 | 0.00 | 0.01 | 0.01 | 0.00 | 0.00 | 0.00 |
| −3.5 | 4.5 | 0.01 | 0.01 | 0.00 | 0.00 | 0.00 | 0.01 | 0.01 | 0.00 | 0.00 | 0.00 | 0.01 | 0.01 | 0.00 | 0.00 | 0.00 | 0.02 | 0.01 | 0.01 | 0.00 | 0.00 | 0.02 | 0.01 | 0.01 | 0.00 | 0.00 |
| −2.5 | 4.5 | 0.01 | 0.01 | 0.01 | 0.00 | 0.00 | 0.01 | 0.01 | 0.01 | 0.00 | 0.00 | 0.02 | 0.01 | 0.01 | 0.00 | 0.00 | 0.02 | 0.02 | 0.01 | 0.01 | 0.00 | 0.03 | 0.02 | 0.01 | 0.01 | 0.00 |
| −1.5 | 4.5 | 0.02 | 0.01 | 0.01 | 0.01 | 0.00 | 0.02 | 0.01 | 0.01 | 0.01 | 0.00 | 0.03 | 0.02 | 0.01 | 0.01 | 0.00 | 0.03 | 0.02 | 0.02 | 0.01 | 0.01 | 0.04 | 0.03 | 0.02 | 0.01 | 0.01 |
| −0.5 | 4.5 | 0.02 | 0.02 | 0.01 | 0.01 | 0.01 | 0.02 | 0.02 | 0.01 | 0.01 | 0.01 | 0.03 | 0.03 | 0.02 | 0.01 | 0.01 | 0.04 | 0.03 | 0.02 | 0.02 | 0.01 | 0.06 | 0.04 | 0.03 | 0.02 | 0.01 |
| 0.5 | 4.5 | 0.02 | 0.02 | 0.02 | 0.01 | 0.01 | 0.02 | 0.02 | 0.02 | 0.01 | 0.01 | 0.03 | 0.03 | 0.03 | 0.02 | 0.01 | 0.05 | 0.04 | 0.03 | 0.02 | 0.02 | 0.06 | 0.06 | 0.04 | 0.03 | 0.02 |
| 1.5 | 4.5 | 0.02 | 0.02 | 0.02 | 0.02 | 0.01 | 0.02 | 0.02 | 0.02 | 0.02 | 0.01 | 0.03 | 0.03 | 0.03 | 0.03 | 0.02 | 0.04 | 0.05 | 0.04 | 0.03 | 0.02 | 0.06 | 0.06 | 0.06 | 0.04 | 0.03 |
| 2.5 | 4.5 | 0.02 | 0.02 | 0.02 | 0.02 | 0.02 | 0.02 | 0.02 | 0.02 | 0.02 | 0.02 | 0.03 | 0.03 | 0.03 | 0.03 | 0.03 | 0.03 | 0.04 | 0.05 | 0.04 | 0.03 | 0.04 | 0.06 | 0.06 | 0.06 | 0.04 |
| 3.5 | 4.5 | 0.01 | 0.02 | 0.02 | 0.02 | 0.02 | 0.01 | 0.02 | 0.02 | 0.02 | 0.02 | 0.02 | 0.03 | 0.03 | 0.03 | 0.03 | 0.02 | 0.03 | 0.04 | 0.05 | 0.04 | 0.03 | 0.04 | 0.06 | 0.06 | 0.06 |
| 4.5 | 4.5 | 0.01 | 0.01 | 0.02 | 0.02 | 0.02 | 0.01 | 0.01 | 0.02 | 0.02 | 0.02 | 0.01 | 0.02 | 0.03 | 0.03 | 0.03 | 0.02 | 0.02 | 0.03 | 0.04 | 0.05 | 0.02 | 0.03 | 0.04 | 0.06 | 0.06 |
| sum | | 2.872 | 2.787 | 2.581 | 2.266 | 1.864 | 4.293 | 4.158 | 3.866 | 3.389 | 2.737 | 8.080 | 7.875 | 7.394 | 6.506 | 5.128 | 17.566 | 17.265 | 16.489 | 14.789 | 11.426 | 47.855 | 47.457 | 46.355 | 43.499 | 34.907 |
| | | 723 | 248 | 142 | 936 | 058 | 587 | 827 | 901 | 774 | 46 | 625 | 645 | 639 | 821 | 288 | 76 | 51 | 26 | 46 | 24 | 48 | 11 | 57 | 85 | 8 |

Simulated irradiance field ($mW/cm^2$) (1 cm increments, upper-right quadrant).

| **$x_r$ (cm) →** | **0.5** | **1.5** | **2.5** | **3.5** | **4.5** |
|---|---|---|---|---|---|
| **$y_r$ (cm) ↓** | | | | | |
| 4.5 | 1.86 | 2.74 | 5.13 | 11.43 | 34.91 |
| 3.5 | 2.27 | 3.39 | 6.51 | 14.79 | 43.50 |
| 2.5 | 2.58 | 3.87 | 7.39 | 16.49 | 46.36 |
| 1.5 | 2.79 | 4.16 | 7.88 | 17.27 | 47.46 |
| 0.5 | 2.87 | 4.29 | 8.08 | 17.57 | 47.86 |

$I_{avg} = 14.54\ mW/cm^2$
$I_{max} = 47.86\ mW/cm^2$
$I_{min} = 1.86\ mW/cm^2$
Count = 100

Entire irradiance field ($mW/cm^2$) (1 cm increments).

| **$x$ (cm) →** | **-4.5** | **-3.5** | **-2.5** | **-1.5** | **-0.5** | **0.5** | **1.5** | **2.5** | **3.5** | **4.5** |
|---|---|---|---|---|---|---|---|---|---|---|
| **$y$ (cm) ↓** | | | | | | | | | | |
| 4.5 | 34.91 | 11.43 | 5.13 | 2.74 | 1.86 | 1.86 | 2.74 | 5.13 | 11.43 | 34.91 |
| 3.5 | 43.50 | 14.79 | 6.51 | 3.39 | 2.27 | 2.27 | 3.39 | 6.51 | 14.79 | 43.50 |
| 2.5 | 46.36 | 16.49 | 7.39 | 3.87 | 2.58 | 2.58 | 3.87 | 7.39 | 16.49 | 46.36 |
| 1.5 | 47.46 | 17.27 | 7.88 | 4.16 | 2.79 | 2.79 | 4.16 | 7.88 | 17.27 | 47.46 |
| 0.5 | 47.86 | 17.57 | 8.08 | 4.29 | 2.87 | 2.87 | 4.29 | 8.08 | 17.57 | 47.86 |
| −0.5 | 47.86 | 17.57 | 8.08 | 4.29 | 2.87 | 2.87 | 4.29 | 8.08 | 17.57 | 47.86 |
| −1.5 | 47.46 | 17.27 | 7.88 | 4.16 | 2.79 | 2.79 | 4.16 | 7.88 | 17.27 | 47.46 |
| −2.5 | 46.36 | 16.49 | 7.39 | 3.87 | 2.58 | 2.58 | 3.87 | 7.39 | 16.49 | 46.36 |
| −3.5 | 43.50 | 14.79 | 6.51 | 3.39 | 2.27 | 2.27 | 3.39 | 6.51 | 14.79 | 43.50 |
| −4.5 | 34.91 | 11.43 | 5.13 | 2.74 | 1.86 | 1.86 | 2.74 | 5.13 | 11.43 | 34.91 |

# 7

# Validation of Photochemical Reactors

## 7.1 Introduction

UV dose is the master variable in photochemical reactor systems. Together with the intrinsic kinetics of the photochemical reactions of interest, the dose (distribution) delivered by a photochemical reactor governs the extent of photochemical reactions within the reactor and the overall performance of the system. The dose distribution delivered by a photochemical reactor depends on a number of system attributes (see Figure 3.22):

- *Reactor geometry*: This defines the physical boundaries of the system and the locations of media interfaces. Solid boundaries within a photoreactor will influence fluid mechanics; they can also influence the spatial distribution of radiant energy as a result of reflection, refraction, or absorbance.
- *Radiation source(s)*: The type, location, output spectrum, and output power of the radiation source(s) will affect the fluence rate field. If the radiation sources or their associated hardware extend into the flow field, they will also affect fluid mechanical behavior.
- *Material properties*: Photoreactors often comprise several materials and are used to treat one or more fluids. The mechanical and optical properties of these materials can influence the fluid flow field and the fluence rate field.
- *Flow rate*: In rough terms, the period of exposure will be inversely related to the flow rate. More broadly, the dose distribution delivered by a photoreactor will be influenced by the flow rate.

At present, methods have not been developed to allow the entire dose distribution delivered by a photoreactor to be monitored in real time. As a result, there is a need to validate photochemical reactors based on measurements that can be related to reactor performance, including flow rate, lamp output power, fluence rate at one or more locations within the irradiated zone of the reactor, and UV transmittance of the fluid being treated. Validation protocols have been developed to allow identification of conditions that yield acceptable performance, relative to specific treatment objectives.

The goal for this chapter is to present common and emerging reactor validation protocols and methods and to present the logic, basic characteristics, and limitations of these protocols. Details of the methods can be found in the original references, and readers are encouraged to read these materials for additional information regarding these methods.

*Photochemical Reactors: Theory, Methods, and Applications of Ultraviolet Radiation*, First Edition. Ernest R. Blatchley III.

Companion website: www.wiley.com/go/blatchley/photochemicalreactors

In a general sense, photochemical reactions are quite fast relative to their chemical and biochemical analogs. Because of this, photochemical reactors tend to be relatively small and sensitive to changes in operating conditions (e.g. lamp failure, changes in optical properties of fluids). For these and other reasons, methods for validation of photoreactors can be more involved, complicated, and conservative than analogous methods for other reactor systems.

The focus of this chapter is on validation methods that are directed at UV disinfection systems for water. Validation protocols for UV disinfection applications involving water and municipal wastewater have been developed by a number of organizations. In principle, validation methods could also be developed for other treatment endpoints (e.g. photolysis), for treatment of other fluids (e.g. air), and for treatment of surfaces.

In the United States, the regulatory approach applied to UV disinfection systems for drinking water shares some features with approaches used for other disinfectants. Specifically, protocols have been developed by which "credit" for inactivation of one or more waterborne microbial pathogens is assigned, based on measurable characteristics of the reactor system. The goal of these approaches is to ensure delivery of water that is safe, in terms of microbial pathogens.

Although several methods are available to accomplish reactor validation, some general characteristics of these methods can be defined. First, most of these methods involve the use of full-scale reactors. Many aspects of the behavior of UV photoreactors are nonlinear, and scaling relationships among these systems are (at present) poorly understood. Because of this, translation of results across reactor scales can introduce unacceptable errors. Second, validation methods often involve quantification of reactor behavior across the range of anticipated operating conditions. For a given reactor, reactor validation may involve the development of empirical relationships to describe reactor performance across the so-called "operating envelope." It is then possible to define conditions that conform to a given regulatory constraint, in terms of reactor performance. Third, reactors are often tested under "worst-case" conditions, such as with aged lamps (or equivalent), upstream and downstream bends in the connecting piping, or other factors. The results of reactor testing are used to define the limits of acceptable operation, as defined by one or more well-defined, measurable endpoints.

## 7.2 Biodosimetry

Most validation protocols employ biodosimetry as the method for characterization of reactor behavior. Biodosimetry was introduced by Qualls and Johnson [1] as a form of biological actinometry in which the response of a population of challenge microorganisms was used to describe the effect of UV exposure. This approach shares some characteristics with chemical actinometry, wherein changes in a population of photosensitive molecules are used to quantify UV exposure. The original application of biodosimetry was based on a small, bench-scale reactor system and was developed as a method to validate numerical models of the "intensity" (more correctly described as fluence rate) field; however, an intermediate result of this test, when applied to a continuous-flow photoreactor, was an estimate of the "equivalent" dose received by the challenge organism. With time, biodosimetry evolved into the default method for characterization of photochemical reactor performance, and the term *reduction equivalent dose* (RED)[a] was defined. Although biodosimetry was originally applied to a small reactor, it can be applied at any scale, another factor that has promoted its widespread use.

In general terms, biodosimetry involves a set of well-defined steps, though many variations of these steps have emerged. The basic steps of biodosimetry are as follows:

1) *Identification of the challenge organism(s)*: One or more microbial or viral species are identified for use as the biological actinometer. Challenge organisms should be easy to grow to high titer in culture and be nonpathogenic to humans. For reasons that will be described later, challenge organisms used in biodosimetry testing will vary with the intended application. Common examples of challenge organisms for wastewater (including reuse) and drinking water applications include bacteria (vegetative cells or spores) and viruses (usually bacterial viruses [phage] are applied). Yeast cultures have been used in some applications [2]. For ballast water treatment, validation based on a marine algal species has been proposed [3]. A well-defined assay must be available to quantify the fraction of the challenge organism population that is inactivated as a result of exposure to UV radiation; the nature of the viability or infectivity assay will depend on the challenge organism. An aqueous suspension of the challenge organism is then prepared for exposure to UV radiation in two reactor types (described below).[b]
2) *Exposure of the challenge organism suspension to a range of well-defined UV doses*: The goal of these experiments is to define the dose–response behavior of the challenge organism (i.e. intrinsic kinetics of inactivation). This is most commonly accomplished using a shallow, well-mixed batch reactor under a collimated beam. A detailed description of this type of experiment is provided by Bolton and Linden [4]. In short, subsamples of the challenge organism suspension are exposed to a range of well-defined, (essentially) single-valued doses under a collimated beam. The exposed samples are then subjected to an appropriate assay to quantify changes in viability or infectivity of the challenge organism. A mathematical model is then fit to the data to allow description of the inactivation response as a function of applied UV dose. Ideally, the model that is fit to the dose–response data should be consistent with the mechanism of inactivation that applies for the organism; however, the primary role of this function is to allow interpolation of microbial responses as a function of applied UV dose. As a result, empirical expressions are often used to describe this behavior.
3) *Exposure of the challenge organism suspension to a range of operating conditions for the reactor of interest*: A matrix of operating conditions is defined for the reactor, usually based on values of flow rate, lamp power, and UV transmittance of the suspension that are anticipated in practical applications. For each operating condition, the challenge organism is introduced to the reactor and maintained at a constant introduction rate until a steady-state condition is reached. At that point, samples of untreated and treated water are collected and subjected to the same viability or infectivity assay as used with samples collected from the collimated beam experiment. The reactor is then adjusted to reach a different steady-state condition within the test matrix, and the processes of sampling and analysis are repeated.
4) *Estimation of RED*: For each operating condition, the fractional inactivation of the challenge organism is quantified. This value is then compared with the UV dose–response curve (developed in step 2) to allow estimation of the RED for that condition. A RED response surface is created for the reactor for the operating conditions applied in biodosimetric testing. As such, it is possible to identify operating conditions that yield acceptable performance, as defined by RED or fractional inactivation of the challenge organism.

Figure 7.1 illustrates a typical "test stand" for validation testing of UV photoreactors. Collectively, the components included in the test stand are intended to allow for meaningful challenges of UV photoreactor performance to be conducted. Characteristics and functions of each component are described briefly below.

- *Delivery pump(s)*: One or more pumps are included to allow controlled, consistent delivery of an aqueous suspension of the challenge agent(s) and matrix modifiers.[c] Downstream valves are used to control flow rate. A backflow prevention valve is often included to protect against flow reversal in the system.
- *UV absorber*: Alternatively referred to as a UV transmittance adjuster, a soluble chemical is added to adjust absorbance (or transmittance) of radiation at the wavelength(s) that characterizes the UV source. Common examples of UV absorbers include lignin sulfonate (LSA) and soluble humic substances, such as the commercial product Super Hume®. LSA is a form of lignin (a complex polymer that is a key component of wood) that is made water soluble by sulfonation. Super Hume® is a concentrated liquid derived from humate. Both of these commonly applied transmittance adjusters are aqueous solutions of natural organic matter that are representative of the organic constituents that are likely to be present in natural waters and treated municipal wastewaters. But in concept, any nonreactive, absorbing chemical could be used for this purpose. Figure 7.2 illustrates UV transmittance spectra for LSA and Super Hume®. In practice, these compounds are applied so that a target value of UV transmittance is reached at a relevant wavelength, often 254 nm. The transmittance spectra illustrated in Figure 7.2 illustrate transmittance behavior across the UV spectrum for aqueous solutions with specified values of $UVT_{254}$. Not surprisingly, these compounds are strong absorbers for wavelengths in the vicinity of the peak output of a low-pressure Hg lamp ($\lambda = 253.7$ nm). Also included in Figure 7.2 are UV transmittance spectra for a commercial humic acid and potassium nitrate ($KNO_3$); these two constituents are responsible for a large fraction of absorbance of UVC radiation in many natural and engineered water supplies. Note that the transmittance spectra for LSA and Super Hume® have shapes that are similar to those that are representative of a combination of humic acid and nitrate, two UV-absorbing compounds that are likely to be present in natural waters and treated municipal wastewater effluents. This is particularly relevant for tests in which a goal is to mimic the absorbance

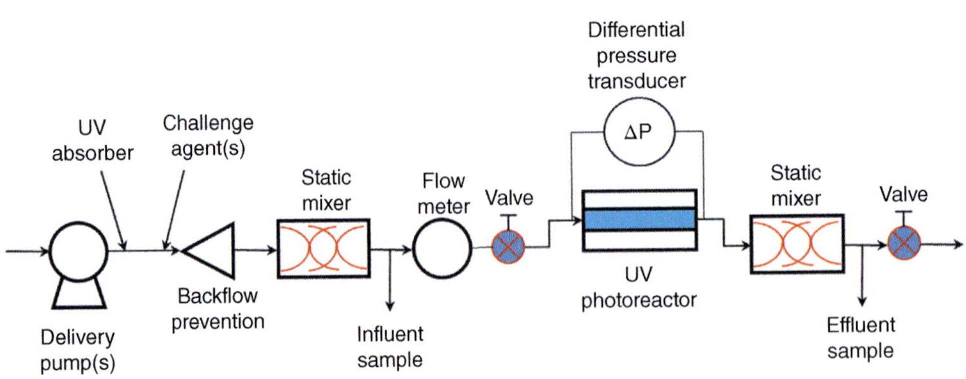


**Figure 7.1** Schematic illustration of typical test stand used for challenge testing and validation of UV photoreactors designed for use in water and wastewater treatment applications.

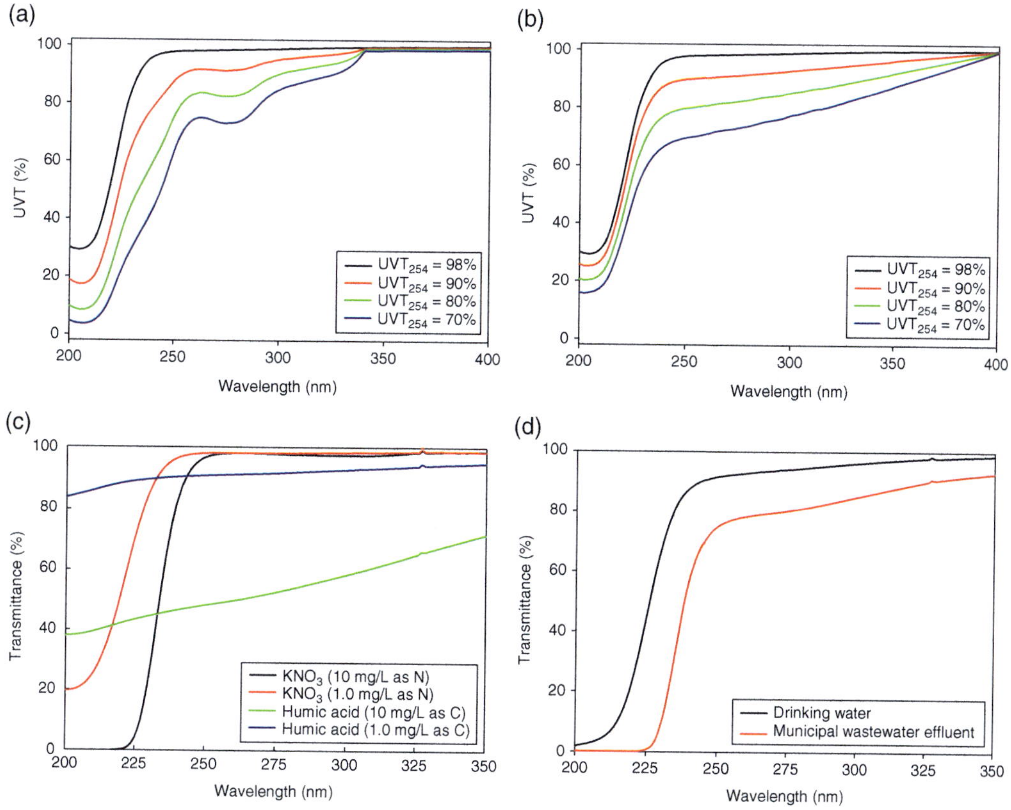


**Figure 7.2** Transmittance spectra for aqueous liquids that are representative of those used in validation testing. (a) UV transmittance spectra of aqueous solutions of LSA for various values of $UVT_{254}$; (b) UV transmittance spectra for Super Hume ® for various values of $UVT_{254}$; UV transmittance spectra for LSA and Super Hume® provided by Harold Wright, Carollo Engineers. (c) UV transmittance spectra for aqueous solutions of $KNO_3$ and humic acid; (d) UV transmittance spectra for drinking water and treated municipal wastewater from communities that share a common groundwater source.

(transmittance) characteristics of natural water supplies, which often contain humic substances and nitrate. For perspective, transmittance spectra for a municipal water supply (from a groundwater source) and a municipal wastewater effluent from a nearby community that gets its drinking water from the same aquifer are illustrated in Figure 7.2. Transmittance spectra for water supplies are highly variable, so the spectra illustrated in Figure 7.2 should be viewed only as examples.

- *Challenge agent(s)*: One or more photoreactive constituents is added to the aqueous suspension that is imposed on the reactor system. It must be possible to relate the change (response) of the challenge agent(s) to the UV dose (distribution) delivered by the reactor under a given operating condition. This is usually accomplished by exposing suspensions of the challenge agent to the output from a collimated beam that houses a UV source with the same (or similar) output spectral characteristics as the reactor undergoing validation. These exposures are repeated over a range of

UV doses. The responses of the challenge agent(s) to UV exposure are then measured as a function of the applied dose. These results define the intrinsic kinetics of the photochemical reaction(s) of interest in the validation process.

- *Static mixers*: These devices are included to promote homogenization of the liquid suspension. At the first location (upstream of the reactor being validated), this device promotes the uniform distribution of the UV absorber and the challenge agent(s). This reduces the likelihood of bias being introduced to the test as a result of nonuniform distribution of these agents. The upstream static mixer also facilitates collection of a representative upstream (untreated) sample, which provides the basis for comparison with corresponding downstream (treated) samples. A sampling port is located immediately downstream of this first static mixer to allow collection of representative influent samples. A second static mixer is placed downstream of the reactor, with a second sampling port, which allows collection of a representative, unbiased effluent (treated) sample. It is important to note, however, that the static mixers that are included in these systems should not affect fluid mechanical behavior within the irradiated zone of the UV photoreactor that is being tested.
- *UV photoreactor*: This is the system that is the subject of the validation test. In some circumstances, it is desirable to plumb the reactor with 90° bends immediately upstream and immediately downstream of the reactor, so as to be representative of (worst-case) nonideal fluid mechanics that may result from space constraints in a full-scale application.
- *Pressure transducer*: Pressure drop (head loss) across the reactor may be of interest in some applications. It is therefore desirable to include instrumentation to allow measurement of the change in pressure across the reactor.

Biodosimetry is a well-established method that provides a tangible measure of reactor performance in the form of microbial inactivation, usually described in the form of RED. Its history of use has promoted familiarity among those in the engineering and regulatory communities that work with UV photoreactors.

However, important drawbacks associated with biodosimetry have been identified. Most important among these is the inability of this method to quantify the UV dose distribution delivered by a photoreactor. In turn, this prevents the direct translation of challenge organism inactivation to the response of another organism, except in cases where the challenge organism and the target organism have identical dose–response behavior. In practice, this shortcoming is overcome through the use of safety factors, which can dramatically increase the capital and operating costs of a UV photoreactor, but are necessary to protect public health when the dose distribution cannot be measured. A related drawback of this method is the use of the term "dose" to describe reactor behavior. Even when included in the broader term reduction equivalent dose, this terminology represents a common source of confusion among users.

## 7.3 Mathematical Descriptions of UV Photoreactor Validation by Biodosimetry

The fundamental goal of biodosimetry is to quantify measured inactivation of a challenge organism and to relate that measurement to an "equivalent" dose (RED). It is instructive to consider a formal mathematical description of this experiment and its interpretation. This formal description will illustrate some of the critical characteristics and limitations of this method of reactor validation.

As described previously, the performance of a photochemical reactor can be described by application of the segregated-flow model (SFM):

$$\left(\frac{[B]}{[B]_0}\right)_{\text{reactor}} = \int_0^{\infty} \left(\frac{[B]}{[B]_0}\right)_{\text{batch}} \cdot E(D) \cdot \mathrm{d}D \tag{7.1}$$

where

| | | |
|---|---|---|
| $[B]$ | = | concentration of photoreactive constituent after UV exposure |
| $[B]_0$ | = | concentration of photoreactive constituent prior to UV exposure |
| $\left(\frac{[B]}{[B]_0}\right)_{\text{reactor}}$ | = | fraction of photoreactive constituent that remains after exposure to reactor system |
| $\left(\frac{[B]}{[B]_0}\right)_{\text{batch}}$ | = | fraction of photoreactive constituent that remains as a function of dose ($D$) in a batch reactor (i.e. an expression of the intrinsic kinetics of the photochemical reaction) |
| $E(D)$ | = | dose distribution function |
| $E(D) \cdot \mathrm{d}D$ | = | fraction of fluid elements that receive a dose between $D$ and $(D + \mathrm{d}D)$. |

For purposes of illustration, we will consider two examples that can be analyzed completely using the SFM. The first example will involve a hypothetical dose distribution and hypothetical reaction kinetics. The second example will be based on dose distribution estimates that were developed for a full-scale reactor based on CFD-E' modeling.

### 7.3.1 Gaussian Dose Distribution, First-Order Kinetics

In the application of the SFM to a photochemical reactor, it is necessary to be able to quantify the dose distribution and the intrinsic kinetics of the reaction of interest. To address these needs, two basic assumptions will be applied in the first simulation:

- *Gaussian dose distribution* – Dose distributions that have been measured or simulated for UV photoreactors often share some characteristics of the Gaussian (Normal) distribution. Specifically, measured or simulated dose distributions (as density functions) often demonstrate a peak near the mean of the distribution, with roughly symmetric tailing on either side of the mean. The Gaussian distribution can be described as a closed-form expression. Despite these characteristics, it is relevant to note that the dose distributions delivered by actual UV photoreactors are likely to deviate from the Gaussian form.
- *First-order, consumptive reaction* – Though not a requirement, we are often interested in consumptive reactions within UV photoreactors. A first-order, consumptive model provides an accurate description of the inactivation responses of some microbes, at least over a limited range of UV doses; however, important deviations from first-order inactivation are observed with most microbes. A first-order, consumptive reaction model also provides a rigorously correct description of a pure photolytic process in dilute solution or on a pointwise basis (see Chapters 2 and 3).

The dose distribution function is a probability density function; as such, it has the following properties:

$$E(D) \cdot \mathrm{d}D = \text{fraction of fluid elements receiving UV dose in the range} : D \leq \text{dose} \leq D + \mathrm{d}D \tag{7.2}$$

$$F(\Delta) = \int_0^{\Delta} E(D) \cdot \mathrm{d}D = \text{fraction of fluid elements receiving UV dose} \leq \Delta \tag{7.3}$$

The function $F(\Delta)$ is a cumulative distribution function which defines the fraction of fluid elements receiving a dose $\leq \Delta$. By extension, Equation 7.3 implies:

$$F(\infty) = \int_0^{\infty} E(D) \cdot \mathrm{d}D \equiv 1 \tag{7.4}$$

In other words, the "area" under a dose distribution density function must equal 1 (dimensionless).

As described above, we will assume a Gaussian dose distribution in this first example. This implies that the dose distribution (density) function takes the form:[d]

$$E(D) = \frac{1}{\sqrt{2\pi\sigma^2}} \exp\left[\frac{-(D-\mu)^2}{2\sigma^2}\right] \tag{7.5}$$

where

$\mu$ = mean dose $[=] \frac{\text{mJ}}{\text{cm}^2}$

$\sigma$ = standard deviation of dose distribution $[=] \frac{\text{mJ}}{\text{cm}^2}$

Figure 7.3 illustrates several examples of Gaussian dose distributions that will be applied with the SFM herein. The mean dose ($\mu$) defines the centroid of the distribution and may be viewed as a location parameter. The standard deviation ($\sigma$) describes the spread of the distribution about the mean. Reactor performance is promoted by a high value of $\mu$ and a low value of $\sigma$.

For simplicity, first-order consumptive kinetics will be assumed for the photochemical reaction of interest. Therefore, for a batch reactor, the fraction of reactant remaining can be defined analytically as:

$$\left(\frac{[B]}{[B]_0}\right)_{\text{batch}} = \exp(-kD) \tag{7.6}$$

where

$k$ = reaction rate constant $[=] \frac{\text{cm}^2}{\text{mJ}}$.

Figure 7.4 illustrates the effect of the rate constant on the extent of photoreaction for UV doses that are relevant in most disinfection applications. Note that for this range of doses, inactivation constants less than approximately $10^{-2} \frac{\text{cm}^2}{\text{mJ}}$ yield only modest changes in reactant concentration for UV doses up to approximately $100 \frac{\text{mJ}}{\text{cm}^2}$. This represents a source of error in the application of biodosimetry, in that minor errors in measurement of challenge agent concentration can lead to large errors in interpretation of reactor performance. Conversely, challenge agents with a rate constant greater than approximately $10^{-2} \frac{\text{cm}^2}{\text{mJ}}$ are quite sensitive to UV doses normally applied in

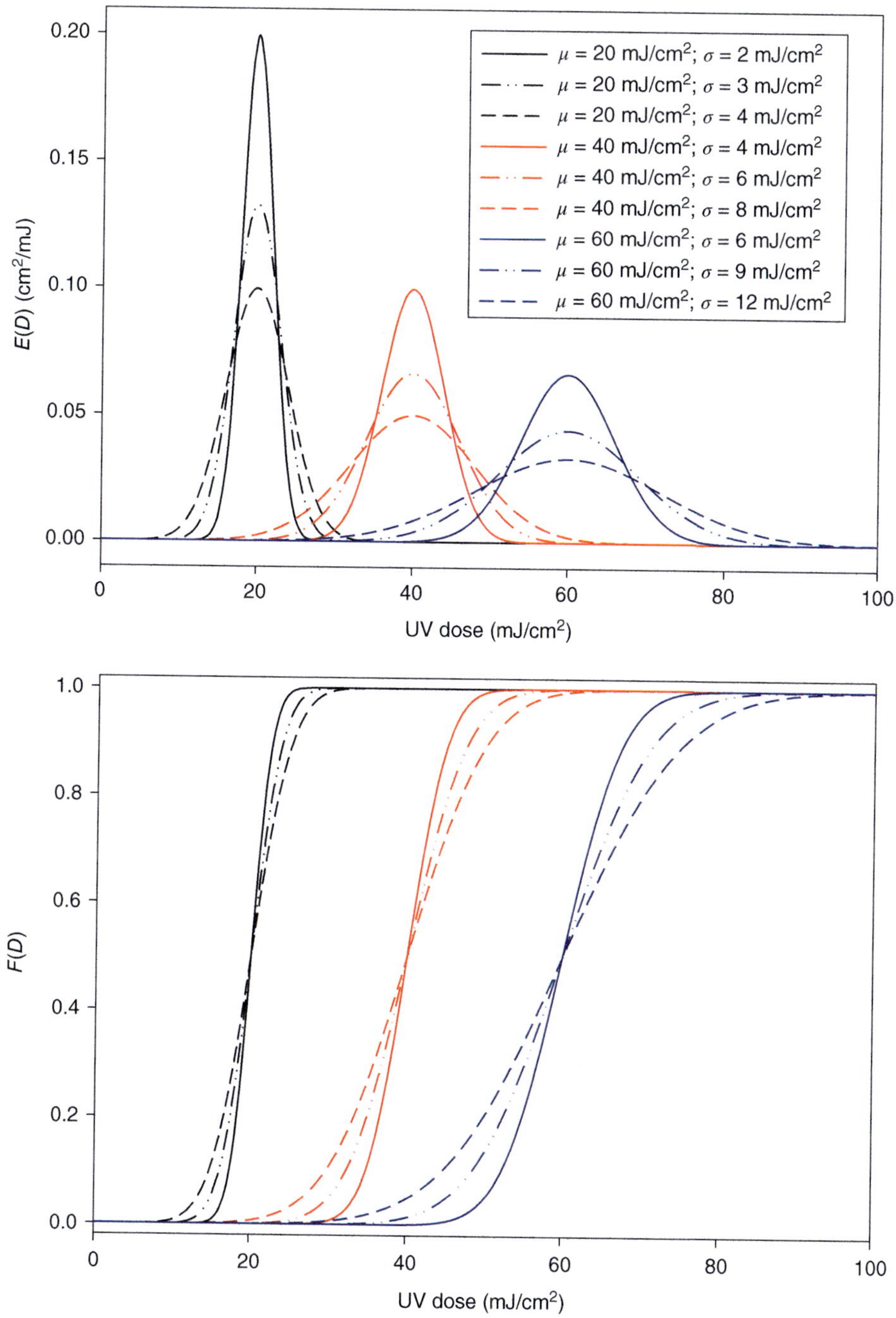


**Figure 7.3** Examples of Gaussian dose distributions in density form (top) and in cumulative form (bottom). Note that the legend in the top panel also applies to the bottom panel.

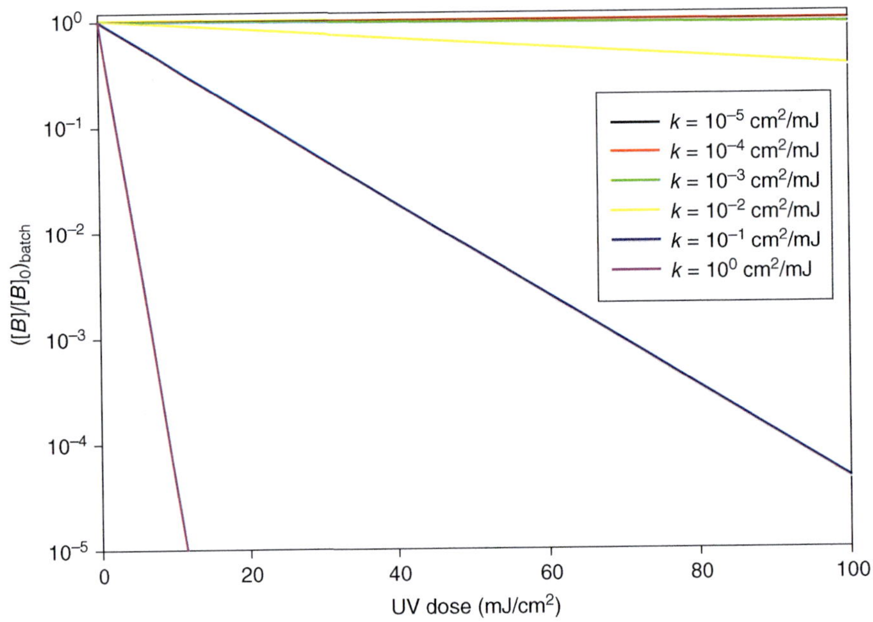


**Figure 7.4** Effect of rate constant ($k$) on the extent of photoreaction in a batch reactor. Note that the vertical axis is presented on a $\log_{10}$ scale.

disinfection applications. For challenge agents with rate constants in this range, analytical methods must exist and be applied that can accurately and precisely quantify large changes (i.e. several $\log_{10}$ units) in their concentration.

Substitution of Equations 7.5 and 7.6 into the SFM yields:

$$\left(\frac{[B]}{[B]_0}\right)_{\text{reactor}} = \frac{1}{\sqrt{2\pi\sigma^2}} \int_0^{\infty} \exp(-kD) \cdot \exp\left[\frac{-(D-\mu)^2}{2\sigma^2}\right] \cdot \mathrm{d}D \tag{7.7}$$

Note that in this simulation of photoreactor performance, the behavior of the reactor system is described entirely by three parameters ($\mu$, $\sigma$, and $k$).

Expansion and combination of the exponential terms yields:

$$\left(\frac{[B]}{[B]_0}\right)_{\text{reactor}} = \frac{1}{\sqrt{2\pi\sigma^2}} \int_0^{\infty} \exp\left[-\left(kD + \frac{D^2 - 2\mu D + \mu^2}{2\sigma^2}\right)\right] \cdot \mathrm{d}D \tag{7.8}$$

which can be rearranged to yield:

$$\left(\frac{[B]}{[B]_0}\right)_{\text{reactor}} = \frac{1}{\sqrt{2\pi\sigma^2}} \int_0^{\infty} \exp\left[-\left(\frac{D^2}{2\sigma^2} + \left(k - \frac{\mu}{\sigma^2}\right)D + \frac{\mu^2}{2\sigma^2}\right)\right] \cdot \mathrm{d}D \tag{7.9}$$

The following is a fundamental identity in calculus:

$$\int_0^{\infty} \exp\left[-\left(ax^2 + bx + c\right)\right] \cdot \mathrm{d}x = \frac{1}{2}\sqrt{\frac{\pi}{a}} \cdot \exp\left(\frac{b^2 - 4ac}{4a}\right) \cdot \mathrm{erfc}\left(\frac{b}{2\sqrt{a}}\right) \tag{7.10}$$

where

$$\text{erfc}(x) = \text{complementary error function} \tag{7.11}$$

A few identities of the error function, erf(*x*), and its complement, erfc(*x*), are as follows:[e]

$$\text{erfc}(x) = 1 - \text{erf}(x) \tag{7.12}$$

$$\text{erfc}(x) = \frac{2}{\sqrt{\pi}}\int_x^{\infty} \mathrm{e}^{-t^2}\,\mathrm{d}t \tag{7.13}$$

$$\text{erf}(x) = \frac{2}{\sqrt{\pi}}\int_0^{x} \mathrm{e}^{-t^2}\,\mathrm{d}t \tag{7.14}$$

Comparison of Equations 7.9 and 7.10 indicates:

$$a = \frac{1}{2\sigma^2}; b = k - \frac{\mu}{\sigma^2}; c = \frac{\mu^2}{2\sigma^2} \tag{7.15}$$

Substitution of these terms into Equation 7.10, and then into Equation 7.9 yields:

$$\left(\frac{[B]}{[B]_0}\right)_{\text{reactor}} = \frac{1}{\sqrt{2\pi\sigma^2}} \cdot \frac{1}{2} \cdot \sqrt{2\pi\sigma^2} \cdot \exp\left(\frac{\left(k - \frac{\mu}{\sigma^2}\right)^2 - 4\left(\frac{1}{2\sigma^2}\right)\left(\frac{\mu^2}{2\sigma^2}\right)}{4 \cdot \frac{1}{2\sigma^2}}\right) \cdot \text{erfc}\left(\frac{k - \frac{\mu}{\sigma^2}}{2\sqrt{\frac{1}{2\sigma^2}}}\right) \tag{7.16}$$

Canceling terms and expanding the square term in the numerator of the exponential argument, we find:

$$\left(\frac{[B]}{[B]_0}\right)_{\text{reactor}} = \frac{1}{2} \cdot \exp\left(\frac{k^2 - 2k\frac{\mu}{\sigma^2}}{\frac{2}{\sigma^2}}\right) \cdot \text{erfc}\left(\frac{k\sigma - \frac{\mu}{\sigma}}{\sqrt{2}}\right) \tag{7.17}$$

Further combining terms, we find:

$$\left(\frac{[B]}{[B]_0}\right)_{\text{reactor}} = \frac{1}{2} \cdot \exp\left(\frac{k^2\sigma^2}{2} - k\mu\right) \cdot \text{erfc}\left(\frac{k\sigma - \frac{\mu}{\sigma}}{\sqrt{2}}\right) \tag{7.18}$$

Equation 7.18 provides an analytical (closed-form) solution to the SFM for the case of a Gaussian dose distribution and first-order, consumptive kinetics. Figure 7.5 illustrates predictions of reactor performance for the dose distributions illustrated in Figure 7.3 and a wide range of first-order reaction rate constants. As with Figure 7.4, simulations based on rate constants less than approximately $10^{-2}\frac{\text{cm}^2}{\text{mJ}}$ demonstrate relatively small changes in reactant concentration. For all cases, the extent of photoreaction increases as the value of the mean dose increases, and as the value of the standard deviation decreases. As alluded to in previous chapters, a narrow dose distribution is desirable in most applications of photoreactors.

In most applications of biodosimetry, the primary goal is to provide an estimate of the RED as a parameter to characterize reactor performance. This is usually accomplished by applying the measured change in challenge agent relative concentration, $\left(\frac{[B]}{[B]_0}\right)_{\text{reactor}}$, to the inverse of the equation

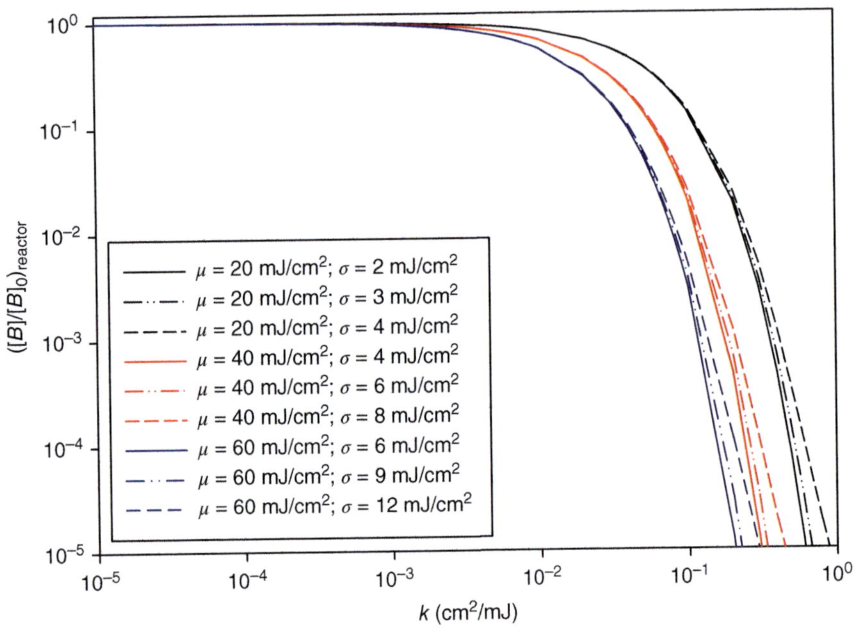


**Figure 7.5** Predictions of reactor performance as a function of rate constant for a range of dose distributions, based on Equation 7.18. Note that both axes are presented using a $\log_{10}$ scale.

that is used to define the intrinsic kinetics of challenge agent photodecay. In the case of a first-order, consumptive reaction, this would be similar to the inverse of Equation 7.6:

$$\text{RED} = \frac{-\ln\left(\frac{[B]}{[B]_0}\right)_{\text{reactor}}}{k} \tag{7.19}$$

Figure 7.6 illustrates the application of Equation 7.19 to the results indicated in Figure 7.5. Several general trends emerge from this analysis:

- The value of RED asymptotically approaches the mean of the dose distribution as the value of $k$ decreases. For the dose distribution examples illustrated in Figure 7.6, the value of RED is essentially constant (RED $\approx \mu$) for values of $k$ less than approximately $10^{-2}\,\frac{\text{cm}^2}{\text{mJ}}$.
- For the dose distribution examples illustrated in Figure 7.6, the value of RED is strongly influenced by the value of the rate constant for $\gtrsim 10^{-2}\,\frac{\text{cm}^2}{\text{mJ}}$. In particular, for a given dose distribution (i.e. fixed values of $\mu$ and $\sigma$), the value of RED decreases as $k$ increases. In other words, the value of RED depends on the challenge organism used in biodosimetry. For a given dose distribution, the value of RED reported from biodosimetry using a UV-resistant microbe (small $k$) will be larger than the value reported from biodosimetry using a UV-sensitive microbe (large $k$). Interestingly, most challenge organisms, as well as most microbial pathogens of interest are characterized by inactivation rate constants that are in the range $k \gtrsim 10^{-2}\,\frac{\text{cm}^2}{\text{mJ}}$. For these organisms, the value of RED is strongly dependent on the challenge agent. And for this reason, any assignment of RED should stipulate the challenge agent used to measure it.

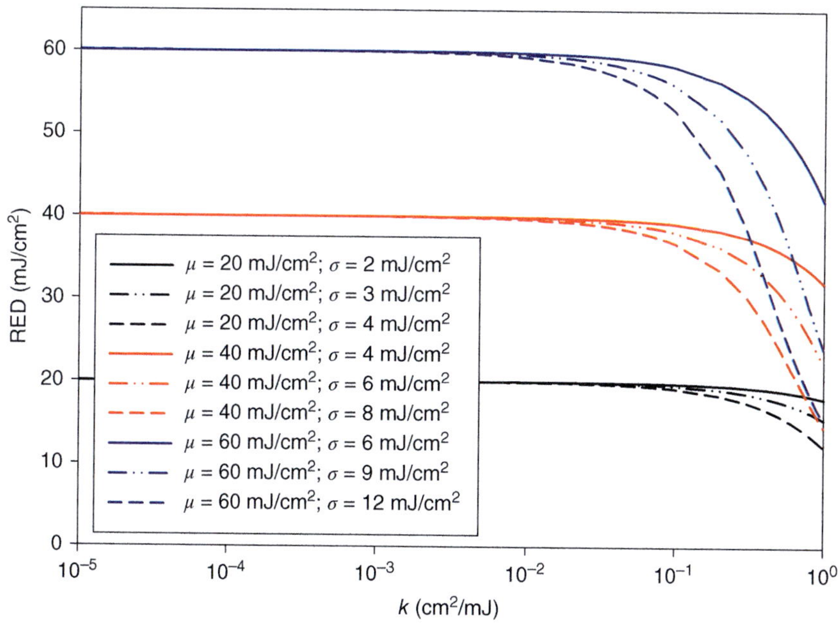


**Figure 7.6** RED as a function of rate constant for a range of Gaussian dose distributions. Note that the horizontal axis is presented using a $\log_{10}$ scale.

- Challenge agents with a relatively large rate constant provide greater sensitivity for quantification of reactor inefficiency, as indicated by a broad dose distribution. In Figure 7.6, this is evident when comparing dose distributions with the same value of mean dose, but different values of the standard deviation. For challenge agents with rate constants greater than about $10^{-2}\ \frac{\text{cm}^2}{\text{mJ}}$, the RED assigned to a given condition decreases as the standard deviation (i.e. spread of the dose distribution about the mean) increases. In other words, inefficient reactors are penalized by the use of challenge agents with large $k$. Conversely, the use of challenge agents with large $k$ also allows identification of reactors with narrow dose distributions.
- Consider a reference challenge agent with a rate constant of $k_1$. For any other agent imposed on the reactor, the results of biodosimetry indicate the following:
  - For challenge agents with rate constant $\geq k_1$, the change in concentration of that agent achieved by the reactor, $\left(\frac{[B]}{[B]_0}\right)_{\text{reactor}}$, will be greater than or equal to the change observed with the challenge agent. However, the magnitude of the differences in $\left(\frac{[B]}{[B]_0}\right)_{\text{reactor}}$ for the two challenge agents cannot be determined unless the dose distribution is known.
  - For challenge agents with rate constant $\leq k_1$, the change in concentration of that agent achieved by the reactor, $\left(\frac{[B]}{[B]_0}\right)_{\text{reactor}}$, will be less than or equal to the change observed with the challenge agent. However, the magnitude of the differences in $\left(\frac{[B]}{[B]_0}\right)_{\text{reactor}}$ for the two challenge agents cannot be determined unless the dose distribution is known.

- For the examples illustrate above, the dose distribution was assumed. However, for most reactor systems, the dose distribution delivered by the system is unknown. Biodosimetry will only yield information regarding the *effect* of the dose distribution; it will not yield a measure of the dose distribution itself. Therefore, in the absence of knowledge of the dose distribution, the results of biodosimetry provide only an upper or lower bound on the performance of the reactor, relative to challenge agents other than the reference challenge agent.

Collectively, these observations indicate benefits of selection of challenge agents that have dose–response behavior that is similar to that of the target of treatment. When this condition is met, the results of reactor testing with the challenge agent will accurately represent the behavior of the target. When this condition is not met, it is not possible to use the inactivation response of a challenge agent to predict (quantitatively) the response of a target organism, though the behavior of the challenge agent can be used to define an upper or lower limit on the inactivation response of the target, depending on how their sensitivities to UV exposure relate to each other. The larger the difference between the UV sensitivities of challenge agent and target microbe, the greater the uncertainty in translating the results of reactor validation experiments to practical application.

These points emerge from the analysis presented above based on assumptions of Gaussian dose distributions and first-order kinetics. However, they apply broadly to other dose distributions and other kinetic expressions.

The last of the points made above is particularly relevant with respect to validation of UV photoreactors for disinfection applications. For example, if we were to consider the use of a challenge agent with an inactivation constant of $0.1\,\frac{\text{cm}^2}{\text{mJ}}$ (nominally the value that corresponds to coliphage MS2), then a biodosimetry test that yields 2 $\log_{10}$ units of inactivation of MS2, corresponding to an MS2 RED of roughly $40\,\frac{\text{mJ}}{\text{cm}^2}$, provides information that organisms with inactivation constants greater than this value will demonstrate inactivation responses of at least 2 $\log_{10}$ units; however, it is not possible to define how much above 2 $\log_{10}$ units the actual inactivation response will be, unless the dose distribution is known. This represents an important limitation of biodosimetry as a validation method, and is the primary reason why generous safety factors are often included in the use of results from biodosimetry.

### 7.3.2 Simulated Dose Distributions, First-Order Kinetics

The preceding illustration was based on a set of contrived dose distributions. The form of the dose distribution was chosen to be similar to dose distributions that have been measured or simulated in actual reactors, and to provide access to a convenient closed-form solution. However, the dose distributions delivered by actual reactors often demonstrate important deviations from a convenient distribution function, such as the Gaussian distribution. As such, there is value in conducting an analogous numerical simulation of a reactor for which dose distributions have been calculated over a range of operating conditions.

Figure 7.7 illustrates dose distributions that were calculated for a UV disinfection system over a range of approach velocities (flow rates); each of these simulations was conducted for a fixed value of UV transmittance and for a fixed value of lamp power. The simulations were conducted using the Reynolds-averaged Navier–Stokes (RANS) form of the $k$–$\varepsilon$ model for CFD simulation and ray tracing to simulate the fluence rate field [5]. Following a Lagrangian approach, a set of 4996 particle trajectories was mapped onto the simulated fluence rate field for each approach velocity. It was then possible to assemble dose distributions for each operating condition based on the trajectory-specific

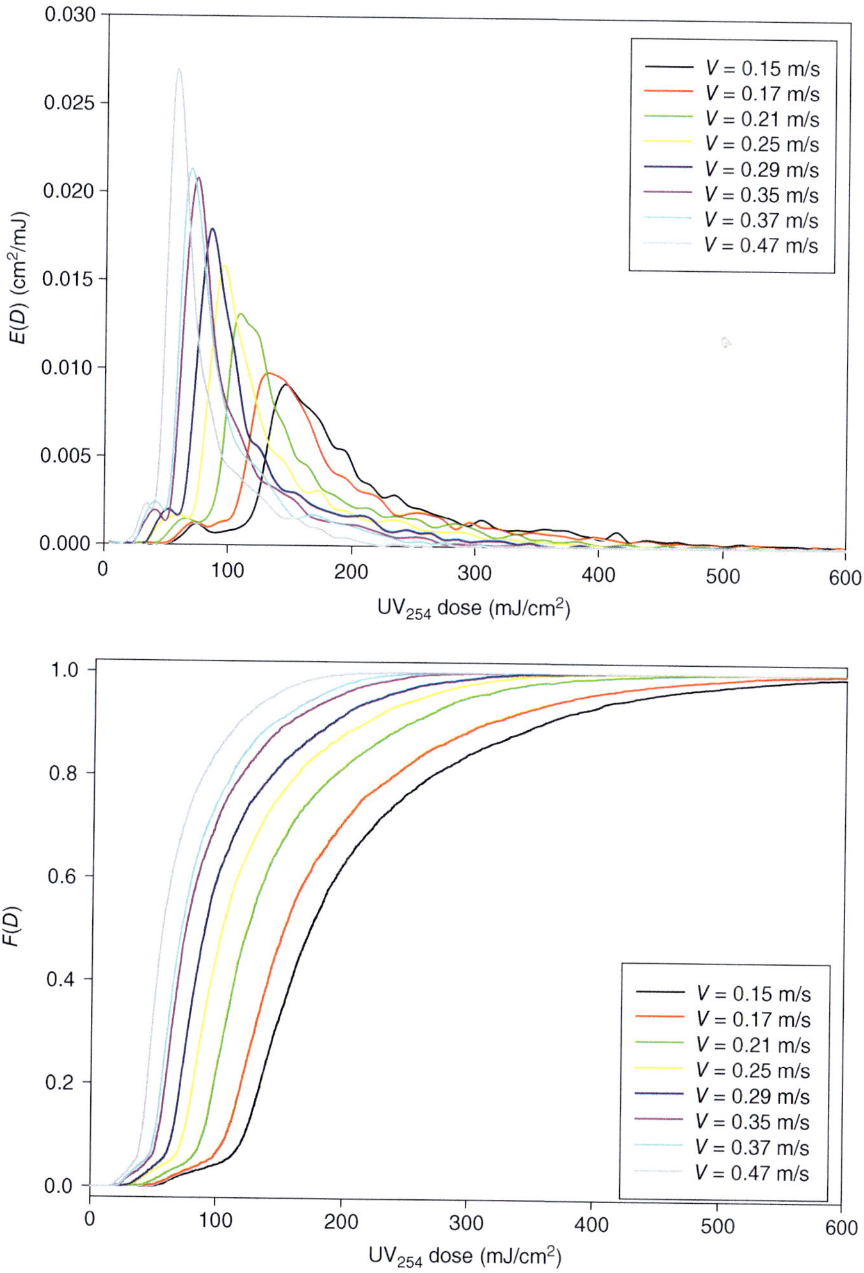


**Figure 7.7** $UV_{254}$ dose distribution estimates in density form (top) and in cumulative form (bottom) based on CFD-E' simulations. For all simulations, the same value of UVT and lamp power was applied [5]. *Source:* Ahmed 2019.

doses that were estimated for each condition. The dose distributions in this reactor (and in many other UV photoreactors) scale roughly linearly with the inverse of flow rate (or approach velocity). It is also evident from these simulated distributions that important deviations from a simple Gaussian form are evident. These deviations are most evident in the "tails" of the dose distributions that exist on both sides of the mean dose. For disinfection applications, the low-dose tail is probably most important, as the low doses within a dose distribution are generally responsible for a large fraction of the organisms that retain viability (or infectivity) following UV exposure.

Simulations of reactor performance were conducted following a Lagrangian approach. Specifically, for each operating condition, the value of $\left(\frac{[B]}{[B]_0}\right)_{\text{batch}}$ was calculated for each particle-specific dose. Because 4996 trajectories were simulated for each operating condition, $E(D) \cdot dD$ was assigned a value of $\frac{1}{4996}$ for each trajectory-specific dose. Following this logic, overall reactor performance, $\left(\frac{[B]}{[B]_0}\right)_{\text{batch}}$, was calculated by the following summation, as an approximation of the SFM:

$$\left(\frac{[B]}{[B]_0}\right)_{\text{reactor}} \approx \sum_{i=1}^{4996} \frac{1}{4996} \cdot \exp\left(-k \cdot D_{\text{trajectory}}\right) \tag{7.20}$$

where

$D_{\text{trajectory}}$ = trajectory-specific dose estimated by Lagrangian simulation.

Figure 7.8 illustrates predictions of challenge agent change resulting from exposure to the dose distributions illustrated in Figure 7.7 as a function of rate constant for a wide range of rate constants, under the assumption of first-order, consumptive reactions. As with the previous example, the extent of change in the concentration of the challenge agent is strongly influenced by the value of the rate constant and the nature of the dose distribution. Small values of the rate constant again lead to circumstances under which only modest changes in the concentration of the challenge agent

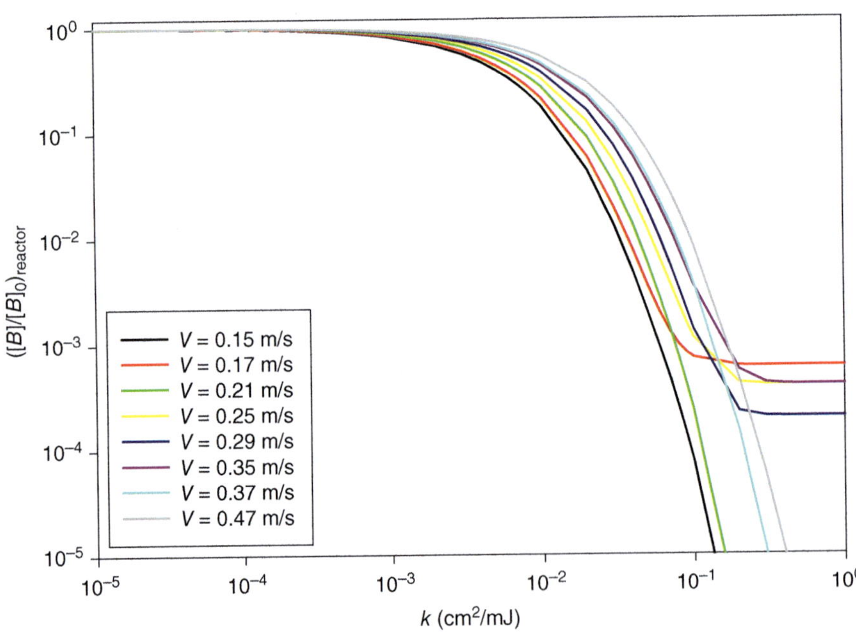


**Figure 7.8** Predictions of reactor performance as a function of rate constant for a range of dose distributions, based on Equation 7.20. Note that both axes are presented using a $\log_{10}$ scale.

are evident. Larger rate constants lead to situations in which the extent of change in the concentration of the challenge agent can be substantially larger. Some anomalous behavior is evident for several of the operating conditions, in that the extent of change in challenge agent concentration does not change with rate constant for values of the rate constant above approximately $0.1\,\frac{cm^2}{mJ}$. The cause of this behavior is unknown, but it provides an indication of the effect of the dose distribution on inactivation. As described with the examples of Gaussian dose distributions, for challenge agents with a rate constant larger than the rate constant for a reference challenge agent with a rate constant of $k_1$, the extent of change in the concentration will be greater than or equal to the value observed for the reference challenge agent.

Following the procedure defined previously (i.e. equation 7.19), it was then possible to define the RED for each operating condition, based on a range of rate constants for the challenge agent (see Figure 7.9). As with the previous example, it is evident that the value of RED asymptotically approaches the mean of the dose distribution ($\mu$, calculated separately for each operating condition based on 4996 trajectory-specific doses) as $k$ approaches zero. Also, as the value of $k$ increases, the value of RED becomes increasingly sensitive to $k$.

The observations that were made from the examples based on Gaussian dose distributions apply for these examples as well. However, additional generalizations may be drawn by comparing the results of these simulations. Specifically, it is evident that the value of the mean of the dose distribution ($\mu$) was 5–10 times larger in the second example than in the first example. Along with this, it is evident that the value of the rate constant ($k$) at which sensitivity of RED to the rate constant becomes evident also changes.

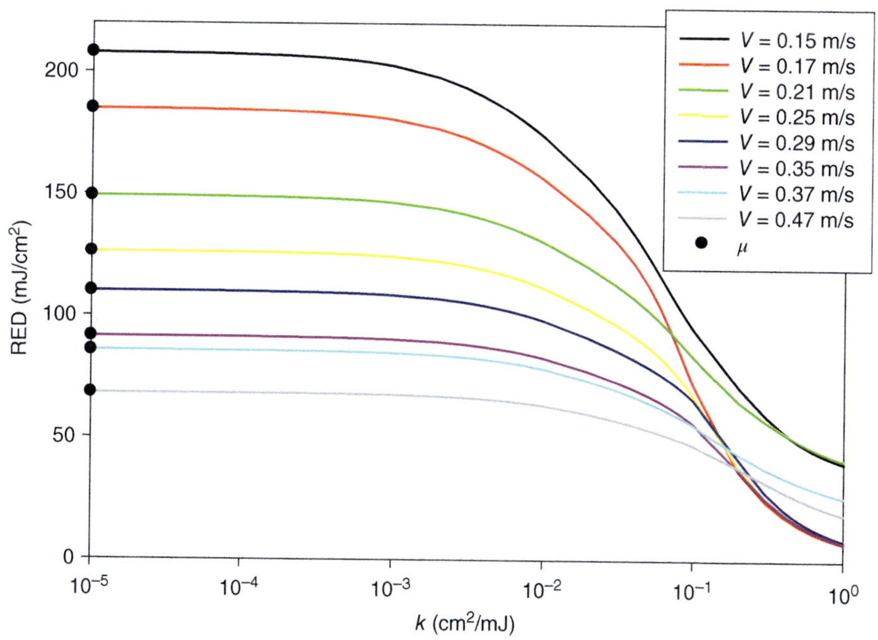


**Figure 7.9** RED as a function of rate constant for a range of simulated dose distributions. The mean of each dose distribution ($\mu$) was calculated based on the entire population of simulated particle trajectories ($n = 4996$). Note that the horizontal axis is presented using a $\log_{10}$ scale.

This behavior can be generalized using the dimensionless variable $k \cdot \mu$. Specifically, in evaluating Figures 7.6 and 7.9, we observe that the value of RED is sensitive to values of $k \cdot \mu \gtrsim 1$. When this condition is met, the test is able to differentiate between reactors that have a narrow dose distribution and those that have a broad dose distribution. Reactors with a narrow dose distribution will yield a value of RED that is closer to the theoretical maximum (i.e. $\text{RED}_{\text{max}} \equiv \mu$) than reactors with a broad dose distribution. Since μ can be calculated independently of a biodosimetry test, it is then possible to combine measured values of RED with calculated values of $\mu$ to assess reactor efficiency.

### 7.3.3 Biodosimetry Experiment

The two preceding examples provided illustrations of the concepts that define biodosimetry and some of the limitations of the results of these experiments. However, these examples were based on assumed or simulated dose distributions and an assumption of first-order kinetics. The following text and figures will illustrate the results of an actual biodosimetry experiment that was used to validate a commercially available UV disinfection system. Because the results of these tests are viewed as proprietary information, the name of the manufacturer and some details of the operating conditions are not included.

The experiment involved the use of coliphage MS2 and coliphage T1 as challenge agents. MS2 is often used as a challenge agent in biodosimetry; in fact, it is viewed as the default challenge organism in some validation protocols (see discussion below). T1 is far more sensitive to UV exposure than MS2 and displays UV dose–response behavior that is similar to that of some common target microbes, including *Cryptosporidium parvum* and *Giardia lamblia*. As such, there is the potential for benefit associated with the inclusion T1 in the validation experiment if the intended application will address inactivation of these common protozoan parasites, or other microbial pathogens that display similar UV dose–response behavior.

The first step in the conduct of this experiment is measurement of the dose–response relationship for both challenge agents. Figure 7.10 illustrates the results of such an experiment for both challenge agents. The data presented in this figure were developed by exposure of aqueous suspensions of the challenge agents to $\text{UV}_{254}$ radiation from a collimated beam that was built around a LP Hg lamp.

For both organisms, it is necessary to define a fitting function. Ideally, the fitting function will have a mathematical form that will be consistent with the mechanism of inactivation. However, the most important role of this fitting function is to allow for interpolation of the dose–response data with minimal error. A common expression used to fit to UV dose–response behavior of phage is a quadratic equation of the following form:

$$D = AI^2 + BI \tag{7.21}$$

where

$D$ = applied UV dose

$I$ = $-\log_{10}\left(\frac{N}{N_0}\right)$

$N$ = concentration of viable challenge organism after UV exposure

$N_0$ = concentration of viable challenge organisms prior to UV exposure

$A, B$ = empirical constants.

Despite the fact that this method of fitting has become common in reactor validation, an argument could be presented that it would be more logical to present an equation that is the inverse of

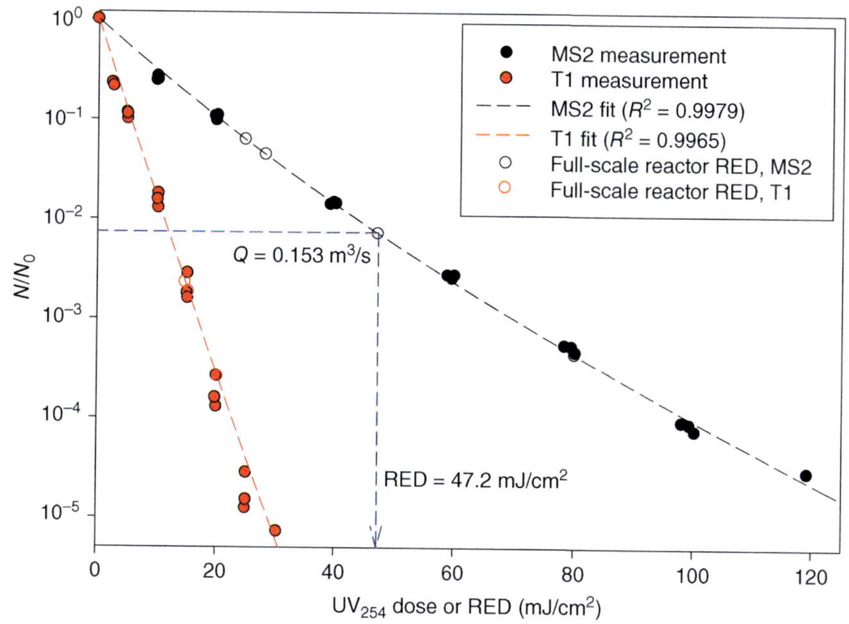


**Figure 7.10** $UV_{254}$ dose–response behavior of coliphage MS2 and coliphage T1 based on exposure of the microbial suspensions in a shallow, well-mixed reactor under a collimated beam. Solid data points indicate measured responses, presented in the form of $N/N_0$ (note that vertical axis is presented on a $\log_{10}$ scale). Dashed lines represent nonlinear fit to data, used for interpolation of RED (see equation 7.21 and Table 7.1). Open circles indicate measured inactivation responses from experiments conducted on a full-scale reactor, with corresponding values of RED calculated from the nonlinear fits to the dose–response data. The dashed arrow indicates an example of an RED calculation for an experiment conducted at a flow rate of $0.153\frac{m^3}{s}$ using MS2 as the challenge agent (see circled data point in Figure 7.11).

this relationship. Equation 7.21 suggests that dose is a function of the extent of microbial inactivation, when in fact the opposite is true.

Table 7.1 provides a summary of the fitting parameters that were developed by nonlinear regression for the data indicated in Figure 7.10. The quadratic function appears to provide a good fit to these data and can be used as an effective tool for interpolation of these data, as will required for estimation of RED.

The next step in reactor validation is to impose suspensions of the same challenge agent(s) on the reactor. This is normally done by applying challenge agent suspensions over a range of operating conditions corresponding to the so-called "operating envelope," or the set of conditions over which the reactor is expected to be operated.

**Table 7.1** Summary of regression parameters for fits of Equation 7.21 to data in Figure 7.10.

| Challenge agent | A | B | $R^2$ |
|---|---|---|---|
| MS2 | 1.5166 | 18.83 | 0.9979 |
| T1 | 0.083983 | 5.351 | 0.9965 |

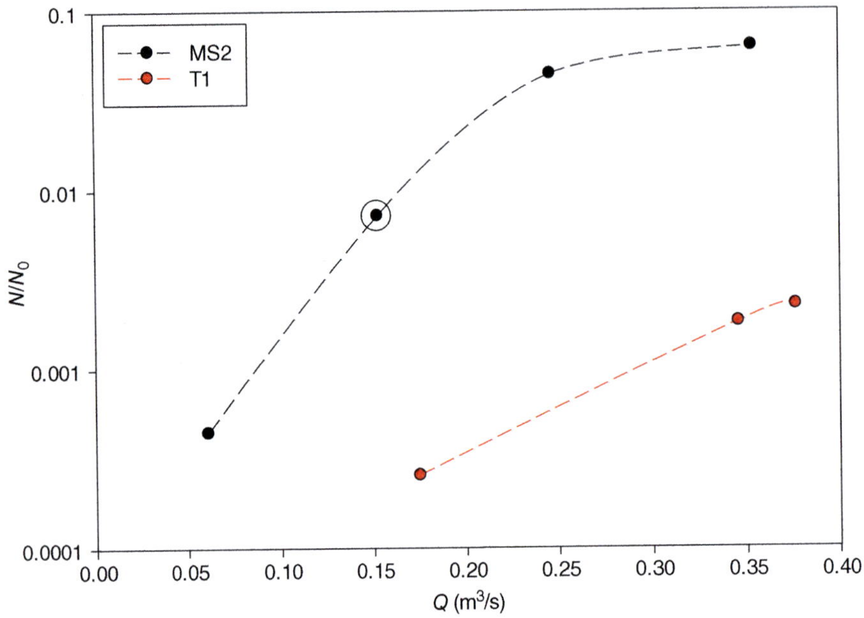


**Figure 7.11** Inactivation of challenge agents as a function of flow rate. The data indicated here illustrate microbial inactivation at fixed values of UV transmittance and lamp power. The circled data point represents an example that is used in Figure 7.10 to illustrate the method by which RED is calculated.

Figure 7.11 illustrates measurements of challenge organism inactivation as a function of flow rate. The data illustrated in Figure 7.11 indicate inactivation responses for MS2 at four flow rates and inactivation of T1 at three flow rates. The value of UV transmittance was fixed for all seven of these experiments, as was lamp power. Therefore, the data presented in Figure 7.11 illustrate the effects of flow rate on microbial inactivation. At a given flow rate, the extent of T1 inactivation always exceeded the extent of MS2 inactivation because of their differences in sensitivity to UV exposure (see Figure 7.10).

The data from Figure 7.11 were then transferred to the dose–response fitting functions in Figure 7.10 to develop estimates of the RED for each operating condition/challenge microbe pair. The circled data point in Figure 7.11 indicates an inactivation response for coliphage MS2 at a flow rate of 0.153 $\frac{m^3}{sJ}$. The estimation of RED for this condition is illustrated in Figure 7.10. Figure 7.12 provides a summary of RED estimates for each of the seven conditions indicated in Figure 7.11.

As described previously, the inactivation response of a relatively resistant challenge agent, such as MS2, can be used to describe a lower bound on the inactivation response of a more sensitive challenge agent. For any value of flow rate (or more precisely, the dose distribution), the extent of T1 inactivation will always be greater than or equal to the extent of MS2 inactivation. Although the MS2 data provide a lower bound on inactivation for more sensitive challenge agents, it is not possible to provide a quantitative estimate of the extent of sensitive agent inactivation with these data. This observation illustrates one of the primary motivations for use of multiple challenge agents in biodosimetry. By following this approach, it is possible to provide quantitative information regarding the inactivation responses of challenge agents with more than one measured sensitivity to UV exposure. In the absence of measurements or estimates of the dose distribution, it is not possible to do this.

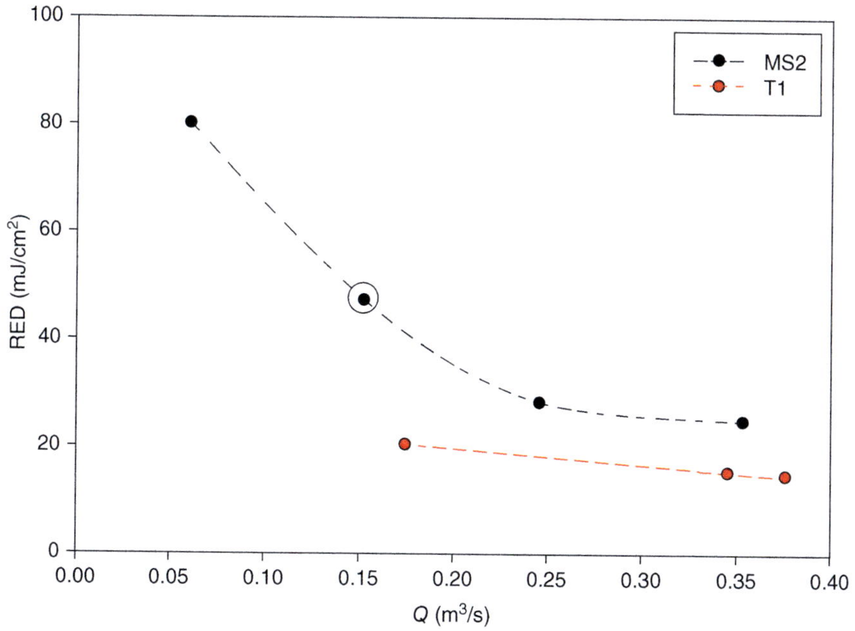


**Figure 7.12** Illustration of RED estimates as a function of volumetric flow rate for experiments based on MS2 and T1 as challenge microbes. The circled data point corresponds to the circled point in Figure 7.11. The method of calculation of RED for that data point is illustrated in Figure 7.10.

Two important trends are evident in Figure 7.12 that were also evident in the previous numerical examples. First, the magnitude of RED is a function of operating condition (here described by flow rate) and the challenge agent. Second, for a fixed operating condition, the magnitude of RED will increase as the value of $k$ decreases. In Figure 7.12, this is observed in that RED estimates based on MS2 are uniformly higher than those based on T1 for the same flow rate condition.

### 7.3.4 Challenge Organisms Commonly Used in Biodosimetry

In principle, any microorganism can be used as a challenge organism in a biodosimetric assay. However, some microbes have attributes that lend themselves to biodosimetry. Desirable characteristics among challenge microbes include the following:

- *Nonpathogenic*: The microbes that are used in biodosimetry must be grown to high titer and will be subjected to an assay that requires them to complete their life cycle and reproduce. As such, the use of nonpathogenic challenge microbes is desirable as a means of protecting those who conduct biodosimetric tests, as well as others who could be exposed to water used in these tests.
- *Mimic UV responses of microbial pathogens*: Ideally, the dose–response behavior of the challenge microbe should be identical, or as close as possible to that of the microbe that limits the performance of the UV disinfection system. By following this approach, the results of the biodosimetry assay can be used to accurately predict the behavior of the target microbe. Ideally, this will be associated with structural and physiological similarity between the challenge and target microbes, as well as similar values of dose–response behavior at the relevant wavelength(s). A related ideal situation is one in which the challenge agent and the target pathogen both display first-order UV dose–response behavior over the range of relevant UV doses. This simplifies the analysis and interpretation of results, but is not a rigorous requirement.

- *Simple and inexpensive to assay*: The response of the challenge organism must be measured before and after exposure to UV radiation, both by a collimated beam reactor (or similar) and the full-scale reactor that is being validated. Simplicity and low cost are both desirable attributes of these assays, in that most validation experiments involve a large number of assays of challenge organism responses to UV exposure.

Figures 7.13–7.17 illustrate replicate measurements of $UV_{254}$ dose–response behavior for challenge organisms that are commonly used in biodosimetry. These figures illustrate several important features of challenge organism dose–response behavior. First, the responses of the microbes are highly variable. Among these challenge organisms, coliphage MS2 is the most resistant of the phage. Second, there are deviations from simple first-order dose–response behavior that are evident in each of these data sets. For the phage data sets, the slopes of their respective dose–response curves decrease as dose increases. This behavior is sometimes referred to as "tailing" and is common to the UV dose–response behavior of many microbes. Tailing has been attributed to population heterogeneity, association of microbes with particles, and aggregation of microbes to form aggregates [6, 7]. Other factors may also contribute to this behavior. For *B. subtilis* spores, there is a characteristic "shoulder" in the dose–response data, indicating little or no inactivation for a finite range of doses in the limit of low dose. The presence of a shoulder has been attributed to the action of repair mechanisms and the requirement for multiple units of photochemical damage to bring about an inactivation response [8]. The presence of a shoulder in the dose–response behavior of a challenge organism is relevant in biodosimetry in that it heavily "weights" the effects of the low-dose end of the dose distribution delivered by a reactor. As such, the use of *B. subtilis* spores (or other challenge microbes that demonstrate a shoulder in their dose–response behavior) can be helpful in identifying reactors with a substantial low-dose tail in their dose distribution.

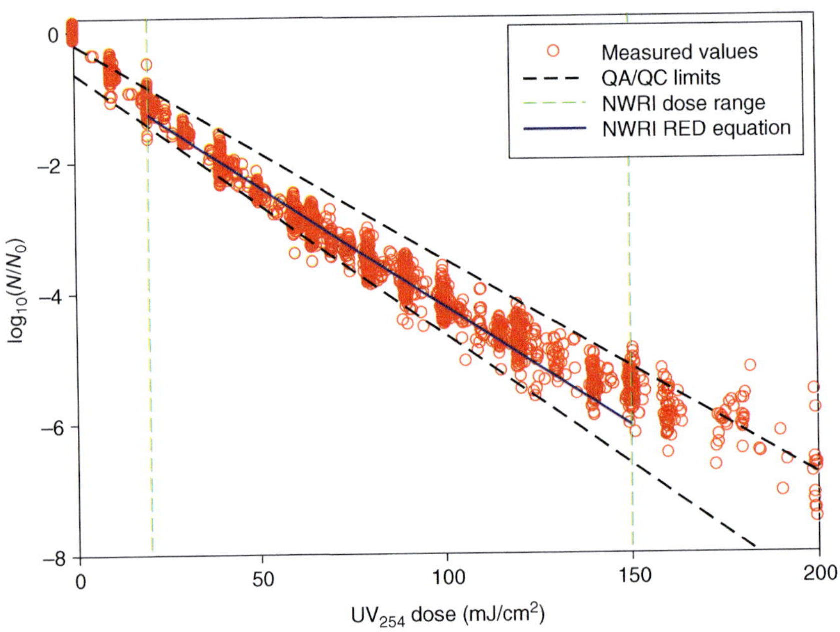


**Figure 7.13** Replicate measurements of $UV_{254}$ dose–response behavior for coliphage MS2. Also included are QA/QC limits for dose–response behavior (as defined by NWRI and NSF/ANSI), the range of accepted doses defined by NWRI, and the NWRI RED equation. *Source:* Data provided by GAP EnviroMicrobial Services, Ltd., London, Ontario, Canada.

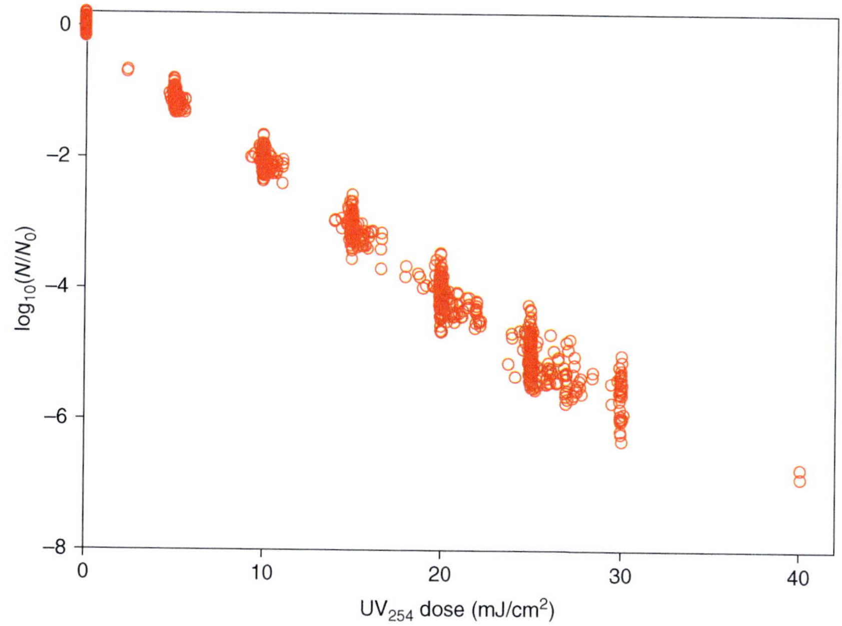


**Figure 7.14** Replicate measurements of $UV_{254}$ dose–response behavior for coliphage T1UV. *Source:* Data provided by GAP EnviroMicrobial Services, Ltd., London, Ontario, Canada.

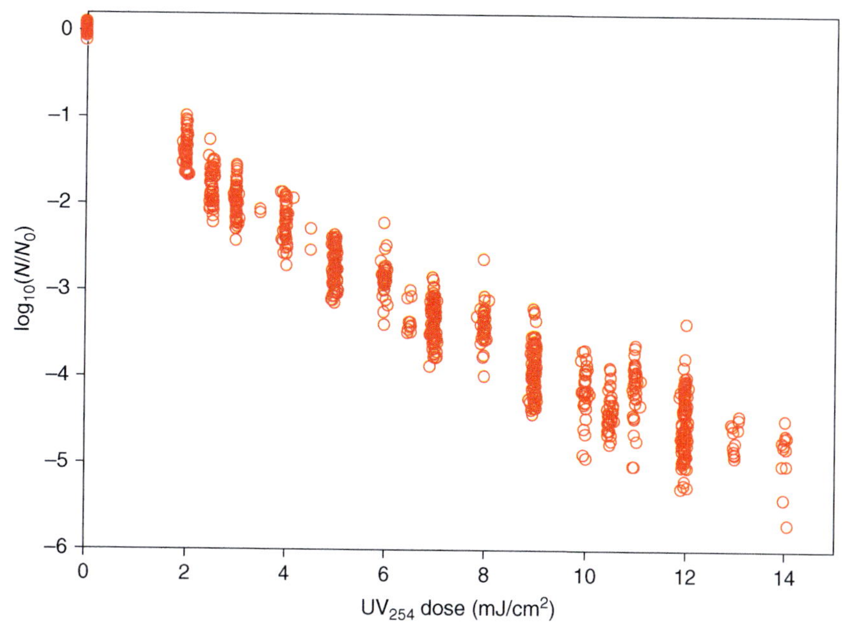


**Figure 7.15** Replicate measurements of $UV_{254}$ dose–response behavior for coliphage T7. *Source:* Data provided by GAP EnviroMicrobial Services, Ltd., London, Ontario, Canada.

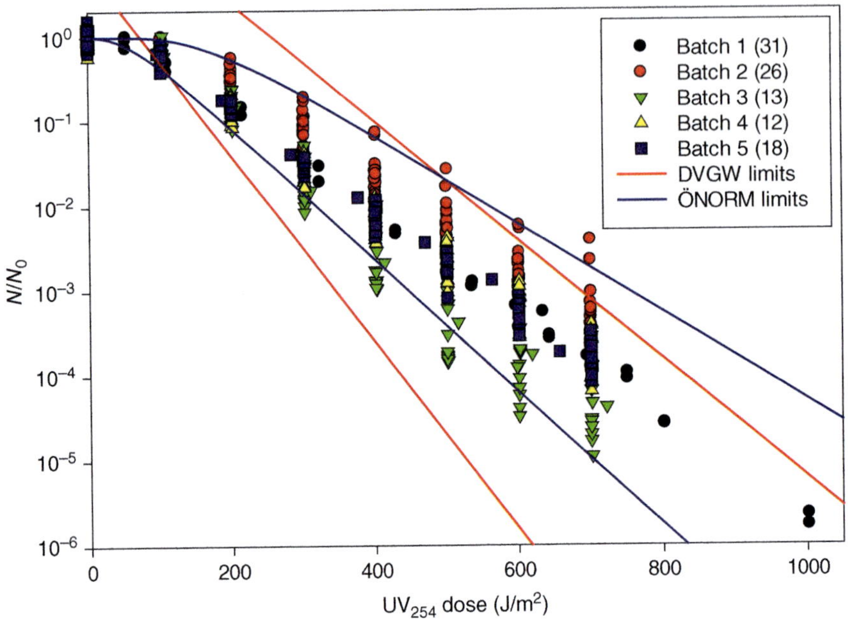


**Figure 7.16** Replicate measurements of $UV_{254}$ dose–response behavior for five batches of *Bacillus subtilis* spores. Parenthetical entries in the legend indicate the number of dose–response curves presented for each batch. Also included are QA/QC limits defined by DVGW and ÖNORM. *Source:* Data provided by TZW: DVGW-Technologiezentrum Wasser.

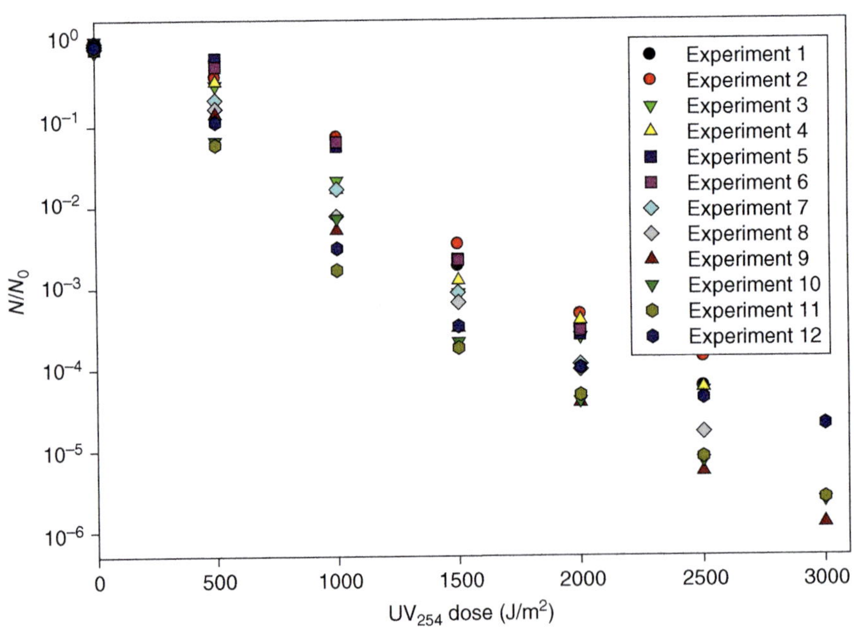


**Figure 7.17** Replicate measurements of $UV_{254}$ dose–response behavior of *Bacillus pumilus* spores. *Source:* Data provided by TZW: DVGW-Technologiezentrum Wasser.

Third, these data sets illustrate the variability that is inherent in the UV dose–response behavior of microorganisms. In considering this variability, it is relevant to note that the data sets presented in Figures 7.13–7.17 were each assembled by single laboratories. Therefore, these data were generated under circumstances where many of the factors that could affect microbial dose–response behavior were controlled (i.e. constant), including methods of culturing, assay methods, interpretation of the data, and possibly the people who conducted the experiments. Variability in reported UV dose–response behavior is likely to increase when data from other laboratories are included, due to increases in variability in these processes.

### 7.3.5 Effects of Variability in Challenge Organism Dose–Response Behavior on Biodosimetry

The variability that is inherent in measurements of UV dose–response behavior of challenge organisms represents a source of error in biodosimetry. The magnitude of this error will depend on the dose distribution delivered by a reactor and the magnitude of the error in challenge microbe dose–response behavior. As an illustrative example, the effects of this variability will be quantified for the range of calculated dose distributions that are illustrated in Figure 7.7. As a reminder, these calculated dose distributions were developed by application of a CFD-E' model to a single reactor at fixed UVT, but with a range of approach velocities (or flow rates).

For each operating condition (flow rate), the inactivation response of coliphage MS2 was calculated using the upper and lower QA/QC limits that have been defined by NWRI [9] and NSF/ANSI [2]. These QA/QC limits are illustrated in Figure 7.13 and are described later in this chapter (see equation 7.66). A graphical summary of these predictions is presented in Figure 7.18. For the range

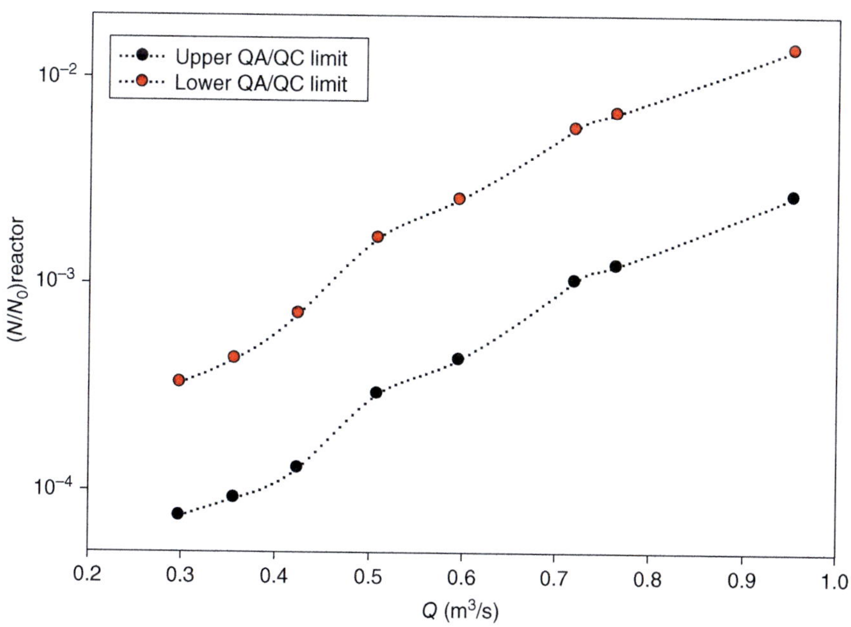


**Figure 7.18** Predicted fractional inactivation of coliphage MS2 for each of the dose distributions defined in Figure 7.7. Calculations were conducted using the upper and lower QA/QC limits for UV dose–response behavior of coliphage MS2, as defined by NWRI and NSF/ANSI in their respective UV reactor validation protocols [2, 9].

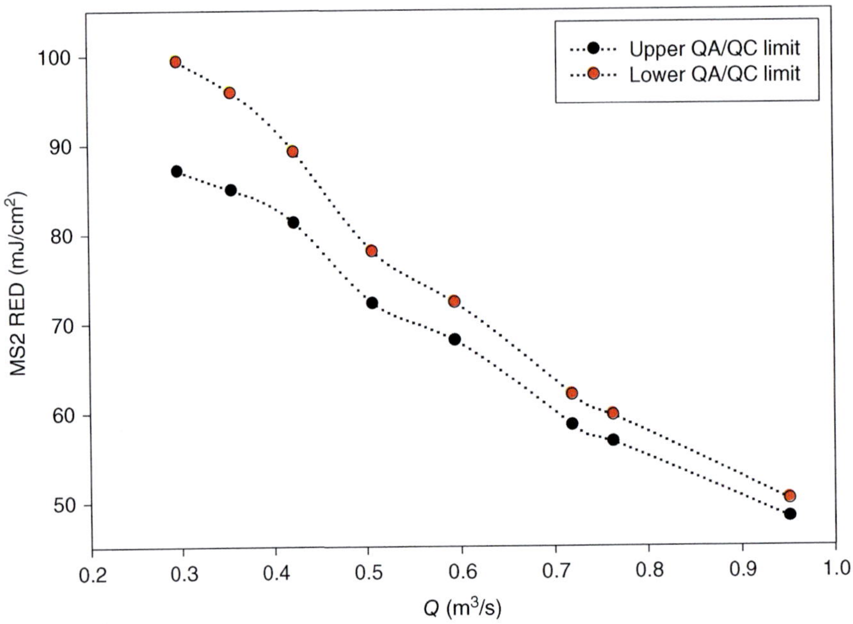


**Figure 7.19** MS2 RED calculated for each of the dose distributions illustrated in Figure 7.7, based on the upper and lower QA/QC limits for UV dose–response behavior of coliphage MS2, as defined by NWRI and NSF/ANSI in their respective UV reactor validation protocols [2, 9].

of operating conditions described in Figure 7.18, the predictions of MS2 inactivation followed qualitatively similar patterns for both QA/QC limits, but differed by a factor of approximately 5×.

The MS2 RED corresponding to each operating condition was then calculated by use of the inverse of the corresponding QA/QC limit (see Figure 7.19). Again, the patterns of predicted behavior were qualitatively similar, but the MS2 RED based on the lower QA/QC limit (corresponding to lower UV sensitivity) were consistently greater than those that were calculated based on the upper QA/QC limit. For the case of this set of dose distributions, the variability in dose–response behavior of MS2 translates to roughly 10% error in RED calculations.

These findings are consistent with the numerical simulations of biodosimetry that were presented previously (see Figures 7.5, 7.6, 7.8, and 7.9). Overall, the information presented in Figures 7.18 and 7.19 illustrates the magnitude of variability (error) that is inherent in biodosimetry, even when all microbiological aspects of the experiment are conducted by a single organization.

### 7.3.6 Use of Chemical Actinometers for Reactor Validation

Validation of UV photoreactors has been conducted by biodosimetry in the vast majority of applications. However, in principle it should be possible to accomplish this objective through the use of a chemical actinometer as the challenge agent, rather than a microbial challenge agent. This approach can present some advantages over conventional biodosimetry. Specifically, there is often less variability in measurements of the concentrations of dissolved chemical constituents (or their associated physical behaviors) than there is in measurements of the concentrations of viable or infective microbes. However, there are important aspects of reactor validation by the use of chemical actinometers that limit the application of this approach; these will be examined in this section.

Desirable characteristics in a chemical actinometer for use as a challenge agent in reactor validation would include the following:

- *Inexpensive*: The challenge agent should be readily available in inexpensive to purchase or produce.
- *Nontoxic*: The water used in a validation process will likely need to be discharged to a treatment facility or to a receiving stream. As such, the challenge agent should not express toxicity toward humans or aquatic organisms.
- *Rate constant (k) similar to that of target*: The extent of photoreaction, often described by the ratio $\left(\frac{[B]}{[B]_0}\right)_{\text{reactor}}$, is governed by the dose distribution and the rate constant. As described above, the rate constant for the challenge agent should mimic that of the target of treatment. For a purely photolytic process, the rate of a photoreaction will be first-order at any point in the reactor, or in dilute solution. As such, first-order kinetics should apply rigorously.
- *Simple, inexpensive, rapid analysis*: The success of any validation process will depend on the availability of analytical procedures that appropriately characterize the effect of UV exposure. The method should yield a measurement of the concentration of the parent compound or a photoproduct, or a physical characteristic of a sample that can be related to one or both of these concentrations.
- *Lack of internal filter*: Some photoproducts absorb strongly at the wavelength(s) used to promote the initial reaction. The products of these reactions can act as an internal optical filter and alter the fluence rate field within the reactor. This could alter the dose distribution delivered by the reactor, which in turn will alter reactor performance and complicate the interpretation of data from a validation experiment.

As with any list of ideal characteristics, it is difficult to find a challenge agent (in the form of a chemical actinometer) that satisfies all of these conditions. However, some chemical actinometers appear to satisfy many of these characteristics.

Table 7.2 provides a summary of several chemical actinometers and their respective values of molar absorption coefficient ($\varepsilon$) and quantum yield ($\Phi$) for $\lambda$ = 253.7 nm, corresponding to the

**Table 7.2** Summary of physical constants and photochemical rate constants ($\lambda$ = 253.7 nm) for common actinometers and other photoreactive compounds.

| Actinometer | $\varepsilon_{253.7}\left(\frac{\text{L}}{\text{mol}\cdot\text{cm}}\right)$ | $\Phi_{253.7}\left(\frac{\text{mole}}{\text{einstein}}\right)$ | $k_{253.7}\left(\frac{\text{cm}^2}{\text{J}}\right)$ |
|---|---|---|---|
| Uranyl oxalate [10] | 3,200 | 0.61 | $9.54\times10^{-3}$ |
| Potassium ferrioxalate [11] | 4,000 | 1.38 | $2.70\times10^{-2}$ |
| 2-Nitrobenzaldehyde [12] | 5,382 | 0.41 | $1.08\times10^{-2}$ |
| Uridine [13] | 9,000 | 0.034 | $1.50\times10^{-3}$ |
| (*E*)-5-[2-(Methoxycarbonyl)ethenyl]cytidine [14] | 7,500 | 0.732 | $2.68\times10^{-2}$ |
| Hydrogen peroxide [15] | 18.6 | 0.5 | $4.55\times10^{-5}$ |
| Hypochlorous acid [16] | 59 | 1 | $2.88\times10^{-4}$ |
| Atrazine [15] | 3,400 | 0.048 | $7.98\times10^{-4}$ |
| $CH_3NCl_2$ | 211 | 0.867 | $8.94\times10^{-4}$ |
| 1,4 Benzoquinone | 16,200 | 0.82 | $6.49\times10^{-2}$ |

dominant line from a LP Hg lamp. These values were then used to calculate corresponding rate constants for their photodegradation based on exposure to monochromatic radiation at a wavelength of 253.7 nm. The rate constants were calculated based on assumptions of a purely photolytic process in which $\varepsilon \cdot [B] \cdot \ell \rightarrow 0$. As described in Chapter 2, this leads to first-order kinetics and will apply in dilute solution or any point in a photochemical reactor. The rate constant for such a process is governed by the product $\varepsilon \cdot \Phi$ as follows (see Chapter 2):

$$k = \frac{2.303 \cdot \varepsilon \cdot \Phi}{U} \tag{7.22}$$

where

$U$ = energy associated with an einstein of photons.

In turn, the energy associated with radiation at a given wavelength is calculated as follows:

$$U = U_p \cdot N_A \tag{7.23}$$

where

$$U_p = h\frac{c_0}{\lambda} \tag{7.24}$$

$U_p$ = photon energy
$h$ = Planck's constant
$c_0$ = speed of light in a vacuum
$\lambda$ = wavelength of radiation
$N_A$ = Avogadro's number.

Substituting Equations 7.23 and 7.24 into Equation 7.22, we find:

$$k = \frac{2.303 \cdot \varepsilon \cdot \Phi \cdot \lambda}{h \cdot c_0 \cdot N_A} \tag{7.25}$$

Equation 7.25 was used to calculate rate constants for photochemical transformation of several chemical actinometers, with unit conversions as needed.

For the chemical actinometers considered in Table 7.2, the rate constants at 253.7 nm are low relative to those of most microorganisms that are used for validation of UV disinfection systems. As such, the use of these chemical actinometers as challenge agents in validation of UV disinfection systems may not be appropriate for validation of photoreactors used in disinfection. However, they may be useful in validation of photoreactors for other treatment endpoints.

Among the actinometers listed in Table 7.2, the compounds 1,4 Benzoquinone and (*E*)-5-[2-(Methoxycarbonyl)ethenyl]cytidine displayed the largest rate constants for radiation at 253.7 nm. The values of these rate constants ($k_{254} = 6.49 \times 10^{-2} \frac{cm^2}{mJ}$ and $k_{254} = 2.60 \times 10^{-2} \frac{cm^2}{mJ}$, respectively) are smaller than the typical value that is observed for common challenge microbes, such as coliphage MS2. However, these values are similar to the rate constants that are representative of the behavior of some relatively resistant microbes that are used as challenge agents or as targets of treatment, including *B. pumilus* spores [17] and Adenovirus [18]. Potential advantages of the use of (*E*)-5-[2-(Methoxycarbonyl)ethenyl]cytidine as a challenge agent for validation of reactors used to inactivate these UV-resistant microbes is that it is a nontoxic, nonpathogenic reagent and its response to UV exposure (fluorescence) is quite repeatable and easy to measure. 1,4-Benzoquinone responds slightly faster to $UV_{254}$ exposure than (*E*)-5-[2-(Methoxycarbonyl)ethenyl]cytidine, but it is a toxic compound and as such its release to the environment would need to be controlled or mitigated.

More broadly, any photoreactive chemical could be used as a challenge agent in photoreactor validation. However, for such a compound to be effective for validation of UV disinfection systems that are used to inactivate common waterborne microbial pathogens, it would need to display a product $\varepsilon \cdot \Phi$ that is larger than those listed in Table 7.2, so that its rate constant was comparable to that of the target pathogen(s).

### 7.3.7 Lagrangian Actinometry

An important limitation of biodosimetry is its inability to measure the UV dose distribution delivered by a photoreactor. Because of this, it is not possible to translate the results of reactor validation with one challenge agent to a response on the part of another challenge agent, unless both agents have the same UV dose–response behavior. This limitation of biodosimetry is associated, in part, with the analytical methods that are used to measure microbial responses to UV radiation. In all cases, these methods yield a binary measure of the response of individual microbes – the microbe is either inactivated or it is not. For a population of microbes, these methods yield a single numerical value that is used to quantify the fraction of microbes within a sample that lose viability or the ability to infect a host. Similar limitations apply to chemical actinometry, when applied for purposes of measuring RED.

Lagrangian Actinometry (LA) was developed as a method to measure the UV dose distribution delivered by a photoreactor. LA requires conjugation of a photochemically active reagent to the surface of microspheres. The reagent chosen for this application allows characterization of the extent of reaction on the surface of each microsphere, preferably through the use of an optical method.

All successful implementations of LA to date have involved conjugation of (*E*)-5-[2-(Methoxycarbonyl)ethenyl]cytidine (hereafter referred to as *S*) to microsphere surfaces through a long-chain linker molecule to prevent quenching of fluorescence on the part of the photoproduct, 3-β-*D*-ribofuranosyl-2,7-dioxopyrido[2,3-*d*]pyrimidine (hereafter referred to as *P*) [19, 20]. Conjugation of *S* to the surface of a microsphere has been accomplished through a biotin:streptavidin linkage. The molecule *S* is biotinylated; the biotinylated form of *S* is then incubated with streptavidin-coated microspheres, which are commercially available. Conjugation via the biotin:streptavidin linkage results in essentially irreversible binding of *S* to the microsphere surface due to near-covalent bond strength [20, 21].

When exposed to UVC radiation, microspheres prepared by this process yield regular, repeatable increases of fluorescence with UV dose (see Figure 7.20). The data presented in Figure 7.20 were collected using flow cytometry based on sample populations of approximately 10,000 microspheres per sample. These data represent the mean of measurements for each sample population. It should be noted that the mean of a population presents no advantage relative to biodosimetry or conventional chemical actinometry, in terms of characterizing UV dose delivery. However, fluorescence measurements tend to be faster and more repeatable than measurements of microbial responses to UV irradiation. It also relevant to note that unirradiated microspheres (i.e. $UV_{254}$ dose = 0) express fluorescence; this behavior is attributable to the fluorescence characteristics of the microsphere material, not fluorescence on the part of *P* on the microsphere surface.

The linkage of *S* to the surface of a microsphere provides a number of important advantages for LA applications, including:

- Microspheres are available commercially and can be produced with narrow ranges of important physical and chemical properties, including shape (usually spherical), size (diameter), density, and surface loading of functional groups, such as streptavidin. These narrow ranges in physical

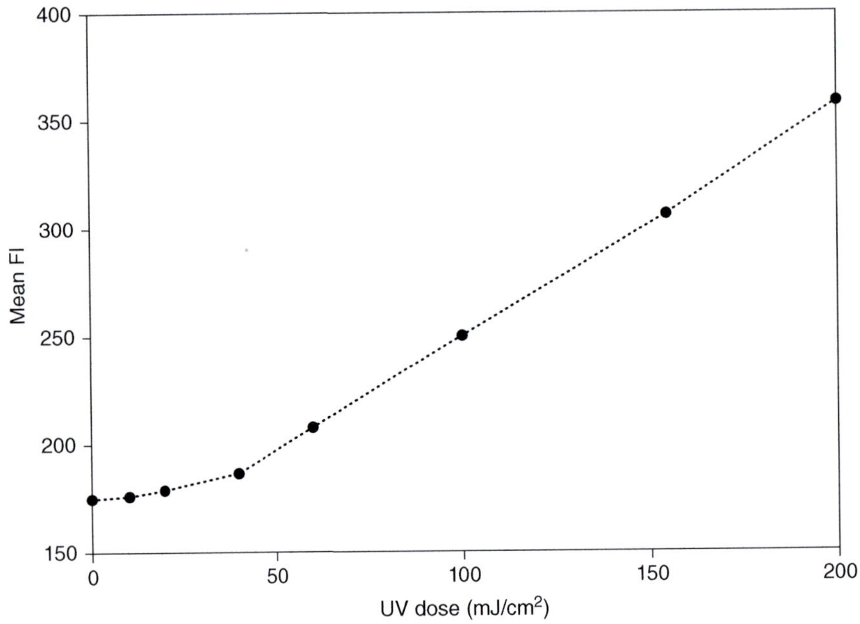


**Figure 7.20** Mean fluorescence intensity (FI) of microspheres as a function of applied $UV_{254}$ dose. Reported FI values represent the mean of measurements collected for roughly 10,000 particles by flow cytometry. *Source:* Data from Fang et al. 2008.

and chemical properties are beneficial in terms of minimizing error in optical measurements of microsphere subpopulations.

- Microspheres can be prepared with size and specific gravity characteristics that are similar to waterborne microbes. As an illustration, consider the data presented in Figure 7.21, which illustrate the terminal settling velocities of spherical particles with densities of 1.05 and 1.10 $\frac{g}{cm^3}$. Terminal settling velocities for these particles were calculated based on Stokes' law (see equation 7.26), which provides a closed-form, analytical description of the settling velocities of spherical particles under conditions of laminar settling. These density values are representative of most waterborne microbes, and microspheres are often produced with density in this range.

$$V_t = \frac{g \cdot \left(\rho_p - \rho\right) \cdot d_p^2}{18 \cdot \mu} \tag{7.26}$$

where

$V_t$ = terminal settling velocity
$g$ = gravitational acceleration
$\rho_p$ = particle density
$\rho$ = fluid density
$d_p$ = particle diameter
$\mu$ = dynamic viscosity of fluid.

Note that most UV photoreactors operate with a fluid approach velocity of 10–100 $\frac{cm}{s}$. Therefore, the terminal settling velocities for microspheres that are used in LA (typical diameter = 5–10 μm) are several orders of magnitude lower than the velocities of the surrounding fluid. As such, the

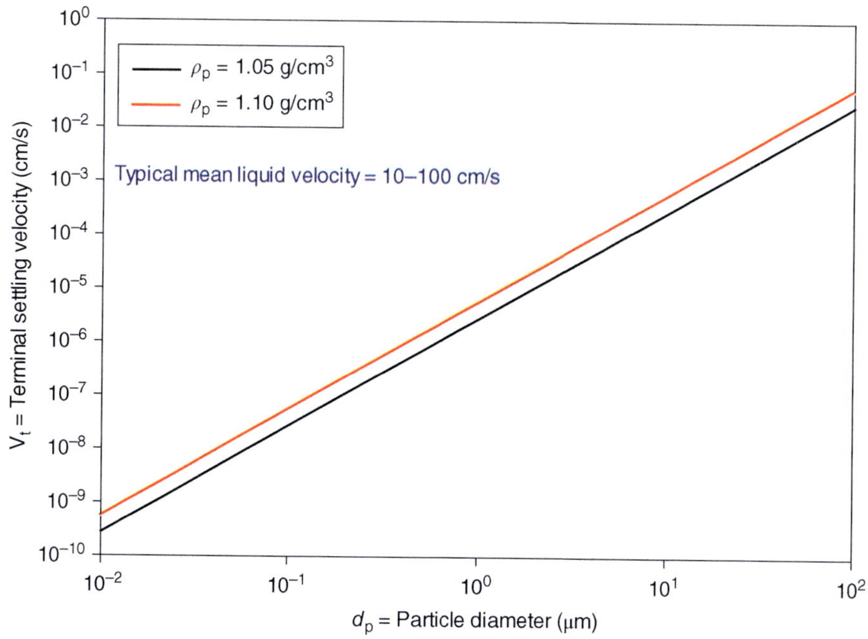


**Figure 7.21** Terminal settling velocity of spherical particles for two different values of particle density. The density values selected for these calculations are representative of the densities of most waterborne microbes. Note that the typical (mean) approach velocity in a UV photoreactor is on the order of 10–100 $\frac{\text{cm}}{\text{s}}$.

movement of nearly-neutrally-buoyant microspheres is similar to that of aquatic microbes, and effectively determined by the velocity of the surrounding fluid.

- Changes in fluorescence of individual microspheres can be measured accurately and rapidly using flow cytometry. And because each microsphere has a large number of *S* molecules conjugated to its surface, the fluorescence intensity (FI) measurement for individual microspheres spans a continuum. In effect, access to a continuum of FI measurements allows the UV dose delivered to individual microspheres to be measured. Of greater significance to UV photoreactor analysis and validation, measurements of FI for a large population of microspheres allow measurement of the dose distribution delivered by the photoreactor.

Figure 7.22 provides a schematic illustration of LA. As described above, microspheres are chosen for these applications to have size (diameter) and specific gravity characteristics that are similar to those of aquatic microorganisms. As such, the trajectories of microspheres and microorganisms will be comparable and governed by the motion of the surrounding fluid. The actinometer used in this method, *S*, is conjugated to the surface of the microsphere through a biotin:streptavidin linkage; however, conjugation also incorporates a long-chain linker molecule as a tether to reduce steric interference, which can result in quenching of fluorescence on the part of the photoproduct, *P*. Exposure of the *S*-conjugated microspheres to UVC radiation causes conversion of *S* to *P* (see Figure 7.23). The FI of the UVC-irradiated microsphere will be governed by the number of *S* molecules that are converted to *P*. As such, the UVC dose delivered to an individual microsphere can be related to the change in FI that results from UVC exposure. Or more broadly, the change in FI distribution among a population of *S*-conjugated microspheres can be used to measure the UVC dose distribution delivered by a photoreactor.

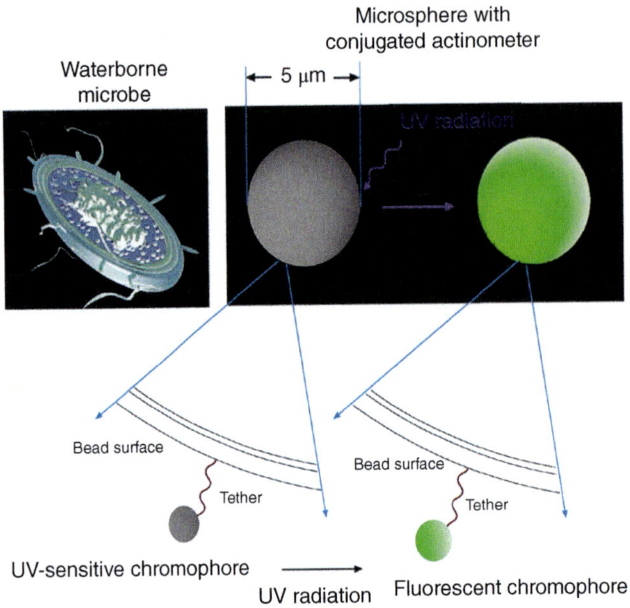


**Figure 7.22** Schematic illustration of Lagrangian Actinometry. Microspheres are selected to have size and specific gravity characteristics that are similar to those of aquatic microorganisms. Actinometer molecules are conjugated to the surface of the microsphere through a long-chain tether molecule to prevent suppression of fluorescence on the part of the photoproduct. Exposure of the microspheres to UVC radiation results in transformation of the starting material, *S*, to a brightly fluorescent photoproduct, *P*. The dose of UVC radiation delivered to an individual microsphere can be related to the strength of the fluorescence response.

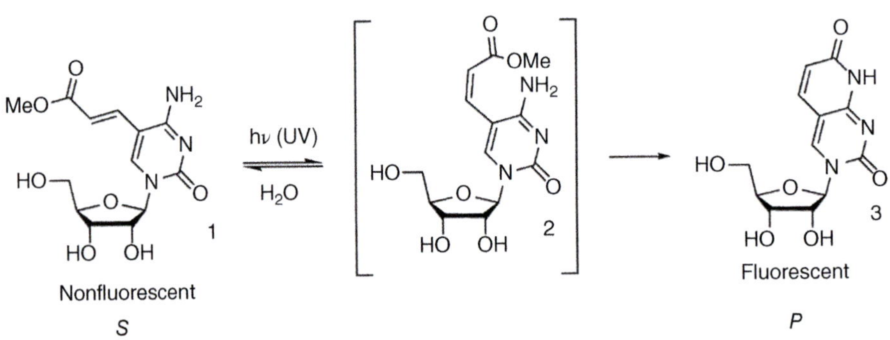


**Figure 7.23** Schematic illustration of the UVC-induced photoconversion of (*E*)-5-[2-(Methoxycarbonyl) ethenyl]cytidine (*S*) to 3-β-*D*-ribofuranosyl-2,7-dioxopyrido[2,3-*d*]pyrimidine (*P*). *Source:* Adapted from [20].

LA involves UVC exposure and subsequent analysis of populations of dyed microspheres. As such, the relevant responses are *distributions* of FI. The FI distribution among a population of UVC-irradiated microspheres is attributable to:

- *Population heterogeneity among the dyed microspheres*: Although current manufacturing processes for microspheres yield narrow distributions of size and surface chemistry, quantifiable variations in these and other attributes of microspheres can be measured.

- *Measurement error associated with flow cytometry*: Flow cytometry represents a family of methods that allow for rapid measurements of optical characteristics of particle populations. The methods are rapid and reproducible, but like all analytical methods, there are sources of error in these measurements.
- *UVC dose distribution delivered to the microsphere population*: Some microspheres within a population will receive a small UVC dose, while others will receive larger doses. The UVC dose distribution governs UVC photoreactor behavior and is the target of LA methods.

The first two factors are unavoidable, but repeatable. The third factor is the primary focus of this method. The effects of these factors on variance of FI among a population of microspheres are independent, and therefore additive. Therefore, the effects of the UV dose distribution on the FI distribution among a population of microspheres can be differentiated from other sources of variability in FI.

### 7.3.8 Convolution Hypothesis – Theoretical Background

A *convolution hypothesis* was developed to describe the behavior of a mixed population of UVC-exposed microspheres. The hypothesis asserts that the FI distribution of a population of microspheres that has been subjected to a UV dose distribution can be described as the convolution of the FI distributions of microspheres that have been subjected to each dose within the dose distribution, weighted by their respective dose fractions. For mathematical convenience, the FI distribution is described as a histogram with a FI increment of 1 FI unit.[f] Similarly, the UV dose distribution is described as a histogram with a dose increment of $1\ \frac{\text{mJ}}{\text{cm}^2}$.

Evaluation of the convolution hypothesis requires the use of linear algebra, which is simplified by the use of vector/index notation. As such, the following definitions are useful in the analysis of the convolution hypothesis and other LA applications; these definitions apply to microsphere samples that have been subjected to a single-valued dose (e.g. UVC exposure in a shallow, well-mixed, batch reactor under a collimated beam):

$$i = \text{index for counting dose } (D) \text{ increments } \left(\text{bin width} = 1\ \frac{\text{mJ}}{\text{cm}^2}\right)$$

$$i = 0, 1, \ldots, m-1 \ (m \text{ dose increments}) \tag{7.27}$$

$$j = \text{index for counting FI increments (bin width} = 1 \text{ FI unit)}$$

$$j = 0, 1, \ldots, n-1 \ (n \text{ FI increments}) \tag{7.28}$$

$$C_{j,i} = \text{number of microspheres that emit } \text{FI}_j \text{ after being exposed to } D_i \tag{7.29}$$

$$C_i = \text{number of microspheres in a sample subjected to } D_i = \sum_{j=0}^{n-1} C_{j,i} \tag{7.30}$$

$$\Gamma_{j,i} = \text{fraction of particles receiving } D_i \text{ that emit } \text{FI}_j = \frac{C_{j,i}}{C_i} \tag{7.31}$$

Two important identities accompany these definitions:

$$\sum_{j=0}^{n-1} \Gamma_{j,i} \equiv 1 \tag{7.32}$$

Equation 7.32 indicates that the sum of all fractions $\frac{C_{j,i}}{C_i}$ across all FI increments must equal 1. In other words, this guarantees that all microspheres receiving a dose of $D_i$ are properly accounted for in the distribution.

$$\sum_{i,j} \Gamma_{j,i} \equiv m + 1 \tag{7.33}$$

Equation 7.33 guarantees that all microspheres within a population are properly normalized across all possible dose increments as well.

When we consider a mixed sample (i.e. a microsphere population that has been subjected to a range of UVC doses), additional definitions are required:

$$A = \text{total number of FI measurements collected for a mixed sample} \tag{7.34}$$

$$A_i = \text{number of particles in a mixed sample that receive dose } D_i \tag{7.35}$$

$$\alpha_i = \text{fraction of particles in a mixed sample that receive dose } D_i = \frac{A_i}{A} \tag{7.36}$$

$$B_j = \text{number of particles in a mixed sample that emit } \mathrm{FI}_j \tag{7.37}$$

$$\beta_j = \text{fraction of particles in a mixed sample that emit } \mathrm{FI}_j = \frac{B_j}{A} \tag{7.38}$$

With these definitions, two additional identities emerge as constraints:

$$\sum_{i=0}^{m-1} \alpha_i \equiv 1 \tag{7.39}$$

$$\sum_{j=0}^{n-1} \beta_j \equiv 1 \tag{7.40}$$

If the convolution hypothesis is correct, then the following should be true:

$$B_j = \alpha_0 A\Gamma_{j,0} + \alpha_1 A\Gamma_{j,1} + \cdots + \alpha_{m-1} A\Gamma_{j,m-1} = \sum_{i=0}^{m-1} \alpha_i A\Gamma_{j,i} \tag{7.41}$$

Dividing both sides of Equation 7.41 by $A$, we find:

$$\beta_j = \sum_{i=0}^{m-1} \alpha_i \Gamma_{j,i} \tag{7.42}$$

### 7.3.9 Convolution Hypothesis – Experimental Verification

An experiment was conducted in which an aqueous suspension of dyed microspheres was prepared. Several equal-volume subsamples of this population were collected and then subjected to a range of $UV_{254}$ doses in a shallow, well-mixed batch reactor under a collimated beam. For each subsample, the FI distribution was measured by flow cytometry. A mixed sample was prepared by blending equal volumes of the subsamples into a single sample; the mixed sample was then analyzed by flow cytometry.

The mixed sample described above was developed to allow preparation of a sample of microspheres that had received a known dose distribution. Since equal volumes of each subsample were

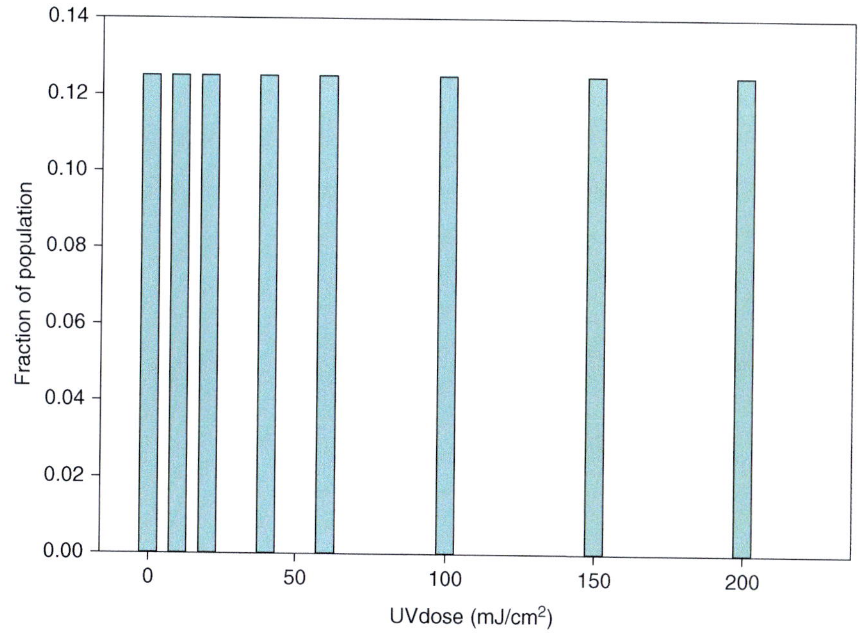


**Figure 7.24** $UV_{254}$ dose distribution for mixed sample in convolution hypothesis experiment.

blended into the mixed sample, then the dose distribution was known. For this specific experiment, equal volumes of microspheres that had been subjected to $UV_{254}$ doses of 0, 10, 20, 40, 60, 100, 150, and 200 $\frac{mJ}{cm^2}$ were blended together to prepare a mixed sample. As such, the dose distribution could be represented as a histogram, with each of these doses representing 1/8 of the total population (see Figure 7.24).

The convolution hypothesis suggests that the FI distribution for this mixed sample should be equal to the convolution of the FI distributions of the samples ($UV_{254}$ doses) that comprised the mixed sample; the "weighting" function in this calculation is the fraction of the population that is represented by each $UV_{254}$ dose. This experiment allowed the FI distributions for the individual subsamples as well as the mixed sample to be measured independently (see Figure 7.25). Also presented in this figure is the convolution of these FI distributions.

The results of this and other, similar experiments supported the validity of the convolution hypothesis. However, the convolution experiment involved an unusual situation, in that it required that a sample be prepared with a known dose distribution. In normal situations, the dose distribution delivered by a photoreactor is unknown; the central goal of LA is to provide a method by which the dose distribution can be measured. The method used to develop an estimate of the dose distribution from a LA experiment involves inversion of Equation 7.42; mathematically, this process is known as deconvolution.

### 7.3.10 Application of LA

The practical application of LA is analogous to biodosimetry in many respects. In both methods, a challenge agent is subjected to UV irradiation under a collimated beam and in a full-scale reactor. The effects of these UV exposures in both reactors are measured by the same method, then

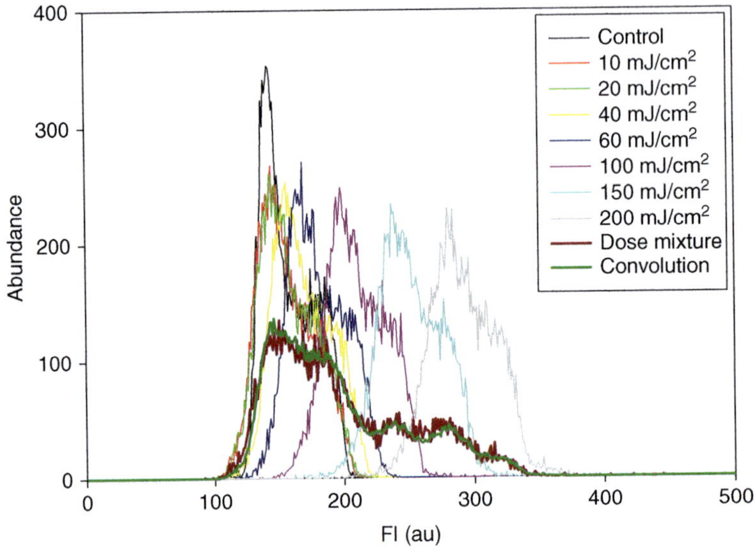


**Figure 7.25** Results of convolution hypothesis experiment. The FI distribution labeled "Dose Mixture" represents the sample that was prepared from equal volumes of each subsample. The FI distribution labeled "Convolution" was calculated as the weighted sum of the contributions of each of the subsamples, with the dose distribution (Figure 7.24) being used as the weighting function.

compared to provide quantitative information about reactor performance. However, the endpoints of treatment, analytical methods, and mathematical analysis are quite different. The specific steps involved in the application of LA have been described in detail as a validation protocol [22]. The protocol document was intentionally developed to parallel the logic and organization of the *UV Disinfection Guidance Manual* (*UVDGM*) [23] which serves as a guidance document in the United States for validation of UV disinfection systems used for treatment of surface waters in production of potable water. Additional information pertaining to the *UVDGM* and other guidance documents is provided later in this chapter. Basic characteristics of the steps involved in application of LA are briefly described below.

### 7.3.11 Microsphere Dose–Response Behavior

As with biodosimetry, it is necessary to define the dose–response behavior of microspheres as a challenge agent. This is usually accomplished by exposure of aqueous suspensions of microspheres in a shallow, well-mixed, batch reactor under a collimated beam. By following this approach, subsamples of microspheres are subjected to a range of single-valued UV doses under a collimated beam, followed by analysis. In the case of microspheres, the analysis is conducted by flow cytometry, yielding FI distributions at each dose.

As with biodosimetry, the data from the collimated beam exposure experiments is used to define the dose–response behavior of the challenge agent. Also as with biodosimetry, it is necessary to identify a mathematical expression to fit to the measured data. Ideally, the fitting function should be based on the known physics and chemistry of the experiments; however, access to such an equation is not required and the most important function of the fitting expression is to allow interpolation of dose–response behavior.

In the case of LA, the measured response for each dose is an FI distribution. Therefore, the fitting function should yield a curved shape that conforms to the shapes of the FI distributions, and that changes in a regular, predictable, repeatable manner as a function of UV dose.

For LA based on the microspheres described above, it has been observed that the Four-Parameter Weibull distribution (equation 7.43) provides a good fit to the measured data. Figure 7.26 illustrates measured FI distributions for several microsphere samples, together with Four-Parameter Weibull fits for each dose. The model provides a consistently good fit across the entire range of doses examined. It is relevant to note that the behavior of these microspheres is regular and predictable across a wide range of UV doses. In fact, the data presented in Figure 7.26 indicate that LA could be used to validate UV photoreactors for applications requiring higher doses than those that are commonly used for disinfection, or that are relevant for applications involving inactivation of UV-resistant microbes, such as Adenovirus.

$$f(x \mid a,b,c,x_0) = a\left(\frac{c-1}{c}\right)\left[\frac{x-x_0}{b} + \left(\frac{c-1}{c}\right)\right]^{c-1} \cdot \exp\left[-\left(\frac{x-x_0}{b} + \left(\frac{c-1}{c}\right)\right)\right]^{c} + \frac{c-1}{c} \tag{7.43}$$

Any function used for fitting (interpolation) of microsphere dose–response behavior must satisfy the characteristics of a density distribution function. Therefore, the Four-Parameter Weibull distribution must satisfy the following condition as a constraint of its application:

$$\sum f(x \mid a,b,c,x_0) \equiv 1 \tag{7.44}$$

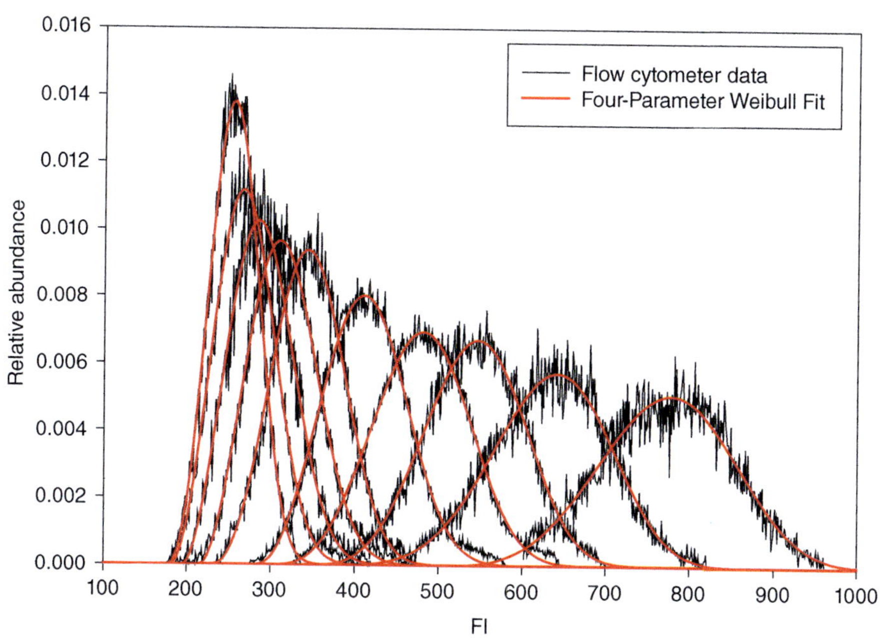


**Figure 7.26** Measured FI distributions for microsphere samples subjected to $UV_{254}$ doses of 0, 10, 20, 40, 60, 100, 150, 200, 300, and 500 $\frac{mJ}{cm^2}$. Also included are fits of the Four-Parameter Weibull function to these FI distributions.

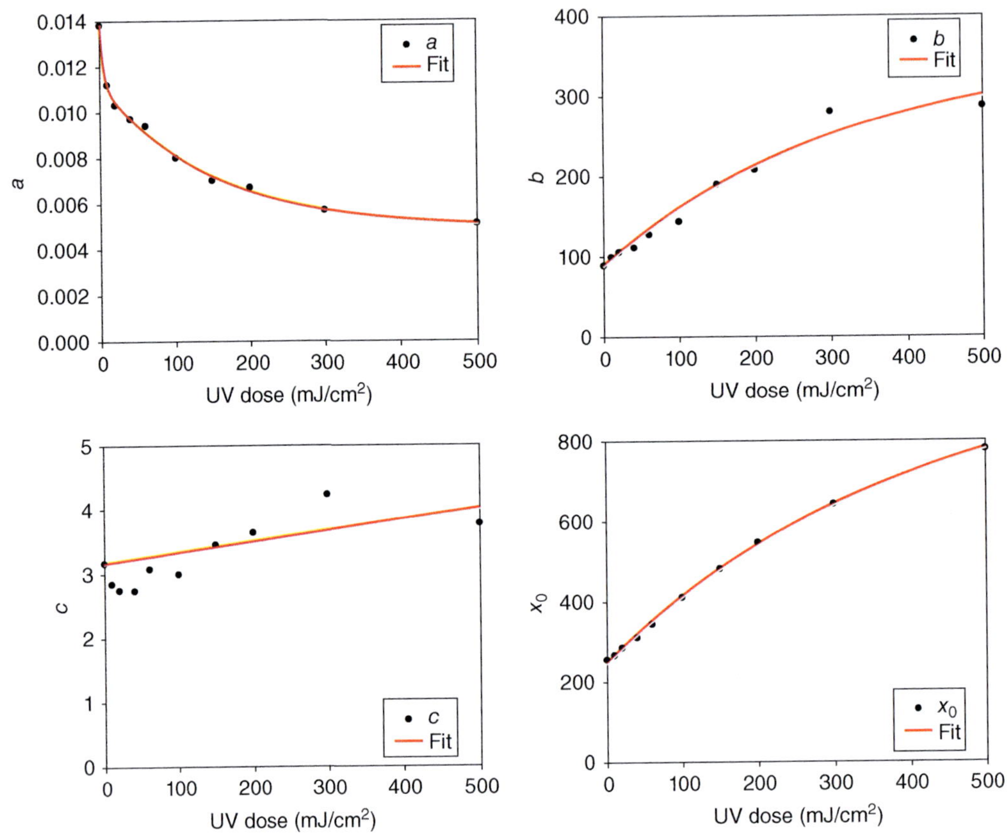


**Figure 7.27** Variation of Four-Parameter Weibull Function parameters as a function of $UV_{254}$ dose.

Figure 7.27 illustrates the dose-dependent behavior of the four parameters that define this model. Among these parameters, $x_0$ is a "location" parameter, which is representative of the mean of each distribution; the parameters $a$, $b$, and $c$ define the shape of the distribution. In general, each of the parameters follows a regular, repeatable pattern as a function of applied $UV_{254}$ dose. The worst of these parameter fits is observed with parameter $c$; sensitivity analysis indicated that the Four-Parameter Weibull function is less sensitive to parameter $c$ than the other three parameters, and the difference between measured and predicted values of this parameter do not appear to have a detrimental effect on the ability of the model to fit the measured FI distributions [19].

### 7.3.12 Reactor Testing

The same suspension of microspheres is then subjected to UV exposure in a full-scale UV photoreactor. This experiment involves the same experimental platform as described for biodosimetry (see Figure 7.1) and also requires that samples be collected when the reactor has reached a steady-state condition. For each steady-state condition evaluated, one or more samples of treated water are collected for analysis. For each of these samples, the microspheres are separated from water by membrane filtration and centrifugation, with subsequent analysis of the microspheres by flow cytometry.

A restatement of the conclusion of the convolution hypothesis (equation 7.42) in matrix notation is:

$$[\beta] = [\Gamma] \times [\alpha] \tag{7.45}$$

where

$$[\beta] = \begin{bmatrix} \beta_0 \\ \vdots \\ \beta_{m-1} \end{bmatrix} \tag{7.46}$$

Effluent sample FI distribution
($m \times 1$ matrix)

$$[\Gamma] = \begin{bmatrix} \Gamma_{0,0} & \cdots & \Gamma_{0,n-1} \\ \vdots & \ddots & \vdots \\ \Gamma_{m-1,0} & \cdots & \Gamma_{m-1,n-1} \end{bmatrix} \tag{7.47}$$

UV dose–response matrix, developed by fitting of Four-Parameter Weibull function to data from a collimated beam experiment

($m \times n$ matrix)

$$[\alpha] = \begin{bmatrix} \alpha_0 \\ \vdots \\ \alpha_{n-1} \end{bmatrix} \tag{7.48}$$

UV dose distribution function
($n \times 1$ matrix)

The goal of LA is to provide a measure of the UV dose distribution function. This can be estimated by deconvolution of the data that result from the collimated beam and full-scale reactor testing as follows:

$$[\Gamma]^{-1} \times [\beta] = [\alpha] \tag{7.49}$$

Equation 7.49 involves the use of linear algebra to solve a series of equations. Solver routines are available in a number of commercial mathematics packages to accomplish this deconvolution.

Many UV reactor validation protocols (described below) call for interpretation of the results of biodosimetry to yield an equation to describe reactor performance, in the form of RED, as a function of the relevant operating parameters, including flow rate, UV transmittance, and sensor readings. This same objective could be accomplished through LA as well; however, the primary benefit of LA is the ability to measure the UV dose distribution delivered by a reactor. When the dose distribution is known, RED becomes superfluous.

One of the primary benefits of measuring the UV dose distribution is that it can be used to accurately predict the performance of a UV reactor relative to any photochemical endpoint. Stated alternatively in the context of disinfection, when the dose distribution is known, it is possible to accurately predict the ability of the reactor to inactivate any microorganism for which accurate dose–response behavior is known. Similarly, knowledge of the dose distribution allows for accurate predictions of the behavior of any photolytic process for which accurate information regarding the molar absorption coefficient ($\varepsilon$) and the quantum yield ($\Phi$) are known.

Integration of dose distribution measurements with dose–response behavior of aquatic microbes through the SFM has been shown to yield accurate predictions of microbial inactivation over a wide range of reactor sizes and for a wide range of challenge organisms [19, 22, 24, 25]. It has been applied to photoreactors based on conventional LP Hg lamps, medium-pressure Hg lamps, and excimer lamps.

### 7.3.13 Integration of MFSD and CFD-E' Simulations

CFD-E' simulations provide a method to estimate the UV dose distribution delivered by a photoreactor over a wide range of conditions. Knowledge of the dose distribution is clearly beneficial in analysis and validation of UV photoreactors; however, CFD-E' models require expertise and computational tools that are not always available. In addition, it is possible to converge on incorrect solutions with these models if proper QA/QC procedures are not followed. Collectively, these attributes have promoted skepticism regarding the appropriateness of CFD-E' models for reactor validation. However, as CFD-E' models and computer hardware improve, the prospects for incorporation of numerical simulations as a part of validation also improve. Moreover, recent improvements in sensor hardware have resulted in measurement tools that can be used to complement numerical simulations for use in reactor monitoring, and by extension in reactor validation.

As an example of this approach, Li et al. [26] integrated measurements of local fluence rate by the use of real-time signals from strategically positioned micro-fluorescent silica detectors (MFSDs), together with real-time signals from flow rate and UV transmittance measurements, as well as CFD-E' simulations to monitor key components of the fluence rate distribution delivered by an operating UV disinfection system. Their method allowed real-time monitoring of reactor performance.

A foundational aspect of this monitoring approach is the linear relationship between lamp output power and the dose distribution. Lamp output can be affected by aging of a lamp, resulting in attenuation of UVC power emitted by the lamp. Fouling of a lamp (sleeve), usually associated with inorganic precipitates [27–31], yields a similar effect on UVC power emitted by a lamp/sleeve assembly. An array of three MFSDs had been previously used to allow real-time measurements of UV transmittance of water, lamp attenuation, and fouling in a simple UV photoreactor [32]. This approach was extended by linking these measurements to CFD-E' simulation results to allow real-time monitoring of important elements of the UV dose distribution in an operating UV disinfection system.

The approach required the use of two basic definitions:

$$\text{LO} = \text{lamp output attentuation coefficient} = \frac{\text{real} - \text{time lamp output power}}{\text{output power of a new lamp}} \tag{7.50}$$

$$\text{SF} = \text{sleeve fouling attenuation coefficient} = \frac{\text{real} - \text{time fluence rate at a fixed location}}{\text{fluence rate at the same location with a clean sleeve}} \tag{7.51}$$

The linear relationships between the dose distribution and lamp output (UVC) power allow the following equations to be defined:

$$D_{\text{RT}}(Q, \text{UVT}) = D_{\text{nom}}(Q, \text{UVT}) \cdot \text{LO} \cdot \text{SF} \tag{7.52}$$

where

$D_{\text{RT}}(Q, \text{UVT})$ = real-time dose distribution at ($Q$, UVT)

$D_{\text{nom}}(Q, \text{UVT})$ = calculated nominal dose distribution at ($Q$, UVT) for LO ≡ 1; SF ≡ 1.

Similarly, because fluence rate (FR) at a given location is linearly dependent on lamp output power, the following relationship can be defined:

$$\frac{\mathrm{FR_{RT}(UVT)}}{\mathrm{FR_{nom}(UVT)}} = \mathrm{LO} \cdot \mathrm{SF} \tag{7.53}$$

where

$\mathrm{FR_{RT}(UVT)}$ = real-time fluence rate at a fixed point, for a given value of UVT
$\mathrm{FR_{nom}(UVT)}$ = nominal (reference) fluence rate at the same location, for the same value of UVT.

The combination of Equations 7.52 and 7.53 yields:

$$D_{\mathrm{RT}}(Q, \mathrm{UVT}) = D_{\mathrm{nom}}(Q, \mathrm{UVT}) \cdot \frac{\mathrm{FR_{RT}(UVT)}}{\mathrm{FR_{nom}(UVT)}} \tag{7.54}$$

$D_{\mathrm{nom}}(Q, \mathrm{UVT})$ was calculated by CFD-E' for a range of flow rates ($Q$) and UV transmittance (UVT) conditions. The terms that represent the quotient $\frac{\mathrm{FR_{RT}(UVT)}}{\mathrm{FR_{nom}(UVT)}}$ can be measured in real time through the use of three strategically placed MFSDs (see Figure 7.28). Therefore, in principle it is possible to use this collection of information to monitor the dose distribution in real time; however, at present this represents an impractical approach. As a simplification, several critical doses that are related to or derived from the dose distribution were measured in real time. These included the RED, $D_{10}$, $D_{\mathrm{mean}}$, and $D_{\mathrm{min}}$. $D_{10}$ represents the dose corresponding to the 10$^{\mathrm{th}}$ percentile of the dose distribution; $D_{\mathrm{mean}}$ represents the mean of the dose distribution; and $D_{\mathrm{min}}$ represents the minimum dose of the dose distribution. Nonlinear regression of the CFD-E' results allowed definition of empirical expressions to estimate the nominal values of these parameters as functions of UVT and $Q$:

$$\mathrm{RED_{nom}}(Q, \mathrm{UVT}) = [2317.3 \cdot \log_{10}(\mathrm{UVT}) + 301.4] \cdot Q^{-0.944} \tag{7.55}$$

$$D_{10,\mathrm{nom}}(Q, \mathrm{UVT}) = [2984.5 \cdot \log_{10}(\mathrm{UVT}) + 386.0] \cdot Q^{-1.046} \tag{7.56}$$

$$D_{\mathrm{mean,nom}}(Q, \mathrm{UVT}) = [5906.6 \cdot \log_{10}(\mathrm{UVT}) + 1009.5] \cdot Q^{-1.070} \tag{7.57}$$

$$D_{\mathrm{min,nom}}(Q, \mathrm{UVT}) = [2225.6 \cdot \log_{10}(\mathrm{UVT}) + 257.9] \cdot Q^{-1.080} \tag{7.58}$$

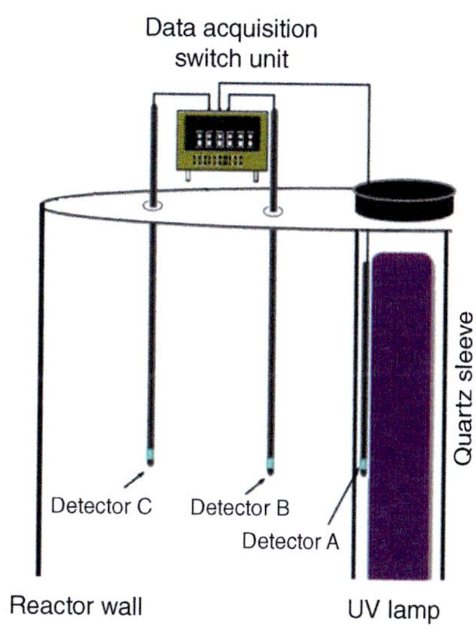


**Figure 7.28** Schematic illustration of positioning of three MFSDs for real-time monitoring of UV photoreactor performance [32]. *Source:* Qiang et. al 2013/with permission of Elsevier.

Although Equations 7.55–7.58 are empirical, their mathematical forms are consistent with the physics that govern the relationships between the monitored parameters (UVT and $Q$) and the predicted doses within the dose distributions. In particular, it is reasonable to expect a $\log_{10}$-dependence of dose on UVT, based on the Beer–Lambert law. Likewise, the exponents on the flow rate term in all four expressions indicate that the dose distribution is (roughly) inversely proportional to the flow rate (i.e. exponent $\approx -1$); this inverse relationship between dose (distribution) and flow rate is reasonable in that residence time is inversely related to flow rate. Figure 7.29 illustrates fits if Equations 7.55–7.58 to values of $RED_{nom}$, $D_{10,nom}$, $D_{mean,nom}$, and $D_{min,nom}$ that were predicted by CFD-E' modeling.

Figure 7.30 illustrates the results of time-course monitoring of fluence rate, flow rate, and UVT in an operating UV disinfection system that was used for secondary disinfection of a public water supply.[g] Several general trends are evident in these data sets. First, there is a steady, gradual decrease in fluence rate within the reactor; this is associated with degradation of lamp output power and the effects of lamp sleeve fouling. Second, a consistent diurnal pattern of water use is evident in the flow rate signal. This is attributable to the normal patterns of water use through the day. Third, the transmittance of the water that was pumped through this reactor appears to have been consistent through the six-month period of this experiment.

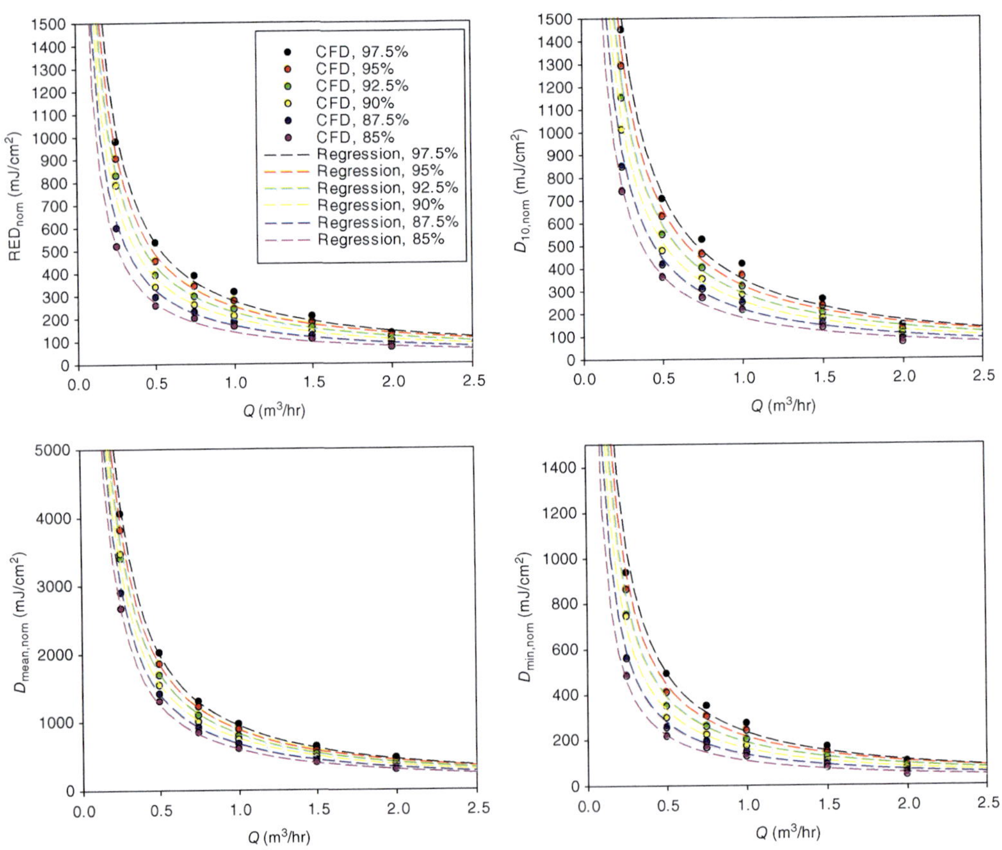


**Figure 7.29** Regression Equations 7.55–7.58 fit to predicted values of $RED_{nom}$ (upper left), $D_{10,nom}$ (upper right), $D_{mean,nom}$ (lower left), and $D_{min,nom}$ (lower right).

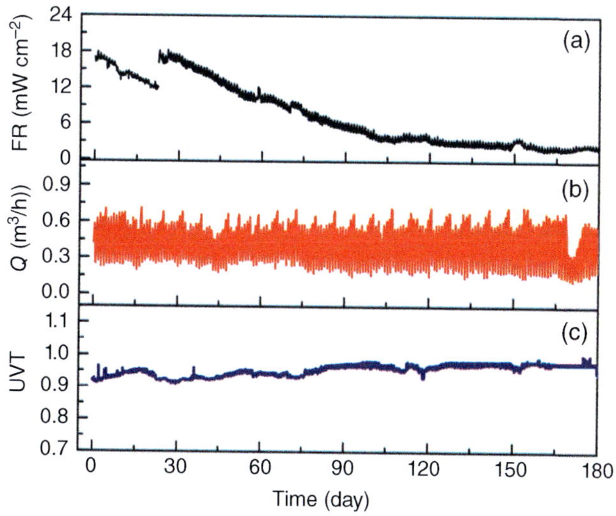


**Figure 7.30** Results of time-course monitoring of (a) fluence rate (FR), (b) flow rate ($Q$), and (c) $UVT_{254}$ over a six-month period for a UV disinfection system used as for secondary disinfection [26]. Source: Li et. al 2017/ with permission of American Chemical Society.

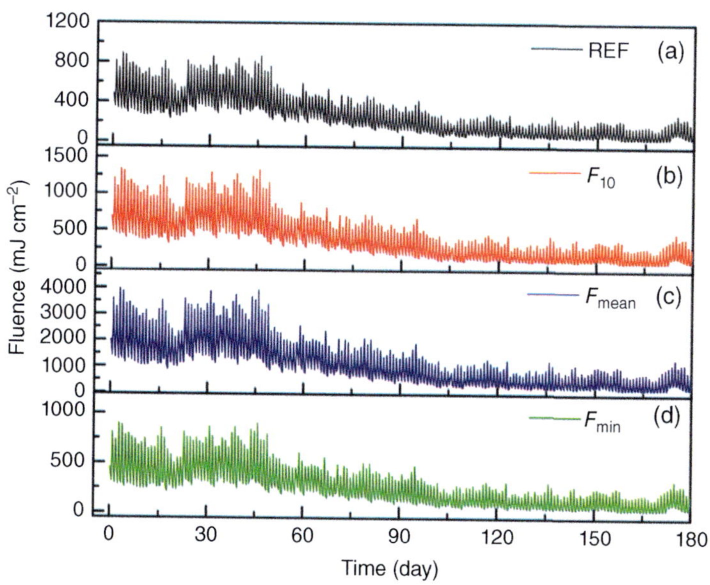


**Figure 7.31** Real-time estimates of $RED_{nom}$ (a), $D_{10}$ (b), $D_{mean}$ (c), and $D_{min}$ (d) over a six-month period for a UV disinfection system used as for secondary disinfection [26]. Source: Li et. al 2017/with permission of American Chemical Society.

Figure 7.31 illustrates the translation of these results to real-time estimates of RED (aka, REF), $D_{10}$, $D_{mean}$, and $D_{min}$ for the same period of time. The effects of the behavior of lamp output and diurnal patterns of water use translate directly to the temporal patterns in these dose estimates for the period of the experiment. Collectively, these results indicate promise for the combined application of physical measurements and numerical modeling as a tool for monitoring of photoreactor performance.

## 7.4 UV Photoreactor Validation Protocols

In municipal applications where UV is employed to accomplish disinfection, current regulations dictate that the ability of the reactor to accomplish a given endpoint of treatment be demonstrated empirically. However, in drinking water applications, the concentration of viable or infective microbes present in treated water is so low that it becomes difficult to measure, at least under normal operating conditions. In most applications, this requires that a UV disinfection system be validated prior to installation.

To address this need, several Federal and private organizations have developed validation protocols for UV disinfection systems. The majority of these validation protocols are based on biodosimetry as the method for reactor analysis.

The objective of this section is to provide brief overviews of the most commonly applied, biodosimetry-based validation protocols. These protocols tend to converge onto similar themes, although the details of implementation and data interpretation may vary. These protocols were developed to address UV disinfection applications in specific countries, geographic regions, or application types. However, they have been adapted by other jurisdictions and for use with other applications. Moreover, the basic principles that inform the design of validation protocols for UV-based water disinfection systems can be expected to inform validation protocols for UV-based reactors used in other settings. For example, a need exists to define standardized protocols for validation of systems that are used to disinfect air and surfaces. These protocols are likely to involve physical measurements that quantify the ability of a system to deliver germicidal UV photons to the target media.

### 7.4.1 Ultraviolet Disinfection Guidance Manual for the Long-Term 2 Enhanced Surface Water Treatment Rule (UVDGM)

In the United States, state regulatory agencies have been given primacy for regulation of public water supplies in their respective states by the US EPA. States are required to implement regulatory approaches that guarantee that public water supplies conform to the *National Primary Drinking Water Standards*, as established by the US EPA. These standards were established under the regulatory authority of the *Safe Drinking Water Act*.

The *Long-Term 2 Enhanced Surface Water Treatment Rule* (*LT2ESWTR*) was enacted to improve protection of public water supplies against contamination by microbial pathogens (especially *Cryptosporidium*, *Giardia*, and enteric viruses) and disinfection byproducts [33]. With respect to microbial quality, the *LT2ESWTR* requires public water utilities to demonstrate that treatment processes achieve the following as minimum levels of treatment:

- *Giardia*: at least 3 $\log_{10}$ removal or inactivation.
- Virus[h]: at least 4 $\log_{10}$ removal or inactivation.
- *Cryptosporidium*: up to 3 $\log_{10}$ removal or inactivation, depending on source water quality.

In recognition of the fact that state regulators often have a wide range of water treatment processes to examine and may lack specific expertise relative to UV disinfection, the US EPA developed the *UVDGM* as a document to provide state agencies with guidance relative to procedures used by engineers in the design, selection, and operation of UV disinfection systems. In addition, the *UVDGM* provides detailed procedures for validation testing and performance standards for UV disinfection systems.

To address compliance with the requirements for control of microbial pathogens, the *LT2ESWTR* established "dose tables" to define the required dose ($D_{req}$) needed to achieve a given level of

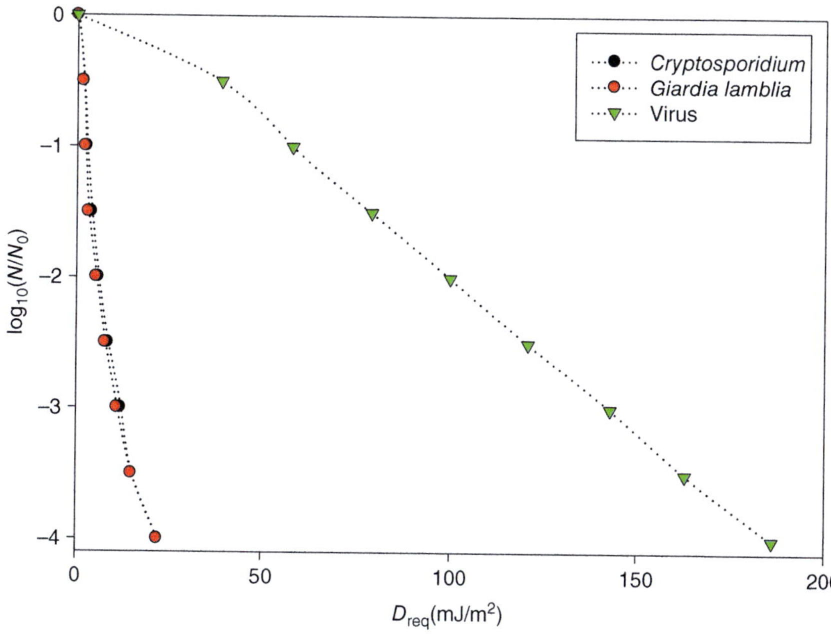


**Figure 7.32** Graphical summary of $UV_{254}$ dose requirements for inactivation or target microbes, as defined by the LT2ESWTR.

inactivation of the three pathogen groups listed above. Figure 7.32 illustrates the "dose" requirements ($D_{req}$) for control of the target pathogens listed in the *LT2ESWTR*. The term $D_{req}$ was defined in the UVDGM as the $UV_{254}$ dose associated with a given extent of inactivation of a target microbe. The data presented in Figure 7.32 indicate that *Cryptosporidium* and *Giardia lamblia* behave similarly in terms of their responses to $UV_{254}$ irradiation; both parasites are quite sensitive to UVC radiation. The "virus" response indicated in Figure 7.32, which is based on Adenovirus, indicates that a substantially larger $UV_{254}$ dose is required to achieve a given extent of inactivation of this microbe, as compared to the protozoan parasites. It should be noted that the apparent "dose–response" behaviors illustrated in Figure 7.32 are conservative, in that they were developed to account for uncertainty that is inherent in the measurement of microbial dose–response behavior.

The term "dose" is associated with a number of different aspects of the behavior of UV photoreactors. Probably because of this, it is a common source of confusion for people who work with these systems. Illustrations of this include the dose-associated definitions that are listed in the *UVDGM*, which are presented in paraphrased form below:

- UV dose $= \int_0^{\tau} E' \cdot dt$
- UV dose distribution = probability density function that describes the doses received by a population of particles that traverses the irradiated zone of a photochemical reactor.
- RED = dose derived by comparison of microbial inactivation in a full-scale reactor with the dose that achieves a similar extent of inactivation under a collimated beam.
- Required dose ($D_{req}$) = dose associated with a given extent of inactivation of target microbes (see Figure 7.32).

- Validated dose ($D_{val}$) = RED estimated by biodosimetry (validation testing).
- Calculated dose = RED calculated from dose-monitoring equation that is developed from biodosimetry testing.

Biodosimetric (validation) testing is used to define operating conditions (flow rate, UVT, lamp status, $E'$ at one or more fixed locations) that will allow delivery of a given value of $D_{req}$. In turn, these conditions can then be associated with a given amount of "credit" for inactivation of one or more target microbes. Validation testing under the *UVDGM* must also account for the effects of lamp aging and fouling, uncertainty associated with sensors, and inlet/outlet configurations. Also, the *UVDGM* requires that validation testing be applied to full-scale reactor systems. In other words, scaling of results among geometrically similar reactors of varying sizes is not allowed.

Within the *UVDGM*, two general dose-monitoring strategies are defined. The first is referred to as the "Intensity (*sic*) Setpoint Approach" in which an "intensity" sensor reading is defined by which a target RED is achieved. A "Single Setpoint" approach is allowed wherein a single sensor reading is defined that can be associated with the target RED. This approach is most appropriate for small systems where the cost of flow measuring equipment may not be justifiable. Another method involves a Variable Setpoint, in which readings from an "intensity" sensor and flow rate meter are used collectively to define the conditions of flow rate and intensity that yield a target RED value.

The second dose-monitoring strategy is referred to as the "Calculated-Dose" approach. In this method, an empirical relationship is developed to describe RED delivered by the reactor as a function of operating conditions. The so-called "dose monitoring equation" is developed based on non-linear regression of the results of biodosimetric testing. A suggested form of the dose-monitoring equation is provided in the *UVDGM*:

$$\text{RED} = 10^a \cdot A_{254}^b \cdot \left(\frac{S}{S_0}\right)^c \cdot \left(\frac{1}{Q}\right)^d \cdot B^e \tag{7.59}$$

where

| | | |
|---|---|---|
| $A_{254}$ | = | absorbance at $\lambda = 254$ nm |
| $S$ | = | measured sensor value |
| $S_0$ | = | sensor value at 100% lamp power |
| $Q$ | = | volumetric flow rate |
| $B$ | = | number of operating banks in series |
| $a, b, c, d, e$ | = | empirical coefficients. |

Note that Equation 7.59 has some similarity to Equation 7.55. By extension of the logic used in the development of Equation 7.55, we may expect that the value of parameter $d$ in Equation 7.59 should be close to 1, which would support an inverse relationship between RED (and other values of dose) and flow rate.

The reading of an "intensity" sensor will be influenced primarily by the output of the lamp(s) in the reactor and the UV transmittance of the water in the reactor. In general terms, readings from a sensor at a fixed location in a reactor will be directly proportional to lamp output power (as affected by lamp aging and fouling of lamp sleeves) and inversely related to UV transmittance.

Viewed another way, the position of a sensor can affect the control strategy for a reactor. If located too close to a lamp, the reading of the sensor is disproportionately affected by lamp power; if too far from the lamp, a relatively long optical path length with result, yielding a situation in which UVT has a disproportionately large effect on the sensor reading [34].

Within the *UVDGM*, the setpoint approach calls for validation of reactors based on at least two test conditions. In the first condition, the reactor is operated under full lamp (ballast) power, while UVT is reduced to yield a desired sensor reading. In the second approach, the reactor is operated at high UVT and lamp power is reduced until the target sensor reading is reached. If the sensor is placed at an "ideal" location, the RED measured under these two test conditions will be essentially identical [35]. This provides motivation for manufacturers to identify an "ideal" sensor location; reactors with "intensity" sensors placed in an ideal location will compete favorably in biodosimetric testing, as compared to otherwise similar reactors with sensors in a nonideal location.

In a general sense, the objective of validation testing (by any method) is to define the operating conditions that yield an acceptable level of treatment, relative to one or more specific treatment objectives. In the case of disinfection, this translates to a definition of operating conditions that guarantee at least a minimum level of inactivation of a target microbe. In the context of biodosimetry, this is interpreted as definition of operating conditions that deliver at least a target value of RED.

The *UVDGM* recommends validation based on biodosimetry. The following steps should be taken in the application of biodosimetry for validation testing, as defined by the *UVDGM*:

- *Bench-scale testing*: One or more challenge organisms are selected for use in biodosimetry. An ideal challenge organism will demonstrate UV dose–response behavior that is identical to that of the target microbe. If validation of a polychromatic reactor is planned, then the action spectrum of the challenge organism should closely match that of the target microbe. The UV dose–response behavior of the challenge microbe is quantified by exposure to a range of well-defined UV doses under a collimated beam.
- *Full-scale reactor testing*: A suspension of the challenge organism is pumped through the full-scale reactor over a range of operating conditions. Common variables in biodosimetry tests include flow rate, lamp power, and UV transmittance. At least three values of each variable are recommended for each reactor. Therefore, if testing is planned to include all three variables, then a minimum of $3^3 = 27$ operating conditions will need to be tested. For each operating condition, samples of influent and effluent are collected to allow measurement of the concentration of viable or infective challenge microbes in the influent and effluent. As such, it is then possible to define $\frac{N}{N_0}$ for the entire range of operating conditions.
- *Estimation of RED*: The measured values of $\frac{N}{N_0}$ for each operating condition are compared with the UV dose–response behavior, as measured based on exposures under a collimated beam, to allow estimation of RED at each operating condition. In other words, RED is calculated across the entire operating envelope. This may be translated into a multiparameter equation to define and RED surface as a function of the operating variables (e.g. Equation 7.59).
- *Adjustment for uncertainty*: The authors of the *UVDGM* recognized many sources of uncertainty (error) that can complicate the interpretation of biodosimetric testing. To account for this, they recommend that the "validated" RED for a reactor be reported based on the following:

$$\text{Validated RED} = \frac{\text{RED}}{\text{VF}} \tag{7.60}$$

where

RED = value of RED calculated in the previous step
VF = validation factor

$$\text{VF} = B_{\text{RED}} \cdot \left(1 + \frac{U_{\text{val}}}{100}\right) \tag{7.61}$$

$B_{\text{RED}}$ = RED bias
$U_{\text{val}}$ = uncertainty of validation.

The value of $B_{RED}$ depends on the UV dose distribution delivered by the reactor as well as the difference in UV sensitivity between the challenge microbe and the target microbe. Ideally, the challenge microbe will display the same UV dose–response behavior as the target microbe; when this is the case, then $B_{RED} \equiv 1$. If the challenge microbe is more resistant than the target microbe, then the value of the RED determined by biodosimetry with the challenge microbe will exceed the RED that would be determined by biodosimetry using the target microbe (see Figures 7.6 and 7.9). Conversely, if the challenge microbe is more sensitive to UV exposure than the target microbe, then biodosimetry using the challenge microbe will be conservative, and a value of $B_{RED} \equiv 1$ can be assigned.

Though not recognized by the *UVDGM*, one of the important advantages of LA is the measurement of the dose distribution. Following the logic described above, measurement of the dose distribution by LA allows for assignment of $B_{RED} \equiv 1$ [22].

$U_{val}$ provides a numerical description of other sources of error (uncertainty) associated with biodosimetry, including uncertainty associated with sensor readings and interpolation of UV dose–response behavior. The value of $U_{val}$ is determined by application of conventional statistical methods, as described in the *UVDGM*.

### 7.4.2 Österreichisches Normungsinstitut (ÖNORM)

UV irradiation is commonly used for drinking water disinfection in Austria. The Austrian government has developed a standardized protocol for reactor validation based on biodosimetry using the spores of *Bacillus subtilis* as the challenge organism. The test evolved from a specification of a minimum RED of 400 $\frac{\text{J}}{\text{m}^2}$ (40 $\frac{\text{mJ}}{\text{cm}^2}$) that was defined in the Austrian Food Act as being sufficient to yield 6 $\log_{10}$ inactivation of health-related transmittable bacteria and 4 $\log_{10}$ inactivation of health-related transmittable viruses.

The ÖNORM regulation also provides a definition of the acceptable range of $UV_{254}$ dose–response behavior for *B. subtilis* spores as a QA/QC check, based on the following equation:

$$\frac{N}{N_0} = 1 - \left(1 - 10^{-kD}\right)^{10^d} \tag{7.62}$$

where

$k$ = slope of the (log)-linear part of the curve (when presented on a semilogarithmic plot)

$D$ = $UV_{254}$ dose $\left(\frac{\text{J}}{\text{m}^2}\right)$

$d$ = shoulder of curve.

The admissible ranges for parameters in Equation 7.62 are as follows:

$$\begin{aligned} k &= 0.0052 - 0.0078\ \frac{\text{m}^2}{\text{J}} \\ d &= 0.49 - 0.91 \end{aligned} \tag{7.63}$$

The QA/QC range for *B. subtilis* spore $UV_{254}$ dose–response behavior based on these limits is illustrated in Figure 7.16. The UV dose–response behavior for this organism is somewhat different than that of many other challenge organisms, in that it is characterized by a "shoulder" in the limit of low doses. In other words, there is a range of $UV_{254}$ doses for which little or no inactivation of these spores is observed. The presence of the shoulder in their $UV_{254}$ dose–response behavior increases the sensitivity of the test to the low part of the $UV_{254}$ dose distribution delivered by a photoreactor.

The objective of biodosimetry under the ÖNORM regulation is to define the range of operating conditions (defined by flow rate, "intensity" sensor reading, and UVT) that yield an RED of at least $400\,\frac{\mathrm{J}}{\mathrm{m}^2}$. The regulation requires the use of lamps that have been in service for at least 100 hours and its output must be adjusted to an "end of lamp life" condition. Lamp output must be constrained to wavelengths greater than 240 nm to prevent ozone formation.

### 7.4.3 Deutsche Vereinigung des Gas- und Wasserfaches (DVGW)

The German standard for biodosimetric testing of UV reactors used for drinking water disinfection is similar to the ÖNORM standard. The same challenge organism (*B. subtilis* spores) is stipulated, lamp output is constrained to wavelengths greater than 240 nm, and the minimum RED for operation is $400\,\frac{\mathrm{J}}{\mathrm{m}^2}$. Biodosimetry is conducted on a full-scale reactor to define operating conditions, defined by maximum flow rate and minimum irradiance (as detected by an internal sensor) that yield an RED of at least $400\,\frac{\mathrm{J}}{\mathrm{m}^2}$.

A QA/QC criterion for the $UV_{254}$ dose–response behavior of the challenge organism has been defined as follows:

$$\log_{10}\left(\frac{N}{N_0}\right) = k \cdot H_0 + d \tag{7.64}$$

where

$k$ = slope of the linear part of the dose–response curve (on a semilogarithmic plot)
$H_0$ = $UV_{254}$ dose
$d$ = distance of the intercept of the straight portion of the curve from the origin of the ordinate.

Parameter limits for this QA/QC criterion are defined as follows:

$$\begin{aligned}
&\text{at } 100\frac{\mathrm{J}}{\mathrm{m}^2},\ \log_{10}\left(\frac{N}{N_0}\right) \geq -0.2\\
&\text{at } 400\frac{\mathrm{J}}{\mathrm{m}^2},\ -1.6 \leq \log_{10}\left(\frac{N}{N_0}\right) \leq -2.6\\
&0.8 \leq d \leq 1.8\\
&0.007\frac{\mathrm{m}^2}{\mathrm{J}} \leq k \leq -0.011\frac{\mathrm{m}^2}{\mathrm{J}}
\end{aligned} \tag{7.65}$$

These QA/QC limits are illustrated in Figure 7.16. Although there is considerable overlap between the QA/QC limits defined by ÖNORM and DVGW, there are also clear differences in these limits as well. As in the case of MS2, these QA/QC limits are related to a source of error in quantification of RED.

### 7.4.4 National Water Research Institute (NWRI)

The National Water Research Institute (NWRI) developed guidelines for validation of UV disinfection systems that are included as part of water reuse and drinking water systems [9]. These guidelines were developed to document procedures to be used for "spot-check" bioassays used in validation of full-scale reactors and to standardize the assignment of "UV dose" in biodosimetric assays based on MS2 as a challenge organism.

For all validation procedures defined under the *NWRI Guidelines*, coliphage MS2 is stipulated as the challenge microbe [9]. QA/QC limits for the $UV_{254}$ dose–response behavior of MS2 were defined such that at least 80% of data points from collimated beam testing were required to fall within the following range:

$$
\begin{aligned}
-\log_{10}\left(\frac{N}{N_0}\right) &= 0.040{\cdot}D + 0.64 \\
-\log_{10}\left(\frac{N}{N_0}\right) &= 0.033{\cdot}D + 0.20
\end{aligned}
\tag{7.66}
$$

These limits are illustrated in Figure 7.13. In addition, the *NWRI Guidelines* constrain the dose–response behavior to the range $20\frac{\text{mJ}}{\text{cm}^2} \le D \le 150\frac{\text{mJ}}{\text{cm}^2}$ (see Figure 7.13). RED values determined by dosimetry are constrained to values greater than $20\frac{\text{mJ}}{\text{cm}^2}$.

The *NWRI Guidelines* include several definitions of the word "dose" that are relevant to the intended applications, but which also contribute to the general confusion that exists around the term. These definitions include (copied verbatim from [9]):

- *Reduction equivalent dose (RED)*: The dose that is assigned to the UV test reactor based on reactor validation testing. The RED is equivalent to that measured with the collimated-beam apparatus for the same degree of inactivation of the target microorganism.
- *Design UV dose*: The RED required for a specific log inactivation of the target microorganism. The design UV dose is used for sizing UV disinfection systems.
- *Operational UV dose*: The UV dose that is established based on the results of the equipment validation testing. The operational UV dose, a useful index of reactor behavior, can be used to make most efficient use of the UV disinfection system (e.g. reduce power demand, reduce the number of reactors or reactor trains on-line) while maintaining the design UV dose.

*RED*, as used in these guidelines conforms to the common definition of this term. The *Design UV dose* is a target of reactor design and operation that is intended to guarantee performance (i.e. acceptable inactivation of microbial pathogens). The *Operational UV dose* is used to describe the value of RED that is assigned to a reactor for a given set of operating conditions, defined by measured values of flow rate, "intensity," and UV transmittance.

The *NWRI Guidelines* provide a description of the procedures to be used in so-called "spot-check" bioassays, which are conducted on installed, full-scale UV disinfection systems. Biodosimetry is conducted following common procedures that have been described elsewhere, in that a suspension of a challenge organism is subjected to $UV_{254}$ irradiation under a collimated beam and in a continuous-flow (full-scale) reactor. The value of RED for a given operating condition is assigned by comparison of the inactivation response of the challenge organism (coliphage MS2) in the full-scale reactor with that observed with a collimated beam. However, in the case of the *NWRI Guidelines*, the assignment of RED is stipulated as follows:

$$
\text{RED} = \frac{\log_{10}\left(\frac{N_0}{N}\right) - 0.5464}{0.0368}
\tag{7.67}
$$

Or rewritten to conform to Figure 7.13:

$$
\log_{10}\left(\frac{N}{N_0}\right) = -0.0368 \cdot \text{RED} - 0.5464
\tag{7.68}
$$

Equation 7.67 is plotted in Figure 7.13 for illustration.

Equations 7.66 and 7.67 allow conversion of observed inactivation response in a full-scale reactor ($\log_{10}\left(\frac{N}{N_0}\right)$) to be translated into an RED value. However, there are at least two problems with this prescribed equation for RED estimation. First, despite the fact that Equation 7.67 conforms to roughly the center of the QA/QC criteria that have been defined for MS2, the equation fails to account for variability in $UV_{254}$ dose–response data that is evident in Figure 7.13 and similar data sets. As described previously, this variability represents a source of error in the assignment of RED and the description of reactor performance. Second, Equations 7.66 and 7.67 imply first-order (log-linear) dose–response behavior on the part of MS2. This assignment may be appropriate as a rough approximation, but it is relevant to note that the data in Figure 7.13 clearly illustrate curvature (tailing) in the limit of large $UV_{254}$ doses.

For wastewater reuse applications, the *NWRI Guidelines* stipulate different values of *Design UV Dose* depending on the nature of upstream treatment:

- *For media filtration*: *Design UV Dose* $\geq 100 \frac{\text{mJ}}{\text{cm}^2}$ under conditions of maximum daily flow and $UVT_{254} \geq 55\%$ (based on a 1.0 cm optical path length). In addition, turbidity is required to be less than 2 NTU always, and less than 5 NTU in 95% of samples.
- *For membrane filtration*: *Design UV Dose* $\geq 80 \frac{\text{mJ}}{\text{cm}^2}$ under conditions of maximum daily flow and $UVT_{254} \geq 65\%$ (based on a 1.0 cm optical path length). In addition, turbidity is required to be less than 0.5 NTU always, and less than 0.2 NTU in 95% of samples.
- *For reverse osmosis*: *Design UV Dose* $\geq 50 \frac{\text{mJ}}{\text{cm}^2}$ under conditions of maximum daily flow and $UVT_{254} \geq 90\%$ (based on a 1.0 cm optical path length). In addition, turbidity is required to be less than 0.5 NTU always, and less than 0.2 NTU in 95% of samples.

### 7.4.5 NSF/ANSI

NSF International and the American National Standards Institute worked together to develop standards for biodosimetric testing of point-of-use and point-of-entry UV disinfection systems [2]. This collaboratively developed standard describes a protocol for application of biodosimetry based on coliphage MS2 to be applied to these small-scale systems.

The protocol stipulates the use of MS2 as the challenge organism and has adopted essentially the same QA/QC standard that was defined in the *NWRI Guidelines* (see Figure 7.13). For Class A reactors, which are intended for use with "visually clear water," the NSF/ANSI protocol calls for identification of operating conditions that yield and MS2-based RED of at least $40 \frac{\text{mJ}}{\text{cm}^2}$, so as to conform to standards that have been established in other, similar protocols (e.g. ÖNORM, DVGW).

## 7.5 Summary

The performance of UV photoreactors is governed by the dose distribution they deliver and the intrinsic kinetics of the photochemical (or photobiological) reactions that take place as a result of UV irradiation. The intrinsic kinetics of photochemical reactions can be measured in a laboratory using well-defined experimental conditions. However, the dose distribution cannot be monitored in real time with currently available tools. As such, it is necessary to conduct experiments to provide empirical data to validate reactor performance.

Most validation methods are based on biodosimetry and are aimed at validation for disinfection applications. Biodosimetry is a well-established method that has been widely applied and is familiar to many in the engineering and regulatory communities. Biodosimetric tests provide a measure of the inactivation response of one or more suspended microbes to UV irradiation under a collimated beam and in a full-scale reactor. The results of these tests are normally reported in the form of a RED. However, biodosimetry suffers from a number of limitations, including the inability of the method to measure the dose distribution delivered by a reactor and the effects of variability in the kinetics (i.e. dose–response behavior) of challenge microbes. These limitations mandate the use of conservative safety factors in the interpretation of biodosimetry results. In addition, the value of the RED that results from a biodosimetry test is strongly influenced by the sensitivity of the challenge microbe (or more generally, challenge agent) to UV exposure. Therefore, reported values of RED should always be accompanied by an indication of the challenge microbe that was applied in biodosimetry.

In principle, similar reactor validation procedures could be conducted using photolabile chemicals (e.g. chemical actinometers). For many actinometers, such an approach could provide benefits of improved repeatability and analytical speed. However, most actinometers are characterized by intrinsic kinetics of photochemical transformation that are much slower than the corresponding rates of UV-induced inactivation of waterborne microbes. Therefore, the use of a chemical actinometer in a validation experiment will be constrained to applications where the kinetics of photoresponse of the target of treatment are similar to those of the actinometer.

Alternatives to conventional biodosimetry have emerged. One such approach involves LA, wherein a UV-sensitive molecule is conjugated to the surface of microspheres. The results of LA experiments yield a measurement of the dose distribution delivered by a photoreactor; LA is the only available method to provide this measurement. Knowledge of the dose distribution delivered by a photoreactor has the important benefit of allowing accurate prediction of the performance of the photoreactor relative to any photochemical endpoint, as long as the dose–response behavior (i.e. intrinsic kinetics) of the target are known.

Another promising alternative involves the coordinated use of CFD-E' modeling and strategic placement of multiple MFSDs. This approach allows real-time monitoring of selected doses within the dose distribution.

## Appendix 7.A Description of the Error Function and Its Complement

The error function, erf($x$), and its complement, erfc($x$), are mathematical functions that often emerge in engineering applications. For some people they are unfamiliar and can be a source of confusion. The objectives of this Appendix are to describe erf($x$) and erfc($x$), to provide their formal mathematical definitions, and to relate them to other, perhaps more familiar functions.

The error function and its complement often appear in analyses that involve integration of the Gaussian (normal) distribution function. Their formal mathematical definitions are as follows:

$$\mathrm{erf}(x) = \frac{2}{\sqrt{\pi}} \int_0^x \mathrm{e}^{-t^2} \cdot \mathrm{d}t \tag{7.A.1}$$

$$\mathrm{erfc}(x) = \frac{2}{\sqrt{\pi}} \int_{x}^{\infty} \mathrm{e}^{-t^2} \cdot \mathrm{d}t \tag{7.A.2}$$

These functions have the following properties:

$$\mathrm{erf}(x) + \mathrm{erfc}(x) \equiv 1 \tag{7.A.3}$$

$$\begin{aligned} &\text{as} \quad x \to \infty, \ \mathrm{erf}(x) \to 1, \ \mathrm{erfc}(x) \to 0 \\ &\text{as} \quad x \to -\infty, \ \mathrm{erf}(x) \to -1, \ \mathrm{erfc}(x) \to 2 \end{aligned} \tag{7.A.4}$$

Figure 7.A.1 illustrates the error function and its complement.

It is generally helpful to have a graphical illustration, such as Figure 7.A.1, to provide a visual reference for the behavior of a function. Given this illustration, it may also be helpful to think of erf($x$) and erfc($x$) as functions that have these shapes and that can be readily calculated for us, as can be done in spreadsheets and in many commercial mathematics packages.

To illustrate this idea, consider the behavior of the functions sin($x$) and cos($x$), which are commonly applied in geometry and trigonometry. Figure 7.A.2 illustrates the well-known behavior of these two functions. These oscillating functions have many applications in engineering and other disciplines and are familiar to most people. In geometry and trignonometry, most are familiar with their definitions as the following ratios of the sides of a right triangle:

$$\sin(\theta) = \frac{\text{opposite}}{\text{hypotenuse}} \tag{7.A.5}$$

$$\cos(\theta) = \frac{\text{adjacent}}{\text{hypotenuse}} \tag{7.A.6}$$

However, in many applications of these functions, people rely on a calculator or computer software to calculate these functions for us. Moreover, most people are unfamiliar with the analytical

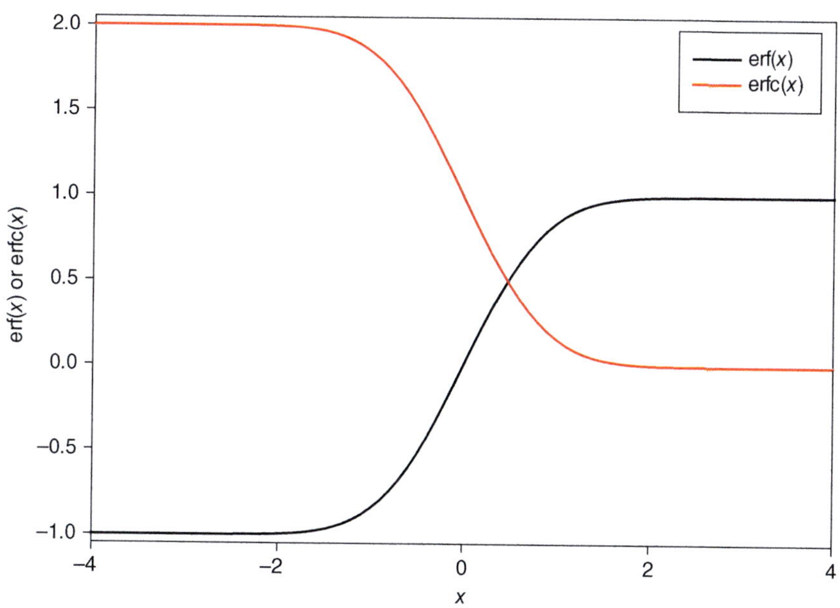


**Figure 7.A.1** Illustration of the error function and its complement.

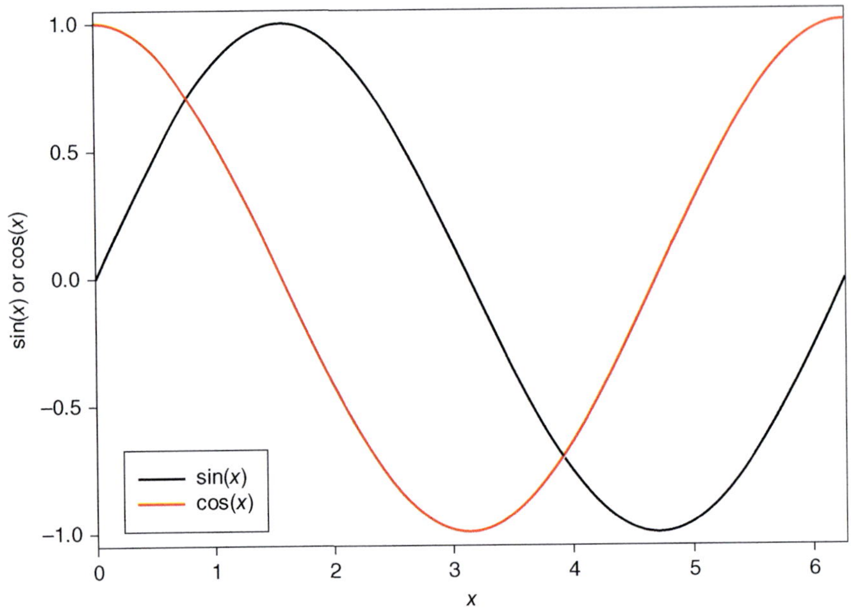


**Figure 7.A.2** Illustration of sine and cosine.

expressions that are used to calculate these two functions. In particular, the following are series expansions that are used to calculate sine and cosine:

$$\sin(x) = \sum_{n=1}^{\infty} \frac{(-1)^{n-1}}{(2n-1)!} x^{2n-1} = x - \frac{x^3}{3!} + \frac{x^5}{5!} - \frac{x^7}{7!} + \cdots \tag{7.A.7}$$

$$\cos(x) = \sum_{n=0}^{\infty} \frac{(-1)^{n}}{(2n)!} x^{2n} = 1 - \frac{x^2}{2!} + \frac{x^4}{4!} - \frac{x^6}{6!} \cdots \tag{7.A.8}$$

When we ask a calculator or computer software to calculate these functions, the algorithms they use are those described in Equations 7.A.7 and 7.A.8.

As in the case of the error function and its complement, there are important identities that are associated with sin and cosine, such as:

$$\sin^2(\theta) + \cos^2(\theta) \equiv 1 \tag{7.A.9}$$

It is perhaps helpful, then, to think of erf($x$) and erfc($x$) in a similar manner. We are now familiar with the general shapes of these functions and how they relate to each other. We also know that there are well-defined analytical functions that can be used to calculate them, and that these functions are available in common software packages.

## Notes

**a** The term *reduction equivalent fluence* (REF) is also used and is equivalent.

**b** Biodosimetry methods described herein will refer to reactors based on monochromatic UVC radiation, such as low-pressure Hg lamps. Closely related methods are available for use with polychromatic reactors.

**c** Reactor testing and validation may also be conducted with soluble reagents (e.g. chemical actinometers) or nonmicrobial particles. The pump(s) are used to deliver the liquid that will carry the challenge agents and matrix modifiers through the reactor system.

**d** An important limitation of this method of photoreactor simulation relates to the low-dose "tail" of the distribution. Specifically, with the Gaussian distribution, it is possible to assign apparent "negative" doses to nontrivial fractions of the dose distribution. A negative UV dose has no physical interpretation (i.e. it is not possible to deliver a "negative" dose). Therefore, the use of this model to simulate the dose distribution can introduce error if a significant fraction of the dose distribution is assigned to negative dose values. In other words, it is important to confirm that Equation 7.4 is satisfied, with the limits of integration described in that equation.

**e** See Appendix 7.A for a description of the error function and its complement.

**f** Note: Fluorescence intensity (FI) is a relative term with no units. The magnitude of FI depends on the fluence rate of radiation used to cause excitation of the fluorophore. FI is presented on a dimensionless, relative scale. However, if the optical characteristics of the instrument used to measure FI are fixed, then FI measurements can be quantitative. This condition is an implicit requirement of LA experiments.

**g** These systems are often used in Asia to ensure delivery of safe water to a residential area or a high-rise apartment building.

**h** Note that the term "virus" here refers to enteric viruses, which comprise a wide range of viruses that originate in the digestive systems of humans and other animals. The responses of these viruses to disinfectants, including UVC irradiation, are highly variable. In many drinking water applications, the UV dose–response behavior of Adenovirus is used as a conservative representation of enteric viruses, as it is among the most UV-resistant of the enteroviruses.

## References


1 Qualls, R.G. and Johnson, J.D. (**1983**). Bioassay and dose measurement in UV disinfection. *Applied and Environmental Microbiology 45* (3): 872–877.

2 NSF/ANSI, NSF/ANSI 55-2017 (**2017**). *Ultraviolet Microbiological Water Treatment Systems*. Ann Arbor, MI: NSF International.

3 Sun, Z. and Blatchley III, E.R. (**2017**). *Tetraselmis* as a challenge organism for validation of ballast water UV systems. *Water Research 121*: 311–319.

4 Bolton, J.R. and Linden, K.G. (**2003**). Standardization of methods for fluence (UV dose) determination in bench-scale UV experiments. *Journal of Environmental Engineering-Asce 129* (3): 209–215.

5 Ahmed, Y.M.H.I.H. (**2017**). Stochastic Evaluation of Disinfection Performance in Large-Scale, Open-Channel UV Photoreactors. West Lafayette, IN: Purdue University.

6 Pennell, K.G., Aronson, A.I., and Blatchley III, E.R. (**2008**). Phenotypic persistence and external shielding ultraviolet radiation inactivation kinetic model. *Journal of Applied Microbiology 104* (4): 1192–1202.

7 Blatchley III, E.R., Dumoutier, N., Halaby, T.N. et al. (**2001**). Bacterial responses to ultraviolet irradiation. *Water Science and Technology 43* (10): 179–186.

8 Severin, B.F., Suidan, M.T., and Engelbrecht, R.S. (**1983**). Kinetic modeling of UV disinfection of water. *Water Research 17* (11): 1669–1678.

9 Emerick, R.W. and Tchobanoglous, G. (**2012**). Ultraviolet Disinfection Guidelines for Drinking Water and Water Reuse. Fountain Valley, CA: National Water Research Institute.

10 Leighton, W.G. and Forbes, G.S. (**1930**). Precision actinometry with uranyl oxalate. *Journal of the American Chemical Society 52*: 3139–3152.

11 Parker, C.A. (**1953**). A new sensitive chemical actinometer. 1. Some trials with potassium ferrioxalate. *Proceedings of the Royal Society of London Series a-Mathematical and Physical Sciences 220* (1140): 104–116.

12 Galbavy, E.S., Ram, K., and Anastasio, C. (**2010**). 2-Nitrobenzaldehyde as a chemical actinometer for solution and ice photochemistry. *Journal of Photochemistry and Photobiology a-Chemistry 209* (2–3): 186–192.

13 Ploeser, J.M. and Loring, H.S. (**1949**). The ultraviolet absorption spectra of the pyrimidine ribonucleosides and ribonucleotides. *Journal of Biological Chemistry 178* (1): 431–437.

14 Shen, C.Y., Fang, S.Y., Bergstrom, D.E., and Blatchley, E.R. (**2005**). (*E*)-5-2-(Methoxycarbonyl)ethenyl – cytidine as a chemical actinometer for germicidal UV radiation. *Environmental Science & Technology 39* (10): 3826–3832.

15 Wols, B.A., Harmsen, D.J.H., Wanders-Dijk, J. et al. (**2015**). Degradation of pharmaceuticals in UV (LP)/$H_2O_2$ reactors simulated by means of kinetic modeling and computational fluid dynamics (CFD). *Water Research 75*: 11–24.

16 Feng, Y.G., Smith, D.W., and Bolton, J.R. (**2007**). Photolysis of aqueous free chlorine species (NOCI and OCI-) with 254 nm ultraviolet light. *Journal of Environmental Engineering and Science 6* (3): 277–284.

17 Link, L., Sawyer, J., Venkateswaran, K., and Nicholson, W. (**2004**). Extreme spore UV resistance of *Bacillus pumilus* isolates obtained from an ultraclean spacecraft assembly facility. *Microbial Ecology 47* (2): 159–163.

18 Beck, S.E., Rodriguez, R.A., Linden, K.G. et al. (**2014**). Wavelength dependent UV inactivation and DNA damage of adenovirus as measured by cell culture infectivity and long range quantitative PCR. *Environmental Science & Technology 48* (1): 591–598.

19 Blatchley III, E.R., Shen, C.Y., Naunovic, Z. et al. (**2006**). Dyed microspheres for quantification of UV dose distributions: photochemical reactor characterization by Lagrangian actinometry. *Journal of Environmental Engineering-Asce 132* (11): 1390–1403.

20 Fang, S.Y., Guan, Y.S., Blatchley III, E.R. et al. (**2008**). Conjugation of (*E*)-5-2-(methoxycarbonyl) ethenyl cytidine to hydrophilic microspheres: development of a mobile microscale UV light actinometer. *Bioconjugate Chemistry 19* (3): 592–597.

21 Fang, S.Y., Guan, Y.S., Blatchley III, E.R. et al. (**2003**). Development of a nucleoside analog UV light sensor. *Nucleosides Nucleotides & Nucleic Acids 22* (5-8): 703–705.

22 Shen, C., Blatchley III, E.R., Cox, E., and Scheible, O.K. (**2016**). Protocol for Validation of UV Disinfection Reactors Using Lagrangian Actinometry. Denver, CO: Water Research Foundation.

23 US EPA (**2006**). Ultraviolet Disinfection Guidance Manual for the Final Long Term 2 Enhanced Surface Water Treatment Rule. Washington, DC: US Environmental Protection Agency (EPA).

24 Shen, C., Scheible, O., Chan, P. et al. (**2009**). Validation of medium-pressure UV disinfection reactors by Lagrangian actinometry using dyed microspheres. *Water Research 43* (5): 1370–1380.

25 Blatchley III, E.R., Shen, C., Scheible, O.K. et al. (**2008**). Validation of large-scale, monochromatic UV disinfection systems for drinking water using dyed microspheres. *Water Research 42* (3): 677–688.

26 Li, M.K., Li, W.T., Qjang, Z.M., and Blatchley III, E.R. (**2017**). On-site determination and monitoring of real-time fluence delivery for an operating UV reactor based on a true fluence rate detector. *Environmental Science & Technology 51* (14): 8094–8100.

27 Lin, L.S., Johnston, C.T., and Blatchley III, E.R. (**1999**). Inorganic fouling at quartz : water interfaces in ultraviolet photoreactors – III. Numerical modelling. *Water Research 33* (15): 3339–3347.

28 Lin, L.S., Johnston, C.T., and Blatchley III, E.R. (**1999**). Inorganic fouling at quartz : water interfaces in ultraviolet photoreactors – II. Temporal and spatial distributions. *Water Research 33* (15): 3330–3338.

**29** Lin, L.S., Johnston, C.T., and Blatchley III, E.R. (**1999**). Inorganic fouling at quartz : water interfaces in ultraviolet photoreactors – I. Chemical characterization. *Water Research 33* (15): 3321–3329.

**30** Wait, I.W. and Blatchley III, E.R. (**2010**). Model of radiation transmittance by inorganic fouling on UV reactor lamp sleeves. *Water Environment Research 82* (11): 2272–2278.

**31** Wait, I.W., Johnston, C.T., and Blatchley III, E.R. (**2007**). The influence of oxidation reduction potential and water treatment processes on quartz lamp sleeve fouling in ultraviolet disinfection reactors. *Water Research 41* (11): 2427–2436.

**32** Qiang, Z.M., Li, M.K., and Bolton, J.R. (**2013**). Development of a tri-parameter online monitoring system for UV disinfection reactors. *Chemical Engineering Journal 222*: 101–107.

**33** US EPA (**2021**). Long Term 2 Enhanced Surface Water Treatment Rule Documents. https://www.epa.gov/dwreginfo/long-term-2-enhanced-surface-water-treatment-rule-documents.

**34** Daniel Schmelling, Christine Cotton, Douglas Owen, Erin Mackey, Harold Wright, Karl Linden, and James Malley, Jr. (**2006**). Ultraviolet Disinfection Guidance Manual for the Final Long Term 2 Enhanced Surface Water Treatment Rule, Washington, DC.

**35** Wright, H.B., Heath, M., and Bandy, J. (**2015**). Yikes! What the UVDGM does not address on UV disinfection. *IUVA News 17* (2): 12–18.

# 8

# Methods for Quantification of Microbial Responses to UVC Irradiation

## 8.1 Introduction

Disinfection represents the most common application of UV radiation. UV-based disinfection processes are used in treatment of municipal wastewater and wastewater reuse applications, in the production of potable water, in industrial water treatment applications, for treatment of swimming pool water, and for treatment of indoor air and surfaces. UV irradiation is also used for antimicrobial applications that may not involve human pathogens, such as in the treatment of ballast water or in the production of pharmaceutical products.

Recent history has provided numerous examples of the consequences of applying analytical methods for quantification of microbial responses to UVC radiation that are inconsistent with the mechanisms of inactivation, the objectives of treatment, or both. The use of these methods can yield fundamentally incorrect or misleading information regarding the effects of UVC irradiation on microbial inactivation. By extension, these methods can also lead to incorrect information pertaining to the design and operation of UV disinfection systems.

The objectives of this chapter are to provide the information needed to identify appropriate analytical methods for quantification of microbial responses to UVC irradiation. First, the mechanisms of inactivation associated with UVC irradiation are summarized. This is followed by a section in which the life cycle and/or infective cycle of relevant microbial groups are presented. The results of tests in which inappropriate methods have been applied are then reviewed, together with a discussion of the implications of the findings of these tests. This is followed by a summary of some of the analytical methods that are appropriate for quantification of responses of microorganisms to UV irradiation, including bacteria, viruses, protozoa, and algae. The chapter concludes with a brief presentation of methods that are emerging as alternatives to conventional assays, often involving the tools of molecular biology.

## 8.2 Mechanisms of Microbial Inactivation Resulting from UVC Irradiation

UV disinfection processes depend on photochemical reactions that take place within critical biomolecules of exposed organisms. Although UV irradiation can lead to structural changes within many biomolecules, the critical targets for microbial inactivation are generally nucleic acids and proteins.

*Photochemical Reactors: Theory, Methods, and Applications of Ultraviolet Radiation*, First Edition. Ernest R. Blatchley III.

Companion website: www.wiley.com/go/blatchley/photochemicalreactors

Nucleic acids are biopolymers that comprise nitrogenous bases, 5-carbon sugars, and one or more phosphate groups. DNA includes four bases: adenine (A), guanine (G), cytosine (C), and thymine (T). RNA also includes four bases, but uracil (U) is substituted for thymine. Structures of purine, pyrimidine, and their derivative bases, which form the building blocks of nucleic acids, are illustrated in Figure 8.1.

As illustrated in Figure 8.2, the pyrimidine and purine bases are effective absorbers of radiation in the UVC and UVB portions of the electromagnetic spectrum. Perhaps not surprisingly, nucleic acids are also effective absorbers of UV radiation, with peak absorption usually being observed in the vicinity of 265 nm (see Figure 8.3). The details of DNA absorption spectra will be influenced by the specific composition of each organisms' DNA, but the general shapes of these spectra are remarkably similar across most DNA samples for the wavelength range 240 nm $\lesssim \lambda \lesssim$ 300 nm.

Proteins and their constituent amino acids can also be strong absorbers of UV radiation, but protein absorption spectra are more variable than those of nucleic acids (depending largely on amino acid composition), and they often demonstrate strongest absorption for wavelengths in the range $\lambda \lesssim$ 240 nm.

When nucleic acids absorb UVC or UVB photons, several photochemical processes are initiated. The most common products of these reactions are cyclopyrimidine dimers (CPDs) and pyrimidine (6-4) pyrimidone photoproducts (*aka* (6-4) photoproducts, or abbreviated (6-4)PP) [1–4]. The efficiency of formation of these products depends largely on the nucleotide sequence within a nucleic acid, with adjacent thymines being generally the most susceptible to this type of damage. Figure 8.4 illustrates the CPD and (6-4)PP that result from UV irradiation of adjacent thymines on a nucleic acid strand.[a]

The formation of CPDs and (6-4)PPs is quite general in most cells, including bacteria and mammalian cells. These photoproducts are also formed within the nucleic acids of multicellular organisms and noncellular organisms, such as viruses. CPDs generally represent 80–90% of the total photoproducts formed within a common cell, with most of the balance being in the form

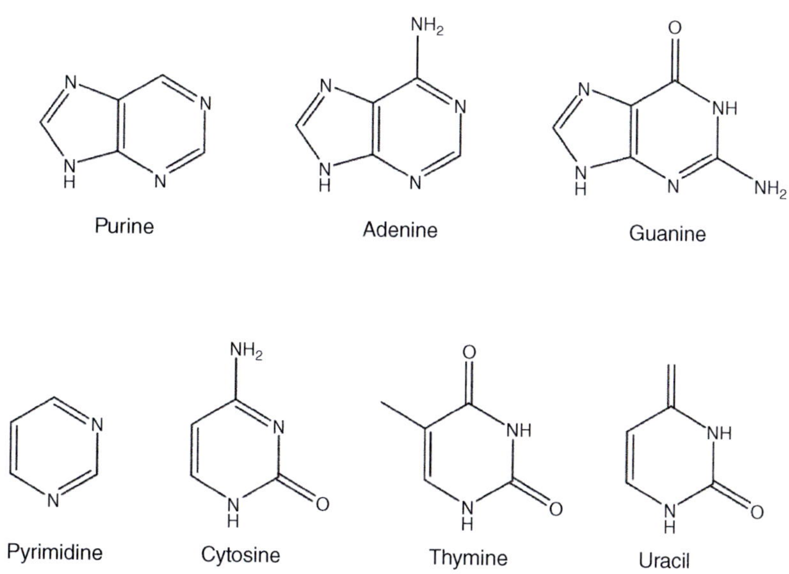


**Figure 8.1** Structures of purine, pyrimidine, and their derivatives, which function as the bases that are used to build nucleic acids.

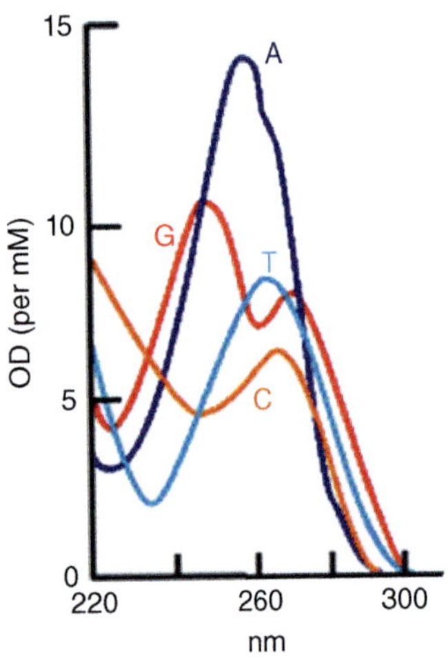


**Figure 8.2** Absorption spectra for adenine (A), guanine (G), cytosine (C), and thymine (T). *Source:* Image from http://www.scienceisart.com/A_DNA/UVspectrum_2.html.

of (6-4)PP [3]. UV irradiation of adjacent pyrimidines at moderate doses results in CPD:(6-4)PP formation at roughly a 9 : 1 molar ratio. Continued irradiation causes the splitting of the pyrimidine dimer by photolysis; therefore, as dose increases, the ratio of CPD:(6-4)PP tends to decrease.

CPDs prevent replication and transcription in cells [4], and as such their formation in microbial cells and viruses represents a key to UV-based disinfection processes. In more general terms, the formation of CPDs and (6-4)PPs in cells is linked to cellular mutation. In mammalian cells, these products are also linked to carcinogenesis [1, 2].

Under conditions in which external stresses are applied, including nutrient limitations (i.e. starvation conditions), some bacteria have the ability to generate a resting stage, known as an endospore (often shortened to "spore"). In the context of UV disinfection processes, spore formation is relevant in that the low water content and altered conformation of spore DNA causes changes in its response to UV irradiation, relative to the DNA of vegetative bacteria from which they were generated [5]. In particular, spores tend to show relatively inefficient formation of CPD and (6-4)PP; instead, the compound 5-thyminyl-5,6-dihydrothymine (*aka* spore photoproduct, abbreviated SP) is formed in its place [6]. SP formation in spores is roughly as efficient as CPD formation in the vegetative cells of the same bacterium; however, the SP-repair process is quite efficient [5]. This behavior leads to relatively high UV-resistance among bacterial spores.

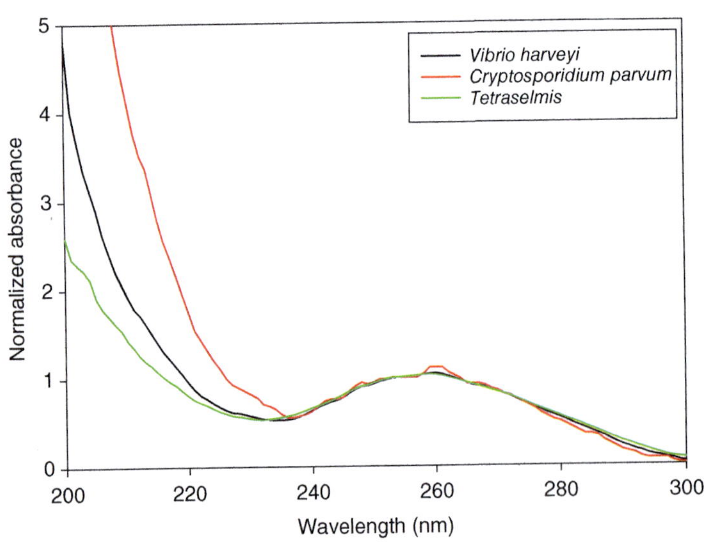


**Figure 8.3** Normalized absorption spectra for DNA extracted from a bacterium (*V. harveyi*), a protozoan parasite (*C. parvum*), and an algal population (*Tetraselmis*). All spectra were normalized to their measured absorption values at $\lambda$ = 254 nm.

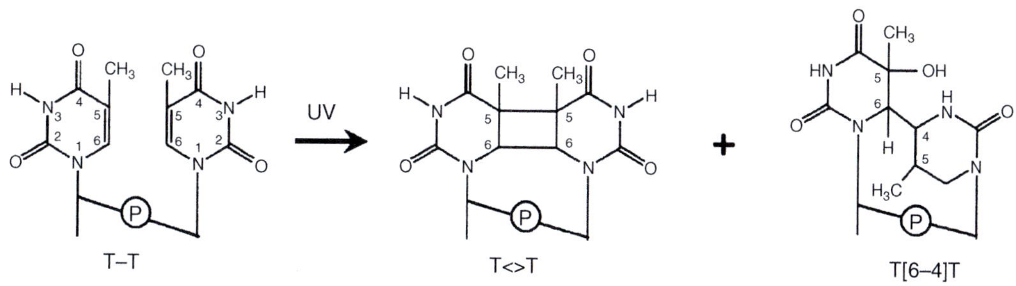


**Figure 8.4** Dominant photoproducts that result from UVC or UVB irradiation of adjacent thymine (T) residues on a DNA strand. The two major products of this reaction are a cyclobutane pyrimidine dimer (middle) and pyrimidine-(6-4)-pyrimidone photoproduct (right). Analogous photoproducts will be generated by other adjacent pyrimidine residue pairs, including T–C and C–T. *Source:* Sancar 2003/American Chemical Society.

**Table 8.1** Summary of culturability and respiration rate responses to $UV_{254}$ irradiation from a collimated beam. Reported culturability ($N$) and respiration rate ($R$) responses are normalized with respect to their values for nonirradiated cultures ($N_0$ and $R_0$, respectively).

| Bacterium | $UV_{254}$ dose $\left(\frac{mJ}{cm^2}\right)$ | $N/N_0$ | $R/R_0$ |
|---|---|---|---|
| *E. coli* | 100 | $10^{-6.8}$ | 0.23 |
| *P. aeruginosa* | 50 | N/A | 0.26 |
| *S. faecalis* | 70 | $10^{-6.0}$ | 0.25 |

*Source:* Data from [9].

Damage to proteins may also play roles in inactivation of UV-exposed microorganisms. Among viruses, there is evidence to indicate that protein damage is of particular relevance when the source of UV radiation applied in the process involves wavelengths of radiation less than approximately 240 nm [7, 8].

In a broad sense, UV-induced damage to nucleic acids and proteins interferes with replication of microorganisms. Interestingly, the doses of UV radiation that prevent microbial replication are generally insufficient to "kill" these organisms; exposed microorganisms may lose the ability to reproduce, but they often retain one or more other physiological capabilities. As an example, consider the data presented in Table 8.1, which summarize responses of laboratory-grown cultures of bacteria to $UV_{254}$ irradiation under a collimated beam. For the three species of bacteria investigated, $UV_{254}$ doses that achieved 6–7 $\log_{10}$ units of reduction in the concentration of viable vegetative bacteria yielded respiration reductions of only about 75% (i.e. roughly 0.6 $\log_{10}$ unit reduction). These vast differences in culturability and respiration indicate that bacterial reproduction is easily disrupted by UVC radiation, but most bacteria retain some basic metabolic functions after UVC exposure, and are probably still "alive" [9].

## 8.3 Reproductive Cycles of Common Microbial Groups

The central goal of disinfection processes is to minimize the risk of disease transmission associated with exposure to microbial pathogens. In related antimicrobial applications, the focus may be on nonpathogenic organisms that present a non-disease threat to human health or the environment.

For example, in ballast water treatment, the goal is to reduce the likelihood of invasions by nonindigenous species.

Regardless of the involvement of pathogenic or nonpathogenic organisms, one of the most common and effective strategies for reducing the likelihood of adverse effects associated with microorganisms is to subject the microbial population to a form of external stress that reduces or eliminates their ability to reproduce. In the case of microbial pathogens, when they are unable to reproduce, they cannot cause infection. Continuing with the ballast water example, if microbes contained within ballast water are unable to reproduce, then their ability to promote an invasion of water to which they are nonindigenous is greatly reduced. Obviously, a dead microorganism will not be able to reproduce, and as such microbial death is an acceptable endpoint (though often difficult to prove). On the other hand, microbial death is not a requirement for disinfection. Rather, the goal is prevention of microbial replication.

Therefore, analytical methods that are used to assess the effectiveness of UV disinfection processes should address microbial reproduction and the degree to which UV irradiation disrupts these processes. With that in mind, brief summaries of the fundamental reproductive processes that are used by major groups of microorganisms are presented below.

### 8.3.1 Bacterial Reproduction

Bacterial reproduction involves a process known as binary fission, in which a single bacterial cell divides to form two identical daughter cells. The basic steps involved in bacterial fission are illustrated in Figure 8.5.

The first step in this process involves duplication of the bacterial genome (i.e. DNA replication). This is followed by growth of the cell wall to allow elongation of the cell. Cell wall growth continues and then involves the formation of an invaginated section to yield the formation of a septum. The

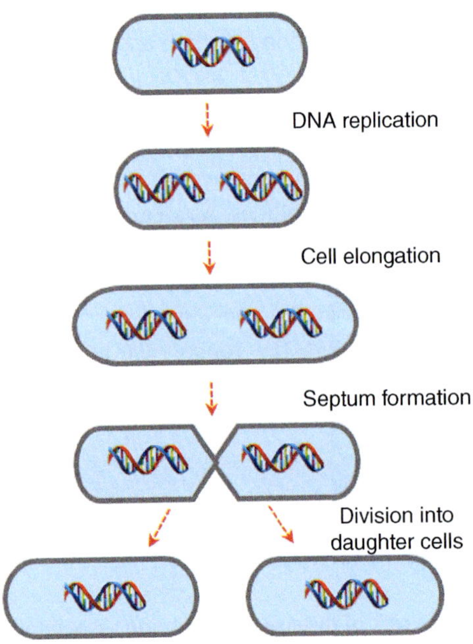


**Figure 8.5** Steps involved in bacterial reproduction by fission. *Source:* http://biology-pictures.blogspot.com/2011/11/binary-fission-in-bacteria.html.

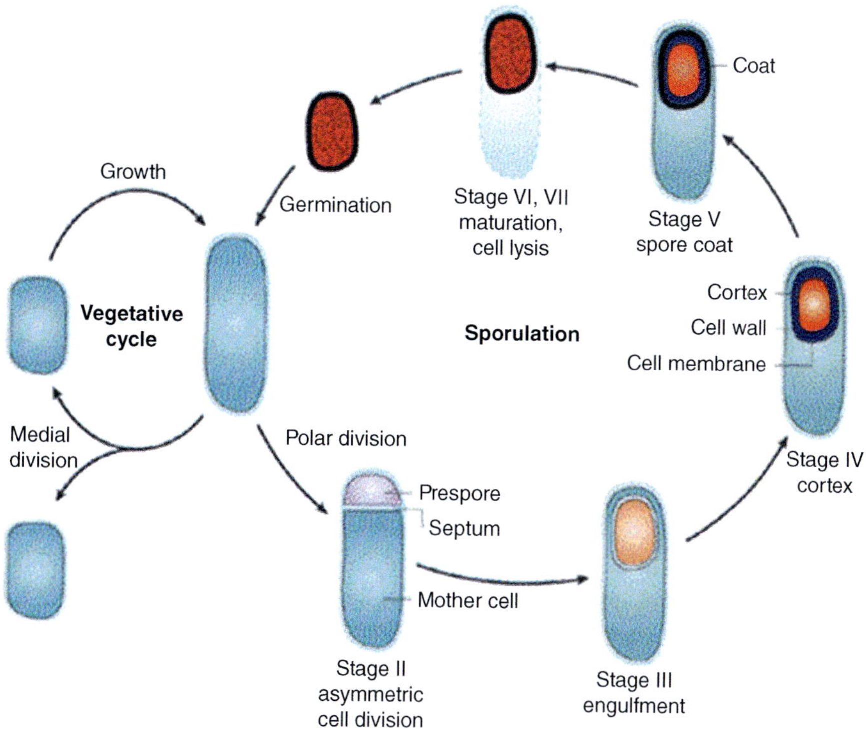


**Figure 8.6** Steps involved in spore formation within a bacterial cell. *Source:* Errington 2003/with permission of Springer Nature. Image from http://www.devbio.biology.gatech.edu/unicellular-model-organisms/bacillus-subtilis-sporulation/.

two halves of the elongated cell are then split to form daughter cells, which share the same DNA as the parent cell.

Bacterial growth by fission is generally a rapid process, with "doubling time" (*aka* generation time) ranging from 10 minutes to weeks, depending on the bacterium and the conditions of growth [10]. The ability of bacterial populations to grow quickly is a key factor in governing the effects of bacterial populations on human health (i.e. with pathogenic bacteria) and in nature.

Under conditions of stress (e.g. starvation), some bacteria have the ability to form a resting stage known as a spore. The steps involved in spore formation are illustrated in Figure 8.6. The spores formed in this process represent a form of the bacterium that is able to germinate to yield a vegetative cell. Germination takes place when environmental conditions are favorable to growth of the vegetative cell.

Spores are dormant and tend to be quite resistant to most forms of external stress. Spores are known to be formed by bacteria from the genera *Bacillus* and *Clostridium*, and have been shown to be 10–50 times more resistant to $UV_{254}$ exposure than their corresponding vegetative cells [11]. Spores do not grow (growth only happens upon germination), but they can remain viable for many years, in some cases as long as thousands of years [10, 12].

### 8.3.2 Replication of Viruses

All viruses are obligate parasites, which means that they cannot reproduce without exploiting a host (cell). Viruses have been defined as noncellular genetic elements that use host cells for replication [10]. Viruses can exist as an extracellular form called a virus particle or virion, which includes one (or sometimes more than one) nucleic acid molecule, proteins, and sometimes additional components. Virions are non-metabolically active and they do not participate in respiration or biosynthetic processes. However, virions provide a mechanism for transfer of viral genetic information from one cell to another.

As a family of microorganisms, viruses are able to infect virtually any type of cell, including animal (including human) cells, bacterial cells, and algae. In historical terms, the relative ease of growing bacteria facilitated the study of bacterial viruses (*aka*, bacteriophage or phage), but improvements in methods of cell culture for animal cells and the emergence of the tools of molecular biology have allowed for dramatic improvements in our collective understanding of animal viruses as well.

The cycles of reproduction for animal and bacterial viruses tend to involve similar steps. Figure 8.7 illustrates the basic steps involved in viral reproduction using a bacteriophage as an example. The process starts with attachment of a virion to a host cell. For any virus, the ability to attach is governed by the surface chemistry of the virion and the host cell, and tends to be highly cell-specific. This is followed by penetration of the virion or its genome into the host cell. The virus then directs the host cell to use its own biosynthetic machinery to replicate viral nucleic acids and proteins. These biomolecules are then assembled into new virions within the host cell. For many virus/host cell combinations, this is followed by lysis of the host cell to release the newly formed virions, which then have the potential to infect additional cells, thereby continuing this reproductive cycle and often killing the host cell.

### 8.3.3 Life Cycle of Protozoa

Although not all protozoa are parasitic, the protozoan organisms that are of greatest interest in the water treatment industry are parasites. A number of protozoan parasites have been identified that

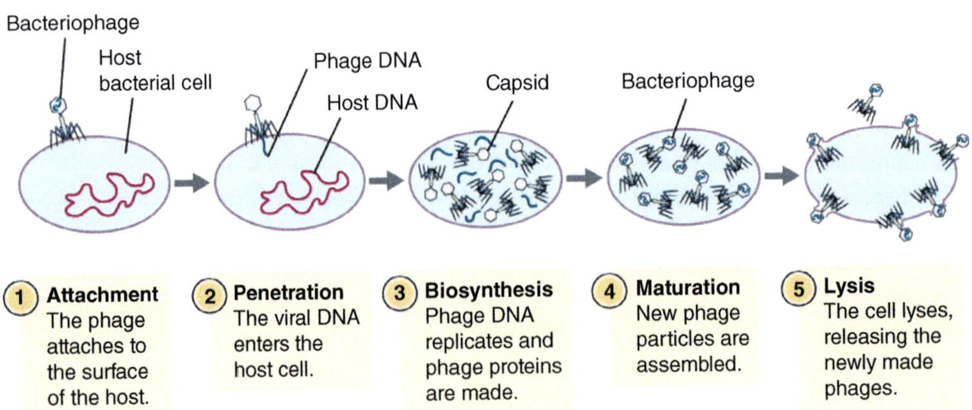


**Figure 8.7** Schematic illustration of steps involved in viral reproduction, using a phage and bacterial host as an example. *Source:* Image from https://courses.lumenlearning.com/microbiology/chapter/the-viral-life-cycle/.

play roles in the public health sector, but the genera *Cryptosporidium* and *Giardia* are probably the most important in water treatment, especially in the context of disinfection processes, including those based on UV radiation.

The reproductive cycles of *Cryptosporidium* and *Giardia* involving human hosts are similar, and are illustrated in Figures 8.8 and 8.9, respectively. In both cases, the parasitic forms of these protozoa are excreted by an infected host, either as the native organism or as a resting stage. From the perspective of disinfection, the resting stages of *Cryptosporidium* and *Giardia* (known as oocysts

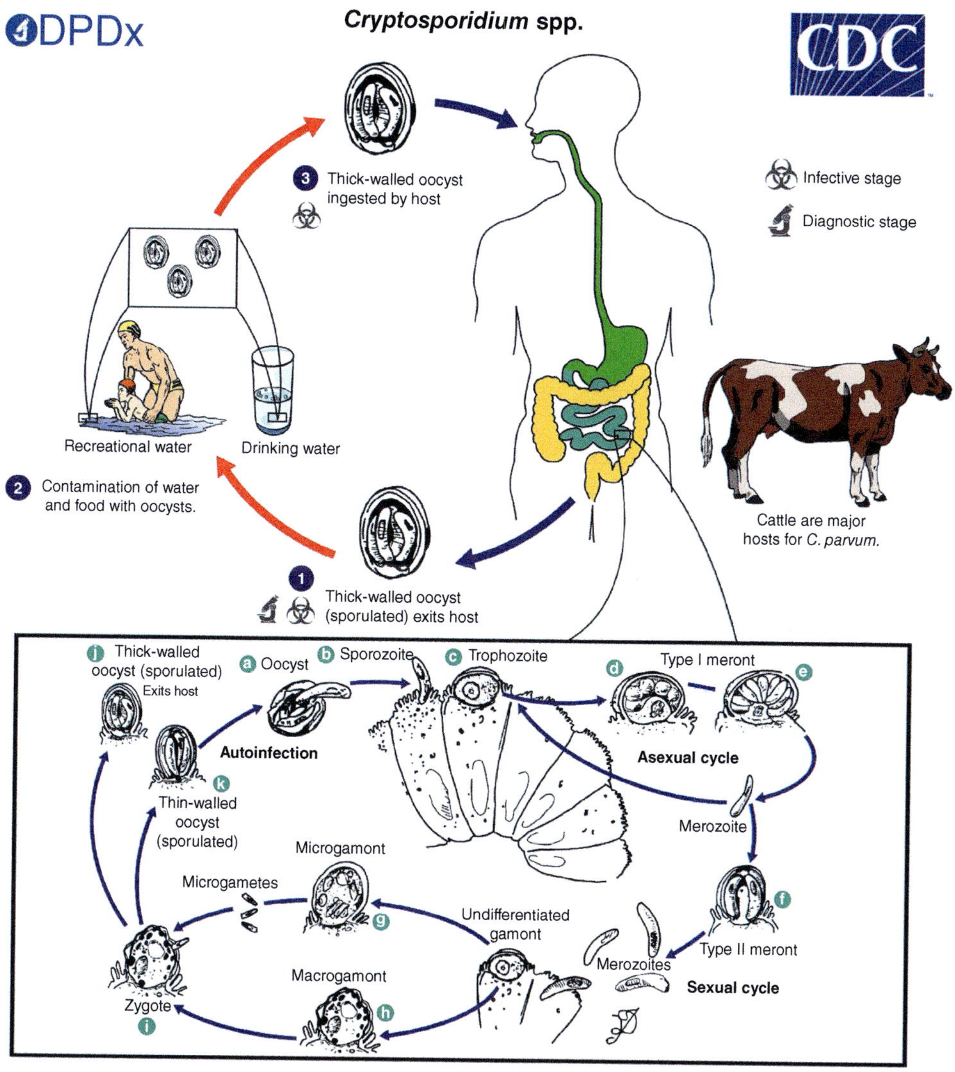


**Figure 8.8** Reproductive cycle of *Cryptosporidium* in humans. Sporulated oocysts, each containing four sporozoites, are excreted by an infected host. Transmission of *Cryptosporidium* often occurs through contact with contaminated water. After entering the host, excystation occurs causing release of the sporozoites, which are then able to attach to and parasitize epithelial cells of the gastrointestinal tract or other tissues. The parasites can then reproduce asexually or sexually to produce microgamonts (male) and macrogamonts (female). After fertilization, oocysts can sporulate in the infected host. *Source:* Image from https://www.cdc.gov/dpdx/cryptosporidiosis/index.html.

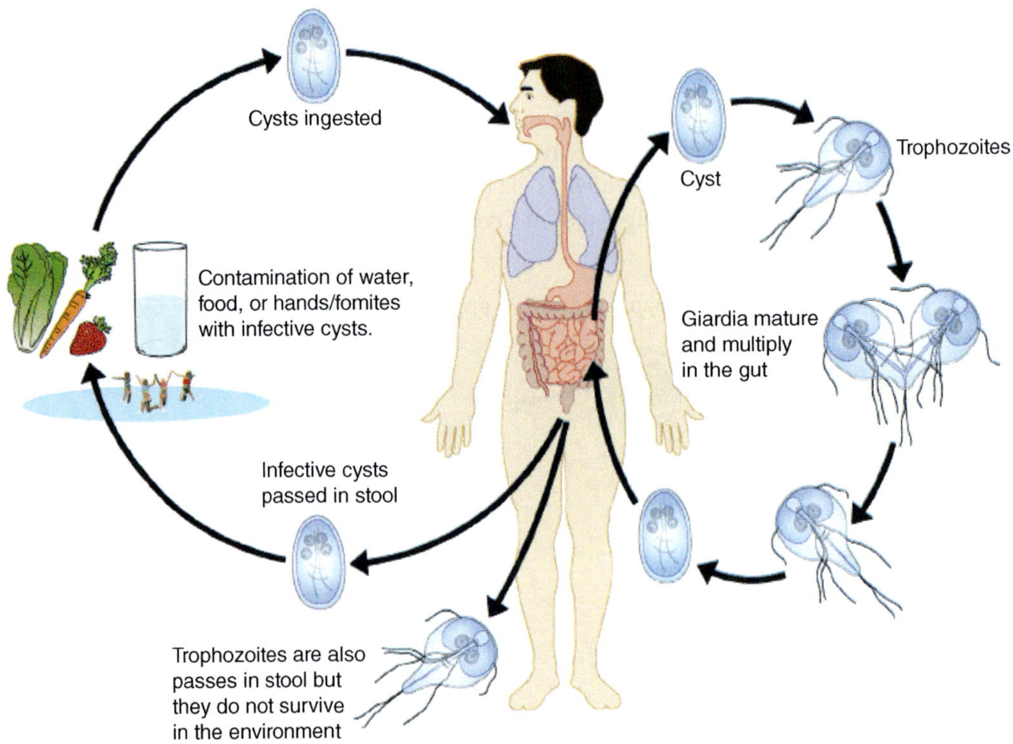


**Figure 8.9** Reproductive cycle of Giardia in humans. Cysts and trophozoites are found in fecal material; fecal excretion provides a pathway for transmission. Ingestion of water containing cysts or trophozoites allows the infective forms to enter the digestive system. Trophozoites reproduce in the small intestine by binary fission. Cyst formation tends to occur as parasites travel toward the colon. Fecal excretion by the infected host allows this cycle to continue. *Source:* Image from: https://www.cdc.gov/parasites/giardia/pathogen.html.

and cysts, respectively) are perhaps most relevant, in that they are hardier and more resistant to external stresses, including disinfectants, than their parent forms. Reproduction of the parasites takes place by asexual and/or sexual processes. The cycle of infection and reproduction continues with fecal excretion.

### 8.3.4 Reproduction of Algae

Algae are eukaryotic organisms that carry out photosynthesis.[b] Algae span a wide range of sizes and morphologies [10]. Algae are critical life forms in aquatic environments, but generally do not function as pathogens toward humans. However, algae (or more generally phytoplankton) are quite important in the design and operation of ballast water treatment systems.

Some algal species reproduce by binary fission (similar to bacteria), while others reproduce via sexual processes. For single-celled algae, the rates of cell division can be similar to those of bacteria. As such, algal populations often demonstrate strongly dynamic behavior in response to changes in water quality, related to temperature, nutrient concentration, pH, and other factors. Conversely, changes in algal populations can also have dramatic effects on water quality, in terms of turbidity, cycling of nutrients, and other factors.

## 8.4 Lessons Learned from use of Inappropriate Methods

Analytical methods that are used to quantify the responses of target organisms to disinfectants, such as UV radiation, need to be consistent with the mechanism(s) of inactivation as well as the endpoint of treatment. In the case of microorganisms that are subjected to UV radiation for purposes of causing their inactivation, there are many examples of the use of assays that have failed to meet one or both of these criteria. The text that follows provides brief descriptions of these assays and the attributes of each that rendered them inappropriate for their intended use.

### 8.4.1 Protozoan Parasites

A wide range of protozoa function as parasites in humans and other mammals. Among them, the organisms belonging to the genera *Cryptosporidium* and *Giardia* have attracted the most attention in water treatment, as they are responsible for a large fraction of diagnosed disease outbreaks associated with drinking water and recreational water. As related to assessments of UV dose–response behavior, *Cryptosporidium* and *Giardia* have also been at the center of the development of analytical methods for quantification of protozoan UV dose–response behavior.

Until the late 1990s, the collective opinion among scientists and engineers who work on water disinfection processes was that UV-based processes were largely ineffective for inactivation of these protozoan parasites. Studies that had been conducted to that date had indicated remarkable resistance on the part of *Cryptosporidum* and *Giardia* to UVC exposure [13]. However, these studies were based largely on *in-vitro* viability assays, including tests involving the use of vital dyes and enhanced excystation.

A common "vital dye" assay involved co-staining of organisms with 4′,6-diamidino-2-phenylindole (DAPI) and propidium iodide (PI) [14]. These dyes are used to provide a measure of the permeability and integrity of the (oo)cyst wall and cytoplasmic membrane. This method is based on the assumptions that intact membranes are permeable to DAPI but impermeable to PI. It was also assumed that (oo)cyst wall integrity correlates strongly with infectivity. If these assumptions apply, then viable (infective) (oo)cysts will selectively accumulate DAPI, while non-infective (i.e. disinfectant-damaged) (oo)cysts will non-selectively accumulate both dyes. Subsequent microscopic examination of the exposed organisms was then used to discriminate "viable" and "nonviable" (oo)cysts. This method was demonstrated to yield results that were strongly correlated with results of *in-vitro* excystation (described below) for environmental samples (i.e. water samples that had not been exposed to any form of water treatment, including disinfection) [15].

As illustrated in Figure 8.8, a critical step in the life/infection cycle of *Cryptosporidium* involves excystation, in which the infective stage of the organism is released from its resting stage. A similar process is illustrated in Figure 8.9 for *Giardia*. Based on knowledge of the importance of this step to the life/infection cycle of these parasites, an analytical method known as *in-vitro* excystation was developed in which the chemistry of the digestive system of the host is mimicked *in-vitro*. Specifically, the target organisms are subjected to conditions of low pH, high $CO_2$ concentration, inclusion of critical enzymes, and elevated temperature. These conditions prevail in mammalian digestive systems and are known to promote excystation. Subsequent microscopic examination of the sample is then used to count the number of excysted and intact (oo)cysts. The idea of this method is that an organism that is unable to complete the excystation process will not be able to infect its host. *In-vitro* excystation had been used successfully to quantify the responses of *Cryptosporidium* to chemical disinfectants, including free chlorine, combined chlorine, ozone, and combinations of these disinfectants [16, 17]. However, the mechanisms of protozoan (oo)cyst inactivation by these chemical disinfectants are fundamentally different than those that govern UV disinfection processes.

In the late 1990s, several groups worked independently to demonstrate that these indirect *in-vitro* methods provided incorrect, misleading information regarding the UV dose–response behavior of *Cryptosporidium* and *Giardia* [14, 18–23]. As alternatives to *in-vitro* assays, methods were developed that involved measurement and confirmation of actual infection (i.e. completion of the entire infection cycle) in mammalian hosts (mice or gerbils) or in tissue culture of host epithelial cells. In general terms, the *in-vitro* methods (i.e. vital dyes and enhanced excystation) indicated minimal inactivation to result from UVC exposure, whereas the methods based on infection of mammalian hosts or epithelial cells grown in tissue culture indicated that these protozoan parasites were quite sensitive to UVC irradiation. In fact, when evaluated using assays that measure actual infection, *Cryptosporidium* and *Giardia* were observed to be roughly as sensitive to UVC radiation as common vegetative bacterial cells. UV-based processes are now the disinfection processes of choice for water when contamination by *Cryptosporidium* and/or *Giardia* is possible.

### 8.4.2 Fish Parasites

*Myxobolus cerebralis* is parasitic toward salmonid fish and is believed to have been introduced from Europe to North American fisheries [24]. It causes whirling disease among salmon and trout, and represents a threat to both wild and hatchery-reared fish. In fishery applications, UV-based treatment can be advantageous to chemical disinfection processes in that there is generally a smaller potential to impart chemical and toxicological change on the water as a result of disinfection. As such, there is interest in understanding the effects of UV-based treatment on fish parasites.

As an initial investigation into the potential for UV to control *M. cerebralis*, the vital stains PI and fluorescein diacetate (FDA) were applied to quantify the responses of this parasite to UVC exposure using a collimated beam and a low-pressure Hg lamp [25]. The results of this investigation suggested that a $UV_{254}$ dose of 1300 $\frac{mJ}{cm^2}$ was required to accomplish effective inactivation of this parasite. However, following lessons learned from examination of the behavior of *Cryptosporidium* and *Giardia*, a second set of experiments was conducted in which actual infection of juvenile rainbow trout was used as the infectivity endpoint of the assay [26]. The results of these tests indicated that a $UV_{254}$ dose of 40 $\frac{mJ}{cm^2}$ was effective for prevention of infection. This was followed by experiments that demonstrated that low doses of $UV_{254}$ radiation allowed the infective stage (triactinomyxon) to invade their trout hosts; however, the UV-irradiated parasite was unable to complete its infective cycle. Interestingly, exposure of trout to $UV_{254}$-irradiated triactinomyxons resulted in immunity of the fish to subsequent exposure to raw (unirradiated) triactinomyxons. Apparently, the $UV_{254}$-irradiated triactinomyxons were able to initiate the infection process in trout hosts. The damage caused within these parasites as a result of $UV_{254}$ exposure prevented them from completing their life/infection cycle in their hosts, but also promoted an immune response in the affected fish, such that they were able to remain unaffected by subsequent exposure to raw triactinomyxons. It is not uncommon for UVC irradiation to be applied in the preparation of vaccines [27–29]. However, it is unclear if analogous immunization benefits will translate to conventional applications of UV-based treatment, such as disinfection of water and wastewater.

### 8.4.3 Algae

Algae comprise a diverse group of organisms that range from unicellular, microscopic organisms to colonial forms that can be more than 30 m long [10]. All algae contain chlorophyll, which they use to carry out oxygenic photosynthesis. Some algae contain other pigments as well. Algae can have profound effects on water quality and can grow quite quickly under favorable conditions.

The treatment of ballast water represents a prominent example of a water treatment application in which UVC irradiation is used for control of algae. In fact, regulations promulgated by the International Maritime Organization (IMO) require that the concentration of *viable* organisms in the (nominal) size range of 10–50 μm be limited to less than 10 viable organisms per mL [30]. This size range is often dominated by algal cells.

Ballast water management regulations have also been passed by individual nations. As an example, the U.S. Coast Guard (USCG) has responsibility for ballast water management in US waters. The USCG has implemented regulations for ships in US waters that are numerically identical to the IMO limits, but stipulate less than 10 *live* organisms per mL, as opposed to 10 *viable* organisms per mL [31].

The distinction between the endpoints *viable* and *live* is critical with respect to the analytical methods that are applied in measurement of the responses of ambient organisms to a ballast water treatment strategy. The word *viable* is defined by the IMO as an organism that is able to reproduce. As such, the assays used to measure viable organisms should provide a measure of the ability of algal cells to reproduce. Evidence of "life" among plankton includes the presence of intact membranes, nucleotide functionality, metabolic competence, or motility [32–35]. By extension, these terms suggest the application of assays that are used to measure these physiological functions among algal populations.

These two endpoints are directly related to analytical methods that are commonly used in the Type Approval testing process that is used to validate and approve treatment technologies for compliance with ballast water management regulations. In particular, the Most Probable Number – Dilution Culture Method (MPN-DCM) is used to quantify viability responses, while differentiation between "live" and "dead" algal cells is often based on the vital dye assay using FDA, sometimes together with 5-chloromethylfluorescein diacetate (CMFDA) [35, 36].

The MPN-DCM is used for estimation of the concentration of viable phytoplankton in a sample based on the following steps: (i) serial dilution of a sample into a series of replicated liquid subcultures, (ii) incubation of the subcultures under favorable conditions, in terms of light, temperature, and nutrients, (iii) scoring each subculture for the presence of one or more viable organisms based on the detection of growth, i.e. accumulation of cells, (iv) probability-based estimation of the most likely number of viable organisms in the original sample, and (v) calculation of concentration in the original sample from that number and dilution of the sample. Although there are important limitations of the MPN-DCM, its critical feature is that the assay itself depends on reproduction of the target organisms during incubation.

FDA is a fluorescein molecule to which two acetate groups are attached. The lipophilic, nonfluorescent FDA molecule readily penetrates the plasma membranes of cells [36]. Once inside the cell, the acetate groups are cleaved by nonspecific esterases to yield a green-fluorescing, hydrophilic fluorescein molecule that is retained by cells with a healthy membrane. CMFDA is mildly thiol reactive and has been reported to yield better cellular retention than FDA. The combined use of FDA + CMFDA has been adopted as a standard method for differentiation of "live/dead" status among phytoplankton cells in ballast water testing.

When applied for testing of UV-based treatment in ballast water testing, the MPN-DCM and FDA/CMFDA assays provide remarkably different measurements of performance. Recent comparisons of these two methods, when applied to the same samples, indicated large differences in phytoplankton responses, with the MPN-DCM assays indicating 1–3 $\log_{10}$ units better inactivation than the FDA + CMFDA assays. In practice, this translates to system sizing requirements that could be beyond the range that is practically applicable if FDA+CMFDA results are applied, but are cost competitive or even favorable if MPN-DCM is applied [37].

### 8.4.4 Viruses

Viruses are simple organisms, often comprising a group of proteins in an external structure known as a capsid, which surrounds one or more nucleic acid molecules. As described previously, viruses are obligate parasites, which means that they require a host (cell) to complete their reproductive cycle. Within the community of microbiologists, viruses are often described as not being "alive," in the sense that they do not participate in respiration or metabolism. However, they can have profound effects on host cells and the organs of affected organisms.

Viral concentrations can be measured by microscopy, but this is a time-consuming process that does not provide information about the ability of a virion to infect. Because of this, it is common to assess the concentration of virus particles using assays that require the infection cycle to be completed. The most common of these methods involve growth of the host cell, with exposure to the virus sample and incubation under conditions that favor growth of the host cell and replication of the virus.

The relative simplicity of virus particles has promoted the use of the tools of molecular biology in the development of methods to measure virus concentrations. Among these methods, variations of the polymerase chain reaction (PCR) are possibly the most important and widely used methods.

PCR is a method whereby a segment of nucleic acid is amplified, yielding a large number of copies of the target segment. When applied to viruses, a small segment of the nucleic acid is selected as the target. Measurements of viral inactivation by PCR can yield substantial underestimates of viral inactivation by UV, if the target segment is not critical for virus reproduction or nonrepresentative of the nucleic acid molecule it is derived from. In these cases, damage to the target segment may not correlate well with actual changes in virus infectivity [38].

Although conventional PCR methods can yield misleading results, recent and emerging methods based on PCR and its derivatives have allowed for improvements in the ability of these methods to provide meaningful measurements of virus inactivation by UV irradiation. Some of these new methods will be presented (briefly) at the end of this chapter.

## 8.5 Quantification of Viable Microorganisms with UV Disinfection Systems

UV irradiation is able to induce damage within critical biomolecules. As described previously, much of this damage is observed within the nucleic acids (DNA and/or RNA); damage to these molecules can prevent replication of microorganisms, but may not "kill" them. In addition, UV irradiation can cause damage to proteins, which appears to be particularly important for viruses. Protein damage can prevent viruses from completing their reproductive cycle [39].

In general terms, UV irradiation is effective as a disinfection process because the damage caused by UV exposure prevents reproduction. As such, any assay used to quantify the effects of UV irradiation should be based on measurements that confirm completion of the reproductive cycle.

In the absence of consideration of ethical issues, the ideal test for measurement of the effectiveness of UV disinfection (and for that matter any other disinfection process) would involve measurements of the ability of such processes to limit disease transmission (i.e. infection) in humans. However, for obvious reasons, such tests are not conducted and are inappropriate. Alternatives to human infection studies include assays that involve growth of target bacteria, or in the case of parasitic pathogens, infection of animal hosts or host cells grown in tissue culture. The use of animal models and tissue culture with parasites is common and provides the advantage of allowing completion of the life/infection cycle of the target pathogen in a relevant host. Animal models provide the benefit of incorporating interactions among organs and tissues within the host, while tissue culture methods are often faster and simpler to conduct.

For the time being, tests that are based on growth (culturing or completion of the reproductive cycle) for target microbes represent the "gold standard" for quantification of the effects of disinfectants, including UV. These tests demonstrate the ability of the target microbe to reproduce and complete its life cycle. The logic supporting the identification of these tests as the "gold standards" is that if these organisms are unable to complete their life cycle (i.e. reproduce), they will not be able to cause infection.

The text below provides brief summaries of the analytical methods that are appropriate and commonly used for quantitative assessments of microbial inactivation by UV irradiation.

### 8.5.1 Bacteria

Historically, the most commonly applied methods for quantification of bacteria have involved assays that are based on bacterial growth (i.e. reproduction). These include conventional plating methods and other *in-vitro* culturing methods. Because these methods provide measures of reproduction and are commonly applied, there is relatively little debate about the appropriateness of these methods for quantification of viable bacterial concentration before or after UV irradiation.

There are many variations on bacterial culturing methods for quantification of bacteria in aqueous samples and other media. Brief summaries of three such methods are provided below; additional details of these methods can be found in the original references that describe these methods.

#### 8.5.1.1 MPN-DCM

Although a number of different methods have been adapted based on the MPN concept, they share several common features and are based on a set of fundamental assumptions [40]:

1) The original sample is well mixed. The well-mixed assumption implies uniform composition of the sample throughout its volume. From the perspective of probability theory, which is applied in estimation of the MPN, this also implies a random distribution of viable organisms within the original sample and sample dilutions.
2) Growth will be observed if one or more viable organisms are present. One of the techniques used in MPN-based methods is serial (often decadic) dilution of the original sample to get to the point where diluted subsamples are expected to contain one or fewer viable microbes. This assumption also implies that after a period of incubation, under conditions that are likely to promote growth of the target microbe, it will be possible to unambiguously discriminate between tubes in which growth is observed and tubes in which no growth is observed.

3) The target organisms are independent. Among other things, this implies that no clumping or agglomeration of target organisms is evident in the sample.

Figure 8.10 provides a schematic illustration of the MPN assay, as applied for estimation of "bacterial density" in a water sample.[c] Serial dilution is applied to an original sample. Subsamples of the original sample are distributed to culture tubes that contain a known volume of "dilution water," which should contain all nutrients needed for growth of the target bacterium. The process of serial dilution is continued until there is confidence that the last set of dilution tubes contains no

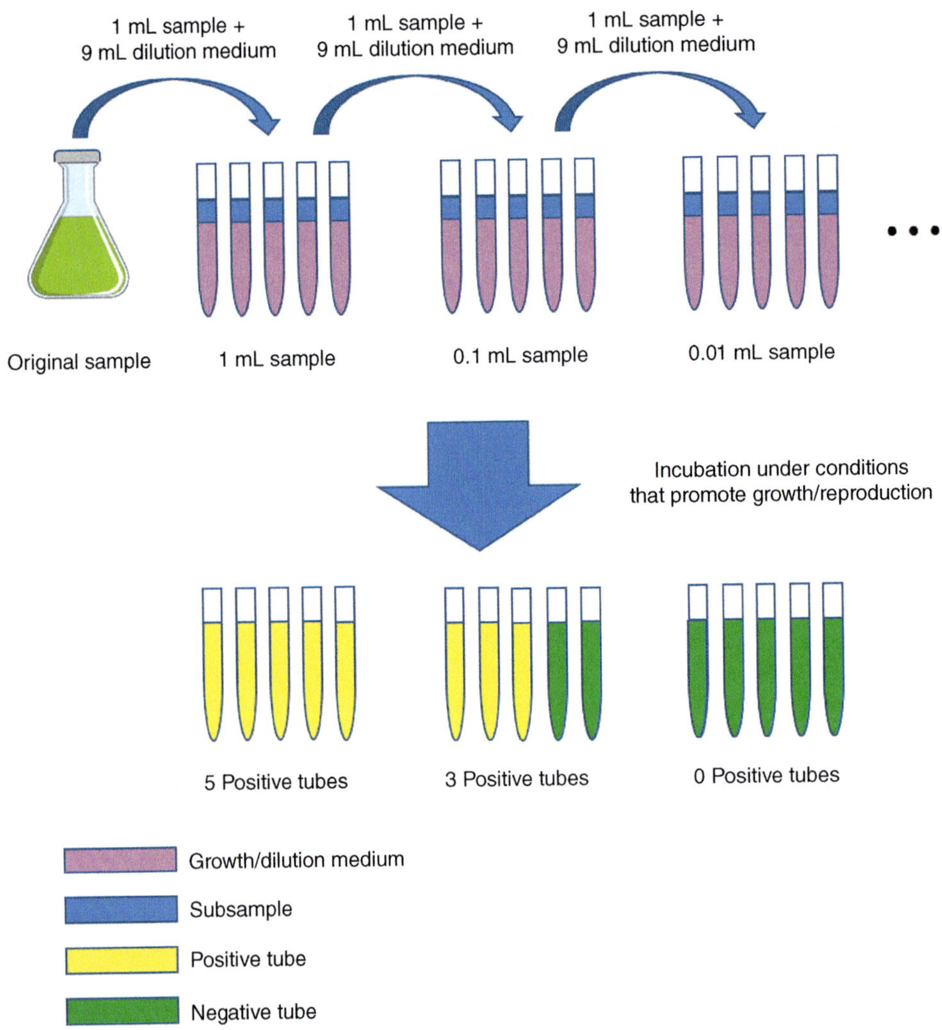


**Figure 8.10** Schematic illustration of MPN-DCM for estimation of the concentration of viable bacteria in a sample. Subsamples of an original sample are distributed among culture tubes that contain a growth medium. Serial (decadic) dilution will continue until it is likely that the most dilute sample is likely to contain no viable bacteria. All diluted subsamples are then incubated under conditions that favor growth of the target bacterium. After incubation, all tubes are "scored" as being positive for growth or negative for growth. The concentration of viable bacteria present in the original sample is estimated based on a statistical estimate of the "most probable number" (MPN) that would yield the pattern of positive and negative tubes, together with data regarding dilution that was applied to the sample.

viable bacteria. The dilution tubes are then incubated under conditions that favor growth of the target bacterium. At the end of the incubation period, each tube is assigned a binary score of positive for growth or negative for growth. This scoring process requires that some aspect of bacterial growth (or a consequence of growth) is used to provide a clear distinction between tubes that are positive and negative for growth. This can be done by production of gas, a colorimetric change, fluorescence, or other signals.

In most applications of MPN-based methods, replicate tubes are prepared at each dilution. The fraction of positive tubes observed among several dilutions is then used to estimate the "most probable number" of microorganisms that would yield the pattern of positive/negative tubes across the dilution series. The concentration in the original sample is then calculated by knowledge of the dilutions that achieved this distribution of scores.

The mathematical details of the MPN method are based on fundamental probability theory. Details of the derivation of these estimates are available in a number of publications [40, 41]. Tabular summaries of MPN estimates, along with corresponding confidence intervals, are also available and are commonly used.

MPN-based methods have a long history of application in analysis of bacterial quality of water samples [42]. The method provides statistically valid, robust estimates of the concentration of viable bacteria in a sample. The method is based on growth (i.e. reproduction) of the target bacterium. However, the method is time-consuming and can involve relatively large amounts of glassware or disposable supplies.

#### 8.5.1.2 Membrane Filtration

Membrane filters can be produced with pore sizes that are highly uniform and evenly distributed across the membrane surface. Moreover, pore sizes for these membranes are commonly in a size range that allows physical separation of bacteria from water. Common pore sizes used to accomplish bacterial separation from water are 0.20–0.45 μm.

Figure 8.11 provides a schematic illustration of the membrane filtration procedure. A water sample is drawn through a sterile membrane by application of differential pressure across the membrane; the pressure differential is often imposed by application of vacuum to the side of the membrane opposite from the sample. This process separates bacteria (and other particles of similar or larger size) in the sample from the water. The filter is then transferred to a sterile Petri dish that contains a growth medium that is specific for the target bacterium. In many membrane filtration methods, the growth medium will be augmented with a chromogenic indicator that yields a specific color or fluorescent character among colonies of the target bacterium that grow during incubation. The Petri dish is then covered, inverted, and transferred to an incubator which is used to maintain conditions (temperature and atmospheric composition) that favor growth of the target bacterium.

A fundamental assumption of membrane filtration assays is that each viable bacterial cell will be able to grow to yield a visible colony after an appropriate period of incubation. Conversely, it is assumed that each visible colony is attributable to one viable bacterial cell (i.e. the effects of bacterial clumping are ignored).

The membrane filters used in these tests are often produced with a grid to facilitate counting. After incubation, the number of colonies on each filter is counted. The concentration of viable bacteria present in the original sample is estimated as the ratio of the number of colonies to the volume of sample drawn through the filter. The standard units of viable bacterial concentration are colony forming units (cfu) per unit volume.

Membrane filtration methods can be applied for quantification of viable bacterial concentration across a broad range of water qualities. For samples that include a high concentration of viable

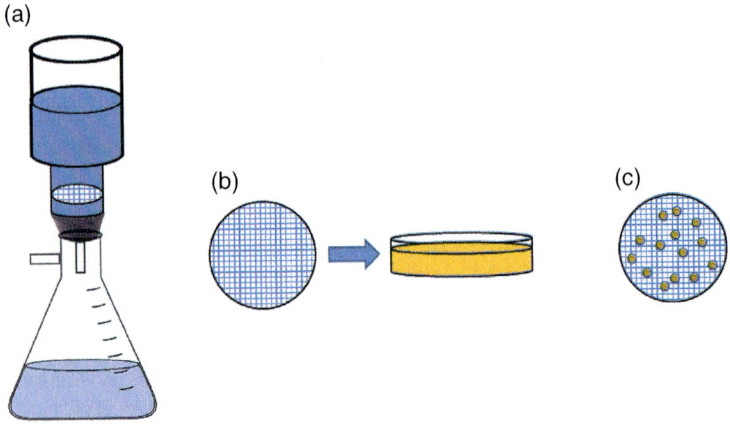


**Figure 8.11** Schematic illustration of steps involved in membrane filtration assay for quantification of the concentration of viable bacteria in a water sample. (a) A water sample of known volume is drawn through a membrane filter, usually by vacuum; the filter is selected to have a pore size that is small enough to capture the target bacteria, while allowing the liquid in the sample to pass through the membrane. (b) The membrane is removed from the filtration assembly and applied to a Petri dish that contains a growth medium that is specific for the target bacterium; the Petri dish is then covered and incubated under conditions that favor growth of the target bacterium. (c) After incubation, bacterial colonies will grow to a visible size, thereby allowing them to be counted by visual inspection. The concentration of viable bacteria in the sample is calculated as the number of colonies divided by the volume of sample drawn through the membrane filter.

bacteria, the volume of sample to be drawn through the filter can be quite small. This issue can be addressed by diluting the original sample in sterile dilution water. This allows the sample to be distributed into a relatively large volume (10–100 mL is common). A large volume of water is beneficial in this test in that it allows complete wetting of the filter and also promotes uniform distribution of the bacteria across the membrane surface.

In samples that contain low concentrations of viable bacteria, it may be necessary to draw a large volume of sample across the filter to achieve colony counts that are statistically reliable. For samples with low bacterial concentrations, this may represent an important limitation of the membrane filtration assay.

#### 8.5.1.3 Compartment Bag Test

UV-based disinfection processes have become common in developed countries. In these settings, water utilities can reliably count on access to laboratory space and equipment needed to conduct routine water quality testing, including tests aimed at defining microbial quality. However, a large fraction of the global population lacks access to potable water. Most people who lack access to safe, affordable water live in economically depressed communities or developing countries. In these settings, economic and infrastructure circumstances often make it difficult to gain access to contemporary water quality testing equipment, methods, and laboratories.

The Compartment Bag Test (CBT) was developed to meet the need for an analytical procedure that is simple, portable, and accessible to people in developing countries [43, 44]. The CBT can be conducted by people who lack formal training in microbiology, and in the absence of a laboratory or electrical power.

The CBT generally involves collection of a 100 mL sample of water from a target source. This could be raw (untreated) water or water that has been subjected to treatment, possibly including

UV disinfection. The sample is added to a bottle that contains a growth medium. The growth medium includes the compound 5-bromo-4-chloro-3-indolyl-beta-D-glucuronic acid [44]. This compound undergoes hydrolysis due to the action of the enzyme β-glucuronidase to yield a blue color; this enzyme is unique to *Escherichia coli*. The sample and growth medium are mixed for 15 minutes to promote uniform distribution of bacteria, growth medium, and the chromogenic indicator.

The sample mixture is then applied to multiple compartments of a sterile compartment bag. These compartments have known volumes; in the commercial version of the CBT assay, the compartment volumes are 1, 3, 10, 30, and 56 mL [44]. The bag is then sealed and incubated for a period of 24–48 hours; the required duration of incubation depends on ambient temperature. One of the basic ideas of this test is that it should be possible to conduct the test without laboratory equipment, including an incubator. After incubation, the compartments are "scored" as positive (blue) for growth or negative (yellow to yellow/brown) for no growth. These results are then used to calculate MPN for the sample.

The mathematical basis for the MPN calculation has its foundations in probability theory. The logic used in defining this estimate for the CBT is described in detail by Gronewold et al. [45]. Specifically, the fundamental assumptions used in MPN assays (listed previously) dictate that the Poisson distribution can be used to accurately describe the number of viable bacteria present in each compartment, based on the volume of each compartment. The joint probability of any pattern of positive/negative compartment signals can be described as the product of individual probabilities.

The CBT has been shown to yield estimates of viable *E. coli* concentration (expressed as MPN per unit volume, including a confidence interval) that are consistent with estimates developed by application of conventional membrane filtration [43, 44]. In its current form, the CBT has been developed for quantification of *E. coli* and $H_2S$-producing bacteria. However, similar principles may allow for extension of the CBT (or related methods) to be applied for in-field measurements of other microorganisms.

### 8.5.2 Viruses

The infective cycle of viruses is illustrated in Figure 8.7. Given that UV-based disinfection processes have the potential to damage the two basic biomolecules that comprise viruses, namely nucleic acids and proteins, it is not always clear how or where the infective cycle of a virus is disrupted by UV irradiation. Moreover, it is possible for UV-induced damage to interrupt this cycle at more than one location for a given virus. Therefore, it is critical that viral assays for UV-disinfected water be based on an endpoint that measures the ability of a viral suspension to complete its infective cycle.

The most common assays that are used for this purpose require that a host cell culture be grown and subjected to exposure to the virus suspension under conditions that favor viral infection. The two most common endpoints for these assays are based on the formation of viral plaques and the development of cytopathic effects (CPEs) within host cells.

#### 8.5.2.1 Plaque Formation

The classic work by Adams and collaborators [46] provides a description of the most basic viral assay based on plaque formation. The assay is known as the "agar layer method," the "double agar layer method," and other names. The basic steps involved in the agar layer method are illustrated in Figure 8.12.

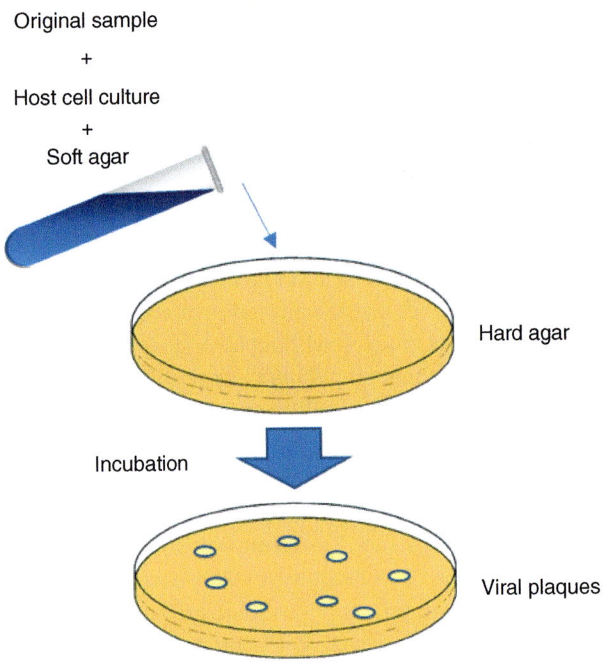


**Figure 8.12** Schematic illustration of steps involved in agar layer method for quantification of infective viruses in a water sample. A sample of water is blended in a suspension with a host cell subculture and a soft agar. The mixture is spread across the surface of a plate containing a preprepared hard agar, then covered and allowed to incubate under conditions that favor viral infection of the host cell. After incubation, visible plaques are counted (reported as plaque forming units, of pfu). This value is divided by the volume of the original sample applied in the test to develop an estimate of the concentration of infective viruses, reported as pfu per unit volume.

A sample of water is blended with a subculture of the viral host cell and a soft agar solution; the soft agar is maintained at a temperature (usually near 45 °C) that is sufficiently high to allow the agar to remain in a molten state, but not cause inactivation of the target virus or its host cells. The sample mixture is then spread across the surface of a preprepared hard agar plate. The plate is covered, inverted, and then incubated under conditions that promote viral infection of the host cells. After incubation, viral plaques are counted.

A plaque is a discolored (often clear or translucent), roughly circular region that is established within a host cell lawn. It results from lysis of the host cell, caused by release of virions, which go on to infect adjacent cells, resulting in additional host cell infection/lysis and the formation of a visible plaque. Each plaque is assumed to be attributable to a single, infective virus. Plaques are counted as plaque forming units (pfu). The concentration of infective virus in the original sample is estimated by dividing the number of plaques by the volume of the sample, and is reported in pfu per unit volume.

The plaque assay is commonly used for estimation of the concentration of infective phage. As such, the host cell culture is developed using bacteria that the phage are able to infect.

With conventional labware and Petri dishes, it is possible to apply sample volumes up to approximately 1 mL with the agar layer method. Larger sample volumes can be accommodated using multiple Petri dishes or containers that allow the development of large host cell lawns.

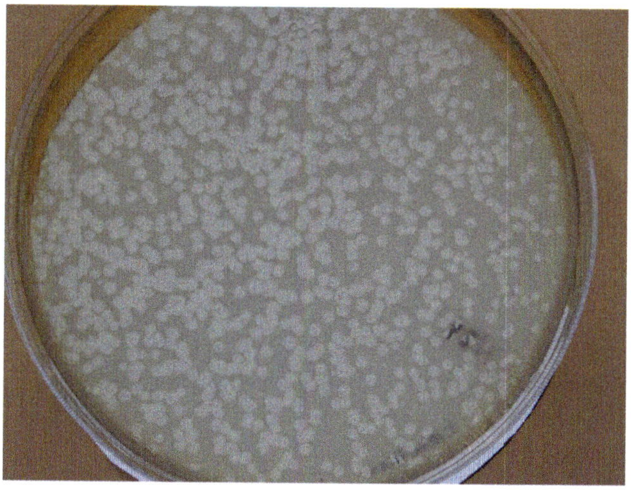

**Figure 8.13** Photographic image of plaques formed in a bacterial cell lawn by coliphage T2. In this case, the plaques are approaching the limit of being too numerous to count, in that several of the plaques have merged. *Source:* Image from: http://w3.marietta.edu/~spilatrs/biol202/labresults/viral_plaques.html.

Figure 8.13 illustrates the endpoint of a plaque assay for coliphage T2. In this image, the plaques are visible to the naked eye as clear/translucent areas that have developed within the bacterial host cell lawn. This image also illustrates an important limitation of the method; specifically, in this case the plaques are so numerous that they are beginning to blend together. Ideally, the number of plaques on a plate should be in the range of 30–300. The lower limit of these guideline values ensures statistical reliability, while the upper limit is based on the ability to differentiate individual plaques within the host cell lawn.

#### 8.5.2.2 Cytopathic Effect (CPE)

Another form of evidence of completion of the virus infection cycle is the development of structural deformations among the host cells, or cytopathic effect (CPE). In assays used to measure viral infectivity, a host cell is grown in a monolayer in an appropriate vessel, such as a Petri dish or a multiwell plate. A range of volumes (or dilutions) of the viral suspension is applied to the host cell lawns and allowed to incubate. After incubation, the host cell lawns are subjected to microscopic examination to identify the presence or absence of CPE, which often shows up in the form of cell rounding or changes to one or more internal components of the host cell. In that sense, tests based on CPE are often binary (i.e. presence vs. absence of infection). When used in quantitative assays, the infective virus concentration can be calculated using the MPN approach described above, based on binary scoring of host cell infection or non-infection (as indicated by the presence or absence of CPE), a series of serial dilutions of the original sample, and replicated measurements.

### 8.5.3 Protozoa

The application of infectivity assays for assessment of the effectiveness of UV irradiation against protozoan parasites represented an important scientific breakthrough. The results of these tests demonstrated the efficacy of UV-based treatment against *Cryptosporidium* and *Giardia*. Perhaps of equal importance, these experiments demonstrated a broader principle with respect to assays that are used for quantification of UV (and some other disinfectants); namely, that it is important to use assays that rely on an endpoint that accurately measures the ability of the target microbe to complete its life/infection cycle in a relevant host.

Two methods are commonly applied for quantification of the concentration of viable *Cryptosporidium* and *Giardia* following exposure to UV radiation (and other disinfectants). These include methods based on animal infection and methods based on completion of the infection cycle within a culture of host cells. Both methods are described briefly below.

#### 8.5.3.1 Animal Infection

Among the analytical methods used to define the concentrations of microbial parasites in water, the methods that involve infection of an animal host as an endpoint are generally considered to represent the "gold standard." Numerous other methods have been developed to quantify the concentrations of these microbial pathogens in water. For these alternative methods, it is desirable to demonstrate equivalence with animal infection assays.

Animal infectivity assays have a number of beneficial features, including involvement of the entire animal host, which allows for interactions among various organs and cells within the host. Importantly, the immune system of the animal host is engaged in the process. Collectively, these attributes of animal infectivity tests allow for true infections to take place and be measured.

Assays based on infection of an animal host have been developed for *Cryptosporidium* and *Giardia*. Various endpoints have been applied in these assays, including counts of resting stage particles (oocysts and cysts, respectively) in fecal matter, or presence/absence of (oo)cysts in the digestive tissues of euthanized animals [14, 19–21].

In broad terms, animal infectivity assays are conducted by forced ingestion of a known number of (oo)cysts, followed by a period of incubation in the animal host. After incubation, evidence of infection is collected, either by examination of fecal material or by euthanizing the animal followed by analysis of the large intestine to identify the presence/absence of evidence of infection [20–22]. These experiments must also be accompanied by experiments in which known quantities of infective (oo)cysts are imposed on the test animals, with subsequent measurement of their response using the same method as is used in the tests with unknown numbers of infective (oo)cysts. The experiments with known quantities of infective particles serve as "calibration" of the animal population against exposure to the target pathogen.

Several procedures have been developed to provide an estimate of the dose of infective parasites imposed on the test animals. In one method, continuous monitoring of fecal material excreted by infected animals indicated a regular pattern of change in the concentration of *Giardia* cysts in fecal material as a function of time [21]. In particular, this pattern was characterized by a latent period, followed by a period of rapid increase. It was demonstrated that the latent period was governed by the dose (i.e. number) of infective cysts. As such, postinfection monitoring of cyst concentration in the fecal material of infected animals could be used to define the latent period, which in turn could be translated into an infective dose.[d] Figure 8.14 illustrates the relationship between the number of infective oocysts in the inoculum and the latent period for multiplication of the *Giardia* cysts in mouse fecal material. This relationship can be used to quantify an unknown number of infective cysts in an inoculum; in turn, this information can be used to calculate the concentration of infective cysts in a water sample before or after exposure to UV radiation.

A method that has been applied for examination of infectivity among *Cryptosporidium* oocysts involves five-day old CD-1 mice inoculated with known numbers of oocysts [20]. After a seven-day incubation period, the mice are euthanized, followed by removal of their large intestine, which is then homogenized. Oocysts are removed from intestinal tissue by centrifugation and washing, then stained with a fluorescein-labeled monoclonal antibody. The resulting suspension is then

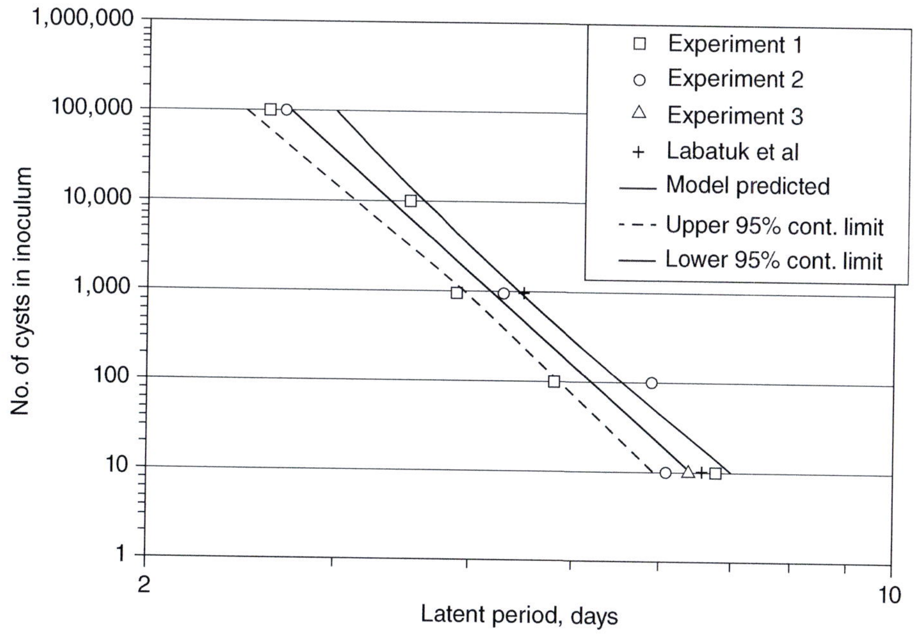


**Figure 8.14** Latent periods for emergence of *Giardia* oocysts in mouse fecal material. *Source:* Craik et al. 2000. Image from [21].

analyzed by flow cytometry to define the presence/absence of oocysts, which is used as evidence of infection/non-infection, respectively. These experiments are preceded by experiments in which mice are subjected to a range of inoculum sizes; the presence/absence of infection is then defined as described above. This preliminary experiment serves as calibration of the mouse population, in that it allows identification of the likelihood of mouse infection as a function of the inoculum size. Estimates of the number of infective oocysts imposed on the mice in the first experiment are then developed by comparison with the "standard curve" for the mouse population, through the use of an appropriate oocyst/mouse dose–response model. Similar methods for mouse infection have been developed in which the infective inoculum size is estimated via an MPN calculator [22]. Many similar protocols are available in the literature to provide estimates of infective inoculum size for protozoan parasites using animal infectivity methods.

#### 8.5.3.2 Cell Culture

Although animal infection assays represent the gold standard for parasite infectivity, these methods have important drawbacks and limitations. Relative to other methods, they are expensive and time-consuming to conduct. In addition, some of the methods involve animal euthanasia, which can be a source of concern. As a result, alternatives to animal infectivity assays have been developed.

Cell culture assays are a common alternative to animal infection because they allow the infection cycle to be completed without the involvement of an animal host. Instead, host cells are grown in cell culture to serve as an artificial host for the parasite.

*Cryptosporidium* oocysts infect the epithelial cells of the human digestive system. Because of this, the cell line that is often used in cell culture assays for *Cryptosporidium* is a human ileocecal adenocarcinoma cell line, HCT-8. These cells are grown in cell culture and function as host cells for infection by *Cryptosporidium*. The cell culture method was initially developed to allow detection of infective oocysts of *Cryptosporidium* [47]; however, the combination of this method with replicate measurements across a dilution series allowed for quantification of infective *Cryptosporidium* in water samples through the use of an MPN calculator [23]. Similar methods have been developed that involve direct, microscopic counting of infective foci across entire cell culture wells.

In the cell culture method, HCT-8 cells are grown in cell culture to approach or reach confluence across multiple culture wells. The host cells are then exposed to a sample that may contain infective *Cryptosporidium* oocysts and allowed to incubate. After incubation, the well contents are exposed to labeled antibodies that are specific for the reproductive stages of *Cryptosporidium*.

In the MPN method, replicate samples of a serially diluted oocyst suspension are scored positive/negative for infection by the presence/absence of infection foci within each well [23]. Scores from the analysis of replicate samples from across a dilution series are then applied to an MPN calculator for estimation of the concentration of infective oocysts in the original sample.

In the direct count method, whole-well counts of infection foci are conducted by microscopy. When combined with the volume of the sample and any dilution of the sample that was applied before analysis, it is then possible to calculate the concentration of infective oocysts in the original sample [48].

The cell culture methods based on MPN and direct counts both yield results that are comparable to results from mouse infectivity assays [23, 48]. Both methods have also been demonstrated to be appropriate for use with UV irradiation and other disinfection processes because the endpoint of the assay is a measurement of actual infection of host cells by the parasite.

The cell culture method has been combined with alternative endpoints for quantification of infective oocysts. Johnson et al. compared the conventional cell culture method that involved detection of infective stages by immunofluorescence microscopy assay (IFA), as described above, with cell culture methods that were combined with PCR or RT-PCR [48]. Among these methods, they found the cell culture/IFA method to be superior to the cell culture/PCR and cell culture/RT-PCR methods because it showed the fewest number of false positives among these methods, was best for detection of low numbers of infective oocysts, and it was the simplest to conduct.

### 8.5.4 Algae

Though generally not relevant in most drinking water or municipal wastewater disinfection applications, phytoplankton, including algae, are relevant to treatment of ballast water in ships. For these applications, UV-based treatment has a number of potential benefits, relative to antimicrobial processes that rely on chemical disinfectants (e.g. chlorine). As with other UV-based disinfection processes, the assays used to measure the responses of algae to UV exposure need to be consistent with the mechanisms of microbial inactivation and the goals (desired endpoints) of treatment.

As described previously, the IMO and the USCG have defined acceptable limits for the concentrations of planktonic organisms in ballast water based on two size classes. For these size-classified plankton, the IMO and USCG limits are numerically identical, but differ in the use of the terms *viable* (IMO) and *live* (USCG) to describe these organisms. This has direct consequences in terms of the assays that have been approved by these two organizations. The IMO has approved a Dilution Culture Method/Most Probable Number (DCM/MPN) assay, while the USCG has approved a method based on inclusion or exclusion of vital dyes.

The lessons learned from work with protozoan parasites and fish pathogens (described above) indicate that any assay that is used to measure a response to UV irradiation should be based on an endpoint that relies on completion of the reproductive cycle. This indicates a clear contradiction with the methods based on vital dyes. In fact, comparisons of algal inactivation based on the use of DCM/MPN and vital dyes have indicated marked differences in apparent responses of algal cells to UV exposure [33, 37, 49–51].

The DCM/MPN assay is similar to other MPN-based assays in some respects; however, important differences are evident for the algal methods, especially as related to the conditions and timing of incubation and the methods used to score samples as positive/negative for growth. Algae are grown and diluted using a medium that provides nutrients needed to support growth of the target algal cells [33, 49, 52]. Samples are subjected to serial dilution across several decades with replication at each dilution. The diluted samples are incubated in an environmental chamber to provide temperature and lighting conditions that promote photosynthetic activity and reproduction of the target algal cells. Incubation periods are typically much longer than will other microorganisms; 14-day incubation periods are common for the algal species that have been examined to date. After incubation, all tubes are scored as positive/negative for algal growth. Scoring can be based on visual inspection (green color) or fluorescence, using excitation and emission wavelengths that are characteristic of chlorophyll. With both of these optical methods, the signal for growth needs to be clearly differentiated from a background signal. Culture tube scores for a dilution series are then applied to an MPN calculator to develop an estimate of the concentration of viable algal cells in the original sample.

## 8.6 Molecular Biology

As described above, microbiological viability/infectivity assays based on growth or completion of the infection cycle represent the current "gold standard" for quantification of viable/infective microbes, and for quantification of the effects of disinfection processes, including those based on UV radiation. However, these methods tend to be time-consuming, resource-intensive, and slow to yield results. In addition, the conditions of incubation used in these assays generally do not allow for growth or reproduction of all organisms. The issue of "viable but not culturable" (VBNC) organisms represents a source of concern for these methods. The VBNC state can be important in applications of all disinfectants, including UV. With UV disinfection applied at conventional doses, the majority of microbial targets are not "killed" but rather lose their ability to reproduce [9].

Taken together, the attributes of growth-based assays point to a need to develop analytical methods that are faster and less expensive, while providing information that can be related broadly and directly to viability or infectivity of these microbes. In concept, these methods could be used for disinfection process control if they can be developed with sufficient speed, simplicity, and economy.

The tools of molecular biology are likely to offer the greatest promise for addressing these needs. In particular, methods that involve PCR, usually coupled with one or more other methods, are being developed as alternatives to conventional culturing or infectivity assays.

### 8.6.1 Polymerase Chain Reaction and Related Methods

The PCR was invented in 1983 and has emerged as perhaps the most commonly applied method in molecular biology. PCR and its derivatives allow for amplification and detection of nucleic acids and genes from organisms. Because UV-based disinfection processes rely heavily on photochemically induced damage to nucleic acids to inactivate microorganisms, the methods based on PCR are a natural fit for investigations of the effects of UV on microbes. Figure 8.15 provides a schematic illustration of the steps involved in PCR, which are summarized briefly below:

1) *Denaturing* – Double-stranded DNA (ds-DNA) is heated to close to the boiling point of water (~95 °C) to rupture the H-bonds that exist between bases in a DNA strand. This results in the formation of two single strands of the original DNA.

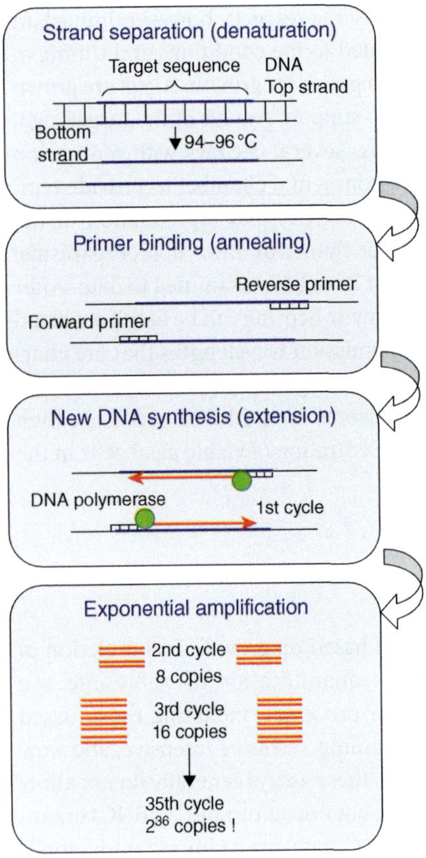


**Figure 8.15** Schematic illustration of steps involved in PCR. *Source:* Image from: https://www.ncbi.nlm.nih.gov/probe/docs/techpcr/.

2) *Annealing* – The solution is cooled to 50–65 °C to allow attachment of primers to the DNA strands. The primers are short nucleic acid segments that include nucleotides that are complementary to each end of the sequence to be copied. The primers function as starting points for synthesis of the new DNA and include a forward and reverse primer to promote strand replication in opposite directions on the two single strands of DNA created by denaturing.
3) *Extending* – Solution temperature is increased to 72 °C, which represents the optimum temperature for function of the enzyme Taq DNA polymerase, which was originally extracted from the thermophilic bacterium *Thermus aquaticus*. Taq DNA polymerase promotes the addition of complementary bases, which will have also been added to solution, in the forward and reverse directions. This leads to the formation of DNA segments that are identical to those of the original target segments.

These steps are repeated as a cycle, with binary replication occurring in each cycle. The process is often carried out for 20–40 cycles, with a typical run time of a few hours. For perspective, it is relevant to note that if binary replication continues for 30 cycles, then a single DNA fragment (amplicon) will be reproduced $2^{30}$ times, resulting in production of roughly $10^9$ copies of the amplicon.

A derivative method is known as Real-Time, or quantitative PCR (q-PCR). In q-PCR, an additional step is included to allow labeling of the amplicon with a molecule that becomes fluorescent as a result of binding to DNA. Because of this, the method allows for quantification of the production of the amplicon.

Another related method is Reverse-Transcription PCR (RT-PCR), which can be used for quantification of RNA. The first step in RT-PCR involves conversion of RNA to complementary DNA (cDNA) by use of the enzyme reverse transcriptase. The cDNA formed in this process then functions as a template for amplification by PCR.

Conventional PCR is based on relatively short amplicons, in part because longer amplicons may increase the duration of the extension step. It is relevant that several types of damage within the amplicon can inhibit DNA polymerase, including the formation of dimers that result from exposure to UVC radiation. In the PCR methods, amplification occurs with undamaged target segments of nucleic acid. However, when the target segment is damaged (e.g. by UVC exposure), amplification is inhibited. As a result, the inhibition of this signal provides a measure of inactivation of the organism from which the nucleic acid was extracted.

Conventional PCR and related methods have been shown to underestimate the extent of inactivation among UV-exposed target microbes [38]. This is probably because the amplicons used in these assays are typically quite small (100 bp or so) and may not be representative of the entire genome of the target organism. As such, methods based on conventional PCR may not be appropriate for quantification of the effects of UVC irradiation in target microbes. However, a number of important modifications to PCR and its derivatives have been developed to overcome this limitation. Some of these are described below.

#### 8.6.1.1 Integrated Cell Culture/PCR

Culture-based methods for evaluation of viral infectivity are slow and time-consuming. By comparison, methods based on PCR are rapid and have greater sensitivity, specificity, and precision than those based on cell culture. The integration of cell culture methods with PCR allows for the strengths of both method types to be combined [53]. Because of this, several so-called Integrated Cell Culture/PCR (ICC/PCR) methods have been developed for quantification of infective viruses, and for quantification of viral UV dose–response behavior. The basic framework for these methods, as applied for evaluation of viral UV dose–response behavior is as follows:

1) Host cells are grown to confluence in appropriate vessels.
2) A virus suspension is subjected to a range of well-defined UV doses, usually using a collimated beam device. These experiments are conducted with a known (i.e. measured separately) concentration of the infective virus in suspension.
3) Serial dilution of the exposed virus suspension is then conducted, with replicate application of dilutions across cell-culture plates with confluent host cells.
4) The virus-exposed host cells are then incubated under conditions that favor infection of the host cell by the target virus. The incubation period needs to be sufficient to allow detection of infection or non-infection by a PCR-based method (see below). Because of the sensitivity of PCR-based methods, this period of incubation can be considerably shorter than the incubation period required for conventional cell culture assays.
5) After incubation, the cells and viruses are removed from the vessel; host cells or viruses are analyzed by an appropriate PCR-based method for evidence of infection or non-infection. This infection/non-infection endpoint is used as an alternative to conventional endpoints, such as CPE or plaque formation.

Ko et al. applied an ICC/PCR method for detection of messenger RNA (mRNA) in host cells. The idea behind this method is that the presence of mRNA can be taken as positive evidence of infection of the host cells [54]. When applied to Adenovirus 41 (Ad41), they observed that Ad41 were apparently more resistant to $UV_{254}$ radiation than when their dose–response behavior was measured using CPE as the endpoint. Similarly, a method based on ICC/qPCR was used for comparison with conventional cell culture methods using CPE as the endpoint for quantifying $UV_{254}$ dose–response behavior of Ad4 [55]. They too found that the ICC/qPCR results indicated apparent differences in UV sensitivity of Ad4 than was indicated by the CPE-based method. A similar study involved a comparison of ICC/RT/q-PCR with a CPE-based assay for measurement of Ad41 $UV_{254}$ dose–response behavior [56]. In this study, the results of the two methods were statistically indistinguishable. It is not clear from these three separate investigations that methods based solely on cell culture are "better" or "worse" than ICC/PCR methods for quantification of virus infectivity, or for quantification of virus UV dose–response behavior. Both method types have merit and are likely to be used in the future.

These integrated methods have been applied for evaluation of other viruses as well. Li et al. compared ICC/real-time RT-PCR with conventional cell culture (using CPE as the endpoint) for quantification of rotavirus $UV_{254}$ dose–response behavior [57]. The PCR-based method indicated apparent UV resistance that was somewhat higher than was quantified using the cell culture method alone. A broader investigation was conducted by Ryu et al. in which they quantified $UV_{254}$ dose–response behavior for four enteroviruses (Coxsackievirus A10, Poliovirus 1, Enterovirus 70, and Echovirus 30) using ICC/RT-qPCR [58]. They applied their method in pure virus suspensions and with mixtures of these viruses; they found the method to be 100% specific for the selected viruses, and to display limits of detection that were two to three orders of magnitude better than those that are possible using cell culture methods. Their measurements of $UV_{254}$ dose–response behavior for all four viruses were similar to results that had been published previously based on cell culture methods.

#### 8.6.1.2 Long Amplicon q-PCR

Microbial inactivation in UV-based disinfection processes is largely attributable to inhibition of DNA production, and to lesser extent the production of RNA and proteins [59]. PCR-based methods are well suited for quantification of microbial responses to UV radiation because they provide measurements of damage to nucleic acids. Many types of UV-induced damage are known to inhibit DNA polymerase, including dimer formation and nucleic acid strand breaks [60–63]. In PCR methods, this inhibition results in a decrease in PCR-product formation, which is analogous to reductions in infectivity or reproduction that are quantifiable in cell culture-based methods.

In principle, UV-induced damage could occur anywhere along the length of a nucleic acid strand. However, PCR methods will generally involve amplification of a segment of the strand, known as an amplicon. As amplicon length increases, the likelihood of detecting a change in a PCR signal also increases. With this in mind, several methods have been developed that involve applications of long amplicons as a means of improving agreement between cell culture-based methods and methods that involve PCR for quantification of damage induced by exposure to UVC radiation. These methods have been developed for a wide range of targets, including mammalian cells [60], bacteria, and viruses.

***8.6.1.2.1 Bacteria*** Süß et al. compared plate counts and qPCR for analysis of *Pseudomonas aeruginosa* [62]. They applied q-PCR using amplicon lengths of 100, 500, and 900 bp as well as conventional plate counting to measure the responses of unirradiated bacteria and bacteria that had been subjected to a $UV_{254}$ dose of 60 $\frac{mJ}{cm^2}$ and for a period of roughly 2.5 days to allow for repair. In general, they observed that agreement between plate counts and qPCR data improved as amplicon length increased; this trend was evident in samples before UV exposure, immediately after UV exposure, and in the period during which repair was monitored.

Banihashemi et al. conducted an experiment with two bacteria, *Salmonella enterica* and *Campylobacter jejuni*, in which plate counts were compared with q-PCR based on short (~120–170 bp) and long (~1500–1600 bp) amplicons. The plate counts and q-PCR data were in good agreement for unirradiated cells; however, the short-amplicon PCR data showed essentially no change among the UV-irradiated bacteria ($UV_{254}$ dose = 50 $\frac{mJ}{cm^2}$), while the q-PCR data based on the long amplicons was in good agreement with plate counts.

Nocker et al. conducted an experiment in which the $UV_{265}$ dose–response behavior of *E. coli* was measured using plate counts and real-time PCR using short (~100 bp) and long (~1500 bp) amplicons. The long-amplicon data yielded good agreement with the plate count data in terms of the trend as a function of applied dose ($UV_{265}$ doses ranging from 0 to 40 $\frac{mJ}{cm^2}$). However, the short-amplicon data showed only minimal change across this dose range.

These studies indicate that PCR-based methods that involve long amplicons may represent an approach that can be applied for quantification of bacterial responses to UVC radiation. As PCR methods evolve, it will be possible to apply them for quantification of reactor performance, or perhaps even for monitoring of reactor behavior.

***8.6.1.2.2 Viruses*** Typical viral genomes are on the order of $10^3$–$10^4$ bp, whereas bacterial genomes typically range from $10^6$ to $10^7$ bp. Because of this, the potential exists to apply amplicons that represent a larger fraction of a viral genome than a bacterial genome. Several examples of long-amplicon PCR methods have been developed for examination of viral responses to UV exposure.

Rodriguez et al. examined qPCR for quantification of the responses of Ad 2 to $UV_{254}$ exposure [64]. Amplicons of 1,100, 6,000, and 10,100 bp were amplified; cell culture-based infectivity was applied in parallel for comparison. As with the bacteria described above, the sensitivity of the q-PCR method increased as amplicon length increased, and the measurements of Ad 2 inactivation based on the long-amplicon q-PCR measurements were in good agreement with measurements based on cell culture. Interestingly, they noted that the increased sensitivity of the long-amplicon q-PCR method yielded a measurement that was effective for relatively low $UV_{254}$ doses, but appeared to reach a saturation condition in the limit of high doses. It was suggested that the short-amplicon q-PCR method might be more appropriate for high doses, which are commonly needed for inactivation of Adenovirus.

This same method was adapted to quantify the action spectrum of Ad 2 for UVC wavelengths [65]. The cell culture and q-PCR data were in good agreement for most wavelengths at or above 254 nm; however, for shorter wavelengths, the cell culture data indicated greater loss of infectivity than the loss of DNA amplification that was displayed by q-PCR. This suggests that inactivation of Ad 2 at these shorter wavelengths may involve damage to biomolecules other than DNA. In particular, it is likely that damage to viral proteins contributed to inactivation of Ad 2 at these short wavelengths.

Ho et al. applied long-amplicon q-PCR for quantification of the $UV_{254}$ dose–response behavior of the bacteriophage ϕX-174; a conventional plaque assay was applied in parallel [61]. As with the other methods described above, inactivation responses generally increased with amplicon length, as did agreement between the q-PCR data and the measurements by plaque assay. However, they observed some tailing in the $UV_{254}$ dose–response behavior of the phage in the limit of large dose. This behavior was attributed to nonspecific UV-induced damage, but this may also be an indication of the "saturation" issue that was described by Rodriguez et al. [64].

The methods described above all support the use of long amplicons for PCR-based measurements of UV dose–response behavior. The logic behind this approach is that a long amplicon will provide a better representation of the target genome than will a small amplicon. This argument is especially valid for viruses, which are characterized by relatively small genomes.

Another perspective on these methods suggests that PCR data based on short amplicons may also provide valuable information regarding microbial inactivation. Specifically, Pecson et al. provided a detailed framework for extension (extrapolation) of short-amplicon data to inactivation responses of viruses [66]. The basic idea behind this method is that PCR-based measurements provide a measurement of damage to an amplicon. If it is assumed that the amplicon is representative of the entire genome, then the measured changes to the amplicon can be extrapolated to the entire genome, which in turn can be used to predict viral inactivation. This method depends on the validity of the assumption described above, but provided reasonable agreement with viral inactivation data based on cell culture.

Pecson et al. also provided some suggestions for how to extend small-amplicon data for predictions of virus inactivation when the assumption listed above was not met. Specifically, they suggested the use of multiple amplicons and calibration of PCR-based results against cell infectivity assay data.

The framework provided by Pecson et al. included a probabilistic model for simulation of viral responses to UVC exposure. Additional information regarding this probabilistic approach and its implications for simulation of microbial inactivation kinetics will be provided in Chapter 9.

#### 8.6.1.3 Molecular Viability Testing

The tools of molecular biology can be used to measure details of microbial growth and reproduction. Ribosomal ribonucleic acid (rRNA) is required for protein synthesis in all living organisms. The production of rRNA in bacteria proceeds through rRNA precursors that are unique to each bacterium. These pre-rRNA molecules can represent as much as 25% of total cellular rRNA, which facilitates their detection. In addition, these pre-rRNA compounds show rapid changes in response to environmental stresses and circumstances that promote bacterial growth, such as exposure to substrates.

With these facts in mind, a bacterial assay was demonstrated based on quantification of rRNA among bacteria [67]. The method, known as "microbial viability testing" (MVT), involves measurements of changes in rRNA among bacteria when subjected to stress or growth conditions. Three different species of bacteria (*E. coli*, *Enterococcus faecalis*, and *Aeromonas hydrophila*) were used in these experiments, wherein the aqueous suspensions of these bacteria were exposed to $UV_{254}$, then split and analyzed by conventional plate counting and by MVT. The MVT measurements were consistent with plate count results, although there was a lag period of about one hour in which rRNA production continued after $UV_{254}$ exposure. Like the other molecular methods described in this section, MVT is faster than conventional plate counting. In addition, it provides a measurement of the effects on UV exposure on bacterial growth, at the molecular level.

#### 8.6.1.4 Intercalating Dyes/PCR

Propidium monoazide (PMA) and ethidium monoazide (EMA) are photoreactive molecules that preferentially form covalent bonds with ds-DNA. PMA and EMA are weak fluorophores; however, when exposed to visible light in the presence of ds-DNA, both compounds bind to DNA and become brightly fluorescent [68].

Bacteria with intact membranes/cell walls are able to exclude PMA and EMA; however, when bacterial membranes/cell walls are compromised, both compounds are able to diffuse into bacterial cells where they can interact with DNA. Assays based on inclusion or exclusion of PMA and EMA, as measured by fluorescence among these bacterial cells, have been demonstrated to be effective for discrimination of "live" and "dead" bacterial cells, where cell killing was accomplished by exposure to isopropyl alcohol or heat [68].

As an extension of this approach, it was noted that intercalation of PMA inhibits PCR amplification. In a similar manner, a PMA/PCR assay was able to discriminate among "live" and "dead" bacterial cells, where cell killing was again accomplished by exposure to isopropyl alcohol or heat [69].

These methods share some characteristics with assays based on so-called "vital dyes" that had been used to assess infectivity of protozoan resting stages. However, as with these protozoan tests, the use of these intercalating dyes for measurements of bacterial responses to UVC radiation are not appropriate, as the doses of UVC radiation that are required to inactivate most bacteria are substantially smaller than the UVC doses that would be required to impart significant damage on bacterial cell walls or membranes.

However, a related method has been applied for examination of the responses of Adenovirus to UVC radiation [39]. In this method, EMA was coupled with qPCR and RT/q-PCR to examine the responses of Adenovirus to UV exposure at three characteristic wavelengths: 224, 254, and 280 nm. All three wavelengths were effective for accomplishing Adenovirus inactivation; however, at the molecular level, the responses of Adenovirus to UVC radiation at these three characteristic wavelengths were quite different. Specifically, radiation at characteristic wavelengths of 254 and 280 nm was effective for imposing nucleic acid damage, but has relatively little effect on the proteins within the viral capsid. In contrast, radiation at a characteristic wavelength of 224 nm was able to induce nucleic acid damage somewhat less efficiently than the longer wavelengths, but was substantially more effective for causing damage to the viral capsid. However, it appeared that none of the damage caused by exposure to UVC radiation at these wavelengths was effective in preventing attachment of virions to the host cell. Instead, the capsid damage caused by relatively short wavelength radiation appeared to cause viral inactivation by inhibiting transfer of the viral genomic material to the host cell nucleus.

## 8.7 Summary

Disinfection and antimicrobial processes represent the most common application for UV photoreactors. The availability of analytical methods that provide accurate, meaningful measurements of the effects of UV radiation on microbial populations is critical for the design and operation of these systems. The assays that are used for this purpose must be designed and implemented in a manner that is consistent with the mechanisms of microbial inactivation and the goals of treatment. Mistakes in the recent history of the application of these methods have taught the importance of adhering to these methodological requirements. The first of these related to the use of improper methods for quantification of the responses of protozoan (oo)cysts to UVC radiation, but similar, more recent mistakes have also been noted with applications aimed at control of algae and fish parasites.

In the case of UV-based processes, inactivation of microbes is largely attributable to photochemically induced damage to nucleic acids (DNA and RNA); for short-wavelength UVC radiation, damage to proteins can also contribute significantly to microbial inactivation. The current "gold standard" methods for measurement of UV-induced inactivation of microorganisms are assays that indicate completion of the reproductive or infective cycle. For bacteria, these methods are based on reproduction; for parasitic organisms, including viruses and protozoa, the methods are based on measurement of the ability of the parasite to infect a host cell or organism.

The recent, rapid evolution of the field of molecular biology, especially involving methods that involve PCR, has allowed the development and demonstration of a number of methods that detect biochemical changes within microbes. These methods have been successfully adapted to allow meaningful measurements of UV-induced damage to nucleic acids and their precursors that can be used as surrogates for actual viability or infectivity assays. The nature of these methods is such that they provide increased speed and specificity, as compared to conventional methods based on cell culture. As these methods continue to evolve, it is possible that they may be adapted for validation of UV photoreactors, or even for their control.

## Notes

**a** Additional details of the mechanisms and kinetics by which UV-induced damage is induced, as well as the mechanisms and kinetics of repair processes will be presented in Chapter 9.

**b** Note than "blue-green algae" (known formally as cyanobacteria) are photosynthetic bacteria; they are not algae.

**c** Note that the term "bacterial density" is a misnomer. Strictly speaking, this term could be interpreted as a measure of the specific gravity of bacterial cells. However, the intended connotation with this term is a measure of the concentration of viable bacterial cells in a sample.

**d** It is worth noting that no alternative to animal infectivity has been developed for *Giardia*. Therefore, animal infectivity represents effectively the only method that can be used to quantify infective *Giardia* cysts, and it is also the only available method for quantifying the effects of disinfectants, including UV, on *Giardia* infectivity.

## References


**1** Brash, D.E. (**1988**). UV mutagenic photoproducts in *Escherichia-coli* and human-cells – a molecular-genetics perspective on human-skin cancer. *Photochemistry and Photobiology 48* (1): 59–66.

**2** Mitchell, D.L. and Nairn, R.S. (**1989**). The biology of the (6-4) photoproduct. *Photochemistry and Photobiology 49* (6): 805–819.

**3** Sancar, A. (**2003**). Structure and function of DNA photolyase and cryptochrome blue-light photoreceptors. *Chemical Reviews 103* (6): 2203–2237.

**4** Sancar, A. (**1994**). Structure and function of DNA photolyase. *Biochemistry 33* (1): 2–9.

**5** Setlow, P. (**1995**). Mechanisms for the prevention of damage to DNA in spores of *Bacillus* species. *Annual Review of Microbiology 49*: 29–54.

**6** Setlow, P. (**2001**). Resistance of spores of *Bacillus* species to ultraviolet light. *Environmental and Molecular Mutagenesis 38* (2-3): 97–104.

**7** Eischeid, A.C., Thurston, J.A., and Linden, K.G. (**2011**). UV disinfection of adenovirus: present state of the research and future directions. *Critical Reviews in Environmental Science and Technology 41* (15): 1375–1396.

**8** Eischeid, A.C. and Linden, K.G. (**2011**). Molecular indications of protein damage in adenoviruses after UV disinfection. *Applied and Environmental Microbiology 77* (3): 1145–1147.

**9** Blatchley III, E.R., Dumoutier, N., Halaby, T.N. et al. (**2001**). Bacterial responses to ultraviolet irradiation. *Water Science and Technology 43* (10): 179–186.

**10** Brock, T.D., Madigan, M.T., Martinko, J.M., and Parker, J. (**1994**). *Biology of Microorganisms*, 7e. Englewood Cliffs, NJ: Prentice Hall.

**11** Moeller, R., Setlow, P., Reitz, G., and Nicholson, W.L. (**2009**). Roles of small, acid-soluble spore proteins and core water content in survival of *Bacillus subtilis* spores exposed to environmental solar UV radiation. *Applied and Environmental Microbiology 75* (16): 5202–5208.

**12** Moeller, R., Stackebrandt, E., Reitz, G. et al. (**2007**). Role of DNA repair by nonhomologous-end joining in *Bacillus subtilis* spore resistance to extreme dryness, mono- and polychromatic UV, and ionizing radiation. *Journal of Bacteriology 189* (8): 3306–3311.

**13** Rice, E.W. and Hoff, J.C. (**1981**). Inactivation of *Giardia lamblia* cysts by ultraviolet irradiation. *Applied and Environmental Microbiology 42* (3): 546–547.

**14** Clancy, J.L., Hargy, T.M., Marshall, M.M., and Dyksen, J.E. (**1998**). UV light inactivation of *Cryptosporidium* oocysts. *Journal American Water Works Association 90* (9): 92–102.

**15** Campbell, A.T., Robertson, L.J., and Smith, H.V. (**1992**). Viability of *Cryptosporidium parvum* oocysts – correlation of *in-vitro* excystation with inclusion or exclusion of fluorogenic vital dyes. *Applied and Environmental Microbiology 58* (11): 3488–3493.

**16** Rennecker, J.L., Driedger, A.M., Rubin, S.A., and Marinas, B.J. (**2000**). Synergy in sequential inactivation of *Cryptosporidium parvum* with ozone/free chlorine and ozone/monochloramine. *Water Research 34* (17): 4121–4130.

**17** Rennecker, J.L., Marinas, B.J., Owens, J.H., and Rice, E.W. (**1999**). Inactivation of *Cryptosporidium parvum* oocysts with ozone. *Water Research 33* (11): 2481–2488.

**18** Clancy, J.L., Bukhari, Z., Hargy, T.M. et al. (**2000**). Using UV to inactivate *Cryptosporidium*. *Journal American Water Works Association 92* (9): 97–104.

**19** Bukhari, Z., Hargy, T.M., Bolton, J.R. et al. (**1999**). Medium-pressure UV for oocyst inactivation. *Journal American Water Works Association 91* (3): 86–94.

**20** Craik, S.A., Weldon, D., Finch, G.R. et al. (**2001**). Inactivation of *Cryptosporidium parvum* oocysts using medium- and low-pressure ultraviolet radiation. *Water Research 35* (6): 1387–1398.

**21** Craik, S.A., Finch, G.R., Bolton, J.R., and Belosevic, M. (**2000**). Inactivation of *Giardia muris* cysts using medium-pressure ultraviolet radiation in filtered drinking water. *Water Research 34* (18): 4325–4332.

**22** Slifko, T.R., Huffman, D.E., Dussert, B. et al. (**2002**). Comparison of tissue culture and animal models for assessment of *Cryptospridium parvum* infection. *Experimental Parasitology 101* (2-3): 97–106.

**23** Slifko, T.R., Huffman, D.E., and Rose, J.B. (**1999**). A most-probable-number assay for enumeration of infectious *Cryptosporidium parvum* oocysts. *Applied and Environmental Microbiology 65* (9): 3936–3941.

**24** Hedrick, R.P., El-Matbouli, M., Adkison, M.A., and MacConnell, E. (**1998**). Whirling disease: re-emergence among wild trout. *Immunological Reviews 166*: 365–376.

**25** Hedrick, R.P., McDowell, T.S., Marty, G.D. et al. (**2000**). Ultraviolet irradiation inactivates the waterborne infective stages of *Myxobolus cerebralis*: a treatment for hatchery water supplies. *Diseases of Aquatic Organisms 42* (1): 53–59.

**26** Hedrick, R.P., Petri, B., McDowell, T.S. et al. (**2007**). Evaluation of a range of doses of ultraviolet irradiation to inactivate waterborne actinospore stages of *Myxobolus cerebralis*. *Diseases of Aquatic Organisms 74* (2): 113–118.

**27** Bachmann, M.F., Bast, C., Hengartner, H., and Zinkernagel, R.M. (**1994**). Immunogenicity of a viral model vaccine after different inactivation procedures. *Medical Microbiology and Immunology 183* (2): 95–104.

**28** Tian, F., Hou, M., Chen, L. et al. (**2013**). Proteomic analysis of schistosomiasis japonica vaccine candidate antigens recognized by UV-attenuated cercariae-immunized porcine serum IgG2. *Parasitology Research 112* (8): 2791–2803.

**29** Schirrmacher, V., Haas, C., Bonifer, R., and Ertel, C. (**1997**). Virus potentiation of tumor vaccine T-cell stimulatory capacity requires cell surface binding but not infection. *Clinical Cancer Research 3* (7): 1135–1148.

**30** International Maritime Organization (**2017**). International Convention for the Control and Management of Ships' Ballast Water and Sediments (BWM). http://www.imo.org/en/About/Conventions/ListOfConventions/Pages/International-Convention-for-the-Control-and-Management-of-Ships'-Ballast-Water-and-Sediments-(BWM).aspx (accessed 17 August 2017).

**31** *Federal Register*. (**2012**). *Standards for Living Organisms in Ships' Ballast Water Discharged in U.S. Waters*, vol. *77*, 17254–17320. Washington, D.C.: Security, D. o. H.

**32** Breeuwer, P. and Abee, T. (**2000**). Assessment of viability of microorganisms employing fluorescence techniques. *International Journal of Food Microbiology 55* (1–3): 193–200.

33 Cullen, J.J. and MacIntyre, H.L. (**2016**). On the use of the serial dilution culture method to enumerate viable phytoplankton in natural communities of plankton subjected to ballast water treatment. *Journal of Applied Phycology 28* (1): 279–298.

34 Emerson, J.B., Adams, R.I., Roman, C.M.B. et al. (**2017**). Schrodinger's microbes: tools for distinguishing the living from the dead in microbial ecosystems. *Microbiome 5*: 86.

35 Steinberg, M.K., Lemieux, E.J., and Drake, L.A. (**2011**). Determining the viability of marine protists using a combination of vital, fluorescent stains. *Marine Biology 158* (6): 1431–1437.

36 Dorsey, J., Yentsch, C.M., Mayo, S., and McKenna, C. (**1989**). Rapid analytical technique for the assessment of cell metabolic-activity in marine microalgae. *Cytometry 10* (5): 622–628.

37 Blatchley III, E.R., Cullen, J.J., Petri, B. et al. (**2018**). The biological basis for ballast water performance standards: "viable/non-viable" or "live/dead"? *Environmental Science & Technology 52* (15): 8075–8086.

38 Sobsey, M.D., Battigelli, D.A., Shin, G.A., and Newland, S. (**1998**). RT-PCR amplification detects inactivated viruses in water and wastewater. *Water Science and Technology 38* (12): 91–94.

39 Vazquez-Bravo, B., Goncalves, K., Shisler, J.L., and Marinas, B.J. (**2018**). Adenovirus replication cycle disruption from exposure to polychromatic ultraviolet irradiation. *Environmental Science & Technology 52* (6): 3652–3659.

40 Cochran, W.G. (**1950**). Estimation of bacterial densities by means of the most probable number. *Biometrics 6* (2): 105–116.

41 Eisenhart, C. and Wilson, P.W. (**1943**). Statistical methods and control in bacteriology. *Bacteriological Reviews 7* (2): 57–137.

42 McGrady, M.H. (**1915**). The numerical interpretation of fermentation-tube results. *The Journal of Infectious Diseases 17* (1): 183–212.

43 Wang, A., McMahan, L., Rutstein, S. et al. (**2017**). Household microbial water quality testing in a peruvian demographic and health survey: evaluation of the compartment bag test for *Escherichia coli*. *American Journal of Tropical Medicine and Hygiene 96* (4): 970–975.

44 Stauber, C., Miller, C., Cantrell, B., and Kroell, K. (**2014**). Evaluation of the compartment bag test for the detection of *Escherichia coli* in water. *Journal of Microbiological Methods 99*: 66–70.

45 Gronewold, A.D., Sobsey, M.D., and McMahan, L. (**2017**). The compartment bag test (CBT) for enumerating fecal indicator bacteria: basis for design and interpretation of results. *Science of the Total Environment 587*: 102–107.

46 Adams, M.H. (**1959**). *Bacteriophages*. New York: Interscience Publishers.

47 Slifko, T.R., Freidman, D., Rose, J.B., and Jakubowski, W. (**1997**). An in vitro method for detecting infectious *Cryptosporidium* oocysts with cell culture. *Applied and Environmental Microbiology 63* (9): 3669–3675.

48 Johnson, A.M., Di Giovanni, G.D., and Rochelle, P.A. (**2012**). Comparison of assays for sensitive and reproducible detection of cell culture-infectious *Cryptosporidium parvum* and *Cryptosporidium hominis* in drinking water. *Applied and Environmental Microbiology 78* (1): 156–162.

49 MacIntyre, H.L., Cullen, J.J., Whitsitt, T.J., and Petri, B. (**2018**). Enumerating viable phytoplankton using a culture-based Most Probable Number assay following ultraviolet-C treatment. *Journal of Applied Phycology 30* (2): 1073–1094.

50 MacIntyre, H.L. and Cullen, J.J. (**2016**). Classification of phytoplankton cells as live or dead using the vital stains fluorescein diacetate and 5-chloromethylfluorescein diacetate. *Journal of Phycology 52* (4): 572–589.

51 Cullen, J.J. (**2018**). Quantitative framework for validating two methodologies that are used to enumerate viable organisms for type approval of ballast water management systems. *Science of the Total Environment 627*: 1602–1626.

52 Sun, Z. and Blatchley III, E.R. (**2017**). *Tetraselmis* as a challenge organism for validation of ballast water UV systems. *Water Research 121*: 311–319.

53 Gerba, C.P., Betancourt, W.Q., Kitajima, M., and Rock, C.M. (**2018**). Reducing uncertainty in estimating virus reduction by advanced water treatment processes. *Water Research 133*: 282–288.

54 Ko, G.P., Cromeans, T.L., and Sobsey, M.D. (**2005**). UV inactivation of adenovirus type 41 measured by cell culture mRNA RT-PCR. *Water Research 39* (15): 3643–3649.

55 Gerrity, D., Ryu, H., Crittenden, J., and Abbaszadegan, M. (**2008**). UV inactivation of adenovirus type 4 measured by integrated cell culture qPCR. *Journal of Environmental Science and Health Part a-Toxic/Hazardous Substances & Environmental Engineering 43* (14): 1628–1638.

56 Ding, N., Craik, S.A., Pang, X.L. et al. (**2017**). Assessing UV inactivation of adenovirus 41 using integrated cell culture real-time qPCR/RT-qPCR. *Water Environment Research 89* (4): 323–329.

57 Li, D., Gu, A.Z., He, M. et al. (**2009**). UV inactivation and resistance of rotavirus evaluated by integrated cell culture and real-time RT-PCR assay. *Water Research 43* (13): 3261–3269.

58 Ryu, H., Schrantz, K.A., Brinkman, N.E., and Boczek, L.A. (**2018**). Applicability of integrated cell culture reverse transcriptase quantitative PCR (ICC-RTqPCR) for the simultaneous detection of the four human enteric enterovirus species in disinfection studies. *Journal of Virological Methods 258*: 35–40.

59 Harm, W. (**1980**). *Biological Effects of Ultraviolet Radiation*. Cambridge: Cambridge University Press.

60 Furda, A., Santos, J.H., Meyer, J.N., and Van Houten, B. (**2014**). Quantitative PCR-based measurement of nuclear and mitochondrial DNA damage and repair in mammalian cells, 2nd edition. *Molecular Toxicology Protocols 1105*: 419–437.

61 Ho, J., Seidel, M., Niessner, R. et al. (**2016**). Long amplicon (LA)-qPCR for the discrimination of infectious and noninfectious phix174 bacteriophages after UV inactivation. *Water Research 103*: 141–148.

62 Suss, J., Volz, S., Obst, U., and Schwartz, T. (**2009**). Application of a molecular biology concept for the detection of DNA damage and repair during UV disinfection. *Water Research 43* (15): 3705–3716.

63 Nocker, A., Shah, M., Dannenmann, B. et al. (**2018**). Assessment of UV-C-induced water disinfection by differential PCR-based quantification of bacterial DNA damage. *Journal of Microbiological Methods 149*: 89–95.

64 Rodriguez, R.A., Bounty, S., and Linden, K.G. (**2013**). Long-range quantitative PCR for determining inactivation of adenovirus 2 by ultraviolet light. *Journal of Applied Microbiology 114* (6): 1854–1865.

65 Beck, S.E., Rodriguez, R.A., Linden, K.G. et al. (**2014**). Wavelength dependent UV inactivation and DNA damage of adenovirus as measured by cell culture infectivity and long range quantitative PCR. *Environmental Science & Technology 48* (1): 591–598.

66 Pecson, B.M., Ackermann, M., and Kohn, T. (**2011**). Framework for using quantitative PCR as a nonculture based method to estimate virus infectivity. *Environmental Science & Technology 45* (6): 2257–2263.

67 Weigel, K.M., Nguyen, F.K., Kearney, M.R. et al. (**2017**). Molecular viability testing of UV-inactivated bacteria. *Applied and Environmental Microbiology 83* (10): e00331-17.

68 Nocker, A., Cheung, C.Y., and Camper, A.K. (**2006**). Comparison of propidium monoazide with ethidium monoazide for differentiation of live vs. dead bacteria by selective removal of DNA from dead cells. *Journal of Microbiological Methods 67* (2): 310–320.

69 Nocker, A., Sossa-Fernandez, P., Burr, M.D., and Camper, A.K. (**2007**). Use of propidium monoazide for live/dead distinction in microbial ecology. *Applied and Environmental Microbiology 73* (16): 5111–5117.

# 9

# UV Disinfection of Drinking Water and Municipal Wastewater

## 9.1 Introduction

Disinfection represents a critical step in the production of potable water and in treatment of municipal wastewater. Disinfection in these settings is applied with a goal of minimizing human exposure to waterborne, microbial pathogens. The inclusion of disinfection processes as part of water treatment has been central to the virtual elimination of several important waterborne, communicable diseases (e.g. cholera and typhoid fever) in communities where these practices have been broadly applied. Conversely, outbreaks of these diseases are generally associated with communities, primarily in economically depressed areas or developing countries where social, economic, political, and cultural constraints often prevent or limit the extent to which disinfection and other forms of treatment are applied.

In general, water and wastewater treatment systems share a number of common features; however, there are also important differences between these systems that relate to water quality, treatment objectives, and individual unit operations. Likewise, there are many similarities between UV disinfection systems that are applied in production of potable water and wastewater treatment; but important differences among these systems are also evident.

The goal for this chapter is to present the basic principles that govern the design and application of UV disinfection systems in water and wastewater treatment. Features that these systems have in common are presented, including variables that affect process performance and the methods of analysis that are applied to these systems. Also included is a presentation of the aspects of water and wastewater UV disinfection systems that are specific to each application.

## 9.2 Primary vs. Secondary Disinfection

Disinfection processes are used to limit human exposure to microbial pathogens. In water treatment, disinfection can be accomplished by physical separation of microbial pathogens from water or by (photo)chemical alteration of aqueous constituents to reduce the concentration of viable or infective pathogenic microbes. Primary disinfection processes are those in which a disinfection process is used to inactivate microbial pathogens. In contrast, secondary disinfection is a process whereby an antimicrobial agent is applied to prevent repair or regrowth of microbial pathogens in water that has been subjected to a primary disinfection process. These terms are often used in the context of drinking water treatment, but may also be relevant in wastewater settings, especially when reuse of the treated water is planned.

*Photochemical Reactors: Theory, Methods, and Applications of Ultraviolet Radiation*, First Edition. Ernest R. Blatchley III.

Companion website: www.wiley.com/go/blatchley/photochemicalreactors

A wide range of disinfection processes can be applied to achieve primary disinfection, including UV irradiation and chemical disinfectants, such as (inorganic) chloramines, free chlorine, chlorine dioxide, and ozone, as well as combinations of these agents. Secondary disinfection generally requires the application of a chemical disinfectant that will provide a stable disinfectant residual, usually with the goal of ensuring the presence of the disinfectant throughout a distribution or storage system. Chlorine-based disinfectants are often applied for this purpose. UV irradiation is generally incompatible with the goals of secondary disinfection.

## 9.3 Motivations for Use of UV-Based Disinfection

UV irradiation is commonly used as a primary disinfectant because of its ability to function as a broad-spectrum antimicrobial agent. As will be discussed in more detail later, most vegetative bacteria, many viruses, and most relevant protozoan parasites are quite sensitive to UV exposure; however, some viruses and bacterial spores are relatively resistant to UV irradiation. But even when considering the extremes in reported sensitivity to UV exposure, the range of UV doses required to accomplish effective inactivation of most microbes is small as compared to the range of chemical disinfectant doses required to meet the same objective (see Figures 9.1 and 9.2). Alternatively, at the nominal doses that are commonly applied in conventional disinfection processes, UV irradiation is effective for control of most microbial pathogens. In general terms, common chemical disinfectants also function as broad-spectrum antimicrobial agents; however, the range of chemical disinfectant doses required to achieve a given extent of inactivation across a broad range of waterborne microbes tends to be wider than observed for UV-based processes.

Chlorination and UV irradiation are sometimes applied together to achieve disinfection. In part, this approach is used because the antimicrobial behaviors of these two agents are complementary. Specifically, UV-based processes tend to be highly effective for control of vegetative bacterial cells and protozoa, but may achieve limited inactivation of some viruses. Chlorine is also effective for control of vegetative bacteria; however, chlorination is relatively ineffective for control of some protozoan parasites, but tends to be highly effective for control of viruses. The ability of chlorine to function as a secondary disinfectant also contributes to the compatibility of disinfection systems that combine the use of UV and chlorine.

The photochemical reactions that lead to microbial inactivation in UV-based disinfection processes tend to be rapid and efficient. As such, the doses of UVC radiation required to achieve target levels of microbial inactivation are small as compared to the doses required to accomplish effective conversion of most other constituents in water, including most dissolved chemicals. One consequence of this fact is that UV disinfection processes tend to generate low or non-detectable concentrations of regulated disinfection byproducts (DBPs). It should be noted, however, that most of the historical focus on DBP formation has been on regulated DBPs. Moreover, the list of regulated DBPs was developed at a time when chemical disinfectants, especially free chlorine, represented essentially the only option for disinfection of water or wastewater. As such, the majority of regulated DBPs are associated with chlorination, with some being associated with disinfectants that are less frequently applied, such as ozone.

As a generalization, UV-based disinfection processes tend to produce (regulated) DBPs less effectively than chlorination and other chemical disinfection processes. It is inaccurate to suggest that there is no DBP formation associated with UV irradiation; however, much less is known about DBP formation in UV-based processes than with chemical disinfection processes.

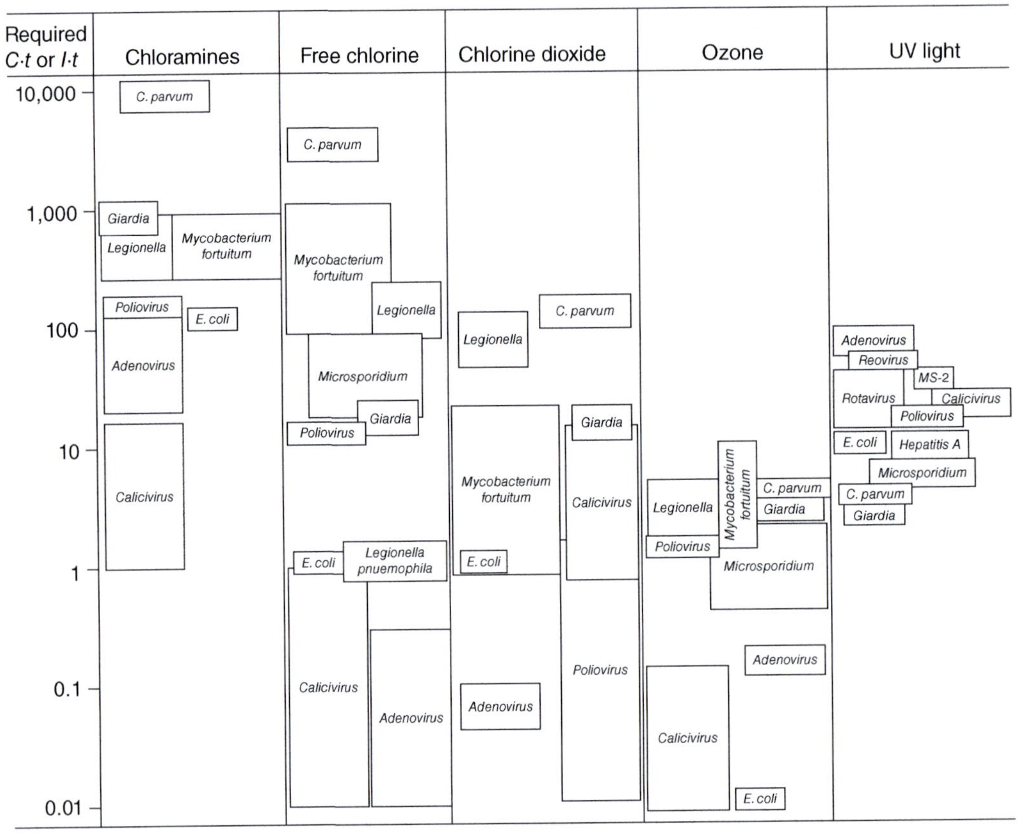


**Figure 9.1** Reported conditions of disinfection required for 99% inactivation of microbial pathogens based on the application of chloramines, free chlorine, chlorine dioxide, ozone, and $UV_{254}$ radiation. *Source:* Adapted from Jacangelo et al. [1].

The combined application of chemical disinfectants with UV irradiation can result in promotion of some DBP formation or decomposition reactions. For example, the combined application of chlorine and UV has been shown to increase the rate of chloramine decay. In the presence of some amine compounds, such as amino acids, creatinine, and uric acid, UV irradiation has been shown to enhance the formation of some DBPs, relative to chlorination alone [3–6]. This behavior is relevant in settings where UV irradiation is applied to water sources where substantial quantities of organic-N are present in water along with free chlorine. Important examples of such settings include swimming pools and water reuse applications.

UV-based disinfection systems present other advantages relative to conventional chemical disinfection systems. Because the reactions that are responsible for disinfection in UV systems are fast, the reactors themselves tend to occupy a small footprint, relative to their chemical analogs. Similarly, the electrical power requirements for UV-based disinfection tend to compare favorably with those of chlorine- or ozone-based disinfection.

As an illustration, consider the power requirements to disinfect water at a flow rate of $10^6 \ \frac{\text{L}}{\text{day}}$ using chlorine or UV irradiation. Electrochemical generation of chlorine is among the most power-intensive industrial processes. A representative power requirement for generation of chlorine by this method is on the order of $2.8 \ \frac{\text{kW} \cdot \text{hr}}{\text{kg} \cdot \text{Cl}_2}$ [7]. If we assume that chlorine is applied a dose of

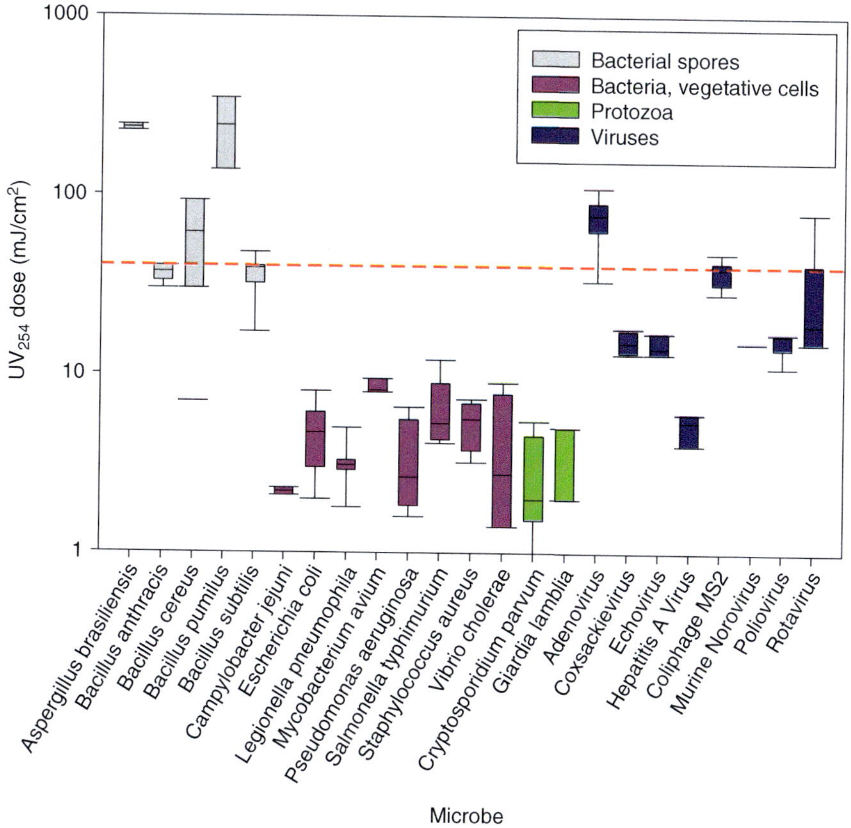


**Figure 9.2** Reported variability in $UV_{254}$ responses of microorganisms. Note that vertical axis is presented on a $\log_{10}$ scale. Data are presented as "box-and-whisker" representations of reported $UV_{254}$ doses required to achieve 2 $\log_{10}$ units (99%) inactivation of each microbe; from top to bottom, the horizontal lines in each box represent the $90^{th}$, $75^{th}$, $50^{th}$ (median), $25^{th}$, and $10^{th}$ percentile of each reported range. The dashed, horizontal red line is included to indicate a $UV_{254}$ dose of 40 $\frac{mJ}{cm^2}$, which is representative of the nominal dose (or RED) delivered in many common water disinfection processes. *Source:* Data from Malayeri et al. [2].

$1.0\ \frac{mg}{L}$ as $Cl_2$, this translates to an electrical power requirement of $2.8\ \frac{kW\cdot hr}{day}$ of operation. For comparison, disinfection of this same flow rate of water based on $UV_{254}$ radiation from conventional low-pressure Hg lamps would require on the order of 750 W of input electrical power, which in turn would translate to roughly 1.8 kW hr of electrical power consumption per day of operation. Of course, this type of calculation will be affected by a wide range of assumptions, as well as location-specific conditions that will influence how chlorine or UV might be applied for disinfection. To reiterate, the point of this calculation is that the electrical power requirements of UV-based disinfection are similar or favorable as compared with conventional chlorine-based disinfection.

Lastly, UV-based disinfection systems are generally perceived favorably by the public, especially relative to chlorination systems. In some communities (or more generally in some countries), chlorination is viewed unfavorably or even banned from public use because of concern over DBP formation, the generation of unpleasant taste and odor, or a combination of these factors.

## 9.4 Traditional View of Drinking Water and Municipal Wastewater as Separate Domains

In historical terms, UV disinfection systems for drinking water and municipal wastewater evolved separately (see Chapter 1). Interestingly, the first municipal-scale UV disinfection systems were applied in drinking water production. These systems were unreliable and emerged roughly coincidentally with the development of disinfection systems that were based on chlorine, which was both inexpensive and reliable. The rapid emergence of chlorine-based disinfection effectively eliminated UV-based disinfection from consideration in most applications at the time, including municipal-scale systems.

Among wastewater applications, UV-based disinfection systems began to emerge and compete with chlorination in the 1980s. As discoveries of drawbacks of chlorination became better understood (e.g. ineffectiveness against protozoan parasites, formation of DBPs, generation of disinfection-associated toxicity, and safety concerns associated with use of gaseous chlorine [8–14]), UV disinfection became more popular. This promoted the development of more efficient UV disinfection systems for wastewater. After the discovery of the effectiveness of UV against protozoan parasites (especially *Cryptosporidium* and *Giardia*) [9–17], UV disinfection of drinking water also emerged as an effective alternative to, or supplement for chlorination.

The communities of engineers and scientists who work on water and wastewater disinfection systems have often operated more-or-less independently of each other. Collectively, these attributes have allowed the principles of design and regulation of these systems to evolve along separate, but largely parallel tracks. For purposes of this chapter, both system types will be presented. However, it is important to recognize that important differences and similarities exist among water and wastewater disinfection.

### 9.4.1 Differences Between Water and Wastewater Disinfection

Perhaps the most obvious differences between water and wastewater disinfection systems relate to the quality of the water being treated. In general, wastewater will contain higher concentrations of particles, dissolved constituents, and microorganisms than source waters used in production of drinking water. The higher particle concentrations are relevant to UV-based disinfection because they negatively affect the optical characteristics of the water, both by scattering and absorption of UV radiation, and because particles can protect microorganisms from exposure to UV radiation (and other disinfectants). The dissolved constituents in wastewater often include unsaturated organic compounds that tend to absorb UVC radiation, thereby reducing UV transmittance (UVT) at the relevant wavelength(s). Other dissolved constituents that can affect UVT include natural organic matter (NOM), proteins, and nitrate. High counts of viable microorganisms translate into a larger (nominal) UV dose requirement to achieve a given target concentration of these microbes in disinfected water.

In addition to being characterized by relatively high concentrations of these constituents, wastewaters often display greater variability in these characteristics than drinking water sources. Water quality characteristics in wastewater are strongly influenced by the water uses that contribute to inflow at a facility. By contrast, water sources used for drinking water production tend to demonstrate less variability in quality than wastewater; this is especially true of groundwater sources.

Flow rates in wastewater systems can demonstrate considerable variability, especially in systems that include combined sewers. In these systems, runoff events from rainfall or snowmelt can dramatically increase volumetric flow rates delivered to a wastewater treatment facility. These changes in flow rate can also have dramatic effects on water quality, particularly at the beginning of a runoff event, when resuspension and transfer of materials from runoff surfaces can be promoted.

Another important distinction between water and wastewater applications of UV disinfection relates to the objectives of treatment; more broadly, UV disinfection system design is governed by the intended uses of the treated water. In potable water production, the goal is to provide water that is both affordable and safe to drink; in the context of disinfection, this means that the risk of exposure to microbial pathogens through direct ingestion should be low in drinking water supplies. In addition, the risks associated with exposure to DBPs must also be acceptably low. It is also important to recognize that in many countries potable water supplies are also used for applications other than drinking water, including bathing, household applications (e.g. washing clothes and dishes), and fire protection.

In contrast, wastewater disinfection systems are designed to minimize risks of human exposure, largely in connection with recreational activities or downstream uses of treated wastewater. As such, the allowable concentrations of (indicator) organisms in disinfected wastewater tend to be higher than those that are observed with drinking water systems.

Increasingly, reuse of treated wastewater is being practiced, especially in areas that experience regular water-stress events. A broad spectrum of reuse applications exists, ranging from irrigation of nonedible crops to direct potabilization. The central goal of disinfection in these applications will always focus on minimization of the risks of human exposure to microbial pathogens, but because the likely paths to human exposure are quite different among these endpoints and generally intermediate to those of drinking water and conventional wastewater treatment, the regulations are also intermediate to those of drinking water and conventional wastewater treatment.

Drinking water regulations are generally mandated by Federal agencies that have responsibility for public health and safety. In the United States, the U.S. EPA has responsibility for drinking water quality and has established the *National Primary Drinking Water Regulations* (NPDWR) to ensure safety of public water supplies. The NPDWRs include legally enforceable requirements for treatment as well as *Maximum Contaminant Levels* (MCLs) for contaminants in several broad categories, including microorganisms, disinfectants, DBPs, inorganic chemicals, organic chemicals, and radionuclides [18].

Systems that rely on surface water sources must also comply with the *Surface Water Treatment Rule* (SWTR). The SWTR directs the design and operation of potable water production from surface water sources, with particular focus on protection against prominent waterborne microbial pathogens, including *Cryptosporidium*, *Giardia*, enteric viruses, and *Legionella*. Filtration and disinfection are commonly applied to remove or inactivate these microbes, although the SWTR allows filtration to be excluded from treatment under some circumstances. Because the protozoan parasites *Cryptosporidium* and *Giardia* are so sensitive to exposure to UVC radiation, UV irradiation has emerged as the disinfection process of choice for inactivation of these microbes, though other disinfection processes (e.g. ozonation) can also be effective. Systems that are effective for control of *Cryptosporidium*, *Giardia*, and enteric viruses are also likely to be effective for control of *Legionella* and most other (nonregulated) bacteria. Drinking water production based on surface water sources is common among large municipalities. Consequently, UV disinfection has been incorporated into the treatment process trains of drinking water utilities in many urban areas.

The concentrations of viable or infective microbial pathogens in source waters can be quite low, often below the limit of detection. For many surface water sources, periods of high pathogen concentration tend to be episodic, and often coincide with hydrologic events that promote runoff and erosion. In addition, the concentrations of the microbes that can cause illness (i.e. microbial pathogens) can be difficult to measure. Rather than regulate drinking water on the basis of microbial quality, the NPDWRs and the SWTR mandate that conditions of treatment be established to provide predefined extents of removal or inactivation of *Cryptosporidium*, *Giardia*, and enteric viruses. Physical separation processes (e.g. filtration or membrane separation devices) and disinfection processes are given "credit" for well-defined quantities of "$\log_{10}$ removal or inactivation" of these microbes. Collectively, these removal and inactivation credits must add up to at least the requirements of the SWTR and the NPDWRs, as required for a given application.

For municipal wastewater systems, the regulatory strategy is quite different. Again using the US EPA's approach as an illustration, each point-source discharge (including all municipal wastewater treatment facilities) is issued a discharge permit under the *National Pollutant Discharge Elimination System* (NPDES). NPDES permits allow controlled discharge of pollutants from point sources, with a goal of protecting downstream water quality. More generally, these regulatory limits are intended to protect human health and the environment from exposure to wastewater-associated pollutants. Most NPDES permits do not stipulate the set of technologies used to meet the limits they describe. Rather, the people and organizations who are responsible for each point-source discharge are given the flexibility to choose the combination of treatment processes they will use to meet discharge permit limitations. Compliance is checked by routine monitoring of effluent water quality by a certified laboratory.

As related specifically to wastewater disinfection, effluent microbial quality is generally controlled through the use of NPDES permit limitations on the concentrations of indicator bacteria. For systems that discharge to lakes and rivers, these limits are usually defined in terms of an upper limit on the concentration of *Escherichia coli*, fecal coliforms, or both. For systems that discharge to coastal waters, *Enterococcus* limits are often included. In recognition of the toxicity associated with residual (+1-valent) chlorine, NPDES permits may also include an upper limit on the residual chlorine concentration in an effluent. The EPA is also considering the inclusion of permit limitations that would allow for control of enteric viruses. These limits are motivated by the fact that enteric viruses are responsible for a large fraction of foodborne and waterborne illness. The human Norovirus (HNV), in particular, has been identified as the causative agent for 95% of all nonbacterial acute gastroenteritis and at least 60% of all food-borne illnesses [19].

UV disinfection is remarkably effective for control of vegetative bacteria, including the common indicator bacteria described above, and most pathogenic bacteria (see Figures 9.1 and 9.2). Some enteric viruses have been shown to be recalcitrant to UVC exposure. Adenoviruses, in particular, are known to be resistant to UVC radiation at wavelengths that are known to be effective for inflicting damage to DNA (i.e. radiation in the vicinity of 260–270 nm, which coincides roughly with the output of low-pressure Hg lamps). However, these same viruses have been shown to be somewhat more sensitive to radiation at wavelengths below 240 nm, probably because of damage to proteins that is caused by these relatively short wavelengths [20–23].

Until recently, there was relatively little information to define the susceptibility of HNV to UV disinfection processes. To a large degree, the lack of information related to HNV was attributable to the fact that no cell-culture-based method has been developed to allow for use of an infectivity assay, as is commonly done with viruses. This situation may change in the future, as a cell-culture assay for HNV has been developed and demonstrated [24]. Although this method has yet been applied for quantification of the responses of HNV to disinfectants, there has been progress in this area to provide quantitative information about the behavior of a closely related virus. Specifically, the murine (mouse) Norovirus (MNV) has been evaluated as a surrogate for HNV. MNV and HNV are closely related, in terms of their morphology and the composition of their genomes. Figure 9.3 illustrates the $UV_{254}$ dose–response behaviors of MNV and Coliphage MS2 in two aquatic

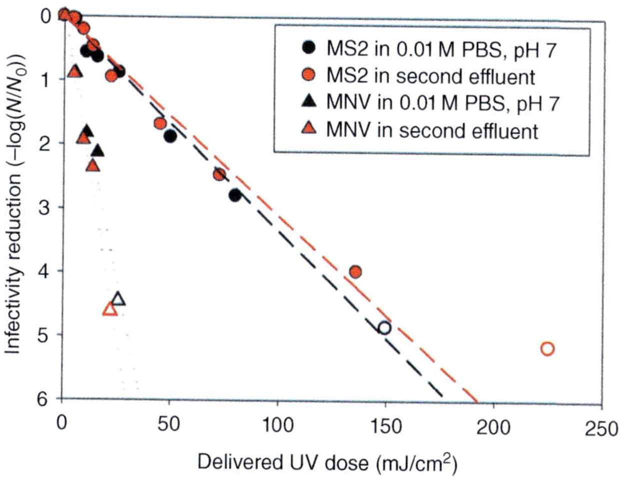


**Figure 9.3** Measured $UV_{254}$ dose–response behaviors of Coliphage MS2 and murine Norovirus (MNV) in phosphate-buffered saline (PBS) solution and in secondary municipal wastewater effluent [25]. *Source:* Weng et al. 2018/with permission of Elsevier.

matrixes [25]. More than 4 $\log_{10}$ units of MNV inactivation was observed at a $UV_{254}$ dose of approximately $30\frac{mJ}{cm^2}$, which corresponds with a nominal dose target that is often used in design and operation of UV disinfection systems that are used in treatment of municipal wastewater. By comparison, MS2 demonstrated only about 1 $\log_{10}$ unit of inactivation at this same $UV_{254}$ dose. These data suggest that UV disinfection should be effective for control of Noroviruses, and that MS2 is not an appropriate surrogate for Noroviruses in UV disinfection applications.

As described above, there is increasing interest in reuse of municipal wastewater. Common reuse applications include irrigation of nonedible or edible crops, but have expanded to include a wide range of other end uses, up to and including direct potabilization. Regulations for these applications are, in some respects, intermediate to those applied for potable water and municipal wastewater applications, largely because the finished water and intended applications also tend to be intermediate to those that apply to potable water production and municipal wastewater treatment.

In the United States, regulation of reuse applications is the responsibility of each state. An example of the regulatory framework for wastewater reuse has been defined by the State of California through regulations in *Titles 17* and *22* of the *California Code of Regulations* [26]. These regulations stipulate treatment requirements for a range of reuse applications, as summarized in Table 9.1.

In California, when a utility plans to use UV as an alternative to conventional chlorination, they are required to demonstrate that disinfection performance equivalent to the requirements of chlorination are met. The National Water Research Institute (NWRI) developed a set of guidelines to assist designers and regulators in validation of UV-based disinfection systems for reuse applications [27]. The NWRI Guidelines for UV-based disinfection in wastewater reuse applications are summarized in Table 9.2. They stipulate conditions used in validation experiments (i.e. biodosimetry) and in operation of these systems.

The State of California has also identified acceptable uses for wastewater effluents that have been treated to conform to the definitions described above. The basic goal of identifying these acceptable uses is to allow use of reclaimed water in a water-stressed areas, while providing an acceptable measure of protection against the risk of communicable disease transmission associated with these reuse applications. Table 9.3 provides a summary of unacceptable and unacceptable uses for reclaimed water in California.

**Table 9.1** Summary of reuse regulations defined by the State Water Resources Control Board in California.

| | Treatment minima | | Disinfection | | |
|---|---|---|---|---|---|
| **Reuse category** | **Oxidation**[a] | **Filtration**[b] | **Total coliforms** | | **Other requirements** |
| Disinfected secondary-23 recycled water | ✓ | | $< 23\ \frac{\text{MPN}}{100\ \text{mL}}$<br>$< 240\ \frac{\text{MPN}}{100\ \text{mL}}$ | 7-day median<br>No more than one sample in a 30-day period | |
| Disinfected secondary-2.2 recycled water | ✓ | | $< 2.2\ \frac{\text{MPN}}{100\ \text{mL}}$<br>$< 23\ \frac{\text{MPN}}{100\ \text{mL}}$ | 7-day median<br>No more than one sample in a 30-day period | |
| Disinfected tertiary recycled water | ✓ | ✓ | $< 2.2\ \frac{\text{MPN}}{100\ \text{mL}}$<br>$< 23\ \frac{\text{MPN}}{100\ \text{mL}}$<br>$< 240\ \frac{\text{MPN}}{100\ \text{mL}}$ | 7-day median<br>No more than one sample in a 30-day period<br>Always | Chlorine[c]<br>Other disinfectants[d] |

*Source:* Adapted from [26].

[a] "Oxidized wastewater" is defined as wastewater in which organic matter has been stabilized, is nonputrescible, and contains dissolved oxygen.

[b] "Filtered wastewater" is oxidized wastewater that has been subjected to a subsequent physical separation process, which may include treatment by the use of a granular media filter and/or by membrane-based processes (i.e. microfiltration, ultrafiltration, nanofiltration, or reverse osmosis). Hydraulic loading limitations are defined for granular media filters, and turbidity limitations are applied for all filtered waters.

[c] Minimum conditions required for chlorine-based disinfection include a minimum disinfectant exposure of $450\ \frac{\text{mg} \cdot \text{min}}{\text{L}}$ and a modal contact time of at least 90 minutes under peak dry-weather flow conditions.

[d] When alternatives to chlorine disinfection are applied, they must be validated to achieve at least 5 $\log_{10}$ units inactivation of coliphage MS2, poliovirus, or an alternative challenge virus that is at least as resistant as poliovirus.

**Table 9.2** Summary of recommended UV disinfection system validation conditions, as defined by the National Water Research Institute [27].

| **Filtration process** | **Design dose**[a] $\left(\frac{\text{mJ}}{\text{cm}^2}\right)$ | **Operating conditions**[b] | **Turbidity requirements** |
|---|---|---|---|
| Non-membrane[c] | 100 | Maximum daily flow rate<br>$UVT_{254} \geq 55\%$ | $\leq$2 NTU (24-hour mean)<br>$\leq$5 NTU (95% of time)<br>$\leq$2 NTU (always) |
| Membrane filtration[d] | 80 | Maximum daily flow rate<br>$UVT_{254} \geq 65\%$ | $\leq$0.2 NTU (95% of time)<br>$\leq$0.5 NTU (always) |
| Reverse osmosis | 50 | Maximum daily flow rate<br>$UVT_{254} \geq 90\%$ | $\leq$0.2 NTU (95% of time)<br>$\leq$0.5 NTU (always) |

[a] Reduction Equivalent Dose (RED), as measured by biodosimetry. Coliphage MS2 is recommended as the challenge organism.

[b] $UVT_{254}$ measured through a 1.0 cm optical path length.

[c] Granular media, cloth, or synthetic media.

[d] Microfiltration (MF) or ultrafiltration (UF).

**Table 9.3** Summary of allowed and non-allowed reuse applications for recycled water in California. Green indicates allowed applications; red indicates non-allowed applications.

| Non-potable recycled water uses | Disinfected tertiary recycled water | Disinfected secondary-2.2 recycled water | Disinfected secondary-23 recycled water | Undisinfected secondary recycled water |
|---|---|---|---|---|
| **Irrigation** | | | | |
| Food crops where recycled water contacts the edible portion of the crop, including all root crops | Green | Red | Red | Red |
| Parks and playgrounds | Green | Red | Red | Red |
| School yards | Green | Red | Red | Red |
| Residential landscaping | Green | Red | Red | Red |
| Unrestricted-access golf courses | Green | Red | Red | Red |
| Any other irrigation uses not prohibited by other provisions of the California Code of Regulations | Green | Red | Red | Red |
| Food crops, surface-irrigated, above-ground edible portion, and not contacted by recycled water | Green | Green | Red | Red |
| Cemeteries | Green | Green | Green | Red |
| Freeway landscaping | Green | Green | Green | Red |
| Restricted-access golf courses | Green | Green | Green | Red |
| Ornamental nursery stock and sod farms with unrestricted public access | Green | Green | Green | Red |
| Pasture for milk animals for human consumption | Green | Green | Green | Red |
| Nonedible vegetation with access control to prevent use as a park, playground, or school yard | Green | Green | Green | Red |
| Orchards with no contact between edible portion and recycled water | Green | Green | Green | Red |
| Vineyards with no contact between edible portion and recycled water | Green | Green | Green | Red |
| Nonfood-bearing trees, including | Green | Green | Green | Red |
| Christmas trees not irrigated less than 14 days before harvest | Green | Green | Red | Red |
| Fodder and fiber crops and pasture for animals not producing milk for human consumption | Green | Green | Red | Red |
| Seed crops not eaten by humans | Green | Green | Green | Green |
| Food crops undergoing commercial pathogen-destroying processing before consumption by humans | Green | Green | Green | Green |
| Ornamental nursery stock, sod farms not irrigated less than 14 day before harvest | Green | Green | Green | Green |

*(Continued)*

**Table 9.3** (Continued)

| Non-potable recycled water uses | Disinfected tertiary recycled water | Disinfected secondary-2.2 recycled water | Disinfected secondary-23 recycled water | Undisinfected secondary recycled water |
|---|---|---|---|---|
| **Impoundment supply** | | | | |
| Nonrestricted recreational impoundments, with supplemental monitoring for pathogenic organisms | | | | |
| Restricted recreational impoundments and publicly accessible fish hatcheries | | | | |
| Restricted recreational impoundments and publicly accessible fish hatcheries | | | | |
| Landscape impoundments without decorative fountains | | | | |
| **Cooling or air conditioning supply** | | | | |
| Industrial or commercial cooling or air conditioning involving cooling tower, evaporative condenser, or spraying that creates a mist | | | | |
| Industrial or commercial cooling or air conditioning not involving cooling tower, evaporative condenser, or spraying that creates a mist | | | | |
| **Other non-potable water applications** | | | | |
| Flushing toilets and urinals | | | | |
| Priming drain traps | | | | |
| Industrial process water that may contact workers | | | | |
| Structural fire fighting | | | | |
| Decorative fountains | | | | |
| Commercial laundries | | | | |
| Consolidation of backfill material around potable water pipelines | | | | |
| Artificial snow making for commercial outdoor uses | | | | |
| Commercial car washes, not heating the water, excluding the general public from washing process | | | | |
| Industrial process water that will not come into contact with workers | | | | |
| Industrial boiler feedwater | | | | |
| Nonstructural fire fighting | | | | |

**Table 9.3** (Continued)

| Non-potable recycled water uses | Disinfected tertiary recycled water | Disinfected secondary-2.2 recycled water | Disinfected secondary-23 recycled water | Undisinfected secondary recycled water |
|---|---|---|---|---|
| Backfill consolidation around non-potable piping | | | | |
| Soil compaction | | | | |
| Mixing concrete | | | | |
| Dust control on roads and streets | | | | |
| Cleaning roads, sidewalks, and outdoor work areas | | | | |
| Flushing sanitary sewers | | | | |

*Source:* Modified from [28].

Another important distinction between UV disinfection systems used in production of potable water and those used for disinfection of wastewater relates to available head (potential) energy to drive flow through these systems. In potable water production facilities, it is common for water to be pumped through a UV disinfection system, perhaps followed by a wet well, then to a distribution system. Although there is always motivation to minimize energy losses (i.e. pumping costs) through treatment systems, the head losses that usually accompany UV disinfection systems tend to be small. This is one of the motivating factors for the use of closed-vessel reactors in drinking water disinfection. Figure 9.4 illustrates several examples of closed-vessel UV disinfection systems used in the production of potable water. These systems operate under positive pressure. They often are built around MP or LP Hg lamps, but alternative UV sources, such as UV LEDs and excimer lamps may also be used for these applications.

In contrast, UV disinfection systems used in municipal wastewater treatment are often constrained by head loss because treated (disinfected) wastewater is often discharged by gravity to a nearby body of water, such as a river, lake, or coastal ocean water. In part because of this, many UV disinfection systems used for disinfection of municipal wastewater are designed as open-channel reactors. Head loss through these systems is generally modest; however, because these systems operate with a free surface, head loss results in a change in free surface elevation. As such, reactor designs must include measures to address the slope of the free surface profile. Common solutions to this problem include the use of a "flap gate," the use of a stepped-channel, or an elongated weir (see section below titled "Hydraulic Behavior of UV Disinfection Systems"). Figure 9.5 illustrates common UV system lamp configurations used in municipal wastewater disinfection applications.

It should be noted that wastewater reuse applications have characteristics that are intermediate to those of conventional potable water production and municipal wastewater treatment. In fact, these applications share aspects of both types of application. Similarly, when it comes to head loss and its effects on UV disinfection process hardware selection, the details of the application will govern viable reactor types. In some reuse applications, open-channel UV disinfection systems are warranted, while in other applications, closed-vessel reactors are appropriate. Moreover, in reuse

**Figure 9.4** Photographic images of UV photoreactors used for disinfection in potable water production across a wide range of application types. (Upper left) Point-of-use device, 52 cm long, 10 cm diameter, power consumption 50 W, nominal flow rate rating 34 $\frac{L}{min}$ to deliver a nominal $UV_{254}$ dose of 40 $\frac{mJ}{cm^2}$. (Upper right) Two LPHO amalgam lamp reactors operating in parallel. Each reactor houses eight LPHO lamps. Reactor length 119 cm, diameter 20 cm. (Lower left) Cross-flow reactor based on medium-pressure Hg lamp technology. System with eight MP lamps is rated for a nominal flow rate of 3950 $\frac{m^3}{hr}$. (Lower right) Large-scale reactors based on LPHO lamp technology. Reactors pictured are located within the Catskill–Delaware Ultraviolet Water Treatment Facility, which provides drinking water for the City of New York, NY. It is the largest UV disinfection facility in the world. *Source:* Images courtesy of Trojan Technologies.

applications, UV systems may also be used to promote changes in water chemistry through direct photolysis or radical-mediated reactions (see Chapter 10).

## 9.5 Disinfection Kinetics

Microbial inactivation in UV disinfection systems results from photochemically induced damage to one or more critical biomolecules [29]. Like all photochemical reactions, the rates of photochemical damage to biomolecules depend on the ability of the target molecule to absorb radiation (formally quantified by the molar absorption coefficient, $\varepsilon$) and the efficiency by which this

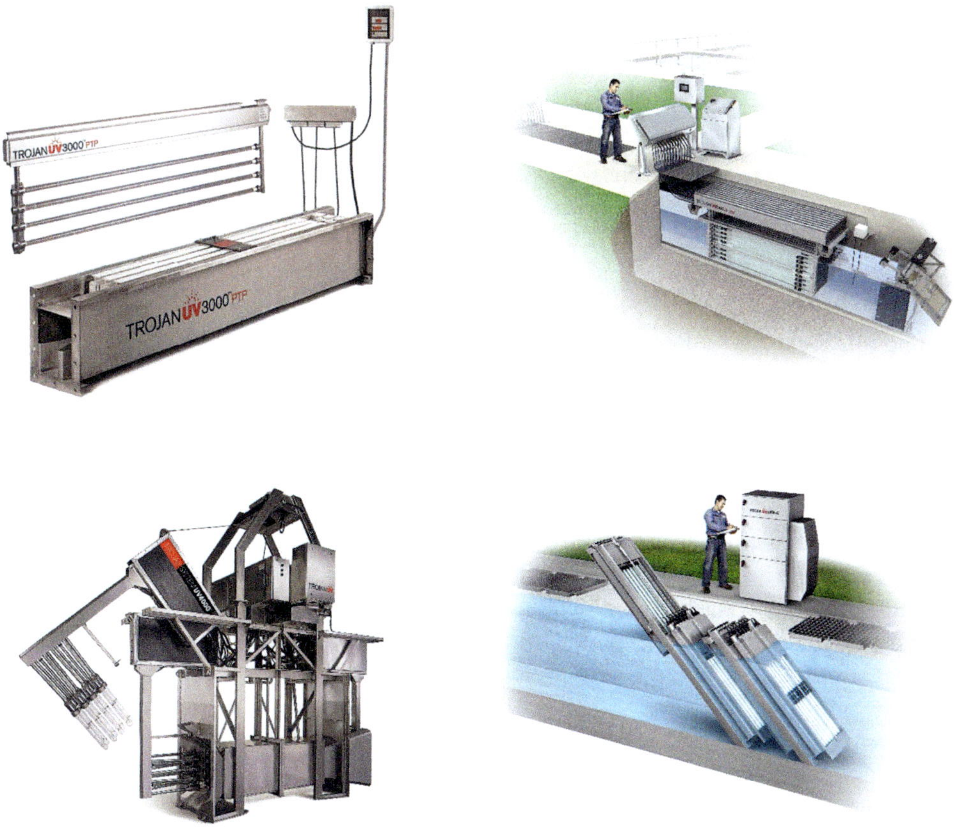


**Figure 9.5** Images of UV photoreactors used for disinfection of municipal wastewater. (Upper left) LPHO reactor designed for applications in "Packaged Treatment Plants" (flow rates in the range of 3–5 MGD [470–790 $\frac{m^3}{hr}$]). (Upper right) Artist's rendering of a conventional open-channel reactor system with LPHO lamps with a "horizontal" lamp orientation. (Lower left) Medium-pressure lamp reactor; open-channel entrance and exit sections not pictured. (Lower right) Artist's rendering of an open-channel LPHO reactor with a diagonal lamp orientation. *Source:* Images courtesy of Trojan Technologies.

absorbed radiation energy is used to bring about chemical change (formally quantified by the quantum yield, $\Phi$).

For most microorganisms, photochemical damage to nucleic acids (DNA and RNA) is largely responsible for their inactivation. Absorption spectra for pyrimidine and purine bases as well as DNA have been presented previously (see Figures 8.2 and 8.3, respectively). These compounds tend to absorb strongly across the UVC spectrum, with peak absorbance being observed in the vicinity of 260 nm. Quantum yields for dimer formation among the purines (adenine [A] and guanine [G]) have been reported to be on the order of $10^{-4}\ \frac{\text{mole}}{\text{einstein}}$, which is roughly one to two orders of magnitude lower than the pyrimidines (thymine [T], cytosine [C], and uracil [U]) [29]. The two most important forms of nucleic acid damage associated with UVC exposure are the formation of cyclobutane pyrimidine dimers (CPD) and pyrimidine pyrimidone photoadducts (6–4 photoproducts). In general, the next most important form of damage involves the production of pyrimidine photohydrates, which tend to form readily in uracil and cytosine, but not in thymine. Because of this, the formation of photohydrates tends to be more important in RNA than in DNA; in fact, recent evidence suggests that photohydrate formation may be the dominant form of UVC-induced damage in RNA [30].

Among the nucleic acids, damage to DNA tends to play a greater role in microbial inactivation than damage to RNA for several reasons [29]. First, DNA is a unique molecule that carries genetic information and is present as one copy per cell. Second, RNA exists in several forms (mRNA, tRNA, and rRNA), which are often present as multiple copies that can be replaced. Third, DNA comprises polymers of adenine (A), guanine (G), cytosine (C), and thymine (T), whereas in RNA, T is replaced by uracil (U). This is relevant in that T is more efficient than the other pyrimidines or the purines in participating in photodimerization reactions. Therefore, replacement of T with another base will generally decrease reaction efficiency.

Proteins can also be damaged by exposure to UVC radiation, but the UV absorption characteristics of proteins are substantially more variable than those of the nucleic acids, largely because of the wide range of structures that exist among the amino acids that comprise them. Also, protein molecules typically exist as multiple copies. In general, protein damage resulting from UVC exposure is less efficient than damage to nucleic acids. An important exception to this generalization has been observed at UVC wavelengths less than about 240 nm, where many proteins often absorb effectively. For radiation at these short wavelengths, damage to proteins can be an important contributor to microbial inactivation, especially among viruses, where the protein coat (capsid) of a virion and other proteins often play critical roles in viral infection and reproduction.

The responses of viruses to germicidal UV radiation (UVC and UVB) are particularly important because they are responsible for a large fraction of waterborne disease. Their simple structure, which generally involves a nucleic acid molecule surrounded by (and to a certain degree protected by) a protein coat (capsid) also facilitates understanding of the basic photobiochemistry that is responsible for their inactivation. Experiments conducted with viral single-stranded (ss) and double-stranded (ds) nucleic acids that had been stripped from their corresponding capsids demonstrated the following order of response to $UV_{254}$ irradiation [30]:

$$\text{ssDNA} > \text{ssRNA} \approx \text{ds DNA} > \text{dsRNA}$$

This order of response is generally consistent with the order of sensitivity of virus types to UVC irradiation and is largely attributable to differences in the structures of these various forms of nucleic acids.

### 9.5.1 Mathematical Models of UV Disinfection Kinetics

The intrinsic kinetics of photoreactions are critical to understanding the behavior of photochemical reactor systems. In a purely photolytic process, the kinetics of the reaction can be monitored by tracking the concentration of the parent molecule or a product of photolysis as a function of the applied dose of radiation. In that sense, kinetic expressions for pure photochemical processes provide a literal description of reaction progress.

In the case of UV disinfection, the intrinsic kinetics of inactivation are a characteristic of the target organism and the wavelength(s)-imposed radiation. Also, the endpoint of interest is microbial inactivation, which is quantified as a loss of the ability of the organism to reproduce, infect a host cell, or otherwise complete its life cycle (see Chapter 8). It is clear that microbial inactivation results from UV-induced damage to one or more critical biomolecules; however, the assays that are used to measure the response of a microbial population to UV exposure do not usually provide a direct measure of this damage. Rather, these assays provide a measure of the *effect* of this damage on the microbe in terms of its ability to reproduce or infect. As a result, the mathematical models that are used to describe UV disinfection kinetics (or for that matter, the kinetics of any disinfection process) include a measure of empiricism, which is used to relate a change in viability/infectivity to a UV dose.

It is also important to note that inactivation of an individual microbial particle can result from one or more photochemical events, but can also be influenced by repair processes, secondary reactions, and other complicating factors. Therefore, it should be apparent that models of disinfection kinetics tend to provide a less rigorous description of the response of a photochemical target than the models that are used to describe pure photochemical reactions. However, the logic used in the development of models to describe UV disinfection kinetics is borrowed from, and in many ways similar to the logic that is used to describe the kinetics of pure photochemical reactions.

#### 9.5.1.1 Single-Event Model

The simplest model used to describe UV disinfection kinetics assumes that a single photochemical event, which results in a single unit of photochemical damage (often referred to as a lesion), will cause inactivation of the microbial target. This model follows the logic of an elementary, bimolecular reaction or an elementary photochemical reaction (see Chapter 2) in that the kinetics of the reaction are described as being functions of terms that describe the availability of each reactant to participate in the reaction. In the case of UV disinfection, the "reactants" of interest are viable/infective target microbes and UV photons; the availabilities of these reactants to participate in photochemical reactions are described by the concentration of viable/infective target microbes and the local fluence rate, respectively. The logic used in developing the single-event model is similar to the logic used to develop the so-called Chick–Watson kinetic model that is often used with chemical disinfectants [31, 32]. As such, the single-event model is sometimes referred to as a Chick–Watson model.

The fundamental form of the single-event model is as follows:

$$\frac{dN}{dD} = -kN \tag{9.1}$$

where

$N$ = concentration of viable target organisms $[=]\ \frac{\text{cfu}}{\text{L}}$ or $\frac{\text{pfu}}{\text{L}}$

$k$ = inactivation constant $[=]\ \frac{\text{m}^2}{\text{J}}$ or $\frac{\text{cm}^2}{\text{mJ}}$

$D$ = UV dose applied to organisms $[=]\ \frac{\text{J}}{\text{m}^2}$ or $\frac{\text{mJ}}{\text{cm}^2}$.

The formal definition of the applied dose is:

$$D = \int_0^{\tau} E'(t) \cdot dt \tag{9.2}$$

where

| | | | | |
|---|---|---|---|---|
| $E'(t)$ | = | fluence rate | [=] | $\frac{W}{m^2}$ or $\frac{mW}{cm^2}$ |
| $t$ | = | time | [=] | s |
| $\tau$ | = | exposure duration (upper limit of integration) | [=] | s. |

Separation of variables and integration allows for the development of an algebraic expression to describe the rate of microbial inactivation by the single-event model:

$$\ln\left(\frac{N}{N_0}\right) = -kD \tag{9.3}$$

Figure 9.6 provides a graphical illustration of the single-event model, as described by Equation 9.3. When plotted in the form of $\ln\left(\frac{N}{N_0}\right)$ vs. $D$, the form that naturally emerges from Equation 9.3, the slope of a best-fit (straight) line through the data, forced through the origin,[a] provides an estimate of the inactivation constant.

The single-event model often provides a reasonable fit to observed UV dose–response data for simple organisms, including many viruses and vegetative bacterial cells. It is also generally applicable for small extents of inactivation, even for higher organisms. However, there are common deviations from single-event behavior that are observed in actual UV dose–response data sets. The most common deviations from single-event (i.e. first-order) behavior are so-called "shoulder" and "tailing." A shoulder will be observed in a microbial population when little or no inactivation is

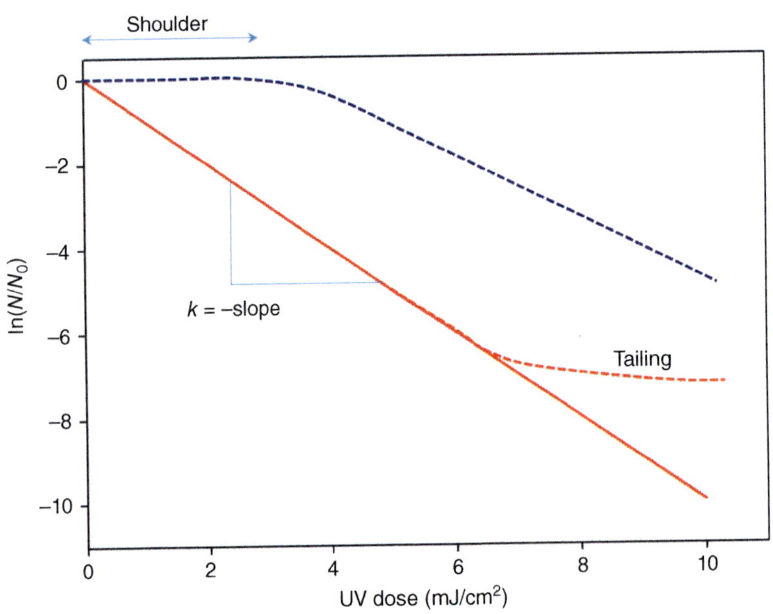


**Figure 9.6** Illustration of the single-event model for the case of $k = 1.0\,\frac{cm^2}{mJ}$ (solid red line: -), an inactivation constant that is representative of many vegetative bacteria. The negative of the slope of this line provides an estimate of the inactivation constant. Common deviations from single-event kinetics are illustrated by the dashed lines. The blue dashed line illustrates a dose–response relationship with a characteristic "shoulder," whereas the red dashed line indicates "tailing" in the dose–response relationship.

observed until a threshold dose is applied, after which the fraction of viable organisms in the population often falls off rapidly, usually in a pattern that is consistent with first-order (i.e. Chick–Watson or single-event) response. The existence of a shoulder has been attributed to the ability of some organisms to retain viability even after experiencing a nominal amount of UV-induced damage. Alternatively, some have suggested that the existence of a shoulder may be attributable to repair mechanisms, which may operate in parallel with reactions that cause damage and are likely to be most effective when relatively small amounts of damage have been imposed on the target microbe. Tailing has been attributed to microbial protection by particles and to phenotypic variation among microbes in a population [33]. In most cases, it is difficult or impossible to provide a quantitative estimate of the fraction of tailing behavior that can be attributed to these two factors.

Although Figure 9.6 illustrates a graphical representation of UV dose–response behavior among a microbial population that is the logical derivative based on the mathematical expressions that define disinfection kinetics, many people are more comfortable with an alternative graphical representation, as illustrated in Figure 9.7. Here, the data from Figure 9.5 are transformed to provide an illustration of UV dose–response behavior in the form of $\log_{10}\left(\frac{N}{N_0}\right)$ vs $D$. Although this graphical form does not emerge directly from the governing equation, it provides a useful graphical summary of dose–response behavior, in that some people are more comfortable with base-10 log interpretation than base-e. Also, many times disinfection designs are based on a target value of $\log_{10}$ units of microbial inactivation.

#### 9.5.1.2 Series-Event Model

The existence of a "shoulder" in a UV dose–response relationship is sometimes attributed to the ability of a microorganism to withstand a limited amount of UV-induced damage before any change in viability is observed. The series-event model was developed from the hypothesis that a threshold level of damage can be withstood by microorganisms [29, 34]. The series-event model is based on the assumption that an individual target, such as a DNA molecule within an organism, may require

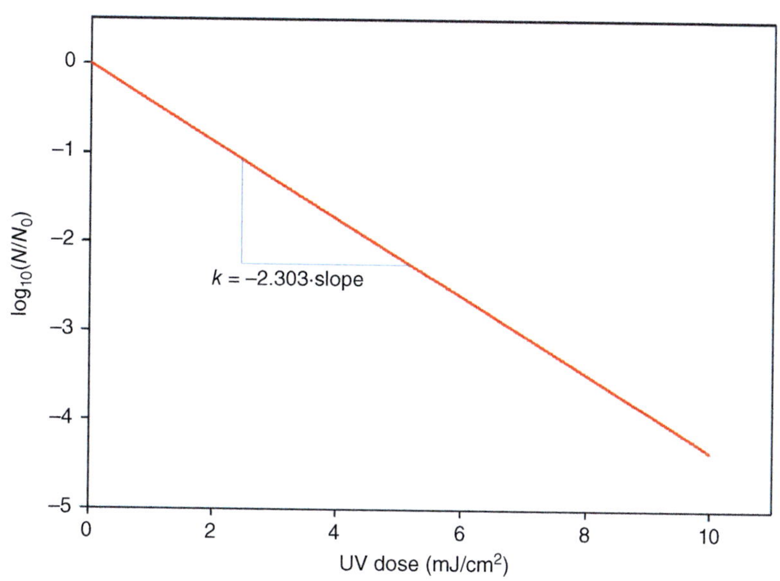


**Figure 9.7** Illustration of the single-event model for the case of $k = 1.0\ \frac{\text{cm}^2}{\text{mJ}}$ in base-10 log form.

more than one unit of damage to cause inactivation of the organism that contains the target. Damage to the target molecule within each microbe is assumed to accumulate in discrete, integer values. In this idealized scenario, microorganisms will only lose viability after they have reached this threshold; organisms that have accumulated less than the damage threshold will retain viability.

The logic of this model is based on serial (step-wise) accumulation of UV-induced damage (lesions) within a microorganism:

$$M_0 \xrightarrow{h\nu} M_1 \xrightarrow{h\nu} M_2 \xrightarrow{h\nu} \cdots \xrightarrow{h\nu} M_{n-1} \Big| \xrightarrow{h\nu} M_n \xrightarrow{h\nu} \cdots \tag{9.4}$$

where

$M_i$ = microorganism with i units of damage

$h\nu$ = energy associated with a single photon.

It is assumed that all photochemical events in this series are identical and that they are characterized by an identical rate constant, $k$. It is further assumed that all microorganisms with damage at event-levels of $n - 1$ or less retain viability (see red dividing line in equation 9.4), while all microorganisms with event-levels of $n$ or higher are inactivated. In that sense, $n$ describes a threshold number of units of damage that are required to initiate microbial inactivation.

The concentration of microbes at each event level can be quantified using the logic of a simple photochemical reaction. For organisms at event-level 0 ($i = 0$), their concentration is defined as:

$$\frac{\mathrm{d}[M_0]}{\mathrm{d}D} = -k[M_0] \tag{9.5}$$

where $[M_i]$ represents the concentration of microbes with $i$ units of accumulated photochemical damage.

For event levels $i \geq 1$, a similar expression can be written to describe the dose-dependent concentration of microbes at a given event level:

$$\frac{\mathrm{d}[M_i]}{\mathrm{d}D} = -k[M_i] + k[M_{i-1}] \tag{9.6}$$

Based on the definitions provided above, we can also define the concentration of viable organisms, $N$, often expressed in units of $\frac{\text{cfu}}{\text{L}}$ or $\frac{\text{pfu}}{\text{L}}$, or similar:

$$N = \sum_{i=0}^{n-1} [M_i] \tag{9.7}$$

In effect, Equations 9.5, 9.6, and 9.7 define a series of differential equations that can be used to quantify the concentration of viable organisms:

$$\begin{aligned}
\frac{\mathrm{d}[M_0]}{\mathrm{d}D} &= -k[M_0] \\
\frac{\mathrm{d}[M_1]}{\mathrm{d}D} &= -k[M_1] + k[M_0] \\
\frac{\mathrm{d}[M_2]}{\mathrm{d}D} &= -k[M_2] + k[M_1] \\
&\vdots \\
\frac{\mathrm{d}[M_{n-1}]}{\mathrm{d}D} &= -k[M_{n-1}] + k[M_{n-2}]
\end{aligned} \tag{9.8}$$

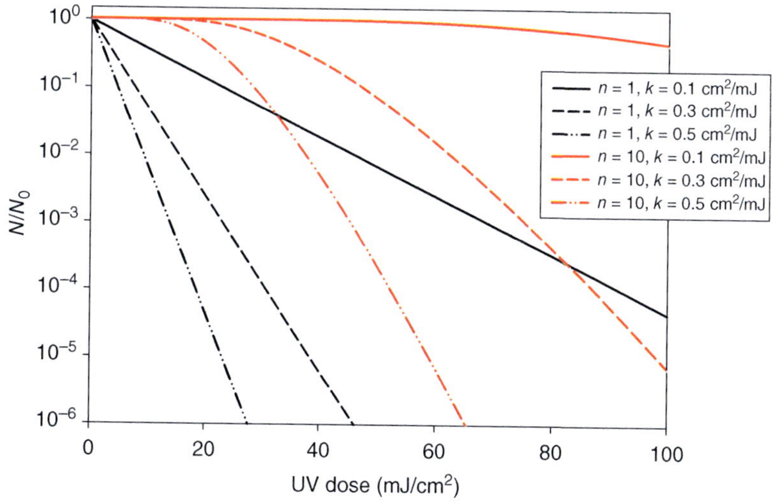


**Figure 9.8** Illustration of series-event kinetics for a range of parameter values. Black lines illustrate examples for $n = 1$ with increasing values of $k$; this model corresponds to the single-event model. Red lines illustrate examples of $n = 10$ with increasing values of $k$.

Solution of this series of differential equations yields the following governing equation:

$$\frac{N}{N_0} = \exp(-kD)\sum_{i=0}^{n-1}\frac{(kD)^i}{i!} \tag{9.9}$$

where

$N_0$ = concentration of viable organisms prior to UV exposure $[=] \frac{\text{cfu}}{\text{L}}$ or $\frac{\text{pfu}}{\text{L}}$.

Figure 9.8 illustrates the series-event model for a range of parameter values. Black lines in Figure 9.8 illustrate the series-event model for $n = 1$, with $k$ ranging from 0.1 to 0.5 $\frac{\text{cm}^2}{\text{mJ}}$. The series-event model for $n = 1$ reduces to the single-event model (i.e. equation 9.9 reduces to equation 9.1 for $n = 1$). The model indicates that the rate of inactivation will increase linearly with the rate constant.

The red lines in Figure 9.8 illustrate examples corresponding to $n = 10$ for the same range of values of $k$ as applied for the single-event model. These curves are characterized by a "shoulder," the width of which is governed by the value of $n$ in the model. The slope of the dose–response relationship increases with dose after the shoulder, eventually becoming parallel to the line for lower values of $n$ at the same dose.

Regression analysis can be applied for estimation of "best-fit" values of model parameters. However, the process is somewhat more complicated with the series-event model than with the single-event model because of the inclusion of two parameters and the mathematical form of Equation 9.9, which governs the shape of these curves. One approach to parameter estimation is to conduct stepwise regression, wherein a value of $n$ is selected, then nonlinear regression is applied to the data set to identify a conditional "best-fit" estimate of $k$. Most regression tools will provide a numerical value of critical equation fitting parameters, including the sum of residual sum of squared errors (RSS).[b] This procedure is repeated over a wide range of values of $n$. The "best-fit" pair of $(n, k)$ may be identified as the pair that yields the lowest value of RSS among the conditional best-fit values.

Figure 9.9 illustrates actual data that have been "fit" to the series-event model for three microorganisms: a virus, a bacterium, and a fungus. For the virus, the single-event model is used to fit to

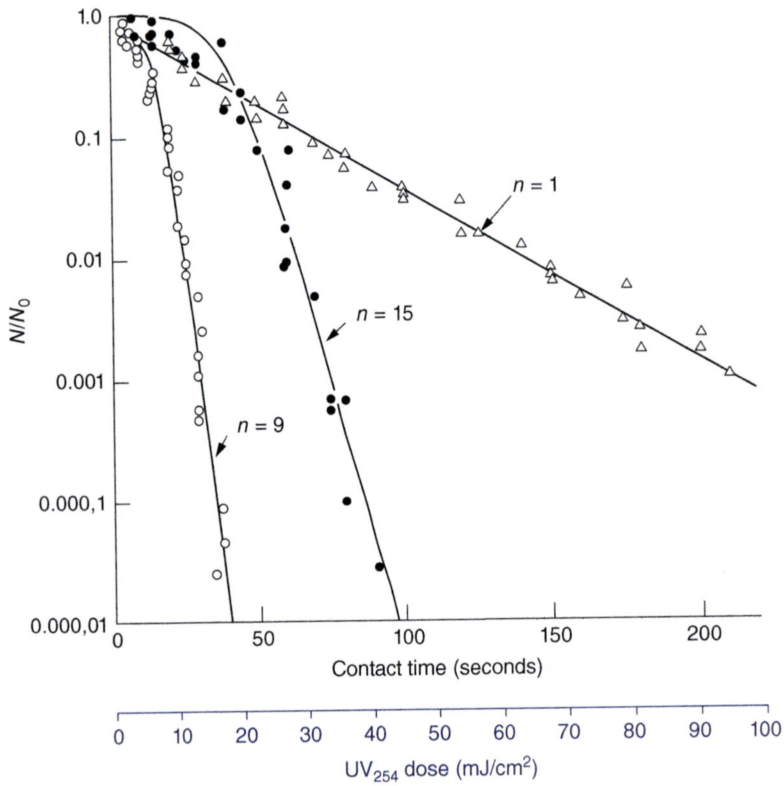


**Figure 9.9** Fits of the series-event model to $UV_{254}$ dose–response behavior of three microorganisms. Open triangles (Δ) represent responses of bacteriophage f2; open circles (○) indicate responses of the bacterium *Escherichia coli*; closed circles (•) indicate responses of the fungus *Candida parapsilosis*. Figure modified from the original version [34] to include $UV_{254}$ doses delivered at each corresponding exposure time (secondary horizontal axis in blue).

the data. As the organisms increase in complexity, the magnitude of the shoulder increases, as reflected in the value of the threshold parameter, $n$.

#### 9.5.1.3 Multi-Target Model

A related model assumes that each microbe may include one or more distinct target molecules, with each target requiring one or more units of damage to cause inactivation of a target. However, microbial inactivation will require that one or more such targets are inactivated.

In the simplest examples of the multi-target model, it is assumed that each target requires one unit of damage to cause its inactivation. Furthermore, it assumed that a critical number of targets, $n_c$, all be inactivated within a microbe to cause its inactivation. For modeling purposes, we will define:

$$[T] = \text{concentration of undamaged targets} \tag{9.10}$$

It is assumed that all targets are identical, in terms of their susceptibility to photochemical damage. Inactivation of targets is assumed to follow first-order kinetics:

$$\frac{d[T]}{dD} = -k[T] \tag{9.11}$$

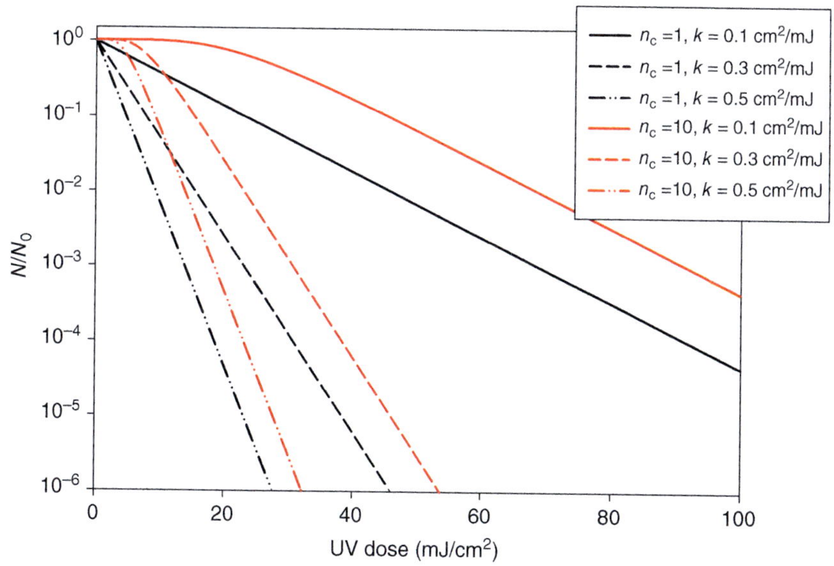


**Figure 9.10** Illustration of multi-target kinetics for a range of parameter values. Black lines illustrate examples for $n_c = 1$ with increasing values of $k$; this model corresponds to the single-event model. Red lines illustrate examples of $n_c = 10$ with increasing values of $k$.

Separation of variables and integration yields:

$$\frac{[T]}{[T]_0} = \exp(-kD) \tag{9.12}$$

Equation 9.12 describes the fraction of targets that will remain undamaged after exposure to a given UV dose. By extension,

$$1 - \exp(-kD) = \text{fraction of targets that are damaged after exposure to a given dose} \tag{9.13}$$

If we assume that $n_c$ targets must be inactivated to cause microbial inactivation, then the fraction of organisms within a population that retains viability is defined as:

$$\frac{N}{N_0} = 1 - (1 - \exp(-kD))^{n_c} \tag{9.14}$$

Figure 9.10 provides a graphical illustration of the implications of the multi-target model. For cases in which $n_c = 1$, the model reduces to the single-event model, corresponding to simple first-order kinetics. For values of $n_c > 1$, the model yields a "shoulder" in the dose–response curve. The slope of the dose–response curve increases with increasing dose, eventually becoming parallel to the slope of the curve at $n_c = 1$ for corresponding values of the inactivation constant, $k$.

The multi-target model provides a graphical illustration of UV dose–response data that illustrates a characteristic shoulder, followed by a region that is roughly log-linear, as was evident with the series-event model. However, the multi-target model is less consistent with the mechanism of inactivation that is commonly associated with microorganisms than the series-event model. As such, the series-event model may be viewed as superior to the multi-target model for descriptions of the UV dose–response behavior of microbial populations.

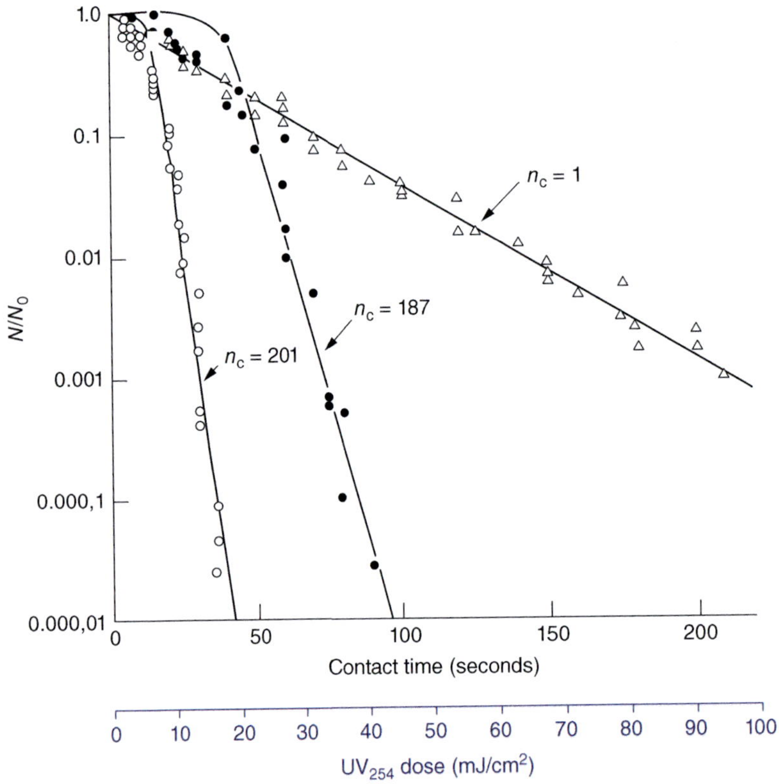


**Figure 9.11** Fits of the multi-target model to $UV_{254}$ dose–response behavior of three microorganisms. Open triangles (Δ) represent responses of bacteriophage f2; open circles (○) indicate responses of the bacterium *Escherichia coli*; closed circles (•) indicate responses of the fungus *Candida parapsilosis*. Figure modified from the original version [34] to include $UV_{254}$ doses delivered at each corresponding exposure time (secondary horizontal axis in blue).

Figure 9.11 illustrates actual data that have been "fit" to the multi-target model for the same three microorganisms indicated in Figure 9.9. For the virus, the single-event model is again used to fit to the data. As the organisms increase in complexity, the magnitude of the shoulder increases, as reflected in the value of the "target" parameter, $n_c$.

#### 9.5.1.4 Two-Population Models

The other common deviation from single-event kinetics is "tailing" behavior, which manifests as a decrease in the kinetics of inactivation with increasing UV dose. Graphically, tailing behavior appears as a declining slope in a UV dose–response plot.

Tailing behavior has been attributed to population heterogeneity (i.e. phenotypic variation) and interactions of microorganisms with particles. In the case of phenotypic variability, the idea is that within a population of microbes, there is likely to be a range of physical and physiological characteristics, such as size, shape, and resistance to sources of external stress.

Waterborne microorganisms often demonstrate an affinity for particles. Inclusion in a particle can provide protection from outside stress and also provide a common location for individuals within a species population. In the context of UV disinfection systems, it has also been demonstrated that exposure to UVC radiation promotes aggregation of bacteria [35]. More broadly, particle-association has been demonstrated to result in apparent resistance to UVC exposure for a wide range of waterborne microbes [36–40].

In many situations, it is not possible to define the specific roles played by phenotypic variability and particle-association in affecting the responses of microorganisms to UVC exposure. However, tailing behavior is evident in many measured dose–response relationships for waterborne microbes and needs to be accounted for in UV disinfection system designs.

The Phenotypic Persistence and External Shielding (PPES) model was developed to account for the existence of a shoulder and tailing behavior in microbial dose–response relationships [33]. The PPES model is a two-population model in which the overall response of a microbial population is represented as the sum of the responses of two subpopulations, one that is sensitive to UVC exposure, the other is resistant to UVC exposure:

$$N_0 = A_0 + B_0 \tag{9.15}$$

where

$A_0$ = initial concentration of viable, susceptible microbes

$B_0$ = initial concentration of viable, persistent microbes.

The responses of each subpopulation to UVC exposure are assumed to be described by the series-event model, or a simplified form (i.e. single-event model). If shouldering behavior is evident in measured dose–response behavior, then a series-event description of the dynamic behavior of the sensitive subpopulation may be warranted. On the other hand, the nature of tailing behavior is such that it is difficult to identify a shape other than log-linear in the tailing portion of most dose–response curves. Therefore, the responses of the two subpopulations can be described by:

$$A = A_0 \cdot \exp\left(-k_A \cdot D\right) \cdot \sum_{i=0}^{n-1} \frac{(k_A \cdot D)^i}{i!} \tag{9.16}$$

$$B = B_0 \cdot \exp\left(-k_B \cdot D\right) \tag{9.17}$$

where

$A$ = concentration of viable microorganisms from the sensitive subpopulation

$B$ = concentration of viable microorganisms from the persistent subpopulation.

Finally, it is assumed that the overall concentration of viable microorganisms ($N$) is the sum of the concentrations of microorganisms from the two subpopulations:

$$N = A + B \tag{9.18}$$

Substitution of Equations 9.16 and 9.17 into Equation 9.18 and rearrangement yields:

$$\frac{N}{N_0} = \frac{A_0}{N_0} \cdot \exp\left(-k_A \cdot D\right) \cdot \sum_{i=0}^{n-1} \frac{(k_A \cdot D)^i}{i!} + \frac{B_0}{N_0} \cdot \exp\left(-k_B \cdot D\right) \tag{9.19}$$

Equations 9.15 and 9.19 represent the PPES model in its basic form. Collectively, the equations that comprise this model include four independent parameters ($A_0$ or $B_0$, $k_A$, $k_B$, and $n$). In general, these parameters cannot be measured independently; this is a common attribute of two-population models. Therefore, it is necessary to apply a logical process to a dose–response data set to develop estimates for each of these parameters. This can be done by piecewise regression, in which an inflection point is identified within a data set that allows identification of $\frac{A_0}{N_0}$ or $\frac{B_0}{N_0}$. Once this is done,

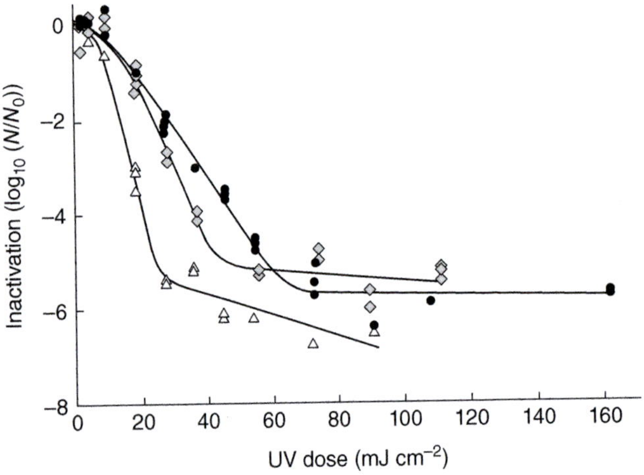


**Figure 9.12** Measured UVC dose–response behavior of *B. subtilis* spores at wavelengths of 222 nm (Δ), 254 nm (♦), and 282 nm (•). *Source:* Data from Pennell et al. [33].

then the regression methods described above can be applied to data that have been assigned to each subpopulation for purposes of estimation of the other model parameters.

Figure 9.12 illustrates fits of the PPES model to measured responses of *Bacillus subtilis* spores to UVC exposure at three wavelengths: 222, 254, and 282 nm. At each of these wavelengths, a well-defined "shoulder" is apparent, followed by nearly log-linear loss of viable spores, then a sharp decline in inactivation efficiency that is identifiable by an inflection point in the dose–response curve at each wavelength. The PPES model provides good fits to each of these data sets. Similar fits of the PPES model to other dose–response data sets are presented by Pennell et al. [33].

## 9.6 Microbial Repair Processes

A fraction of the ambient solar spectrum is able to cause damage to critical biomolecules, including nucleic acids and proteins. The UVB portion of the solar spectrum causes damage by mechanisms that are similar to those that are associated with UVC radiation, but UVA radiation is also known to cause damage to biomolecules. As a result of this, organisms across a wide range of taxonomic orders have evolved to develop processes that accomplish repair of UV-induced damage. Among these processes, those that are responsible for repair of damage to nucleic acids, especially DNA, are critical and quite common.

Some basic terminology is applied in the descriptions of these processes. Specifically, the term "reactivation" (aka, "recovery") is used to describe macroscopically observable regaining of the ability of an organism to reproduce or infect a host, while the term "repair" is used to describe processes that take place at the molecular scale that lead to reactivation [29].

Focusing on the molecular scale, at least three basic categories of repair processes have been identified:

1) Photochemical reversal of UV-induced damage: Photochemical reactions lead to reversal of UV-induced damage to DNA. Examples include photoenzymatic repair and photolysis of pyrimidine dimers to (re)form their respective monomers.

2) Replacement of UV-damaged nucleotides: This is an example of a "dark repair" process in which enzymatic processes promote removal of damaged nucleotides and replacement with undamaged molecules.
3) Recombination repair: This is a second example of "dark repair" in which undamaged regions of DNA are replicated to allow re-formation of ds DNA.

Brief descriptions of each process are presented below.

### 9.6.1 Photochemical Reversal of UV-Induced Damage

These processes are sometimes referred to as "photoreactivation" processes, in that they involve one or more photochemical reactions that result in reversal or repair of UV-induced damage to DNA. The most common of these processes, photoenzymatic repair, was first reported by Kelner [41, 42], but has been observed subsequently among many prokaryotes (bacteria, cyanobacteria) and eukaryotes (algae, fungi, protozoa, arthropods, vertebrates, and some plants) [29]. However, not all species within these groups demonstrate this capability.

Photoreactivation enzymes (E), or photolyases, are monomeric proteins that comprise 450–550 amino acids and two non-covalently bound chromophores [43, 44]. Photoreactivation generally follows a mechanism that can be described by Michaelis–Menten kinetics, wherein a photoreactivation enzyme will recognize and bind to a substrate, in the form of cyclopyrimidine dimer (CPD) or other UV-induced damage. The enzyme–substrate complex will absorb radiation, usually in the range of 310 nm $\lesssim \lambda \lesssim$ 480 nm (near UV to violet-blue) to yield the enzyme molecule and a repaired DNA segment:

$$E + S \underset{k_2}{\overset{k_1}{\rightleftarrows}} ES \overset{h\nu}{\rightarrow} E + P \tag{9.20}$$

The first step in this process is a "dark" reaction that leads to the formation of an enzyme/lesion complex (ES). The rate constants for the reversible binding of E and S are such that the equilibrium for this process tends to lie far to the right, meaning most of the enzyme will be bound to UV-damaged lesions. ES will absorb radiation in the range 310 nm $\lesssim \lambda \lesssim$ 480 nm effectively; the quantum yield for photolysis of ES is thought to range between $0.1 \lesssim \Phi \lesssim 1.0 \ \frac{\text{mole}}{\text{einstein}}$ [29]. As described in Chapter 2, the rate of a photochemical reaction is governed by the product of the molar absorption coefficient ($\varepsilon$) and the quantum yield ($\Phi$). Figure 9.13 illustrates this product for several photolyases as a function of wavelength; such a plot is referred to as an "absolute action spectrum," in that it illustrates the basic factors that govern a photochemical reaction rate as a function of wavelength.

A second type of photoreactivation process involves re-formation of pyrimidine monomers by photolysis of their dimers. The Thymine : Thymine dimer $\left(\widehat{TT}\right)$ and its monomer will be used to illustrate this behavior. It is known that the formation of $\widehat{TT}$ is relatively efficient in DNA, as compared to the reactions that lead to the formation of other pyrimidine dimers. In experiments involving Thymine/Thymine dinucleotides as a simplification of a DNA molecule, it was shown that the absorption spectrum of $\widehat{TT}$ is quite different than the absorption spectrum for Thymine alone. These spectra are illustrated in Figure 9.14.

The differences in the absorption spectra for Thymine and its dimer are largely attributable to the formation of a cyclobutane ring, which eliminates the 5-6 double-bond of the nucleotide. This double-bond is responsible for strong absorption in the UVC range. Also included in Figure 9.14 is the ratio of these absorption coefficients, $\epsilon_{\widehat{TT}} : \epsilon_T$.

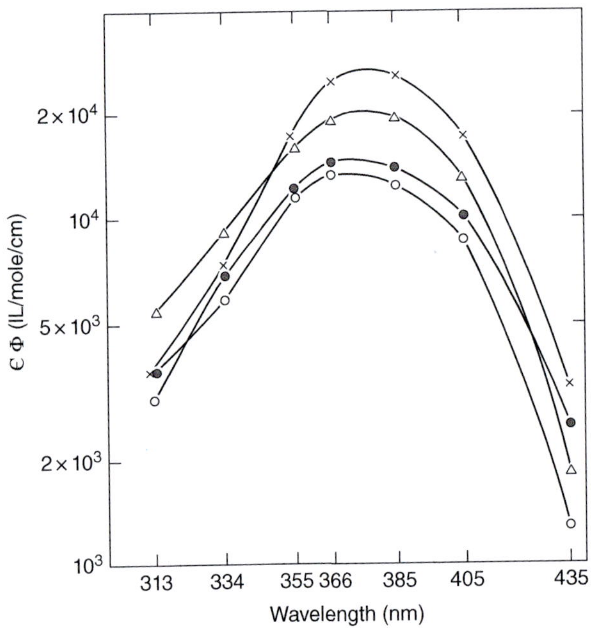


**Figure 9.13** Absolute action spectra for photoenzymatic repair of several aquatic microbes [29]. Stationary-phase *E. coli* $B_{s-1}$-160 cells (*x*); stationary-phase *Saccharomyces cerevisiae* uvrl-2 (•); *Haemophilus* transforming DNA in-vitro with yeast photoreactivation enzyme at low purification (○) and at high purification (Δ).

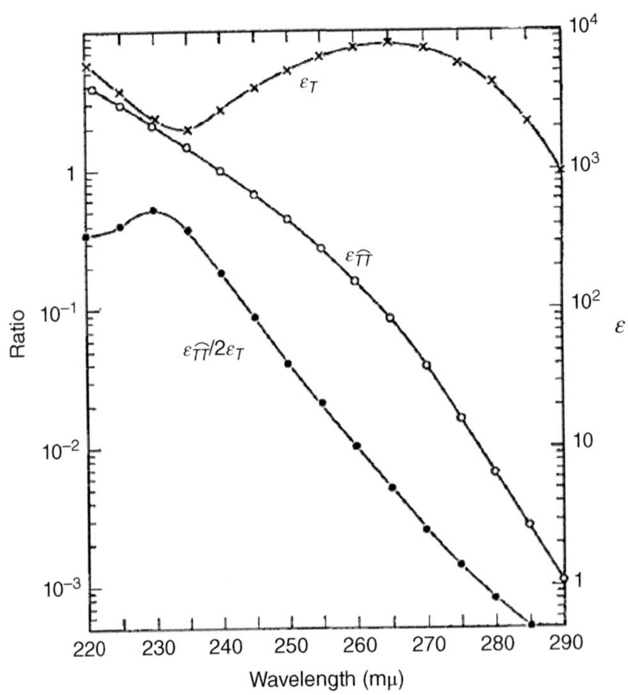


**Figure 9.14** Absorption spectra for Thymine/Thymine dimer $\left(\widehat{TT}\right)$ and Thymine (*T*) alone [45]. The molar absorption coefficients for the dimer $\left(\epsilon_{\widehat{TT}}\right)$ and the monomer ($\epsilon_T$) refer to the right, vertical axis. The left, vertical axis represents the ratio $\epsilon_{\widehat{TT}} : \epsilon_T$.

Because pyrimidine dimers can be formed *and* re-monomerized by exposure to UVC radiation, it is possible to establish a photosteady-state condition between $T$ and $\widehat{TT}$. Equation 9.21 illustrates this process, along with the factors that govern the rates of the forward and reverse processes. Since the rates of these reactions are governed by the products of their molar absorption coefficients and their quantum yields, $\epsilon_T \cdot \Phi_T$ and $\epsilon_{\widehat{TT}} \cdot \epsilon\Phi_{\widehat{TT}}$ for the forward and reverse reactions are illustrated in Equation 9.21, respectively.

$$T + T \underset{\epsilon_{\widehat{TT}} \cdot \epsilon\Phi_{\widehat{TT}}}{\overset{\epsilon_T \cdot \Phi_T}{\rightleftarrows}} \widehat{TT} \tag{9.21}$$

Not surprisingly, the photosteady-state condition is strongly influenced by the ratio of their respective molar absorption coefficients, which is illustrated in Figure 9.14 (left, vertical axis) as a function of wavelength [45]. The data illustrated in this figure illustrate that the Thymine monomer is always a more effective absorber than the dimer; however, for wavelengths less than roughly 235 nm, the difference between these absorption coefficients is small. On the other hand, at longer wavelengths, that dimer becomes substantially less effective at absorbing UV radiation than the monomer. In relative terms, this means that the reverse reaction becomes slower than the forward reaction for these longer wavelengths. Consequently, the photosteady-state condition at 280 nm is strongly shifted toward dimer formation, whereas this condition is shifted more toward the monomer at shorter wavelengths.

### 9.6.2 Excision-Resynthesis Repair (aka Excision Repair)

This process involves four discrete steps, each of which is promoted by a separate enzyme [29]. The process is illustrated in schematic form in Figure 9.15.

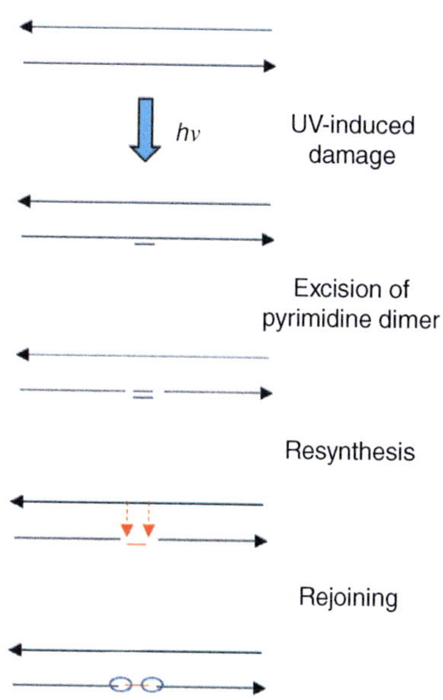


**Figure 9.15** Schematic illustration of steps involved in excision repair (adapted from Harm [29]). From the top: exposure to UV radiation causes the formation of a pyrimidine dimer; the damaged segment is cleaved from the DNA strand; resynthesis of the original (undamaged) segment is promoted using the complementary DNA strand as a template; the resynthesized material is covalently bound to the DNA strand to complete the repair.

In the first step, the repair endonuclease promotes cleavage of the DNA backbone adjacent to the lesion on a DNA strand. This is followed by excision of the damaged section. Resynthesis of the original (undamaged) nucleotide sequence then follows, using the intact, complementary strand as the template for its formation. Lastly, the resynthesized nucleotide sequence is covalently bound to the DNA strand to complete the repair process.

Excision repair is common among many microbial cells, but has also been demonstrated in mammalian cells, including human cells. It appears to be a common defense mechanism among organisms that are subjected to solar UVB radiation, which causes DNA damage. Excision repair allows repair of pyrimidine dimers as well as spore photoproduct, and in that sense it is less specific than photorepair as a defense mechanism against UV-induced damage.

### 9.6.3 Recombination Repair

This dark repair process is similar, though distinct from and complementary to excision repair and it requires the *recA* gene. Following UV irradiation, cells that are able to replicate will generate polynucleotide sequences that are much shorter than those that are formed among similar, unirradiated cells. These short strands are formed because pyrimidine dimers within the DNA strand prevent complete replication, and lead to the formation of gaps within the daughter DNA strands. These short strands are joined into higher molecular weight strands, which then serve as templates for DNA synthesis. The UV photoproducts (i.e. pyrimidine dimers) remain in the original, parent DNA material but are absent in the replicated DNA.

### 9.6.4 Effects of Repair Processes

The repair processes serve to reverse damage caused by exposure to UV radiation among many cell types. In effect, these processes reduce the *effective* UV dose delivered to these cells by repairing damage or replacing damaged nucleic acid segments.

This behavior is critical to public health, in the sense that it increases the nominal UV dose requirement for production of safe water or water that provides an acceptably low risk of disease transmission. The repair processes are most relevant in the limit of low UV doses, where the extent of nucleic acid damage is modest. As the extent of nucleic acid damage increases, it becomes increasingly difficult for cells to repair the damage. Consequently, UV disinfection processes require that the UV dose (distribution) delivered in the treatment process yields damage to critical biomolecules that exceeds the ability of the target microbes to repair UV-induced damage.

## 9.7 Hydraulic Behavior of UV Disinfection Systems

As with most reactor systems, mass and momentum transport play several crucial roles in governing the behavior of UV photoreactors, including those used for disinfection. Microscale mixing is critical to UV dose delivery in a photoreactor. On a larger (macro) scale, fluid mechanics also affects process performance and the design of UV photoreactors. Relevant macroscale hydraulic behavior in UV disinfection systems includes head loss and flow conditioning.

### 9.7.1 Head Loss

Generally speaking, head losses across UV photoreactors used in water treatment tend to be small relative to those that are observed in other processes, such as physical separation processes like filtration and membrane-based separations. Despite the relatively small magnitude of these losses, they can play important roles in UV disinfection system performance. Also, system designs need to account for these energy losses so as to minimize their adverse effects on process performance.

Head losses in systems that involve fluid motion past solid objects can be attributed to two groups of phenomena. First, there are frictional losses, that are attributable to the effects of fluid shear (viscosity) expressed at the solid:fluid boundary and in its immediate vicinity. Second, there are losses associated with pressure differentials that are associated with bluff bodies in flowing fluid. Table 9.4 provides schematic illustrations of fluid behavior in the immediate vicinity of some basic solid:fluid boundary shapes and the associated trends in form drag and skin friction.

For the case of a flat surface oriented nominally parallel to the direction of flow (Table 9.4A), skin friction will dominate. Momentum boundary layers will develop in the vicinity of the solid:fluid interface. Table 9.4B illustrates a solid:fluid boundary that corresponds to a "streamlined" shape, which tends to reduce, but not eliminate form drag. The streamlined shape will reduce velocity gradients on the downstream side of the object, as well as reducing the drop of pressure in this same area, relative to a cylindrical object (Table 9.4C). A cylindrical object will experience considerable form drag, largely in connection with the formation of a low-pressure recirculation zone that will be established in the area immediately downstream of the cylinder. The circulation patterns that emerge in these areas tend to be unstable and to display oscillatory behavior, sometimes referred to as "vortex shedding." The example illustrated in Table 9.4D is a flat plate oriented perpendicular to the nominal flow direction. Head loss across this object will be almost entirely attributable to form drag and will be accompanied by a recirculation zone with vortex shedding. The energy (head) loss that is observed with a fluid moving past this object is largely attributable to the difference in pressure that will be established between the regions located immediately upstream and downstream of the object.

Head losses present differently in open-channel and closed-vessel reactors. Therefore, the effects of these energy losses on each system type will be presented separately.

#### 9.7.1.1 Open-Channel Systems

Head loss in open-channel systems coincides with the change in free surface elevation as a function of downstream distance. Regardless of the lamp orientation, this change in liquid level (depth) can adversely affect the performance of UV disinfection systems if there is a significant deviation either above or below the design liquid depth. Actual liquid level in an open-channel reactor will be determined by the flow rate, the hydraulic capacity of the open channel, and any downstream liquid-level control device that is used to regulate liquid depth. Figure 9.16 illustrates the possible effects of inadequate liquid-level control for a system with lamps oriented parallel to the nominal flow direction (i.e. horizontal lamps, parallel to flow).

The effluent liquid-level measurements illustrated in Figure 9.16 were conducted on an open-channel UV disinfection system for which liquid-level control was accomplished using an elongated weir. At the highest flow rates, effluent liquid level was such that a substantial fraction of

**Table 9.4** Schematic illustrations of fluid behavior in the vicinity of basic solid:fluid boundaries and the associated trends in form drag and skin friction. Arrows in C and D indicate nominal directions of recirculation in the vortex shedding regions of these flow fields.

| Example | Solid : fluid boundary shape and corresponding streamlines | Form drag | Skin friction |
|---|---|---|---|
| A | 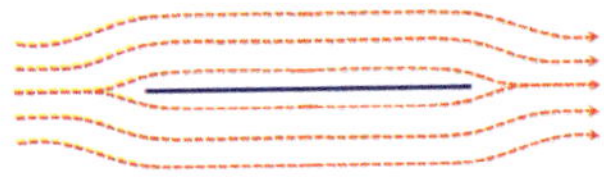 |   |   |
| B | 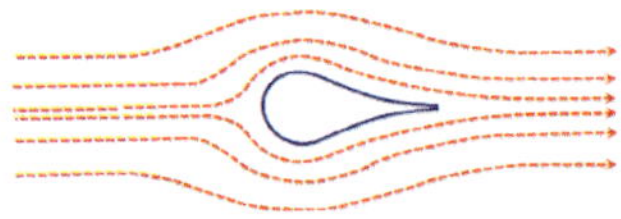 | | |
| C | 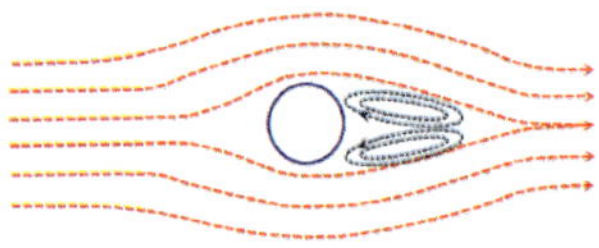 | | |
| D | 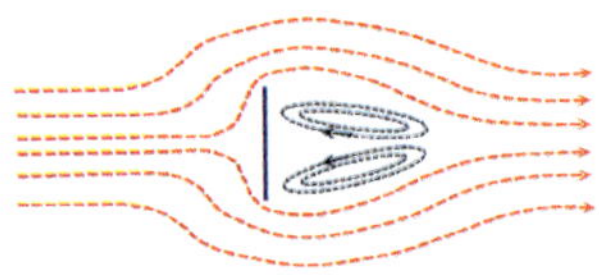 | | |

the flow was exposed to a low fluence rate. In addition, at these high flow rates, the exposure time was relatively short. These conditions will lead to a dose distribution that is characterized by a large fraction of microbial targets receiving a low UV dose; in other words, this reactor is not likely to perform well under these circumstances. As flow rate decreased, liquid level also decreased; system performance will improve as flow rate decreases because of reductions in the fraction of fluid that is poorly irradiated and longer residence time in the irradiated zone, both of which will shift the UV dose distribution toward larger values.

In some systems, it is possible for the liquid level to drop to a point where some or all of a horizontally oriented lamp, and possibly the lamp electrodes, could be unsubmerged (note: this behavior is not evident in the results illustrated in Figure 9.16). This can promote overheating of the lamp as well as inorganic foulant accumulation, both of which can adversely affect process performance. The need for liquid-level control will be evident in all open-channel reactor systems, irrespective of the lamp orientation.

In open-channel systems, the free surface profile will change along the entire length of the irradiated zone. The pattern of change in the free surface profile will be influenced by lamp orientation. For reactors with a horizontal lamp orientation (parallel to the direction of flow), energy losses will be attributable to two general mechanisms, as described above. The first is associated with the development of a momentum boundary layer adjacent to all stationary, solid surfaces (skin friction). These losses will be attributable to viscous losses along channel walls and the lamp/sleeve assemblies themselves. The second mechanism will be attributable to the support structures (form drag), which are likely to be vertically oriented, perpendicular to the nominal

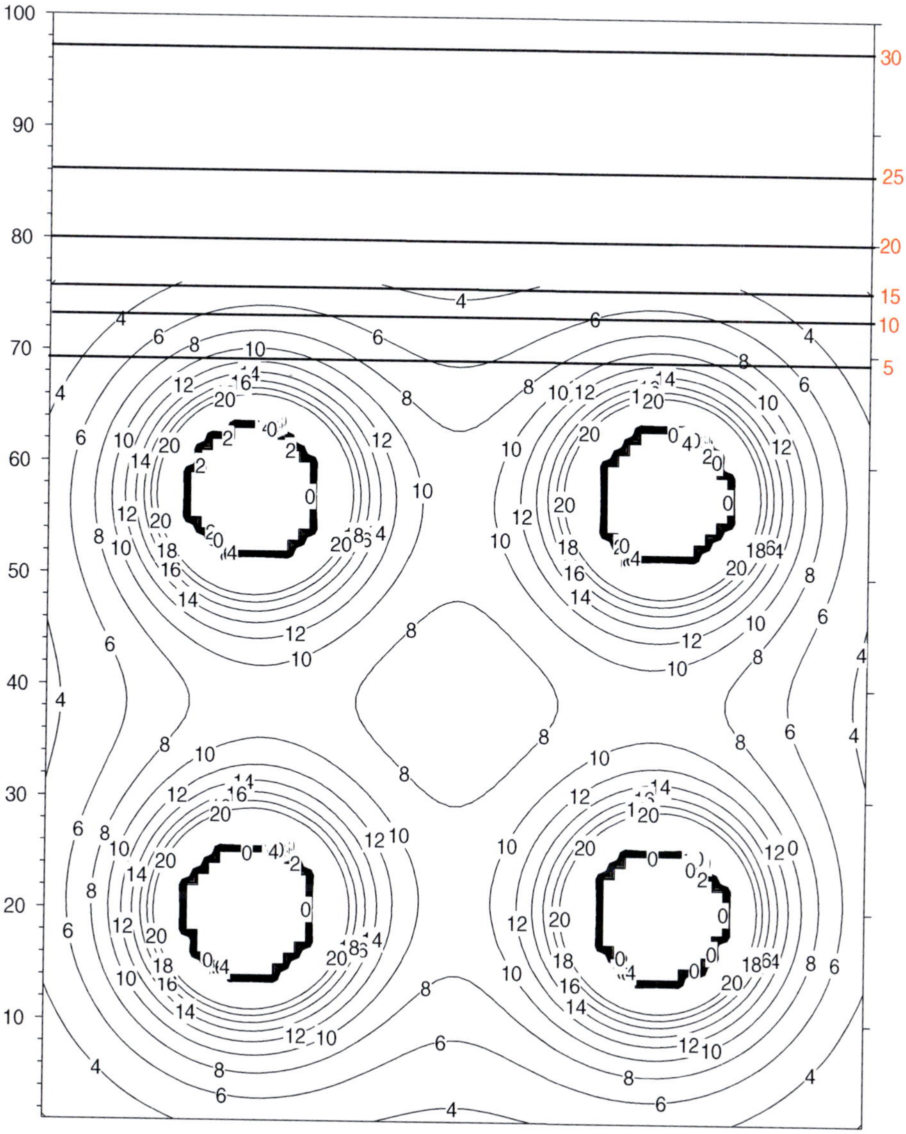


**Figure 9.16** Measured values of effluent liquid depth in a 4-lamp, open-channel UV photoreactor a flow rates ranging from 5 to 30 $\frac{m^3}{hr}$ (numerical values in red font [upper right] indicate flow rate $\left(\frac{m^3}{hr}\right)$ for each liquid level measurement. Liquid depth measurements are superimposed on a numerical simulation of the fluence rate field developed using the Line-Source Integration method; *Source:* Redrawn from Blatchley 1997.

direction of flow. These structures will act as "bluff bodies" in the flow field, resulting in the formation of an oscillating vortex structure (i.e. recirculation zone) on the downstream side of each bluff body; this behavior represents a classic phenomenon of fluid mechanics that has been studied extensively [47–49].

For open-channel UV disinfection systems with a vertical lamp orientation, the same energy loss mechanisms will contribute to overall head loss. However, because each lamp/sleeve assembly will function as a bluff body, along with the vertical support structures, the losses associated with bluff bodies will tend to dominate overall head loss.

Some recently developed open-channel UV disinfection systems are characterized by lamps with a diagonal lamp orientation (see Figure 9.5). These systems will be characterized by fluid mechanical behavior that shares features with horizontal and vertical systems.

At the approach velocities used in open-channel UV disinfection systems, the head losses associated with bluff bodies tend to be larger than those associated with momentum boundary layers near solid surfaces. In "horizontal" open-channel UV disinfection systems, the free surface profile is characterized by sharp changes in free surface elevation immediately downstream of support structures, which are attributable to form drag imposed by each support structure. By comparison, the free surface profile for channel segments corresponding to the irradiated zone (where losses due to momentum boundary layers will dominate) will display modest, sometimes imperceptible slope. For "vertical" open-channel UV disinfection systems, the free surface profile will show sharp drops in free surface elevation immediately downstream of each row of lamps and each set of support structures; these changes in free surface profile are caused by form drag losses across each row of lamps and support structures. The slope of the free surface at other locations in the system will be modest.

Head loss in any system will be a strong function of approach velocity or flow rate. As a generalization, head loss can be expressed as the sum of energy losses due to viscous and inertial forces (skin friction and form drag, respectively). Viscous losses will dominate under laminar flow conditions, while inertial losses tend to dominate for systems that operate in the turbulent regime. Viscous losses will scale linearly with approach velocity, while inertial losses are proportional to the square of approach velocity. As indicated by Equation 9.22, overall head loss can be described as the sum of viscous and inertial losses:

$$\frac{\Delta H}{L} = \underbrace{aV}_{\text{viscous}} + \underbrace{b\rho V^2}_{\text{inertial}} \tag{9.22}$$

where

$\Delta H$ = head loss observed across the system
$L$ = length of flow path across the system
$V$ = fluid approach velocity to system
$\rho$ = fluid density
$a, b$ = empirical constants.

Figure 9.17 (top) illustrates a typical lamp configuration for a system with a vertical lamp orientation. Each lamp bank will include several rows of lamps, typically arranged in a staggered pattern to promote lateral mixing. Figure 9.17 (bottom) illustrates measured values of head loss (normalized per bank of lamps) as a function of approach velocity for systems with vertical lamp orientations. Also included in this figure are fits of Equation 9.22 to these data sets. For the range of hydraulic loadings applied in these systems, the relationship between head loss and approach velocity is dominated by the second term in Equation 9.22. This and other evidence suggest that these systems almost always operate in the turbulent regime. Similar trends in head loss have been reported for systems with horizontal lamp orientations (i.e. lamp axis parallel to nominal flow direction); however, the magnitude of the head loss in these systems tends to be lower than with a vertical lamp orientation, largely because there are fewer bluff bodies for the flow to encounter.

Head losses observed under these operating conditions are typically on the order of a few cm per bank. The change in free surface elevation can be managed by the use of a sloped or stepped channel. Downstream liquid-level control is also beneficial, usually accomplished by inclusion of an

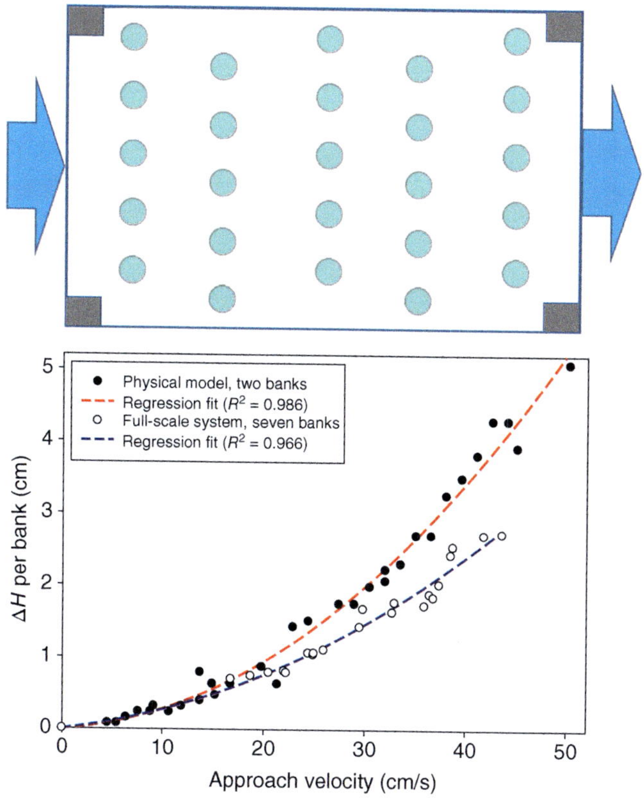


**Figure 9.17** (Top) Plan view of a typical lamp configuration in an open-channel UV disinfection system with a vertical lamp orientation. (Bottom) Measured values of head loss in systems with vertical lamp orientations. Data are included for measurements based on a physical model of a UV disinfection system that involved two lamp banks as well as a full-scale system that included seven lamp banks. Also included are fits of Equation 9.22 to these data sets. *Source:* Data from Blatchley et al. [50].

elongated weir or a flap gate. Figure 9.18 illustrates a stepped-channel as well as common liquid level control devices.

When medium-pressure (MP) Hg lamps are used as the source of UV radiation in a disinfection system, it is critical that the entire lamp/sleeve assembly remain submerged whenever the lamp is illuminated. The water that flows around the lamp/sleeve assembly is critical for cooling of the lamp. For this and other reasons, UV disinfection systems that operate with MP sources must be designed to keep the lamps submerged at all times. However, in wastewater settings, it is often desirable to avoid pumping through the disinfection system.

Figure 9.19 illustrates a MP UV disinfection system for wastewater. The hydraulics of this system have some features of closed-vessel and open-channel systems. Specifically, the lamps are maintained in a condition that guarantees that they are submerged when illuminated. The lamps swing into a tightly controlled location within the cross section of a stainless steel channel. Water entering and leaving the irradiated zone flows through an open channel.

Head loss in these systems is manifested as a change in free surface elevation from the influent side of the reactor to the effluent side. Figure 9.20 illustrates head loss as a function of flow rate for a MP UV disinfection system for municipal wastewater. Also included in this figure is a fit of

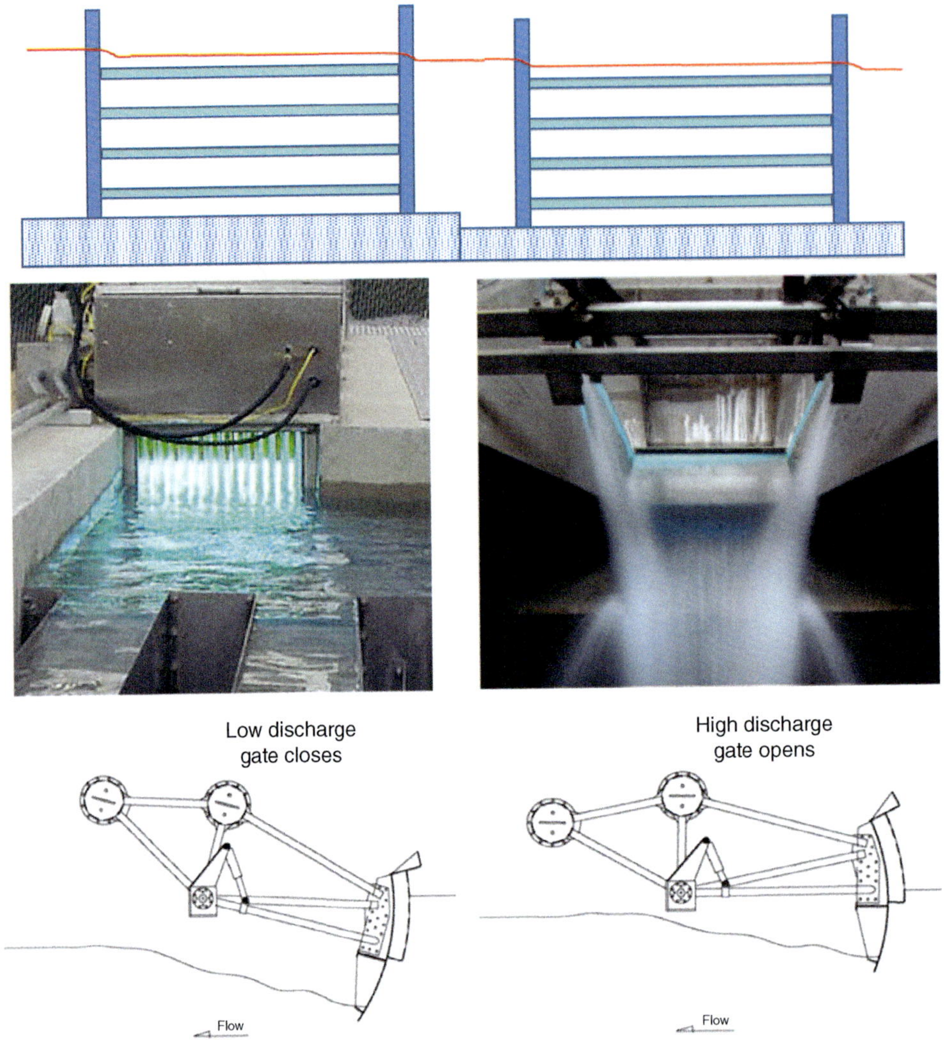


**Figure 9.18** (Top) Schematic illustration of stepped channel with an open-channel system employing horizontal lamps; the red line illustrates a characteristic free surface profile. (Left center) Open channel UV disinfection system with an elongated weir used for liquid level control. Image from: https://www.americanairandwater.com/water/wastewater-uv.htm. (Right center) Mechanical-level control gate (Photo courtesy of Golden Harvest, Inc.). (Lower) Schematic illustrations of the function of a mechanical-level control gate at low discharge (left) and high discharge (right).

Equation 9.22 to the data. Although the data set is sparse, the general form of the equation is consistent with the observed head loss as a function of flow rate. Also, the relationship between head loss and flow rate indicates that head loss is dominated by the inertial term, thereby suggesting that the system operates in the turbulent regime across the entire range of operating flow rates. Note that the head loss observed across this system is substantially larger than the head losses observed with systems based on conventional LP Hg lamps, which generally operate at much lower approach velocity.

**Figure 9.19** Schematic illustration of medium-pressure UV disinfection system used for wastewater treatment. *Source:* Image courtesy of Trojan Technologies.

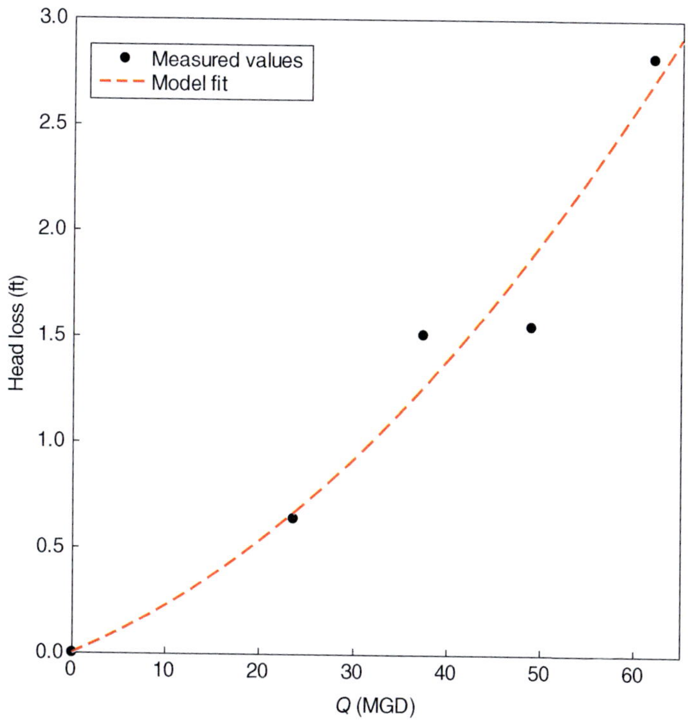


**Figure 9.20** Head loss as a function of flow rate for a Trojan UV 4000 Plus installed at the Decatur, IL Sanitary District. Also included is a fit of Equation 9.22 to the data.

#### 9.7.1.2 Closed-Vessel Systems

In contrast with open-channel reactors, closed-vessel reactors operate under positive gauge pressure. This guarantees complete flooding of the irradiated zone of the reactor and complete submergence of the lamp/sleeve assemblies. Head loss in these systems manifests as a change in pressure across the reactor system. These systems are often designed to include lamps that are oriented perpendicular to the direction of flow, sometimes referred to as a "cross-flow" orientation. As such, head losses attributable to fluid mechanics around bluff bodies represent a substantial contribution to overall head loss.

Because these systems are often fed by pumped water, it is likely that the head losses across the UV system will be small relative to available head. Figure 9.21 illustrates head loss as a function of flow rate for a closed-vessel reactor. These graphs illustrate trends of head loss measurements for each system, based on fits of Equation 9.22 to each data set; actual head loss measurements are not included in this image. As with the other systems, the turbulent (second-order) term dominates over the entire operating range, thereby suggesting that these systems operate in the turbulent regime.

### 9.7.2 Flow Conditioning

The lamp/sleeve assemblies and their associated support structures cause disturbances in the flow field, relative to the open channels that are used to convey water to and from these areas of a UV photoreactor. In addition to contributing to head loss, these flow field disturbances also dampen velocity gradients, thereby promoting uniformity in the velocity field.

Figure 9.22 illustrates two simple open-channel UV disinfection systems, one with a vertical lamp orientation, the other with a horizontal lamp orientation. For these systems, a series of measurements was conducted to quantify the effects of the lamp modules on the axial (downstream) and transverse (cross-stream) velocity profiles. The inlet structure to this system,

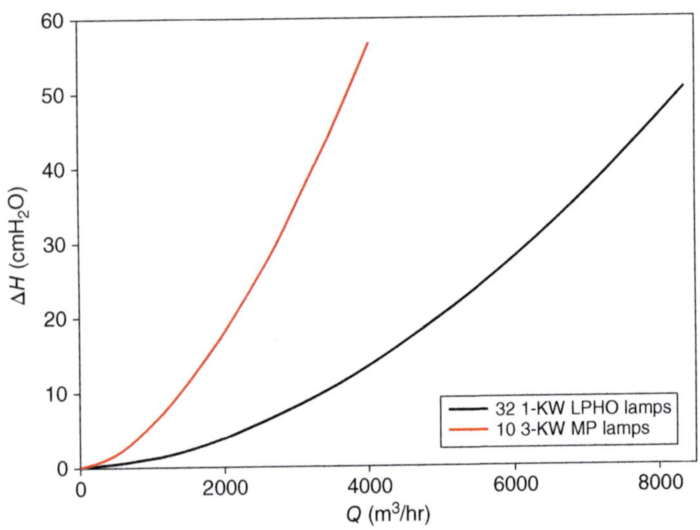


**Figure 9.21** Head loss as a function of volumetric flow rate for two closed-vessel, cross-flow reactors with similar nominal power ratings. Trends in head loss are presented in the form of a fit of Equation 9.22 to measurements of head loss as a function of flow rate for a cross-flow reactor that included 32 1-kW LPHO lamps and for a different cross-flow reactor with 10 3-kW MP lamps.

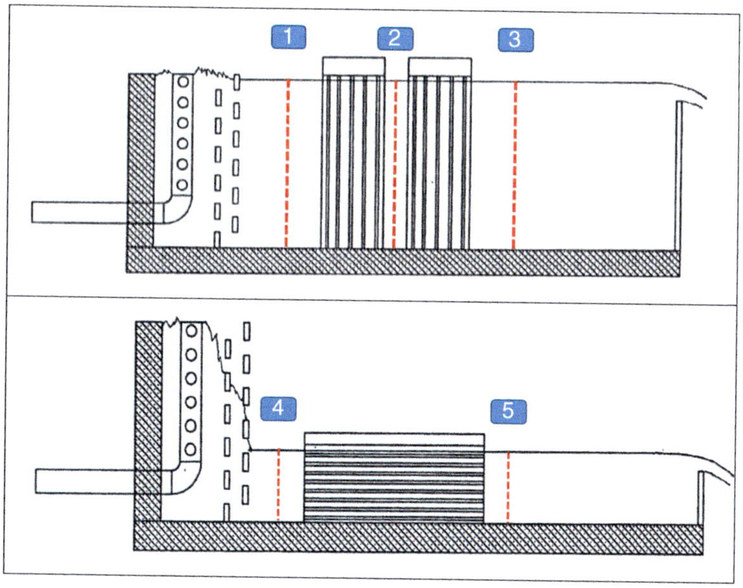


**Figure 9.22** Schematic illustration of pilot-scale reactor system configured with vertical lamp modules (top) and horizontal lamp modules (bottom). Dashed, red, vertical lines indicated locations of velocity profile measurements. Numbers indicate cross-section labels.

which was built for pilot-scale experiments to house lamps with either a vertical or a horizontal lamp orientation, yielded a highly nonuniform velocity field upstream of the modules.

Local measurements of velocity were conducted at the centerline of each of the cross sections illustrated in Figure 9.22. For the system with the vertical lamp configuration, measurements of transverse (*X*) and longitudinal (*Y*) velocity were conducted upstream of the first module (location 1), between the two modules (location 2), and downstream of the second module (location 3). Figure 9.23 provides a summary of these velocity measurements. Similar measurements were conducted on the system with the horizontal lamp configuration at cross-section locations immediately upstream and immediately downstream of the lamp bank (locations 4 and 5 in Figure 9.22; see Figure 9.24).

An ideal upstream velocity profile would take the form of a uniform longitudinal (axial) velocity and zero transverse velocity. For both lamp orientations, the upstream velocity profile imposed on the lamp banks was quite irregular, with a notable "jet" being evident in the axial velocity profile at roughly 1.5–2 ft (45–60 cm) above the channel bottom. The transverse velocity profiles also showed considerable deviation from the ideal case. However, after passing through a lamp array, the velocity profile approached an ideal condition, with the exception of a momentum boundary layer with a thickness of 4–6 (10–15 cm) inches being evident in the axial velocity profile in both systems. This boundary layer is a characteristic of flow in a channel and is essentially unavoidable. In the case of the system with the vertical lamp orientation, flow conditioning was further improved after passing through a second lamp bank.

The fluid mechanical behavior within these lamp arrays is quite complicated. However, in simplistic terms, the flow conditioning behavior within these systems is attributable to energy losses around the lamp/sleeve assemblies and the support structures. Local energy losses are roughly proportional to the square of local velocity. Therefore, the resistance to flow offered by the lamp

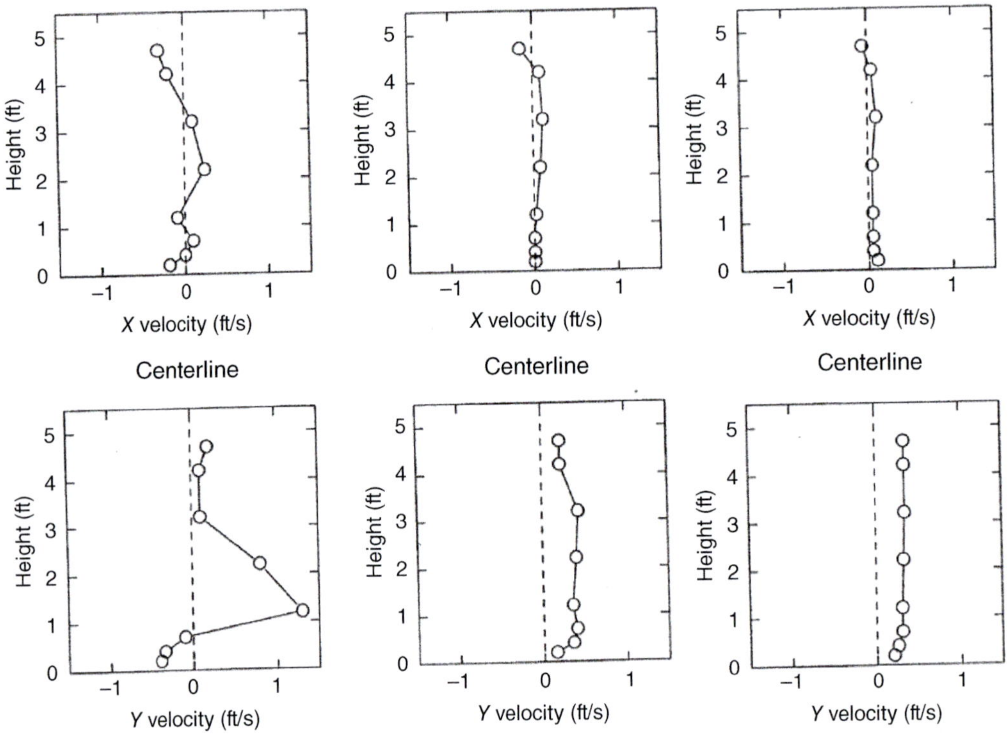


**Figure 9.23** Measured values of transverse (*X*, top row) and longitudinal (*Y*, bottom row) velocity for open-channel UV disinfection system with vertical lamp orientation at a fixed flow rate, at the channel centerline for cross-sections 1 (left), 2 (center), and 3 (right) from Figure 9.22.

modules in areas of locally high velocity will be higher than the resistance offered in areas of locally low velocity. This results in dampening of velocity gradients within the system. This behavior is conceptually similar to the approach used in wind tunnels, where a "honeycomb" section is often included upstream of the "test section" to straighten the flow and reduce local velocity gradients.

Flow conditioning is also relevant among large systems that require two or more operating channels within a system. In addition to the target flow conditions described above for a single-channel system, an ideal upstream hydraulic condition for a multichannel system will would involve equal distribution of flow among all operating channels. However, these systems are often constrained in terms of space allowed for the footprint of the system and associated inlet and outlet hydraulic structures.

As an illustration, consider the UV disinfection system that was installed at the Belmont Advanced Wastewater Treatment Plant (AWTP) in Indianapolis, IN. UV disinfection was installed as a retrofit at this facility, which had previously employed ozone-based disinfection or chlorine-based disinfection. Space limitations dictated that the inlet hydraulic structures involved a number of sharp turns for water, immediately upstream of the UV disinfection system's inlet structure. The system comprises seven channels, which are all expected to be operational under the design flow condition of 170 MGD $\left(27,000\ \frac{m^3}{hr}\right)$. In the absence of any supplemental flow conditioning (i.e. original design condition), the flow was unevenly distributed among the seven channels, and the velocity profile imposed on the first bank of UV lamps was nonideal, in that several of the channels experienced a significant cross-channel flow pattern, as well as deviations from the condition of

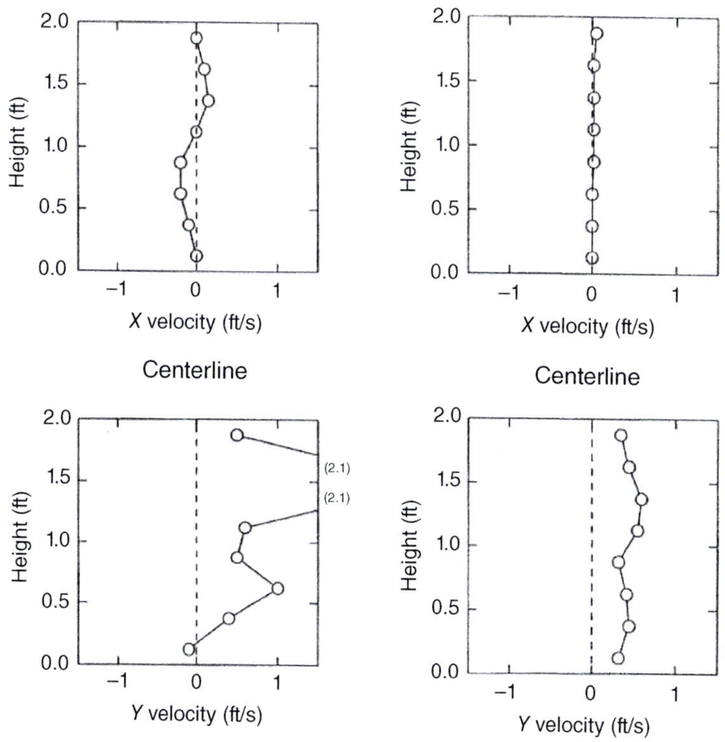


**Figure 9.24** Measured values of transverse (*X*, top row) and longitudinal (*Y*, bottom row) velocity for open-channel UV disinfection system with horizontal lamp orientation at a fixed flow rate, at the channel centerline for cross-sections 4 (left), and 5 (right) from Figure 9.22.

a uniform axial velocity profile (see Figure 9.25, top). The upstream end of each channel was modified to include a transition section, which allowed a gradual increase in the channel cross-sectional area, followed by two stilling plates, which were fabricated by cutting a series of circular holes into two aluminum plates (see Figure 9.25, center). These changes in the upstream geometry of the channel allowed for substantial improvements in flow distribution among the seven channels, as well as promoting velocity profile uniformity across each channel (see Figure 9.25, bottom). Similar approaches have been applied in other large-scale UV photoreactors, usually in municipal-scale applications.

## 9.8 Fouling in UV Photoreactors

Fouling can occur in any application of UV radiation, but the effects of fouling on water quality are perhaps more acute with disinfection systems than in other processes. In addition, most of the research that has been conducted on fouling in UV photoreactors has involved reactors that are designed for disinfection applications.

UV radiation functions as a broad-spectrum antimicrobial agent because the mechanisms of inactivation involve damage to biomolecules that are common to all forms of life (i.e. nucleic acids and proteins). The ability of UV disinfection systems to meet this function depends on their ability to deliver UVC photons to microbial targets. In some settings, fouling of the quartz sleeves that

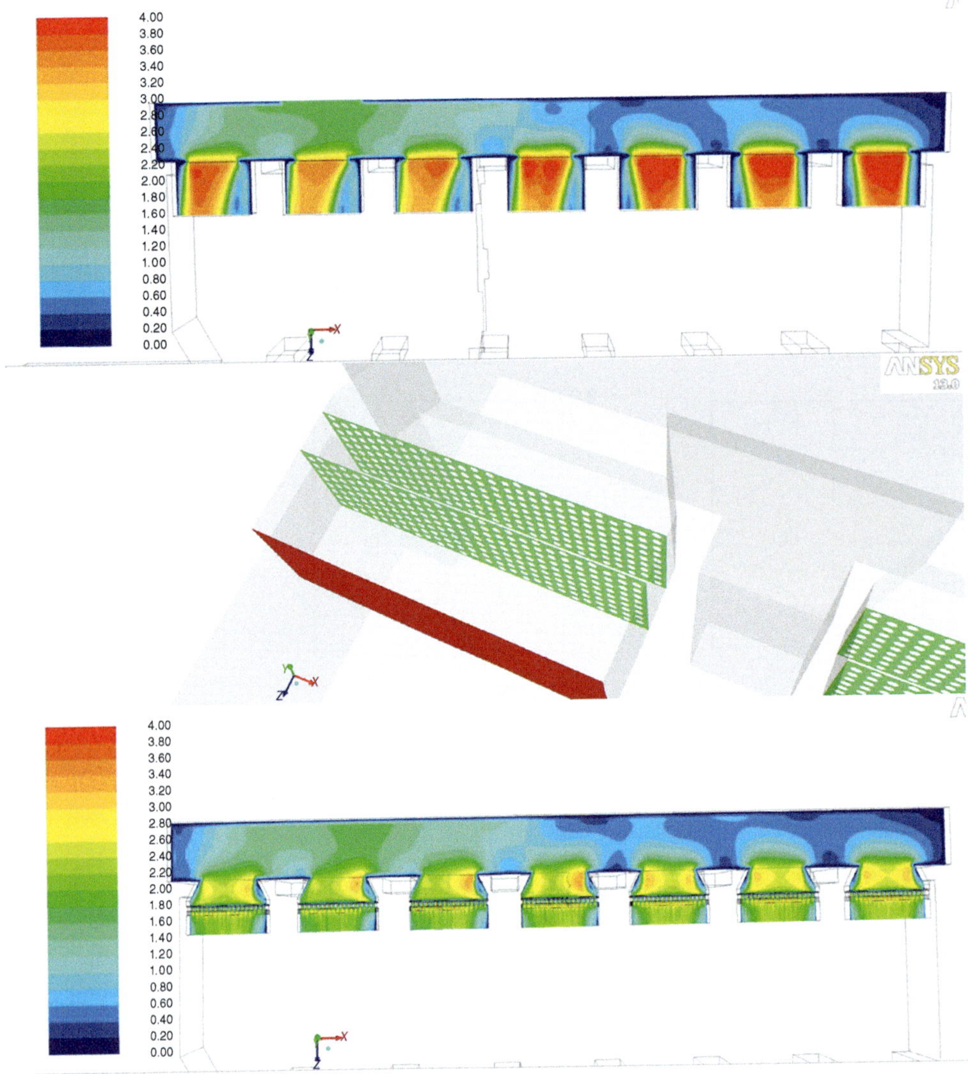


**Figure 9.25** (Top) CFD simulation of upstream flow distribution and velocity fields among seven channels in Belmont AWTP (Indianapolis, IN) UV disinfection system, as originally installed. Substantial deviations from optimum flow distribution were evident in all seven channels that would operate under the design flow condition of 170 MGD $\left(27{,}000\ \frac{\text{m}^3}{\text{hr}}\right)$. Inclusion of an upstream flow conditioning structure (center) within each channel resulted in promotion of a more uniform flow distribution among the seven channels, as well as improved uniformity of the velocity profile upstream of each lamp bank (bottom). *Source:* Images courtesy of Malcolm Pirnie/ARCADIS, Indianapolis, IN.

surround UV lamps can represent an important barrier to photon transfer from the radiation source to the target.

Quartz fouling can occur on the interior or exterior surfaces of quartz sleeves, which can involve the actual lamp or the sleeve that is used to separate the lamp from water. Interior fouling can be caused within the lamp itself by sputtering of tungsten from the lamp electrodes onto the quartz

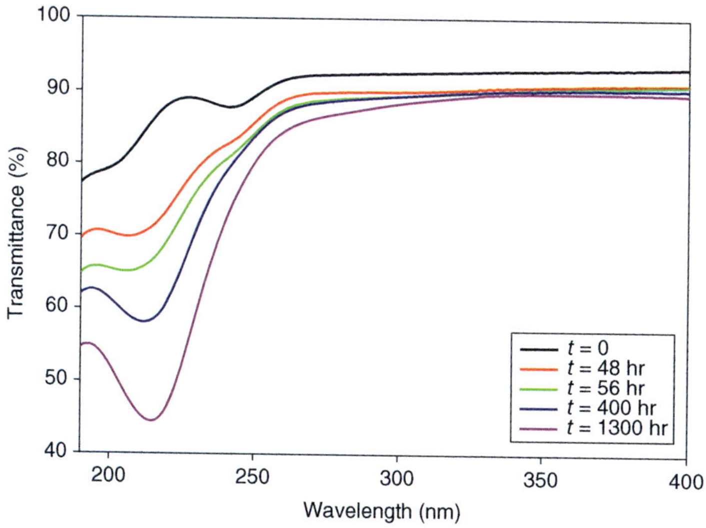


**Figure 9.26** Time-course changes in UVT for quartz containing SiOH at a concentration below 1 ppm. Time refers to the period of exposure of the quartz sample to VUV radiation. *Source:* Data from Arnold and Adam [51].

walls of the lamp enclosure. In addition, UV radiation can cause changes to the chemical composition of the quartz itself.

The quartz used in UV photoreactors is often produced by melting pegmatite (quartz-containing igneous rock) under reducing conditions [51]. The resulting material is largely quartz ($SiO_2$) with low SiOH content and a high concentration of SiH groups. Both of these groups within the quartz matrix will respond to UV radiation at wavelengths below about 230 nm. Two specific solid-phase photochemical reactions have been identified that affect the optical properties of quartz [51]:

$$\mathrm{SiH} + h\nu \xrightarrow{\lambda \approx 215\ \mathrm{nm}} \mathrm{Si}^* + \mathrm{H} \tag{9.23}$$

$$\mathrm{SiOH} + h\nu \xrightarrow{\lambda \approx 265\ \mathrm{nm}} \mathrm{SiO}^* + \mathrm{H} \tag{9.24}$$

Aging experiments conducted with quartz materials using a vacuum UV (VUV) source have been observed to cause changes to the UVT spectra of these materials; these changes are sometimes referred to as "solarization" of quartz. Figure 9.26 illustrates the response of a quartz material with low SiOH content (i.e. [SiOH] < 1 ppm) to VUV exposure. UVT of this material demonstrated a monotonic decrease with time for the wavelength range 190 nm $\lesssim \lambda \lesssim$ 300 nm over the entire period of exposure.

Figure 9.27 illustrates the results of a similar experiment in which a quartz sample with OH content of approximately 200 ppm was exposed to VUV radiation. The sample showed an initial ($t \leq 48$ hr) decline in UVT across the wavelength range 190 nm $\lesssim \lambda \lesssim$ 340 nm, followed by recovery of transmittance. The gradual recovery of transmittance is due to recombination within the quartz matrix [51]:

$$\mathrm{SiO}^* + \mathrm{Si}^* \rightarrow \mathrm{Si-O-Si} \tag{9.25}$$

The quartz surfaces that surround common UV sources, such as mercury lamps, are also susceptible to exterior fouling. When the lamps are not illuminated, biofilms and other organic materials

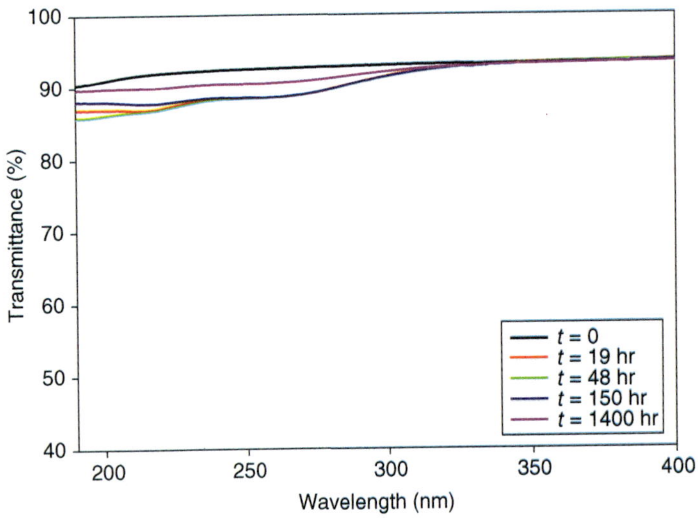


**Figure 9.27** Time-course changes in UVT for quartz containing SiOH at a concentration of approximately 200 ppm. Time refers to the period of exposure of the quartz sample to VUV radiation. *Source:* Data from Arnold and Adam [51].

can rapidly accumulate on these quartz surfaces. Illumination of these lamps leads to rapid removal of organic foulants [52]. However, lamp illumination can lead to the formation of inorganic foulants. Although lamps in UV systems may not always be illuminated, the primary concern with fouling potential in UV systems relates to the inorganic foulants that can accumulate when the lamps are on. As such, the focus of this section will be on inorganic (exterior) foulants.

### 9.8.1 Mechanisms of Inorganic (Exterior) Fouling

Conventional UV disinfection systems share many common features with tube/bundle heat exchangers. In both system types, a fluid (often water or air) will flow past an array of heated tubes. In the case of water-based systems, these tubes offer opportunities for the formation of foulants by thermally enhanced precipitation and accumulation of particles by inertial impaction. As such, it is reasonable to expect that some similarities in fouling behavior will be evident among UV disinfection systems and tube/bundle heat exchangers. However, there are also important differences between heat exchangers and UV lamp arrays. These include the rates of transport for mass, heat, and momentum, differences in chemistry of the fouling surfaces, and the presence (absence) of UV radiation [53].

Building on the knowledge base that had been established for fouling in heat exchangers, at least two mechanisms have been identified that can contribute to inorganic fouling in UV lamp arrays. The first mechanism involves thermally enhanced precipitation. In addition to the UV radiation that is emitted by UV sources, they all release heat as well. The heat released by these lamps will result in localized heating of water in the thermal boundary layer that surrounds each lamp/sleeve structure.

The potential for a system to yield a precipitate is governed, in part, by the difference between the actual, local condition and a relevant, local equilibrium condition. To examine this behavior, it is necessary to define equilibrium for a precipitation/dissolution reaction.

By convention, these reactions are written as dissolution reactions in the forward direction, with precipitation appearing as the corresponding reverse reaction. Using a metal hydroxide as an example, the reversible reaction will be written as follows:

$$Me(OH)_{n,s} \rightleftharpoons Me^{n+} + nOH^{-} \tag{9.26}$$

where $Me(OH)_{n,\,s}$ represents the crystalline, solid form of the metal:hydroxide precipitate, $Me^{n+}$ represents the metal ion that participates in the process (or is generated as a result of dissolution), and $OH^{-}$ represents the hydroxide ion, which serves as the (anionic) ligand that associates with the metal ion to form the precipitate. The equilibrium expression for this reversible reaction can be written following the general rules for chemical kinetics and equilibria [54]:

$$K = \frac{\{Me^{n+}\}_{eq}\{OH^{-}\}_{eq}^{n}}{\{Me(OH)_{n,s}\}_{eq}} \tag{9.27}$$

where $K$ represents the equilibrium constant for this process and $\{i\}_{eq}$ represents the activity of constituent $i$ in this reaction at an equilibrium condition. By convention, the activity of a pure crystalline solid is defined as 1. Therefore, the equilibrium constant for precipitation/dissolution reactions usually takes on the following form:

$$K_{sp} = \{Me^{n+}\}_{eq}\{OH^{-}\}_{eq}^{n} \tag{9.28}$$

where $K_{sp}$ represents the solubility product for the reversible reaction described in Equation 9.26. It is important to note that the activities indicated in Equation 9.28 represent an equilibrium condition. Most systems operate in a nonequilibrium condition.

We can quantify the status of the system relative to equilibrium by the use of the *ion activity product (IAP)*, which is formally defined as:

$$IAP = \{Me^{n+}\}\{OH^{-}\}^{n} \tag{9.29}$$

where $\{i\}$ represents the *actual* activity of constituent $i$ in this system. Three fundamental conditions can be identified that relate IAP to $K_{sp}$:

- When IAP $< K_{sp}$, the actual activities (~molar concentrations) of the constituent ions in the system are below those that would bring the system to equilibrium. This would be considered a sub-saturation (sub-equilibrium) condition, such that we would expect the rate of dissolution to exceed the rate of precipitation. As a result, we expect net dissolution of the solid to take place.
- When IAP $= K_{sp}$, the actual activities (~molar concentrations) of the constituent ions in the system are equal to those that represent an equilibrium condition. Therefore, the rate of dissolution will be equal to the rate of precipitation, and we should expect the system's composition to remain constant.
- When IAP $> K_{sp}$, the actual activities (~molar concentrations) of the constituent ions in the system exceed those that define an equilibrium condition. This would be considered a supersaturation (super-equilibrium) condition. Under this condition, the rate of the precipitation (reverse) reaction will exceed the rate of the dissolution (forward) reaction, resulting in net precipitation of the crystalline solid.

An alternative method used to define the relationship between the actual condition and the equilibrium condition involves the use of a saturation index (SI):

$$SI = \log_{10}\left(\frac{IAP}{K_{sp}}\right) \tag{9.30}$$

When SI = 0, the system is in an equilibrium condition. When SI < 0, the system is subsaturated; when SI > 0, the system is supersaturated and we expect that precipitation will take place. In general terms, the larger the value of SI, the faster the rate of precipitation. However, this generalization is largely qualitative. It is difficult to predict precipitation rates based solely on calculated values of SI. This is partly attributable to the fact that equilibrium constants are generally defined for pure, crystalline solids, whereas in many precipitation processes, the initial solid formed by the process is amorphous. Further complicating the quantitative application of *SI* for prediction of inorganic fouling rates in UV photoreactors is the fact that attempts to characterize the solids that form in these processes have indicated that they tend to be amorphous co-precipitates that are difficult to characterize, and for which little or no thermodynamic data are available [53, 55–59].

The rates of the forward and reverse reactions in these processes are temperature-dependent. As such, the values of the solubility product will also change with temperature. For many metal:anion pairs, the thermodynamics of their precipitation/dissolution reactions are such that their solubility product decreases as temperature increases. A consequence of this behavior is that the metal:anion pairs become less soluble as temperature increases. This condition, sometimes referred to as *inverse solubility*, will exist when the enthalpy of dissolution is negative. Metal:anion pairs that demonstrate this behavior include: $CaCO_3$, $CaSO_4$, $MgCO_3$, $MgSO_4$, and $FePO_4$, though many other examples exist [53].

As the temperature of water in the thermal boundary layer increases, these metal:anion pairs can change from a condition of slight subsaturation to a condition of slight supersaturation, which promotes precipitation in the vicinity of the lamp sleeve surface [60]. At near-neutral pH, the quartz surface will express net negative surface charge, which may provide a mechanism for attraction of metal ions to the quartz surface. However, measured values of metal loadings within inorganic foulants that accumulate on these quartz surfaces are orders of magnitude higher than the surface charge density for quartz [53]. Therefore, ion exchange is believed to play no more than a minor role in fouling of quartz sleeves in UV photoreactors.

Thermally promoted precipitates have been observed to form in a pattern that includes an induction (lag) period, in which only minor accumulation is observed, followed by a longer period in which the metal:anion pairs accumulate following zero-order kinetics. One possible explanation of this behavior is that the surface chemistry of the quartz sleeve changes from a surface that is dominated by silane groups to a surface that is dominated by mixed-metal precipitates. After this induction period, the affinity of the surface for the metal:anion pairs appears to increase. Because the chemistry of the water that passes the sleeve is likely to remain essentially constant, the driving force for precipitation will also remain constant, as well the rate of precipitation (i.e. zero-order kinetics of precipitation) [52, 53, 60].

Several circumstances can be identified under which thermally enhanced precipitation will be promoted. For example, in systems where groundwater is treated in a UV reactor, the water that enters the system will probably have been in direct contact with aquifer minerals for a long period of time. As such, it is likely that the water will be in equilibrium with aquifer minerals at the local groundwater temperature. Localized heating of this water will move it to a condition of slight supersaturation, resulting in precipitation. A second example may involve chemical addition to water from a surface water source, or perhaps municipal wastewater. Common chemical additives that could promote precipitation could include salts of $Al^{3+}$ or $Fe^{3+}$ that may be added to as coagulant chemicals or to promote precipitation of phosphate. Aluminum- and iron-based precipitates are commonly observed in UV systems that are located downstream of points of metal salt addition [55, 59, 61, 62].

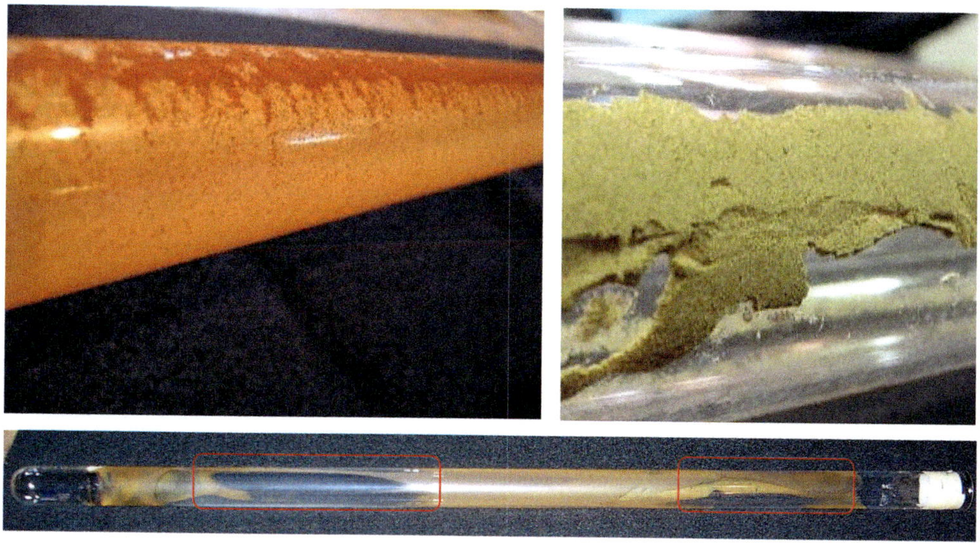

Figure 9.28 Photos of fouled quartz sleeves [61, 63]. (Top left) Quartz sleeve from a system with a horizontal lamp orientation in which iron-based fouling dominated. Preformed particles tended to settle on the top surface of this sleeve. (Top right) Particle-based fouling in a water supply that was characterized by a high concentration of suspended particles, but little potential to support thermally enhanced precipitation. Particles in the influent settled on the top surface of this sleeve, which was rotated to bring the top surface into clear view of the camera. (Bottom) Sleeve from a reactor system in which aggressive fluid shear prevented precipitate accumulation in at least two locations (see red boxes).

A second mechanism that contributes to fouling in UV disinfection systems (and other UV photoreactors) is particle impaction. This mechanism is similar to particle collection in granular media filters. In effect, the quartz sleeves function as a collector surface for pre-existing particles. Under this mechanism, the top surfaces of lamp sleeves tend to be most heavily fouled, when the lamp/sleeve orientation is parallel to the nominal flow direction and horizontal. In systems where the lamps are oriented perpendicular to the direction of flow, particle fouling tends to dominate on the upstream side of the sleeve. Figure 9.28 illustrates several examples of fouling behavior on quartz sleeves, including fouling by the two mechanisms described above.

Fouling is also affected by characteristics of water treatment and fluid mechanics that are often site-specific. For example, oxidation–reduction (redox) potential has been shown to have a profound effect on fouling dynamics. One illustration of this behavior emerged from an experiment in which fouling was allowed to proceed in a small-scale UV reactor that was installed at several locations within the process train of a potable water production facility that was used to treat groundwater. Figure 9.29 provides a schematic illustration of the process train at this facility and the locations where the UV photoreactor was installed.

The small-scale UV system, which operated using three LPHO (mercury amalgam) lamps, was allowed to run continuously over a period of three to eight days at each location to allow for fouling to proceed. An irradiance sensor was installed close to one of the three lamps to allow real-time, continuous measurement of irradiance delivered from one lamp to a nearby location. The water chemistry at each location was essentially constant over the entire period of the experiment. Table 9.5 provides a summary of several relevant water quality measurements at each of the four locations where the UV photoreactor was positioned.

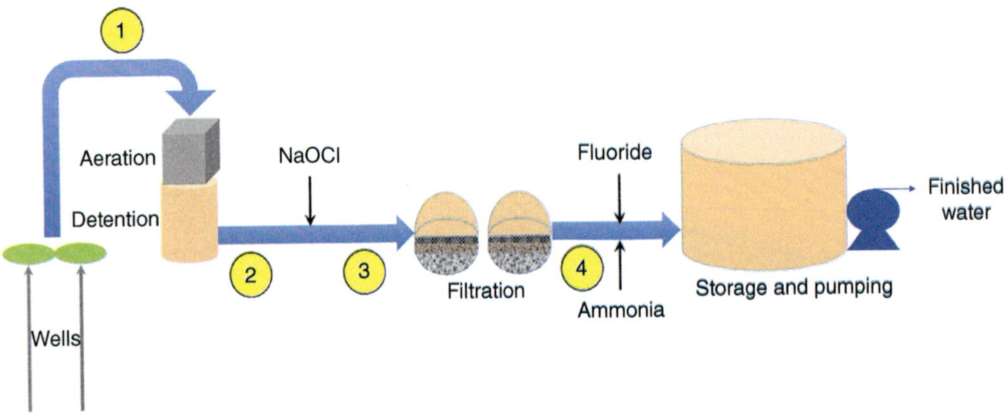


**Figure 9.29** Schematic illustration of South Wellfield Water Treatment Facility, Indianapolis, IN. Water is extracted from a series of wells in the South Wellfield area. Water is then treated by aeration, followed by detention to allow for $O_2$-induced oxidation of reduced iron and manganese. This is followed by chlorination to further promote these oxidation reactions. Water then passes through multimedia filters, after which it is ammonified, to allow chloramination, and fluoridated. After storage, finished water is pumped to the distribution system. Schematic courtesy of Citizens Energy Group, Indianapolis, IN. Fouling experiments were conducted at four locations within this system; 1: raw water; 2: aerated water; 3: chlorinated water; 4: filtered water.

**Table 9.5** Measured values of concentrations of inorganic constituents and other water quality parameters in water that may have contributed to fouling in the South Wellfield experiments.

| Parameter | Units | Raw water | Aerated water | Chlorinated water | Filtered water |
|---|---|---|---|---|---|
| Ca | $\frac{mg}{L}$ | 103 | 101 | 102.35 | 102.8 |
| Fe | $\frac{mg}{L}$ | 1.888 | 1.64 | 1.032 | 0.005 |
| K | $\frac{mg}{L}$ | <2.0 | <2.0 | <2.0 | 2 |
| Mg | $\frac{mg}{L}$ | 30.54 | 30.3 | 30.26 | 30.56 |
| Mn | $\frac{mg}{L}$ | 0.104 | 0.123 | 0.095 | <0.0004 |
| Na | $\frac{mg}{L}$ | 24.86 | 26.5 | 28.37 | 31.61 |
| Fluoride | $\frac{mg}{L}$ | 0.09 | 0.11 | 0.1 | 0.1 |
| Chloride | $\frac{mg}{L}$ | 58.87 | 62.13 | 63 | 65.04 |
| Phosphate | $\frac{mg}{L}$ | <0.10 | <0.10 | <0.10 | <0.10 |
| Sulfate | $\frac{mg}{L}$ | 69 | 69.49 | 72 | 75.05 |
| Electron activity, $E_h$ | mV, NHE | 149 | 150 | 870 | 808 |
| Dissolved oxygen | $\frac{mg}{L}$ | 0.09 | 7.87 | 7.55 | 7.43 |
| pH | | 7.4 | 7.6 | 7.6 | 7.9 |
| Temperature | °C | 13.3 | 13.6 | 13.4 | 13.1 |
| Alkalinity | $\frac{mg}{L}$ as $CaCO_3$ | 300 | 304 | 304 | 306 |

Among the dissolved constituents that are listed in Table 9.5, only the concentrations of iron, manganese, and dissolved oxygen changed across this system. Iron and manganese are both known to form recalcitrant fouling materials in UV systems; in addition, the inclusion of iron and manganese in foulant materials is known to lead to strong absorption of UV radiation. Another important change in water quality that is evident in Table 9.5 is electron activity ($E_h$), which is closely related to oxidation/reduction (redox) potential. A high value of $E_h$ is indicative of conditions that favor oxidation.

The South Wellfield treatment facility was designed to accomplish removal of iron and manganese. Removal of these metals is promoted by oxidation as a result of addition of two oxidants. The first of these is molecular oxygen, which is added to water through an aeration system. The aeration system is followed by a detention basin which allows for roughly 20 minutes of reaction time to allow for oxidation of these reduced metals. This is followed by chlorination to promote further oxidation of these reduced metals. In their oxidized forms, iron and manganese for low-solubility precipitates that are easily removed by filtration.

Figure 9.30 illustrates the effects of foulant accumulation at each of four locations within the South Wellfield treatment system. In the raw (untreated) water (location 1 in Figure 9.29), irradiance was reduced to roughly 25% of its original value (i.e. in the unfouled reactor) within 8 days of exposure. In the aerated water (location 2 in Figure 9.29), the foulant became essentially opaque to $UV_{254}$ radiation within three days. With the chlorinated water (location 3 in Figure 9.29), the rate of fouling was intermediate to the previous two locations. In the filtered water, the rate of fouling was apparently quite slow.

The addition of oxidizing chemicals promotes oxidation and subsequent precipitation of iron- and manganese-containing foulants. The quartz sleeves, which express net negative surface charge at near-neutral pH, probably functioned as nucleation sites for these precipitation reactions. Downstream of the filtration system, the potential for fouling was dramatically diminished, as compared

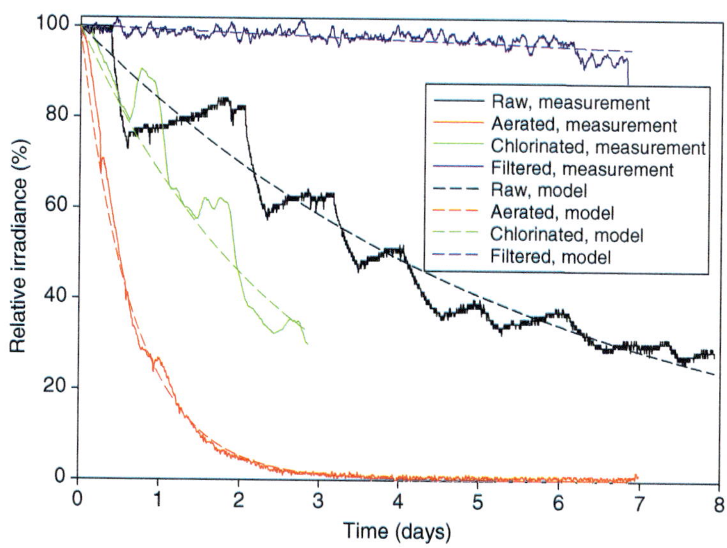


**Figure 9.30** Measured and modeled values of relative irradiance at a fixed location in a UV reactor that experienced fouling. The reactor was installed at each of the four location indicated in Figure 9.31. Relative irradiance is defined as the irradiance at any time divided by the initial irradiance at a given location. *Source:* Data from Wait et al. [59].

to the other locations where experiments were conducted within this system, despite the fact that electron activity ($E_h$) was quite high in this finished water. The fact that iron and manganese were almost completely removed by the combination of oxidation and filtration probably explains this behavior.

As described previously, the time-course behavior of fouling has been observed to follow zero-order kinetics (i.e. constant rate of accumulation). The Beer–Lambert law teaches that irradiance will decrease in an exponential fashion as the concentration of absorbing compounds increases. The combined effects of these two phenomena suggest an exponential decrease in irradiance at a given location in a photoreactor that experiences fouling. As an illustration of this behavior, best-fit lines to the following equation were fit to each of the data sets indicated in Figure 9.30.

$$\text{Relative Irradiance} = 100 * e^{-kt} \quad (9.31)$$

where

$k$ = rate constant for irradiance decline

$t$ = time.

### 9.8.2 Methods for Mitigation of Fouling

Foulant materials that accumulate on the interior surfaces of quartz sleeves or lamps are difficult to remove because of their physical/chemical characteristics and because they are difficult to gain access to. As a result, a common strategy to counteract the effects of interior fouling is periodic replacement of the quartz sleeve or lamp, as part of routine maintenance.

Exterior foulant materials are readily accessible and span a range of physical and chemical characteristics. Exterior foulants have textures that range from hardened surface precipitates, to granular, or gelatinous coatings. Mitigation measures include physical abrasion and chemical dissolution of precipitated materials.

Abrasion methods involve erosion of surface foulants by friction. These methods can be effective for removal of foulants, but also have the potential to damage the surface of a quartz sleeve. Areas of damage may emerge from repeated abrasion in the form of grooves or pits on the surface of the sleeve. These areas of localized damage can then function and harbor subsequent foulant materials, thereby allowing for permanent fouling of a quartz sleeve [56].

Exterior foulant materials can also be removed by dissolution. This usually involves local or general immersion of the fouled sleeve or a sleeve section in a dilute solution of a strong acid, often phosphoric acid [64]. The low pH of this acid solution causes dissolution of the metal:ligand precipitates, and is generally effective in removing foulants and restoring the UV transmissive features of the quartz sleeve.

## 9.9 Selection and Design of UV Disinfection Systems

For many common applications of UV disinfection, hardware can be selected "off-the-shelf." For many reactors, manufacturers of UV photoreactors will have conducted the actual design of the reactor based on *numerical prototyping* methods that may employ computational fluid dynamics and fluence rate field modeling (CFD-E models, see Chapter 6). The primary benefits of numerical prototyping are that it allows for relatively rapid and inexpensive quantification of the strengths and weaknesses of a given reactor design. Moreover, it is relatively simple and inexpensive to numerically "modify" reactor geometry so as to seek improvements in the aspects of reactor design that limit its performance. However, characterizations of reactor performance based on CFD-E simulations are generally insufficient. In many applications, regulatory or guidance documents call for empirical validation of reactor performance based on biodosimetry (see Chapter 7).

An important limitation of biodosimetry relates to its inability to quantify the dose distribution delivered by a reactor under a given operating condition. Because of this, it is not possible to develop quantitatively accurate predictions of reactor performance for any target microbe, unless the target has the same UV dose–response behavior as the challenge organism used in biodosimetry. The effects of this shortcoming can be mitigated by the use of multiple challenge organisms in biodosimetry (e.g. selection of two challenge organisms that "bracket" the dose–response behavior of the target microbe). An important consequence of all biodosimetric validation methods is the need for inclusion of one or more safety factors. The use of safety factors is warranted as a measure of engineering conservatism, so as to protect public health. However, they often yield system configurations that are overdesigned and inefficient.

Another factor that affects overall system capital cost is the need for the system to be able to effectively treat a "worst-case" operating condition, often defined as a combination of high flow rate and low UVT that collectively yield poor system performance. Again, design to deal with "worst-case" conditions represents a responsible approach, and one that works to protect public health. However, this also results in system configurations that exceed treatment requirements for most operating conditions. And since the simplest method of operation often involves continuous use of all lamps in a system, this can result in energy costs that far exceed the requirements for a given set of treatment objectives.

Among large-scale UV disinfection systems, such as those that are used in many municipal applications for production of potable water or wastewater/reuse applications, the need for redundancy within the system will be a component of design. In these applications, systems must be able to operate continuously, even when maintenance is required on the system. This capability is often met by the use of multiple channels that operate in parallel. This introduces requirements for inclusion of upstream and downstream hydraulic structures that are used to provide uniform distribution of flow among operating channels. For open-channel systems, hydraulic control structures will also be required to maintain an appropriate liquid level within the irradiated zone of the reactor (see section on "Hydraulic Behavior of UV Disinfection Systems").

Optimization of system performance may be warranted in some systems; this is most likely to be the case with systems that have large numbers of UV lamps, either because of high flow rate, a high treatment standard (e.g. wastewater reuse), or both. For these systems, it may be worthwhile to refine operating conditions so as to optimize energy use and reactor efficiency.

### 9.9.1 Stochastic Methods of Reactor Analysis

Traditionally, CFD-E models have been applied for deterministic simulations of reactor performance. In other words, conventional CFD-E model applications have been conducted to provide a well-defined function to relate disinfection efficacy to operating conditions. However, actual performance of UV disinfection systems demonstrates variability. In a general sense, this variability is attributable to at least two factors. The first is error in the assays that are used to measure reactor operating conditions and performance. These errors can be minimized by the careful, consistent application of analytical methods, but will always be a part of any measurement of reactor behavior (see Figures 7.13–7.17). The second source of variability is inherent in the operating parameters that govern reactor performance (Figure 9.31). For any given application, these input parameters (flow

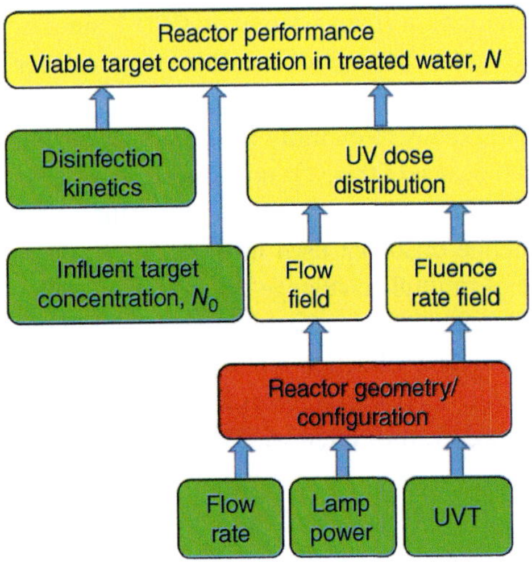


**Figure 9.31** Schematic illustration of factors that affect disinfection efficacy in a UV disinfection system. Parameters identified in green boxes are variable; parameters identified in the red box are fixed for a given reactor and measureable; parameters identified in yellow boxes are dependent on other parameters; the direction of dependence among the variables is indicated by the arrows in the figure. *Source:* Modified from Figure 3.22 to conform to disinfection applications.

rate, lamp power, UVT, $N_0$, and UV dose–response behavior) demonstrate variability, which translates into observed variability in reactor performance.

Recent research has demonstrated the potential benefits of nondeterministic (i.e. stochastic) simulations of reactor behavior based on CFD-I modeling [65, 66]. This work was conducted under the hypothesis that variability observed in the performance (i.e. disinfection efficacy) of UV disinfection systems is attributable to variability in the input parameters that govern their performance. Since these input parameters can be measured independently, it is possible to quantify their variability, as well as correlations that exist among these parameters.

The general philosophy that has been applied in these stochastic simulations to date is that the UV dose distribution delivered under any given operating condition (as defined by reactor configuration, lamp power, UVT, and flow rate) is deterministic. By extension, this implies that variability in disinfection efficacy ($N/N_0$) or effluent microbial quality ($N$) is largely attributable to variability in the concentration of viable target organisms in the influent water ($N_0$) and variability in UV dose–response behavior.

Figures 9.32–9.36 illustrate measured values of UV dose–response behavior for three common indicator bacterial groups (*E. coli*, fecal coliforms, and *Enterococcus*) from replicate experiments with undisinfected effluent samples from three municipal wastewater treatment facilities. These graphs illustrate variability in $N_0$ and UV dose–response behavior. For each of these data sets, measured values of $N_0$ varied by one to three orders of magnitude (see values of $N$ corresponding to $UV_{254}$ dose = 0 for each data set). The trends in $UV_{254}$ dose–response behavior for each data set are largely similar; however, close inspection of these data indicate variations in the slopes of these curves.

Qualitatively similar behavior was illustrated previously for $UV_{254}$ dose–response behavior for challenge organisms that are used in reactor validation by biodosimetry (see Figures 7.13–7.17). It is relevant to compare the trends in dose–response behavior that are evident among these data sets. The challenge organisms that are used in biodosimetry are generally grown in a laboratory under well-defined, controlled conditions of incubation. In contrast, the "wild" indicator organisms that are naturally present in water samples will often have experienced harsher, less well-controlled conditions of growth prior to exposure to UV radiation. As a result, it is logical to expect that variability in UV dose–response behavior will be greater among "wild" samples than with samples

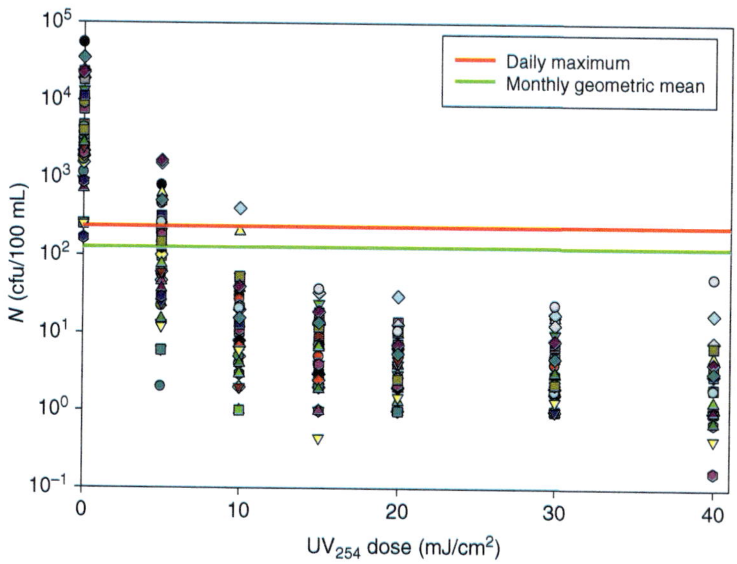


**Figure 9.32** $UV_{254}$ dose–response behavior for *E. coli* in effluent samples received from the Belmont WRP in Indianapolis, IN, from bench-scale collimated beam experiments. The red and green lines illustrate the permit limits based on *E. coli* for the Belmont facility. *Source:* Data from Ortiz [67].

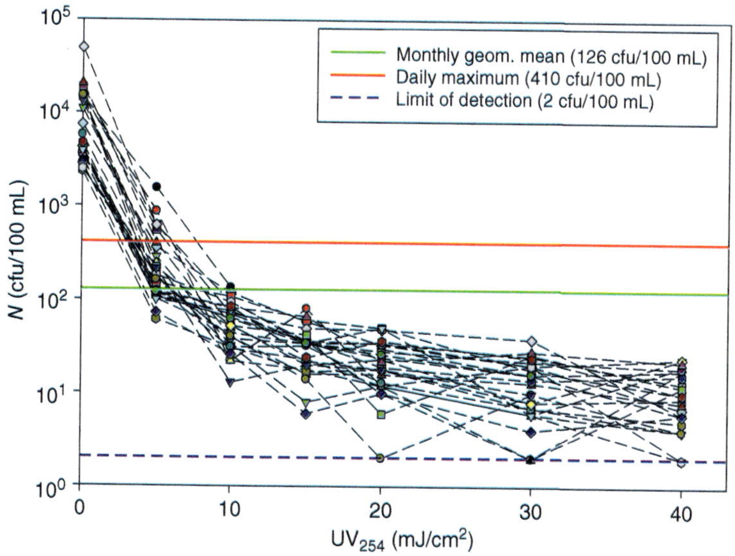


**Figure 9.33** $UV_{254}$ dose–response behavior for *E. coli* in effluent samples received from the O'Brien WRP in Chicago, IL, from bench-scale collimated beam experiments. The red and green lines illustrate the permit limits based on *E. coli* for the O'Brien facility; the dashed horizontal line indicates the limit of detection for the analytical method. *Source:* Data from Li [68].

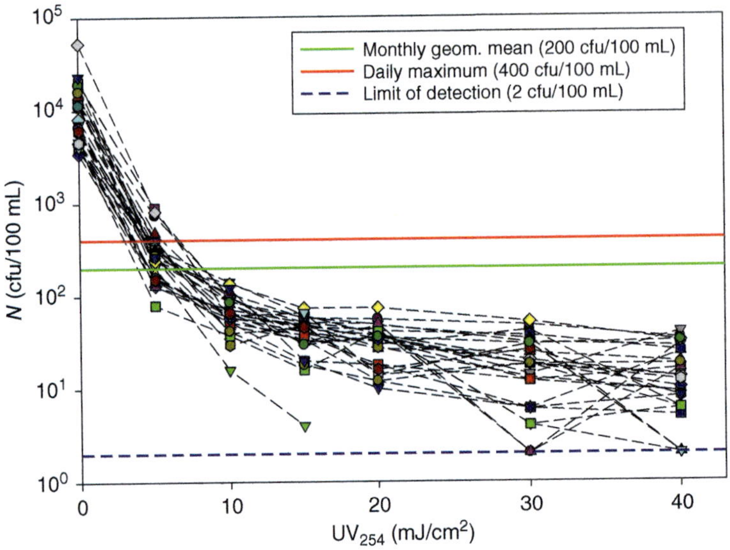


**Figure 9.34** $UV_{254}$ dose–response behavior for fecal coliform bacteria in effluent samples received from the O'Brien facility in Chicago, IL, from bench-scale collimated beam experiments. The red and green lines illustrate the NPDES permit limit for the O'Brien facility. The dashed horizontal line indicates the limit of detection for the analytical method. *Source:* Data from Li [68].

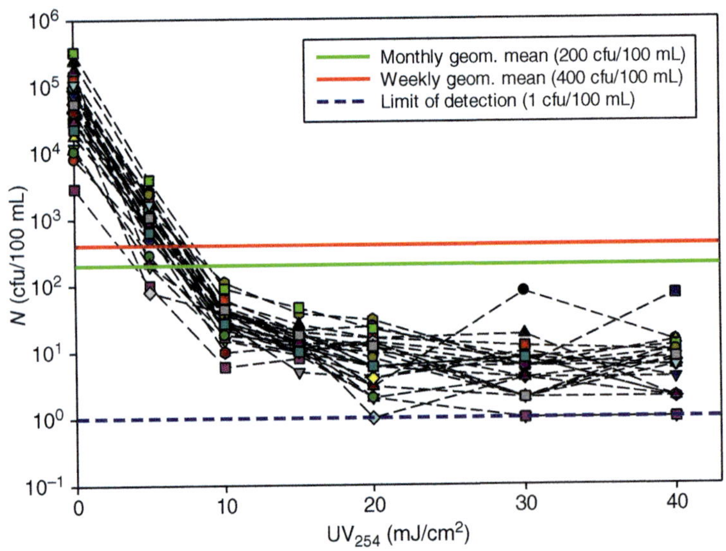


**Figure 9.35** $UV_{254}$ dose–response behavior for fecal coliform bacteria in effluent samples received from the 26th Ward WWTP in New York, NY, from bench-scale collimated beam experiments. The red and green lines illustrate the NPDES permit limit for the 26th Ward facility. The dashed horizontal line indicates the limit of detection for the analytical method. *Source:* Data from Li [68].

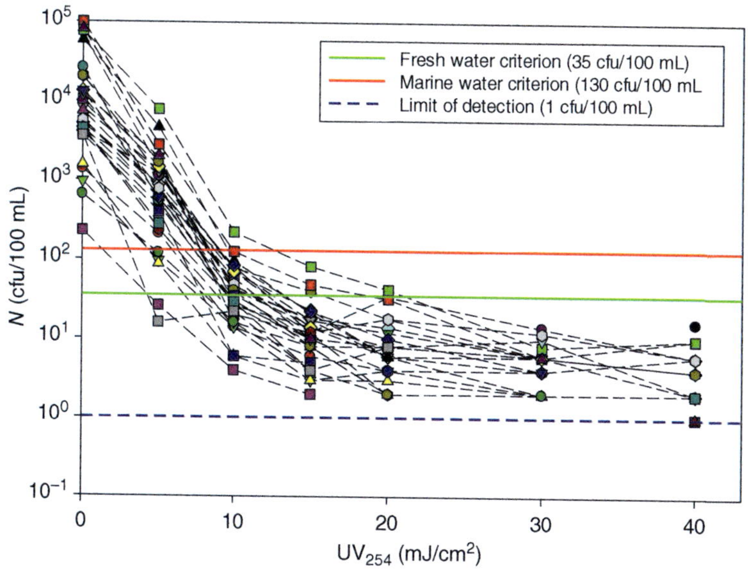


**Figure 9.36** $UV_{254}$ dose–response behavior for *Enterococcus* in effluent samples received from the 26th Ward WWTP in New York, NY, from bench-scale collimated beam experiments. The red and green lines illustrate the NPDES permit limit for the 26th Ward facility. The dashed horizontal line indicates the limit of detection for the analytical method. *Source:* Data from Li [68].

grown in a laboratory. However, significant variation is evident in all of these data sets. By extension, it is logical to expect that variation will be evident in values of $N_0$ and $UV_{254}$ dose–response behavior, which will translate directly into variability in measured values of effluent water quality and reactor performance.

Given the variability in $N_0$ and $UV_{254}$ dose–response behavior that is common to virtually all sources of water, it is relevant to consider the effects of these sources of variability on disinfection efficacy. Stochastic simulation approaches were applied to address that topic specifically by applying CFD-E models with variable inputs. Two types of stochastic analysis were applied for analysis of disinfection efficacy, including variability, for the open-channel UV disinfection system that is in use at the Belmont AWT facility in Indianapolis, IN [65, 66]. Figure 9.37 illustrates this system.

The first of these methods is known as "one-at-a-time" (OAT) analysis, wherein a single parameter is allowed to vary, with all other input parameter values being held constant at a representative value. Two forms of OAT analysis were applied; for both methods, the PPES model was applied to describe and interpolate $UV_{254}$ dose–response behavior for the target microbe [69], which was *E. coli*. Like most kinetic models, the PPES model can be applied to disinfection kinetic data (i.e. from a collimated beam experiment with batch samples) that has been normalized. Figure 9.38 provides an illustration of $N/N_0$ as a function of $UV_{254}$ dose (data from Figure 9.32 are presented in normalized form in Figure 9.38). The raw data (Figure 9.32) provide a clear indication of variability in $N_0$, but the density of data points in the graph masks some of the variability in UV dose–response behavior. In contrast, the variability in UV dose–response behavior is clearly evident in the normalized form of this graph (see Figure 9.38).

The OAT analyses were used to simulate reactor performance over a range of operating conditions. Table 9.6 provides a summary of these conditions. Ambient biodosimetry (AB) experiments

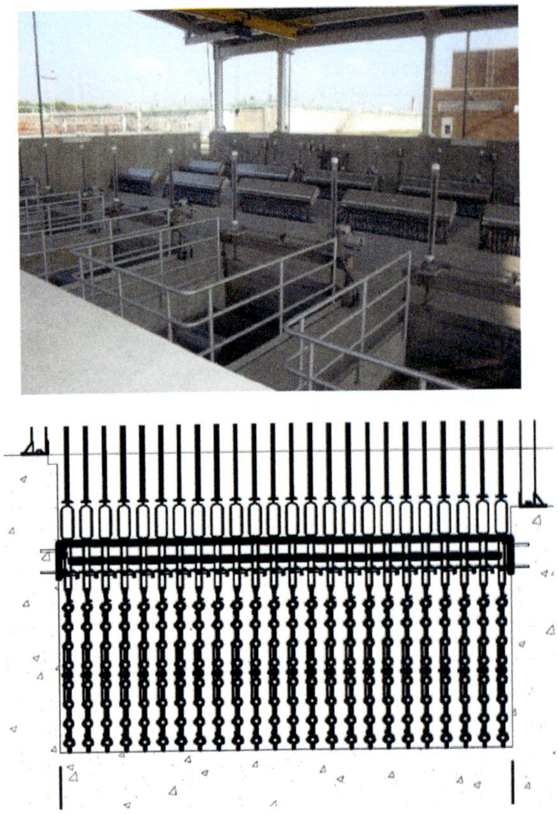

Figure 9.37 (Top) Photographic image of multichannel UV disinfection system located at the Belmont AWT facility in Indianapolis, IN. (Bottom) Schematic illustration of reactor cross section, which involves an 8x24 array of LPHO lamps [66].

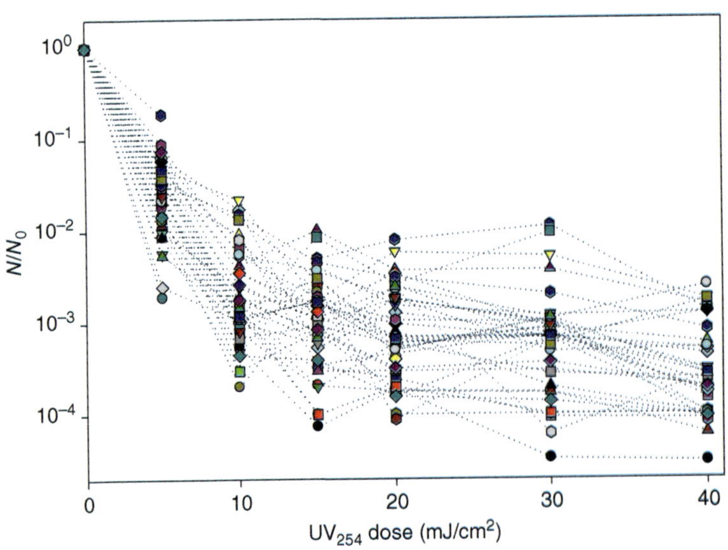


Figure 9.38 Measured $UV_{254}$ dose–response data for *E. coli* in replicate, undisinfected effluent samples from the Belmont AWT facility in Indianapolis, IN. *Source:* Data from Ortiz [67].

**Table 9.6** Summary of operating conditions for which AB experiments were conducted and for which OAT stochastic simulations were applied.

| Operating condition | Flow rate $\left(\frac{m^3}{hr}\right)$ | $N_{Ch}$ | NB | UVT $\left(\frac{\%}{cm}\right)$ | Lamp power (%) | Approach velocity $\left(\frac{m}{s}\right)$ |
|---|---|---|---|---|---|---|
| 1 | 13,200 | 4 | 1 | 73 | 80 | 0.47 |
| 2 | 19,300 | 7 | 2 | 68 | 67 | 0.37 |
| 3 | 12,900 | 6 | 1 | 70 | 72 | 0.29 |
| 4 | 11,000 | 6 | 1 | 71 | 60 | 0.25 |
| 5 | 10,600 | 7 | 1 | 70 | 60 | 0.21 |
| 6 | 10,800 | 7 | 1 | 70 | 60 | 0.21 |
| 7 | 7,510 | 7 | 1 | 69 | 82.5 | 0.15 |
| 8 | 8,980 | 7 | 1 | 73 | 62 | 0.17 |
| 9 | 12,300 | 7 | 1 | 72 | 62 | 0.24 |
| 10 | 18,100 | 7 | 2 | 63 | 80 | 0.35 |
| 11 | 10,900 | 7 | 1 | 62 | 76.5 | 0.21 |
| 12 | 10,600 | 7 | 1 | 71 | 76.5 | 0.20 |
| 13 | 9,060 | 7 | 1 | 71 | 76.5 | 0.18 |
| 14 | 9,140 | 7 | 1 | 72 | 76.5 | 0.18 |

[a] Acronym definitions: $N_{Ch}$ = number of operating channels; NB = number of operating lamp banks; UVT = transmittance of water at 254 nm through a 1.0 cm optical path.
*Source:* Data from Ahmed et al. [65].

were conducted at each of these operating conditions. In AB experiments, replicate samples of undisinfected and disinfected effluent are collected and analyzed for viable target organism concentration. As such, the data that result from these measurements provide a measure of variability in reactor performance for a given operating condition.

The first OAT analysis involved the use of variable PPES model parameters, with a fixed, representative value of $N_0$. For this simulation, the mean of measured values of $N_0$ from across the 14 operating conditions listed in Table 9.6 was applied. Values of PPES model parameters were allowed to vary across the ranges of parameter estimates that were generated by fitting of the PPES model to each of the 46 data sets illustrated in Figure 9.38. The second OAT analysis involved a fixed set of the PPES model parameters and variable $N_0$ values. The PPES model parameters applied in this study were the means of each parameter from the regression analysis that was applied to the 46 normalized data sets illustrated in Figure 9.38, while the variable values of $N_0$ were randomly selected from across the range of measured values of this parameter in all 14 AB experiments.

Figure 9.39 provides an illustration of the results of this analysis. Measured values and stochastic simulation results were in good agreement across all operating conditions. Measurements and simulation results were similar in terms of their respective means, and in terms of the range of values that were reported. Moreover, the results from all operating conditions indicate that reactor performance was consistently in compliance with discharge permit limitations, with measured and predicted values consistently at least an order of magnitude below the regulatory limits.

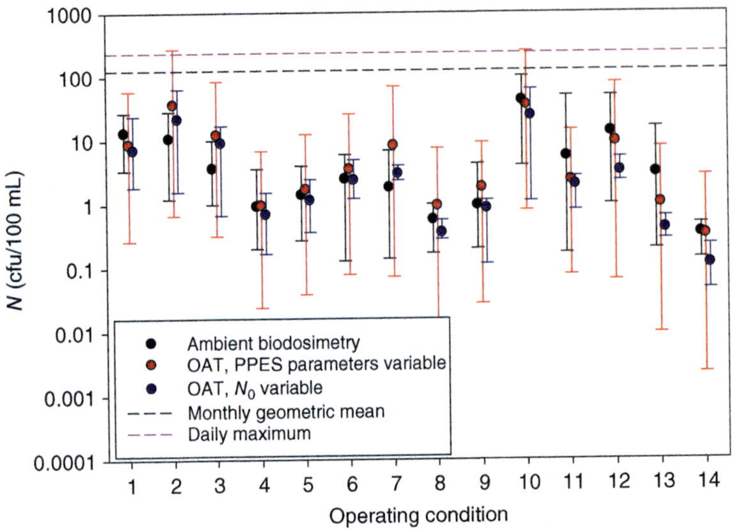


**Figure 9.39** Viable *E. coli* in the effluent of the Belmont AWT facility. Values reported include measurements collected during ambient biodosimetry, predictions based on stochastic applications of CFD-I using the OAT method with variable PPES model parameters, and predictions based on stochastic applications of CFD-I using the OAT method with variable $N_0$. For each data set, the symbol represents the mean of the reported values, while the error bars represent minimum and maximum values. Also included for reference are NPDES permit limitations for viable *E. coli* based on a monthly geometric mean and a daily maximum.

The second method of stochastic simulation was based on a Monte Carlo (MC) approach, wherein a large number of input parameter sets was selected for use in simulations of reactor performance over a wide range of conditions. In a MC simulation, a large population of input data sets is applied to a governing model to simulate the performance of a system, including its variability. A key feature of the MC approach is the incorporation of a method of data selection that allows for identification of data sets that are representative of the actual input parameters that govern system performance. This can be accomplished through the use of a *copula* function, which represents an algorithm that accounts for correlations among input parameters. In effect, the inclusion of a copula allows for guided random sampling of input parameters so as to guarantee a sample population that is representative of measured data and correlations among them.

Application of the copula is initiated by random selection of one input parameter. The selection of subsequent input parameters within an input parameter set will be influenced by the correlations that exist among the input parameters. Each input data set is then applied to the governing relationship, which in the case of UV disinfection systems is the segregated-flow model, to yield a prediction of system performance. The process is repeated a large number of times to develop a large population of process performance estimates for a given operating condition. Collectively, this process allows for probabilistic predictions of process performance for essentially any operating condition.

Figure 9.40 illustrates the outcome of such simulations for four operating conditions for the open-channel UV disinfection system at the Belmont AWT facility. Key features of these four operating conditions are summarized in Table 9.7. All four operating conditions were simulated for UVT = 60% and lamp power at 60% of capacity.

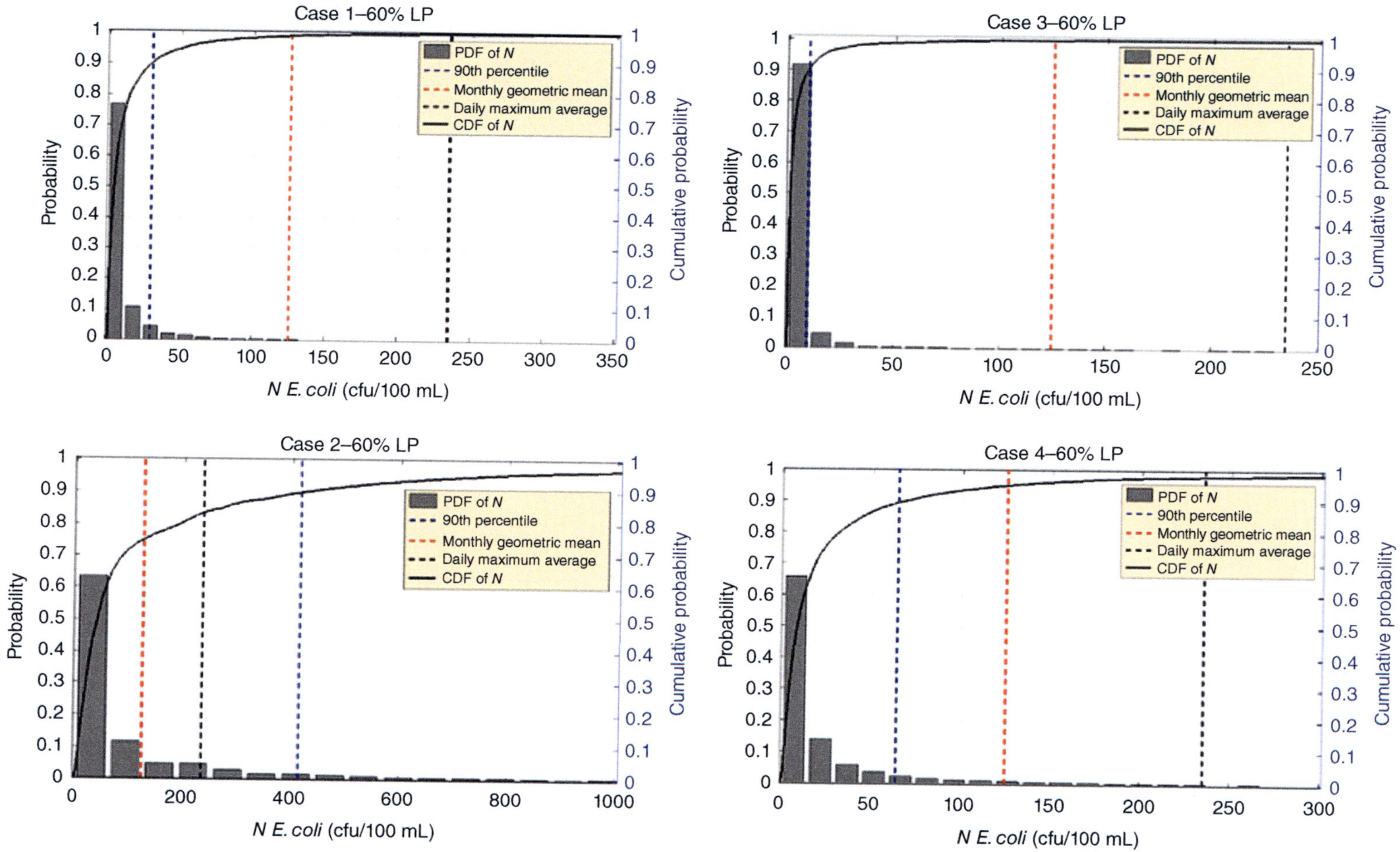


**Figure 9.40** Distributions of predicted effluent *N* (*E. coli*) from Monte Carlo simulations for the Belmont UV disinfection system for four distinct operating conditions. For each case, the bars indicate the probability density function of N as a histogram (left vertical axis), while the solid black line represents the cumulative density function (right vertical axis). The blue, vertical, dashed line represents the 90th percentile of the model predictions, while the red and black vertical dashed lines indicate the NPDES permit limitations for effluent viable *E. coli* for the Belmont AWT facility. Note that the horizontal axis scale is not identical in each panel.

**Table 9.7** List of the operating conditions tested in MC simulations of Belmont system.

| Case# | Flow $\left(\frac{m^3}{hr}\right)$ | $N_{Ch}$ | NB/channel | Flow/channel ($m^3$/hr/channel) | Velocity $\left(\frac{m}{s}\right)$ |
|---|---|---|---|---|---|
| 1 | 7,880 | 7 | 1 | 1120 | 0.15 |
| 2 | 23,600 | 7 | 1 | 3370 | 0.47 |
| 3 | 7,880 | 7 | 2 | 1120 | 0.15 |
| 4 | 23,600 | 7 | 2 | 3370 | 0.47 |

For cases 1 and 3, the 90th percentile of predicted performance was well below the regulatory limits for this system, as defined in their NPDES permit limitations as a daily maximum and a monthly geometric mean of viable *E. coli* concentration. For these two operating conditions, it appears that the system could be reliably operated by reducing the number of operating channels, thereby allowing a fraction of the lamps in the system to be turned off. In contrast, for cases 2 and 4, the 90th percentile of the MC simulations exceeded (case 2) or was close to these regulatory limits. Operationally, this suggests that for these operating conditions, more lamp banks should be turned on (case 2) or lamp power should be increased (case 4) to improve disinfection efficacy in the system.

Overall, these stochastic approaches to operation of UV disinfection systems offer the potential to optimize system performance. This optimization approach has the potential to improve energy efficiency, while simultaneously improving process reliability. These stochastic approaches are time-consuming and require computational expertise and software to conduct the simulations on which these probabilistic characterizations are based. As such, they are best suited for applications to systems that involve large numbers of lamps, either because of high flow rate, a high treatment standard (e.g. reuse), or both.

## 9.10 Summary

Disinfection of potable water and municipal wastewater is practiced with a goal of reducing the risks of human exposure to microbial pathogens, thereby reducing the risk of communicable disease transmission. UV-based disinfection systems have been implemented extensively in these public works applications and have been demonstrated to be effective for meeting the general goals of disinfection described above.

UV systems designed for disinfection of potable water and municipal wastewater share many common features, but also demonstrate some important differences, specifically in terms of the goals of treatment, the methods used to measure process performance, and the geometries of the respective systems. In addition, the regulatory approaches that are used to dictate performance standards are also different among these applications. Despite these differences, the same principles of physics, chemistry, and microbiology govern the behavior of these systems.

Process behavior in UV disinfection systems is governed by the concentration of viable or infective pathogens that enter the system, the nature of the pathogens (which influences their intrinsic kinetics of inactivation by exposure to UV radiation), and the dose distribution delivered by the reactor. These factors can be accurately represented using mathematical models and computational tools, thereby allowing accurate simulations of process performance using tools like CFD-E models.

The fundamental characteristics of UV disinfection systems and the fluids they treat also affect practical aspects of their performance, including fouling of quartz surfaces and hydraulic behavior. These phenomena are understood at a largely qualitative level, but existing knowledge is sufficient to guide practice.

For large-scale UV disinfection systems, the use of stochastic models to simulate process performance, including inherent variability, may be warranted. These stochastic methods allow optimization of system performance, while at the same time improving process reliability and the ability to comply with regulatory constraints.

## Notes

**a** The single-event model suggests that a plot of $\ln\left(\frac{N}{N_0}\right)$ vs. $D$ should yield a straight line that passes through the origin with a slope of $-k$. Linear regression of data that are expected to conform to this model implies that the data from this function will be fit to the general equation of a line: $y = mx + b$. In this form, the dependent variable ($y$) is $\ln\left(\frac{N}{N_0}\right)$, the independent variable ($x$) is dose, and the slope of the line ($m$) is the negative of the inactivation constant. The "$y$-intercept" term ($b$) has no physical interpretation in this model, and as such should be fixed at a value of zero (i.e. best-fit line should be forced through the origin).

**b** For any individual point in a data set, it is possible to identify the vertical distance from the measured value of the dependent variable ($N/N_0$, in this case) to the value that corresponds to the model fit at a given value of the independent variable (dose, in this case). This vertical distance is squared to provide an unbiased estimate of model fit at a given point. RSS represents the sum of these squared error terms for all values of dose.

## References


1 Jacangelo, J.G., Patania, N.L., Trussell, R.R. et al. (**2002**). *Inactivation of Waterborne Emerging Pathogens by Selected Disinfectants*. Denver, CO: AWWA Research Foundation.

2 Malayeri, A.H., Mohseni, M., Cairns, B. et al. (**2016**). *Fluence (UV Dose) Required to Achieve Incremental Log Inactivation of Bacteria, Protozoa, Viruses and Algae*. IUVA News.

3 Li, J. and Blatchley III, E.R. (**2009**). UV photodegradation of inorganic chloramines. *Environmental Science & Technology 43* (1): 60–65.

4 Zhang, X.R., Li, W.G., Blatchley III, E.R. et al. (**2015**). UV/chlorine process for ammonia removal and disinfection by-product reduction: Comparison with chlorination. *Water Research 68*: 804–811.

5 Weng, S.C., Li, J., Wood, K.V. et al. (**2013**). UV-induced effects on chlorination of creatinine. *Water Research 47* (14): 4948–4956.

6 Weng, S.C. and Blatchley III, E.R. (**2013**). Ultraviolet-induced effects on chloramine and cyanogen chloride formation from chlorination of amino acids. *Environmental Science & Technology 47* (9): 4269–4276.

7 Brinkmann, T., Santonja, G.G., Schorcht, F. et al. (**2014**). *Best Available Techniques (BAT) Reference Document for the Production of Chlor-Alkali*, 344. Seville: European Commission, Joint Research Centre, Institute for Prospective Technological Studies.

8 Blatchley III, E.R., Hunt, B.A., Duggirala, R. et al. (**1997**). Effects of disinfectants on wastewater effluent toxicity. *Water Research 31* (7): 1581–1588.

9 Craik, S.A., Weldon, D., Finch, G.R. et al. (**2001**). Inactivation of *Cryptosporidium parvum* oocysts using medium- and low-pressure ultraviolet radiation. *Water Research 35* (6): 1387–1398.

10 Craik, S.A., Finch, G.R., Bolton, J.R., and Belosevic, M. (**2000**). Inactivation of *Giardia muris* cysts using medium-pressure ultraviolet radiation in filtered drinking water. *Water Research 34* (18): 4325–4332.

11 Clancy, J.L., Bukhari, Z., Hargy, T.M. et al. (**2000**). Using UV to inactivate *Cryptosporidium*. *Journal American Water Works Association 92* (9): 97–104.

12 Clancy, J.L., Hargy, T.M., Marshall, M.M., and Dyksen, J.E. (**1998**). UV light inactivation of *Cryptosporidium* oocysts. *Journal American Water Works Association 90* (9): 92–102.

13 Bukhari, Z., Hargy, T.M., Bolton, J.R. et al. (**1999**). Medium-pressure UV for oocyst inactivation. *Journal American Water Works Association 91* (3): 86–94.

14 Mofidi, A.A., Meyer, E.A., Wallis, P.M. et al. (**2002**). The effect of UV light on the inactivation of *Giardia lamblia* and *Giardia muris* cysts as determined by animal infectivity assay (P-2951-01). *Water Research 36* (8): 2098–2108.

15 Rochelle, P.A., Marshall, M.M., Mead, J.R. et al. (**2002**). Comparison of *in vitro* cell culture and a mouse assay for measuring infectivity of *Cryptosporidium parvum*. *Applied and Environmental Microbiology 68* (8): 3809–3817.

16 Slifko, T.R., Huffman, D.E., Dussert, B. et al. (**2002**). Comparison of tissue culture and animal models for assessment of *Cryptospridium parvum* infection. *Experimental Parasitology 101* (2-3): 97–106.

17 Slifko, T.R., Huffman, D.E., and Rose, J.B. (**1999**). A most-probable-number assay for enumeration of infectious *Cryptosporidium parvum* oocysts. *Applied and Environmental Microbiology 65* (9): 3936–3941.

18 US EPA (**2021**). National Primary Drinking Water Regulations. https://www.epa.gov/ground-water-and-drinking-water/national-primary-drinking-water-regulations#Microorganisms (accessed 2021).

19 DiCaprio, E., Ma, Y.M., Purgianto, A. et al. (**2012**). Internalization and dissemination of human norovirus and animal caliciviruses in hydroponically grown romaine lettuce. *Applied and Environmental Microbiology 78* (17): 6143–6152.

20 Rodriguez, R.A., Bounty, S., and Linden, K.G. (**2013**). Long-range quantitative PCR for determining inactivation of adenovirus 2 by ultraviolet light. *Journal of Applied Microbiology 114* (6): 1854–1865.

21 Linden, K.G., Thurston, J., Schaefer, R., and Malley, J.P. (**2007**). Enhanced UV inactivation of adenoviruses under polychromatic UV lamps. *Applied and Environmental Microbiology 73* (23): 7571–7574.

22 Beck, S.E., Hull, N.M., Poepping, C., and Linden, K.G. (**2018**). Wavelength-dependent damage to adenoviral proteins across the germicidal UV spectrum. *Environmental Science & Technology 52* (1): 223–229.

23 Vazquez-Bravo, B., Goncalves, K., Shisler, J.L., and Marinas, B.J. (**2018**). Adenovirus replication cycle disruption from exposure to polychromatic ultraviolet irradiation. *Environmental Science & Technology 52* (6): 3652–3659.

24 Ettayebi, K., Crawford, S.E., Murakami, K. et al. (**2016**). Replication of human noroviruses in stem cell-derived human enteroids. *Science 353* (6306): 1387–1393.

25 Weng, S., Dunkin, N., Schwab, K.J. et al. (**2018**). Infectivity reduction efficacy of UV irradiation and peracetic acid-UV combined treatment on MS2 bacteriophage and murine norovirus in secondary wastewater effluent. *Journal of Environmental Management 221*: 1–9.

26 California State Water Resource Control Board (**2015**). Regulations Related to Recycled Water. Sacramento, CA.

27 Emerick, R.W. and Tchobanoglous, G. (**2012**). *Ultraviolet Disinfection Guidelines for Drinking Water and Water Reuse*. National Water Research Institute: Fountain Valley, CA.

28 Sheikh, B. (**2013**). *Recycled Water Uses Allowed in California* (ed. East Bay Municipal Utilities District (EBMUD) Office of Water Recycling). Oakland, CA.

29 Harm, W. (**1980**). *Biological Effects of Ultraviolet Radiation*. Cambridge: Cambridge University Press.

30 Qiao, Z., Ye, Y.Y., Chang, P.H. et al. (**2018**). Nucleic acid photolysis by $UV_{254}$ and the impact of virus encapsidation. *Environmental Science & Technology 52* (18): 10408–10415.

31 Chick, H. (**1908**). An investigation of the laws of disinfection. *Journal of Hygiene 8* (1): 92–158.

32 Watson, H.E. (**1908**). A note on the variation of the rate of disinfection with change in the concentration of the disinfectant. *Journal of Hygiene 8* (4): 536–542.

33 Pennell, K.G., Aronson, A.I., and Blatchley III, E.R. (**2008**). Phenotypic persistence and external shielding ultraviolet radiation inactivation kinetic model. *Journal of Applied Microbiology 104* (4): 1192–1202.

34 Severin, B.F., Suidan, M.T., and Engelbrecht, R.S. (**1983**). Kinetic modeling of UV disinfection of water. *Water Research 17* (11): 1669–1678.

35 Blatchley III, E.R., Dumoutier, N., Halaby, T.N. et al. (**2001**). Bacterial responses to ultraviolet irradiation. *Water Science and Technology 43* (10): 179–186.

36 Jolis, D., Lam, C., and Pitt, P. (**2001**). Particle effects on ultraviolet disinfection of coliform bacteria in recycled water. *Water Environment Research 73* (2): 233–236.

37 Templeton, M.R., Andrews, R.C., and Hofmann, R. (**2005**). Inactivation of particle-associated viral surrogates by ultraviolet light. *Water Research 39* (15): 3487–3500.

38 Loge, F.J., Emerick, R.W., Thompson, D.E. et al. (**1999**). Factors influencing ultraviolet disinfection performance part I: light penetration to wastewater particles. *Water Environment Research 71* (3): 377–381.

39 Emerick, R.W., Loge, F.J., Ginn, T., and Darby, J.L. (**2000**). Modeling the inactivation of particle-associated coliform bacteria. *Water Environment Research 72* (4): 432–438.

40 Emerick, R.W., Loge, F.J., Thompson, D., and Darby, J.L. (**1999**). Factors influencing ultraviolet disinfection performance part II: association of coliform bacteria with wastewater particles. *Water Environment Research 71* (6): 1178–1187.

41 Kelner, A. (**1949**). Photoreactivation of ultraviolet-irradiated *Escherichia-coli*, with special reference to the dose-reduction principle and to ultraviolet-induced mutation. *Journal of Bacteriology 58* (4): 511–522.

42 Kelner, A. (**1949**). Effect of visible light on the recovery of *Streptomyces griseus conidia* from ultraviolet irradiation injury. *Proceedings of the National Academy of Sciences of the United States of America 35* (2): 73–79.

43 Sancar, A. (**2003**). Structure and function of DNA photolyase and cryptochrome blue-light photoreceptors. *Chemical Reviews 103* (6): 2203–2237.

44 Sancar, A. (**1994**). Structure and function of DNA photolyase. *Biochemistry 33* (1): 2–9.

45 Johns, H.E., Rappaport, S.A., and Delbruck, M. (**1962**). Photochemistry of thymine dimers. *Journal of Molecular Biology 4*: 104–114.

46 Blatchley III, E.R. (**1997**). Numerical modelling of UV intensity: application to collimated-beam reactors and continuous-flow systems. *Water Research 31* (9): 2205–2218.

47 Roshko, A. (**1993**). Perspectives on bluff-body aerodynamics. *Journal of Wind Engineering and Industrial Aerodynamics 49* (1–3): 79–100.

48 Zdravkovic, M.M. (**2003**). *Flow Around Circular Cylinders, Applications*, vol. *2*, 592. Oxford University Press.

49 Zdravkovic, M.M. (**1997**). *Flow Around Circular Cylinders, Fundamentals*, vol. *1*, 694. Oxford University Press.

50 Blatchley III, E. R., Duggirala, R., Chiu, K.P. et al. (**1994**). Macro-scale hydraulic behavior in open channel UV disinfection systems. *Proc. WEFTEC '94, WEF*, vol. VI, 473–484, Alexandria, VA.

51 Arnold, E. and Adam, V. (**2013**). Influence of quartz glass properties on UV-performance of sleeves for systems: working with medium-pressure lamps. *IUVA News 15* (2): 14–16.

52 Lin, L.S., Johnston, C.T., and Blatchley III, E.R. (**1999**). Inorganic fouling at quartz : water interfaces in ultraviolet photoreactors – II. Temporal and spatial distributions. *Water Research 33* (15): 3330–3338.

53 Lin, L.S., Johnston, C.T., and Blatchley III, E.R. (**1999**). Inorganic fouling at quartz : water interfaces in ultraviolet photoreactors – I. Chemical characterization. *Water Research 33* (15): 3321–3329.

54 Benjamin, M.M. (**2015**). *Water Chemistry*, 2e. Long Grove, IL: Waveland Press.

55 Sheriff, M. and Gehr, R. (**2001**). Laboratory investigation of inorganic fouling of low pressure UV disinfection lamps. *Water Quality Research Journal of Canada 36* (1): 71–92.

56 Peng, J., Qiu, Y., and Gehr, R. (**2005**). Characterization of permanent fouling on the surfaces of UV lamps used for wastewater disinfection. *Water Environment Research 77* (4): 309-+.

57 Nessim, Y. and Gehr, R. (**2006**). Fouling mechanisms in a laboratory-scale UV disinfection system. *Water Environment Research 78* (12): 2311–2323.

58 Wait, I.W. and Blatchley III, E.R. (**2010**). Model of radiation transmittance by inorganic fouling on UV reactor lamp sleeves. *Water Environment Research 82* (11): 2272–2278.

59 Wait, I.W., Johnston, C.T., and Blatchley III, E.R. (**2007**). The influence of oxidation reduction potential and water treatment processes on quartz lamp sleeve fouling in ultraviolet disinfection reactors. *Water Research 41* (11): 2427–2436.

60 Lin, L.S., Johnston, C.T., and Blatchley III, E.R. (**1999**). Inorganic fouling at quartz : water interfaces in ultraviolet photoreactors – III. Numerical modelling. *Water Research 33* (15): 3339–3347.

61 Wait, I.W. (**2005**). *Fouling of Quartz Surfaces in Potable Water Ultraviolet Disinfection Reactors*. West Lafayette, IN: Purdue University.

62 Gehr, R. and Wright, H. (**1998**). UV disinfection of wastewater coagulated with ferric chloride: recalcitrance and fouling problems. *Water Science and Technology 38* (3): 15–23.

63 Derrick, B. (**2005**). *Inorganic Fouling and Effects on Intensity in Groundwater Applications of Ultraviolet Photoreactors*. West Lafayette, IN: Purdue University.

64 Blatchley III, E.R., Bastian, K.C., Duggirala, R.K. et al. (**1996**). Ultraviolet irradiation and chlorination dechlorination for municipal wastewater disinfection: assessment of performance limitations. *Water Environment Research 68* (2): 194–204.

65 Ahmed, Y.M., Ortiz, A.P., and Blatchley III, E.R. (**2019**). Stochastic evaluation of disinfection performance in large-scale, open-channel UV photoreactors. *Journal of Environmental Engineering, ASCE 145* (10): 04019071.

66 Ahmed, Y.M.H.I.H. (**2017**). *Stochastic Evaluation of Disinfection Performance in Large-Scale, Open-Channel UV Photoreactors*. West Lafayette, IN: Purdue University.

67 Ortiz, A.P. (**2014**). *Variability in UV Disinfection of Municipal Wastewater*. West Lafayette, IN: Purdue University.

68 Li, X. (**2017**). *Variability in UV Disinfection System Performance in Municipal Wastewater Treatment*. West Lafayette, IN: Purdue University.

69 Pennell, K.G., Aronson, A.I., and Blatchley III, E.R. (**2008**). Phenotypic persistence and external shielding (PPES) ultraviolet radiation inactivation kinetic model. *Journal of Applied Microbiology 104* (4): 1192–1202.

# 10

# Photolysis and Advanced Reaction Processes for Control of Trace Contaminants

## 10.1 Introduction

Historically, water and wastewater treatment have focused on conventional, bulk characteristics of water quality. Treatment processes have evolved to provide reliable, robust measures for control of these contaminants based on physical separation, as well as chemical and biochemical transformations, often implemented in systems that involve combinations of these approaches. In the drinking water realm, the pollutants that are targeted for treatment often include colloidal particles, microbial pathogens, and nuisance solutes, such as reduced iron and manganese. For municipal wastewater systems, the focus of treatment has traditionally been on colloidal particles, reduced organics, nutrients, and microbial pathogens.

Since roughly the 1990s, there has been increasing recognition of the importance and effects of micropollutants as trace contaminants in water treatment. These contaminants are often released directly to the environment or are formed as by-products of treatment. Most of these compounds are attributable to human activity, but some are produced or altered by natural processes. Common sources of trace contaminants include urban and agricultural runoff, municipal wastewater effluents, and industrial releases [1]. Important categories of micropollutants include: pharmaceuticals and personal care products (PPCPs) [2], endocrine disrupting chemicals (EDCs), antibiotics, and disinfection byproducts (DBPs). Typical concentrations of these contaminants are in the range of $\frac{\text{ng}}{\text{L}} - \frac{\mu\text{g}}{\text{L}}$, but their presence at concentrations outside this range is certainly possible.

The compounds that fit the definitions of trace contaminants (micropollutants) have a number of features that motivate attention to their behavior:

- Among the compounds that have medicinal applications, their effects on human health are typically observed at low concentration (dose), by design.
- They can have adverse effects on human health and/or the environment.
- They are often present as dilute mixtures, yet the effects of mixtures are largely unknown.
- Some are targets of new or future regulations. For example, several of these compounds are listed on the US EPA's existing Contaminant Candidate Lists (e.g. CCL3 and CCL4), or are being considered for CCL5.
- Many of these compounds are ineffectively removed or degraded by conventional treatment processes.

Because of this list of characteristics, a need exists to implement alternative processes to degrade or separate these compounds to yield less harmful or non-harmful products in water treatment.

*Photochemical Reactors: Theory, Methods, and Applications of Ultraviolet Radiation*, First Edition. Ernest R. Blatchley III.

Companion website: www.wiley.com/go/blatchley/photochemicalreactors

UV-based photochemical processes represent alternatives to conventional treatment processes. The most basic of these involve direct photolysis, in which UV radiation is used to cause transformation of target chemicals by photochemical cleavage of one or more bonds. These direct photochemical processes tend to be selective, but are effective for a relatively small number of trace contaminants. Direct photolysis achieves chemical transformation through reaction mechanisms that are fundamentally different from those of conventional processes, and as such, direct photolysis can be used to augment the potential for chemical transformation in treatment.

A more common approach is to use UV radiation to promote reactions that yield reactive intermediates as photolysis products. These reactive intermediates, which often include free radicals, are able to initiate reactions that may not have taken place otherwise or that would otherwise proceed slowly. As with the direct processes, they open up reaction pathways that are often different from those that would result from conventional thermal or biochemical reactions. Consequently, they complement conventional treatment processes.

UV radiation is used to promote so-called advanced oxidation processes (AOPs) and advanced reduction processes (ARPs). AOPs and ARPs proceed through radical intermediates that are produced as the primary products of a photochemical reaction. In many UV/AOPs and UV/ARPs, the reactive intermediates are produced by photolysis of a promoter chemical that is usually added immediately upstream of the UV system. The radical intermediates that are produced tend to be highly reactive, with reaction rates that often proceed at rates that approach diffusional limitation. Some of these radicals are essentially nonselective, while others tend to be selective for molecules that contain functional groups or moieties that are structurally consistent with their mechanisms of reaction.

The nominal UV dose required to operate these processes is often one to two orders of magnitude higher than that required for disinfection. Therefore, UV/AOPs and UV/ARPs are likely to yield effective disinfection, along with the chemical transformations they are designed to accomplish. The large dose requirements of UV/AOP and UV/ARP systems translate to relatively high capital and operating costs, as well as high energy demand, as compared with the costs of UV disinfection systems. Therefore, there is a need for accurate, predictive models of process dynamics to allow optimization of these systems.

The chemical environment within a UV/AOP or UV/ARP system tends to be complex and aggressive; reactions within these systems may include direct photolysis, reactions involving the promoter chemicals directly, and reactions involving the reactive intermediates that are formed from photolysis of the promoter. Predictive models of process behavior in these systems need to include terms to account for this complex range of reaction processes. In addition, because the reactions tend to be rapid and because strong spatial gradients in local reaction rates are observed in these systems, predictive models will often require incorporation of sub-models to account for local transport behavior (i.e. fluid mechanics) and the spatial distribution of radiant energy (i.e. fluence rate field model).

Lastly, the complex chemical environment that characterizes UV/AOP and UV/ARP applications results in systems where a wide range of chemical transformations can occur. As a result, there are likely to be intended and unintended chemical transformations that take place. The designers and users of these systems need to be aware of the potential for this wide range of chemical transformations to accompany these process applications.

The central goal of this chapter is to present the basic principles of direct and indirect photochemical processes. This should allow the reader to conduct or interpret the results of laboratory or bench-scale experiments and translate them to full-scale applications. Similarly, readers should be able to apply the principles that are presented in this chapter to simulate the behavior of UV-based processes for transformation of known contaminants or contaminants that emerge in the future. Lastly, readers should recognize that these processes are likely to have applications in settings that include water and wastewater treatment, as well as a number of other settings. Examples of locations where these processes could be applied include: biomedical/medical applications, analytical chemistry, industrial operations, pharmaceutical production, beverages, and food production.

## 10.2 Direct Photolysis

Photolytic processes can be initiated by radiation from across a wide range of the electromagnetic spectrum, including vacuum UV (VUV), UVC, UVB, UVA, and visible radiation. However, most applications are based on UVC radiation.

The basic principles that govern the kinetics of direct photochemical reactions and photochemical reactor dynamics were presented in Chapters 2 and 3, and are briefly restated here. In the most general sense, the rate of a photochemical process is governed by the volumetric rate of photon absorption and the quantum yield for the reaction. For a system in which monochromatic radiation is used to promote a pure photochemical reaction, the (local) kinetics of the reaction can be described as follows (see Chapter 2 for additional information about these relationships):

$$r_B = -\frac{d[B]}{dt} = \frac{2.303 \cdot E' \cdot \varepsilon \cdot [B]}{U} \cdot \Phi \tag{10.1}$$

Or equivalently:

$$\frac{d[B]}{E' \cdot dt} = -\frac{2.303 \cdot \varepsilon \cdot [B]}{U} \cdot \Phi \tag{10.2}$$

where

| | | | | |
|---|---|---|---|---|
| $r_B$ | = | (local) rate of photochemical reaction | [=] | $\frac{\text{mole}}{\text{L} \cdot \text{s}}$ |
| $[B]$ | = | (local) molar concentration of target compound | [=] | $\frac{\text{mole}}{\text{L}}$ |
| $t$ | = | time | [=] | seconds |
| $E'$ | = | (local) fluence rate | [=] | $\frac{\text{W}}{\text{m}^2}$ |
| $\varepsilon$ | = | molar absorption coefficient of target compound | [=] | $\frac{\text{L}}{\text{mole} \cdot \text{cm}}$ |
| $\Phi$ | = | quantum yield for reaction of interest | [=] | $\frac{\text{mole}}{\text{einstein}}$ |
| $U$ | = | energy per einstein at wavelength of interest | [=] | $\frac{\text{J}}{\text{einstein}}$. |

The denominator on the left-hand side of Equation 10.2 represents an infinitesimally small dose increment, $dD$, that the solution containing the target molecule is exposed to. Therefore, Equation 10.2 can be rewritten equivalently as follows:

$$\frac{d[B]}{dD} = -\frac{2.303 \cdot \varepsilon \cdot [B]}{U} \cdot \Phi \tag{10.3}$$

Equations 10.1–10.3 indicate that the *local* kinetics of a pure photochemical reaction are first-order with respect to the *local* target reactant concentration and the *local* fluence rate. In that sense, the kinetics of a photochemical reaction can be described using a conventional first-order rate expression:

$$\frac{d[B]}{dD} = -k \cdot [B] \tag{10.4}$$

where

$k$ = first-order reaction rate constant $[=] \dfrac{cm^2}{mJ}$

The rate constant for Equation 10.4 can be defined by inspection (see equations 2.1, 2.2, 10.3, and 10.4):[a]

$$k = \frac{2.303 \cdot \varepsilon \cdot \Phi \cdot \lambda}{h \cdot c_0 \cdot N_A} \tag{10.5}$$

where

$U_p$ = photon energy $[=] \dfrac{J}{photon}$

$h$ = Planck constant $= 6.626 \times 10^{-34} \dfrac{J \cdot s}{photon}$

$c_0$ = speed of light in vacuum $= 2.998 \times 10^8 \dfrac{m}{s}$

$\lambda$ = wavelength of radiation $[=]$ nm

$U$ = energy per einstein at wavelength $\lambda$ $[=] \dfrac{J}{einstein}$

$N_A$ = Avogadro's number $= 6.023 \times 10^{23} \dfrac{photon}{einstein}$.

It is important to reiterate that Equations 10.1–10.4 are fundamental rate expressions that apply rigorously at a location or region where the fluence rate and the reactant concentration are single-valued, such as at a point in a photoreactor. In most photoreactors, the fluence rate demonstrates strong spatial variation, which dictates that equally strong spatial variation in the local photochemical reaction rate will be evident in the system. Dose is the master variable that governs the performance of photochemical reactor systems. Therefore, any model that is used to predict photochemical reactor performance must provide a detailed, accurate representation of the dose distribution that is delivered to the solution that contains the targets of treatment.

For pure photochemical reactions, the effects of the dose distribution on reactor performance can be accounted for by use of the segregated-flow model (see Chapter 3):

$$\left(\frac{[B]}{[B]_0}\right)_{reactor} = \int_0^{\infty} \left(\frac{[B]}{[B]_0}\right)_{batch} \cdot E(D) \cdot dD \tag{10.6}$$

where

$[B]$ = concentration of photoreactive constituent after UV exposure

$[B]_0$ = concentration of photoreactive constituent prior to UV exposure

$\left(\dfrac{[B]}{[B]_0}\right)_{reactor}$ = fraction of photoreactive constituent that remains after UV exposure in a reactor system

$\left(\frac{[B]}{[B]_0}\right)_{batch}$ = fraction of photoreactive constituent that remains as a function of dose ($D$) in a batch reactor (i.e. the intrinsic kinetics of the photoreaction of interest)
$E(D)$ = dose distribution function
$E(D) \cdot dD$ = fraction of fluid elements that receive a dose between $D$ and $(D + dD)$.

The kinetics of a pure photochemical process can be described by any of the rate expressions listed above (i.e. equations 10.1–10.4); however, for use in the segregated-flow model, as described in Equation 10.6, it is probably best to use a rate expression that is written using dose ($D$) as the independent variable (i.e. equation 10.3 or 10.4).

Equation 10.6 implies explicit, detailed knowledge of the dose distribution delivered by a photoreactor, $E(D)$. If we assume a Gaussian dose distribution and first-order kinetics (as described by equations 10.3 and 10.4), then Equation 10.6 has a closed-form solution (see derivation in Chapter 7):

$$\left(\frac{[B]}{[B]_0}\right)_{reactor} = \frac{1}{2} \cdot \exp\left(\frac{k^2\sigma^2}{2} - k\mu\right) \cdot \operatorname{erfc}\left(\frac{k\sigma - \frac{\mu}{\sigma}}{\sqrt{2}}\right) \tag{10.7}$$

where

$\mu$ = mean of dose distribution $[=] \frac{mJ}{cm^2}$
$\sigma$ = standard deviation of dose distribution $[=] \frac{mJ}{cm^2}$.

Few, if any photoreactors deliver a dose distribution that is rigorously Gaussian, but many dose distributions are close to Gaussian in shape, so the analysis provided by Equation 10.7 is instructive. However, there are circumstances where it may be possible to simplify the analysis of photoreactor performance, while reaching a similar or acceptable solution. In particular, it may be useful to calculate photoreactor performance based on average dose, rather than by accounting for the entire dose distribution.

A calculation of reactor performance based on average dose is equivalent to an assumption that all fluid elements receive the average dose. Under these circumstances, the performance of a photolytic reactor is defined as:

$$\left(\frac{[B]}{[B]_0}\right)_{reactor} = \exp(-k\mu) \tag{10.8}$$

Equation 10.8 is also equivalent to the assumption of an ideal plug-flow reactor, which yields a theoretical upper limit on reactor performance.

A comparison of the predictions of Equations 10.7 and 10.8 is instructive as a demonstration of the importance of the dose distribution for calculation of reactor performance. To illustrate, the two equations were applied for analysis of several hypothetical reactors that were assumed to be characterized by Gaussian dose distributions. Additional assumptions are that the reactor uses UV sources that are monochromatic at a characteristic wavelength of 254 nm, that mean UV doses delivered by these reactors range from 100 to 1000 $\frac{mJ}{cm^2}$, and the dose distributions are characterized by standard deviations that range from 10 to 20% of the mean. These assumptions are representative of well-designed photoreactors that are used for photolytic applications. The dose distributions used

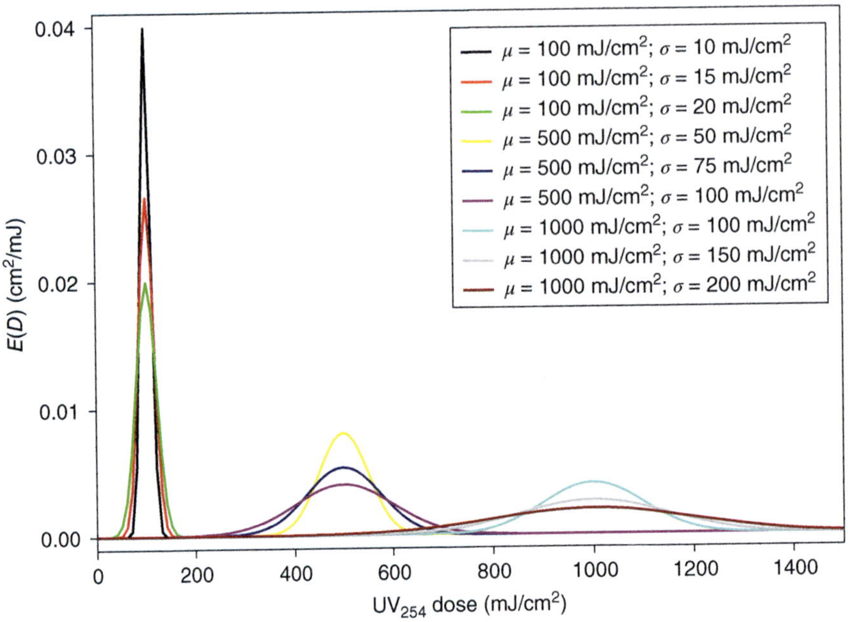


**Figure 10.1** Gaussian dose distributions (density form) used to simulate performance of UV reactors used for photolysis. Mean and standard deviation for each distribution are indicated in the legend.

for these simulations are illustrated in Figure 10.1. The results of these calculations, which are based on Equations 10.7 and (10.8), are illustrated in Figure 10.2.

The results of these simulations indicate that for modest conversion of the target compound, say 1–3 $\log_{10}$ units of conversion, the error associated with an assumption of an ideal dose distribution (i.e. application of equation 10.8) to predict reactor performance is less than 20–30%. If error of this magnitude is acceptable for a design application, then the use of Equation 10.8 to simulate reactor performance may be justified. However, if more accurate calculations are required, then a more detailed analysis (i.e. the segregated-flow model) may be required.

The logic used in this calculation provides justification for predictions of photoreactor performance based on an average dose under some circumstances. However, this also implies that the value of average dose is truly representative of the dose distribution delivered by the reactor. As a first cut, it is probably justifiable to calculate the average dose as the product of the average fluence rate within the irradiated zone and the mean hydraulic detention time in the irradiated zone. These values can be calculated fairly easily using a simple fluence rate field model (applied to the geometry of the reactor, the lamp characteristics, and the optical properties of the fluid being treated), as well as knowledge of the reactor geometry and flow rate through the reactor.

The approach applied in this analysis is analogous to the approach used in Chapter 7, where the focus was on validation of UV photoreactors that are used for disinfection of water. The same principles of physics and chemistry apply for UV systems that are used to reach photolytic treatment endpoints; however, several important differences exist among these reactor systems:

- In disinfection applications, it is common to design and operate a system to achieve 3–4 $\log_{10}$ units of conversion, or more. For photolytic processes, it is common to design for more modest changes in target concentration, often in the range of 1–3 $\log_{10}$ units.

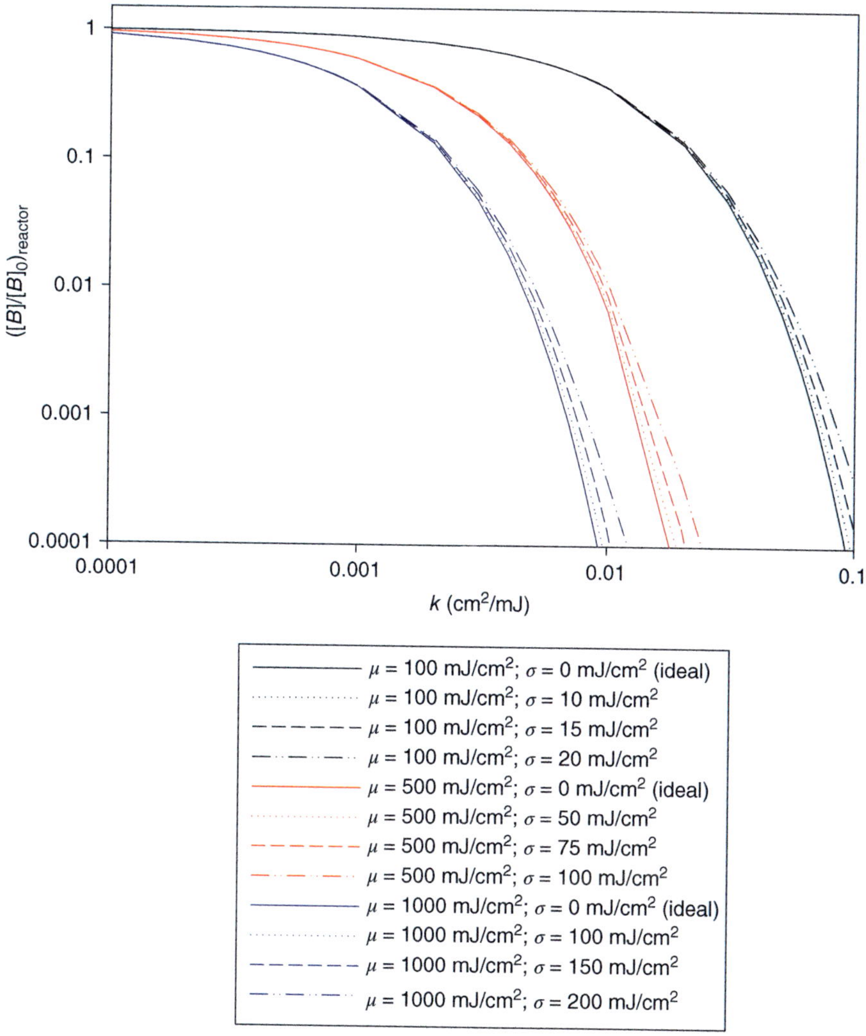


**Figure 10.2** Simulations of UV photoreactor performance for photolysis of a target compound, B. Solid lines represent simulations of reactor performance as a function of photoreaction rate constant for a given value of the mean dose ($\mu$), which corresponds to ideal photoreactor behavior (i.e. $\sigma = 0$). Remaining lines indicate predictions of reactor performance for dose distributions with finite values of the standard deviation for the dose distribution, based on an assumption of a Gaussian dose distribution.

- Pure photolytic reactions are often characterized by smaller rate constants than those that are representative of most disinfection applications. In other words, pure photolysis tends to be much slower than disinfection, in terms of intrinsic kinetics.
- A consequence of the two previous points is that the nominal dose requirements of UV reactors that are used for pure photolysis are generally much higher than those required for disinfection. Nominal (average) UV doses of $1000 \frac{\text{mJ}}{\text{cm}^2}$ or more are not uncommon in UV photoreactors that are used for photolysis.
- When a UV photoreactor is installed for pure photolysis, it is likely that disinfection requirements will also be met by this system, although this should be verified in any application.

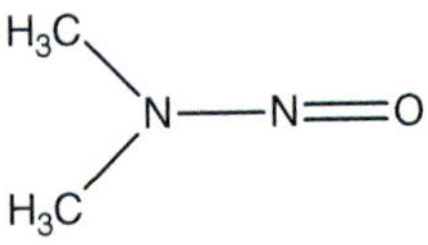


Figure 10.3 Structure of N-nitrosodimethylamine (NDMA).

**Example 10.1 Photolysis of NDMA**

Photolysis is the process of choice for chemical transformation of some pollutants. Prime examples of this are the compound *N*-nitrosodimethylamine (NMDA), and many of the other *N*-dialkyl-nitrosamines (i.e. nitrosamines). The *N*-nitrosamines, including NDMA, are extremely potent carcinogens [3]. As an indication of the risks posed by these compounds, the US EPA identified an NDMA concentration of $0.42\frac{\text{ng}}{\text{L}}$ in tap water as a screening level, based on a $10^{-6}$ lifetime excess cancer risk [3]. NDMA is present as a waste product from the production of rocket fuels based on unsymmetrical dimethylhydrazine (UDMH), from combustion of tobacco, and in cooked meats. NDMA is present in beer and is found in a number of toiletry items and cosmetics. It is also formed as a disinfection by-product from chlorination, which is at least partially responsible for its presence in drinking water supplies and swimming pools [4, 5]. As such, NDMA (and other nitrosamines) are ubiquitous, but are usually present at low concentrations.

NDMA is a relatively low molecular weight, polar molecule (see structure in Figure 10.3). It is slow to biodegrade and responds poorly to most conventional physico/chemical treatment processes, including air stripping and adsorption to activated carbon. On the other hand, it is known to degrade when exposed to ambient solar radiation, which probably contributed to examination of its behavior in response to UVC irradiation.

The absorption spectra for NDMA and other *N*-nitrosamines are characterized by two bands, one with a peak near 230 nm and the second, smaller band with a peak centered near 330 nm (see Figure 10.4). For each of the *N*-nitrosamines illustrated in Figure 10.4, the primary peak corresponds with a $\pi \rightarrow \pi^*$ electronic transition, which leads to cleavage of the *N–N* bond that is observed to result from nitrosamine exposure to UVC radiation. The secondary peak corresponds to an $n \rightarrow \pi^*$ electronic transition, and is responsible for degradation of nitrosamines as a result of exposure to ambient solar radiation, especially in the UVA range; however, it should be noted that despite the fact that the electronic transitions associated with these two bands both lead to cleavage of the *N–N* bond and degradation of nitrosamines, the reaction mechanisms associated with these two bands are not identical. Specifically, the $\pi \rightarrow \pi^*$ transition yields products that react with molecular oxygen ($O_2$), while the products of the $n \rightarrow \pi^*$ transition do not [6]. Ultimately, the degradation products from these two mechanisms are similar, and are dominated by methylamine, dimethylamine (DMA), and nitrite, with smaller quantities of several compounds, including formaldehyde, formate, and nitrate.

Exposure of NDMA to UVC radiation has been shown to promote its decay through homolytic cleavage of the *N–N* bond [7]. This results in the formation of a dimethylaminium radical and a nitroso radical (see Figure 10.5) [7, 8]. These radicals go on to participate in other reactions that yield dimethyl amine (DMA) and nitrite as the primary products, with trace concentrations of nitrate, formaldehyde, formate, methylamine, and other compounds; these products are similar to those that result from irradiation of nitrosamines under ambient solar radiation [6]. For radiation at 254 nm, the quantum yield for this reaction has been reported to be $0.30\frac{\text{mole}}{\text{einstein}}$ [9], with a molar absorption coefficient of $1650\frac{\text{L}}{\text{mole}\cdot\text{cm}}$ [10]. The quantum yield for NDMA photolysis for wavelengths corresponding to the first absorption band (200 nm $\lesssim \lambda \lesssim$ 280 nm) has been reported to be essentially constant across this wavelength range, presumably because this range of wavelengths is associated with the $\pi \rightarrow \pi^*$ transition [9]. By contrast, the second absorption band

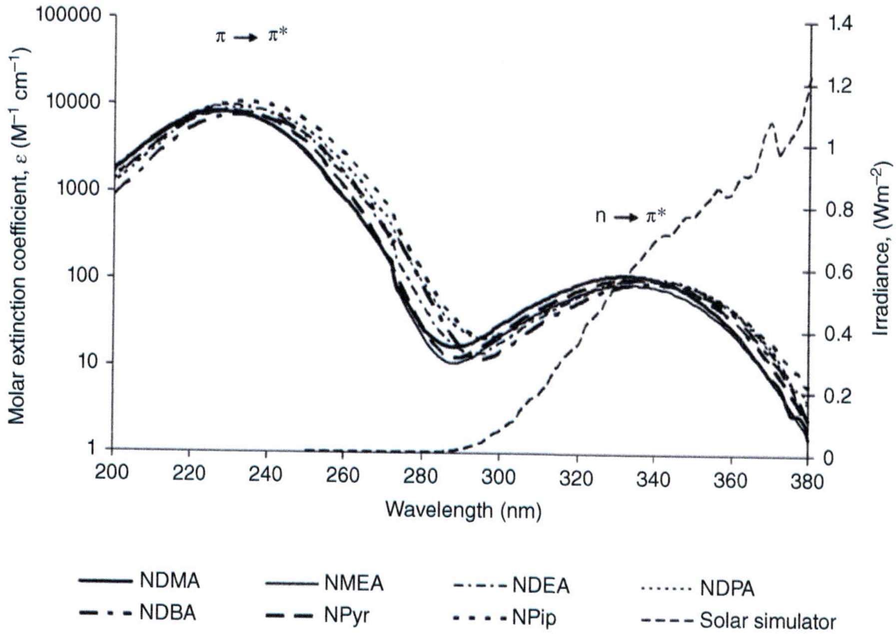


**Figure 10.4** Absorption spectra for NDMA and six other *N*-nitrosamine compounds. Each of the seven *N*-nitrosamines illustrated (NDMA = *N*-nitrosodimethylamine; NMEA = *N*-nitrosodiethylamine; NDEA = *N*-nitrosomethylethylamine; NDPA = *N*-nitrosodi-*n*-propylamine; NDBA = *N*-nitrosodi-*n*-butylamine; NPip = *N*-nitrosopiperidine; NPyr = *N*-nitrosopyrrolidine) share similar bands, centered at nominally 230 nm ($\pi \rightarrow \pi^*$ transition) and 330 nm ($n \rightarrow \pi^*$ transition). Also included is a spectrum of simulated solar irradiance, which illustrates the potential for phototransformation of nitrosamines by exposure to ambient solar radiation. *Source:* Plumlee and Reinhard 2007/with permission of American Chemical Society.

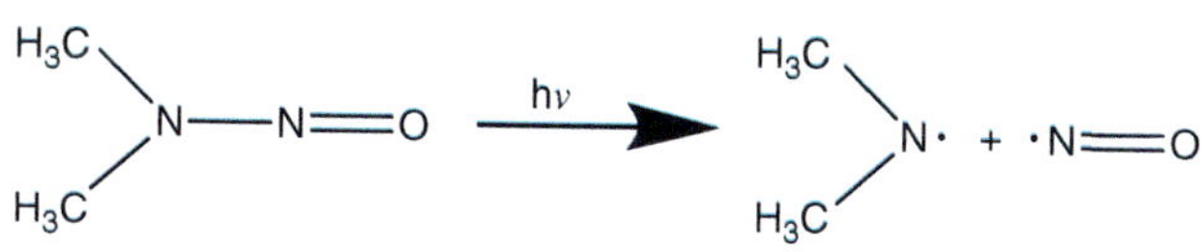

**Figure 10.5** Primary photochemical process associated with UVC photolysis of NDMA.

(290 nm $\lesssim \lambda \lesssim$ 380 nm) has been associated with a quantum yield of approximately $0.41 \frac{\text{mole}}{\text{einstein}}$; quantum yields for other *N*-nitrosamines are similar or slightly larger for these wavelengths in the UVB and UVA ranges [6].

As a rough approximation, a $UV_{254}$ dose of roughly $1000 \frac{\text{mJ}}{\text{cm}^2}$ is commonly used in applications where NDMA photolysis is a treatment objective; applications with dose delivery on this order have been reported to yield about 1 $\log_{10}$ unit of NDMA transformation [8].

Using this information as background, we will estimate NDMA degradation resulting from exposure to $UV_{254}$ radiation in a photoreactor that yields a (hypothetical) Gaussian dose distribution with a mean dose ($\mu$) of $1000 \frac{\text{mJ}}{\text{cm}^2}$ and a standard deviation ($\sigma$) of $200 \frac{\text{mJ}}{\text{cm}^2}$. Estimates of process performance will be conducted based on the closed-form solution presented above (equation 10.7), the simplified solution based on the assumption of an ideal (uniform) dose distribution (equation 10.8), and by numerical integration of the segregated-flow model (an approximation of equation 10.6).

For each of these calculations, it will be necessary to calculate the rate constant for NDMA photolysis based on the reaction parameters defined above. This is accomplished by use of Equation 10.5:[b]

$$k = \frac{2.303 \cdot \varepsilon \cdot \Phi \cdot \lambda}{h \cdot c_0 \cdot N_A} = \frac{2.303 \cdot \left(1650 \frac{\mathrm{L}}{\mathrm{mole \cdot cm}}\right) \cdot \left(0.30 \frac{\mathrm{mole}}{\mathrm{einstein}}\right) \cdot (254\,\mathrm{nm})}{\left(6.626 \times 10^{-34} \frac{\mathrm{J \cdot s}}{\mathrm{photon}}\right) \cdot \left(2.998 \times 10^{8} \frac{\mathrm{m}}{\mathrm{s}}\right) \cdot \left(6.023 \times 10^{23} \frac{\mathrm{photon}}{\mathrm{einstein}}\right)}$$
$$\cdot \frac{J}{10^3\mathrm{mJ}} \cdot \frac{10^3\mathrm{cm}^3}{\mathrm{L}} \cdot \frac{\mathrm{m}}{10^9\mathrm{nm}} = 2.420 \times 10^{-3} \frac{\mathrm{cm}^2}{\mathrm{mJ}} \tag{10.E.1.1.}$$

The closed-form solution accounts for the entire dose distribution and provides an exact solution to describe NDMA transformation by this system. Substituting terms into Equation 10.7, we find:

$$\left(\frac{[\mathrm{NDMA}]}{[\mathrm{NDMA}]_0}\right)_{\mathrm{reactor}} = \frac{1}{2} \cdot \exp\left(\frac{k^2\sigma^2}{2} - k\mu\right) \cdot \mathrm{erfc}\left(\frac{k\sigma - \frac{\mu}{\sigma}}{\sqrt{2}}\right) = \frac{1}{2}$$
$$\cdot \exp\left(\frac{\left(2.420 \times 10^{-3} \frac{\mathrm{cm}^2}{\mathrm{mJ}}\right)^2 \left(200 \frac{\mathrm{mJ}}{\mathrm{cm}^2}\right)^2}{2} - \left(2.420 \times 10^{-3} \frac{\mathrm{cm}^2}{\mathrm{mJ}}\right) \cdot \left(1000 \frac{\mathrm{mJ}}{\mathrm{cm}^2}\right)\right)$$
$$\cdot \mathrm{erfc}\left(\frac{\left(2.420 \times 10^{-3} \frac{\mathrm{cm}^2}{\mathrm{mJ}}\right) \cdot \left(200 \frac{\mathrm{mJ}}{\mathrm{cm}^2}\right) - \frac{1000 \frac{\mathrm{mJ}}{\mathrm{cm}^2}}{200 \frac{\mathrm{mJ}}{\mathrm{cm}^2}}}{\sqrt{2}}\right) = 0.100 \tag{10.E.1.2}$$

The simplified solution based on the assumption of an ideal (uniform) dose distribution (equation 10.8) represents the theoretical upper limit on NDMA conversion in this reactor system. This upper limit is calculated as:

$$\left(\frac{[\mathrm{NDMA}]}{[\mathrm{NDMA}]_0}\right)_{\mathrm{reactor}} = \exp(-k\mu) = \exp\left(-\left(2.420 \times 10^{-3} \frac{\mathrm{cm}^2}{\mathrm{mJ}}\right) \cdot \left(1000 \frac{\mathrm{mJ}}{\mathrm{cm}^2}\right)\right) = 0.089 \tag{10.E.1.3}$$

The results from the segregated flow model (exact solution) and the single-valued dose model (theoretical upper limit on performance) differ by about 11%. This suggests that estimates of reactor performance based on the average dose can provide a good approximation of actual reactor performance for UV reactors that are used to promote NDMA photolysis, where extents of reactant conversion tend to be modest; however, users need to be aware of the error associated with these calculations.

In most situations, the dose distribution will not conform to a well-defined shape, such as the Gaussian distribution. However, if the dose distribution is known, it is possible to accurately calculate process performance in a reactor that is being used to reach a purely photolytic endpoint because the kinetics of the reaction are identically first-order. This calculation is conducted by applying the trapezoidal rule as a method of performing numerical integration (details of this calculation are presented in an attached spreadsheet):

$$\left(\frac{[\text{NDMA}]}{[\text{NDMA}]_0}\right)_{\text{reactor}} = \int_0^{\infty} \left(\frac{[\text{NDMA}]}{[\text{NDMA}]_0}\right)_{\text{batch}} \cdot E(D) \cdot \text{d}D \approx \sum_i \left(\frac{[\text{NDMA}]}{[\text{NDMA}]_0}\right)_{\text{batch},i} \cdot E(D_i) \cdot \Delta D_i = 0.100 \tag{10.E.1.4}$$

For this calculation, the values $\left(\frac{[\text{NDMA}]}{[\text{NDMA}]_0}\right)_{\text{batch},i}$, which represent the extent of reaction for a dose of $D_i$, were calculated for the dose corresponding to the central value of each dose increment. The value of $E(D_i)$ was the average value of the dose distribution function for each dose increment, and the value $\cdot\Delta D_i$, which represents the width of each dose increment, was chosen to be $10\frac{\text{mJ}}{\text{cm}^2}$. Note that the prediction of process performance based on numerical integration is identical to that of the closed-form solution, when reported to three decimal places. More generally, all three model predictions yield similar results.

This example suggests that calculations of process performance in photoreactors that are used to bring about pure photolytic reactions can be conducted using simplifications, such as the use of a single-valued dose distribution, without introducing substantial error, as long as the extent of photoconversion is relatively small (i.e. 1–3 $\log_{10}$ units of conversion). The magnitude of the error associated with use of such a simplified expression in this case was modest. For reactor systems that require larger extents of conversion or that are characterized by a broader dose distribution, the magnitude of this error will increase, and may warrant the application of a more detailed method of prediction of reactor performance. However, in most cases, it appears that simplified models will yield reasonably accurate process simulation results.

## 10.3 Electrical Power Consumption

As compared with disinfection applications, UV photoreactors used to reach treatment targets by direct photolysis generally require much more power per unit volume of fluid to reach their treatment objectives. UV photoreactors used for direct photolysis may require UV doses on the order of $1000\ \frac{\text{mJ}}{\text{cm}^2}$, whereas those used for disinfection generally require less than 10% of this value. For well-designed UV photoreactors, the nominal or average dose delivered is roughly linearly related to the lamp output power, which in turn is linearly related to input electrical power. Therefore, reactors used for direct photolytic processes require roughly an order of magnitude more input power per unit volume of fluid treated than those used for disinfection.

Bolton et al. [11] introduced several figures-of-merit that have been adopted for quantification and comparison of electrical power requirements for UV photoreactors. Several such figures-of-merit were suggested, with the choice depending on the nature of the application.

The selection of a figure-of-merit to use in analysis of energy requirements for a photochemical process depends on the intrinsic kinetics of the reaction of interest. In other words, the term $\left(\frac{[B]}{[B]_0}\right)_{\text{batch}}$ from the segregated flow model (equation 10.6) influences the selection of a figure-of-merit.

### 10.3.1 Electrical Energy per Mass

In systems where the pollutant concentration is sufficiently high to allow all incident photons to be absorbed by the target molecule (i.e. an opaque solution), the rate of a direct photochemical reaction will be zero-order. This was demonstrated previously for the case of a shallow, well-mixed, batch reactor under a collimated beam (see Chapter 2). Under these circumstances, all incident photons are absorbed by the target molecule, and the intrinsic kinetics are defined as:

$$r'_B = -\frac{d[B]}{dD} = \frac{\Phi}{\ell \cdot U} \tag{10.9}$$

where $\ell$ represents the depth of the reactor. This kinetic expression indicates that the mass (or molar) rate of contaminant decay is directly related to the rate at which photons are imposed on the system, which in turn is directly related to electrical power. Based on this logic, the term Electrical Energy per Mass ($E_{EM}$) was proposed. For a batch process, $E_{EM}$ is defined as:[c]

$$E_{EM} = \frac{P \cdot t}{\forall \cdot MW \cdot ([B]_i - [B]_f)} \tag{10.10}$$

where

$P$ = electrical power used by system [=] kW
$t$ = period of exposure [=] hr
$\forall$ = volume of fluid being irradiated [=] $m^3$
MW = molecular weight of target compound [=] $\frac{g}{mole}$
$[B]_i$ = initial concentration of target compound [=] $\frac{mole}{L}$
$[B]_f$ = final concentration of target compound [=] $\frac{mole}{L}$.

In a similar manner, $E_{EM}$ can be defined for continuous-flow processes as follows:

$$E_{EM} = \frac{P}{Q \cdot MW \cdot ([B]_i - [B]_f)} \tag{10.11}$$

where

$Q$ = volumetric flow rate of fluid being treated [=] $\frac{m^3}{hr}$.

### 10.3.2 Electrical Energy per Order

A more common circumstance for photolysis applications in water (or air) treatment is one in which the contaminant is present at low concentration. Under these circumstances, the intrinsic reaction rate is defined as (see derivation in Chapter 2):

$$r'_B = -\frac{d[B]}{dD} = \frac{2.303 \cdot \varepsilon \cdot [B]}{U} \cdot \Phi \tag{10.12}$$

This kinetic expression indicates that the rate of photodecay of the target compound is first-order with respect to contaminant concentration. Under these circumstances, the amount of energy required to reduce the contaminant concentration by 1 $\log_{10}$ unit is independent of contaminant concentration. In such a situation, it is logical to consider the amount of electrical power required to reduce the contaminant concentration by 1 $\log_{10}$ unit. This is defined as the Electrical Efficiency per Order ($E_{EO}$). For a batch reactor, $E_{EO}$ is defined as:

$$E_{EO} = \frac{P \cdot t}{\forall \cdot \log_{10}\left(\frac{[B]_i}{[B]_f}\right)} \tag{10.13}$$

Following similar logic, $E_{EO}$ is defined for continuous-flow systems as:

$$E_{EO} = \frac{P}{Q \cdot \log_{10}\left(\frac{[B]_i}{[B]_f}\right)} \tag{10.14}$$

In practical applications, $E_{EO}$ for continuous-flow reactors (Equation 10.14) is likely to be used most frequently, as most photoreactors operate as continuous-flow systems, and the targets of photolysis are often present at low concentration. In any reactor setting, the relationship between $P$ and $Q$ will depend on reactor efficiency (i.e. dose distribution) and reaction efficiency (i.e. the product $\varepsilon \cdot \Phi$). If these attributes of a reactor are known and can be related to operational features of the system, such as lamp output power and UV transmittance of water, then it may be possible to define $E_{EO}$ based on these basic parameters.

## 10.4 Physical/Chemical Properties of Photosensitive Contaminants

Equations 10.1–10.5 indicate that for dilute solutions, the local rate of a photochemical reaction is first-order with respect to the concentration of the target contaminant and the local fluence rate. Moreover, the rate constant for this reaction at a given wavelength is governed by two properties of the target compound: the molar absorption coefficient ($\varepsilon$) and the primary quantum yield ($\Phi$). Therefore, it is possible to compare direct photochemical reaction rates of compounds that are subjected to the same fluence rate field in a reactor by comparing the product of these parameters ($\varepsilon \cdot \Phi$) for the wavelength of interest.

Appendix 10.A provides a summary of reported values of these parameters for radiation at 254 nm, which coincides with the output of a conventional low-pressure Hg lamp. Although other sources of radiation are being used with increasing frequency, values of $\varepsilon$ and $\Phi$ have been reported more frequently for 254 nm than for other wavelengths.

## 10.5 Laboratory Experiments for Estimation of Photochemical Kinetic Parameters

The fundamental governing equation to define the rate of a photochemical process in an arbitrary reactor system was presented in Chapter 2. It is repeated below in slightly paraphrased form for clarity, here being written for a specific target compound, $T$:

$$r_T = -\frac{d[T]}{dt} = I_a \cdot \Phi_\lambda \tag{10.15}$$

where

| | | | | |
|---|---|---|---|---|
| $r_T$ | = | rate of reaction of the target molecule, $T$ | [=] | $\frac{\text{mole}}{\text{L}\cdot\text{s}}$ |
| $[T]$ | = | molar concentration (activity) of $T$ in solution | [=] | $\frac{\text{mole}}{\text{L}}$ |
| $t$ | = | time | [=] | s |
| $I_a$ | = | volumetric rate of photon absorption by $T$ | [=] | $\frac{\text{einstein}}{\text{L}\cdot\text{s}}$ |
| $\Phi_\lambda$ | = | quantum yield at wavelength $\lambda$ | [=] | $\frac{\text{mole}}{\text{einstein}}$. |

Equation 10.15 is quite general and can be applied to any reactor system that is subjected to radiation at the wavelength of interest. Figure 10.6 provides a schematic illustration of a system with an arbitrary geometry that is being subjected to monochromatic UV radiation at wavelength, $\lambda$. For this system, the external surface area across which radiation is transmitted is defined by $A$, while the fluid volume in the system is defined as $\forall$. The product of the incident fluence rate, $E_{0,\lambda}'$, and the surface area defines the rate at which photons are imposed on the reactor system. In a more general sense, we could allow the applied fluence rate to vary across the surface of the reactor. In such a situation, the rate at which photons are imposed on the reactor system would be defined by the integral of the applied fluence rate over the surface area.

The volumetric rate of photon absorption by the target molecule will depend on the molar absorption coefficient ($\varepsilon_\lambda$), the presence of other absorbing compounds in the system, and the geometry of the system. For experiments designed to allow accurate estimation of the quantum yield, it is often desirable to use a simple reactor geometry.

There are several methods that have been applied for this purpose. Three of the most commonly used approaches are summarized below. In all cases, it is necessary to be able to be able to conduct time-course measurements of the concentration of the target compound (or a photoproduct) and to be able to quantify the volumetric rate of photon absorption by the target compound.

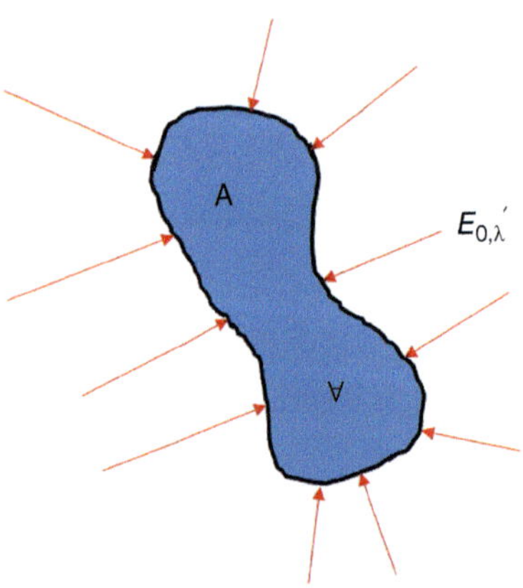


Figure 10.6 Schematic illustration of arbitrary reactor being subjected to monochromatic irradiation at an incident fluence rate of $E_{0,\lambda}'$. The reactor is characterized by an external surface area A, and a fluid volume ∀.

### 10.5.1 Collimated Beam

In this method, a well-mixed, batch reactor is placed under a (quasi) collimated source of radiation, which is imposed normal (perpendicular) to the surface of the reacting solution. This approach is well adapted for quantification of the incident fluence rate by the use of a radiometer; most radiometer detectors are designed to accurately quantify irradiance for radiation imposed normal to the detector surface.

In principle, any concentration of the target compound can be used in an experiment to quantify the quantum yield; it is only necessary to be able to quantify the volumetric rate of photon absorption by the target compound and the rate of compound photodecay. However, the easiest methods involve the use of a solution that allows all incident photons to be absorbed by the target compound (large value of $\varepsilon_\lambda[T]\ell$) or a situation in which the solution approaches a condition of transparency: $\varepsilon_\lambda[T]\ell \to 0$. The first case defines an opaque solution, for which photodecay will follow zero-order kinetics. The second case, a transparent solution, is one in which photodecay will be first-order. Details of the derivations of the governing equations for both special cases are presented in Chapter 2.

### 10.5.2 Effective Path Length Approach

In a classic paper, Zepp [12] described a method that involves two separate, but linked experiments that are used for estimation of the quantum yield of a direct photochemical reaction. The first of these involves an arbitrary reactor system being subjected to spatially uniform, monochromatic fluence rate.[d]

The basic governing equation that defines progress in a photochemical reaction (Equation 10.15) can be decomposed as follows:

$$r_T = -\frac{d[T]}{dt} = E_{0,\lambda}{}' \cdot \left(\frac{A}{\forall}\right) \cdot \Phi_\lambda \cdot F_{S,\lambda} \cdot F_{T,\lambda} \tag{10.16}$$

where

| | | |
|---|---|---|
| $r_T$ | = | rate of target compound photodecay |
| $[T]$ | = | (molar) concentration of target compound |
| $t$ | = | time |
| $E_{0,\lambda}{}'$ | = | incident fluence rate |
| $A$ | = | surface area exposed to incident radiation |
| $\forall$ | = | volume of fluid being irradiated |
| $\Phi_\lambda$ | = | quantum yield for photodecay of target compound |
| $F_{S,\lambda}$ | = | fraction of radiation at wavelength λ that is absorbed by the solution |
| $F_{T,\lambda}$ | = | fraction of absorbed radiation at wavelength λ that is absorbed by the target compound. |

The Beer–Lambert law can be used to quantify $F_{S,\lambda}$:

$$F_{S,\lambda} = 1 - 10^{-(\alpha_\lambda + \varepsilon_{T,\lambda}[T])\ell} \tag{10.17}$$

where

| | | | | |
|---|---|---|---|---|
| $\alpha_\lambda$ | = | absorption coefficient of solvent | [=] | $\frac{1}{\text{cm}}$ |
| $\varepsilon_{T,\lambda}$ | = | molar absorption coefficient of target compound | [=] | $\frac{\text{L}}{\text{mole}\cdot\text{cm}}$ |
| $[T]$ | = | molar concentration of target compound | [=] | $\frac{\text{mole}}{\text{L}}$ |
| $\ell$ | = | effective path length of reactor | [=] | cm. |

In most situations where photochemical reactions take place in the environment, the concentration of the target compound is small, such that the condition of a dilute solution can be assumed. In practical terms, this assumption is valid when $F_{S,\lambda} \lesssim 0.1$, but the error associated with this assumption decreases as $F_{S,\lambda} \to 0$. When this condition is met, and when $F_{T,\lambda} \approx 1$ (i.e. when all absorbed photons are absorbed by the target compound), Equation 10.16 reduces to:

$$r_T = -\frac{d[T]}{dt} = 2.303 \cdot E_{0,\lambda}{}' \cdot \left(\frac{A}{\forall}\right) \cdot \varepsilon_\lambda \cdot \ell \cdot \Phi_\lambda \cdot [T] \tag{10.18}$$

For the design of an experiment to allow accurate estimation of the quantum yield, it is necessary to conduct the experiment under conditions such that $E_{0,\lambda}' \cdot \left(\frac{A}{\forall}\right)$, $\varepsilon_\lambda$, and $\ell$ are known or quantifiable. Specifically, separation of variables and integration of Equation 10.18 yields the following:

$$\ln\left(\frac{[T]}{[T_0]}\right) = -2.303 \cdot E_{0,\lambda}' \cdot \left(\frac{A}{\forall}\right) \cdot \varepsilon_\lambda \cdot \ell \cdot \Phi_\lambda \cdot t \tag{10.19}$$

where $[T_0]$ is the molar concentration of the target compound at the beginning of the experiment. Linear regression of the relationship $\ln\left(\frac{[T]}{[T_0]}\right)$ vs. $t$ (forced through the origin) will yield a line with slope of $-2.303 \cdot E_{0,\lambda}' \cdot \left(\frac{A}{\forall}\right) \cdot \varepsilon_\lambda \cdot \ell \cdot \Phi_\lambda$. Therefore, the quantum yield can be estimated by:

$$\Phi_\lambda = \frac{-\text{slope}}{2.303 \cdot E_{0,\lambda}' \cdot \left(\frac{A}{\forall}\right) \cdot \varepsilon_\lambda \cdot \ell} \tag{10.20}$$

The product $E_{0,\lambda}' \cdot \left(\frac{A}{\forall}\right)$ represents the volumetric rate at which photons at the wavelength of interest are applied to the system; this product can be quantified by the use of a chemical actinometer,[e] as a reference compound, $R$. For this approach, it is convenient to apply the actinometer under conditions such that all incident photons are absorbed by the actinometer (i.e. $F_{S,\lambda} \equiv 1$ and $F_{T,\lambda} \equiv 1$).[f] Under these conditions, Equation 10.16 is simplified and can be rearranged to:

$$E_{0,\lambda}' \cdot \left(\frac{A}{\forall}\right) = \frac{-r_R}{\Phi_\lambda} = \frac{-\frac{\mathrm{d}[R]}{\mathrm{d}t}}{\Phi_\lambda} \tag{10.21}$$

where

$r_R$ = molar rate of change of reference compound concentration [=] $\frac{\text{mole}}{\text{L} \cdot \text{s}}$

$[R]$ = molar concentration of reference compound (actinometer) [=] $\frac{\text{mole}}{\text{L}}$.

Therefore, estimation of the product $E_{0,\lambda}' \cdot \left(\frac{A}{\forall}\right)$ requires that time-course measurements of the reference compound concentration be conducted and that quantum yield is known. In addition, the experiment with the reference compound must be conducted with the same solution volume and conditions of irradiation as used in experiments with the target compound.

The effective path length, $\ell$, is a relevant parameter only for experiments that are conducted using a dilute solution. However, estimation of $\ell$ requires that the target compound be sufficiently soluble in the solvent to allow development of an opaque solution (i.e. a solution in which all incident photons are absorbed by the target molecule). Under this condition, the rate of photolysis of the target molecule will be maximized in the reactor system. Mathematically, this can be defined as follows:

$$r_{\max} = \left(\frac{\mathrm{d}[T]}{\mathrm{d}t}\right)_{\max} = -\Phi_\lambda \cdot E_{0,\lambda}' \cdot \left(\frac{A}{\forall}\right) \tag{10.22}$$

We then define a dummy variable $X$:

$$X = \frac{r_T}{r_{max}} = 1 - 10^{-\varepsilon_\lambda \cdot [T] \cdot \ell} \quad (10.23)$$

A plot of $-\log_{10}(1 - X)$ vs. $\varepsilon_\lambda \cdot [T]$ gives a straight line through the origin with a slope of $\ell$. Collectively, Equations 10.16 through 10.23, along with the experimental approaches described above, allow for accurate estimation of the quantum yield for a direct photochemical reaction. However, this approach requires that the target compound be sufficiently soluble to allow generation of an opaque solution.

This approach has a benefit of being applicable to essentially any reactor geometry. It has been used for evaluation of photochemical kinetics in natural systems and in controlled laboratory experiments.

### 10.5.3 Parallel Photolysis

In this method, separate, dilute solutions of a reference compound and the target compound are prepared such that the following conditions are met:

$$\begin{aligned} &F_{S,\lambda} \lesssim 0.1, \mathrm{F}_{\mathrm{R},\lambda} \approx 1 \text{ for the reference compound solution and} \\ &F_{S,\lambda} \lesssim 0.1, \mathrm{F}_{\mathrm{T},\lambda} \approx 1 \text{ for the target compound solution} \end{aligned} \quad (10.24)$$

The containers used to hold these solutions during irradiation must be identical, in terms of geometry and material of construction. Under these conditions, the rates of photolysis of the two compounds can be described as follows:

$$r_R = \left(\frac{\mathrm{d}[R]}{\mathrm{d}t}\right)_\lambda = -2.303 \cdot E_{0,\lambda}{}' \cdot \left(\frac{A}{\forall}\right) \cdot \varepsilon_{R,\lambda} \cdot \ell \cdot \Phi_{R,\lambda} \cdot [R] \quad (10.25)$$

$$r_T = \left(\frac{\mathrm{d}[T]}{\mathrm{d}t}\right)_\lambda = -2.303 \cdot E_{0,\lambda}{}' \cdot \left(\frac{A}{\forall}\right) \cdot \varepsilon_{T,\lambda} \cdot \ell \cdot \Phi_{T,\lambda} \cdot [T] \quad (10.26)$$

Integration of Equations 10.25 and 10.26 yields:

$$\ln\left(\frac{[R]}{[R_0]}\right) = -2.303 \cdot E_{0,\lambda}{}' \cdot \left(\frac{A}{\forall}\right) \cdot \varepsilon_{R,\lambda} \cdot \ell \cdot \Phi_{R,\lambda} \cdot t \quad (10.27)$$

$$\ln\left(\frac{[T]}{[T_0]}\right) = -2.303 \cdot E_{0,\lambda}{}' \cdot \left(\frac{A}{\forall}\right) \cdot \varepsilon_{T,\lambda} \cdot \ell \cdot \Phi_{T,\lambda} \cdot t \quad (10.28)$$

Linear regression of measured values of $\ln\left(\frac{[R]}{[R_0]}\right)$ vs. $t$ yields a straight line with an intercept of zero (i.e. the regression must be forced through the origin) and a slope of:

$$\text{Slope}_R = -2.303 \cdot E_{0,\lambda}{}' \cdot \left(\frac{A}{\forall}\right) \cdot \varepsilon_{R,\lambda} \cdot \ell \cdot \Phi_{R,\lambda} \quad (10.29)$$

Similarly, linear regression of measured values of $\ln\left(\frac{[T]}{[T_0]}\right)$ vs. $t$ yields a straight line with an intercept of zero (i.e. the regression must be forced through the origin) and a slope of:

$$\text{Slope}_T = -2.303 \cdot E_{0,\lambda}{}' \cdot \left(\frac{A}{\forall}\right) \cdot \varepsilon_{T,\lambda} \cdot \ell \cdot \Phi_{T,\lambda} \quad (10.30)$$

By algebra, the following equalities can be identified from Equations 10.29 and 10.30:

$$-2.303 \cdot E'_{0,\lambda} \cdot \left(\frac{A}{\forall}\right) \cdot \ell = \frac{\text{Slope}_R}{\varepsilon_{R,\lambda} \cdot \Phi_{R,\lambda}} = \frac{\text{Slope}_T}{\varepsilon_{T,\lambda} \cdot \Phi_{T,\lambda}} \tag{10.31}$$

Algebraic rearrangement of the second equality in 10.31 results in an expression that can be used for estimation of the quantum yield of the target compound:

$$\Phi_{T,\lambda} = \frac{\text{Slope}_T}{\text{Slope}_R} \cdot \frac{\varepsilon_{R,\lambda}}{\varepsilon_{T,\lambda}} \cdot \Phi_{R,\lambda} \tag{10.32}$$

In theory, this method could be used with any batch reactor system, as long as the exposure cells and conditions of the experiment guarantee the same exposure history in each cell. This approach is commonly used with a merry-go-round reactor (see Chapter 5). Essentially any reference compound can be used in these experiments, as long as the conditions defined in 10.24 are met.

## 10.6 Effects of Solution Composition on Rates of Direct Photolysis

The rate of a direct photochemical process can be influenced by the composition of a solution, both optically and chemically. In terms of optical behavior, the presence of compounds that do not interact with the target compound, but that absorb radiation at the wavelength(s) of interest function as *inner filters*. This effect is made evident by examination of the Beer–Lambert law to a solution that contains multiple absorbing compounds:

$$A_\lambda = \left(\alpha_\lambda + \sum_i \varepsilon_i \cdot [i]\right) \cdot \ell \tag{10.33}$$

where

$\alpha_\lambda$ = absorption coefficient for solvent [=] $\frac{1}{\text{cm}}$

$\varepsilon_i$ = molar absorption coefficient for $i$th constituent [=] $\frac{\text{L}}{\text{mole} \cdot \text{cm}}$

$[i]$ = molar concentration of $i$th constituent [=] $\frac{\text{mole}}{\text{L}}$

$\ell$ = optical path length [=] cm.

For most photochemical reactions of interest, water or air represent the solvent fluid. For these two fluids, $\alpha_\lambda \approx 0$ for radiation in the visible, UVA, UVB, and UVC portions of the electromagnetic spectrum. However, for VUV, water and air will both absorb strongly. The implications of this behavior for VUV process applications will be addressed later in this chapter.

The chemical composition of a solution that contains a target compound can also influence the photochemical behavior of the target by altering the chemical structure of the target. A prime example of this behavior is associated with solution pH. Many compounds demonstrate acid/base behavior. As a result, the chemical form of such a compound will change with pH. The behavior of free chlorine will be used to illustrate this point.

A practical definition of the *free available chlorine* (FAC) concentration is:[g]

$$\text{FAC} = [\text{HOCl}] + [\text{OCl}^-] \tag{10.34}$$

Hypochlorous acid (HOCl) functions as a weak acid, as defined by the following equilibrium (for room temperature conditions):

$$HOCl \rightleftharpoons H^{+} + OCl^{-} \quad K_a = \frac{[H^{+}][OCl^{-}]}{[HOCl]} \quad pK_a = 7.47 \tag{10.35}$$

The distribution of HOCl and $OCl^-$ is governed by pH, as illustrated by Figure 10.7. At pH values below the $pK_a$ of HOCl, the acid form (HOCl) dominates, while at pH values above the $pK_a$ of HOCl, its conjugate base ($OCl^-$) dominates. As pH moves away from the $pK_a$, the extent to which the acid or base form dominates will increase. For example, at pH ≈ 6.5 (roughly 1 pH unit below the $pK_a$ of HOCl), more than 90% of FAC is present as HOCl.

HOCl and $OCl^-$ both undergo photolysis when exposed to UVC radiation, but their behaviors are distinct, both in terms of their ability to absorb UVC radiation and in photolytic reactions that result from absorption of UVC photons.

The absorption spectra of HOCl and $OCl^-$ are illustrated in Figure 10.8. Both forms of FAC are fairly weak absorbers of UVA, UVB, and UVC radiation. The absorption behavior of these two forms of FAC plays a critical role in the chlorine/UV process. A common configuration for this process is to use a low-pressure Hg lamp as a source of UV radiation. Both of the common forms of FAC are weak absorbers at the wavelength that characterizes this radiation source ($\lambda \approx 254$ nm). Coincidentally, there are isosbestic points[h] in these spectra at wavelengths of approximately 215 and 255 nm. As a result, UV absorption of FAC is relatively insensitive to pH at the wavelength that characterizes low-pressure Hg lamps (254 nm). It is also worth noting that peak absorption for $OCl^-$ occurs at a wavelength of approximately 292 nm; at this wavelength, the molar absorption coefficient for $OCl^-$ is roughly 10 times greater than at 254 nm. Therefore, photolysis of $OCl^-$ is likely to be much more efficient with a UV source that emits in the vicinity of 292 nm than with a low-pressure Hg lamp [14].

HOCl and $OCl^-$ photolyze differently when exposed to UV radiation. For $\lambda \lesssim 400$ nm, HOCl undergoes homolytic cleavage to yield two radicals:

$$HOCl + h\nu \xrightarrow{\lambda \lesssim 400\ \text{nm}} HO\cdot + \cdot Cl \tag{10.36}$$

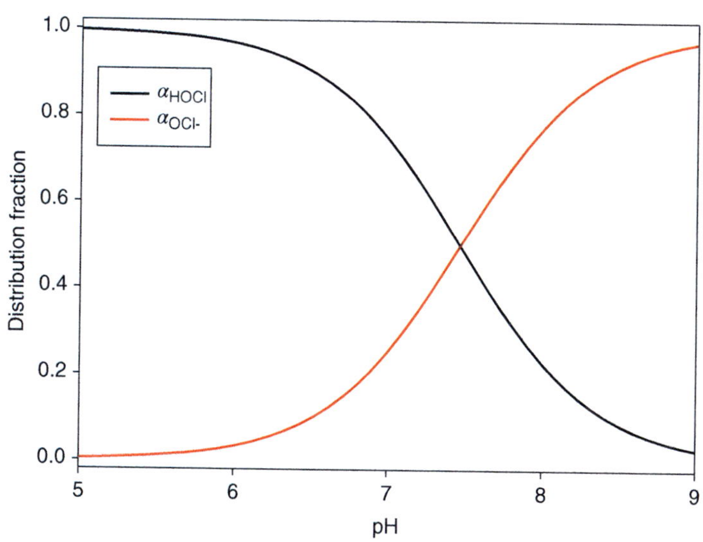


**Figure 10.7** Distribution fractions ($\alpha$) for HOCl and $OCl^-$ as a function of pH at room temperature.

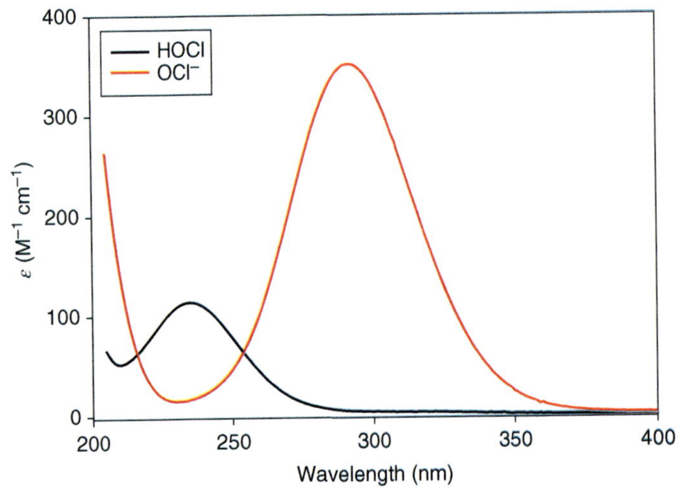


**Figure 10.8** UV absorption spectra for HOCl and $OCl^-$.

As will be discussed later in this chapter, HO· and ·Cl are highly reactive and can contribute to degradation of other compounds in solution.

In contrast, the photolysis of $OCl^-$ follows distinct pathways that are governed by the wavelength of radiation. For $\lambda \lesssim 320$ nm, $OCl^-$ photodegrades via two different pathways:

$$OCl^- + h\nu \xrightarrow{\lambda \lesssim 320\ \text{nm}} O^{\cdot -} + Cl^- \tag{10.37a}$$

$$OCl^- + h\nu \xrightarrow{\lambda \lesssim 320\ \text{nm}} O(^1D) + Cl^- \tag{10.37b}$$

For longer wavelengths, a different reaction pathway predominates:

$$OCl^- + h\nu \xrightarrow{\lambda \gtrsim 320\ \text{nm}} O(^3P) + Cl^- \tag{10.38}$$

The products of reactions 10.37 and 10.38 tend to be highly reactive, but their reactivities are different and can involve the formation of other reactive intermediates. As a result, the effects of pH and wavelength on photolysis of free chlorine, and subsequent reactions, can be quite dramatic.

From an analytical perspective, the pH-dependent absorption behavior of compounds that display acid/base behavior can be used to estimate the $pK_a$ of such a compound. This is done by algebraic manipulation of the relationship that defines acid/base equilibria and the Beer–Lambert law.

We will evaluate this behavior by considering a monoprotic acid.[i] The acid will undergo deprotonation to yield its conjugate base.

$$HA \rightleftharpoons H^+ + A^- \tag{10.39}$$

The forward and reverse reactions are fast, such that this reaction can always be assumed to be in equilibrium, as defined by its acid dissociation constant, $K_a$.

$$K_a = \frac{[H^+][A^-]}{[HA]} \tag{10.40}$$

Application of the "p" operator (i.e. $pX = -\log_{10}(X)$) to both sides of this equation yields:

$$pK_a = pH + \log_{10}\frac{[A^-]}{[HA]} \tag{10.41}$$

Equation 10.41 is sometimes referred to as the Henderson–Hasselbalch equation, which relates the pH of a solution to the p$K_a$ and the molar concentrations of an acid and its conjugate base. Algebraic manipulation of Equation 10.41 yields:

$$\frac{[A^-]}{[HA]} = 10^{\mathrm{pH}-\mathrm{p}K_a} \tag{10.42}$$

It is often useful to define the total concentration of the acid and its conjugate base ($C_T$); for many compounds that display acid/base behavior, analytical methods will yield a measure of the total concentration, rather than the concentration of the acid or its conjugate base.

$$C_T = [HA] + [A^-] \tag{10.43}$$

Combining Equations 10.42 and 10.43, we get:

$$C_T = [HA] + [HA] \cdot 10^{\mathrm{pH}-\mathrm{p}K_a} \tag{10.44}$$

Further rearrangement yields:

$$[HA] = \frac{C_T}{1 + 10^{\mathrm{pH}-\mathrm{p}K_a}} \tag{10.45}$$

The Beer–Lambert law can be used to describe absorption at a given wavelength ($A_\lambda$) by a mixture of an acid and its conjugate base:

$$A_\lambda = \varepsilon_{0,\lambda}[HA]\ell + \varepsilon_{1,\lambda}[A^-]\ell \tag{10.46}$$

where

$\varepsilon_{0,\lambda}$ = molar absorption coefficient of protonated form at wavelength $\lambda$ $[=]$ $\frac{\mathrm{L}}{\mathrm{mole \cdot cm}}$

$\varepsilon_{1,\lambda}$ = molar absorption coefficient of deprotonated form at wavelength $\lambda$ $[=]$ $\frac{\mathrm{L}}{\mathrm{mole \cdot cm}}$.

Substitution of Equation 10.43 into Equation 10.46 yields:

$$A_\lambda = \varepsilon_{0,\lambda}[HA]\ell + \varepsilon_{1,\lambda}(C_T - [HA])\ell \tag{10.47}$$

which can be rearranged to yield:

$$A_\lambda = (\varepsilon_{0,\lambda} - \varepsilon_{1,\lambda})[HA]\ell + \varepsilon_{1,\lambda}C_T\ell \tag{10.48}$$

Substitution of Equation 10.45 into Equation 10.48, along with some minor algebraic manipulation yields an equation that relates absorption by a solution containing both the acid and its conjugate base to $C_T$, the molar absorption coefficients of the constituent forms, pH, p$K_a$, and the optical path length, all of which can be measured independently:

$$A_\lambda = (\varepsilon_{0,\lambda} - \varepsilon_{1,\lambda})\frac{C_T}{1 + 10^{\mathrm{pH}-\mathrm{p}K_a}}\ell + \varepsilon_{1,\lambda}C_T\ell \tag{10.49}$$

A plot of $A_\lambda$ vs. pH yields a characteristic sigmoidal shape, with an inflection point at pH = p$K_a$. As such, an experiment in which $A_\lambda$ is measured at a range of pH values allows estimation of the p$K_a$ for a target ionizable compound. Figure 10.9 illustrates the application of this method for estimation of the p$K_a$ of acetaminophen.

Accuracy in application of this method depends on selection of a wavelength for which there is a large difference in the values of $\varepsilon_{0,\lambda}$ and $\varepsilon_{1,\lambda}$. If an isosbestic point was selected for this experiment,

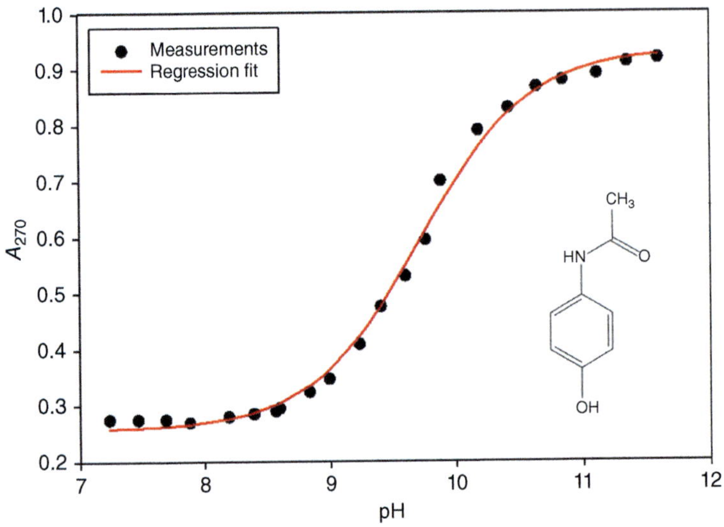


**Figure 10.9** UV absorbance ($\lambda$ = 270 nm) of acetaminophen as a function of pH. Structure of the acid form of acetaminophen is included in the inset. Symbols represent measurements, while the solid red line represents a fit of Equation 10.49 to the data. The inflection point in this fit, which is evaluated by nonlinear regression of Equation 10.49, provides an estimate of the p$K_a$ of acetaminophen (9.72 ± 0.03). *Source:* Data from Carlson et al. [15].

there would be no variation in the value of $A_\lambda$ with $pH$ and it would not be possible to develop an estimate of p$K_a$.

Beyond estimation of p$K_a$, this method of analysis is useful for evaluation of the pH-dependence of absorption of compounds that display acid/base behavior. Also, if the p$K_a$ of an acid/base pair is known, regression analysis can also be applied to Equation 10.49 to develop estimates of the molar absorption coefficients of the protonated and deprotonated forms of the compound of interest.

Solution pH affects the distribution of ionizable compounds among their protonated and deprotonated forms, which are distinct, but related compounds; these forms often have chemical characteristics that are quite different from each other, including reactivity toward other chemicals, molar absorption coefficients, and quantum yields. All of these factors can affect the fate of these compounds in the environment.

As illustrated previously, the kinetics of a pure photochemical reaction can be defined as follows:

$$r_B = -\frac{\mathrm{d}[B]}{\mathrm{d}t} = k \cdot E_i' \cdot [B] \tag{10.50}$$

$$k = \frac{2.303 \cdot \varepsilon \cdot \Phi}{U} \tag{10.51}$$

Following the approach developed by Canonica et al. [16], we will examine the dynamic behavior of a system that includes a target chemical that displays acid/base behavior. We will assume that the compound is a monoprotic acid, although the logic used here could be extended to polyprotic molecules. The acid/base behavior of this chemical can be described using fundamental principles of equilibrium chemistry (see equations 10.39–10.45). We expect that both forms of this chemical, HA and $A^-$, will have distinct physical and chemical properties. In a general sense, we may write a rate expression that is based on the total concentration of the chemical:

$$r_{\text{overall},\lambda} = -k_\lambda^{\text{app}} \cdot E_i' \cdot C_T \tag{10.52}$$

where

$$k_\lambda^{app} = \frac{2.303 \cdot \varepsilon_\lambda^{app} \cdot \Phi_\lambda^{app}}{U_\lambda} \tag{10.53}$$

where

$k_\lambda^{app}$ = apparent rate parameter for compound (at $\lambda$) [=] $\frac{L \cdot cm^2}{mW \cdot mole}$

$\varepsilon_\lambda^{app}$ = apparent molar absorption coefficient for compound (at $\lambda$) [=] $\frac{L}{mole \cdot cm}$

$\Phi_\lambda^{app}$ = apparent quantum yield for compound (at $\lambda$) [=] $\frac{mole}{einstein}$.

This apparent rate parameter is not a constant, in that it depends on the distribution of the target chemical between its protonated and deprotonated forms. The rate parameter can be decomposed into a set of more fundamental parameters:

$$k_\lambda^{app} = \sum_i \alpha_i \cdot k_i \tag{10.54}$$

The terms $\alpha_i$ (distribution fractions) represent the fraction of the compound that exists in its protonated and deprotonated forms.[j] For the case of a monoprotic acid, $\alpha_0$ = fraction of compound present in the protonated form, while $\alpha_1$ = fraction of compound present in the deprotonated form. This approach is constrained by the following identity:

$$\sum_i \alpha_i \equiv 1 \tag{10.55}$$

The terms $k_i$ in Equation 10.54 are true rate constants for each of the forms of the chemical. For example, for the protonated form:

$$k_{0,\lambda} = \frac{2.303 \cdot \varepsilon_{0,\lambda} \cdot \Phi_{0,\lambda}}{U_\lambda} \tag{10.56}$$

where

$\varepsilon_{0,\lambda}$ = (true) molar extinction coefficient for protonated form (at $\lambda$) [=] $\frac{L}{mole \cdot cm}$

$\Phi_{0,\lambda}$ = (true) quantum yield for protonated form (at $\lambda$) [=] $\frac{mole}{einstein}$.

If the value of p$K_a$ is known, the values of $\varepsilon_{0,\lambda}$ and $\varepsilon_{1,\lambda}$ can be estimated by measurement of absorbance (at wavelength $\lambda$) as a function of pH, following an approach similar to that illustrated in Figure 10.9. The values of $\Phi_{0,\lambda}$ and $\Phi_{1,\lambda}$ can then be estimated from an experiment in which the rate of photodecay of the compound (at wavelength $\lambda$) is quantified as a function of pH.

Table 10.1 provides a summary of data that were generated for three compounds that demonstrate acid/base behavior. The parameters that are summarized in this table represent chemical characteristics of the distinct forms of each compound (i.e. the protonated and deprotonated forms of each chemical). By decomposing the problem in this manner, it is possible to develop accurate estimates of the rate of direct photolysis at essentially any pH. This fundamental approach allows generalization to a wide range of scenarios, because it is based on distinct chemical characteristics of the participating chemical forms.

**Table 10.1** Summary of pH-dependent photochemical parameters for three pharmaceutical contaminants found in water.

| Compound | Structure | p$K_a$ | $\varepsilon_0\varepsilon_1$ $\left(\frac{L}{mole \cdot cm}\right)$ | $\Phi_0\Phi_1$ $\left(\frac{mole}{einstein}\right)$ | $\varepsilon_1$ $\left(\frac{L}{mole \cdot cm}\right)$ | $\Phi_1\Phi_0$ $\left(\frac{mole}{einstein}\right)$ |
|---|---|---|---|---|---|---|
| 17α-ethinylestradiol | 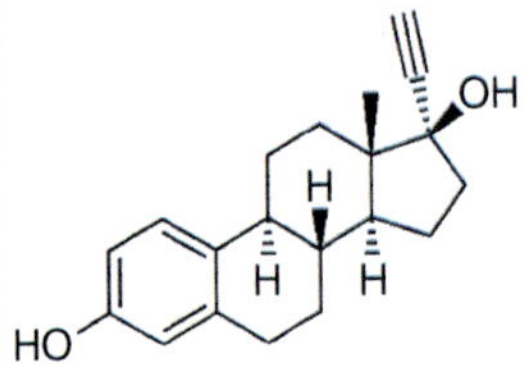 | 10.4 | 750 | 0.026 | 3,060 | 0.057 |
| Diclofenac | 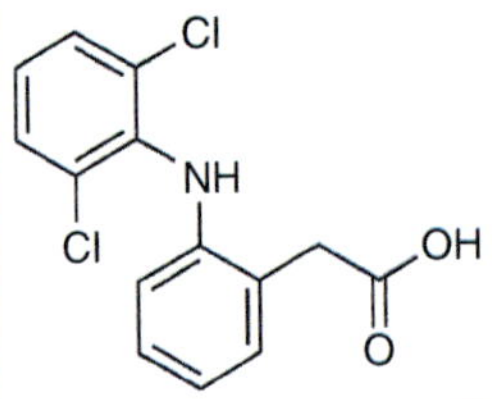 | 4.2 | 4,260 | 0.272 | 4,260 | 0.384 |
| Sulfamethoxazole | 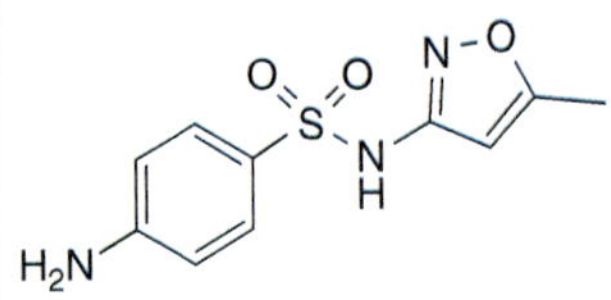 | 5.7 | 11,890 | 0.212 | 16,760 | 0.046 |

*Source:* Data from Canonica et al. [16].

### Example 10.2 pH-Dependence of Direct Photolysis of Pharmaceuticals

Prepare a graph to illustrate $k_{254}^{app}$ as a function of pH for $0 \leq pH \leq 14$ for the three contaminants listed in Table 10.1; express the values of $k_{254}^{app}$ in units of $\frac{cm^2}{mJ}$. Also prepare a graph to illustrate the $UV_{254}$ dose required to accomplish 1 $\log_{10}$ unit of photodecay of each of these compounds for the same pH range. This $UV_{254}$ dose is sometimes referred to as $D_{10}$; express values of $D_{10}$ in units of $\frac{mJ}{cm^2}$.

To develop answers to these questions, we will begin by evaluating the distribution fractions of the protonated and deprotonated forms. This begins with the fundamental equilibrium relationship for an acid/base system:

$$HA \rightleftharpoons H^+ + A^- \quad K_a = \frac{[H^+][A^-]}{[HA]} \tag{10.E.2.1}$$

We define the total concentration of both forms as:

$$C_T = [HA] + [A^-] \tag{10.E.2.2}$$

We can rearrange Equation 10.E.2.1 to yield expressions that describe the concentration of the protonated form, $[HA]$, as a function of $K_a$, $[H^+]$, and $[A^-]$:

$$[HA] = \frac{[H^+][A^-]}{K_a} \tag{10.E.2.3}$$

Similarly, we can define the concentration of the deprotonated form as a function of $K_a$, $[H^+]$ and $[HA]$:

$$[A^-] = \frac{K_a[HA]}{[H^+]} \tag{10.E.2.4}$$

Combining Equations 10.E.2.2 and 10.E.2.3) with some algebraic manipulations, we derive a general expression for the distribution fraction of the deprotonated form, $\alpha_1$:

$$\alpha_1 = \frac{[A^-]}{C_T} = \left(1 + \frac{[H^+]}{K_a}\right)^{-1} \tag{10.E.2.5}$$

Similarly, we can combine Equations 10.E.2.2 and 10.E.2.4 to develop a general expression for the distribution fraction of the protonated form, $\alpha_0$:

$$\alpha_0 = \frac{[HA]}{C_T} = \left(1 + \frac{K_a}{[H^+]}\right)^{-1} \tag{10.E.2.6}$$

The overall rate of photodecay of the protonated and deprotonated forms (collectively) is defined by:

$$r_{C_T} = -\frac{dC_T}{dt} = k_{254}^{app} \cdot E_i' \cdot C_T \tag{10.E.2.7}$$

Decomposition of the rate constant in this expression allows quantification of the contributions of both forms to the overall photodegradation rate:

$$\frac{dC_T}{dt} = -\left(k_{0,254} \cdot E_{254}' \cdot \alpha_0 \cdot C_T + k_{1,254} \cdot E_{254}' \cdot \alpha_1 \cdot C_T\right) \tag{10.E.2.8}$$

But the first-order rate constants for the two forms can be further decomposed:

$$\frac{dC_T}{dt} = -\left(\frac{2.303 \cdot \varepsilon_{0,254} \cdot \Phi_{0,254}}{U_{254}} \cdot \alpha_0 + \frac{2.303 \cdot \varepsilon_{1,254} \cdot \Phi_{1,254}}{U_{254}} \cdot \alpha_1\right) \cdot E_{254}' \cdot C_T \tag{10.E.2.9}$$

Inspection of Equations 10.E.2.7 and 10.E.2.9 yields an expression for $k_{254}^{app}$:

$$k_{254}^{app} = \frac{2.303 \cdot \varepsilon_{0,254} \cdot \Phi_{0,254}}{U_{254}} \cdot \alpha_0 + \frac{2.303 \cdot \varepsilon_{1,254} \cdot \Phi_{1,254}}{U_{254}} \cdot \alpha_1 \tag{10.E.2.10}$$

For purposes of illustration, a sample calculation is presented below, with appropriate unit conversions, for estimation of $k_{254}^{app}$ for sulfamethoxazole at pH = 6. Slight rearrangement of Equation 10.E.2.10 yields:

$$k_{254}^{app} = \frac{2.303}{U_{254}} \cdot \left[\varepsilon_{0,254} \cdot \Phi_{0,254} \cdot \alpha_0 + \varepsilon_{1,254} \cdot \Phi_{1,254} \cdot \alpha_1\right] \tag{10.E.2.11}$$

The parameter $U_{254}$ represents the amount of energy per einstein for radiation at $\lambda = 254$ nm:

$$U = h \cdot \frac{c_0}{\lambda} \cdot N_A = \left(6.626 \times 10^{-34} \frac{\text{J} \cdot \text{s}}{\text{photon}}\right) \cdot \frac{2.998 \times 10^8 \frac{\text{m}}{\text{s}}}{254 \text{ nm}} \cdot \left(6.023 \times 10^{23} \frac{\text{photon}}{\text{einstein}}\right) \cdot \frac{10^9 \text{nm}}{\text{m}}$$
$$= 4.71 \times 10^5 \frac{\text{J}}{\text{einstein}} \tag{10.E.2.12}$$

We also need to calculate the distribution fractions for the protonated and deprotonated forms:

$$\alpha_0 = \left(1 + \frac{K_a}{[H^+]}\right)^{-1} = \left(1 + \frac{10^{-5.7}}{10^{-6}}\right)^{-1} = 0.3339 \tag{10.E.2.13}$$

$$\alpha_1 = \left(1 + \frac{[H^+]}{K_a}\right)^{-1} = \left(1 + \frac{10^{-6}}{10^{-5.7}}\right)^{-1} = 0.6661 \tag{10.E.2.14}$$

We now have numerical values for all parameters that are included in Equation 10.E.2.11. Substitution of corresponding values of these parameters yields:

$$k_{254}^{app} = \frac{2.303}{\left(4.71\times10^5 \frac{J}{einstein}\right)} \cdot \left[\left(11{,}890\frac{L}{mole\cdot cm}\right)\cdot\left(0.212\frac{mole}{einstein}\right)\cdot(0.3339) + \left(16{,}760\frac{L}{mole\cdot cm}\right)\cdot\left(0.046\frac{mole}{einstein}\right)\cdot(0.6661)\right] \cdot \frac{10^3 cm^3}{L}\cdot\frac{J}{10^3 mJ} = 6.63\times10^{-3}\frac{cm^2}{mJ} \tag{10.E.2.15}$$

This calculation was repeated for pH values ranging from 0 to 14 for all three compounds identified in the problem statement using a spreadsheet. Figure 10.10 illustrates calculated values of $k_{254}^{app}$ for all three compounds across the pH range $0 \rightarrow 14$.

For all three compounds, the plot of $k_{254}^{app}$ as a function of pH has a sigmoidal shape, with an inflection point at pH = p$K_a$ and limits that are defined by the values of rate constants for the protonated and deprotonated forms. The direction of the sigmoidal curve (i.e. up or down as pH increases from $0 \rightarrow 14$) depends on the relative magnitudes of the products $\varepsilon_{0,254}\cdot\Phi_{0,254}\cdot\alpha_0$ and $\varepsilon_{1,254}\cdot\Phi_{1,254}\cdot\alpha_1$. For sulfamethoxazole, $\varepsilon_{0,254}\cdot\Phi_{0,254}\cdot\alpha_0 > \varepsilon_{1,254}\cdot\Phi_{1,254}\cdot\alpha_1$ for pH < p$K_a$, which indicates that the protonated form is more photoreactive than the deprotonated form. Therefore, we expect

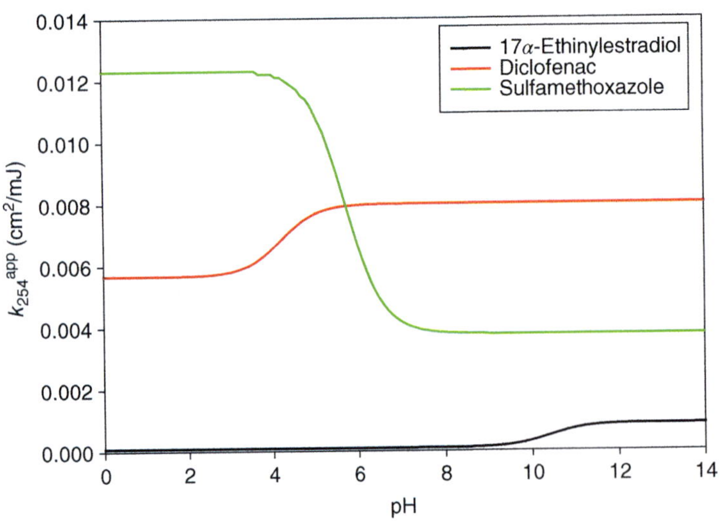


**Figure 10.10** Apparent rate constants ($k_{254}^{app}$) for photolysis of 17α-ethinylestradiol, diclofenac, and sulfamethoxazole at a wavelength of 254, as a function of pH.

sulfamethoxazole to photodecay faster at pH < p$K_a$ than under other conditions. For 17α-ethinylestradiol and diclofenac, the opposite trends are evident, so their photolysis is favored under conditions where pH > p$K_a$. This analysis also indicates that among these chemicals, 17α-ethinylestradiol is much less sensitive to radiation at 254 nm than either diclofenac or sulfamethoxazole across the entire pH range.

For calculation of $D_{10}$, we first use the definition of UV dose:

$$\mathrm{d}D_{254} = E'_{254} \cdot \mathrm{d}t \tag{10.E.2.16}$$

Applying this definition together with some algebraic manipulation of Equation 10.E.2.9 yields:

$$\frac{\mathrm{d}C_T}{\mathrm{d}D_{254}} = -\left(\frac{2.303 \cdot \varepsilon_{0,254} \cdot \Phi_{0,254}}{U_{254}} \cdot \alpha_0 + \frac{2.303 \cdot \varepsilon_{1,254} \cdot \Phi_{1,254}}{U_{254}} \cdot \alpha_1\right) \cdot C_T \tag{10.E.2.17}$$

Separation of variables and integration yields:

$$\ln\left(\frac{C_T}{C_{T,0}}\right) = -\frac{2.303}{U_{254}}\left(\varepsilon_{0,254} \cdot \Phi_{0,254} \cdot \alpha_0 + \varepsilon_{1,254} \cdot \Phi_{1,254} \cdot \alpha_1\right) \cdot D_{254} \tag{10.E.2.18}$$

For 1 $\log_{10}$ unit of transformation of the parent compound, $\frac{C_T}{C_{T,0}} = 0.1$. Rearrangement of Equation 10.E.2.18 allows calculation of $D_{10}$:

$$D_{10} = -\frac{\ln(0.1) \cdot U_{254}}{2.303 \cdot \left(\varepsilon_{0,254} \cdot \Phi_{0,254} \cdot \alpha_0 + \varepsilon_{1,254} \cdot \Phi_{1,254} \cdot \alpha_1\right)} \tag{10.E.2.19}$$

or more simply:

$$D_{10} = -\frac{\ln(0.1)}{k_{254}^{\mathrm{app}}} \tag{10.E.2.20}$$

As an example, we will apply the result of the previous calculation for the apparent rate constant of sulfamethoxazole at pH = 6 to estimate the value of $D_{10}$ for the same condition:

$$D_{10} = -\frac{\ln(0.1)}{6.63 \times 10^{-3}\,\frac{\mathrm{cm}^2}{\mathrm{mJ}}} = 347\,\frac{\mathrm{mJ}}{\mathrm{cm}^2} \tag{10.E.2.21}$$

Similar calculations were performed for all three target compounds over the pH range 0 → 14. The results of these calculations are presented in Figure 10.11. Note that the vertical axis in this graph is presented using a $\log_{10}$ scale to facilitate comparisons of the values of $D_{10}$ for the three compounds across this pH range.

Again, we see plots that take on a sigmoidal shape, with an inflection point at pH = p$K_a$ and limits (asymptotic values) that correspond to the values of $D_{10}$ for the protonated and deprotonated forms of the three target compounds.

The values of $D_{10}$ for 17α-ethinylestradiol are two to three orders or magnitude greater than those required for most disinfection applications. The capital and operating costs associated with implementation of a $UV_{254}$ system used to cause degradation of 17α-ethinylestradiol are likely to be unacceptably high. It is likely that other measures will be needed for reduction of the concentration of 17α-ethinylestradiol in water.

For diclofenac and sulfamethoxazole, the $UV_{254}$ doses required to achieve 1 $\log_{10}$ unit degradation are also greater than those required for most disinfection applications, but correspond to

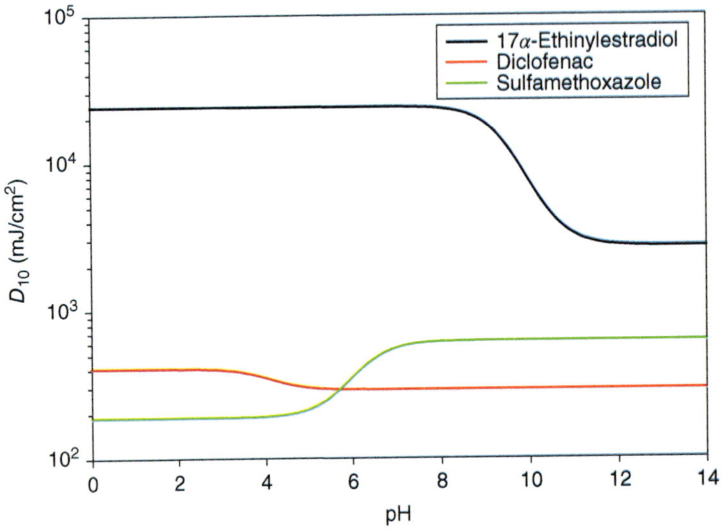


**Figure 10.11** $UV_{254}$ doses required for 1 $\log_{10}$ unit photodegradation ($D_{10}$) of 17α-ethinylestradiol, diclofenac, and sulfamethoxazole at a wavelength of 254, as a function of pH.

systems that are in use today for direct photolysis or AOP applications. Even for $UV_{254}$ systems that are designed for disinfection applications, there will be measurable conversion of diclofenac and sulfamethoxazole (and other compounds); direct photochemical conversion of compounds in water should be considered when designing UV photoreactors.

In broad terms, the extent of reactant conversion by direct photolysis is likely to be modest for most chemicals in water. Typical chemical conversion by photolysis will be less than 1–2 $\log_{10}$ units. By comparison, conversions of 3–5 $\log_{10}$ units are common in disinfection applications. As described above, the relatively modest reactant conversions that are typical of direct photochemical applications allow for simplification of reactor analysis, in the sense that calculations of reactor performance based on the average dose delivered by the system are likely to introduce only small error when compared with an exact solution that accounts for the dose distribution, as is accomplished with the segregated-flow model.

## 10.7 Indirect Processes

Direct photochemical processes result in chemical transformation of compounds through elementary photochemical reactions, which result from electronic excitation of the target molecule caused by absorption of a photon. In turn, this may result in cleavage of a chemical bond, thereby promoting transformation of the target compound.

Another group of processes involve direct photolysis of a promoter compound in solution to yield one or more reactive intermediates. In turn, these intermediates can participate in thermal chemical reactions with target compounds. These processes are sometimes referred to as *indirect photolysis*, and can be important, even critical to the ultimate fate of compounds in the environment.

In many situations, direct and indirect processes may occur simultaneously, such that distinct degradation pathways may operate in parallel. It is generally desirable to be able to predict the rates and effects of these elementary processes independently, because the net effect of these parallel processes is additive. Also, by having quantitative information about the kinetics and mechanisms of the constituent processes, it is possible to develop models that have general applicability.

In natural settings, dissolved organic matter (DOM, including humic and fulvic acids), may function as a photosensitizer. In other words, DOM may absorb ambient solar radiation to yield reactive intermediates, including hydroxyl radicals (HO·), singlet oxygen ($^1O_2$), and triplet state DOM ($^3$DOM) [17]. Direct and indirect processes play important roles in governing the fate and behavior of naturally occurring compounds, as well as trace compounds (of anthropogenic origin). However, similar processes have been adapted into engineering applications wherein so-called "artificial" sources of UV radiation are applied immediately downstream of the point at which an activating chemical is added to the system. The activating (promoter) chemical(s) are chosen to be photoactive, and to yield one or more reactive intermediates as primary products of photolysis. In turn, these reactive intermediates can go on to react with trace contaminants, thereby offering the potential to improve water quality.

Most engineered, indirect processes involve injection of a promoter chemical to a liquid stream, which is then introduced to a UV photoreactor. Exposure to UV radiation results in cleavage of a bond within the promoter, thereby leading to production of one or more reactive intermediates, often in the form of free radicals. The free radicals are then available to react via thermal chemical reactions with constituents that are present in solution. At least five reaction types may proceed under these circumstances:

- direct photolysis of the target compound,
- direct thermal chemical reactions between the activating agent and the target compound,
- thermal chemical reactions between the reactive intermediates that are generated from photolysis of the activating agent and the target compound,
- thermal chemical reactions between reactive intermediates, often resulting in the formation of a relatively stable (i.e. less reactive) product,
- thermal chemical reactions between reactive intermediates and nontarget compounds in solution; these reactions "compete" with the target compound for the reactive intermediate; sometimes these reactions may "quench" reactive intermediates.

These reaction processes proceed in parallel and can be treated as independent processes, in the sense that the effects of these processes are additive. Therefore, definition of kinetic expressions for the fundamental, elementary processes allows description of the effects of the combined processes.

## 10.8 Experimental Methods Used to Estimate Rate Constants for Reactive Intermediates

The reactive intermediates that are generated in indirect processes are often extremely reactive, with rate constants that may approach the diffusional limitation.[k] At room temperature, rate constants for diffusion-limited, bimolecular, thermal chemical reactions are on the order of $10^{10}\frac{L}{\text{mole}\cdot\text{s}}$. For reactions that proceed at such high rates, it is necessary to use specialized experimental methods to develop accurate estimates of reaction rate constants. A number of these methods have been developed, several of which are described briefly below.

### 10.8.1 Direct Measurement

A common direct method is *flash photolysis*, a process in which a beam of (monochromatic) radiation, usually from a pulsed laser, is imposed on a sample. The beam is designed with a wavelength that allows production of a reactive intermediate in solution. Ideally, the target chemical will not absorb the incident radiation, so that direct photolysis of the target does not play a role in the process [18]. Also, the absorption of radiation by the activating chemical in solution should produce only a single reactive intermediate; it may be acceptable to produce more than one reactive intermediate if conditions are such that only one of the intermediates reacts with the target compound. The solution that is exposed to the beam will generally contain the target chemical at a high concentration, such that the concentration of the target chemical remains essentially constant. Under these conditions, the concentration of the reactive intermediate will decay following pseudo-first-order kinetics. Therefore, if methods are available to monitor the decay of the reactive intermediate, it is then possible to measure reaction kinetics directly, thereby allowing estimation of a fundamental reaction rate constant. It is also possible to develop an accurate estimate of a reaction rate constant by monitoring the rate of reaction product generation, which will also follow pseudo-first-order kinetics if the promoter concentration and the applied fluence rate remain constant.

Regardless of the focus of measurement (i.e. decay of the reactive intermediate or formation of a reaction product), it is necessary to use an analytical method that is fast enough to describe true reaction progress. In most cases, these measurements are conducted optically, using instruments that are designed with fast optical components. With this in mind, the experiment is often conducted by following absorbance as a function of time at a fixed wavelength that is at or near a local absorption maximum for a reactive intermediate or the reaction product. Interference from other compounds in solution that absorb at the same wavelength must be accounted for; ideally, only the reactive intermediate or the reaction product will absorb at the selected wavelength.

Establishment of reaction conditions that yield pseudo-first-order kinetics is critical to these experiments. Therefore, it is necessary to check the validity of the pseudo-first-order assumption by conducting the experiment at several concentrations of the target compound.

A related experimental method is *pulse radiolysis*, in which a short, intense pulse of ionizing radiation is imposed on a sample. The radiation will cause the formation of one or more reactive intermediates, which then go on to participate in thermal chemical reactions. The conditions of experiment and experimental design constraints described above for flash photolysis generally translate well to pulse radiolysis. Likewise, the analytical methods that are used to develop estimates of reaction rate constants are similar as well.

### 10.8.2 Competition Kinetics

For many chemicals, it is not possible to accurately measure the decay of a reactive intermediate or formation of a reaction product on a timescale that is relevant to the reactions that take place with the reactive intermediates that are relevant with indirect photochemical processes. Under these circumstances, it common to examine the decay of a target chemical and a reference chemical simultaneously, in the same solution. This approach, often referred to as a *competition kinetics* experiment, allows the target and reference chemicals to be exposed to the same reaction conditions (i.e. be subjected to the same history of exposure to the reactive intermediate) [19–21]. Under these conditions, it is possible to compare the rates of decay of the target and reference chemical as a means to develop an estimate of the reaction rate constant for the target chemical.

Competition kinetics experiments have been applied for estimation of bimolecular rate constants for many target chemicals. The logic behind this approach is summarized below. First, we will assume that a reactive intermediate (*I*) is formed by photolysis of a promoter chemical (*P*):

$$P \overset{h\nu}{\rightarrow} I \tag{10.57}$$

The reactive intermediate is likely to react with a number of compounds in solution, including background chemicals (e.g. pH buffers), the target chemical, and the reference chemical:

$$I + S_i \overset{k_{S_i}}{\rightarrow} \text{Products} \tag{10.58}$$

$$I + T \overset{k_T}{\rightarrow} \text{Products} \tag{10.59}$$

$$I + R \overset{k_R}{\rightarrow} \text{Products} \tag{10.60}$$

where

$S_i$ represents dissolved chemicals that react with $I$, excluding the target chemical and reference chemical

$i$ = number of chemicals that react with $I$, excluding the target chemical and reference chemical

$k_{S_i}$ = bimolecular reaction rate constant for reaction between $I$ and the $i$th constituent in solution (other than the target chemical and reference chemical)

$k_T$ = bimolecular reaction rate constant for reaction between $I$ and the target chemical, $T$

$k_R$ = bimolecular reaction rate constant for reaction between $I$ and the reference chemical, $R$.

The time-dependence of the concentration of the reactive intermediate will be governed by its rate of formation and the reactions that it participates in with dissolved chemicals in solution. For the reaction sequence described by Equations 10.57–10.60, this yields:

$$\frac{\mathrm{d}[I]}{\mathrm{d}t} = \Phi \cdot \frac{GF}{\forall} - \left(\sum_i k_{S_i}[S_i][I] + k_T[T][I] + k_R[R][I]\right) \tag{10.61}$$

where

$\Phi$ = quantum yield for the formation of $I$ from $P$ [=] $\dfrac{\text{mole}}{\text{einstein}}$

$G$ = total photon emission rate from the radiation source [=] $\dfrac{\text{einstein}}{\text{s}}$

$F$ = fraction of emitted photons that are absorbed by $P$

$\forall$ = volume of solution being irradiated [=] L.

The first term on the right-hand side of Equation 10.61 represents the rate of formation of $I$ by photolysis of $P$.[1] The parenthetical term on the right-hand side of Equation 10.61 represents the sum of the rates of all reactions in solution that involve $I$.

Ideally, a competition kinetics experiment will be conducted under conditions such that degradation of the target chemical and the reference chemical will be solely attributable to bimolecular reactions with the reactive intermediate. When this is true, then their rates of decay can be described as:

$$\frac{\mathrm{d}[T]}{\mathrm{d}t} = -k_T[T][I] \tag{10.62}$$

$$\frac{d[R]}{dt} = -k_R[R][I] \tag{10.63}$$

A critical criterion for selection of a reference chemical is that the value of the bimolecular rate constant, $k_R$, should be accurately defined. Assuming this is the case, then the relative reactivity, $\beta$, becomes relevant:

$$\beta = \frac{k_T}{k_R} \tag{10.64}$$

Separation of variables and integration of Equations 10.62 and 10.63 yields two general equations that describe the time-dependent behavior of the concentrations of the target chemical and reference chemical:

$$\ln\left(\frac{[T]}{[T]_0}\right) = -k_T\int_0^t [I] \cdot dt \tag{10.65}$$

$$\ln\left(\frac{[R]}{[R]_0}\right) = -k_R\int_0^t [I] \cdot dt \tag{10.66}$$

Inspection of Equations 10.64–10.66 indicates the following:

$$\beta = \frac{\ln\left(\frac{[T]}{[T]_0}\right)}{\ln\left(\frac{[R]}{[R]_0}\right)} \tag{10.67}$$

Therefore, linear regression of a plot of $\ln\left(\frac{[T]}{[T]_0}\right)$ vs. $\ln\left(\frac{[R]}{[R]_0}\right)$, forced through the origin, will yield a line with a slope of $\beta$.

A special case of this approach involves conditions under which the concentration of the reactive intermediate is maintained at a constant, steady-state value. This condition will be met when the rates of formation and decay of the reactive intermediate are equal. Under these conditions, $\frac{d[I]}{dt} = 0$. Substitution of this information into Equation 10.61 yields an estimate of the steady-state concentration of the reactive intermediate, $[I]_{SS}$:

$$[I]_{SS} = \frac{\Phi \cdot \frac{GF}{\forall}}{\sum_i k_{S_i}[S_i] + k_T[T] + k_R[R]} \tag{10.68}$$

If the sum $k_T[T] + k_R[R]$ is much less than the sum $\sum_i k_{S_i}[S_i]$, then the composition of the solution becomes essentially constant, and $[I]_{SS}$ becomes essentially constant. When this occurs, the rates of decomposition of $T$ and $R$ become pseudo-first-order:

$$\ln\left(\frac{[T]}{[T]_0}\right) = -k_T \cdot [I]_{SS} \cdot t = -k_T' \cdot t \tag{10.69}$$

$$\ln\left(\frac{[R]}{[R]_0}\right) = -k_R \cdot [I]_{SS} \cdot t = -k_R' \cdot t \tag{10.70}$$

where

$k_T'$ = pseudo-first-order reaction rate constant for degradation of $T$ by I [=] $s^{-1}$
$k_R'$ = pseudo-first-order reaction rate constant for degradation of $R$ by I [=] $s^{-1}$.

$k_T'$ and $k_R'$ can be estimated by linear regression of plots of $\ln\left(\frac{[T]}{[T]_0}\right)$ vs. $t$ and $\ln\left(\frac{[R]}{[R]_0}\right)$ vs. $t$, based on the slopes of "best-fit" lines through these data that are forced through the origin. The relative reactivity can be estimated from:

$$\beta = \frac{\text{slope}_T}{\text{slope}_R} = \frac{k_T'}{k_R'} \tag{10.71}$$

where

$\text{slope}_T$ = slope of linear regression of $\ln\left(\frac{[T]}{[T]_0}\right)$ vs. $t$ forced through the origin
$\text{slope}_R$ = slope of linear regression of $\ln\left(\frac{[R]}{[R]_0}\right)$ vs. $t$ forced through the origin.

The more general approach described by Equation 10.67 tends to yield regression lines that are straighter than those that are developed by the approach described by Equation 10.71 [21]. This is probably because the approach described by Equation 10.71 requires that the concentration of the reactive intermediate reaches a true steady-state value, and that this value does not change throughout the course of the experiment. Also, when possible, it is helpful to select a reference chemical that displays similar rate of decay as the target chemical. This will dictate that the extents of conversion of the target and reference chemical are similar over the course of the experiment, which tends to reduce error in estimation of the rate constant of interest.

## 10.9 UV Advanced Oxidation Processes (UV/AOPs)

Conventional water and wastewater treatment processes are effective for control of conventional contaminants, including suspended particles, compounds that are readily biodegradable, and for inactivation of most waterborne microbial pathogens. However, the combined effects of increasing human population and climate change have resulted in increased pressure to use water supplies that had previously been viewed as having marginal or undesirable quality. Simultaneously, there has been increasing scrutiny placed on the fate and effects of chemical contaminants that are present at trace concentrations (typically $\frac{\mu g}{L}-\frac{ng}{L}$). By their very nature, and sometimes by design, these chemicals often exert adverse effects on human health and the environment at low concentration. Also, it is common for these chemicals to coexist; the effects of mixtures of these contaminants are largely unknown.

For these and other reasons, interest in the development of "advanced" treatment processes has been increasing since the 1990s, coincident with the first reports of the potential adverse effects of these compounds on human health and the environment [2]. Among the most important of these processes are the so-called AOPs.

A general definition for an AOP is any process that results in the formation of one or more highly reactive, oxidizing, intermediate compounds. These intermediates often react rapidly and indiscriminately to oxidize compounds in solution, including trace contaminants. As such, AOPs often represent a possible approach for degradation of trace contaminants and improvement of water quality.

UV photoreactions tend to promote cleavage of covalent bonds, resulting in the formation of one or more reactive intermediates, often in the form of free radicals. Many UV AOPs involve the introduction of a promoter chemical that is photolabile, followed immediately by UV irradiation. This sequence results in the formation of reactive intermediates.

It is important to note that for most of the oxidizing intermediates that are formed in UV AOPs, there are other processes that can lead to their formation. Once formed, the subsequent thermal chemical reactions proceed at their intrinsic rates. As such, it is common to use reaction rate constants that have been estimated through the use of experimental methods that rely on other forms of radical formation, such as application of ionizing radiation or heat.

For many AOPs, the hydroxyl radical ($HO^{\cdot}$) is a primary reaction product; in some senses, $HO^{\cdot}$ is viewed as a reference oxidant for AOPs, including those that are initiated by UV irradiation. $HO^{\cdot}$ reacts with many compounds at rates that approach the diffusion limitation. It is also among the most powerful oxidizing agents known, with a standard electrode potential of approximately 2.73 V.

The standard electrode potential, $E^0$ (also known as standard reduction potential [SRP], standard oxidation potential, standard redox potential, and sometimes just standard potential), represents the potential for a molecule (atom, ion, complex, or radical) to be reduced. A large, positive value of $E^0$ indicates a molecule that has a strong tendency to accept electrons, thereby being reduced. In other words, molecules that have a large, positive value of $E^0$ function as strong oxidants. Conversely, molecules with large, negative values of $E^0$ tend to function as strong reducing agents (i.e. they are easily oxidized or lose electrons easily, thereby promoting reduction of another molecule).

Like all potentials, $E^0$ provides a relative measurement, and must be reported relative to a standardized value. By convention, values of $E^0$ are reported relative to the following reaction, which is assigned a value of $E^0 = 0\ V$.

$$2H^+ + 2e^- \rightleftharpoons H_2 \tag{10.72}$$

Table 10.2 provides a list of standard electrode potentials for a number of inorganic radicals that may participate in reactions that are promoted through AOPs. For each radical, the corresponding half-reaction (written as a reduction, following standard practice) is included. Radicals and their corresponding half-reactions are presented in decreasing order of $E^0$. Therefore, the strongest oxidizing agents (i.e. those compounds with the greatest potential to be reduced) are found at the top of the table, while the strongest reducing agents (i.e. those with the greatest potential to be oxidized) are found at the bottom of the table.

Many AOPs involve introduction of a promoter chemical into a UV photoreactor, or immediately upstream of the photoreactor. As such, target chemicals will be exposed to the promoter chemical followed (often immediately) by UV. Using this logic, the naming convention that will be applied for these systems will be to identify the promoter chemical (usually in its neutral or general form), followed by UV. This naming convention will be used to identify several specific UV-based AOPs that are used in practice today, and are discussed below.

Free radicals and other reactive intermediates can be generated by a wide range of processes. Beyond UV-based processes, there are processes in which ionizing radiation, heat, ultrasound, or thermal chemical reactions are used to generate radicals. These processes vary widely in terms of the efficiency of radical generation and energy required to degrade target organic compounds. As described previously, the term electrical efficiency per order ($E_{EO}$) represents a standardized figure-of-merit for comparison of energy costs associated with AOPs and other processes [11, 23]. $E_{EO}$ is based on the concept that the kinetics of these processes tends toward first-order behavior in the limit of dilute solution; most contaminants that are the target of AOPs are present at low (trace) concentration. Figure 10.12 provides a graphical comparison of $E_{EO}$ values that have been reported in the refereed literature for a range of AOPs. Among these, the UV-based AOPs tend to be on the low end of the $E_{EO}$ spectrum, which provides motivation to pursue their use in treatment of water containing trace contaminants.

**Table 10.2** Reduction half-reactions and corresponding standard electrode potentials for inorganic radicals that are likely to participate in advanced oxidation processes and advanced reduction processes.

| | Half-cell reaction | Standard electrode potential (V) | |
|---|---|---|---|
| Stronger oxidizing agent → | $H_2PO_4^{\cdot} + e^- \rightleftharpoons H_2PO_4^-$ | $2.75 \pm 0.01$ | Stronger reducing agent → |
| | $ClOH^{\cdot-} + e^- + H^+ \rightleftharpoons Cl^- + H_2O$ | $2.740 \pm 0.018$ | |
| | $HO^{\cdot} + e^- + H^+ \rightleftharpoons H_2O$ | $2.730 \pm 0.017$ | |
| | $NO_3^{\cdot} + e^- \rightleftharpoons NO_3^-$ | $2.466 \pm 0.019$ | |
| | $SO_4^{\cdot-} + e^- \rightleftharpoons SO_4^{2-}$ | $2.437 \pm 0.019$ | |
| | $Cl^{\cdot} + e^- \rightleftharpoons Cl^-$ | $2.432 \pm 0.018$ | |
| | $ClO_3^{\cdot} + e^- \rightleftharpoons ClO_3^-$ | $2.38 \pm 0.03$ | |
| | $O(^3P) + H^+ + e^- \rightleftharpoons HO^{\cdot}$ | $2.3 \pm 0.1$ | |
| | $Cl_2^{\cdot-} + e^- \rightleftharpoons 2Cl^-$ | $2.126 \pm 0.017$ | |
| | $Br^{\cdot} + e^- \rightleftharpoons Br^-$ | $1.96 \pm 0.02$ | |
| | $ClOH^{\cdot-} + e^- \rightleftharpoons Cl^- + OH^-$ | $1.912 \pm 0.018$ | |
| | $OH^{\cdot} + e^- \rightleftharpoons OH^-$ | $1.902 \pm 0.017$ | |
| | $Br_2^{\cdot-} + e^- \rightleftharpoons 2Br^-$ | $1.63 \pm 0.02$ | |
| | $O(^3P) + e^- \rightleftharpoons O^{\cdot-}$ | $1.6 \pm 0.1$ | |
| | $CO_3^{\cdot-} + e^- \rightleftharpoons CO_3^{2-}$ | $1.57 \pm 0.03$ | |
| | $PO_3^{\cdot 2-} + H^+ + e^- \rightleftharpoons HPO_3^{2-}$ | $1.54 \pm 0.04$ | |
| | $HO_2^{\cdot} + e^- + H^+ \rightleftharpoons H_2O_2$ | $1.46 \pm 0.01$ | |
| | $S_2O_8^{2-} + e^- \rightleftharpoons SO_4^{\cdot-} + SO_4^{2-}$ | $1.44 \pm 0.08$ | |
| | $ClO^{\cdot} + e^- \rightleftharpoons ClO^-$ | $1.39 \pm 0.03$ | |
| | $I^{\cdot} + e^- \rightleftharpoons I^-$ | $1.35 \pm 0.02$ | |
| | $BrO_2^{\cdot} + e^- \rightleftharpoons BrO_2^-$ | $1.290 \pm 0.005$ | |
| | $S_4O_6^{\cdot 3-} + e^- \rightleftharpoons 2S_2O_3^{2-}$ | $1.10 \pm 0.01$ | |
| | $I_2^{\cdot-} + e^- \rightleftharpoons 2I^-$ | $1.05 \pm 0.02$ | |
| | $NO_2^{\cdot} + e^- \rightleftharpoons NO_2^-$ | $1.04 \pm 0.02$ | |
| | $ClO_2^{\cdot} + e^- \rightleftharpoons ClO_2^-$ | $0.935 \pm 0.003$ | |
| | $SO_5^{\cdot-} + e^- \rightleftharpoons SO_5^{2-}$ | $0.81 \pm 0.01$ | |

(Continued)

**Table 10.2** (Continued)

| | Half-cell reaction | Standard electrode potential (V) | |
|---|---|---|---|
| | $SO_3^{\cdot -} + e^- \rightleftharpoons SO_3^{2-}$ | $0.73 \pm 0.02$ | |
| | $Cl_2 + e^- \rightleftharpoons Cl_2^{\cdot -}$ | $0.666 \pm 0.017$ | |
| | $Br_2 + e^- \rightleftharpoons Br_2^{\cdot -}$ | $0.55 \pm 0.02$ | |
| | $HOCl + e^- \rightleftharpoons ClOH^{\cdot -}$ | $0.25 \pm 0.08$ | |
| | $I_2 + e^- \rightleftharpoons I_2^{\cdot -}$ | $0.19 \pm 0.02$ | |
| | $O_2 + H^+ + e^- \rightleftharpoons HO_2^{\cdot}$ | $0.10 \pm 0.02$ | |
| | $NO^{\cdot} + H^+ + e^- \rightleftharpoons HNO$ | $-0.15 \pm 0.02$ | |
| | $O_2 + e^- \rightleftharpoons O_2^{\cdot -}$ | $-0.18 \pm 0.02$ | |
| | $CO_2 + e^- \rightleftharpoons CO_2^{\cdot -}$ | $-1.90 \pm 0.02$ | |
| | $H^+ + e^- \rightleftharpoons H^{\cdot}$ | $-2.31 \pm 0.03$ | |
| | $e^- \rightleftharpoons e^{\cdot -}$ | $-2.88 \pm 0.02$ | |

*Source:* Data from Armstrong et al. [22].

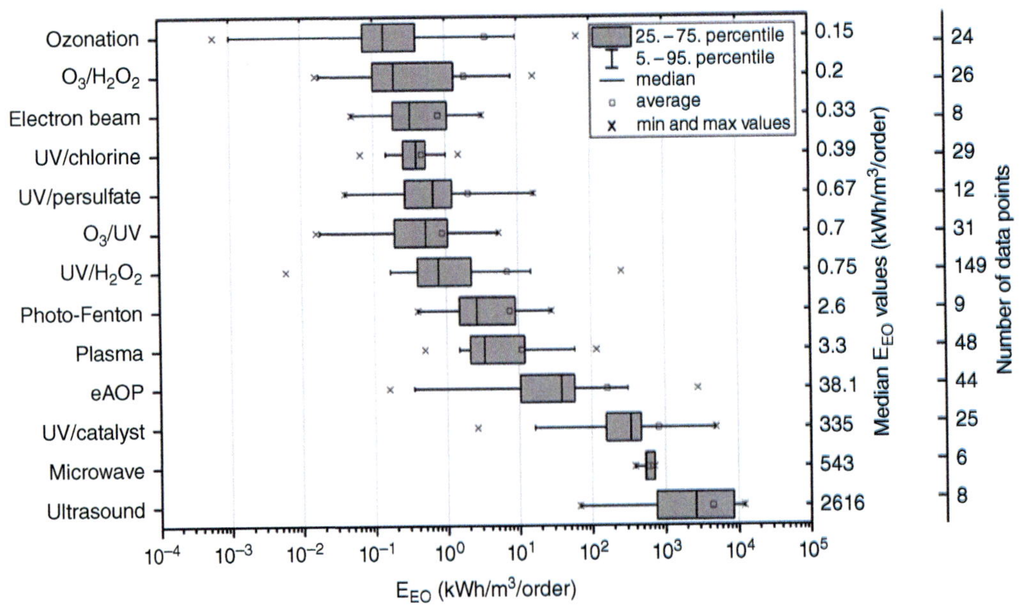


**Figure 10.12** Summary and comparison of published $E_{EO}$ values for AOPs. Vertical bars in the boxes represent the 25th, 50th, and 75th percentiles of reported values, respectively. Whiskers illustrate 5th and 95th percentiles. Median values and number of data points are reported on the second and third *y*-axis, respectively. *Source:* Miklos et. al 2018/with permission of Elsevier.

## 10.10 $H_2O_2$/UV Process

Among UV AOPs, the $H_2O_2$/UV process is one of the most commonly applied, and perhaps the process with the longest history of application. Like most UV AOPs, the $H_2O_2$/UV process is strongly influenced by solution composition and the wavelength(s) of UV radiation imposed on the reacting solution.

Hydrogen peroxide displays acid/base behavior, which affects process efficiency:

$$H_2O_2 \rightleftharpoons H^+ + HO_2^- \quad K_a = \frac{[H^+][HO_2^-]}{[H_2O_2]} \quad pK_a = 11.62 \tag{10.73}$$

In neutral or near-neutral solutions, the dominant form is the fully protonated, neutral molecule $H_2O_2$, which is a weak absorber of UVC radiation $\left(\varepsilon_{H_2O_2,254} = 19.6\frac{L}{\text{mole}\cdot\text{cm}}\right)$ [24]. $H_2O_2$ is photolyzed by UVC radiation to yield two hydroxyl radicals:

$$H_2O_2 \xrightarrow{h\nu} 2HO^{\cdot} \tag{10.74}$$

At 254 nm, this reaction has been demonstrated to have a quantum yield (based on $H_2O_2$ decay) of approximately $0.50\frac{\text{mole}}{\text{einstein}}$ for concentrations ranging from $2\times10^{-5}$ M $\leq [H_2O_2] \leq 0.1$ M ($0.7\frac{\text{mg}}{L}$ to $3400\ \frac{\text{mg}}{L}$) [24]. Note that the quantum yield based on $HO^{\cdot}$ formation is approximately $1.0\frac{\text{mole}}{\text{einstein}}$ for this same concentration range (see stoichiometry defined in reaction 10.74).

The conjugate base of $H_2O_2$ (the hydroperoxyl ion, $HO_2^-$) undergoes a similar reaction, with a similar quantum yield and similar reaction products:

$$HO_2^- + H_2O \xrightarrow{h\nu} 2HO^{\cdot} + OH^- \tag{10.75}$$

An advantage of this reaction is that $HO_2^-$ is a much stronger absorber of UVC radiation than $H_2O_2$ $\left(\varepsilon_{HO_2^-,254} = 229\frac{L}{\text{mole}\cdot\text{cm}}\right)$ [24]. Therefore, the formation of hydroxyl radicals is expected to be roughly 10 times more efficient at pH values above the $pK_a$ of $H_2O_2$ than at near-neutral pH. However, the high pH required to reach this efficiency is often difficult to achieve in practical applications. Also, acidification will probably be required after treatment to bring the pH back to a near-neutral condition. It is important to note also that pH changes may affect the chemical behavior of other compounds in solution that display acid/base behavior.

Sauer et al. [25] reported that photolysis of the hydroperoxyl ion also yields the formation of triplet oxygen (see reaction 10.76) [25]. The quantum yield for this reaction was reported to be $0.08\frac{\text{mole}}{\text{einstein}}$ at a wavelength of 248 nm. Reaction 10.76 does not appear to play an important role in the dynamics of the $H_2O_2$/UV process and will not be discussed further.

$$HO_2^- \xrightarrow{h\nu} O(^3P) + OH^- \tag{10.76}$$

Figure 10.13 illustrates UV absorption spectra for $H_2O_2$ and $HO_2^-$. Their absorption behaviors are largely independent of temperature for $1\,°C \leq T \leq 25\,°C$. Absorption is quite weak in the UVA range, but increases steadily as wavelength decreases through the UVB and UVC ranges. For wavelengths less than approximately 300 nm, $HO_2^-$ is roughly 10× more effective at absorbing radiation than $H_2O_2$; however, absorption is modest for both forms across the entire UV range.

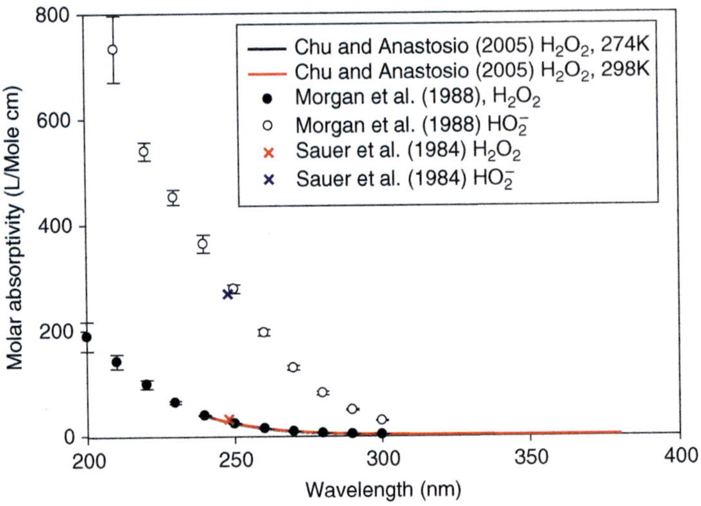


**Figure 10.13** UV absorption spectra for $H_2O_2$ and $HO_2^-$. Solid red and black lines represent smooth curves through data presented by Chu and Anastasio [26] for $H_2O_2$ at 1 and 25 °C, respectively [26]. Circles (error bars) represent mean (95% confidence interval) for multiple measurements at specific wavelengths by Morgan et al. [27], with closed circles representing $H_2O_2$, while open circles represent $HO_2^-$ [27]. The red cross and the blue cross represent values for $H_2O_2$ and $HO_2^-$, respectively reported by Sauer et al. at 248 nm [25].

The low efficiency of photon absorption by $HO_2^-$ and $H_2O_2$ is somewhat counteracted by the high quantum yield for the process. Nonetheless, low absorption efficiency represents an important limitation of this process. This behavior requires use of relatively high concentrations of $H_2O_2$, relatively high UVC doses, or both to ensure sufficient formation of $HO^{\cdot}$ to achieve a given treatment endpoint. In many situations, the low molar absorption coefficient is compensated by addition of $H_2O_2$ concentrations on the order of 10's of $\frac{mg}{L}$. In turn, this generally requires that the $H_2O_2$ residual that remains after UVC exposure be quenched prior to use of the treated water.

Like many UV-based advanced reaction processes, the $H_2O_2$/UV process depends on the formation of reactive intermediates to accomplish degradation of target pollutants in water. For the $H_2O_2$/UV process, the dominant intermediates, in terms of degradation of target contaminants, are the hydroxyl radical and carbonate radical. With many reaction processes that involve radical intermediates, process dynamics will depend on the rate of direct photolysis of the target compound(s), the rates at which participating radicals are formed (i.e. initiation), the rates at which the participating radicals react with ambient constituents in solution to form other, related reactive intermediates (i.e. propagation), and the rates at which reactive intermediates react with each other to form nonreactive compounds (i.e. termination).

The $H_2O_2$/UV process is initiated by photolysis of $H_2O_2$ and its conjugate base, $HO_2^-$, as described by reactions 10.74–10.76. Reactions that govern propagation and termination are summarized in Tables 10.2 and 10.3, respectively.

It is also relevant to consider acid/base behavior among the compounds listed in Tables 10.2 and 10.3. Beyond the equilibrium described in Equation 10.73, there are two other equilibria involving reactive intermediates that must be accounted for:

**Table 10.3** Summary of propagation reactions that govern the dynamics of the $H_2O_2$/UV process.

| Reaction | Recommended[a] rate constant $\left(\frac{L}{mole \cdot s}\right)$ | Reference | Reaction number |
|---|---|---|---|
| $HO^{\cdot} + H_2O_2 \rightarrow H_2O + HO_2^{\cdot}$ | $(3.2 \pm 1.5) \times 10^7$ | Yu [28] | (10.77) |
| $HO^{\cdot} + HO_2^- \rightarrow OH^- + O_2^{\cdot -} + H^+$ | $(7.5 \pm 1.0) \times 10^9$ | Christensen et al. [29] | (10.78) |
| $O^{\cdot -} + H_2O_2 \rightarrow OH^- + O_2^{\cdot -} + H^+$ | $(4.2 \pm 0.2) \times 10^9$ | Yu [28] | (10.79) |
| $O^{\cdot -} + HO_2^- \rightarrow OH^- + O_2^{\cdot -}$ | $(7.8 \pm 0.8) \times 10^9$ | Yu [28] | (10.80) |
| $HO_2^{\cdot} + H_2O_2 \rightarrow HO^{\cdot} + H_2O + O_2$ | 3.7 | Farhataziz and Ross [30] | (10.81) |
| $O_2^{\cdot -} + H_2O_2 \rightarrow HO^{\cdot} + OH^- + O_2$ | $(0.13 \pm 0.07)$ | Weinstein and Bielski [31] | (10.82) |

[a] Rate constants listed here are those recommended by Stefan [32].

$$HO_2^{\cdot} \rightleftharpoons H^+ + O_2^{\cdot -} \quad K_a = \frac{[H^+][O_2^{\cdot -}]}{[HO_2^{\cdot}]} \quad pK_a = 4.7 \tag{10.89}$$

$$HO^{\cdot} \rightleftharpoons H^+ + O^{\cdot -} \quad K_a = \frac{[H^+][O^{\cdot -}]}{[HO^{\cdot}]} \quad pK_a = 11.9 \tag{10.90}$$

Like most acid/base reactions, the forward and reverse reactions that contribute to these processes are extremely fast in both directions, such that these reactions can be considered to be in equilibrium. This provides another illustration of the important roles played by pH in governing reaction dynamics in advanced reaction processes.

Other reactions to consider in the $H_2O_2$/UV process include so-called "scavenging" reactions between reactive intermediates and compounds that are present in solution. The scavenging reactions lead to consumption of the reactive intermediates that are responsible for degradation of target pollutants. Therefore, these reactions decrease the efficiency of the $H_2O_2$/UV process (Tables 10.4 and 10.5).

**Table 10.4** Summary of termination reactions that govern the dynamics of the $H_2O_2$/UV process.

| Reaction | Recommended[a] rate constant $\left(\frac{L}{mole \cdot s}\right)$ | Reference | Reaction number |
|---|---|---|---|
| $HO_2^{\cdot} + HO_2^{\cdot} \rightarrow H_2O_2 + O_2$ | $8.3 \times 10^5$ | Bielski et al. [33] | (10.83) |
| $HO_2^{\cdot} + O_2^{\cdot -} + H_2O \rightarrow H_2O_2 + O_2 + OH^-$ | $9.7 \times 10^7$ | Bielski et al. [33] | (10.84) |
| $O_2^{\cdot -} + O_2^{\cdot -} + 2H_2O \rightarrow H_2O_2 + O_2 + 2OH^-$ | <0.35 | Weinstein and Bielski [31] | (10.85) |
| $HO^{\cdot} + HO^{\cdot} \rightarrow H_2O_2$ | $5.5 \times 10^9$ | Buxton et al. [18] | (10.86) |
| $HO^{\cdot} + HO_2^{\cdot} \rightarrow H_2O + O_2$ | $6 \times 10^9$ | Buxton et al. [18] | (10.87) |
| $HO^{\cdot} + O_2^{\cdot -} \rightarrow OH^- + O_2$ | $8 \times 10^9$ | Buxton et al. [18] | (10.88) |

[a] Rate constants listed here are those recommended by Stefan [32].

**Table 10.5** Common scavenging reactions that contribute to the dynamic behavior of the $H_2O_2$/UV process.

| Reaction | Recommended[a] rate constant $\left(\frac{L}{mole\ s}\right)$ | Reaction number |
|---|---|---|
| $HO^{\cdot} + HCO_3^- \rightarrow CO_3^{\cdot -} + H_2O$ | $8.5 \times 10^6$ | (10.91) |
| $HO^{\cdot} + CO_3^{2-} \rightarrow CO_3^{\cdot -} + OH^-$ | $3.9 \times 10^8$ | (10.92) |
| $HO^{\cdot} + PO_4^{3-} \rightarrow$ n. d.[b] | $<10^7$ | (10.93) |
| $HO^{\cdot} + HPO_4^{2-} \rightarrow$ n. d.[b] | $1.5 \times 10^5$ | (10.94) |
| $HO^{\cdot} + H_2PO_4^- \rightarrow H_2PO_4^{\cdot} + OH^-$ | $2 \times 10^4$ | (10.95) |
| $HO^{\cdot} + H_3PO_4 \rightarrow H_2PO_4^{\cdot} + H_2O$ | $2.7 \times 10^6$ | (10.96) |
| $HO^{\cdot} + NO_2^- \rightarrow NO_2^{\cdot} + OH^-$ | $1.0 \times 10^{10}$ | (10.97) |
| $HO^{\cdot} + Br^- \rightarrow BrOH^-$ | $1.1 \times 10^{10}$ | (10.98) |

[a] Rate constants listed here are those recommended by Buxton et al. [18].
[b] n.d. = not defined.

There are many other reactions that function to scavenge reactive intermediates, including reactions that involve natural organic matter (NOM, e.g. humic and fulvic materials). These constituents can contribute significantly to decay of reactive intermediates, such as the hydroxyl radical, and "rate constants" for these reactions have been reported. However, because of the undefined structures of compounds that comprise NOM, these rate constants are generally reported using units that allow normalization based on the molar (or mass) quantity of carbon involved, rather than a molar quantity of a specific reactant.

Collectively, the reactions that define initiation, propagation, termination, the relevant equilibria, and scavenging reactions will determine the forms of the reactive intermediates as well as their local concentrations. The reactive intermediates will react with target chemicals following conventional thermal reaction processes, with most reactions being elementary and second-order. Rate constants for these processes have been estimated in experiments and reported widely in the literature. Buxton et al. [18] and Neta et al. [34] have assembled comprehensive critical reviews and summaries of measured rate constants for reactions involving the hydroxyl radical and many other radicals with a wide range of compounds that are commonly found in aqueous systems, or that may exist in low concentrations.

The factors that govern the performance of UV-AOPs are similar to those that govern the behavior of other UV-based photoreactor systems. Specifically, process performance will be governed by the fluence rate field, fluid mechanics, and the intrinsic kinetics of the reactions of interest. Because of this, simulations of the $H_2O_2$/UV process often involve integrated applications of computational fluid dynamics (CFD), a fluence rate field model, and summaries of the kinetics of the fundamental reactions that contribute to process performance.

Figure 10.14 provides an illustration of the application of this numerical modeling approach for simulation of atrazine degradation in a $H_2O_2$/UV process [35]. Atrazine is relevant to consider, in that it is known that direct photolysis, as well as reactions with radical intermediates will contribute significantly to its degradation. The simulation was conducted for a reactor that included four lamps oriented perpendicular to the nominal flow direction. The flow field (Figure 10.14 top) indicated that the fastest fluid flow was through the central region of the reactor, while areas above and

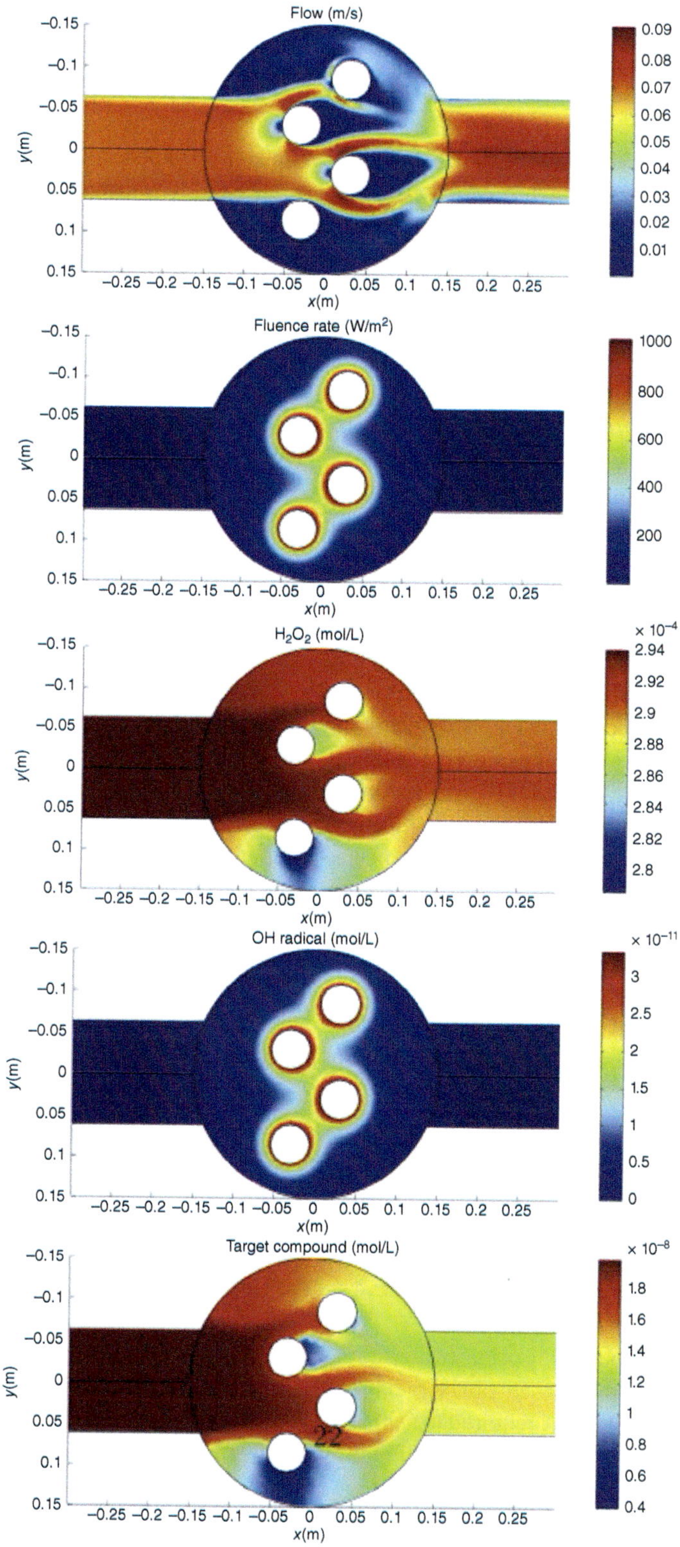


**Figure 10.14** Results of numerical simulation of degradation of atrazine in an $H_2O_2$/UV process by Wols and Hofman-Caris [35]/with permission of Elsevier, for a four-lamp reactor. Top panel illustrates the simulated flow field; second panel illustrates the simulated fluence rate field; third panel illustrates the $H_2O_2$ field; fourth panel illustrates the hydroxyl radical field; bottom panel illustrates the concentration profile for the target compound, atrazine.

below the central region were characterized by a horizontal velocity that was roughly 10% of the velocity through the central section of the reactor. The fluence rate field was characterized by areas of high fluence rate in the immediate vicinity of the lamps, but substantially lower fluence rate in areas more distant from the lamps (Figure 10.14, second panel). The simulated steady-state concentration field for hydrogen peroxide (Figure 10.14, third panel) indicated steep gradients in the concentration of $H_2O_2$ close to the lamps, and also in an area that was characterized by low velocity below the lamps. Loss of $H_2O_2$ in near-lamp regions is largely attributable to the high fluence rate in these parts of the reactor. The relatively low concentration of $H_2O_2$ near the bottom of the reactor cross section was probably attributable to the relatively low velocity in this area, which would allow for relatively long exposure, thereby allowing exposure of $H_2O_2$ to a relatively large UV dose. The profile of hydroxyl radical concentration (Figure 10.13, fourth panel) was quite similar to the fluence rate field map; this provides a clear indication that $H_2O_2$ photolysis, which yields $HO^{\cdot}$ as a product, takes places at a rate that is directly proportional to the local fluence rate. As such, the steady-state concentration of $HO^{\cdot}$ is likely to follow a spatial pattern that is quite similar to the fluence rate field for reactors in which the $H_2O_2$/UV process is applied. Lastly, Figure 10.14 (bottom) illustrates the spatial distribution of atrazine in this reactor system. The steepest gradients in atrazine concentration are evident in the areas where the gradients in $HO^{\cdot}$ concentration are highest. This modeling approach yielded predictions of atrazine degradation that were close to measured values of atrazine degradation for the same operating conditions [35]. More broadly, similar modeling approaches have been shown to yield accurate simulations of the performance of $H_2O_2$/UV systems for degradation of a wide range of trace contaminants [36].

Process simulations based on these integrated modeling methods have a number of advantages over other, simpler approaches. First, they are able to capture much of the relevant physics and chemistry that govern the behavior of the process. Second, they can be used to identify aspects of reactor design that limit overall reactor performance. Third, with contemporary software and computer hardware systems, it is possible to develop detailed, accurate simulations of process performance in a relatively short period of time (i.e. hours to days). As such, it is then possible to use these modeling approaches as the basis of reactor design and for reactor optimization. Moreover, these modeling approaches can be used to predict the ability of an existing reactor system to bring about changes in aquatic constituents that may emerge as targets after reactor implementation.

### 10.10.1 Chlorine/UV Process

The $H_2O_2$/UV process is limited by small values of $\varepsilon$ and the need to quench residual $H_2O_2$. Both of these characteristics reduce process efficiency and add to operating costs. As a result, there is interest in development and application of processes that may achieve similar treatment endpoints, but at lower cost.

The chlorine/UV process addresses both of the primary limitations of the $H_2O_2$/UV process. Free chlorine compounds are generally better absorbers of UV radiation than $HO_2^-$ and $H_2O_2$. Because of this, the molar concentration of free chlorine required to achieve a given level of treatment in the chlorine/UV process tends to be lower than the corresponding molar concentration of hydrogen peroxide required in the $H_2O_2$/UV process. Also, residual chlorine may not need to be quenched, as it is commonly applied in water treatment and can serve a beneficial function as a secondary oxidant or disinfectant in finished water.

In its most common form, the chlorine/UV process involves photolysis of free chlorine compounds to generate radical intermediates. Therefore, to understand the behavior of the chlorine/

UV process, it is necessary to examine the chemistry and photochemistry of the various forms of free chlorine that can exist in solution.

The term "free chlorine" refers to forms of oxidized (+1-valent or zero-valent) chlorine that exist in solution; these are the forms of chlorine that can accept electrons (i.e. function as oxidizing agents). Several forms of free chlorine exist, and well-defined equilibria govern their distribution. Equations 10.99–10.102 summarize these equilibria and their respective equilibrium constants, all reported at 25 °C. Values of equilibrium constants were taken from Odeh et al. [37] and Deborde and von Gunten [38].

$$Cl_2 + H_2O \rightleftharpoons HOCl + H^+ + Cl^- \quad K_h = \frac{[HOCl][H^+][Cl^-]}{[Cl_2]} = 1.04 \times 10^{-3}\ M^2 \tag{10.99}$$

$$HOCl \rightleftharpoons H^+ + OCl^- \quad K_a = \frac{[H^+][OCl^-]}{[HOCl]} = 3.39 \times 10^{-8}\ M \tag{10.100}$$

$$HOCl + HOCl \rightleftharpoons Cl_2O + H_2O \quad K_{Cl_2O} = \frac{[Cl_2O]}{[HOCl]^2} = 8.7 \times 10^{-3}\ M^{-1} \tag{10.101}$$

$$Cl_2 + Cl^- \rightleftharpoons Cl_3^- \quad K_{Cl_3^-} = \frac{[Cl_3^-]}{[Cl_2][Cl^-]} = 0.18\ M^{-1} \tag{10.102}$$

These equilibrium expressions can be used to illustrate the speciation of free chlorine among its various forms. Figure 10.15 illustrates the speciation of free chlorine as a function of pH in a sample of water that contains $5.0\frac{mg}{L}$ ($7.0 \times 10^{-5}$ M) of free chlorine and chloride ($Cl^-$) at a molar concentration of $10^{-4}$ M ($3.6\frac{mg}{L}$), which is representative of chlorinated tap water. For comparison, Figure 10.16 illustrates free chlorine speciation as a function of pH in a water sample that contains $7.0 \times 10^{-5}$ M free chlorine and $Cl^-$ at a concentration of 0.545 M, which is representative of chlorinated sea water [39]. Note that the vertical axis scales in Figures 10.15 and 10.16 are logarithmic (base 10).

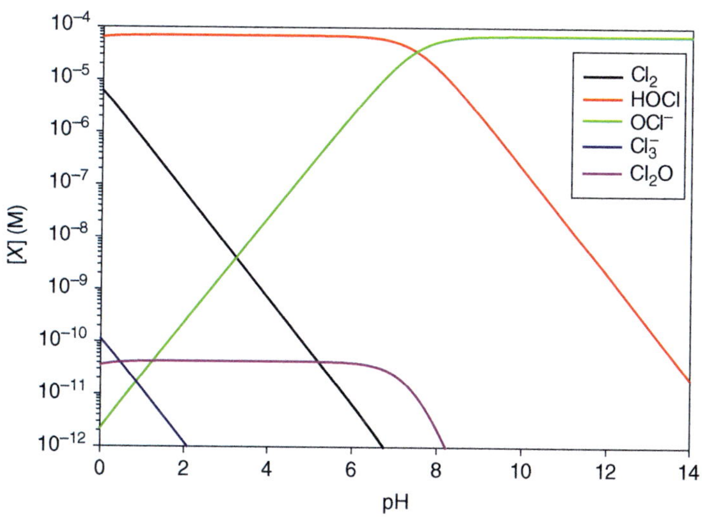


**Figure 10.15** Distribution diagram for free chlorine compounds in a sample that contains $7.0 \times 10^{-5}$ M free chlorine concentration and chloride ion at a concentration of $10^{-4}$ M. These conditions are representative of chlorinated tap water. Temperature is assumed to be 25 °C.

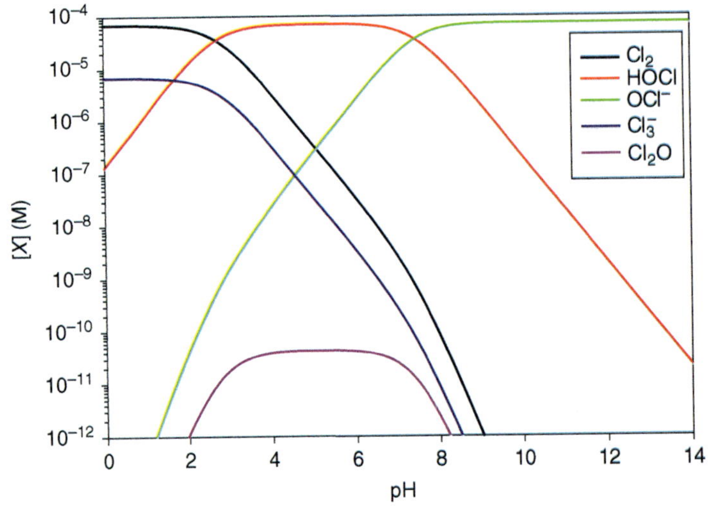


**Figure 10.16** Distribution diagram for free chlorine compounds in a sample that contains $7.0 \times 10^{-5}$ M free chlorine concentration and chloride ion at a concentration of 0.545 M. These conditions are representative of chlorinated sea water. Temperature is assumed to be 25 °C.

For the case of a low chloride ion concentration, as would be typical of tap water (Figure 10.15), free chlorine is dominated by HOCl and $OCl^-$, with the distribution between these two forms being governed by pH (relative to $pK_a \approx 7.47$). At near-neutral pH, the molar concentrations of HOCl and $OCl^-$ are at least five to six orders of magnitude higher than those of the other forms of free chlorine. Despite these large differences, the other forms of free chlorine can play important roles in the behavior of chlorine. This is because molecular chlorine ($Cl_2$) and chlorine monoxide ($Cl_2O$) are better substituting agents than HOCl and $OCl^-$. Rate constants for reactions involving $Cl_2$ and $Cl_2O$ can be many orders of magnitude greater than their corresponding rate constants based on HOCl and $OCl^-$ [40, 41]. Therefore, $Cl_2$ and $Cl_2O$ can contribute significantly to chemical dynamics in reactions that involve free chlorine.

When the chloride ion concentration is high, as it will be in sea water (see Figure 10.16), we observe that HOCl and $OCl^-$ still dominate at most pH values. However, the concentrations of $Cl_2$ and the trichloride ion ($Cl_3^-$) approach the concentrations of HOCl and $Cl_2$ at low pH. Therefore, it is possible for $Cl_2$ and $Cl_3^-$ to contribute significantly to solution chemistry, especially when the chloride ion concentration is high.

Quantification of free chlorine is based on the electrochemical behavior of the contributing species and the concept of electrochemical equivalence. Equations 10.103–10.107 summarize the electrochemical behavior of the compounds that comprise free chlorine, as well as reported values of the SRP for each compound:[m]

$$Cl_{2,aq} + 2e^- \longrightarrow 2Cl^- \qquad SRP = 1.36\ V \tag{10.103}$$

$$HOCl + H^+ + 2e^- \longrightarrow H_2O + Cl^- \qquad SRP = 1.49\ V \tag{10.104}$$

$$OCl^- + H_2O + 2e^- \longrightarrow Cl^- + 2OH^- \qquad SRP = 0.90\ V \tag{10.105}$$

$$2H^+ + Cl_2O + 4e^- \longrightarrow H_2O + 2Cl^- \qquad SRP = n.a. \tag{10.106}$$

$$Cl_3^- + 2e^- \longrightarrow 3Cl^- \qquad SRP = n.a. \tag{10.107}$$

The half-cell reactions that are presented in Equations 10.103–10.107 also form the basis of electrochemical equivalence for these compounds. Specifically, these reactions indicate that all of the forms of free chlorine indicated in these equations have the ability to accept two electrons per molecule, except chlorine monoxide, which has the ability to accept four electrons per molecule. This information can be used to provide a formal definition of free chlorine:

$$\text{Free Chlorine} = [Cl_2] + [HOCl] + [OCl^-] + 2[Cl_2O] + [Cl_3^-] \tag{10.108}$$

Equation 10.108 is a more detailed expression of free chlorine than is normally presented in the literature, in that it accounts for the electrochemical contributions of free chlorine forms that are often present at trace concentrations (i.e. $Cl_2$, $Cl_2O$, and $Cl_3^-$). Common definitions of free chlorine often include only contributions from HOCl and $OCl^-$ because of the dominant roles they play in terms of molar concentration.

The distribution diagrams illustrated in Figures 10.15 and 10.16 indicate that many forms of free chlorine can contribute to the dynamic behavior of chlorinated solutions. However, it is also relevant to consider the effects of solution composition (i.e. pH and chloride ion concentration) on the photochemistry of free chlorine.

Figure 10.17 illustrates the absorption spectra of the compounds that comprise free chlorine. The data for HOCl and $OCl^-$ are identical to those that are presented in Figure 10.8 (note that the vertical axis in Figure 10.17 is presented on a $\log_{10}$ scale) and are included for comparison.

As described previously, HOCl and $OCl^-$ (i.e. the dominant forms of free chlorine on a molar basis for most conditions) are weak to modest absorbers of UV radiation, with $OCl^-$ being a stronger absorber for wavelengths longer than about 255 nm. For UV systems based on LP Hg lamps (i.e. $\lambda = 254$ nm), absorption by the hypochlorite ion is modest $\left(\varepsilon_{254} = 70 \frac{\text{L}}{\text{mole} \cdot \text{cm}}\right)$. It is worth noting that the maximum value of the molar absorption coefficient for the hypochlorite ion is observed at a wavelength of approximately 292 nm $\left(\varepsilon_{254} = 350 \frac{\text{L}}{\text{mole} \cdot \text{cm}}\right)$, which is roughly 10 times higher than the value observed at 254 nm. Because the quantum yield for photolysis of the hypochlorite ion shows only

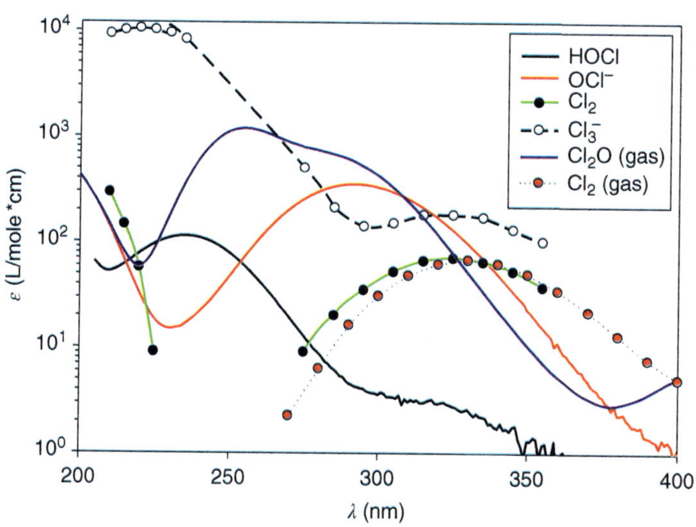


**Figure 10.17** Absorption spectra for compounds that comprise free chlorine. Absorption spectra for HOCl and $OCl^-$ were measured in solutions that contained low concentrations of chloride. The data points that are included for the absorption spectra for aqueous $Cl_2$ and $Cl_3^-$ are reported values from Wang et al. [43]. Cubic splines are drawn through these data to illustrate trends. Note that the reported values of $\varepsilon$ for aqueous $Cl_2$ were zero for wavelengths from 225 to 275 nm and are therefore excluded here. The spectrum for $Cl_2O$ was derived from a gas-phase absorption cross section that was reported by Papanastasiou et al. [44]. For comparison, a measured absorption spectrum for gas-phase $Cl_2$ is also included; these data were converted from an absorption cross section for $Cl_2$ reported by Ganske et al. [45]. A cubic spline was fit to the data to illustrate spectral trends.

slight wavelength-dependence for wavelengths in the range of 255 nm $\lesssim \lambda \lesssim$ 305 nm, it is likely that efficiency in the chlorine/UV process will be promoted by the use of a UV source that emits in the vicinity of the absorbance peak, and at a pH well above the p$K_a$ of HOCl [14]. Among currently available sources of UV radiation, this suggests that UV LEDs or excimer lamps may represent viable alternatives to conventional LP Hg lamps for the chlorine/UV process.

Molecular chlorine ($Cl_2$) is a weak absorber at all UVC, UVB, and UVA wavelengths. Because it also represents a small molar fraction of free chlorine under most practical conditions, it is likely that photolysis of $Cl_2$ is a negligible contributor to the chlorine/UV process. On the other hand, $Cl_3^-$ has a molar absorption coefficient that is roughly two orders of magnitude higher than that of HOCl across almost the entire UV spectrum. Although its low mole fraction will limit the contributions of $Cl_3^-$ to the chlorine/UV process, it is possible that it could be a significant contributor under some conditions (e.g. solutions with high [$Cl^-$]); to date, the contributions of $Cl_3^-$ to the chlorine/UV process remain largely unknown. Lastly, chlorine monoxide ($Cl_2O$) is a stronger absorber than HOCl and $OCl^-$ for radiation in the range 225 nm $\lesssim \lambda \lesssim$ 300 nm. Therefore, $Cl_2O$ may also contribute to the chemistry of the chlorine/UV process, but like $Cl_3^-$ , its contributions are likely to be minor relative to those of HOCl and $OCl^-$ and are largely unknown.

Note that the absorption spectrum for $Cl_2O$ that is presented in Figure 10.17 was measured in the gas phase [44]. An absorption spectrum for gas-phase $Cl_2$ is also included for comparison. The $Cl_2$ spectra that are illustrated in Figure 10.17 indicate that solvation may play a minor role in photon absorption for this molecule.

This gas-phase $Cl_2O$ spectrum is similar to another spectrum for $Cl_2O$ that was measured in solution using carbon tetrachloride as the solvent [46]. No absorption data were found in the literature for aqueous $Cl_2O$, but it is likely that its absorption spectrum in aqueous solution is similar to the spectrum presented in Figure 10.17. As such, it is likely that $Cl_2O$ plays a minor role in the chlorine/UV process, for reasons described above.

The chemistry of the chlorine/UV process is more complex than that of the $H_2O_2$/UV process, largely because the parent molecules that used to initiate the chlorine/UV process undergo photolysis to yield two distinct radical types, which in turn participate in a broad array of reactions that involve transformation to secondary radicals, quenching by dissolved constituents, and other reactions. Table 10.6 lists fundamental reactions and their corresponding rate parameters for the chlorine/UV process [49, 50].

The reactions presented in Table 10.6 are divided into four categories. The first category includes primary photochemical events that initiate the Cl/$UV_{254}$ AOP by yielding reactive intermediates, in the form of free radicals. Values of the molar absorption coefficient and quantum yield are given for these compounds at 254 nm. Values of these parameters may vary for other wavelengths of UV radiation.

The second group includes rapid, reversible reactions that are often assumed to be in equilibrium. Rate constants are indicated for the forward and reverse reactions. Many of these are acid/base reactions, where proton transfer between forms tends to be rapid.

The third group includes reactions that are responsible for radical speciation. These reactions are important in that they result in changes in the forms of free radicals as a function of solution chemistry. Specific aspects of solution chemistry that can affects radical speciation include pH, the concentrations of pH buffer compounds (e.g. carbonate and phosphate), and the presence of halide ions, such as $Cl^-$. Speciation is important among free radicals because they display wide variations of reactivity and selectivity. For example, $HO^{\cdot}$ is an extremely powerful oxidant and tends to be nonselective. By comparison, the chlorine radicals tend to be less powerful oxidants, but are often more selective.

The fourth group includes reactions that result in quenching of free radicals, often by reacting with another free radical. These reactions tend to be rapid and result in the formation of compounds that are less reactive than the radicals that caused them. However, some of the products of these reactions are precursors for the formation of other radicals. For example, reaction 71 from

**Table 10.6** Fundamental reactions and reaction rate parameters for chlorine/$UV_{254}$ system.

| Reaction type | Reaction number | Reaction | Reaction rate parameter[a] | References |
|---|---|---|---|---|
| Primary photochemical event (initiation) | 1 | $HOCl \xrightarrow{h\nu} HO^{\cdot} + Cl^{\cdot}$ | $\varepsilon_{254} = 59 \frac{L}{mole \cdot cm}$ $\Phi_{254} = 1.0 \frac{mole}{einstein}$ | Feng et al. [47] |
| | 2 | $OCl^- \xrightarrow{h\nu} O^{\cdot -} + Cl^{\cdot}$ | $\varepsilon_{254} = 66 \frac{L}{mole \cdot cm}$ $\Phi_{254} = 0.9 \frac{mole}{einstein}$ | |
| | 3 | $H_2O_2 \xrightarrow{h\nu} 2HO^{\cdot}$ | $\varepsilon_{254} = 19.6 \frac{L}{mole \cdot cm}$ $\Phi_{254} = 0.5 \frac{mole}{einstein}$ | Baxendale and Wilson [24] |
| | 4 | $HO_2^- \xrightarrow{h\nu} O^{\cdot -} + Cl^{\cdot}$ | $\varepsilon_{254} = 229 \frac{L}{mole \cdot cm}$ $\Phi_{254} = 0.5 \frac{mole}{einstein}$ | |
| Rapid reversible reactions (equilibria) | 5 | $H^+ + OCl^- \rightarrow HOCl$ | $5.00 \times 10^{10}$ $M^{-1}$ $s^{-1}$ | Fang et al. [48] Guo et al. [49] |
| | 6 | $HOCl \rightarrow H^+ + OCl^-$ | 1600 $s^{-1}$ | |
| | 7 | $H^+ + HO_2^- \rightarrow H_2O_2$ | $5.00 \times 10^{10}$ $M^{-1}$ $s^{-1}$ | |
| | 8 | $H_2O_2 \rightarrow H^+ + HO_2^-$ | 0.13 $s^{-1}$ | |
| | 9 | $H^+ + OH^- \rightarrow H_2O$ | $1.00 \times 10^{11}$ $M^{-1}$ $s^{-1}$ | |
| | 10 | $H_2O \rightarrow H^+ + OH^-$ | $1.00 \times 10^{-3}$ $s^{-1}$ | |
| | 11 | $H^+ + O_2^{\cdot -} \rightarrow HO_2^{\cdot}$ | $5.00 \times 10^{10}$ $M^{-1}$ $s^{-1}$ | |
| | 12 | $HO_2^{\cdot} \rightarrow H^+ + O_2^{\cdot -}$ | $7.00 \times 10^5$ $s^{-1}$ | |
| | 13 | $H_2PO_4^- + H^+ \rightarrow H_3PO_4$ | $5.00 \times 10^{10}$ $M^{-1}$ $s^{-1}$ | |
| | 14 | $H_3PO_4 \rightarrow H_2PO_4^- + H^+$ | $3.87 \times 10^8$ $s^{-1}$ | |
| | 15 | $HPO_4^{2-} + H^+ \rightarrow H_2PO_4^-$ | $5.00 \times 10^{10}$ $M^{-1}$ $s^{-1}$ | |
| | 16 | $H_2PO_4^- \rightarrow HPO_4^{2-} + H^+$ | $3.15 \times 10^3$ $s^{-1}$ | |
| | 17 | $PO_4^{3-} + H^+ \rightarrow HPO_4^{2-}$ | $5.00 \times 10^{10}$ $M^{-1}$ $s^{-1}$ | |
| | 18 | $HPO_4^{2-} \rightarrow PO_4^{3-} + H^+$ | $2.50 \times 10^{-2}$ $s^{-1}$ | |
| | 19 | $HCO_3^- + H^+ \rightarrow H_2CO_3$ | $5.00 \times 10^{10}$ $M^{-1}$ $s^{-1}$ | |
| | 20 | $H_2CO_3 \rightarrow HCO_3^- + H^+$ | $5.00 \times 10^5$ $s^{-1}$ | |
| | 21 | $CO_3^{2-} + H^+ \rightarrow HCO_3^-$ | $5.00 \times 10^{10}$ $M^{-1}$ $s^{-1}$ | |

*(Continued)*

**Table 10.6** (Continued)

| Reaction type | Reaction number | Reaction | Reaction rate parameter[a] | References |
|---|---|---|---|---|
| | 22 | $HCO_3^- \rightarrow CO_3^{2-} + H^+$ | $2.50\ s^{-1}$ | |
| | 23 | $Cl_2 + Cl^- \rightarrow Cl_3^-$ | $2.00 \times 10^4\ M^{-1}\ s^{-1}$ | |
| | 24 | $Cl_3^- \rightarrow Cl_2 + Cl^-$ | $1.10 \times 10^5\ s^{-1}$ | |
| Radical speciation | 25 | $HO^{\cdot} + HPO_4^{2-} \rightarrow HPO_4^{\cdot-} + OH^-$ | $1.50 \times 10^5\ M^{-1}\ s^{-1}$ | Fang et al. [48] Guo et al. [49] |
| | 26 | $HO^{\cdot} + H_2PO_4^- \rightarrow HPO_4^{\cdot-} + H_2O$ | $2.00 \times 10^4\ M^{-1}\ s^{-1}$ | |
| | 27 | $H_2O_2 + HPO_4^{\cdot-} \rightarrow H_2PO_4^- + HO_2^{\cdot}$ | $2.70 \times 10^7\ M^{-1}\ s^{-1}$ | |
| | 28 | $HO^{\cdot} + H_2O_2 \rightarrow HO_2^{\cdot} + H_2O$ | $2.70 \times 10^7\ M^{-1}\ s^{-1}$ | |
| | 29 | $HO^{\cdot} + HO_2^- \rightarrow HO_2^{\cdot} + OH^-$ | $7.50 \times 10^9\ M^{-1}\ s^{-1}$ | |
| | 30 | $HO_2^{\cdot} + H_2O_2 \rightarrow O_2 + HO^{\cdot} + H_2O$ | $3.00\ M^{-1}\ s^{-1}$ | |
| | 31 | $O_2^{\cdot-} + H_2O_2 \rightarrow O_2 + HO^{\cdot} + OH^-$ | $0.130\ M^{-1}\ s^{-1}$ | |
| | 32 | $O^{\cdot-} + H_2O \rightarrow HO^{\cdot} + OH^-$ | $1.80 \times 10^6\ M^{-1}\ s^{-1}$ | |
| | 33 | $HO^{\cdot} + CO_3^{2-} \rightarrow CO_3^{\cdot-} + OH^-$ | $3.90 \times 10^8\ M^{-1}\ s^{-1}$ | |
| | 34 | $HO^{\cdot} + HCO_3^- \rightarrow CO_3^{\cdot-} + H_2O$ | $8.50 \times 10^6\ M^{-1}\ s^{-1}$ | |
| | 35 | $HO^{\cdot} + H_2CO_3 \rightarrow CO_3^{\cdot-} + H_3O^+$ | $1.00 \times 10^6\ M^{-1}\ s^{-1}$ | |
| | 36 | $H_2O_2 + CO_3^{\cdot-} \rightarrow HCO_3^- + HO_2^{\cdot}$ | $4.30 \times 10^5\ M^{-1}\ s^{-1}$ | |
| | 37 | $HO^{\cdot} + Cl^- \rightarrow ClOH^{\cdot-}$ | $4.30 \times 10^9\ M^{-1}\ s^{-1}$ | |
| | 38 | $ClOH^{\cdot-} \rightarrow HO^{\cdot} + Cl^-$ | $6.10 \times 10^9\ s^{-1}$ | |
| | 39 | $ClOH^{\cdot-} + H^+ \rightarrow Cl^{\cdot} + OH^-$ | $2.10 \times 10^{10}\ M^{-1}\ s^{-1}$ | |
| | 40 | $ClOH^{\cdot-} + Cl^- \rightarrow Cl_2^{\cdot-} + OH^-$ | $1.00 \times 10^5\ M^{-1}\ s^{-1}$ | |
| | 41 | $Cl^{\cdot} + H_2O \rightarrow ClOH^{\cdot-} + H^+$ | $4.50 \times 10^3\ M^{-1}\ s^{-1}$ | |
| | 42 | $Cl^{\cdot} + OH^- \rightarrow ClOH^{\cdot-}$ | $1.80 \times 10^{10}\ M^{-1}\ s^{-1}$ | |
| | 43 | $Cl^{\cdot} + H_2O_2 \rightarrow HO_2^{\cdot} + Cl^- + H^+$ | $2.00 \times 10^9\ M^{-1}\ s^{-1}$ | |
| | 44 | $Cl^{\cdot} + Cl^- \rightarrow Cl_2^{\cdot-}$ | $8.50 \times 10^9\ M^{-1}\ s^{-1}$ | |
| | 45 | $Cl^{\cdot} + HOCl \rightarrow ClO^{\cdot} + Cl^- + H^+$ | $3.00 \times 10^9\ M^{-1}\ s^{-1}$ | |
| | 46 | $Cl^{\cdot} + OCl^- \rightarrow ClO^{\cdot} + Cl^-$ | $8.30 \times 10^9\ M^{-1}\ s^{-1}$ | |
| | 47 | $Cl_2^{\cdot-} \rightarrow Cl^{\cdot} + Cl^-$ | $6.00 \times 10^4\ s^{-1}$ | |

**Table 10.6** (Continued)

| Reaction type | Reaction number | Reaction | Reaction rate parameter[a] | References |
|---|---|---|---|---|
| | 48 | $Cl_2^{\cdot-} + H_2O_2 \rightarrow HO_2^{\cdot} + 2Cl^- + H^+$ | $1.40 \times 10^5$ $M^{-1}$ $s^{-1}$ | |
| | 49 | $Cl_2^{\cdot-} + OH^- \rightarrow Cl^- + ClOH^{\cdot-}$ | $4.50 \times 10^7$ $M^{-1}$ $s^{-1}$ | |
| | 50 | $HClOH \rightarrow ClOH^{\cdot-} + H^+$ | $1.00 \times 10^8$ $M^{-1}$ $s^{-1}$ | |
| | 51 | $HClOH \rightarrow Cl^{\cdot} + H_2O$ | $1.00 \times 10^2$ $M^{-1}$ $s^{-1}$ | |
| | 52 | $HClOH + Cl^- \rightarrow Cl_2^{\cdot-} + H_2O$ | $5.00 \times 10^9$ $M^{-1}$ $s^{-1}$ | |
| | 53 | $Cl_3^- + HO_2^{\cdot} \rightarrow Cl_2^{\cdot-} + HCl + O_2$ | $1.00 \times 10^9$ $M^{-1}$ $s^{-1}$ | |
| | 54 | $Cl_3^- + O_2^{\cdot-} \rightarrow Cl_2^{\cdot-} + Cl^- + O_2$ | $3.80 \times 10^9$ $M^{-1}$ $s^{-1}$ | |
| | 55 | $Cl_2 + O_2^{\cdot-} \rightarrow O_2 + Cl_2^{\cdot-}$ | $1.00 \times 10^9$ $M^{-1}$ $s^{-1}$ | |
| | 56 | $Cl_2 + HO_2^{\cdot} \rightarrow H^+ + O_2 + Cl_2^{\cdot-}$ | $1.00 \times 10^9$ $M^{-1}$ $s^{-1}$ | |
| | 57 | $HOCl + HO^{\cdot} \rightarrow ClO^{\cdot} + H_2O$ | $2.00 \times 10^9$ $M^{-1}$ $s^{-1}$ | |
| | 58 | $HOCl + O_2^{\cdot-} \rightarrow Cl^{\cdot} + OH^- + O_2$ | $7.50 \times 10^6$ $M^{-1}$ $s^{-1}$ | |
| | 59 | $HOCl + HO_2^{\cdot} \rightarrow Cl^{\cdot} + H_2O + O_2$ | $7.50 \times 10^6$ $M^{-1}$ $s^{-1}$ | |
| | 60 | $OCl^- + HO^{\cdot} \rightarrow ClO^{\cdot} + OH^-$ | $8.80 \times 10^9$ $M^{-1}$ $s^{-1}$ | |
| | 61 | $OCl^- + O_2^{\cdot-} + H_2O \rightarrow Cl^{\cdot} + 2OH^- + O_2$ | $2.00 \times 10^8$ $M^{-1}$ $s^{-1}$ | |
| | 62 | $OCl^- + CO_3^{\cdot-} \rightarrow ClO^{\cdot} + CO_3^{2-}$ | $5.70 \times 10^5$ $M^{-1}$ $s^{-1}$ | |
| | 63 | $Cl^{\cdot} + CO_3^{2-} \rightarrow Cl^- + CO_3^{\cdot-}$ | $5.00 \times 10^8$ $M^{-1}$ $s^{-1}$ | |
| | 64 | $Cl^{\cdot} + HCO_3^- \rightarrow Cl^- + CO_3^{\cdot-} + H^+$ | $2.20 \times 10^8$ $M^{-1}$ $s^{-1}$ | |
| | 65 | $Cl_2^{\cdot-} + CO_3^{2-} \rightarrow 2Cl^- + CO_3^{\cdot-}$ | $1.60 \times 10^8$ $M^{-1}$ $s^{-1}$ | |
| | 66 | $ClO^{\cdot} + CO_3^{2-} \rightarrow OCl^- + CO_3^{\cdot-}$ | 600 $M^{-1}$ $s^{-1}$ | |
| | 67 | $Cl_2^{\cdot-} + HCO_3^- \rightarrow 2Cl^- + CO_3^{\cdot-} + H^+$ | $8.00 \times 10^7$ $M^{-1}$ $s^{-1}$ | |
| | 68 | $HO^{\cdot} + ClO_2^- \rightarrow ClO_2^{\cdot} + OH^-$ | $6.30 \times 10^9$ $M^{-1}$ $s^{-1}$ | |
| | 69 | $Cl_2^{\cdot-} + ClO_2^- \rightarrow ClO_2^{\cdot} + 2Cl^-$ | $1.30 \times 10^8$ $M^{-1}$ $s^{-1}$ | |
| | 70 | $ClO^{\cdot} + ClO_2^- \rightarrow OCl^- + ClO_2^{\cdot}$ | $9.40 \times 10^8$ $M^{-1}$ $s^{-1}$ | |

(*Continued*)

**Table 10.6** (Continued)

| Reaction type | Reaction number | Reaction | Reaction rate parameter[a] | References |
|---|---|---|---|---|
| Radical combination (quenching) | 71 | $HO^{\cdot} + HO^{\cdot} \rightarrow H_2O_2$ | $5.50 \times 10^9\ M^{-1}\ s^{-1}$ | Fang et al. [48] Guo et al. [49] |
| | 72 | $HO^{\cdot} + HO_2^{\cdot} \rightarrow O_2 + H_2O$ | $7.10 \times 10^9\ M^{-1}\ s^{-1}$ | |
| | 73 | $HO^{\cdot} + O_2^{\cdot} \rightarrow O_2 + OH^-$ | $1.00 \times 10^{10}\ M^{-1}\ s^{-1}$ | |
| | 74 | $HO_2^{\cdot} + HO_2^{\cdot} \rightarrow H_2O_2 + O_2$ | $8.30 \times 10^9\ M^{-1}\ s^{-1}$ | |
| | 75 | $HO_2^{\cdot} + O_2^{\cdot-} \rightarrow HO_2^- + O_2$ | $9.70 \times 10^7\ M^{-1}\ s^{-1}$ | |
| | 76 | $HO_2^- + CO_3^{\cdot-} \rightarrow HCO_3^- + O_2$ | $3.00 \times 10^7\ M^{-1}\ s^{-1}$ | |
| | 77 | $HO^{\cdot} + CO_3^{\cdot-} \rightarrow X$ | $3.00 \times 10^9\ M^{-1}\ s^{-1}$ | |
| | 78 | $O_2^{\cdot-} + CO_3^{\cdot-} \rightarrow CO_3^{2-} + O_2$ | $6.00 \times 10^8\ M^{-1}\ s^{-1}$ | |
| | 79 | $CO_3^{\cdot-} + CO_3^{\cdot-} \rightarrow X$ | $3.00 \times 10^7\ M^{-1}\ s^{-1}$ | |
| | 80 | $Cl^{\cdot} + Cl^{\cdot} \rightarrow Cl_2$ | $8.80 \times 10^7\ M^{-1}\ s^{-1}$ | |
| | 81 | $Cl_2^{\cdot-} + Cl_2^{\cdot-} \rightarrow Cl_2 + 2Cl^-$ | $8.30 \times 10^8\ M^{-1}\ s^{-1}$ | |
| | 82 | $Cl_2^{\cdot-} + Cl^{\cdot} \rightarrow Cl_2 + Cl^-$ | $2.10 \times 10^9\ M^{-1}\ s^{-1}$ | |
| | 83 | $Cl_2^{\cdot-} + HO_2^{\cdot} \rightarrow O_2 + 2Cl^- + H^+$ | $3.00 \times 10^9\ M^{-1}\ s^{-1}$ | |
| | 84 | $Cl_2^{\cdot-} + O_2^{\cdot-} \rightarrow O_2 + 2Cl^-$ | $1.00 \times 10^9\ M^{-1}\ s^{-1}$ | |
| | 85 | $Cl_2^{\cdot-} + H_2O \rightarrow Cl^- + HClOH$ | $23.4\ M^{-1}\ s^{-1}$ | |
| | 86 | $ClO^{\cdot} + ClO^{\cdot} \rightarrow Cl_2O_2$ | $2.50 \times 10^9\ M^{-1}\ s^{-1}$ | |
| | 87 | $HO^{\cdot} + ClO^{\cdot} \rightarrow ClO_2^- + H^+$ | $1.00 \times 10^9\ M^{-1}\ s^{-1}$ | |
| | 88 | $HO^{\cdot} + ClO_2^{\cdot} \rightarrow ClO_3^- + H^+$ | $4.00 \times 10^9\ M^{-1}\ s^{-1}$ | |

[a] Reaction parameters listed in this column are rate constants for elementary reactions, except as indicated.

Table 10.6 involves self-reaction of $HO^{\cdot}$ to form $H_2O_2$, which can then go on to photolyze to re-form the $HO^{\cdot}$ radicals (reaction 3 from Table 10.6).

The reactions described in Table 10.6 are elementary. Therefore, the kinetic expressions and reaction rate parameters can be incorporated into a model to simulate the behavior of one or more target compounds in a batch reactor. Application of this fundamental approach can be cumbersome, but offers the important benefit of providing detailed, accurate predictions of process behavior over a wide range of operating conditions and solution chemistries.

An illustration of this approach was presented by Fang et al. for prediction of the degradation of benzoic acid (BA) by the chlorine/UV process [48]. A model was developed to account for BA degradation by $HO^{\cdot}$, $Cl^{\cdot}$, $O^{\cdot-}$, and $Cl_2^{\cdot-}$. The contributions of direct photolysis and reactions with free chlorine (HOCl and $OCl^-$) were shown to be minor contributors to BA degradation, and were neglected. Similarly, reactions involving other secondary radicals (e.g. $ClOH^{\cdot}$ and $ClO^{\cdot}$) were assumed to provide negligible contributions to BA decay, and were therefore not included in the model. Their model was used to simulate decay of BA in a well-mixed, batch reactor subjected to irradiation under a pseudo-collimated $UV_{254}$ source.

Using the assumptions listed above, an overall rate expression for BA decay was defined as follows:

$$\frac{d[BA]}{dt} = -\left(k_{HO^\cdot}[HO^\cdot]_{SS} + k_{Cl^\cdot}[Cl^\cdot]_{SS} + k_{Cl_2^{\cdot-}}[Cl_2^{\cdot-}]_{SS} + k_{O^{\cdot-}}[O^{\cdot-}]_{SS}\right) \cdot [BA] \tag{10.109}$$

where

$k_i$ = second-order rate constant for reaction between radical $i$ and BA [=] $\frac{L}{\text{mole} \cdot s}$

$[i]_{SS}$ = steady-state concentration of radical $i$ [=] $\frac{\text{mole}}{L}$.

Collectively, the terms in Equation 10.104 indicate that the rate of BA decay in this system is pseudo-first-order. Therefore, Equation 10.109 can be simplified to:

$$\frac{d[BA]}{dt} = -k_{O,BA} \cdot [BA]$$

$$k_{O,BA} = k_{HO^\cdot}[HO^\cdot]_{SS} + k_{Cl^\cdot}[Cl^\cdot]_{SS} + k_{Cl_2^{\cdot-}}[Cl_2^{\cdot-}]_{SS} + k_{O^{\cdot-}}[O^{\cdot-}]_{SS} \tag{10.110}$$

where

$k_{O,\,BA}$ = overall pseudo-first-order rate constant for decay of BA [=] $s^{-1}$.

The radicals that participate in this process are formed rapidly and decay rapidly. Using the hydroxyl radical as an example, its dynamic behavior is defined by a balance between reactions that yield $HO^\cdot$ as a product and those that consume it:

$$\frac{d[HO^\cdot]}{dt} = r_{formation} - r_{consumption} \tag{10.111}$$

For the work conducted by Fang et al., a well-mixed, batch reactor was subjected to pseudo-collimated $UV_{254}$ irradiation using a device illustrated in Figure 10.18. This device was developed

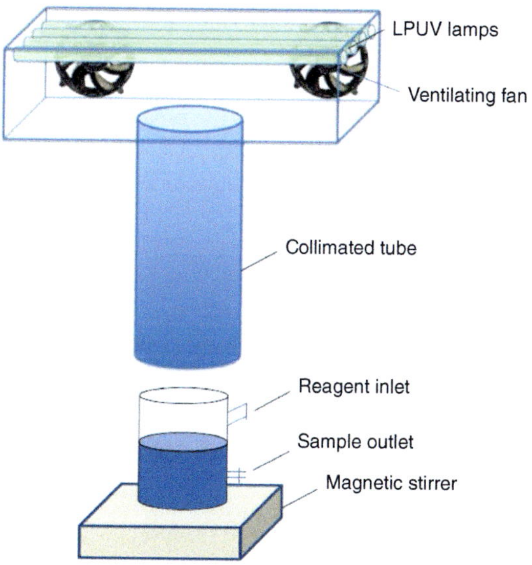


**Figure 10.18** Schematic illustration of well-mixed, batch reactor and collimated beam used for analysis of benzoic acid decay in the chlorine/UV process by Fang et al. [48]/with permission of American Chemical Society.

to isolate the effects of reactions (only) on the dynamic behavior of BA in solution. In other words, transport behavior did not play a role in this process. However, it is important to recognize that mixing behavior within the batch reactor was assumed to be ideal; the well-mixed assumption is not likely to apply rigorously to such a system, at least not with respect to photolysis and radical-mediated reactions. However, for modest extents of reaction, the error introduced by the well-mixed assumption is likely to be small.

For the system above, the rates of hydroxyl radical formation and consumption can be described as follows:

$$r_{\text{formation}} = \Phi_{\text{HOCl}} \cdot \frac{E_0}{\forall} \cdot \ell \cdot f_{\text{HOCl}} \cdot \left(1 - 10^{-A}\right) + k_4[\text{ClOH}^{\cdot-}] + k_{15}[\text{O}^{\cdot-}][\text{H}_2\text{O}] \tag{10.112}$$

where

$\Phi_{\text{HOCl}}$ = quantum yield for HOCl photolysis
$E_0$ = incident fluence rate
$\forall$ = liquid volume in reactor
$\ell$ = effective path length for reactor (see Zepp [12])
$f_{\text{HOCl}}$ = fraction of incident radiation absorbed by HOCl
= $\dfrac{\varepsilon_{\text{HOCl}} \cdot [\text{HOCl}] \cdot \text{L}}{A}$
$A$ = absorbance of reacting solution at wavelength of interest
= $\sum_i \varepsilon_i \cdot [i] \cdot \ell$
$i$ = any compound in solution that absorbs incident radiation at the wavelength of interest
$k_4$ = rate constant for autodecay of $\text{ClOH}^{\cdot-}$
$[\text{ClOH}^{\cdot-}]$ = concentration of $\text{ClOH}^{\cdot-}$
$k_{15}$ = second-order rate constant for reaction between $\text{O}^{\cdot-}$ and water
$[\text{O}^{\cdot-}]$ = concentration of $\text{O}^{\cdot-}$.

The first term in Equation 10.112 describes the rate of $\text{HO}^{\cdot}$ formation by HOCl photolysis. The remaining terms describe $\text{HO}^{\cdot}$ formation by autodecay of other radicals in solution.

The rate of $\text{HO}^{\cdot}$ consumption will be described by the sum of all (bimolecular) reactions that involve $\text{HO}^{\cdot}$ as a reactant:

$$r_{\text{consumption}} = \sum_j k_j \cdot [j] \cdot [\text{HO}^{\cdot}] \tag{10.113}$$

where

$j$ = any compound that reacts with $\text{HO}^{\cdot}$
$k_j$ = second-order rate constant for reaction between $j$ and $\text{HO}^{\cdot}$
$[\text{HO}^{\cdot}]$ = concentration of $\text{HO}^{\cdot}$.

At steady state, the rates of $\text{HO}^{\cdot}$ formation and decay will be equal. Therefore, the time-derivative in Equation 10.111 becomes zero and the value of $[\text{HO}^{\cdot}]_{\text{SS}}$ can be estimated by algebra:

$$[\text{HO}^{\cdot}]_{\text{SS}} = \frac{\Phi_{\text{HOCl}} \cdot \frac{E_0}{\forall} \cdot \ell \cdot f_{\text{HOCl}} \cdot \left(1 - 10^{-A}\right) + k_4[\text{ClOH}^{\cdot-}]_{\text{SS}} + k_{15}[\text{O}^{\cdot-}]_{\text{SS}}[\text{H}_2\text{O}]}{\sum_j k_j \cdot [j]} \tag{10.114}$$

Similar approaches are used to estimate the steady-state concentrations of other free radicals in solution.

Equations 10.109 and 10.110 can be used to describe the dynamic behavior of BA as a function of exposure to four reactive free radicals: $\text{HO}^{\cdot}$, $\text{Cl}^{\cdot}$, $\text{Cl}_2^{\cdot-}$, and $\text{O}^{\cdot-}$. But since the model explicitly

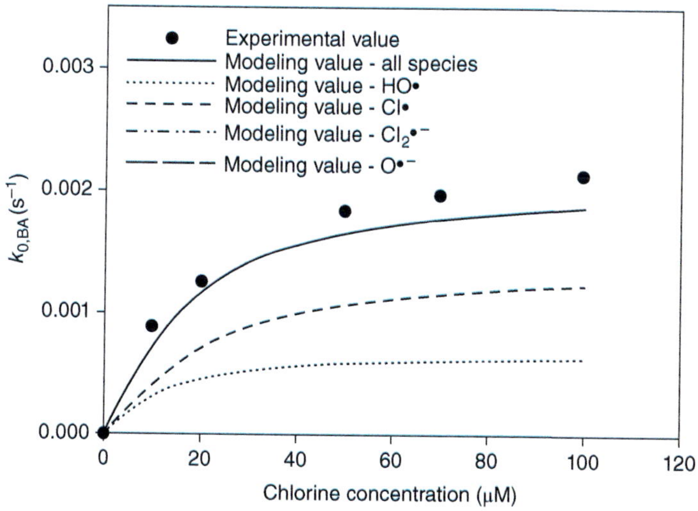


**Figure 10.19** Overall and individual pseudo-first-order rate constants (*k*) for BA degradation by $HO^{\cdot}$, $Cl^{\cdot}$, $Cl_2^{\cdot-}$, and $O^{\cdot-}$ under a range of free chlorine concentrations in the chlorine/UV process. Initial reaction conditions: pH 6.0, $[BA]_0$ = 5 μM [48]. *Source:* Fang et. al 2014/with permission of American Chemical Society.

includes elementary reactions that are responsible for the formation and decay of these free radicals, it is able to accurately simulate BA dynamics over a wide range of conditions. As an example, Figure 10.19 illustrates the effect of initial free chlorine concentration on $k_{O,\ BA}$ [48]. For all conditions of this experiment, degradation of BA was dominated by reactions between BA and either $HO^{\cdot}$ or $Cl^{\cdot}$; the contributions of BA reactions with $Cl_2^{\cdot-}$ and $O^{\cdot-}$ were negligible, as were direct photolysis and reactions with free chlorine (not shown in Figure 10.19). For the entire range of operating conditions used in this experiment, decay by BA reaction with $Cl^{\cdot}$ was roughly twice as fast as decay by reaction between BA and $HO^{\cdot}$. The model consistently underestimated the rate of BA decay by about 10% over the entire range of operating conditions used in this experiment. Some of this error was probably due to a lack of inclusion of the effects of reactions between BA and secondary radicals, as well as direct photolysis and direct reactions between BA and free chlorine.

Figure 10.20 illustrates measured and simulated behavior of BA over a range of carbonate alkalinity concentrations [48]. This graph is relevant in that it accounts for the effects of $HCO_3^-$ and $CO_3^{2-}$ as radical scavengers. Both forms of carbonate alkalinity will react with $HO^{\cdot}$ and $Cl^{\cdot}$, but the effect of this scavenging mechanism appears to be more important for $Cl^{\cdot}$ than for $HO^{\cdot}$ for the conditions of this experiment.

Figure 10.21 provides a third example of the ability of the model of Fang et al. [48] to account for dynamic behavior of BA in the chlorine/UV process. The chloride ion is able to affect speciation of reactive free radicals via the reversible reaction:

$$Cl^{\cdot} + Cl^- \rightleftharpoons Cl_2^{\cdot-} \tag{10.115}$$

Therefore, we expect that the steady-state concentration of $Cl_2^{\cdot-}$ will increase with $[Cl^-]$; consequently, the rate of BA decay by reaction with $Cl_2^{\cdot-}$ will increase with $[Cl^-]$, as indicated in Figure 10.21. However, even under the condition of $[Cl^-] = 20$ mM (the highest chloride ion concentration used in this experiment), overall degradation was still dominated by reactions that involve $HO^{\cdot}$ and $Cl^{\cdot}$, despite the fact that the behavior of these two radicals is essentially unaffected by the concentration of $Cl^-$.

A similar, but somewhat expanded modeling approach was developed by Guo et al. to simulate the responses of a wide range of PPCPs to the chlorine/UV process [49]. Their model included a

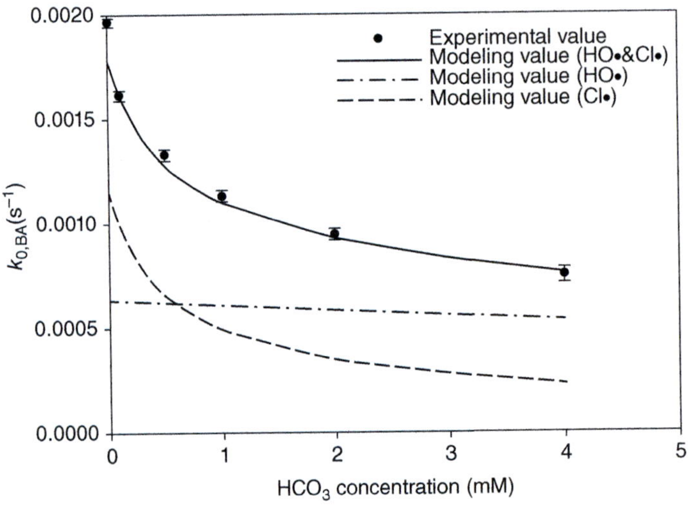


**Figure 10.20** Overall and individual pseudo-first-order rate constants for BA degradation by OH˙ and Cl˙ under different concentrations of bicarbonate in the chlorine/UV process. Initial conditions: pH 6, $[BA]_0 = 5\ \mu M$, free chlorine dosage = 70 μM [48]. *Source:* Fang et. al 2014/with permission of American Chemical Society.

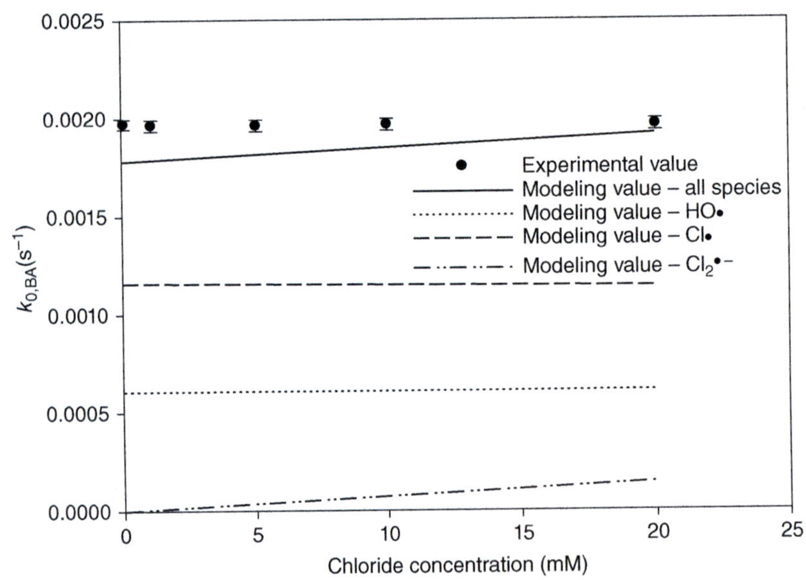


**Figure 10.21** Overall and individual pseudo-first-order rate constants (*k*) for BA degradation by HO˙, Cl˙, $Cl_2^{\cdot-}$, and $O^{\cdot-}$ as a function of chloride ion concentration in the chlorine/UV process. Initial conditions: pH 6.0, $[BA]_0 = 5\ \mu M$, free chlorine = 70 μM.

wider range of contaminants and also accounted for other reaction processes that were ignored by Fang et al. [48]. Specifically, they observed that direct photolysis, bimolecular reactions with free chlorine, and bimolecular reactions with the free radicals HO˙, Cl˙, $Cl_2^{\cdot-}$, and ClO˙ all contributed to PPCP degradation, but the contributions of these reaction processes varied widely across the 27 compounds they included in their investigation. For purposes of this paper, the term "Reactive Chlorine Species" (*RCS*) was used to describe the collective contributions of chlorine-containing free radicals in their experiments. For these experiments, the effects of *RCS* comprised contributions from Cl˙, $Cl_2^{\cdot-}$, and ClO˙.

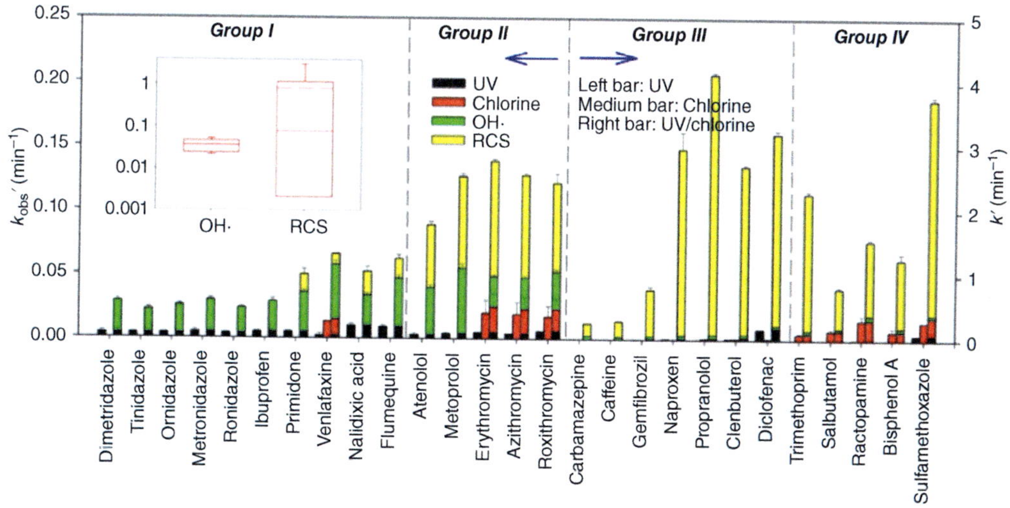


**Figure 10.22** Comparison of the first-order rate constants ($k'_{obs}$) of 27 PPCPs by direct UV photolysis, direct reactions with free chlorine, and radicals generated in the chlorine/UV process. The inset shows ranges of the pseudo-first-order rate constant attributable to $HO^{\cdot}$ and that attributable to all reactive chlorine species (RCS) ($k_{RCS}'$) for all 27 PPCPs. Initial conditions: pH = 7, $[PPCP]_0$ = 1 μg $L^{-1}$, $[chlorine]_0$ = 10 μM, average UV fluence rate = 0.78 mW $cm^{-2}$.

Figure 10.22 illustrates the results of their modeling effort to simulate the contributions of each of the reaction processes described above. The inset for Figure 10.22 illustrates the range of pseudo-first-order rate constants for reactions involving $HO^{\cdot}$ and RCS. The pseudo-first-order rate constants for reactions between $HO^{\cdot}$ and each of these 27 compounds were quite consistent, while those of RCS spanned roughly three orders of magnitude. This is in agreement with the general observation that $HO^{\cdot}$ tends to be nonselective as an oxidant, while *RCS* are often far more selective.

Figure 10.22 also includes an empirical division of PPCPs among four different groups, depending on how they reacted to the chlorine/UV process in these experiments. Table 10.7 provides a summary of the behavior of these groups. In general, $HO^{\cdot}$ tends to be a nonselective oxidant that tends to react rapidly with most organic compounds. On the other hand, RCS tend to be more selective. $Cl^{\cdot}$ reacts more rapidly than $HO^{\cdot}$ with BA, chlorobenzene, and phenol; $Cl_2^{\cdot-}$ tends to be selective for olefins and aromatic compounds that include hydroxyl, methoxy, or amino functional groups; $ClO^{\cdot}$ reacts rapidly with methoxy-substituted aromatic compounds [49].

Figure 10.23 illustrates pH-dependence for reaction kinetic parameters for each of the 27 PPCP compounds included in the study of Guo et al. [49]. Among compounds in group I, where degradation by reaction with $HO^{\cdot}$ dominated, a consistent pattern of decreasing rate of decay with increasing pH is evident. This pH-dependence is probably attributable to several pH-dependent factors, including the rate of $HO^{\cdot}$ formation, the efficiency of scavenging, and acid/base behavior of the target compounds.

The studies by Fang et al. [48] and Guo et al. [49] illustrate many of the critical reactions that contribute to contaminant decay in the chlorine/UV process. However, there are other compounds that contribute to the behavior of these systems. For example, the carbonate radical ($CO_3^{\cdot-}$) has been shown to be reactive toward some micropollutants, including algal toxins, paroxetine, and sulfachloropyridazine [50].

These studies demonstrate that the dynamic behavior of contaminants in a chlorine/UV process can be accurately simulated by a model that accounts for relevant reaction kinetics. However, it is also important to note that these studies were developed to describe the behavior of the chlorine/UV process in a well-mixed, batch reactor, where dynamic behavior depends only on reaction

**Table 10.7** Compound groupings used by Guo et al. [49] to differentiate PPCP reactivity toward oxidants that are available in the chlorine/UV process. Two-dimensional compound structures are from https://pubchem.ncbi.nlm.nih.gov/.

| Group | Compound | Structure | Rate constant range | Dominant oxidant |
|---|---|---|---|---|
| I | Dimetridazole | 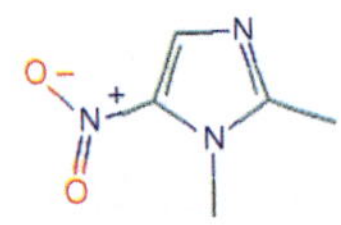 | $\frac{k_{HO^{\cdot}}}{k_{RCS}} > 1$ | $HO^{\cdot}$ |
| | Tinidazole | 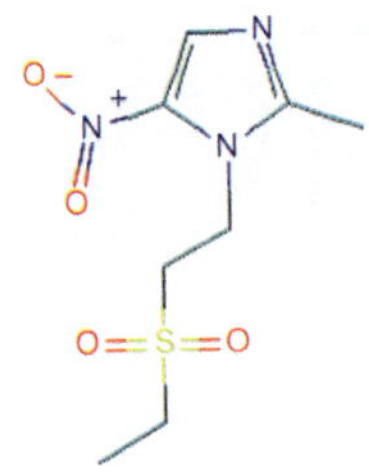 | | |
| | Ornidazole | 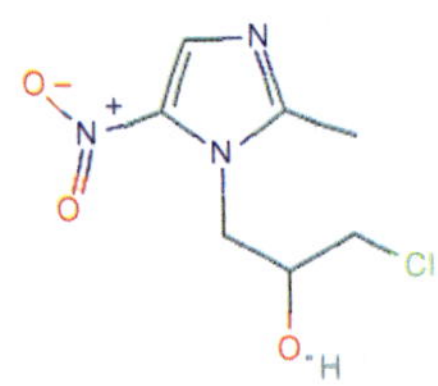 | | |
| | Ronidazole | 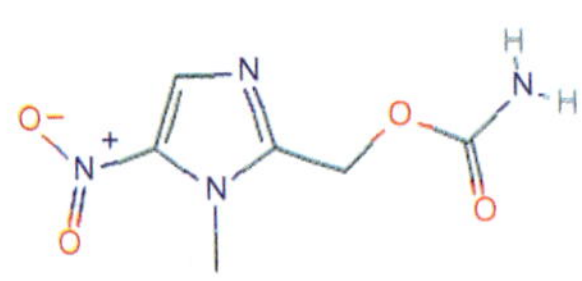 | | |
| | Metronidazole | 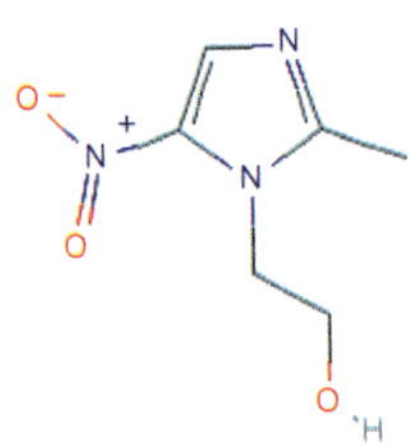 | | |
| | Ibuprofen | 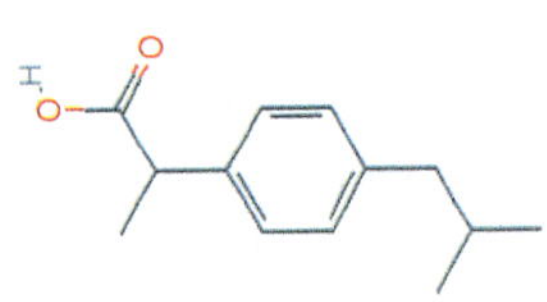 | | |
| | Primidone | 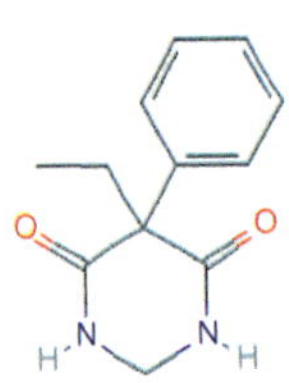 | | |

Table 10.7 (Continued)

| Group | Compound | Structure | Rate constant range | Dominant oxidant |
|---|---|---|---|---|
| | Venlafaxine | 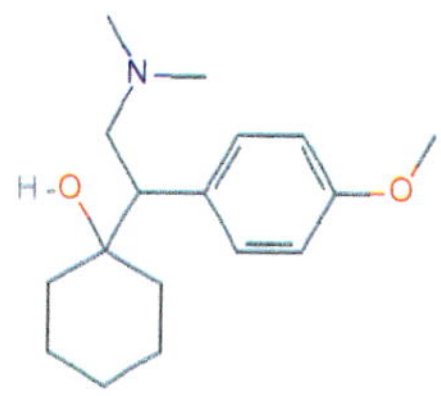 | | |
| | Nalidixic acid | 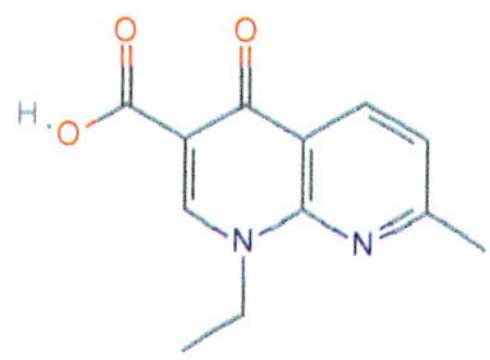 | | |
| | Flumequine | 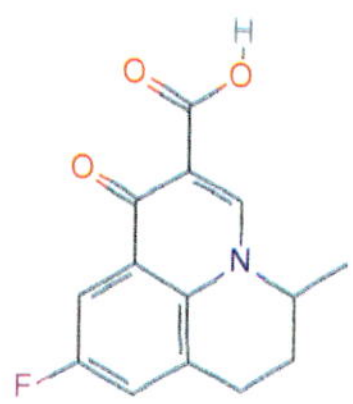 | | |
| II | Atenolol | 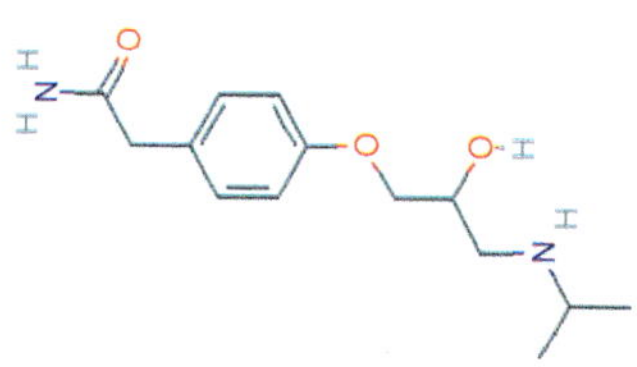 | $0.2 < \frac{k_{HO^\cdot}}{k_{RCS}} < 1$ | $HO^\cdot$ and RCS |
| | Metoprolol | 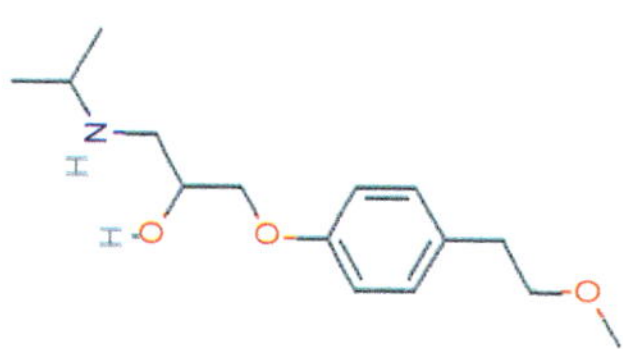 | | |
| | Erythromycin | 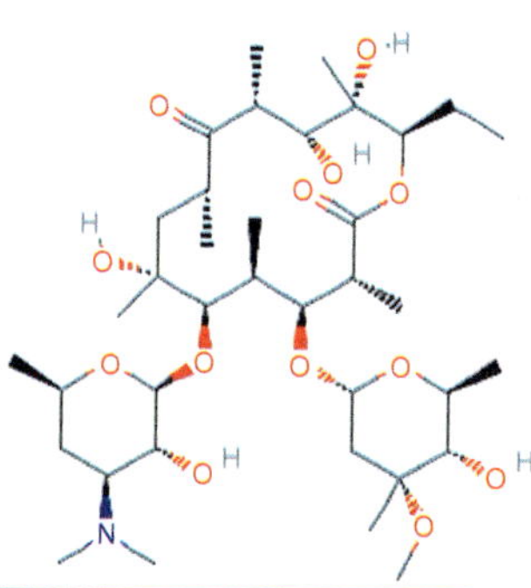 | | |

(Continued)

**Table 10.7** (Continued)

| Group | Compound | Structure | Rate constant range | Dominant oxidant |
|---|---|---|---|---|
| | Azithromycin | 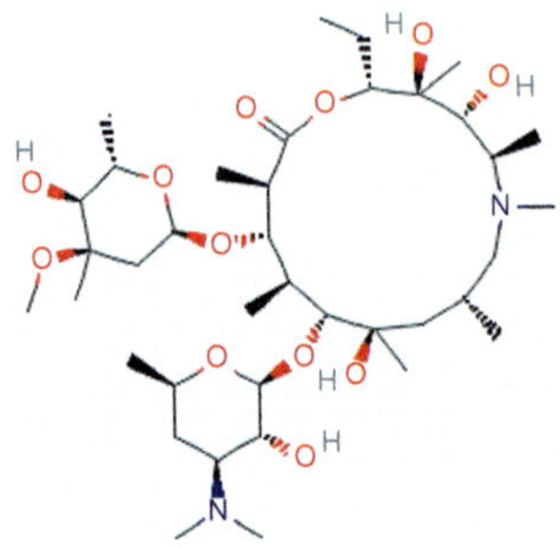 | | |
| | Roxithromycin | 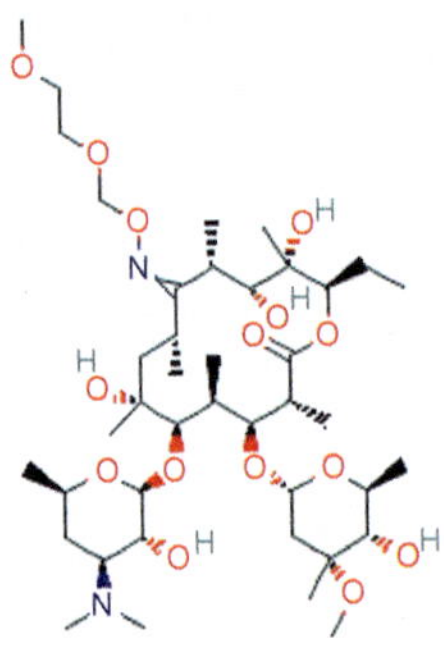 | | |
| III | Carbamazepine |  | $\frac{k_{HO^\cdot}}{k_{RCS}} < 0.05$ | RCS |
| | Caffeine | 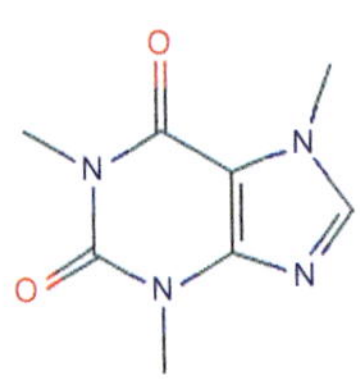 | | |
| | Gemfibrozil | 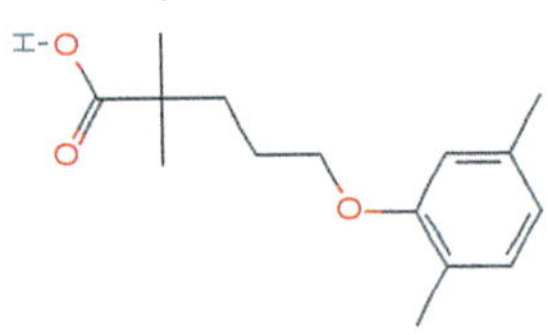 | | |
| | Naproxen | 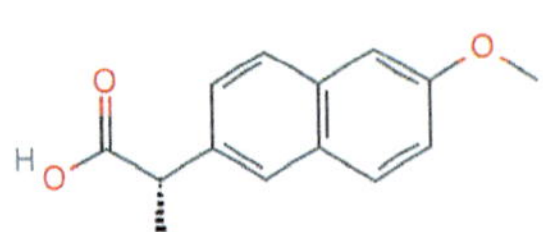 | | |

**Table 10.7** (Continued)

| Group | Compound | Structure | Rate constant range | Dominant oxidant |
|---|---|---|---|---|
| | Propranolol | 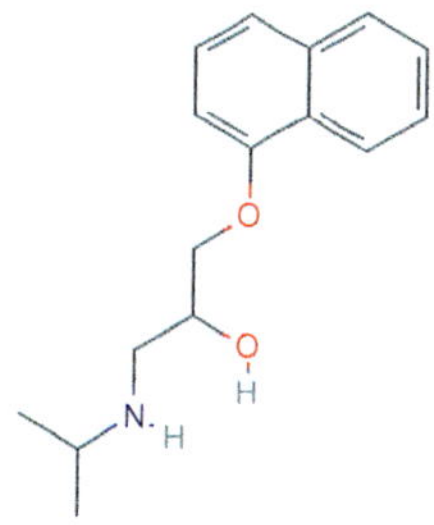 | | |
| | Clenbuterol | 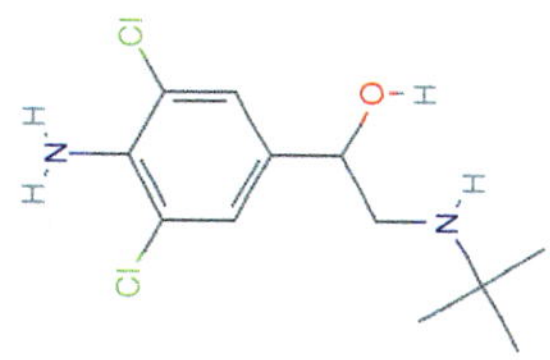 | | |
| | Diclofenac | 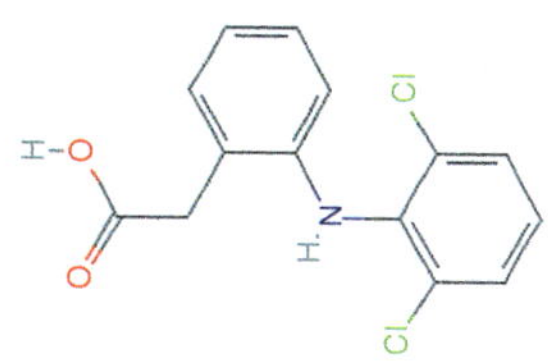 | | |
| IV | Trimethoprim | 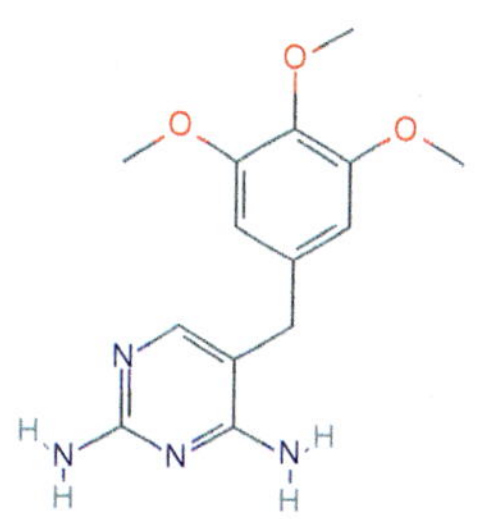 | $\frac{k_{HO^\cdot}}{k_{RCS}} < 0.05$ and $k_{FC} > 0.02$ | RCS and $HOCl/OCl^-$ |
| | Salbutamol | 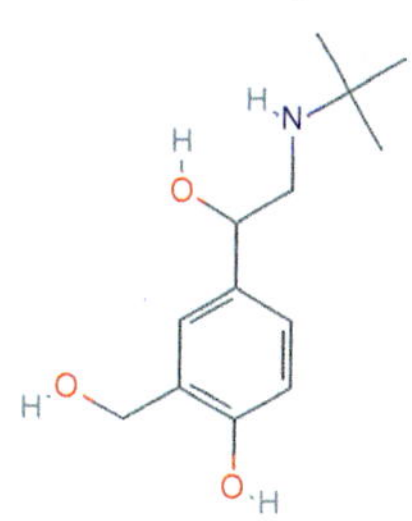 | | |
| | Ractopamine | 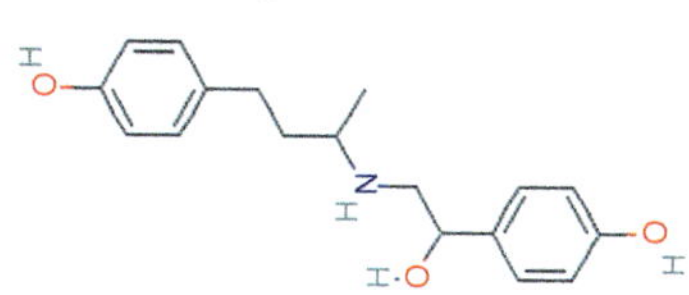 | | |

(*Continued*)

**Table 10.7** (Continued)

| Group | Compound | Structure | Rate constant range | Dominant oxidant |
|---|---|---|---|---|
| | Bisphenol A | 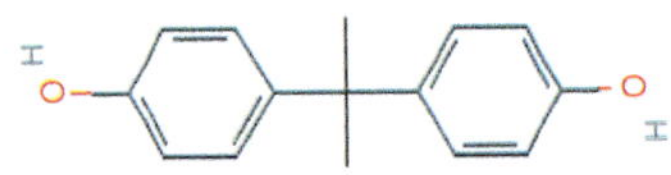 | | |
| | Sulfamethoxazole | 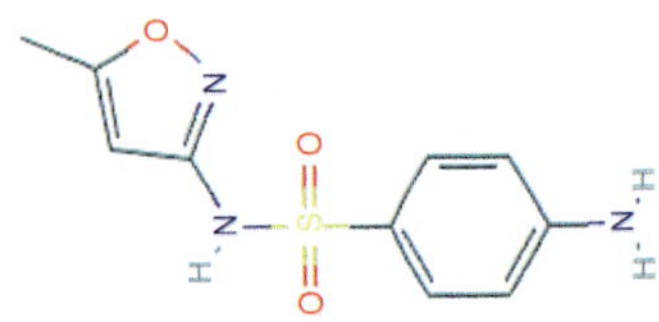 | | |

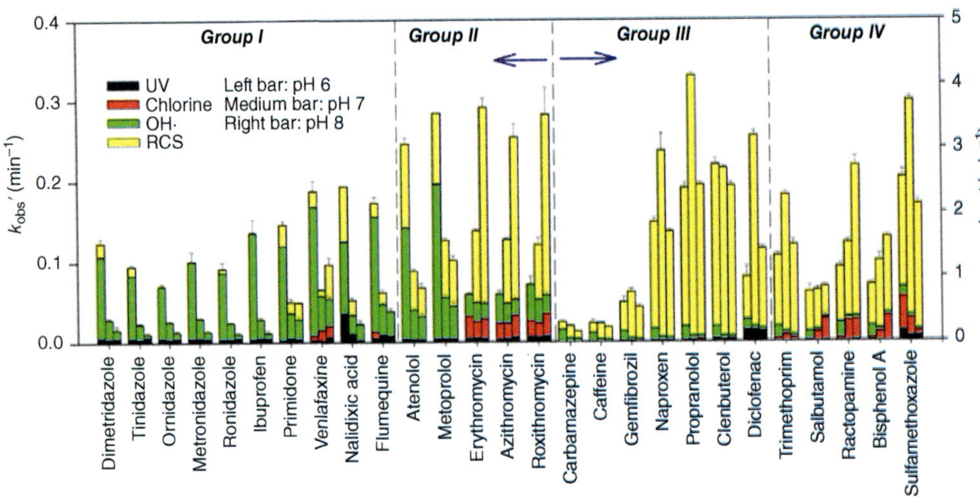


**Figure 10.23** Effect of pH on observed pseudo-first-order rate constants for decay of 27 PPCPs. Also included are pseudo-first-order rate constants for each of the participating reaction mechanisms. Initial conditions: $[\text{PPCP}]_0 = 1\ \mu g\ L^{-1}$, $[\text{chlorine}]_0 = 10\ \mu M$, average UV fluence rate = 0.78 mW cm$^{-2}$. *Source:* Guo et. al 2017/with permission of American Chemical Society.

kinetics. For descriptions of process performance in actual continuous-flow reactor systems, it would be necessary to incorporate model components to account for transport behavior. This can be accomplished by linking the reaction kinetics to a CFD model, as described previously for the $H_2O_2$/UV process.

### 10.10.2 Chloramine/UV Process

The thermal chemistry that governs the formation and decay of inorganic chloramines ($NH_2Cl$, $NHCl_2$, and $NCl_3$) is well understood [51, 52]. A series of reversible substitution (and disproportionation) reactions is responsible for interconversion of ammonia and the inorganic chloramines.

$$NH_3 + HOCl \rightleftharpoons NH_2Cl + H_2O \tag{10.116}$$

$$NH_2Cl + HOCl \rightleftharpoons NHCl_2 + H_2O \tag{10.117}$$

$$NHCl_2 + HOCl \rightleftharpoons NCl_3 + H_2O \tag{10.118}$$

$$NH_2Cl + NH_2Cl \rightleftharpoons NHCl_2 + NH_3 \tag{10.119}$$

Together with a series of oxidation/reduction reactions, these reactions govern the dynamic behavior of the inorganic chloramines.

The chemistry of the inorganic chloramines is relevant in a wide range of water treatment applications. In some potable water applications, ammonia is added to water along with (or prior to) chlorination to promote the formation of the inorganic chloramines as a stable disinfectant residual. In other settings, the use of the chloramines in potable water production is undesirable, or perhaps even forbidden. The presence of even low concentrations of ammonia-N can result in essentially complete transformation of free chlorine to inorganic chloramines; this phenomenon is often observed in chlorination of municipal wastewater effluents, often without being a formal treatment objective. Swimming pools are another common application in which chlorination results in the formation of the inorganic chloramines, despite the fact that swimming pool water contains little or no free ammonia. In these settings, inorganic chloramine formation results from reactions between free chlorine and organic-N compounds, including urea, uric acid, creatinine, and amino acids that are introduced to swimming pools via excretion of human body fluids, mostly urine and sweat [40, 53–55]. These reactions are known to result in the formation of $NCl_3$, which allows subsequent formation of $NHCl_2$ and $NH_2Cl$ by hydrolysis (i.e. the reverse of reactions 10.117–10.118).

Chlorination of ammonia yields one or more N–Cl bonds. These bonds tend to absorb radiation in the UVC portion of the electromagnetic spectrum quite well, as compared with the N–H bond. This behavior is illustrated in Figure 10.24, which illustrates the absorption spectra of $NH_3$ and the inorganic chloramines. $NH_3$ and $NH_4^+$ absorb essentially no radiation across the UVC, UVB, and UVA portions of the spectrum. In contrast, each of the inorganic chloramines demonstrate modest absorption across the UV spectrum. At 254 nm, the molar absorption coefficients for these compounds follow the trend: $\varepsilon_{NCl_3} > \varepsilon_{NHCl_2} > \varepsilon_{NH_2Cl}$, but at other wavelengths the trend is variable.

The N–Cl bond is photolabile; absorption of UV photons leads to homolytic cleavage of the N–Cl bond in the inorganic chloramines. This results in the formation of an amine (*aka* aminyl) radical

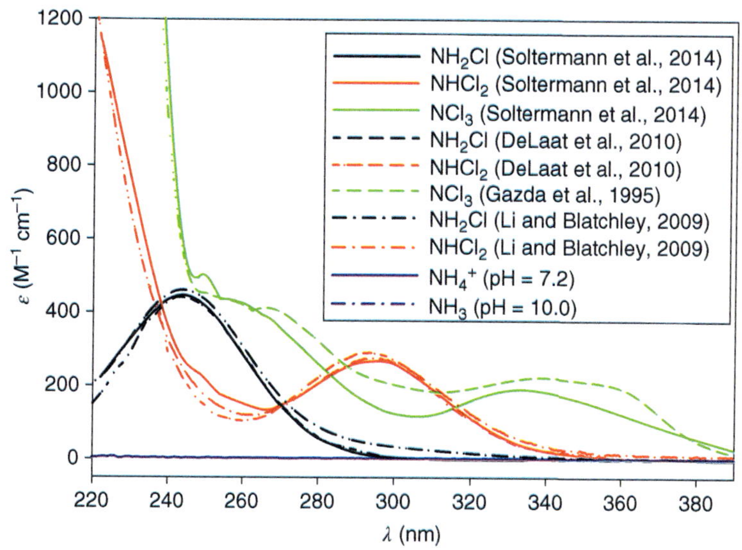


**Figure 10.24** UV absorption spectra for ammonia-N and the inorganic chloramines.

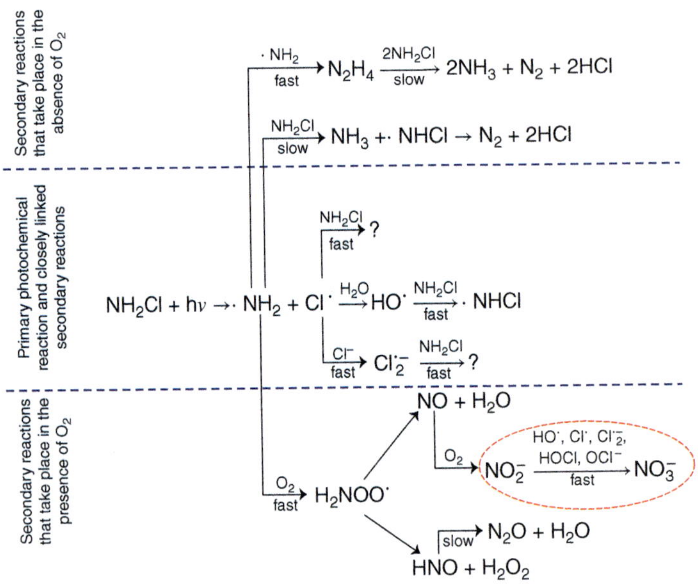


**Figure 10.25** Schematic illustration of the mechanism responsible for degradation of $NH_2Cl$ as a result of exposure to UVC radiation. The schematic was synthesized from the work of several investigations [56–59]. The presence of molecular oxygen ($O_2$) plays a key role in the rate of $NH_2Cl$ decay and in the formation of reaction products. Reactions within the red dashed oval are believed to be largely responsible for inhibition of $NH_2Cl$ decay in the presence of $O_2$.

and a chlorine radical. In turn, these radicals will participate in a number of thermal chemical reactions that result in the formation of secondary radicals and stable products. UV-induced degradation of inorganic chloramines has been studied most extensively with $NH_2Cl$, and is illustrated in schematic form in Figure 10.25.

After the initial photochemical event, the chlorine radical participates in a series of rapid parallel reactions that result in degradation of $NH_2Cl$ and formation of secondary radicals, $HO^{\cdot}$ and $Cl_2^{\cdot-}$. In turn, these radicals will also react rapidly with $NH_2Cl$. The aminyl radical ($\cdot NH_2$) will react with itself to form hydrazine ($N_2H_4$), which will then react slowly with $NH_2Cl$ to form $NH_3$, $N_2$, and HCl. The aminyl radical is likely to react directly with $NH_2Cl$, but it appears that this reaction will be slow [59, 60].

Molecular oxygen ($O_2$) will react rapidly with $\cdot NH_2$ to yield $\cdot NH_2OO$, which in turn will decay to yield NO and HNO. Importantly, NO will react with $O_2$ to yield nitrite ($NO_2^-$) which reacts rapidly with many oxidants to form nitrate ($NO_3^-$) as a stable form of nitrogen in solution. HNO will slowly decay to yield nitrous oxide ($N_2O$).

The formation of $NO_2^-$ through the $O_2$-mediated reaction pathways appears to be a critical step in this process because it represents a source of oxidant demand. In particular, the oxidants that exist in solution (i.e. $HO^{\cdot}$, $Cl^{\cdot}$, $Cl_2^{\cdot-}$, HOCl, $OCl^-$, and $NH_2Cl$) will all react rapidly with $NO_2^-$ to form $NO_3^-$. Interestingly, each of these oxidants will also react with $NH_2Cl$. Therefore, the formation of $NO_2^-$ opens reaction pathways that compete with $NH_2Cl$ for these oxidants. Consequently, the presence of $O_2$ in solution has the effect of slowing the rate of $NH_2Cl$ decay. More broadly, the rate of UV-induced decay of $NH_2Cl$ is strongly influenced by solution composition. This probably explains why reported values of the effective quantum yield for $NH_2Cl$ photolysis are so variable in the literature [56–59].

Figure 10.26 illustrates the influence of $O_2$ on the rate of $NH_2Cl$ decay in experiments that were conducted in well-mixed, batch reactors under a collimated beam, as a function of wavelength.

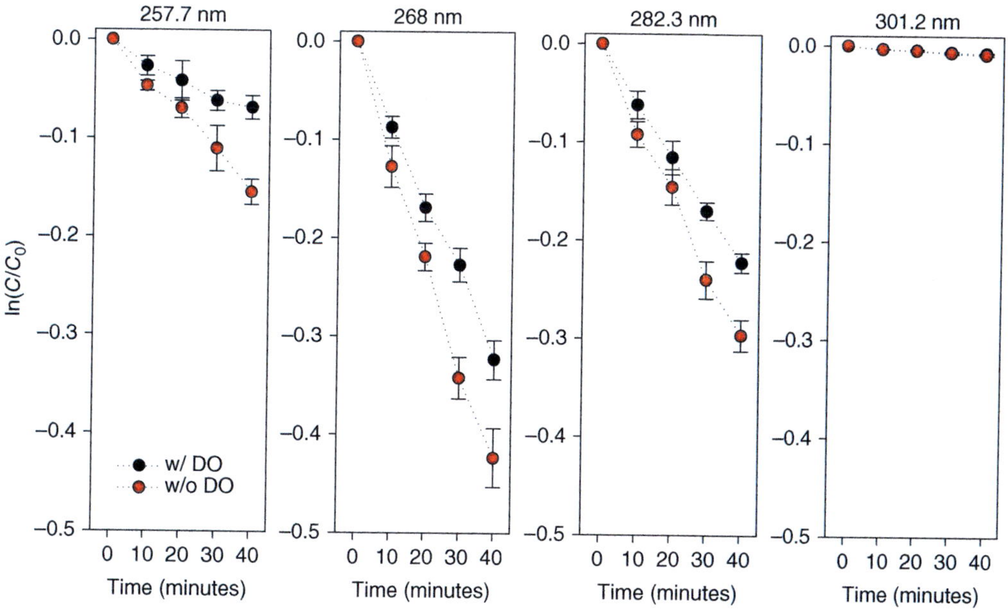


**Figure 10.26** Time-course UV photodecay of $NH_2Cl$ in well-mixed, batch reactors under collimated UV radiation at four wavelengths. Experiments at each wavelength were conducted in the presence of dissolved oxygen (•) and in the absence of dissolved oxygen (•). Error bars represent difference between the mean of duplicate measurements and a given measurement. *Source:* Data from Yin [59] and Yin et al. [61].

At each wavelength, the presence of $O_2$ in solution was associated with a notable reduction in the rate of $NH_2Cl$ decay.

Figure 10.26 also illustrates that the photodecay of $NH_2Cl$ shows strong wavelength-dependence. This is attributable to wavelength-dependence in both the molar absorption coefficient of $NH_2Cl$ and the primary quantum yield for its photodecay ($\Phi_{NH_2Cl}$). Values of $\Phi_{NH_2Cl}$ were estimated by conducting photodecay experiments in the presence of radical scavengers. Table 10.8 summarizes values of the molar absorption coefficient and primary quantum yields for $NH_2Cl$ and $NHCl_2$ photodecay at several wavelengths. These data indicate that $NH_2Cl$ photodecay is promoted effectively by radiation at wavelengths that are close to the characteristic wavelength of a low-pressure Hg lamp. However, $NHCl_2$ decay appears to be favored by longer wavelengths, such as those that are available from UV LEDs and excimer lamps.

**Table 10.8** Summary of molar absorption coefficients and primary quantum yields for photodecay of $NH_2Cl$ and $NHCl_2$ over a range of UV wavelengths. Also included are the products of these parameters at each wavelength.

| $\lambda$ (nm) | $\varepsilon_{NH_2Cl}$ $\left(\frac{L}{mole \cdot cm}\right)$ | $\Phi_{NH_2Cl}$ $\left(\frac{mole}{einstein}\right)$ | $\varepsilon_{NH_2Cl} \cdot \Phi_{NH_2Cl}$ $\left(\frac{L}{cm \cdot einstein}\right)$ | $\varepsilon_{NHCl_2}$ $\left(\frac{L}{mole \cdot cm}\right)$ | $\Phi_{NHCl_2}$ $\left(\frac{mole}{einstein}\right)$ | $\varepsilon_{NHCl_2} \cdot \Phi_{NHCl_2}$ $\left(\frac{L}{cm \cdot einstein}\right)$ |
|---|---|---|---|---|---|---|
| 257.7 | 345 | 0.34 | 117 | 96 | 0.93 | 89 |
| 268 | 174 | 0.32 | 56 | 111 | 0.84 | 93 |
| 282.3 | 75 | 0.31 | 23 | 214 | 0.65 | 139 |
| 301.2 | 4 | 0.27 | 1.1 | 261 | 0.52 | 136 |

*Source:* Data from Yin [59] and Yin et al. [61].

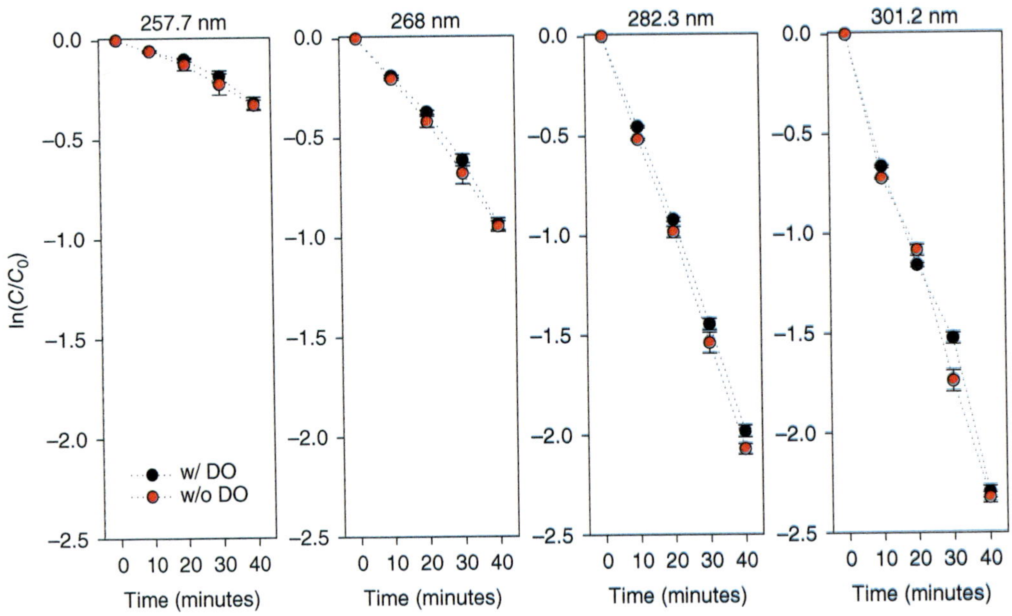


**Figure 10.27** Time-course UV photodecay of $NHCl_2$ in well-mixed, batch reactors under collimated UV radiation at four wavelengths. Experiments at each wavelength were conducted in the presence of dissolved oxygen (•) and in the absence of dissolved oxygen (•). Error bars represent difference between the mean of duplicate measurements and a given measurement. *Source:* Data from Yin [59] and Yin et al. [61].

Figure 10.27 illustrates the effect of dissolved oxygen on photodecay of $NHCl_2$ for a range of UVC wavelengths. Unlike $NH_2Cl$, it appears that $O_2$ has little or no effect on the photodecay of $NHCl_2$. This suggests that the mechanism of $NHCl_2$ photodecay does not involve reactions with $O_2$ to the same extent as the photodecay of $NH_2Cl$ does. As with the other inorganic chloramines, it is likely that $NHCl_2$ photodecay is initiated by homolytic cleavage of an N–Cl bond:

$$NHCl_2 + h\nu \rightarrow NHCl^{\cdot} + Cl^{\cdot} \tag{10.120}$$

However, it appears that $NHCl^{\cdot}$ is less reactive toward $O_2$ than is $NH_2^{\cdot}$.

The chloramine/UV process yields several reactive intermediates that function as oxidants toward a wide range of compounds. A simple example relates to the use of this process to control ammonia-nitrogen ($NH_3$–N). Common control strategies for management of $NH_3$–N include biochemical nitrification, ion exchange, and breakpoint chlorination. The chemistry of breakpoint chlorination is well understood and represents a specific application of the general reaction mechanism that describes chlorination of ammoniacal nitrogen, as described previously [51, 52].

UV-enhanced degradation of $NH_3$–N relies largely on the formation of reactive radicals as primary photoproducts of $NH_2Cl$ photolysis ($\cdot NH_2$ or $Cl^{\cdot}$), or as secondary radicals that are formed by reactions between $Cl^{\cdot}$ and aqueous constituents; $\cdot NH_2$ tends to react slowly with primary amines [62]. Secondary radical formation reactions include:

$$Cl^{\cdot} + Cl^{-} \rightarrow Cl_2^{\cdot -} \qquad k = 8.0 \times 10^9 \ \frac{\text{L}}{\text{mole} \cdot \text{s}} \tag{10.121}$$

$$\begin{aligned} &Cl^{\cdot} + H_2O \rightarrow ClO^{\cdot -} + H^{+} && k = 2.5 \times 10^5 \ \text{s}^{-1} \\ &ClO^{\cdot -} + HO^{\cdot} + Cl^{-} && k = 6.1 \times 10^9 \ \text{s}^{-1} \end{aligned} \tag{10.122}$$

Ammonia-N participates in an acid/base reaction that affects its form in solution:

$$NH_4^+ \rightleftharpoons NH_3 + H^+ \quad pK_a = 9.25 \tag{10.123}$$

As such, the majority of ammonia-N will exist in the form of $NH_4^+$ at pH conditions that are common in water supplies. $NH_4^+$ is essentially inert to direct photolysis, but will undergo reactions with the radicals that are produced from $NH_2Cl$ photolysis [63]:

$$NH_4^+ + Cl^{\cdot} \rightarrow \text{products} \qquad k = 2.6 \times 10^8 \frac{L}{\text{mole} \cdot s} \tag{10.124}$$

$$NH_4^+ + Cl_2^{\cdot -} \rightarrow \text{products} \qquad k = 3.5 \times 10^5 \frac{L}{\text{mole} \cdot s} \tag{10.125}$$

$$NH_4^+ + HO^{\cdot} \rightarrow \cdot NH_2 + H_2O + H^+ \qquad k = 1.4 \times 10^7 \frac{L}{\text{mole} \cdot s} \tag{10.126}$$

These radicals have been shown to be important contributors to oxidation of $NH_4^+$.

Figure 10.28 illustrates a hypothesized mechanism to describe the UV-assisted oxidation of $NH_3$–N when +1-valent chlorine is present in solution. As described previously, photolysis of $NH_2Cl$ results in the formation of $\cdot NH_2$ and $Cl^{\cdot}$. In turn, $Cl^{\cdot}$ reacts with other aqueous constituents to yield other reactive radicals: $Cl_2^{\cdot -}$ and $HO^{\cdot}$. Reduced nitrogen, in the forms of $NH_3$–N and the inorganic chloramines, will decay based on the reactions that define the classic breakpoint chlorination experiment. The radical-based reactions that are described by Equations 10.124–10.126 will also contribute to this decay process, resulting in enhanced degradation of $NH_3$–N.

Figure 10.29 illustrates the effect of inclusion of $UV_{254}$ irradiation on the breakpoint chlorination experiment. The inclusion of $UV_{254}$ irradiation ($UV_{254}$ dose = 800 $\frac{mJ}{cm^2}$) resulted in faster oxidation of N(III) and faster degradation of +1-valent chlorine (largely in the form of $NH_2Cl$) than in an experiment without $UV_{254}$ exposure. This same process was shown to result in reductions in the formation of trihalomethanes and haloacetic acids, while promoting the formation of haloacetonitriles, relative to the process of chlorination only [64].

The contributions of reactions 10.124–10.126 to $NH_4^+$ degradation will be governed by the products of their respective rate constants and the steady-state concentrations of the radicals that participate in each of these reactions. It has been shown that for near-neutral pH conditions that are typical of most water treatment applications, reaction 10.124 dominates the decay of $NH_4^+$, with minor contributions from reactions 10.125 and 10.126 [63].

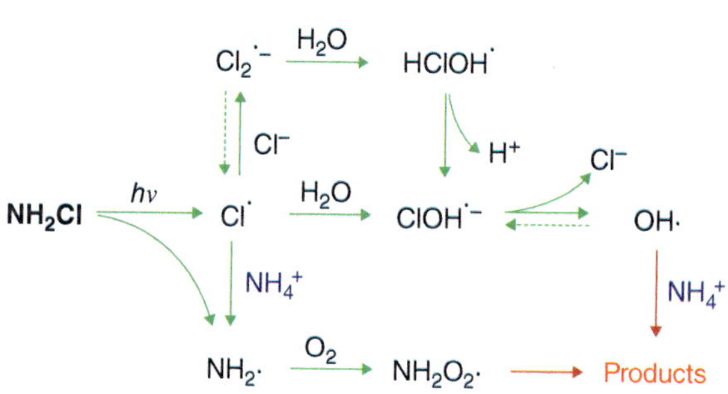


**Figure 10.28** Illustration of proposed radical reaction pathways that contribute to $NH_4^+$ degradation in the UV/$NH_2Cl$ process. *Source:* Zhang et. al 2019/with permission of Elsevier.

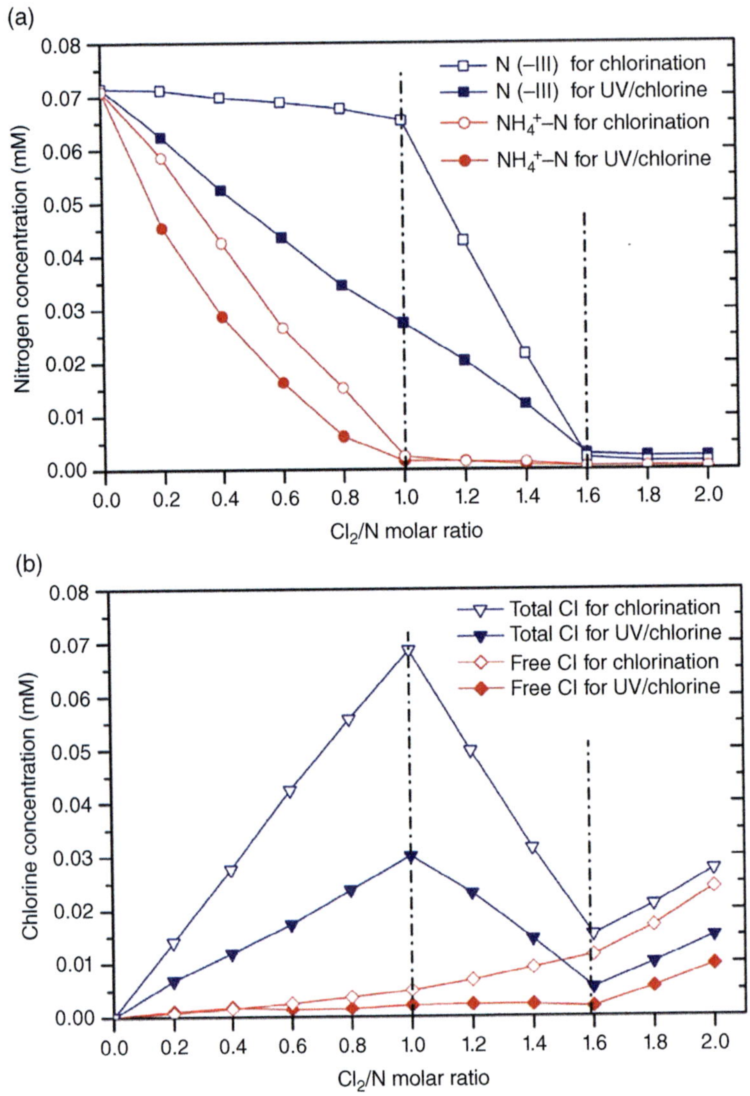


**Figure 10.29** Comparison of (a) nitrogen removal and (b) residual chlorine concentration as a function of $Cl_2$/N molar ratio for the UV/chlorine process and chlorination process ([Ammonia-N]$_0$ = 0.07 mM, pH = 7.20, $UV_{254}$ dose = 800 $\frac{mJ}{cm^2}$, reaction time = 200 seconds for chlorination). *Source:* Zhang et al. 2015/with permission of Elsevier.

The dynamic behavior of $NH_4^+$ (and other reduced compounds) in solution will be governed by the combined effects of reactions that lead to radical formation, secondary reactions that result in radical propagation and termination, the radical-based reactions (10.124–10.126), and other thermal chemical reactions that involve $NH_4^+$. Kinetic parameters are available for each of these processes; when combined in a numerical simulation, this modeling approach provides an accurate prediction of the dynamic behavior of $NH_4^+$ in solution (see Figure 10.30).

The $NH_2Cl$/UV process may be included intentionally or unintentionally as part of treatment. An example of unintentional use is water reuse operations. In these settings, inorganic chloramines

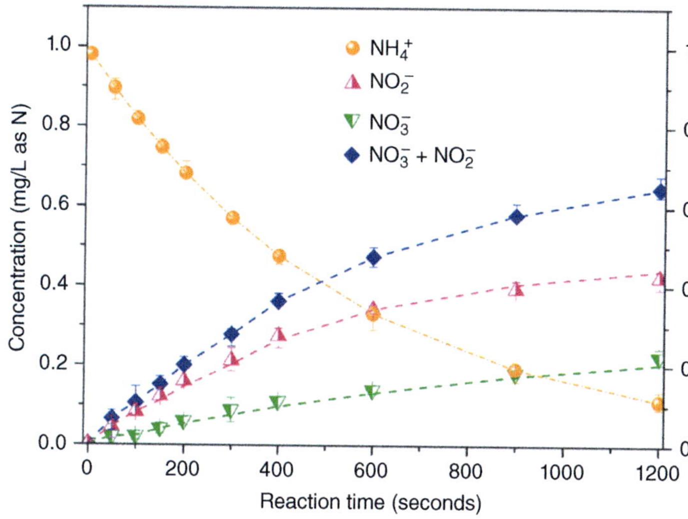


**Figure 10.30** Measured and simulated behavior of $NH_4^+$ and degradation products in the chloramine/UV process. Initial conditions: $[NH_4^+]$ = 70 μM, $[NH_2Cl]$ = 3 mM, pH = 7.5 with 10 mM phosphate buffer, UV fluence rate = $2.00 \frac{mW}{cm^2}$; symbols illustrate measured values; dashed lines illustrate predicted results from a numerical model implemented using Kintecus. *Source:* [63]. Zhang et al. 2019/with permission of Elsevier.

(largely in the form of $NH_2Cl$) are formed or added upstream of membrane-based processes (e.g. microfiltration [MF] and reverse osmosis [RO]) to mitigate fouling. $NH_2Cl$ will permeate RO membranes and contribute to chemistry in downstream UV-based AOPs. Low molecular weight, neutral contaminants are also likely to be present in RO permeate. As such, these contaminants are often targets of treatment in water reuse plants and may represent design or regulatory constraints for operation of AOPs.

As described previously, $NH_2Cl$ photolysis yields $\cdot NH_2$ and $Cl^\cdot$ as primary photoproducts. $\cdot NH_2$ is generally not an important contributor to degradation of other compounds in solution. However, $Cl^\cdot$ reacts with water and $Cl^-$ to yield $HO^\cdot$ and $Cl_2^{\cdot -}$, respectively; $Cl^\cdot$, $HO^\cdot$, and $Cl_2^{\cdot -}$ all are important contributors to oxidation of aqueous contaminants.

The $NH_2Cl$/UV process has been shown to be effective for oxidation of a wide range of aqueous contaminants. For example, the $NH_2Cl$/UV process has been demonstrated to be roughly equivalent to the free chlorine/UV and $H_2O_2$/UV processes for control of benzoate (aromatic), carbamazepine (alkene), and 1,4-dioxane (alkane) [62, 65].

The use of the $NH_2Cl$/UV process for control of micropollutants has been observed as an artifact of treatment (a *de facto* treatment process) and in these circumstances, performance of this UV-AOP was not optimized. However, with recognition of the role this process can play in improving water quality, and with the emergence of a deep, quantitative understanding of the elementary reactions that are responsible for this process, it is now possible to optimize the process using mathematical models [62, 65].

The foundation of the $NH_2Cl$/UV process is the fact that the N–Cl bond is quite photolabile. This allows conversion of $NH_3$–N, a form of nitrogen that is essentially inert to UV photochemistry, to become photoactive. A key aspect of this behavior is the alteration of the UV absorption spectrum of $NH_3$-N that results from the formation of one or more N–Cl bonds (see Figure 10.24). This behavior is not limited to $NH_3$–N; in fact, it is demonstrated with essentially any compound that includes one or more amine groups. Similar behavior has been demonstrated with urea (an amide), but the process is generally less efficient than with amines [66].

HOCl tends to react rapidly with amines to yield their N-chlorinated derivatives. The rates of N-chlorination of primary and secondary amines tends to be rapid, with N-chlorination of tertiary amines tending to proceed more slowly [38]. However, the bond energy associated with the N–Cl bond tends to coincide (roughly) with the output spectrum of common UVC sources, such as low-pressure Hg lamps or UVC LEDs. Examples of the effects of N-chlorination on the absorption spectra of amines are presented in Figures 10.31 and 10.32.

Figure 10.31 illustrates the absorption spectra for DMA and *N*-chlorodimethylamine (CDMA). DMA is a simple secondary amine that is essentially a non-absorber of UV radiation. It undergoes rapid N-chlorination to yield CDMA, which displays notable absorption in the UVC range.

The effects of N-chlorination on absorption spectra are also evident among compounds that display measurable absorption by UV radiation. Figure 10.32 illustrates the absorption spectrum for Anatoxin-a (ANTX-a), a bicyclic secondary amine within the family of cyanotoxins. ANTX-a is a potent neurotoxin, sometimes referred to as "very fast death factor." ANTX-a is a fairly strong absorber of UV radiation; N-chlorination of ANTX-a to yield Cl-ANTX-a results in an increase in absorption across a wide range of wavelengths. Increases in $\varepsilon_{254}$ attributable to N-chlorination are indicated by the inset tables in Figures 10.31 and 10.32.

Free chlorine reacts with DMA to yield CDMA; similarly, free chlorine reacts with ANTX-a to yield *N*-chloro-ANTX-a (Cl-ANTX-a). In both cases, the inclusion of the N–Cl bond increases the susceptibility of an amine to photodegradation. Figure 10.33 illustrates this behavior in quantitative terms.

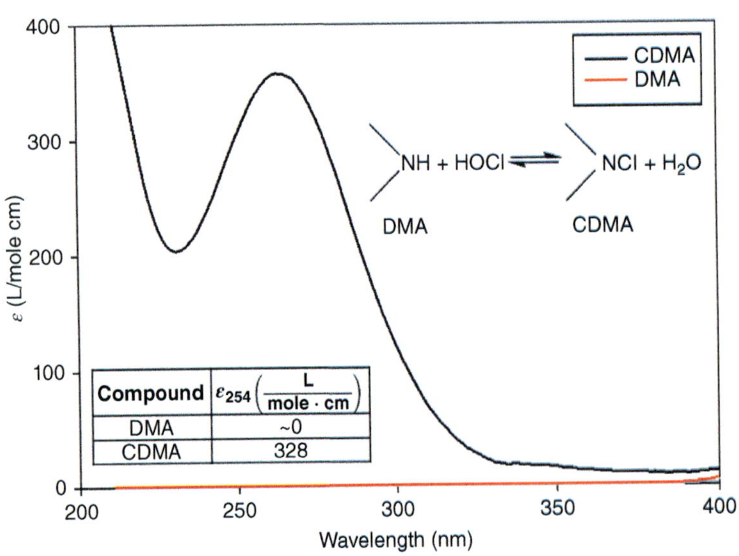


**Figure 10.31** UV absorption spectra for dimethylamine (DMA) and *N*-chlorodimethylamine (CDMA) at pH = 7.0. The reaction that is responsible for CDMA formation from DMA is illustrated in the inset upper right. Also included as an inset (lower left) are values of $\varepsilon_{254}$ for DMA and CDMA. *Source:* modified from Chen [67] and Chen and Blatchley [68].

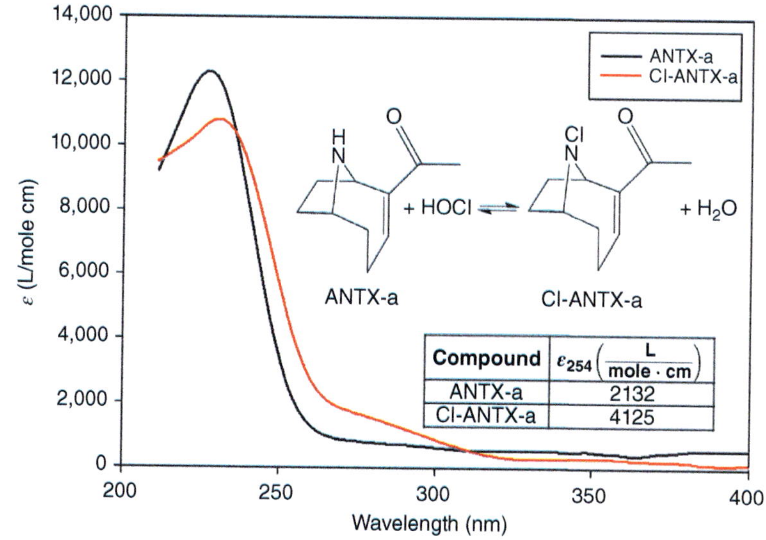


**Figure 10.32** UV absorption spectra for Anatoxin-a (ANTX-a) and *N*-chloro-Anatoxin-a (Cl-ANTX-a) at pH = 7.0. The reaction that is responsible for Cl-ANTX-a formation from ANTX-a is illustrated in the inset (upper right). Also included as an inset (lower right) are values of $\varepsilon_{254}$ for ANTX-a and Cl-ANTX-a. *Source:* modified from Chen [67] and Chen and Blatchley [68].

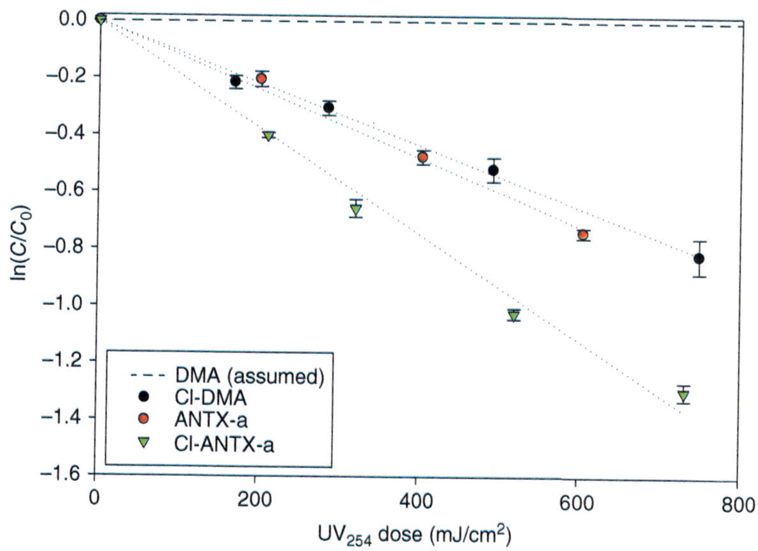


**Figure 10.33** Pseudo-first-order $UV_{254}$ photodecay of DMA, ANTX-a, and their respective N-chlorinated forms. Photodegradation of DMA was assumed to be negligible because it does not absorb $UV_{254}$ radiation. *Source:* Data from Chen [67] and Chen and Blatchley [68].

DMA does not absorb radiation at 254 nm, so no photodegradation is possible. However, CDMA absorbs radiation at this wavelength effectively, and undergoes photodegradation at a rate that follows pseudo-first-order kinetics when present in dilute solution.

ANTX-a absorbs radiation at 254 nm. In fact, both the protonated and deprotonated forms of ANTX-a ($pK_a = 9.36$) absorb at 254 nm. Both forms also undergo direct photolysis. However,

N-chlorination of ANTX-a increases the molar absorption coefficient of the parent molecule (Figure 10.32) and also increases its photosensitivity.

ANTX-a expresses chlorine demand. Specifically, a chlorine:precursor (Cl:P) molar ratio of 1.4:1 appears to be just sufficient to accomplish complete N-chlorination of ANTX-a [67]. When the Cl:P molar ratio is increased to above this value, free chlorine will remain in solution. The presence of free chlorine allows for photolysis of $HOCl/OCl^-$ to yield reactive intermediates, which may also react with ANTX-a and its N-chlorinated derivative. As such, the rate of photodegradation of Cl-ANTX-a increases as the Cl:P ratio increases (see Figure 10.34).

The behavior illustrated above for DMA and ANTX-a is not limited to secondary. In fact, it appears to be observed generally among amines. Similar observations have been made with a wide range of amine compounds, including amino acids, creatinine, uric acid, urea, and other cyanotoxins [54, 55, 66, 69, 70]. As such, it appears that chlorination followed by UV irradiation represents a process that selectively and preferentially attacks amine compounds (and amides) via the mechanism of formation of one or more N–Cl bonds, with subsequent photolysis to open up radical-mediated reaction pathways.

The ability of prechlorination to promote selective degradation of amines represents a useful strategy for water treatment. However, it is important to recognize that the products of these reactions may also be of concern. For example, chlorination of creatinine, a common metabolite and constituent of human urine, yields *N*-chloro-creatinine. UV irradiation of *N*-chloro-creatinine results in cleavage of the N–Cl bond, which in turn leads to cleavage of the heterocyclic ring structure, which on subsequent exposure to free chlorine results in the formation of dichloromethylamine ($CH_3NCl_2$) and inorganic chloramines. In turn, these compounds are also photolabile. $CH_3NCl_2$ photolyzes to yield HCN as an intermediate, which then undergoes rapid chlorination to yield cyanogen chloride (ClCN), as illustrated in Figure 10.35 [54].

A separate investigation allowed examination of the effects of chlorination and UV exposure on amino acids, which are also found as common constituents in water. The nature of the reactions

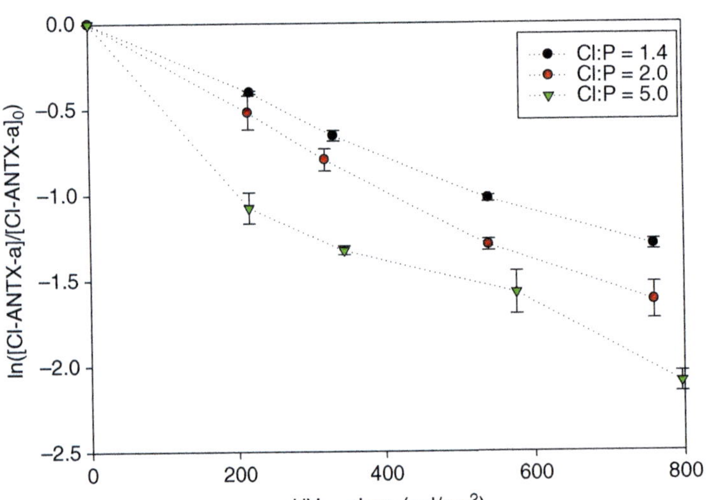


**Figure 10.34** Effect of Cl:P ratio on photodegradation of Cl-ANTX-a. A Cl:P ratio of 1.4:1 is sufficient to complete N-chlorination of ANTX-a. Therefore, application of free chlorine above this molar ratio yields free chlorine in solution after N-chlorination. *Source:* Data from Chen [67] and Chen and Blatchley [68].

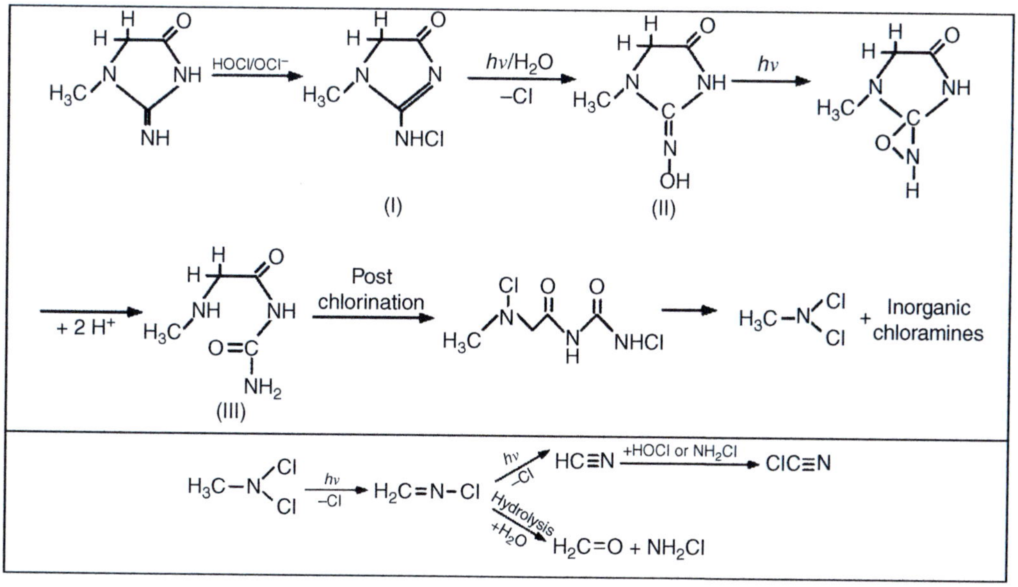


**Figure 10.35** Mechanisms of decay of creatinine and derivative compounds as a result of chlorination followed by UV irradiation. Top panel illustrates the response of creatinine to chlorination, followed by UV irradiation in the presence of free chlorine. Bottom panel illustrates the response of dichloromethylamine, which appears as an intermediate in this process. *Source:* Weng et al. 2013/with permission of Elsevier.

that are observed in this process depends on the extent of N-chlorination. Using L-arginine as an illustration of the behavior of aliphatic amino acids, Figure 10.36 indicates that the location of N-chlorination depends on the Cl:P molar ratio. Each of the N–Cl bonds that is formed in this process is susceptible to subsequent UV-induced cleavage, which allows for different reaction pathways to be opened. The products of the subsequent thermal chemical reactions include the inorganic chloramines and ClCN. Similarly for L-histidine, which includes a heterocyclic group, the Cl:P molar ratio determines the extent and location of N-chlorination, which again influences the thermal chemical reactions that follow photodechlorination of these N–Cl bonds. Products of these reactions again include a blend of inorganic chloramines and ClCN [55].

### 10.10.3 Persulfate/UV Process

The persulfate/UV process incorporates two persulfate compounds: peroxodisulfate (PDS) and peroxomonosulfate (PMS). Both compounds function as precursors to the formation of sulfate radicals $\left(SO_4^{\cdot -}\right)$ and other radicals, including $HO^{\cdot}$.

PDS ($S_2O_8^{2-}$) and hydrogen peroxide both include a central peroxide group (–O – O–). The structures of both compounds are illustrated in Figure 10.37. This peroxide group plays a critical role in the absorption behavior of PDS and $H_2O_2$, resulting in similar absorption behavior for these two compounds [71]. When subjected to UV radiation, these central peroxide bonds in PDS and $H_2O_2$ undergo homolytic cleavage to yield the sulfate and hydroxyl radicals, respectively. Photolysis of $H_2O_2$ has been described previously; the photolysis of PDS was described by Mark et al. [71]. Although the absorption coefficients for $H_2O_2$ and $S_2O_8^{2-}$ are quite similar, the quantum yields for their homolytic cleavage are different, with $H_2O_2$ having a photolysis quantum yield of

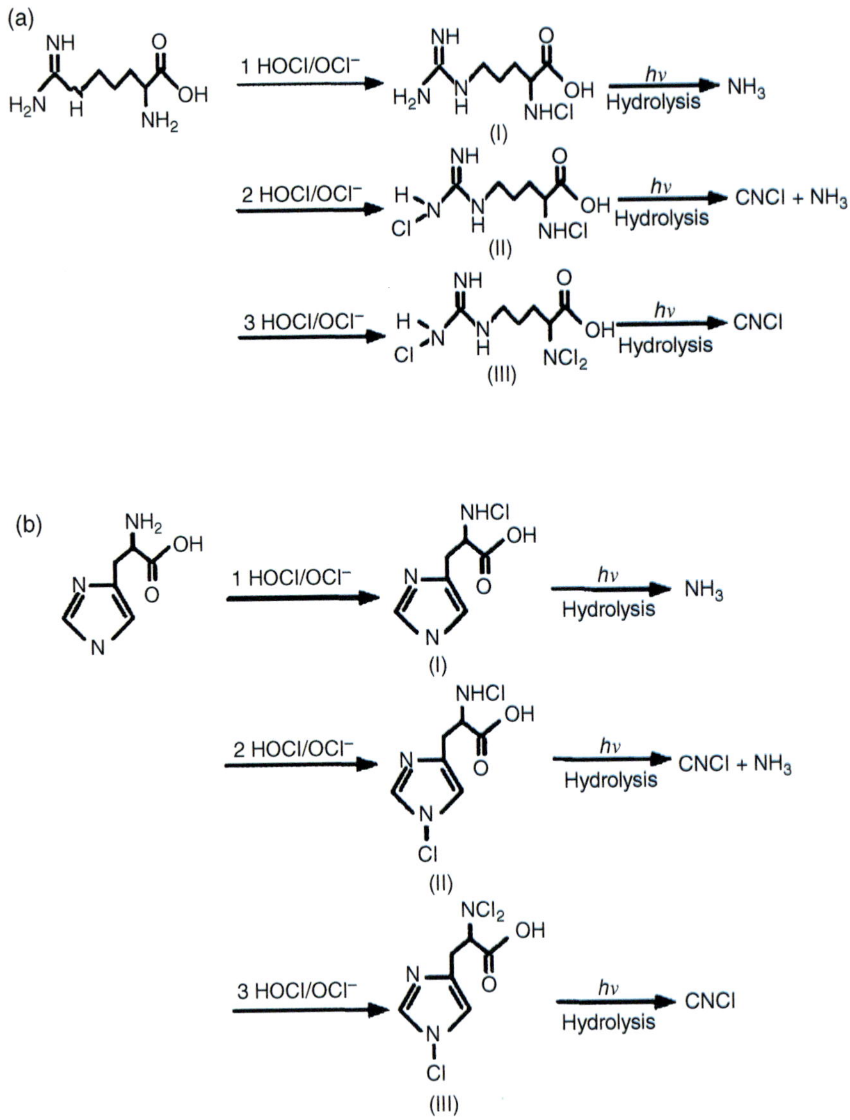


**Figure 10.36** Hypothesized mechanisms to describe responses of amino acids to chlorination and UV exposure. (a) Responses of L-arginine; (b) responses of L-histidine. *Source:* Weng and Blatchley 2013/with permission of American Chemical Society.

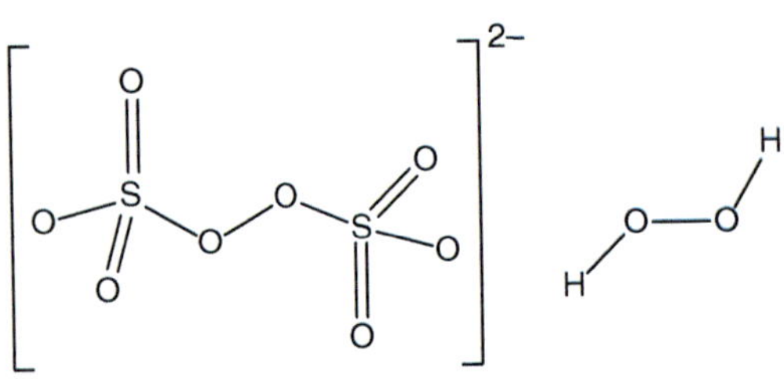


**Figure 10.37** Structures of peroxodisulfate anion (left) and hydrogen peroxide (right), both of which include a central peroxide bond (–O–O–) that is largely responsible for the absorption behavior and photochemistry of these molecules. Both molecules undergo homolytic cleavage of this bond to yield radicals (see equations 10.74 and 10.127).

$0.5\frac{\text{mole}}{\text{einstein}}$ and $S_2O_8{}^{2-}$ having a photolysis quantum yield of $0.7\frac{\text{mole}}{\text{einstein}}$ [72]. These translate to a formation quantum yield for $HO^{\cdot}$ of $1.0\frac{\text{mole}}{\text{einstein}}$ from photolysis of $H_2O_2$ and a formation quantum yield for $SO_4{}^{\cdot-}$ of $1.4\frac{\text{mole}}{\text{einstein}}$ from photolysis of $S_2O_8{}^{2-}$.

$$S_2O_8{}^{2-} \xrightarrow{h\nu} 2SO_4^{\cdot-} \quad \Phi_{254} = 1.4\frac{\text{mole}}{\text{einstein}} \text{ for } SO_4{}^{\cdot-} \text{ formation} \tag{10.127}$$

The photochemical behavior of PMS is somewhat more complicated than that of PDS because it displays acid/base behavior that is relevant at pH conditions that are common in water treatment. Specifically:

$$HSO_5{}^{-} \rightarrow H^{+} + SO_5{}^{2-} \quad pK_a = 9.3 \tag{10.128}$$

This acid/base behavior is particularly relevant because $HSO_5{}^{-}$ and $SO_5{}^{2-}$ display very different absorption behavior for UVC radiation. At 254 nm, $\varepsilon_{HSO_5{}^{-}} = 13.8\frac{\text{L}}{\text{mole} \cdot \text{cm}}$, while $\varepsilon_{SO_5{}^{2-}} = 149.5\frac{\text{L}}{\text{mole} \cdot \text{cm}}$ [69]. At practical pH values, PMS exists as a mixture of the protonated and deprotonated forms. As such, the effective absorption behavior of these solutions at 254 nm is a strong function of pH (see Figure 10.38). In fact, the absorption behavior of PMS and is deprotonated form at 254 nm are quantitatively similar to $H_2O_2$ and its conjugate base, the hydroperoxyl ion, $HO_2{}^{-}$ (see Figure 10.13). Both forms of PMS undergo photolysis to yield two reactive radicals, and it has been suggested that the quantum yield for this reaction is essentially the same for the protonated and deprotonated forms [73]:

$$\frac{HSO_5{}^{-}}{SO_5{}^{2-}} \xrightarrow{h\nu} SO_4{}^{\cdot-} + HO^{\cdot} \quad \Phi_{254} = 0.52\frac{\text{mole}}{\text{einstein}} \tag{10.129}$$

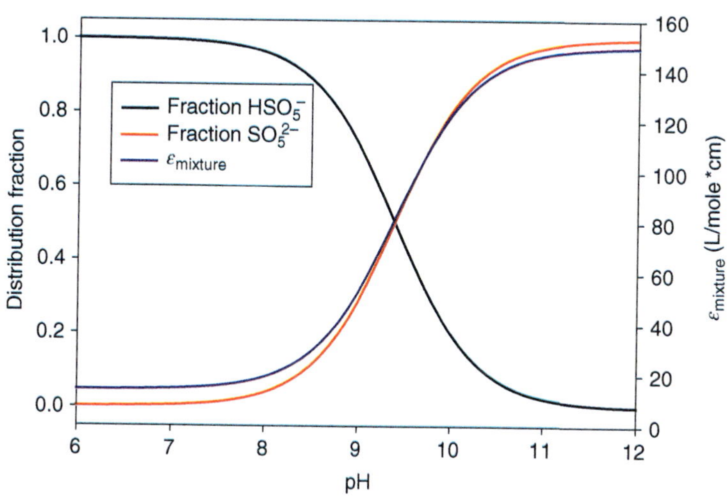


**Figure 10.38** Distribution of PMS between is protonated and deprotonated forms (left vertical axis). Right vertical axis refers to the molar absorption coefficient of the mixture of these two forms.

The data presented in Figure 10.38 indicate that the PMS/UV process will be more effective at pH above the p$K_a$ of PMS than at pH values below this p$K_a$. This behavior has been confirmed experimentally [73]. However, it is often impractical to operate at a pH above 9.3 in water treatment applications, so this represents an important barrier to use of the PMS/UV process.

Table 10.9 provides a partial list of reactions that are fundamental to the persulfate/UV process in most applications; the reaction list presented in Table 10.9 does not include reactions that focus on the behavior of HO$^{\cdot}$, as these have been listed previously (see Table 10.6). The sulfate radical is characterized by a SRP of 2.5–3.1 V, indicating that $SO_4^{\cdot-}$ functions as a powerful oxidant. $SO_4^{\cdot-}$ participates in a wide range of reaction processes, including direct electron transfer, addition, and H-abstraction [74]. Though $SO_4^{\cdot-}$ is highly reactive, it tends to be more selective than HO$^{\cdot}$.

$SO_4^{\cdot-}$ and HO$^{\cdot}$ show important similarities and differences. Both react rapidly with common anions, including $HCO_3^-$, $Cl^-$, and $Br^-$, which function as scavengers of the primary radicals, and which also allow the formation of secondary radicals, including $CO_3^{\cdot-}$, $Cl^{\cdot}$, $Cl_2^{\cdot-}$, and others

**Table 10.9** Partial list of reactions that govern the persulfate/UV$_{254}$ process. Reactions included in this table are those that involve PDS, PMS, or their derivative radicals. Other reactions that will influence the behavior of this process, such as those that involve other radicals, have been listed previously (see Table 10.6). Reactions and reaction rate parameters are from Mark et al. [71], Guan et al. [73], and Sharma et al. [74].

| Reaction type | Reaction number | Reaction | Reaction rate parameter[a] |
|---|---|---|---|
| Primary photochemical event (initiation) | 1 | $S_2O_8^{2-} \xrightarrow{h\nu} 2SO_4^{\cdot-}$ | $\varepsilon_{254} = 20 \frac{L}{mole \cdot cm}$ $\Phi_{254} = 0.7 \frac{mole}{einstein}$ |
| | 2 | $HSO_5^- \xrightarrow{h\nu} SO_4^{\cdot-} + HO^{\cdot}$ | $\varepsilon_{254} = 13.8 \frac{L}{mole \cdot cm}$ $\Phi_{254} = 0.52 \frac{mole}{einstein}$ |
| | 3 | $SO_5^{2-} \xrightarrow{h\nu} SO_4^{\cdot-} + O^{\cdot-}$ | $\varepsilon_{254} = 149.5 \frac{L}{mole \cdot cm}$ $\Phi_{254} = 0.52 \frac{mole}{einstein}$ |
| Rapid reversible reactions (equilibria) | 4 | $HSO_5^- \rightleftharpoons H^+ + SO_5^{2-}$ | p$K_a$ = 9.3 |
| Radical speciation | 5 | $HO^{\cdot} + HSO_5^- \rightarrow SO_5^{\cdot-} + H_2O$ | $1.7 \times 10^7\ M^{-1}\ s^{-1}$ |
| | 6 | $HO^{\cdot} + SO_5^{2-} \rightarrow SO_5^{\cdot-} + OH^-$ | $2.1 \times 10^9\ M^{-1}\ s^{-1}$ |
| | 7 | $SO_4^{\cdot-} + HSO_5^- \rightarrow SO_5^{\cdot-} + HSO_4^-$ | $<10^5\ M^{-1}\ s^{-1}$ |
| | 8 | $SO_4^{\cdot-} + OH^- \rightarrow HO^{\cdot} + SO_4^{2-}$ | $6.5 \times 10^7\ M^{-1}\ s^{-1}$ |
| | 9 | $HO^{\cdot} + HSO_4^- \rightarrow SO_4^{\cdot-} + H_2O$ | $6.9 \times 10^5\ M^{-1}\ s^{-1}$ |
| Radical combination | 10 | $SO_4^{\cdot-} + SO_4^{\cdot-} \rightarrow S_2O_8^{2-}$ | $3.1 \times 10^8\ M^{-1}\ s^{-1}$ |
| | 11 | $HO^{\cdot} + SO_4^{\cdot-} \rightarrow HSO_5^-$ | $1 \times 10^{10}\ M^{-1}\ s^{-1}$ |
| | 12 | $SO_4^{\cdot-} + SO_4^{\cdot-} \rightarrow SO_5^{\cdot-} + SO_5^{\cdot-} + O_2$ | $1 \times 10^8\ M^{-1}\ s^{-1}$ |

[a] Reaction parameters listed in this column are rate constants for elementary reactions, except as indicated.

that tend to be more selective than their parent radicals, but may also contribute significantly to decay of target contaminants [72, 75]. The extent of the contributions of these secondary radicals to contaminant degradation depends on the specific reaction matrix. In general, the contributions of secondary radicals to contaminant degradation will become more important as the concentrations of their precursor anions (e.g. $HCO_3^-$, $Cl^-$, and $Br^-$) increase. For example, the secondary radicals will contribute significantly to contaminant degradation in brines and seawater, where the concentrations of the precursor anions are highly relative to most other applications of this process [72, 74, 75].

### 10.10.4 Vacuum UV

By definition, VUV includes electromagnetic radiation in the range 10 nm $\leq \lambda \leq$ 200 nm. Radiation in this range gets its name from the fact that gas-phase (and liquid-phase) molecular oxygen ($O_2$) and molecular nitrogen ($N_2$) both absorb strongly in this region. As such, effective penetration of VUV radiation requires that it be transmitted through a vacuum.

Liquid water ($H_2O$) is also an effective absorber of VUV radiation. Among the longer VUV wavelengths, the numerical value of the molar absorption coefficient for $H_2O$ is quite small; however, because the molar concentration of liquid $H_2O$ is so high ($[H_2O] \approx 55.6$ M), absorbance by $H_2O$ is actually quite high. As a result, VUV radiation penetrates only small distances, even in the absence of dissolved $O_2$ and $N_2$. Figure 10.39 illustrates the molar absorption coefficient for liquid $H_2O$ in the

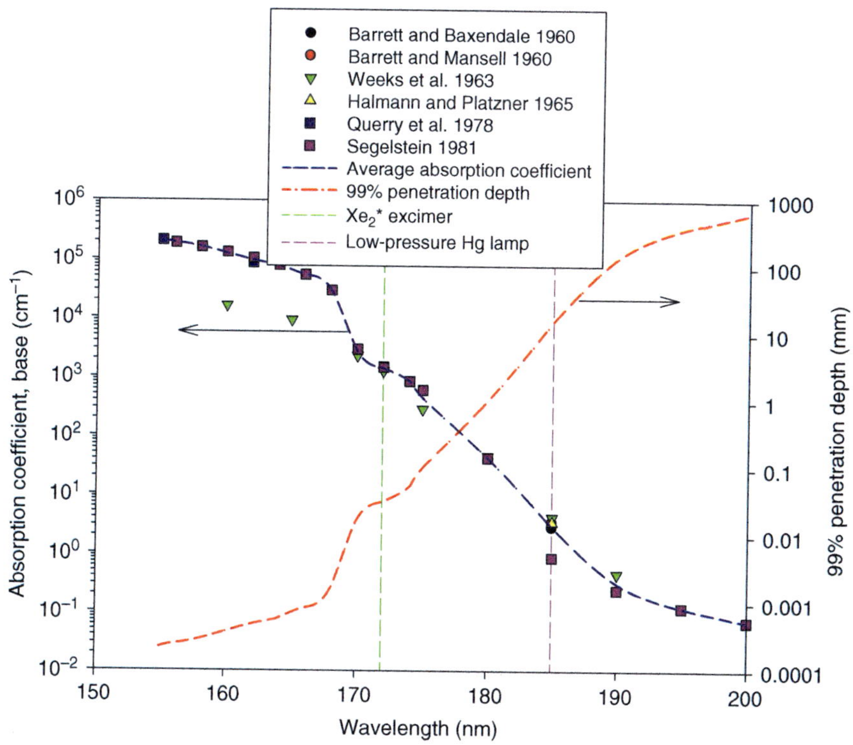


**Figure 10.39** VUV absorption spectrum for liquid $H_2O$, as reported by various authors (left vertical axis) [76–81]. Data summary is from Crapulli et al. [82]. Right vertical axis refers to depth of penetration corresponding to 99% loss of photons by $H_2O$. Also included are characteristic wavelengths corresponding to the output from two common VUV sources: a $Xe_2*$ excimer lamp and a low-pressure Hg lamp.

VUV range, as reported by several investigators. Also included in this figure is the depth corresponding to 99% absorption of radiation as a function of wavelength. The nominal output wavelengths of two commonly used sources of VUV radiation are also included: 172 nm, which represents the peak output of $Xe_2*$ excimer lamps, and 185 nm, corresponding to a line that comprises roughly 10% of the output power of conventional low-pressure Hg lamps [83].

Absorption of VUV radiation by $H_2O$ results in at least two important photochemical reactions [84]:

$$H_2O \xrightarrow{h\nu} HO + H \qquad \begin{array}{c} \Phi_{123.6} \approx 1\dfrac{\text{mole}}{\text{einstein}} \\ \Phi_{172} = 0.42\dfrac{\text{mole}}{\text{einstein}} \\ \Phi_{185} = 0.33\dfrac{\text{mole}}{\text{einstein}} \\ \Phi_{193} \approx 0 \end{array} \tag{10.130}$$

The quantum yield for this reaction varies roughly linearly between $\Phi_{193} \approx 0$ and $\Phi_{123.6} \approx 1\dfrac{\text{mole}}{\text{einstein}}$ [84].

A second, somewhat less-efficient photochemical reaction also takes place as a two-step process:

$$\begin{array}{l} H_2O \xrightarrow{h\nu} \left[e_{aq}^{\;-} + H_2O^+\right] + (H_2O) \\ \left[e_{aq}^{\;-} + H_2O^+\right] + (H_2O) \rightarrow e_{aq}^{\;-} + HO^{\cdot} + H_3O^+ \quad \Phi = 0.05 \pm 0.01\dfrac{\text{mole}}{\text{einstein}} \end{array} \tag{10.131}$$

In reaction 10.131, the first step involves the formation of a hydrated electron in a solvent (water) cage. In the second step, a fraction of these caged electrons is able to escape the cage to become free, hydrated electrons, $e_{aq}^{\;-}$. The quantum yield for this reaction is largely independent of wavelength across the VUV range.

Collectively, reactions 10.130 and 10.131 are responsible for creation of a highly reactive environment, in which strong oxidants and strong reductants may both be present. As such, these processes have the potential to open a wide range of reaction pathways for degradation of aqueous contaminants.

Direct photolysis of $H_2O$ and the resulting thermal chemical reactions that are driven by the reactive intermediates that form from reactions 10.130 and 10.131 are largely confined to the photochemical boundary layer that exists immediately adjacent to interface between the irradiated liquid and the surface of the VUV source. The thickness of this photochemical boundary layer will correspond roughly with the 99% penetration depth illustrated in Figure 10.39. For radiation at 185 nm, the 99% penetration depth is roughly 16 mm. For 172 nm radiation, the 99% penetration depth is only about 30 μm. Within this photochemical boundary layer, steep gradients in the VUV fluence rate field will be observed, along with similarly steep spatial gradients in the concentrations of the reactive intermediates.

Figure 10.40 illustrates simulated radial profiles of fluence rates at 185 and 254 nm within an annular VUV reactor based on a low-pressure Hg lamp. The annular space between the lamp sleeve and the reactor wall for this reactor was only about 0.5 cm thick, yet most of the radiation at 185 was absorbed before reaching the outside wall. In contrast, the fluence rate at 254 nm was more uniform, with the fluence rate reaching the outside wall being slightly more than 50% of the value at the inside surface of the annular irradiated zone.

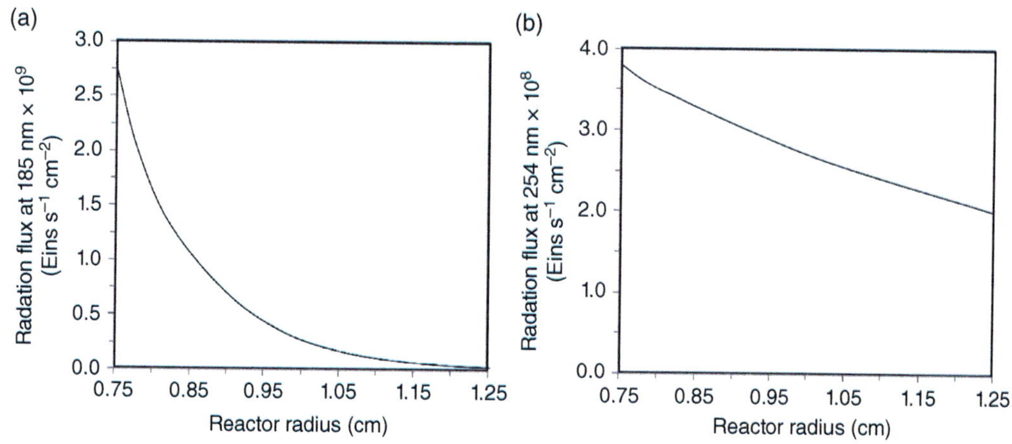


**Figure 10.40** Results of a numerical simulation of fluence rate as a function of radial position in an annular reactor at 185 nm (top) and 254 nm for a VUV reactor based on a low-pressure Hg lamp. *Source:* Imoberdorf and Mohseni 2011/with permission of Elsevier.

The thermal chemistry that results from the formation of the reactive intermediates that appear in reactions 10.130 and 10.131 will be strongly influenced by the chemistry of the solution. Of particular importance are the roles played by molecular oxygen ($O_2$) in solution. In particular, $O_2$ will scavenge $H^{\cdot}$ and $e_{aq}^-$ as follows:

$$H^{\cdot} + O_2 \rightarrow HO_2^{\cdot} \qquad k = 1 \times 10^{10} \frac{L}{mole \cdot s} \tag{10.132}$$

$$e_{aq}^- + O_2 \rightarrow O_2^{\cdot -} \qquad k = 2 \times 10^{10} \frac{L}{mole \cdot s} \tag{10.133}$$

In solutions where $O_2$ is in equilibrium with atmospheric oxygen ($[O_2]_{aq} \approx 0.25 - 0.29$ mM), $O_2$ will compete effectively with other substrates for $H^{\cdot}$ and $e_{aq}^-$. As a result, the roles of these reducing agents are greatly diminished in the presence of dissolved oxygen, under conditions that approach equilibrium with the ambient atmosphere, and the active chemistry that takes place within and near the photochemical boundary layer is dominated by the effects of $HO^{\cdot}$ [85].

An example of a common aqueous constituent that will be affected by the presence or absence of $O_2$ is nitrate ($NO_3^-$). In the presence of $O_2$, $NO_3^-$ is stable. However, in the absence of $O_2$, $NO_3^-$ will be reduced by diffusion-controlled reactions with $e_{aq}^-$ and $H^{\cdot}$ [85]. Products of these reactions will include nitrite ($NO_2^-$) and nitrous oxide ($N_2O$).

The rate of $HO^{\cdot}$ formation by VUV photolysis of $H_2O$ can be orders of magnitude faster than by $UV_{254}$ irradiation of $H_2O_2$. However, the formation of $HO^{\cdot}$ is confined to the photochemical boundary layer and its immediate vicinity. In a broader sense, the reactions that govern the dynamic behavior of VUV photoreactors, which include direct photolysis and radical-mediated thermal chemical reactions, tend to be extremely fast. As such, the dynamic behavior of VUV photoreactors is limited by mass transport behavior, rather than reaction kinetics. Because of this, accurate simulations of VUV reactor dynamics require the use of models that account for transport behavior and the spatial distributions of the governing reaction processes.

Crapulli et al. [82] presented a numerical simulation of VUV reactor that involved annular reactor systems (see Figure 10.41). The annular water layer that existed in the annular space between the lamp surface and the outside wall of the reactor was roughly 0.5 cm thick. VUV radiation did

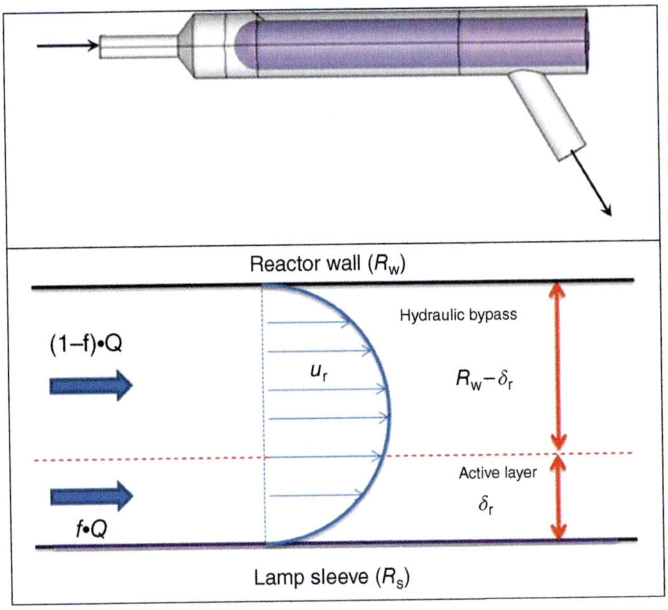


**Figure 10.41** (Top) Schematic illustration of annular VUV reactor simulated by model of Crapulli et al. [82]. (Bottom) Schematic illustration of the photochemical boundary layer ("active layer") and the non-photochemically active layer ("hydraulic bypass") in this reactor. The steady-state velocity field in this annular layer is also indicated. *Source:* Crapulli et al. 2014/with permission of Elsevier.

not penetrate the full thickness of this water layer, resulting in a situation in which part of the flow field was essentially unirradiated (i.e. non-photochemically active).

A CFD model was linked to a model to describe the kinetics of the participating reactions in this reactor system for two VUV sources: a $Xe_2*$ excimer lamp and a low-pressure Hg lamp. The results of simulations for these two VUV sources are illustrated in Figure 10.42. The top panels in Figure 10.42 illustrate the deeper penetration of radiation at 185 nm (low-pressure Hg lamp) as compared with radiation at 172 nm ($Xe_2*$ lamp). Local concentrations of $H_2O_2$ and $HO^{\cdot}$ are governed by a complex set of reactions that include direct photolysis and thermal chemical reactions. The lack of penetration of VUV radiation for both sources also constrains the reach of $H_2O_2$ and $HO^{\cdot}$ into the reacting fluid, but it is important to note that the space in which $H_2O_2$ and $HO^{\cdot}$ are found extends beyond the space that is irradiated by these two sources. This behavior has been attributed to the nonlinear behavior of the photochemical and thermal chemical reactions that govern the dynamic behavior of these two reactive species [82].

## 10.11 UV Advanced Reduction Processes (UV-ARPs)

A wide range of processes is available to promote the formation of free radicals and other reactive compounds that have a high (positive) SRP. By definition, these compounds have a high affinity for electrons, and as such have the ability to accept electrons from substrate molecules, resulting in their oxidation. As described in the previous section, the so-called UV-AOPs have emerged as an important class of treatment processes for oxidation of trace contaminants, many of which

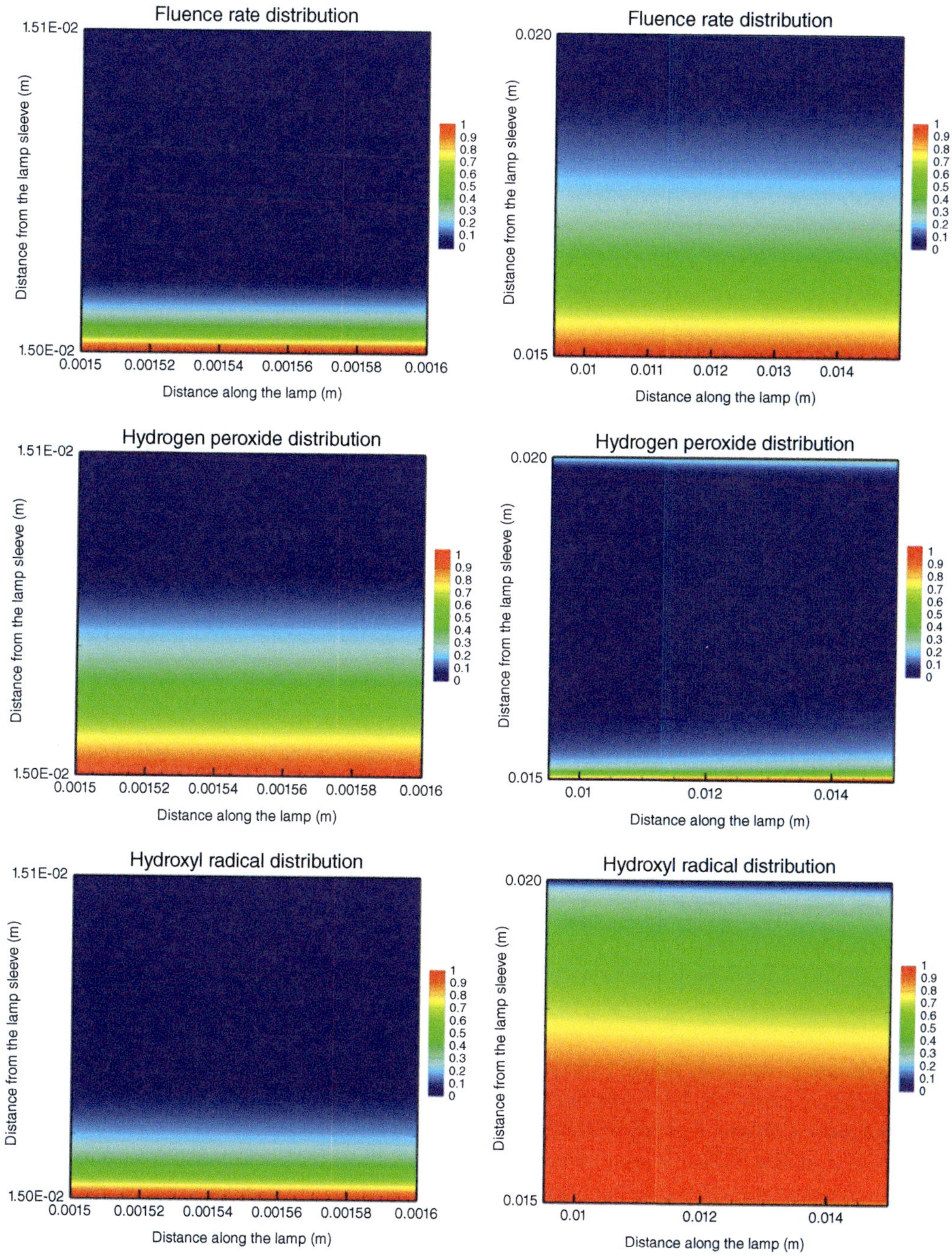


**Figure 10.42** Spatial distributions of distributions of fluence rate (top row), $H_2O_2$ (middle row), and $HO^{\cdot}$, scaled and normalized between their maximum and minimum values for a $Xe_2*$ excimer VUV source (left panels) and a low-pressure Hg lamp (right panels).

are not effectively removed or degraded by conventional water treatment processes and are often present at trace concentrations (μM or lower).

However, not all contaminant compounds are amenable to oxidation. Some compounds have structural characteristics and functional groups that are electron-poor, or that themselves have a high electron affinity. Important examples of these compounds, which are also often present at trace concentrations, include halogenated organics. Within this broad array of compounds are chlorinated solvents (e.g. trichloroethylene [TCE] and polychlorinated biphenyls [PCBs]), brominated flame retardants, and per- and polyfluorinated substances (PFAS). Collectively and historically, these halogenated organic compounds have presented some of the greatest challenges to water treatment, in part because they are so difficult to treat by conventional processes, or more broadly by processes that promote oxidation. Beyond the halogenated organics, there are some highly oxidized inorganic compounds that can be problematic in water supplies, such as perchlorate ($ClO_4^-$), bromate ($BrO_3^-$), or nitrate ($NO_3^-$). For these compounds, transformation via chemical reduction may represent a desirable approach.

ARPs have emerged as a group of treatment processes that are focused on compounds that are amenable to reduction. UV-ARPs are based on photochemical formation of highly reactive reducing agents in solution that will promote the intended chemical reduction processes. Specifically, an objective of UV-ARPs is to use UV radiation to promote the formation of reactive intermediates that have a high (negative) SRP. These compounds have a large potential to donate electrons, which results in chemical reduction of target substrates. As with UV-AOPs, these reactive intermediates will react with other (nontarget) constituents in solution, and these competitive reaction processes must be accounted for in design or simulation of UV-ARPs.

Among the most important of these highly reactive reducing intermediates are the hydrated electron ($e_{aq}^-$) and the hydrogen atom or hydrogen radical ($H^\cdot$). These two species are linked by their acid–base behavior; the hydrated electron is the conjugate base of the hydrogen radical [18]:

$$H^\cdot \rightleftharpoons H^+ + e_{aq}^- \quad pK_a = 9.6 \tag{10.134}$$

Therefore, any solution that contains hydrated electrons will also contain hydrogen radicals and the distribution between these forms will be governed by solution pH.

The hydrated electron and the hydrogen radical are both strong reducing agents, with SRPs of −2.9 and −2.3 V, respectively. The hydrated electron is a strong nucleophile. Its reactivity is influenced by the presence of some functional groups, especially those that have the ability to withdraw electrons from the rest of the molecule. In solutions where the pH is significantly below its $pK_a$ (see Equation 10.134), it tends to be the dominant reducing compound in solution [18].

There are many methods available for the formation of reactive intermediates that function as strong reducing agents. In UV-ARPs, these reducing intermediates are formed through primary photochemical reactions that involve precursors to their formation. Many of the precursors to the formation of these reducing agents are inorganic anions. Since the formation of these intermediates is initiated by primary photochemical processes, it is useful to be familiar with the absorption spectra of these precursors.

The absorption spectra of anions are often characterized by bands that are attributable to intramolecular transitions and charge-transfer-to-solvent (CTTS) bands. Intramolecular transitions are those that involve promotion of an electron from a molecular ground state to a molecular excited state. CTTS bands tend to appear in the UV range and are usually smooth and structureless. They involve the release of an electron to a surrounding solvent (water) cage and formation of a radical [86].

$$X_{aq}^- \xrightarrow{h\nu} (X^\cdot + e^-)_{aq} \tag{10.135}$$

CTTS bands are commonly observed with halide ions, but are also observed among other inorganic anions [87].

Among anions, the behavior of iodide ($I^-$) as a precursor to the formation of the hydrated electron is probably the most studied. This chemistry forms the basis of the iodide actinometer, often paired with an electron scavenger, such as nitrous oxide or iodate [88, 89]. In this process, $I^-$ absorbs a UVC photon to cause the release of an electron to a surrounding water cage. This process has been described as a CTTS reaction:

$$I_{aq}^{-} \xrightarrow{h\nu} (I^{\cdot} + e^{-})_{aq} \tag{10.136}$$

In the absence of compounds in solution that will react with the electron that is released in this process, the reverse reaction can represent the dominant fate of the products of reaction 10.136. However, when compounds that are susceptible to chemical reduction are present in solution, the hydrated electron formed in this process can play an important role. With that in mind, it is possible to use iodide as a precursor to $e_{aq}^{-}$ formation in a UV-ARP. However, this approach has not been widely applied.

### 10.11.1 $SO_3^{2-}$/UV Process

Perhaps the most commonly applied UV-ARP involves the use of tetravalent sulfur (S[IV]) as a precursor to $e_{aq}^{-}$ formation. The chemistry of this process is also central to the environmental fate of $SO_2$ and related compounds that are released to the atmosphere as a result of combustion of fossil fuels that contain sulfur (e.g. coal and petroleum) [90].

The behavior of this system depends on a number of chemical reactions that govern the speciation of S[IV] compounds. These reactions tend to be fast and reversible. As such, it is useful to consider their respective equilibria in evaluating the distribution of S[IV] compounds.

The first of these processes involves partitioning of $SO_2$ between the gas and liquid phases, which is governed by Henry's law [90–92]:

$$SO_{2(g)} \overset{H_2O}{\leftrightarrow} SO_{2(aq)} \quad H = 0.76\frac{L \cdot atm}{mole} \text{ at } 25^{o}C \tag{10.137}$$

Once dissolved in the liquid phase, $SO_{2(aq)}$ reacts $H_2O$ to yield an $SO_2$·water complex (hydrated $SO_2$):

$$SO_{2(aq)} + H_2O \rightleftharpoons SO_2 \cdot H_2O \tag{10.138}$$

Note that some people identify hydrated $SO_2$ as sulfurous acid ($H_2SO_3$); but in fact, $H_2SO_3$ does not exist in solution [92]. Hydrated $SO_2$ will behave as a diprotic acid. Loss of the first proton yields the bisulfite ion:

$$SO_2 \cdot H_2O \rightleftharpoons H^{+} + HSO_3^{-} \quad pK_{a,1} = 1.88 \tag{10.139}$$

The second deprotonation reaction yields the sulfite ion, $SO_3^{2-}$:

$$HSO_3^{-} \rightleftharpoons H^{+} + SO_3^{2-} \quad pK_{a,2} = 7.2 \tag{10.140}$$

The bisulfite ion also participates in a dimerization reaction to form pyrosulfite (disulfite) ions:

$$HSO_3^{-} + HSO_3^{-} \rightleftharpoons S_2O_5^{2-} + H_2O \quad K_D = 0.045\,M^{-1} \tag{10.141}$$

Reaction 10.141 indicates that $S_2O_5^{2-}$ will always be present at a molar concentration that is a small fraction of the molar concentration of $HSO_3^-$. Lastly, the pyrosulfite ion will also display acid/base behavior:

$$HS_2O_5^- \rightleftharpoons H^+ + S_2O_5^{2-} \quad pK_{a,p} = 2.9 \tag{10.142}$$

These equilibrium expressions can be used to illustrate the distribution of S(IV) compounds among these various forms as a function of pH. Figure 10.43 illustrates this behavior for total S(IV) concentrations of 0.1 and 1.0 M.

The distribution diagrams presented in Figure 10.43 were developed from the equilibrium expressions that are presented in Equations 10.139–10.142, but at different total S[IV] molar concentrations. For both conditions, $SO_3^{2-}$ is the dominant form at pH values above $pK_{a,2} = 7.2$. At pH values below $pK_{a,2}$, other forms of S(IV) are present at relevant concentrations. Of immediate interest is the presence of $HSO_3^-$, which is present at high concentrations at all practical pH values below neutral. It is also relevant to note that as the total S[IV] concentration increases, $S_2O_5^{2-}$ becomes relevant. The concentration of $S_2O_5^{2-}$ varies with the square of the concentration of $HSO_3^-$ because it is formed from $HSO_3^-$ by the dimerization reaction described in Equation 10.141.

The distribution of S[IV] among these forms will play an important role in determining the absorbing characteristics of S[IV] solutions. Figure 10.44 illustrates absorption spectra for the forms of S[IV] that are described above.

Speciation of S[IV] compounds affects the responses of S[IV] solutions to UV irradiation, as these compounds respond differently to UV exposure. This is illustrated by the primary photochemical processes that have been identified for the two S[IV] forms that are usually present at the highest concentration [93, 94]:

$$SO_3^{2-} \xrightarrow{h\nu} SO_3^{\cdot -} + e_{aq}^- \tag{10.143}$$

$$HSO_3^- \xrightarrow{h\nu} HSO_3^{\cdot} + e_{aq}^- \tag{10.144}$$

$$HSO_3^- \xrightarrow{h\nu} SO_3^{\cdot -} + H^{\cdot} \tag{10.145}$$

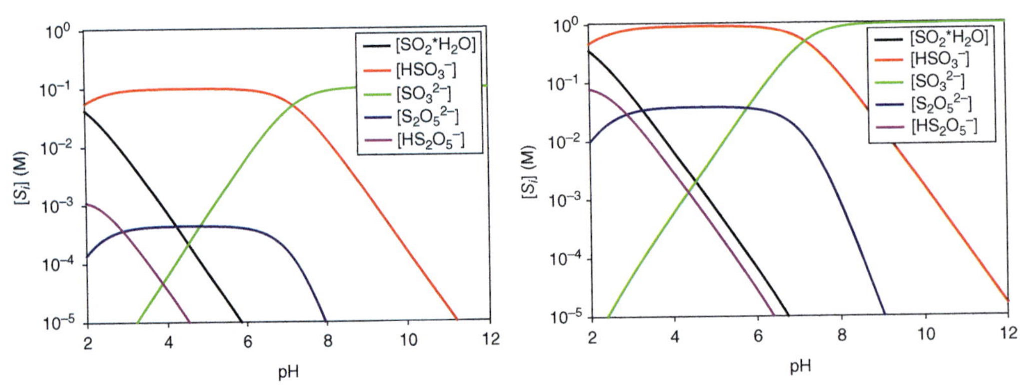


Figure 10.43 Distribution of S[IV] compounds as a function of pH. Left panel illustrates behavior in an aqueous solution with a total S(IV) concentration of 0.1 M; right panel illustrates behavior in an aqueous solution with a total S[IV] concentration of 1.0 M.

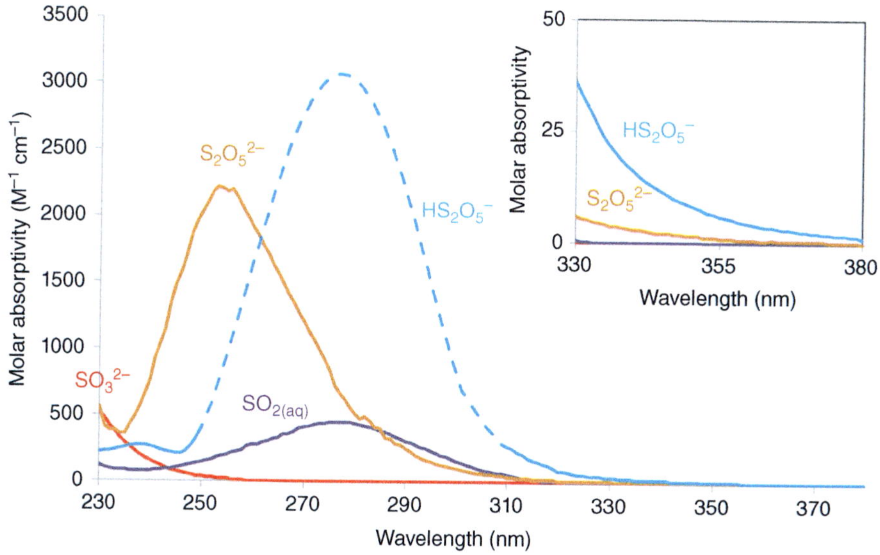


**Figure 10.44** Calculated absorption spectra for S[IV] species. Spectra for each compound were calculated from spectra of mixtures of these compounds. Note that $HSO_3^-$ was described as having "very low and relatively broad absorbance" in this wavelength range, and was not included in this graph for reasons of clarity of presentation. *Source:* Beyad et al. 2014/with permission of Royal Society of Chemistry.

These processes are effective for generation of reactive intermediates that function as strong reducing agents, specifically $e_{aq}^-$ and $H^{\cdot}$. As with many other photochemical processes, the efficiency of these photoreactions is influenced by pH and the wavelength(s) of radiation imposed on the sample.

The two forms of S[IV] that appear to play the greatest role in the formation of these radical reducing compounds are $SO_3^{2-}$ and $HSO_3^-$, largely because they are usually present at higher molar concentration than any of the other forms. However, it is possible that $S_2O_5^{2-}$ could play a relevant role in the photochemistry of this system under some conditions, as its molar absorptivity is orders of magnitude greater than that of $SO_3^{2-}$ or $HSO_3^-$, and it can be present at concentrations that are on the order of 5% of the total S[IV] in the system (see Figure 10.43, right panel). However, the photochemistry of $S_2O_5^{2-}$ is largely undefined at this point.

For chlorinated compounds, $e_{aq}^-$ tends to react rapidly, often close to the diffusional limit. These reactions tend to involve reductive dechlorination, which results in release of $Cl^-$ to solution and replacement of +1-valent chlorine on the target compound by $H^+$. This approach has been demonstrated to be effective for dechlorination of a wide range of chlorinated organic compounds, including TCE, diclofenac, 2,4-dichlorophenol (2,4-DCP), and 1,2-dichloroethane (1,2-DCA) [28, 94–97]. In addition, the hydrated electron has been demonstrated to be effective for reduction with a wide range of highly oxidized inorganic contaminant compounds, including $NO_3^-$, $N_2O$, $ClO_4^-$, $BrO_3^-$, and $IO_3^-$. In fact, these compounds are sometimes added as electron scavengers in experiments where hydrated electrons are known to be produced [88, 89].

Halogenated organic compounds represent important pollutants in the environment and in water supplies because they often express toxicity toward humans and aquatic organisms and they also tend to be quite stable (i.e. persistent). Therefore, dehalogenation processes can represent important tools for protection of water supplies and the environment.

Historically, the focus of attention on halogenated compounds started with chlorinated compounds, including dichloro-diphenyl-trichloroethane (DDT), PCBs, and a wide range of pesticides. Brominated compounds have also received considerable attention, especially as related to the behavior of brominated flame retardants. More recently, significant attention has also been given to per- and polyfluoroalkyl substances (PFAS), a broad group of chemicals that include multiple C–F groups.

PFAS compounds are used in a wide range of applications, including aqueous film-forming foams (AFFF) for firefighting, anti-stick coatings for cookware, and high-performance waxes for snow skis. Figure 10.45 illustrates representative structures of common PFAS compounds. These compounds are fully fluorinated, but with different "headgroups" as the terminus of each molecule.

Aqueous electrons generated by the $SO_3^{2-}/UV_{254}$ process are effective for partial defluorination of these compounds. The speed of defluorination and the location of the C–F bonds that are affected by this process depend on alkyl chain length and the structure of the headgroup (see Figure 10.46) [98]. For example, trifluoroacetate (TFA) is slow to defluorinate, but can be completely defluorinated by this process. Compounds with a carboxylic headgroup tend to undergo rapid defluorination, but the extent of defluorination appears to be limited. This behavior was shown to be consistent with calculated bond dissociation energies for the C–F bonds in these structures [98]. As such, processes that are able to promote the formation of the hydrated electron appear to represent rational strategies for defluorination of PFAS and related compounds.

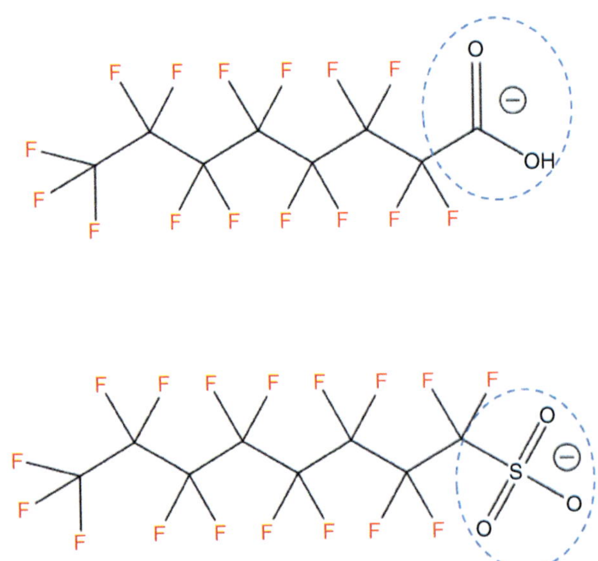

**Figure 10.45** Structures of two common PFAS compounds: top, perfluorooctanoic acid (C8); bottom, perfluorooctanesulfonic acid. Both compounds are indicated in their respective anionic forms, with their headgroups (encircled with dashed, blue line) having lost a proton.

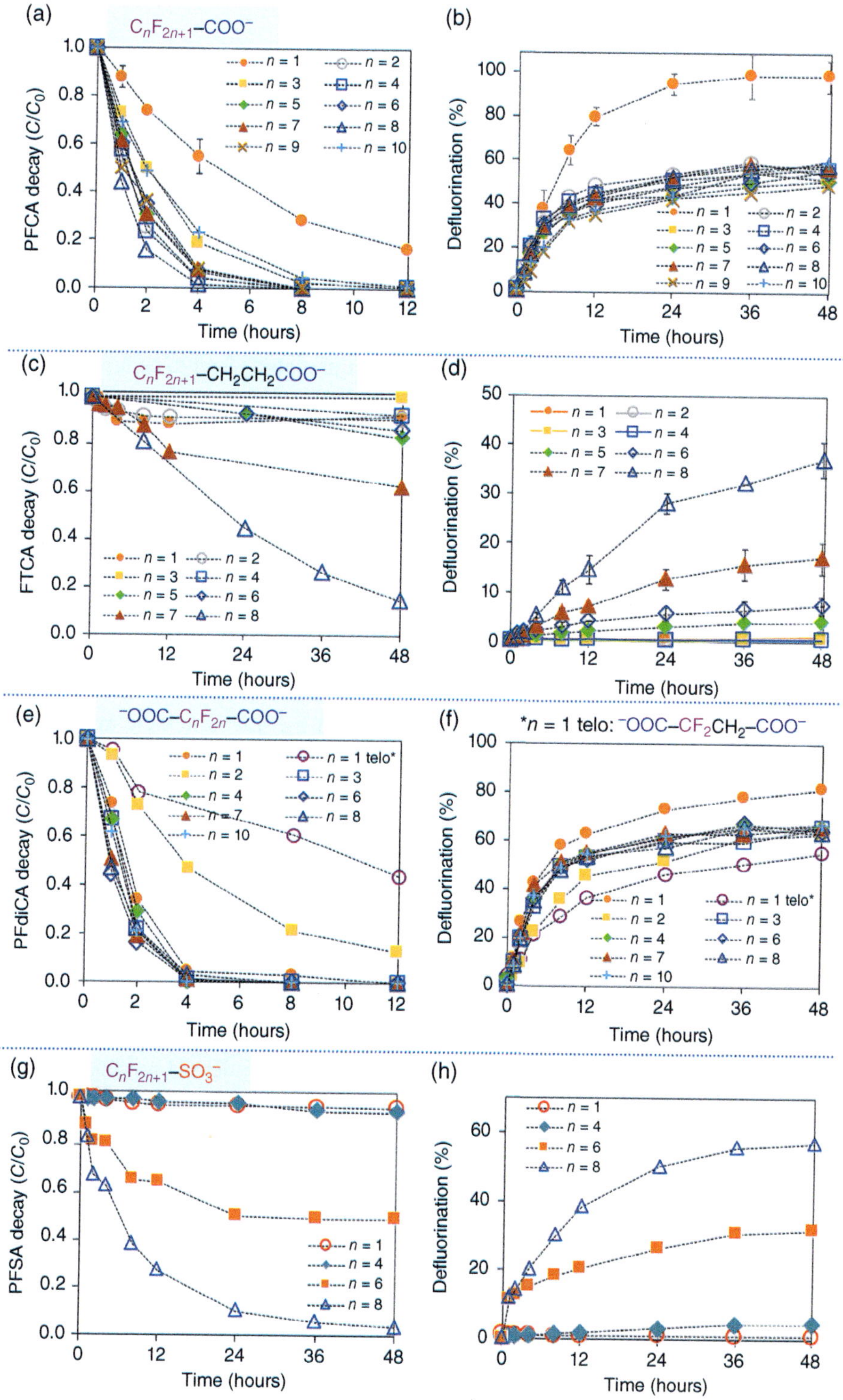


**Figure 10.46** Rate and extent of defluorination for PFAS compounds. (a, b) The rate of parent compound degradation and the extent of overall defluorination, respectively, for perfluorocarboxylic acids (PFCAs); (c, d) similar behavior for fluorotelomer carboxylic acids (FTCAs); (e, f) analogous behavior for polyfluoro dicarboxylic acids (PFdiCAs); (g, h) analogous behavior for perfluoroalkanesulfonic acids (PFSAs). *Source:* Bentel et al. 2019/with permission of American Chemical Society.

## 10.12 Chapter Summary

Direct and indirect photochemical processes can be used to promote reactions with chemical compounds that may not respond to conventional water treatment processes. As such, these processes, which are usually initiated using UV radiation, can be used to complement conventional treatment and provide measures to diminish the concentrations of trace contaminants. Some of these processes are selective and therefore have the potential to influence the behavior of only a narrow range of compounds. However, other processes are essentially nonselective, so their application tends to be aimed at situations where the concentrations of other compounds that may participate in these processes (and represent a source of "demand" for the reacting agents) have been diminished by the use of other processes.

The reactions that promote degradation in direct and indirect photolysis tend to be fast. Consequently, the overall dynamics of these processes tends to be limited by transport processes. Therefore, accurate simulations of these processes often require integrated simulations of fluid mechanics, the fluence rate field, and intrinsic reaction kinetics. Numerical models that follow this approach can provide accurate simulations of photoreactor dynamics, and as such these models represent a potentially important tool for design and optimization of photoreactors that are used for direct or indirect photochemical processes.

## 10.A Appendix

Photochemical properties of trace water contaminants. Values of the quantum yield and molar absorption coefficient are reported at 254 nm. Quantum yield estimates listed below include both observed (apparent) and innate (primary) quantum yield values. The data indicated from Wols and Hofman-Caris [99] are from a critical review of the literature. Readers are encouraged to read the original references cited within that paper for additional details on the experimental methods used and the values of the parameters listed below. When not reported from the reference listed, values of $pK_a$ were taken from https://pubchem.ncbi.nlm.nih.gov, http://foodb.ca/compounds or https://sitem.herts.ac.uk/aeru/iupac/index.htm. Values of $pK_a$ listed as "n.a." are not available, often because compounds show no measurable potential for ionization in the practical pH range. Values of pH represent the conditions of experiments that were reported in the corresponding reference; pH values listed as "n.p." were not provided in the reference. Entries of "Protonated" or "Deprotonated" in the pH column indicate that an experiment was conducted under pH conditions such that either the Protonated or Deprotonated dominated in solution.

| Compound | p$K_a$ | pH | $\Phi_{254}$ $\left(\frac{\text{mole}}{\text{einstein}}\right)$ | $\varepsilon_{254}$ $\left(\frac{\text{L}}{\text{mole}\cdot\text{cm}}\right)$ | $k_{254}$ $\left(\frac{\text{cm}^2}{\text{mJ}}\right)$ | Reference |
|---|---|---|---|---|---|---|
| 1,2,3-Trichlorobenzene | n.a. | 3–9 | 0.190 | 308 | $2.86 \times 10^{-4}$ | Wols and Hofman-Caris [99] |
| 1,4-Dichlorobenzene | 12.39 | 3–9 | 0.600 | 118 | $3.46 \times 10^{-4}$ | Wols and Hofman-Caris [99] |
| 2,4-Dichlorophe noxyacetic acid | 2.73 | 7.0 | 0.009,5 | 173 | $8.04 \times 10^{-6}$ | Wols and Hofman-Caris [99] |
| 2,6-Dinitrotoluene | 1.80 | 8.5 | 0.022,0 | 6,640 | $7.14 \times 10^{-4}$ | Wols and Hofman-Caris [99] |
| 4-*t*-Octylphenol | 10.23 | 7.0 | 0.016,0 | 675 | $5.28 \times 10^{-5}$ | Wols and Hofman-Caris [99] |
| 4-*n*-Nonylphenol | 10.31 | 7.2 | 0.32 | 400 | $6.26 \times 10^{-4}$ | Carlson et al. [15] |
| Acenaphthene | n.a. | 7.0 | 0.052,0 | 1,330 | $3.38 \times 10^{-4}$ | Wols and Hofman-Caris [99] |
| Acetaminophen | 9.76 | 7.2 | 0.006 | 7,800 | $2.29 \times 10^{-4}$ | Carlson et al. [15] |
| | 9.46 | 5.5–6 | 0.001,8 | 8,095 | $7.12 \times 10^{-5}$ | Rivas et al. [100] |
| Alachlor | 0.62 | n.p. | 0.210 | 543 | $5.58 \times 10^{-4}$ | Rivas et al. [100] |
| | | n.p. | 0.140 | 467 | $3.20 \times 10^{-4}$ | Wols and Hofman-Caris [99] |
| | | 6.0 | 0.095,0 | 428 | $1.99 \times 10^{-4}$ | |
| Amoxicillin | 3.2, 11.7 | 3–9 | 0.173 | 1,200 | $1.01 \times 10^{-3}$ | Wols and Hofman-Caris [99] |
| Anthracene | n.a. | 7.6 | 0.201 | 1,040 | $1.01 \times 10^{-3}$ | Wols and Hofman-Caris [99] |
| | | n.p. | 0.096,0 | 1,090 | $5.12 \times 10^{-4}$ | |
| Antipyrine | 1.4 | 5.5–6 | 0.004,1 | 8,386 | $1.68 \times 10^{-4}$ | Rivas et al. [100] |
| Atrazine | 1.68 | n.p. | 0.033,0 | 3,680 | $5.94 \times 10^{-4}$ | Wols and Hofman-Caris [99] |
| | | n.p. | 0.050,0 | 3,860 | $9.44 \times 10^{-4}$ | |
| | | n.p. | 0.060 | 2,647 | $7.76 \times 10^{-4}$ | Sanches et al. [101] |
| Benzene | n.a. | 3–9 | 0.880 | 250 | $1.08 \times 10^{-3}$ | Wols and Hofman-Caris [99] |
| Benzo[a]anthracene | n.a. | 7.6 | 0.001,21 | 69,400 | $4.11 \times 10^{-4}$ | Wols and Hofman-Caris [99] |
| Benzo[a]pyrene | n.a. | n.p. | 0.314 | 1,080 | $1.66 \times 10^{-3}$ | Wols and Hofman-Caris [99] |
| Bisphenol A | 9.78, 10.53 | 7.85 | 0.004,60 | 750 | $1.69 \times 10^{-5}$ | Baeza and Knappe [102] |
| Boldenone | 18.86 | 7.0 | 0.610 | 14,600 | $4.35 \times 10^{-2}$ | Wols and Hofman-Caris [99] |
| Bromoxynil | 3.86 | 7.0 | 0.040 | 4,970 | $9.72 \times 10^{-4}$ | Wols and Hofman-Caris [99] |
| Butachlor | n.a. | 7.0 | 0.820 | 410 | $1.64 \times 10^{-3}$ | Wols and Hofman-Caris [99] |
| Caffeine | 14.0 | 5.5–8.0 | 0.001,8 | 3,920 | $3.45 \times 10^{-5}$ | Wols and Hofman-Caris [99] |
| | | 5.5–6 | 0.001,8 | 4,807 | $4.23 \times 10^{-5}$ | Rivas et al. [100] |
| Carbamazepine | 13.9 | n.p. | 0.000,6 | 6,070 | $1.78 \times 10^{-5}$ | Pereira et al. [103] |
| Carbendazime | 4.29 | 8.4 | 0.002,30 | 4,470 | $5.03 \times 10^{-5}$ | Wols and Hofman-Caris [99] |
| Carbofuran | 3.78 | 5–9 | 0.016,6 | 800 | $6.49 \times 10^{-5}$ | Wols and Hofman-Caris [99] |
| Carbomethoxy fenitrothion | n.a. | 5.6 | 0.000,840 | 5,020 | $2.06 \times 10^{-5}$ | Wols and Hofman-Caris [99] |

(*Continued*)

| Compound | $pK_a$ | pH | $\Phi_{254}$ $\left(\frac{\text{mole}}{\text{einstein}}\right)$ | $\varepsilon_{254}$ $\left(\frac{\text{L}}{\text{mole} \cdot \text{cm}}\right)$ | $k_{254}$ $\left(\frac{\text{cm}^2}{\text{mJ}}\right)$ | Reference |
|---|---|---|---|---|---|---|
| Chloramphenicol | 7.49 | 7.5 | 0.084,0 | 4,330 | $1.78 \times 10^{-3}$ | Wols and Hofman-Caris [99] |
| Chlorfenvinphos | n.a. | 7.0 | 0.089,0 | 7,290 | $3.17 \times 10^{-3}$ | Wols and Hofman-Caris [99] |
| | | n.p. | 0.056,0 | 8,656 | $2.37 \times 10^{-3}$ | Sanches et al. [101] |
| Chlorotetracyclene | 3.58 | 8.0 | 0.029,5 | 14,800 | $2.13 \times 10^{-3}$ | Wols and Hofman-Caris [99] |
| Chlorotoluren | n.a. | 2–9 | 0.031,9 | 6,080 | $9.48 \times 10^{-4}$ | Wols and Hofman-Caris [99] |
| Chlorpyrifos | n.a. | 5.6 | 0.016,0 | 650 | $5.08 \times 10^{-5}$ | Wols and Hofman-Caris [99] |
| Ciprofloxacin | 6.09, 8.74 | 8.0 | 0.013,4 | 22,000 | $1.44 \times 10^{-3}$ | Wols and Hofman-Caris [99] |
| | | n.p. | 0.010,3 | 12,400 | $6.24 \times 10^{-4}$ | Pereira et al. [103] |
| Clofibric acid | 3.18 | 5.5 | 0.010,8 | 380 | $2.01 \times 10^{-5}$ | Wols and Hofman-Caris [99] |
| | | n.p. | 0.539 | 400 | $1.05 \times 10^{-3}$ | Pereira et al. [103] |
| Coumafos | n.a. | 5.6 | 0.027,0 | 2,400 | $3.17 \times 10^{-4}$ | Wols and Hofman-Caris [99] |
| 1,2-Dibromo-3-chloropropane | n.a. | 8.0 | 0.049,0 | 14.7 | $3.52 \times 10^{-6}$ | Wols and Hofman-Caris [99] |
| 2-Methyl-4,6-dinitrophenol | 4.23 | 5.6 | 0.000,480 | 6,830 | $1.60 \times 10^{-5}$ | Wols and Hofman-Caris [99] |
| Desethylatrazine | n.a. | n.p. | 0.059,0 | 3,440 | $9.92 \times 10^{-4}$ | Wols and Hofman-Caris [99] |
| Desethyldesisopropylatrazine | n.a. | n.p. | 0.018,0 | 2,200 | $1.94 \times 10^{-4}$ | Wols and Hofman-Caris [99] |
| Desisopropylatrazine | n.a. | n.p. | 0.059,0 | 3,600 | $1.04 \times 10^{-3}$ | Wols and Hofman-Caris [99] |
| Diatrizoic acid | 1.13, 7.95 | 7.0 | 0.035,0 | 31,200 | $5.34 \times 10^{-3}$ | Wols and Hofman-Caris [99] |
| Diazinon | 2.6 | 5.6 | 0.082,0 | 3,890 | $1.56 \times 10^{-3}$ | Wols and Hofman-Caris [99] |
| | | 7.0 | 0.086,0 | 2,000 | $3.41 \times 10^{-4}$ | |
| Dibenz[a,b]anthracene | n.a. | 7.6 | 0.002,22 | 12,600 | $1.37 \times 10^{-4}$ | Wols and Hofman-Caris [99] |
| Diclofenac | 4.2 | Protonated | 0.272 | 4,260 | $5.67 \times 10^{-3}$ | Canonica et al. [16] |
| | | Deprotonated | 0.384 | 4,260 | $8.00 \times 10^{-3}$ | |
| | 4.0–4.2 | 5.5–6 | 0.28 | 6,140 | $8.41 \times 10^{-3}$ | Meite et al. [104] |
| | 4.15 | 7.85 | 0.213 | 5,202 | $5.42 \times 10^{-3}$ | Baeza and Knappe [102] |
| Dimetridazole | 2.38 | 7.0 | 0.003,20 | 2,240 | $3.50 \times 10^{-5}$ | Wols and Hofman-Caris [99] |
| Diphenhydramine | 8.98 | 7.0 | 0.125 | 388 | $2.37 \times 10^{-4}$ | Wols and Hofman-Caris [99] |
| Disulfoton | n.a. | 8.0 | 0.160 | 160 | $1.25 \times 10^{-4}$ | Wols and Hofman-Caris [99] |
| Diuron | 13.18 | 2–9 | 0.011,5 | 15,700 | $8.83 \times 10^{-4}$ | Wols and Hofman-Caris [99] |
| | | 2–8.5 | 0.012,5 | 16,500 | $1.01 \times 10^{-3}$ | |
| | | n.p. | 0.019,0 | 16,162 | $1.50 \times 10^{-3}$ | Rivas et al. [100] |
| Doxycycline | 3.09 | 5–8 | 0.001,10 | 13,500 | $7.26 \times 10^{-5}$ | Wols and Hofman-Caris [99] |
| | | 8.0 | 0.021,9 | 11,100 | $1.19 \times 10^{-3}$ | |
| | 3.4, 7.7, 9.3 | 5.5–6 | 0.001,1 | 12,129 | $6.52 \times 10^{-5}$ | Rivas et al. [100] |

| | | | | | | |
|---|---|---|---|---|---|---|
| *O*-Ethyl *O*-(4-nitrophenyl) phenylphosphonothioate | n.a. | 5.6 | 0.008,10 | 4,360 | $1.73 \times 10^{-4}$ | Wols and Hofman-Caris [99] |
| Estradiol | 10.4 | 5.5–6 | 0.067 | 420 | $1.38 \times 10^{-4}$ | Meite et al. [104] |
| Estrone | 10.33 | 7.2 | 0.35 | 344 | $5.89 \times 10^{-4}$ | Carlson et al. [15] |
| 17α-Ethinylestradiol | 10.4 | Protonated | 0.026 | 750 | $9.53 \times 10^{-5}$ | Canonica et al. [16] |
| | | Deprotonated | 0.057 | 3,060 | $8.53 \times 10^{-4}$ | |
| | 10.4 | 5.5–6 | 0.062 | 440 | $1.33 \times 10^{-4}$ | Meite et al. [104] |
| | 10.33 | 7.2 | 0.08 | 508 | $1.99 \times 10^{-4}$ | Carlson et al. [15] |
| Etridiazole | 2.77 | 7.0 | 0.460 | 720 | $1.62 \times 10^{-3}$ | Wols and Hofman-Caris [99] |
| Fenchlorphos | n.a. | 5.6 | 0.710 | 635 | $2.20 \times 10^{-3}$ | Wols and Hofman-Caris [99] |
| Fenitrothion | n.a. | 5.6 | 0.009,10 | 4,660 | $2.07 \times 10^{-4}$ | Wols and Hofman-Caris [99] |
| Fensulfothion | n.a. | 5.6 | 0.049,0 | 1,740 | $4.17 \times 10^{-4}$ | Wols and Hofman-Caris [99] |
| Fenthion | n.a. | 5.6 | 0.092,0 | 10,300 | $4.63 \times 10^{-3}$ | Wols and Hofman-Caris [99] |
| Fluoranthene | n.a. | n.p. | 0.447 | 311 | $6.80 \times 10^{-4}$ | Wols and Hofman-Caris [99] |
| Fluorene | n.a. | 7.0 | 0.007,50 | 16,700 | $6.12 \times 10^{-4}$ | Wols and Hofman-Caris [99] |
| Hydrochlorothiazide | 7.9, 9.2 | 3–7 | 0.041,0 | 6,650 | $1.33 \times 10^{-3}$ | Wols and Hofman-Caris [99] |
| Ibuprofen | 4.85 | 7.0 | 0.192 | 256 | $2.40 \times 10^{-4}$ | Wols and Hofman-Caris [99] |
| | | 7.2 | 0.086 | 282 | $1.19 \times 10^{-4}$ | Carlson et al. [15] |
| Iohexol | >13 | n.p. | 0.040,3 | 27,620 | $5.44 \times 10^{-3}$ | Pereira et al. [103] |
| Iopromide | 11.09 | 7.0 | 0.039 | 21,040 | $4.01 \times 10^{-3}$ | Canonica et al. [16] |
| Isazofos | 4.0 | 8.0 | 0.027,0 | 70 | $9.24 \times 10^{-6}$ | Wols and Hofman-Caris [99] |
| Isofenfos | 9.76 | 5.6 | 0.068,0 | 1,260 | $4.19 \times 10^{-4}$ | Wols and Hofman-Caris [99] |
| Isoproturon | n.a. | 8.0 | 0.031,0 | 850 | $1.29 \times 10^{-4}$ | Wols and Hofman-Caris [99] |
| | | 2–9 | 0.003,70 | 6,070 | $1.10 \times 10^{-4}$ | |
| | | n.p. | 0.002,0 | 5,944 | $5.81 \times 10^{-5}$ | Sanches et al. [101] |
| Ketoprofen | 4.45 | n.p. | 0.236,4 | 15,450 | $1.79 \times 10^{-2}$ | Pereira et al. [103] |
| | | 9.0, 7.0 | 0.360 | 15,200 | $2.68 \times 10^{-2}$ | Wols and Hofman-Caris [99] |
| Ketorolac | 3.84 | 5.5–6, 8.0 | 0.006,0 | 6,540 | $1.92 \times 10^{-4}$ | Wols and Hofman-Caris [99] |
| | 3.5 | 5.5–6 | 0.006,0 | 6,368 | $1.87 \times 10^{-4}$ | Rivas et al. [100] |
| Linuron | n.a. | 2–9 | 0.003,60 | 13,400 | $2.36 \times 10^{-4}$ | Wols and Hofman-Caris [99] |
| 4-Chloro-2-methylphenoxyacetic acid | 3.13 | 7.0 | 0.150 | 352 | $2.58 \times 10^{-4}$ | Wols and Hofman-Caris [99] |
| Mefenamic acid | 4.2 | 10.0 | 0.004,70 | 5,500 | $1.26 \times 10^{-4}$ | Wols and Hofman-Caris [99] |
| Methyl chlorpyrifos | n.a. | 5.6 | 0.013,0 | 610 | $3.88 \times 10^{-5}$ | Wols and Hofman-Caris [99] |
| Methyl parathion | 7.15 | 5.6 | 0.000,430 | 4,580 | $9.63 \times 10^{-6}$ | Wols and Hofman-Caris [99] |
| Metolachlor | n.a. | 7.0 | 0.300 | 503 | $7.38 \times 10^{-4}$ | Wols and Hofman-Caris [99] |
| Metoprolol | 9.7 | 7.0 | 0.063,8 | 900 | $2.81 \times 10^{-4}$ | Wols and Hofman-Caris [99] |
| Metronidazole | 2.55 | 7.0 | 0.005,60 | 560 | $1.53 \times 10^{-5}$ | Wols and Hofman-Caris [99] |
| | | 7.0 | 0.003,50 | 2,100 | $3.59 \times 10^{-5}$ | |

(*Continued*)

| Compound | p$K_a$ | pH | $\Phi_{254}$ $\left(\frac{\text{mole}}{\text{einstein}}\right)$ | $\varepsilon_{254}$ $\left(\frac{\text{L}}{\text{mole} \cdot \text{cm}}\right)$ | $k_{254}$ $\left(\frac{\text{cm}^2}{\text{mJ}}\right)$ | Reference |
|---|---|---|---|---|---|---|
| | | 5.5 | 0.09 | 114 | $5.02 \times 10^{-5}$ | Dantas et al. [105] |
| *N*-nitrosodimethylamine | n.a. | 8.1 | 0.30 | 1,650 | $2.42 \times 10^{-3}$ | Sharpless and Linden [9] |
| | | | | | | Minakata and Coscarelli [10] |
| | n.a. | 3 | 0.25 | 1,650 | $2.02 \times 10^{-3}$ | Stefan and Bolton [7] |
| | | 7 | 0.13 | 1,650 | $1.05 \times 10^{-3}$ | |
| Naproxen | 4.15–4.70 | 5.5–6 | 0.024,0 | 3,946 | $4.63 \times 10^{-4}$ | Meite et al. [104] |
| | | | 0.050,0 | 3,200 | $7.82 \times 10^{-4}$ | Wols and Hofman-Caris [99] |
| | | n.p. | 0.009,30 | 4,900 | $2.23 \times 10^{-4}$ | Pereira et al. [103] |
| Nitrobenzene | 3.98 | 8.5 | 0.007,00 | 5,560 | $1.90 \times 10^{-4}$ | Wols and Hofman-Caris [99] |
| Norfloxacin | 6.34, 8.75 | 5.5, 8.8 | 0.003,40 | 15,400 | $2.56 \times 10^{-4}$ | Wols and Hofman-Caris [99] |
| Oxamyl | -2.11 | 6.0 | 0.550 | 5,320 | $1.43 \times 10^{-2}$ | Wols and Hofman-Caris [99] |
| Oxytetracycline | 3.27, 9.5 | 8.0 | 0.004,20 | 14,300 | $2.94 \times 10^{-4}$ | Wols and Hofman-Caris [99] |
| | | 8.0 | 0.018,7 | 13,400 | $1.23 \times 10^{-3}$ | |
| Acetaminophen (Paracetamol) | 9.38 | 5.5–6, 8.0 | 0.001,80 | 8,280 | $7.29 \times 10^{-5}$ | Wols and Hofman-Caris [99] |
| Pentachlorophenol | 4.70 | n.p. | 0.025,0 | 7,400 | $9.04 \times 10^{-4}$ | Rivas et al. [100] |
| Phenacetin | 14.98 | 3–9 | 0.004,60 | 9,100 | $2.05 \times 10^{-4}$ | Wols and Hofman-Caris [99] |
| Phenanthrene | n.a. | 7.0 | 0.006,90 | 40,500 | $1.37 \times 10^{-3}$ | Wols and Hofman-Caris [99] |
| Phenazone (antipyrine) | 1.4 | 5.5–6, 8.0 | 0.004,10 | 8,290 | $1.66 \times 10^{-4}$ | Wols and Hofman-Caris [99] |
| | | 7.0 | 0.006,32 | 8,910 | $2.75 \times 10^{-4}$ | |
| Phenol | 9.99 | 5.6 | 0.021,0 | 980 | $1.01 \times 10^{-4}$ | Wols and Hofman-Caris [99] |
| | | 8.0 | 0.030,0 | 520 | $7.63 \times 10^{-5}$ | |
| Phenytoin | 8.33 | 7.0 | 0.279 | 1,260 | $1.72 \times 10^{-3}$ | Wols and Hofman-Caris [99] |
| Primidone | 12.3 | 7.0 | 0.082,0 | 220 | $8.82 \times 10^{-5}$ | Wols and Hofman-Caris [99] |
| Profenofos | n.a. | 8.0 | 0.026,0 | 460 | $5.85 \times 10^{-5}$ | Wols and Hofman-Caris [99] |
| Progesterone | 18.92 | 5.5–6 | 0.022 | 17,020 | $1.82 \times 10^{-3}$ | Meite et al. [104] |
| Propachlor | n.a. | 7.0 | 0.127 | 421 | $2.51 \times 10^{-4}$ | Wols and Hofman-Caris [99] |
| Propanalol | 9.14 | 3.9 | 0.07 | 28.9 | $9.89 \times 10^{-6}$ | Dantas et al. [105] |
| Propazine | 1.7 | n.p. | 0.099,0 | 3,370 | $1.63 \times 10^{-3}$ | Wols and Hofman-Caris [99] |
| Prothiofos | n.a. | 5.6 | 1.10 | 121 | $6.51 \times 10^{-4}$ | Wols and Hofman-Caris [99] |
| Pyrene | n.a. | 7.6 | 0.003,85 | 18,200 | $3.43 \times 10^{-4}$ | Wols and Hofman-Caris [99] |
| Pyridaphenthion | -2.09 | 5.6 | 0.000,320 | 1,870 | $2.93 \times 10^{-6}$ | Wols and Hofman-Caris [99] |
| Ronidazole | 1.2 | 7.0 | 0.002,21 | 2,260 | $2.44 \times 10^{-5}$ | Wols and Hofman-Caris [99] |
| | | n.p. | 0.010,2 | 2,180 | $1.09 \times 10^{-4}$ | Qin et al. [106] |
| Simazine | 1.62 | n.p. | 0.083,0 | 3,330 | $1.35 \times 10^{-3}$ | Wols and Hofman-Caris [99] |
| Sulfachloropyridazine | 1.67 | 7.2 | 0.003,5 | 21,400 | $3.66 \times 10^{-4}$ | Carlson et al. [15] |

| | | | | | | |
|---|---|---|---|---|---|---|
| Sulfadiazene | 2.02,6.43 | 3.6 | 0.004,30 | 13,590 | $2.86 \times 10^{-4}$ | Baeza and Knappe [102] |
| | | 7.85 | 0.005,81 | 20,150 | $5.72 \times 10^{-4}$ | |
| | | 9.7 | 0.003,78 | 20,660 | $3.82 \times 10^{-4}$ | |
| Sulfamethazine | 2.26, 7.65 | 3.6 | 0.002,82 | 16,196 | $2.23 \times 10^{-4}$ | Baeza and Knappe [102] |
| | | 7.85 | 0.008,70 | 18,525 | $7.88 \times 10^{-4}$ | |
| | | 9.7 | 0.008,49 | 20,538 | $8.53 \times 10^{-4}$ | |
| Sulfamethoxazole | 5.7 | Protonated | 0.212 | 11,890 | $1.23 \times 10^{-2}$ | Canonica et al. [16] |
| | | Deprotonated | 0.046 | 16,760 | $3.77 \times 10^{-3}$ | |
| | 1.74, 5.65 | 3.6 | 0.180 | 11,130 | $9.79 \times 10^{-3}$ | Baeza and Knappe [102] |
| | | 7.85 | 0.029,7 | 16,580 | $2.41 \times 10^{-3}$ | |
| | 1.4 | 7.2 | 0.033 | 17,100 | $2.76 \times 10^{-3}$ | Carlson et al. [15] |
| Terbuthylazine | 2.0 | n.p. | 0.094,0 | 3,830 | $1.76 \times 10^{-3}$ | Wols and Hofman-Caris [99] |
| Testosterone | 17.4 | 5.5–6 | 0.033 | 15,140 | $2.44 \times 10^{-3}$ | Meite et al. [104] |
| Tetracycline | 3.30 | 8.0 | 0.003,80 | 13,500 | $2.51 \times 10^{-4}$ | Wols and Hofman-Caris [99] |
| Tinidazole | 4.7 | 7.0 | 0.001,96 | 2,340 | $2.24 \times 10^{-5}$ | Wols and Hofman-Caris [99] |
| Tolclofos | n.a. | 5.6 | 0.017,0 | 774 | $6.43 \times 10^{-5}$ | Wols and Hofman-Caris [99] |
| Trenbolone | 18.4 | 7.0 | 0.002,90 | 6,300 | $8.93 \times 10^{-5}$ | Wols and Hofman-Caris [99] |
| Triclosan | 8.21 | 4.2 | 0.34 | 1,440 | $2.39 \times 10^{-3}$ | Carlson et al. [99] |
| | | 7.2 | 0.34 | 1,795 | $2.98 \times 10^{-3}$ | |
| | | 10.8 | 0.25 | 4,590 | $5.61 \times 10^{-3}$ | |
| Trifluralin | -1 | 7.0 | 0.600 | 4,970 | $1.46 \times 10^{-2}$ | Wols and Hofman-Caris [99] |
| Trimethoprim | 7.12 | 7.8 | 0.001,18 | 2,940 | $1.70 \times 10^{-5}$ | Wols and Hofman-Caris [99] |
| | 3.23, 6.76 | 3.6 | 0.000,59 | 4,956 | $1.43 \times 10^{-5}$ | Baeza and Knappe [102] |
| | | 7.85 | 0.001,8 | 2,942 | $2.59 \times 10^{-5}$ | |
| | | 9.7 | 0.001,49 | 2,635 | $1.92 \times 10^{-5}$ | |
| Triphenyl phosphate | n.a. | 5.6 | 0.290 | 644 | $9.13 \times 10^{-4}$ | Wols and Hofman-Caris [99] |
| Triphenyltin hydroxide | 5.20 | n.p. | 1.25 | 582 | $3.56 \times 10^{-3}$ | Wols and Hofman-Caris [99] |
| *m*-cresol | 10.1 | 5.6 | 0.057,0 | 302 | $8.42 \times 10^{-5}$ | Wols and Hofman-Caris [99] |
| *m*-nitrophenol | 8.36 | 5.6 | 0.000,190 | 3,410 | $3.17 \times 10^{-6}$ | Wols and Hofman-Caris [99] |
| Butylparaben | 8.47 | 7.0 | 0.003,30 | 15,400 | $2.48 \times 10^{-4}$ | Wols and Hofman-Caris [99] |
| *o*-nitrophenol | 7.23 | 5.6 | 0.002,00 | 4,310 | $4.21 \times 10^{-5}$ | Wols and Hofman-Caris [99] |

General Observations

As described in the text, the rate of a photochemical reaction is governed by the product of the quantum yield and the molar absorption coefficient. Under the assumption of an elementary photochemical reaction (i.e. pure photolysis), the rate of the reaction can be described as follows:

$$\frac{d[B]}{dD} = -k \cdot [B] \tag{10.A.1}$$

where

$$k = \frac{2.303 \cdot \varepsilon \cdot \Phi \cdot \lambda}{h \cdot c_0 \cdot N_A} \tag{10.A.2}$$

Values of this rate constant corresponding to monochromatic radiation at a wavelength of 254 nm are compiled in the sixth column of the table above. Rearrangement and integration of Equation 10.1 allows estimation of the $UV_{254}$ dose required to achieve a given fractional photodegradation of a target compound. For purposes of this analysis, we will develop estimates of the $UV_{254}$ dose required to achieve 1 $\log_{10}$ unit of photodegradation of each compound ($D_{10}$):

$$D_{10} = -\frac{\ln(0.1)}{k_{254}} \tag{10.A.3}$$

In the table above, the calculated values of the rate constant ($k_{254}$) range from roughly $3 \times 10^{-6} \frac{cm^2}{mJ} \lesssim k_{254} \lesssim 5 \times 10^{-2} \frac{cm^2}{mJ}$. Using these values as a representation of the range of responses to be expected for trace contaminants in water, we can identify the range of $UV_{254}$ doses required to achieve 1 $\log_{10}$ unit of photodegradation of these compounds using equation 10.3. A plot of this range of $UV_{254}$ doses is presented in Figure 10.A.1.

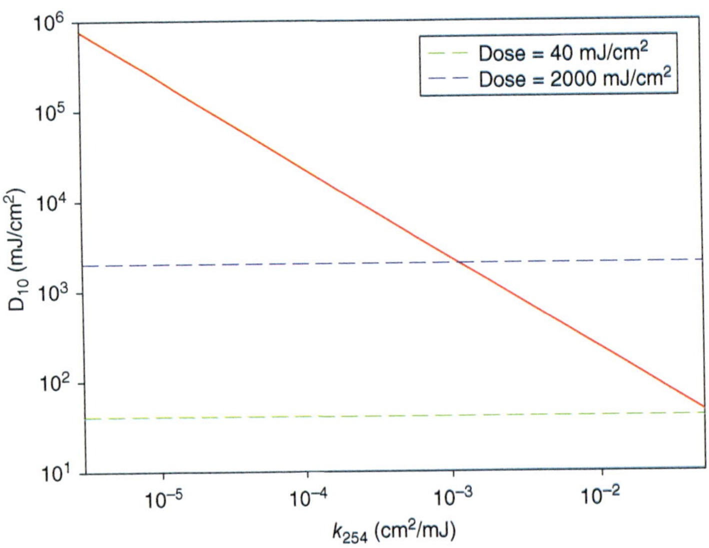


**Figure 10.A.1** Calculated values of the $UV_{254}$ dose required to accomplish 1 $\log_{10}$ unit of photodegradation as a function of the first-order photolysis rate constant. Also included in this graph are horizontal reference lines to indicate $UV_{254}$ doses of 40 and 2000 $\frac{mJ}{cm^2}$, which represent rough indications of the range of $UV_{254}$ doses that can be applied in practical, full-scale applications for disinfection and photolysis applications, respectively.

The results presented in Figure 10.A.1 indicate that photodegradation of dissolved compounds will only be relevant for compounds that have a first-order photolysis rate constant of roughly $10^{-3}\frac{\text{cm}^2}{\text{mJ}}$ or larger. For compounds that have first-order photolysis rate constants below this value, the extent of photodegradation achieved in most practical UV photoreactors will be small or negligible.

## Notes

**a** Equation 10.5 is dimensionally correct, but will require unit conversions to reduce $k$ to units of $\frac{\text{cm}^2}{\text{mJ}}$.

**b** As described previously, unit conversions are implied in the equations given in this text, and are included explicitly in these calculations.

**c** Note that $E_{EM}$ values are normally reported in units of kW hr/kg. As such, it will be necessary to apply some unit conversions to Equation 10.10 to yield an $E_{EM}$ value that has these units.

**d** All of the methods described in this section are presented under the assumption of monochromatic radiation. However, it is possible for any of these methods to be applied with a polychromatic source; the output of these processes would be an estimate of the quantum yield spectrum for the reaction, and would require integration of the polychromatic effects on the target molecule. Methods to account for polychromatic behavior are presented in Chapter 2. In addition, it is often necessary to account for reflection at the reactor surface (and other losses). Again, details of the physics and math required to complete these calculations are presented in Chapter 2.

**e** When a chemical actinometer is used, it is assumed that $\varepsilon_\lambda$ and $\Phi_\lambda$ are known for the actinometer.

**f** Note that if the photochemical products that are generated from the actinometer also absorb an appreciable fraction of the incident radiation, it may be necessary to constrain the analysis of data to only the early portion of this experiment when the concentration of the absorbing photoproduct is low or negligible.

**g** A more rigorous and complete definition of free chlorine is presented later in this chapter as part of the section that addresses the chlorine/UV advanced oxidation process.

**h** An isosbestic point is defined as a "wavelength at which the total absorbance of a sample does not change during a chemical reaction or a physical change of the sample" [13] One way this condition can be met (and is illustrated here) is for a reactant and product to have the same molar absorption coefficient and 1:1 stoichiometry. If the sum of the molar concentrations of the reactant and product remain constant, then absorbance at this wavelength does not change with reaction progress.

**i** It is possible to develop this relationship for diprotic or triprotic acids, but the algebra is more complicated. Also, even when a multiprotic acid is present, there is usually only one of the protonation/deprotonation reactions that is relevant for a given pH range, such as circumneutral pH values that are commonly observed in water treatment.

**j** The subscript index "$i$" refers to the number of protons that have been lost from the fully-protonated form of the molecule.

**k** For a thermal chemical reaction to occur, the reactant molecules must be brought into contact (i.e. their electronic orbitals must overlap) and there must be sufficient time for one or more bonds to be broken or formed. In a general sense, this indicates that the observed rate at which a reaction proceeds can be affected by the rate at which the molecules are brought together, the rate at which the molecules react with each other once this condition has been met, or both. In the extreme condition when the intrinsic rate of the reaction is so fast that (in effect) the reaction proceeds immediately when molecules are brought into contact, the observed reaction rate is then limited by the rate at which molecules are brought into contact. On a small scale, this process will be limited by diffusive transport. Such reactions are governed by this diffusional limitation, which represents a theoretical upper limit on a reaction rate.

**l** This method can be adapted to experimental conditions wherein reactive intermediates are generated by other processes. For such experiments, it is necessary to include terms to describe the rate of formation of the intermediate by other processes.

**m** SRP values are from [42].

## References


1 Miklos, D.B., Remy, C., Jekel, M. et al. (2018). Evaluation of advanced oxidation processes for water and wastewater treatment – a critical review. *Water Research 139*: 118–131.

2 Daughton, C.G. and Ternes, T.A. (1999). Pharmaceuticals and personal care products in the environment: Agents of subtle change? *Environmental Health Perspectives 107*: 907–938.

3 United States Environmental Protection Agency (US EPA) (2014). Technical Fact Sheet: *N*-Nitrosodimethylamine (NDMA). https://www.epa.gov/sites/production/files/2014-03/documents/ffrrofactsheet_contaminant_ndma_january2014_final.pdf.

4 Walse, S.S. and Mitch, W.A. (2008). Nitrosamine carcinogens also swim in chlorinated pools. *Environmental Science & Technology 42* (4): 1032–1037.

5 Soltermann, F., Lee, M., Canonica, S., and von Gunten, U. (2013). Enhanced *N*-nitrosamine formation in pool water by UV irradiation of chlorinated secondary amines in the presence of monochloramine. *Water Research 47* (1): 79–90.

6 Plumlee, M.H. and Reinhard, M. (2007). Photochemical attenuation of *N*-nitrosodimethylamine (NDMA) and other nitrosamines in surface water. *Environmental Science & Technology 41* (17): 6170–6176.

7 Stefan, M.I. and Bolton, J.R. (2002). UV direct photolysis of *N*-nitrosodimethylamine (NDMA): kinetic and product study. *Helvetica Chimica Acta 85* (5): 1416–1426.

8 Mitch, W.A., Sharp, J.O., Trussell, R.R. et al. (2003). *N*-Nitrosodimethylamine (NDMA) as a drinking water contaminant: a review. *Environmental Engineering Science 20* (5): 389–404.

9 Sharpless, C.M. and Linden, K.G. (2003). Experimental and model comparisons of low- and medium-pressure Hg lamps for the direct and $H_2O_2$ assisted UV photodegradation of *N*-nitrosodimethylamine in simulated drinking water. *Environmental Science & Technology 37* (9): 1933–1940.

10 Minakata, D. and Coscarelli, E. (2018). Mechanistic insight into the degradation of nitrosamines via aqueous-phase UV photolysis or a UV-based advanced oxidation process: quantum mechanical calculations. *Molecules 23* (3): 539.

11 Bolton, J.R., Bircher, K.G., Tumas, W., and Tolman, C.A. (2001). Figures-of-merit for the technical development and application of advanced oxidation technologies for both electric- and solar-driven systems – (IUPAC Technical Report). *Pure and Applied Chemistry 73* (4): 627–637.

12 Zepp, R.G. (1978). Quantum yields for reaction of pollutants in dilute aqueous solution. *Environmental Science & Technology 12* (3): 327–329.

13 Braslavsky, S.E. (2007). Glossary of terms used in photochemistry 3(rd) edition (IUPAC recommendations 2006). *Pure and Applied Chemistry 79* (3): 293–465.

14 Yin, R., Ling, L., and Shang, C. (2018). Wavelength-dependent chlorine photolysis and subsequent radical production using UV-LEDs as light sources. *Water Research 142*: 452–458.

15 Carlson, J.C., Stefan, M.I., Parnis, J.M., and Metcalfe, C.D. (2015). Direct UV photolysis of selected pharmaceuticals, personal care products and endocrine disruptors in aqueous solution. *Water Research 84*: 350–361.

16 Canonica, S., Meunier, L., and Von Gunten, U. (2008). Phototransformation of selected pharmaceuticals during UV treatment of drinking water. *Water Research 42* (1–2): 121–128.

17 Wang, Y.F., Roddick, F.A., and Fan, L.H. (2017). Direct and indirect photolysis of seven micropollutants in secondary effluent from a wastewater lagoon. *Chemosphere 185*: 297–308.

18 Buxton, G.V., Greenstock, C.L., Helman, W.P., and Ross, A.B. (1988). Critical-review of rate constants for reactions of hydrated electrons, hydrogen-atoms and hydroxyl radicals (.OH/.O-) in aqueous-solution. *Journal of Physical and Chemical Reference Data 17* (2): 513–886.

19 Cater, S.R., Stefan, M.I., Bolton, J.R., and Safarzadeh-Amiri, A. (2000). UV/$H_2O_2$ treatment of methyl tert-butyl ether in contaminated waters. *Environmental Science & Technology 34* (4): 659–662.

20 Lopez, J.L., Einschlag, F.S.G., Gonzalez, M.C. et al. (2000). Hydroxyl radical initiated photodegradation of 4-chloro-3,5-dinitrobenzoic acid in aqueous solution. *Journal of Photochemistry and Photobiology a-Chemistry 137* (2–3): 177–184.

21 Einschlag, F.S.G., Carlos, L., and Capparelli, A.L. (2003). Competition kinetics using the UV/$H_2O_2$ process: a structure reactivity correlation for the rate constants of hydroxyl radicals toward nitroaromatic compounds. *Chemosphere 53* (1): 1–7.

22 Armstrong, D.A., Huie, R.E., Koppenol, W.H. et al. (2015). Standard electrode potentials involving radicals in aqueous solution: inorganic radicals (IUPAC technical report). *Pure and Applied Chemistry 87* (11-12): 1139–1150.

23 Bolton, J.R. and Stefan, M.I. (2002). Fundamental photochemical approach to the concepts of fluence (UV dose) and electrical energy efficiency in photochemical degradation reactions. *Research on Chemical Intermediates 28* (7–9): 857–870.

24 Baxendale, J.H. and Wilson, J.A. (1957). The photolysis of hydrogen peroxide at high light intensities. *Transactions of the Faraday Society 53* (3): 344–356.

25 Sauer, M.C., Brown, W.G., and Hart, E.J. (1984). O(P-3) atom formation by the photolysis of hydrogen-peroxide in alkaline aqueous-solutions. *Journal of Physical Chemistry 88* (7): 1398–1400.

26 Chu, L. and Anastasio, C. (2005). Formation of hydroxyl radical from the photolysis of frozen hydrogen peroxide. *Journal of Physical Chemistry A 109* (28): 6264–6271.

27 Morgan, M.S., Vantrieste, P.F., Garlick, S.M. et al. (1988). Ultraviolet molar absorptivities of aqueous hydrogen-peroxide and hydroperoxyl ion. *Analytica Chimica Acta 215* (1–2): 325–329.

28 Yu, H., Nie, E., Xu, J. et al. (2013). Degradation of diclofenac by advanced oxidation and reduction processes: kinetic studies, degradation pathways and toxicity assessments. *Water Research 47* (5): 1909–1918.

29 Christensen, H., Sehested, K., and Corftizen, H. (1982). Reactions of hydroxyl radicals with hydrogen peroxide at ambient and elevated temperatures. *Journal of Physical Chemistry 86* (9): 1588–1590.

30 Farhataziz, P.C. and Ross, A.B. (1977). *Selected Specific Rates of Reactions of Transients from Water in Aqueous Solution. III. Hydroxyl Radical and Perhydroxyl Radical and their Radical Ions*, National Standards and Reference Data Series, 59. Washington, DC: U.S. National Bureau of Standards.

31 Weinstein, J. and Bielski, B.H.J. (1979). Kinetics of the interaction of $HO_2$ and $O_2$ – radicals with hydrogen peroxide. The Haber-Weiss reaction. *Journal of the American Chemical Society 101* (1): 58–62.

32 Stefan, M.I. (2017). *Advanced Oxidation Processes for Water Treatment: Fundamentals and Applications*. IWA Publishing, ISBN13: 9781780407180, eISBN: 9781780407197, 710 pp.

33 Bielski, B.H.J., Cabelli, D.E., Arudi, R.L., and Ross, A.B. (1985). Reactivity of $HO_2/O_2$ – radicals in aqueous solution. *Journal of Physical Chemistry Reference Data 14* (4): 1041–1100.

34 Neta, P., Huie, R.E., and Ross, A.B. (1988). Rate constants for reactions of inorganic radicals in aqueous-solution. *Journal of Physical and Chemical Reference Data 17* (3): 1027–1284.

35 Wols, B.A. and Hofman-Caris, C.H.M. (2012). Modelling micropollutant degradation in UV/H2O2 systems: Lagrangian versus Eulerian method. *Chemical Engineering Journal 210*: 289–297.

36 Wols, B.A., Harmsen, D.J.H., Wanders-Dijk, J. et al. (2015). Degradation of pharmaceuticals in UV (LP)/$H_2O_2$ reactors simulated by means of kinetic modeling and computational fluid dynamics (CFD). *Water Research 75*: 11–24.

37 Odeh, I.N., Nicoson, J.S., Hartz, K.E.H., and Margerum, D.W. (2004). Kinetics and mechanisms of bromine chloride reactions with bromite and chlorite ions. *Inorganic Chemistry 43* (23): 7412–7420.

38 Deborde, M. and von Gunten, U. (2008). Reactions of chlorine with inorganic and organic compounds during water treatment – kinetics and mechanisms: a critical review. *Water Research 42* (1-2): 13–51.

39 Stumm, W. and Morgan, J.J. (2013). *Aquatic Chemistry Chemical Equilibria and Rates in Natural Waters*. American Chemical Society.

40 Blatchley III, E.R. and Cheng, M. (2010). Reaction mechanism for chlorination of urea. *Environmental Science & Technology 44* (22): 8529–8534.

41 Sivey, J.D., McCullough, C.E., and Roberts, A.L. (2010). Chlorine monoxide ($Cl_2O$) and molecular chlorine ($Cl_2$) as active chlorinating agents in reaction of dimethenamid with aqueous free chlorine. *Environmental Science & Technology 44* (9): 3357–3362.

42 White, G.C. (1992). *The Handbook of Chlorination and Alternative Disinfectants*, 3e. New York: Van Nostrand Reinhold p xiv, 1308 p.

43 Wang, T.X., Kelley, M.D., Cooper, J.N. et al. (1994). Equilibrium, kinetic, and UV-spectral characteristics of aqueous bromine chloride, bromine and chlorine species. *Inorganic Chemistry 33* (25): 5872–5878.

44 Papanastasiou, D.K., Feierabend, K.J., and Burkholder, J.B. (2011). $Cl_2O$ photochemistry: ultraviolet/vis absorption spectrum temperature dependence and O(P-3) quantum yield at 193 and 248 nm. *Journal of Chemical Physics 134* (20): 204310.

45 Ganske, J.A., Berko, H.N., and Finlaysonpitts, B.J. (1992). Absorption cross-sections for gaseous $ClNO_2$ and $Cl_2$ at 298-K – potential organic oxidant source in the marine troposphere. *Journal of Geophysical Research-Atmospheres 97* (D7): 7651–7656.

46 Cooksey, C.C. and Reid, P.J. (2003). Femtosecond pump-probe studies of dichlorine monoxide in solution. *Journal of Physical Chemistry A 107* (29): 5508–5514.

47 Feng, Y.G., Smith, D.W., and Bolton, J.R. (2007). Photolysis of aqueous free chlorine species (NOCl and OCl-) with 254 nm ultraviolet light. *Journal of Environmental Engineering and Science 6* (3): 277–284.

48 Fang, J.Y., Fu, Y., and Shang, C. (2014). The roles of reactive species in micropollutant degradation in the UV/free chlorine system. *Environmental Science & Technology 48* (3): 1859–1868.

49 Guo, K.H., Wu, Z.H., Shang, C. et al. (2017). Radical chemistry and structural relationships of PPCP degradation by UV/chlorine treatment in simulated drinking water. *Environmental Science & Technology 51* (18): 10431–10439.

50 Wu, Z.H., Guo, K.H., Fang, J.Y. et al. (2017). Factors affecting the roles of reactive species in the degradation of micropollutants by the UV/chlorine process. *Water Research 126*: 351–360.

51 Jafvert, C.T. and Valentine, R.L. (1992). Reaction scheme for the chlorination of ammoniacal water. *Environmental Science & Technology 26* (3): 577–586.

52 Wahman, D.G. (2018). Web-based applications to simulate drinking water inorganic chloramine chemistry. *Journal American Water Works Association 110* (11): E43–E61.

53 Li, J. and Blatchley III, E.R. (2007). Volatile disinfection byproduct formation resulting from chlorination of organic-nitrogen precursors in swimming pools. *Environmental Science & Technology 41* (19): 6732–6739.

54 Weng, S.C., Li, J., Wood, K.V. et al. (2013). UV-induced effects on chlorination of creatinine. *Water Research 47* (14): 4948–4956.

55 Weng, S.C. and Blatchley III, E.R. (2013). Ultraviolet-induced effects on chloramine and cyanogen chloride formation from chlorination of amino acids. *Environmental Science & Technology 47* (9): 4269–4276.

56 Li, J. and Blatchley III, E.R. (2009). UV photodegradation of inorganic chloramines. *Environmental Science & Technology 43* (1): 60–65.

57 De Laat, J., Boudiaf, N., and Dossier-Berne, F. (2010). Effect of dissolved oxygen on the photodecomposition of monochloramine and dichloramine in aqueous solution by UV irradiation at 253.7 nm. *Water Research 44* (10): 3261–3269.

58 Soltermann, F., Widler, T., Canonica, S., and von Gunten, U. (2014). Photolysis of inorganic chloramines and efficiency of trichloramine abatement by UV treatment of swimming pool water. *Water Research 56*: 280–291.

59 Yin, R. (2019). *Investigation of the Optimal UV-LED Photolysis of Chlorine, Inorganic Chloramines and Chlorine Dioxide for Water Advanced Treatment.* Hong Kong University of Science and Technology.

60 Neta, P., Maruthamuthu, P., Carton, P.M., and Fessenden, R.W. (1978). Formation and reactivity of amino radical. *Journal of Physical Chemistry 82* (17): 1875–1878.

61 Yin, R., Blatchley III, E.R., and Shang, C. (2020). UV photolysis of mono- and dichloramine using UV-LEDs as radiation sources: photodecay rates and radical concentrations. *Environmental Science & Technology 54* (13): 8420–8429.

62 Patton, S., Li, W., Couch, K.D. et al. (2017). Impact of the ultraviolet photolysis of monochloramine on 1,4-dioxane removal: new insights into potable water reuse. *Environmental Science & Technology Letters 4* (1): 26–30.

63 Zhang, X.R., Ren, P.F., Li, W.G. et al. (2019). Synergistic removal of ammonium by monochloramine photolysis. *Water Research 152*: 226–233.

64 Zhang, X., Li, W., Blatchley III, E.R. et al. (2015). UV/chlorine process for ammonia removal and disinfection by-product reduction: comparison with chlorination. *Water Research 68*: 804–811.

65 Chuang, Y.H., Chen, S., Chinn, C.J., and Mitch, W.A. (2017). Comparing the UV/monochloramine and UV/free chlorine advanced oxidation processes (AOPs) to the UV/hydrogen peroxide AOP under scenarios relevant to potable reuse. *Environmental Science & Technology 51* (23): 13859–13868.

66 Yang, Q., Guo, Y., Xu, J. et al. (2021). Photolysis of *N*-chlorourea and its effect on urea removal in a combined pre-chlorination and UV254 process. *Journal of Hazardous Materials 411*: 125111.

67 Chen, M. (2018). *Chlorine/UV Treatment of Anatoxin-A.* West Lafayette, IN: Purdue University.

68 Chen, M. and Blatchley III, E.R. (2020). Chlorine/UV treatment of anatoxin-a by activation of the secondary amine functional group. *Environmental Science-Water Research & Technology 6* (5): 1412–1420.

69 Weng, S., Li, J., and Blatchley III, E.R. (2012). Effects of $UV_{254}$ irradiation on residual chlorine and DBPs in chlorination of model organic-N precursors in swimming pools. *Water Research 46* (8): 2674–2682.

70 Zhang, X., Li, J., Yang, J.-Y. et al. (2016). Chlorine/UV process for decomposition and detoxification of microcystin-LR. *Environmental Science & Technology 50* (14): 7671–7678.

71 Mark, G., Schuchmann, M.N., Schuchmann, H.-P., and von Sonntag, C. (1990). The photolysis of potassium peroxodisulfate in aqueous solution in the presence of tert-butanol: a simple actinometer for 254 nm radiation. *Journal of Photochemistry and Photobiology A: Chemistry 55*: 157–168.

72 Yang, Y., Pignatello, J.J., Ma, J., and Mitch, W.A. (2014). Comparison of halide impacts on the efficiency of contaminant degradation by sulfate and hydroxyl radical-based advanced oxidation processes (AOPs). *Environmental Science & Technology 48* (4): 2344–2351.

73 Guan, Y.H., Ma, J., Li, X.C. et al. (2011). Influence of pH on the formation of sulfate and hydroxyl radicals in the UV/peroxymonosulfate system. *Environmental Science & Technology 45* (21): 9308–9314.

74 Sharma, J., Mishra, I.M., Dionysiou, D.D., and Kumar, V. (2015). Oxidative removal of bisphenol A by UV-C/peroxymonosulfate (PMS): kinetics, influence of co-existing chemicals and degradation pathway. *Chemical Engineering Journal 276*: 193–204.

75 Yang, Y., Pignatello, J.J., Ma, J., and Mitch, W.A. (2016). Effect of matrix components on UV/$H_2O_2$ and UV/$S_2O_{82}$-advanced oxidation processes for trace organic degradation in reverse osmosis brines from municipal wastewater reuse facilities. *Water Research 89*: 192–200.

76 Barrett, J. and Mansell, A.L. (1960). Ultra-violet absorption spectra of the molecules $H_2O$, HDO and $D_2O$. *Nature 187* (4732): 138–138.

77 Barrett, J. and Baxendale, J.H. (1960). The photolysis of liquid water. *Transactions of the Faraday Society 56* (1): 37–43.

78 Weeks, J.L., Maeburn, G.M.A.C., and Gordon, S. (1963). Absorption coefficients of liquid water and aqueous solutions in the far ultraviolet. *Radiation Research 19* (3): 559–567.

79 Halmann, M. and Platzner, I. (1965). The photochemistry of phosphorus compounds. Part III. Photolysis of ethyl dihydrogen phosphate in aqueous solution. *Journal of the Chemical Society 1440*: 5380–5385.

80 Querry, M.R., Cary, P.G., and Waring, R.C. (1978). Split-pulse laser method for measuring attenuation coefficients of transparent liquids: application to deionized filtered water in the visible region. *Applied Optics 17* (22): 3587–3592.

81 Segelstein, D.J. (1981). *The Complex Refractive Index of Water*. Kansas City: University of Missouri.

82 Crapulli, F., Santoro, D., Sasges, M.R., and Ray, A.K. (2014). Mechanistic modeling of vacuum UV advanced oxidation process in an annular photoreactor. *Water Research 64*: 209–225.

83 Imoberdorf, G. and Mohseni, M. (2011). Modeling and experimental evaluation of vacuum-UV photoreactors for water treatment. *Chemical Engineering Science 66* (6): 1159–1167.

84 Gonzalez, M.G., Oliveros, E., Worner, M., and Braun, A.M. (2004). Vacuum-ultraviolet photolysis of aqueous reaction systems. *Journal of Photochemistry and Photobiology C-Photochemistry Reviews 5* (3): 225–246.

85 Gonzalez, M.C. and Braun, A.M. (1995). VUV photolysis of aqueous-solutions of nitrate and nitrite. *Research on Chemical Intermediates 21* (8-9): 837–859.

86 Blandamer, M.J. and Fox, M.F. (1970). Theory and applications of charge-transfer-to-solvent spectra. *Chemical Reviews 70* (1): 59.

87 Fox, M.F. and Hunter, T.F. (1969). Charge-transfer-to-solvent spectra. *Nature 223* (5202): 177.

88 Rahn, R.O. (1997). Potassium iodide as a chemical actinometer for 254 nm radiation: Use of iodate as an electron scavenger. *Photochemistry and Photobiology 66* (4): 450–455.

89 Rahn, R.O. (1993). Use of potassium iodide as a chemical actinometer. *Photochemistry and Photobiology 58* (6): 874–880.

90 Beyad, Y., Burns, R., Puxty, G., and Maeder, M. (2014). A speciation study of sulfur(IV) in aqueous solution. *Dalton Transactions 43* (5): 2147–2152.

**91** Sander, R. (2021). Compilation of Henry's Law Constants for Inorganic and Organic Species of Potential Importance in Environmental Chemistry. http://satellite.mpic.de/henry/henry/henry-3.0.pdf (accessed 2021).

**92** Townsend, T.M., Allanic, A., Noonan, C., and Sodeau, J.R. (2012). Characterization of sulfurous acid, sulfite, and bisulfite aerosol systems. *Journal of Physical Chemistry A 116* (16): 4035–4046.

**93** Hayon, E., Treinin, A., and Wilf, J. (1972). Electronic-spectra, photochemistry, and autoxidation mechanism of sulfite-bisulfite-pyrosulfite systems – $SO_2$-, $SO_3$-, $SO_4$-, and $SO_5$-radicals. *Journal of the American Chemical Society 94* (1): 47.

**94** Jung, B., Farzaneh, H., Khodary, A., and Abdel-Wahab, A. (2015). Photochemical degradation of trichloroethylene by sulfite-mediated UV irradiation. *Journal of Environmental Chemical Engineering 3* (3): 2194–2202.

**95** Vellanki, B.P. and Batchelor, B. (2013). Perchlorate reduction by the sulfite/ultraviolet light advanced reduction process. *Journal of Hazardous Materials 262*: 348–356.

**96** Vellanki, B.P., Batchelor, B., and Abdel-Wahab, A. (2013). Advanced reduction processes: a new class of treatment processes. *Environmental Engineering Science 30* (5): 264–271.

**97** Liu, X., Vellanki, B.P., Batchelor, B., and Abdel-Wahab, A. (2014). Degradation of 1,2-dichloroethane with advanced reduction processes (ARPs): effects of process variables and mechanisms. *Chemical Engineering Journal 237*: 300–307.

**98** Bentel, M.J., Yu, Y.C., Xu, L.H. et al. (2019). Defluorination of per- and polyfluoroalkyl substances (PFASs) with hydrated electrons: structural dependence and implications to PFAS remediation and management. *Environmental Science & Technology 53* (7): 3718–3728.

**99** Wols, B.A. and Hofman-Caris, C.H.M. (2012). Review of photochemical reaction constants of organic micropollutants required for UV advanced oxidation processes in water. *Water Research 46* (9): 2815–2827.

**100** Rivas, J., Gimeno, O., Borralho, T., and Sagasti, J. (2011). UV-C and UV-C/peroxide elimination of selected pharmaceuticals in secondary effluents. *Desalination 279* (1–3): 115–120.

**101** Sanches, S., Crespo, M.T.B., and Pereira, V.J. (2010). Drinking water treatment of priority pesticides using low pressure UV photolysis and advanced oxidation processes. *Water Research 44* (6): 1809–1818.

**102** Baeza, C. and Knappe, D.R.U. (2011). Transformation kinetics of biochemically active compounds in low-pressure UV photolysis and UV/$H_2O_2$ advanced oxidation processes. *Water Research 45* (15): 4531–4543.

**103** Pereira, V.J., Weinberg, H.S., Linden, K.G., and Singer, P.C. (2007). UV degradation kinetics and modeling of pharmaceutical compounds in laboratory grade and surface water via direct and indirect photolysis at 254 nm. *Environmental Science & Technology 41* (5): 1682–1688.

**104** Meite, L., Szabo, R., Mazellier, P., and DeLaat, J. (2010). Cinétique de phototransformation de polluants organiques émergents en solution aqueuse diluée. *Revue des Sciences de l'Eau 23* (1): 31–39.

**105** Dantas, R.F., Rossiter, O., Teixeira, A.K.R. et al. (2010). Direct UV photolysis of propranolol and metronidazole in aqueous solution. *Chemical Engineering Journal 158* (2): 143–147.

**106** Qin, L., Lin, Y.L., Xu, B. et al. (2014). Kinetic models and pathways of ronidazole degradation by chlorination, UV irradiation and UV/chlorine processes. *Water Research 65*: 271–281.

# 11

# UV Disinfection of Air and Surfaces

## 11.1 Introduction

As described previously, UVC and UVB radiation promote photochemical reactions within critical biomolecules that lead to inactivation of microorganisms and viruses, including pathogens. It is important to recognize that UVC and UVB radiation can also promote damage to cells and biomolecules that comprise higher organisms, such as plants and animals. In addition, UVC and UVB radiation can also cause damage to building materials, including plastics, paints, and coatings. As such, applications of UV radiation for disinfection and other objectives need to be implemented in a manner that mitigates UV-induced damage to plants, animals, and materials that are not the focus of these applications.

In most practical applications of UV radiation for disinfection, UVC radiation is applied because it tends to be far more effective and efficient than UVB radiation for meeting the objective of pathogen inactivation. In a broader sense, the most common applications of UV for disinfection to date have been in water and wastewater treatment; however, there has been a long-standing recognition of the potential for UV (especially UVC) radiation to function as a disinfectant for air [1].

William F. Wells has been credited with the discovery that some bacteria can be transmitted by airborne particles and coined the term "droplet nuclei" to describe airborne droplets that have been linked to transmission of several diseases. Droplet nuclei are defined as airborne particles with an aerodynamic diameter of less than 10 μm.[a]

These small particles can remain suspended in air for long periods of time, thereby allowing disease transmission between individuals who may never come into close contact [2]. Small particles also have the ability to penetrate deep into the respiratory system, even as far as the alveoli. Table 11.1 provides a summary of several diseases that are believed to be transmitted by droplet nuclei.

Disease transmission via droplet nuclei is sometimes referred to as "airborne transmission." It should be noted that particles larger than droplet nuclei can also cause disease transmission, but the transmission distance for these larger particles tends to be short, owing to the relatively rapid settling of these larger particles. Among the pathogens that have been associated with airborne transmission routes, there may be other routes of transmission as well. Disease severity is sometimes linked to the route of transmission, with airborne transmission often being linked to the most severe cases.

*Photochemical Reactors: Theory, Methods, and Applications of Ultraviolet Radiation*, First Edition. Ernest R. Blatchley III.

Companion website: www.wiley.com/go/blatchley/photochemicalreactors

**Table 11.1** Summary of diseases for which transmission includes an airborne pathway. Reported values of inactivation constant(s) based on exposure to UV radiation at 254 nm ($k_{254}$), or other wavelengths, are also included. Reported inactivation constants were taken from the associated references or developed by linear regression of ln-transformed data reported for each pathogen. Regressions were limited to data that appeared to conform to a single-event (i.e. first-order) inactivation response and were based on concentrations of infective or viable targets that were above the limit of detection.

| Disease(s) or symptoms | Etiologic agent(s) | $k_{254}$ $\left(\frac{m^2}{J}\right)$ for air (a) or liquid water (w) suspension | Reference(s) |
|---|---|---|---|
| Anthrax | *Bacillus anthracis* (spores) | 0.0164 (w)<br>0.0766 (w)[a] | [3]<br>[4] |
| Aspergillosis | *Aspergillus fumigatus*<br>*Aspergillus flavus*<br>*Aspergillus flavus* | 0.0549 (w)[b]<br>0.040 (a)[c]<br>0.0189 (w)[d] | [5]<br>[6]<br>[7] |
| Cold or flu<br>Acute bronchitis<br>Pneumonia<br>Conjunctivitis | Adenovirus | 0.0023 (w) | [8] |
| Severe diarrhea and vomiting | Rotavirus | 0.030 (w) | [9] |
| Influenza | Influenza viruses<br>(Melbourne strain)<br>(NIB-4 strain)<br>(NIB-6 strain)<br>Influenza virus H1N1 | <br>0.098 (w)<br>0.091 (w)<br>0.104 (w)<br>0.18 (a)[e] | <br>[10]<br>[11]<br>[11]<br>[12] |
| Common cold | Rhinovirus<br>Coronavirus OC43<br>Coronavirus OC43<br>Coronavirus 229E | 0.034 (w)[f]<br>0.077 (w)[g]<br>0.59 (a)[e]<br>0.41 (a)[e] | [13]<br>[14]<br>[15]<br>[15] |
| Legionellosis (Legionnaires' disease, Pontiac fever) | *Legionella pneumophila* | 0.065 (w) | [16] |
| Measles | Measles virus (*Morbillivirus*) | 0.105 (w) | [17] |
| Smallpox | Vaccinia virus<br>Variola virus | 0.13 (w) | [18] |
| Cryptococcosis | *Cryptococcus neoformans* | 0.0047 (w) | [19] |
| Tuberculosis | *Mycobacterium tuberculosis* | 0.048 (w)<br>0.47 (a) | [17] |
| Ebola | Ebola virus | 0.14 (w) | [18] |
| Severe acute respiratory syndrome (SARS) | SARS-CoV-1 | 0.19 (w)[f]<br>0.096 (w)[h]<br>0.33 (w)[i] | [13]<br>[20]<br>[21] |
| Middle East respiratory syndrome (MERS) | MERS-CoV | 0.21 (w)[f]<br>0.082 (w)[h]<br>0.33 (w)[i] | [13]<br>[20]<br>[21] |
| Coronavirus disease 2019 (COVID-19) | SARS-CoV-2 | 0.20 (w)[f]<br>0.107 (w)[h]<br>0.33 (w)[i] | [13]<br>[20]<br>[21] |

[a] A significant shoulder was evident in the dose–response data. The value of $k_{254}$ reported here is for a fit of the series-event model, with $n = 13$.
[b] Based on data from an experiment involving output from UV LEDs with nominal peak output at $\lambda = 255$ nm.
[c] Based on data from an experiment involving output from UV LEDs with nominal peak output at $\lambda = 280$ nm.
[d] Experiments were conducted using a flow-through reactor operated at steady state.
[e] Experiments were performed using aerosolized virus particles irradiated with an optically filtered KrCl* excimer lamp, with nominal peak output at $\lambda = 222$ nm.
[f] Predicted using the genomic model of Rockey et al. [13].
[g] Based on data from an experiment involving output from UV LEDs with nominal peak output at $\lambda = 267$ nm.
[h] Predicted using the genomic model of Pendyala et al. [20].
[i] Predicted using the genomic model of Sagripanti and Lytle [21].

Droplet-associated pathogens can also adhere to surfaces, either by settling or inertial impaction. This provides another important route of transmission, namely serial contact with contaminated surfaces (fomites).

Disease transmission via airborne and fomite-based pathways is particularly relevant in hospital settings. Nosocomial (hospital acquired) infections have become increasingly important with the emergence of antibiotic-resistant bacteria. Important examples of antibiotic-resistant bacteria include methicillin-resistant strains of *Staphylococccus aureus* (MRSA), *Clostridium difficile*, vancomycin-resistant *Enterococci* (VRE), vancomycin intermediate *S. aureus* (VISA), and extensively drug-resistant tuberculosis (XDR-TB).

Nosocomial infections are common in current medical practice in developed countries. For example, the Centers for Disease Control and Prevention indicate that roughly 4% of hospital patients in the United States will be diagnosed with at least one infection related to hospital care [22, 23]. In the United Kingdom, it has been reported that roughly 10% of hospital patients experience a Hospital Acquired Infection (HAI) [24].

Many of the pathogens that are responsible for nosocomial infections are susceptible to inactivation by exposure to UVC radiation. Examples of airborne and surface-associated pathogens that can be effectively inactivated with UVC radiation include: *Mycobacterium tuberculosis* [25], fungal spores [26], *C. difficile*, VRE [23], airborne viruses that are associated with the common cold and measles [27, 28], as well as airborne viruses that have been associated with epidemics and pandemics, including SARS-CoV-1, MERS, and SARS-CoV-2 (i.e. coronaviruses) [20, 21, 29]. As such, UVC-based interventions have the potential to reduce the incidence of HAIs and disease outbreaks. However, because of the potential for UVC-based damage to external tissues, especially the skin and eyes, implementation of UVC-based devices in indoor settings needs to be conducted in a manner that limits human exposure below relevant exposure thresholds.

In healthcare facilities, transmission of infectious agents is exacerbated by close proximity of infected and noninfected individuals, as well as people with compromised immune systems. In other settings, infected individuals (symptomatic and asymptomatic) can also come in close contact with large numbers of uninfected individuals; many documented "superspreader" events have taken place in indoor settings where people have been in close contact for extended periods of time. It should be noted that superspreader events have been documented at outdoor events as well. For example, the final match of the 2020 European Football Championship was held at Wembley Stadium in London, England. Contact tracing results identified this as a superspreader event for COVID-19, despite a requirement that all attendees demonstrate proof of vaccination or a negative lateral flow device (LFD) test result for COVID-19 within 48 hours of attending the event [30].

Although HAIs represent an important opportunity for implementation of UV radiation, airborne and surface-associated pathogens can be relevant in many non-healthcare settings, including food processing operations, offices, educational institutions, places of worship, elevators, entertainment venues, indoor sports venues, travel venues and vehicles (airports, airplanes, busses, trains), hotels, and dining establishments. More generally, the potential for disease transmission tends to be greatest in indoor settings where human populations congregate at high density.

The outbreak of Coronavirus Disease 2019 (COVID-19) heightened interest in the use of UV-based interventions to control transmission of the disease. The etiologic agent of the disease is a

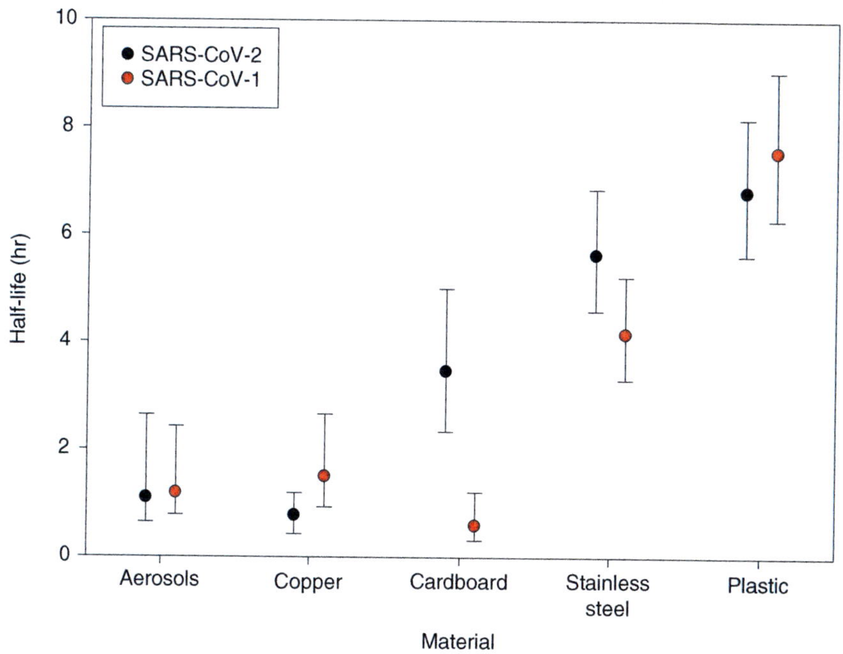


**Figure 11.1** Reported half-lives of infective SARS-CoV-2 and SARS-CoV-1 in aerosols and on surfaces. Data points represent median values; error bars represent the 95% credible interval (2.5%–97.5% quartile range). *Source:* Data adapted from Van Doremalen et al. 2020.

novel coronavirus, severe acute respiratory syndrome coronavirus 2 (SARS-CoV-2), which is closely related to the virus that was responsible for the outbreak of severe acute respiratory syndrome (SARS), SARS-CoV. Both viruses belong to the β-coronavirus genera, which also includes viruses that cause the common cold, and the virus that causes middle-east respiratory syndrome (MERS). SARS-CoV and SARS-CoV-2 have been shown to behave similarly in the environment. Specifically, the decay of the infective forms of these viruses was shown to follow similar patterns in aerosols and on solid surfaces (see Figure 11.1) [31]. Both viruses demonstrate relatively rapid decay in aerosols and on copper surfaces; however, on cardboard, stainless steel, and plastic surfaces, the viruses are generally more stable. Collectively, these data indicate the potential for infective forms of the coronaviruses to persist on or in many media for as long as several days. Moreover, these data indicate the potential for transmission of SARS and COVID-19 by airborne and fomite-based pathways. In other words, contaminated air and contaminated surfaces can both contribute to transmission of these diseases.

The data presented in Figure 11.1 indicate the need for interventions that actively inactivate coronaviruses (and other disease agents) in air and on surfaces. These data also illustrate the benefits of other measures of protection against airborne and surface-associated pathogens, including the use of personal protective equipment (PPE) and frequent hand-washing.

The basic structure and genomic composition of SARS-CoV-2 and other coronaviruses suggests that they should be quite susceptible to inactivation by exposure to UVC radiation [13, 20, 21]. More broadly, it is likely that UVC radiation will be effective for inactivation of most viral and bacterial pathogens that are associated with airborne and fomite-based disease transmission.

Given the consensus opinion that epidemics and pandemics will occur with greater frequency in the future [32], along with the potential for UVC radiation to inactivate pathogens, it is likely that UVC-based interventions will be applied with increasing regularity in the future for disinfection of air and surfaces.

Ultraviolet Germicidal Irradiation (UVGI) is a term that is used to describe applications of UVC and UVB radiation (200 nm $\leq \lambda \leq$ 320 nm) to disinfect air and surfaces [17]. UVGI systems have been predominantly based on low-pressure Hg lamps as the source of germicidal radiation, but have evolved to include systems based on other sources, including UV LEDs, excimer lamps, and Xenon lamps.

The goal of this chapter is to present fundamental aspects of UVGI systems. Common examples of UVGI will be described and illustrated. Recommended human exposure limits for UVC radiation are then presented, along with information focused specifically on Far UVC radiation. The effects of germicidal UV radiation on materials are discussed, followed by a summary of the intrinsic kinetics of UV-based inactivation of pathogens in air and on surfaces. This will be followed by a discussion of current standards for testing of UVGI systems that are used for in-duct disinfection applications, along with simple models to simulate the performance of UVGI systems for air disinfection applications. The chapter concludes with a presentation of the behavior of devices that are used to disinfect solid objects and surfaces.

### 11.1.1 UVGI Devices

The effectiveness of any UV disinfection device depends on its ability to deliver UV photons to target pathogens. For pathogens that are not surrounded by an absorbing medium, UV-based inactivation tends to be rapid and effective. However, airborne and surface-associated pathogens may be sheltered from UV exposure by a surrounding medium. This complicates the application of UV-based disinfection systems and needs to be accounted for in all such applications.

Given the broad definition of UVGI, there are many different modes of application for these devices. In a general sense, the aim of these devices is to introduce germicidal UV radiation in a manner that promotes inactivation of airborne or surface-associated pathogens, while controlling exposure of human tissues and UV-sensitive materials to radiation from these sources. Within this broad scope of application, there are many different types of UVGI systems. Brief summaries and illustrations of common forms of UVGI applications are presented below.

### 11.1.2 Upper-Room UVGI (UR-UVGI)

Upper-room devices are installed in manner that allows effective irradiation of the air space that lies above the height of most humans. Inactivation of airborne pathogens can be effective if the fluence rate within the irradiated zone is sufficiently high and if air circulation and mechanical mixing within the room ensures (at least) intermittent exposure of airborne pathogens to UVC radiation.

With conventional upper-room UVGI (UR-UVGI) installations, the devices are usually wall-mounted (as illustrated in Figure 11.2) or suspended from the ceiling. The devices involve one or more lamps, along with reflecting devices and louvers to constrain output UVC radiation to the upper-room area, above the height of most humans.

Although international standards for the design and operation of UR-UVGI systems are generally lacking, some guidance has emerged through observation. Specifically, systems based on conventional

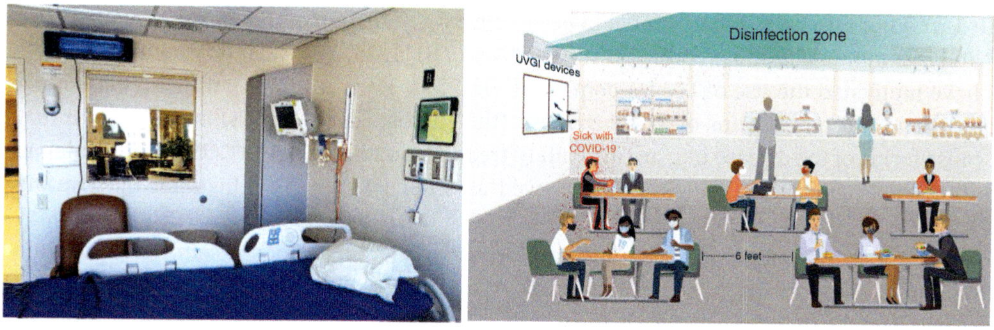


**Figure 11.2** (Left) Photographic image of an UR-UVGI fixture in a hospital room. *Source:* https://www.nytimes.com/2020/05/07/science/ultraviolet-light-coronavirus.html. (Right) Schematic illustration of an UR-UVGI system in a common area. *Source:* https://www.cdc.gov/coronavirus/2019-ncov/community/ventilation/uvgi.html. In both images, UVC fixtures are located on a wall above the height of most people in the room and include reflecting surfaces and louvers to direct output UVC radiation to the upper room region, thereby limiting human exposure to output from the UVC sources. *Source:* Images from ASHRAE [33].

low-pressure Hg lamps (essentially monochromatic output at $\lambda = 254$ nm) should be implemented to achieve a spatially averaged fluence rate of roughly $30-50\frac{\mu W}{cm^2}$ in the upper-room volume, while the spatially averaged fluence rate in the occupied part of the room should be less than $0.2\frac{\mu W}{cm^2}$ [34]. The ceiling height in the room should be at least 2.5 m to allow a sufficient fraction of the air space in the room to be irradiated. Ideally, the air space in the room should be well mixed; the use of ceiling fans and other forms of mechanical mixing has been shown to improve the performance of these systems. This combination of characteristics allows effective delivery of germicidal UV radiation to accomplish inactivation of airborne pathogens, while protecting room occupants from exposure to UV radiation. This set of goals is a requirement of all UVC-based systems that are intended for disinfection of air and surfaces.

When installed with sufficient UVC power, appropriate spacing and locations of UVC sources, and appropriate in-room air circulation, UR-UVGI systems can be effective for inactivation of airborne pathogens (i.e. pathogens associated with droplet nuclei). On the other hand, these systems will be less effective against pathogens associated with relatively large droplets because these particles tend to settle to surfaces within close proximity to their place of introduction to the air space; in other words, many of these particles and their associated pathogens end up on the floor or other horizontal surfaces (e.g. table tops, counters, and desk tops) in the occupied air space of the room, where they will receive little or no exposure to germicidal UV radiation.

Wells et al. [27] conducted one of the first controlled experiments with UR-UVGI systems. Their study involved application of UR-UVGI in selected rooms of a boarding school. Students in irradiated rooms had substantially lower rates of infection of measles, chicken pox, and mumps; frequency of colds was essentially the same in irradiated vs. unirradiated rooms. Subsequent experiments at other locations yielded less-favorable results; however, it is likely that these later studies were confounded by opportunities for disease transmission outside of the settings where UR-UVGI was applied. This led to a general conclusion that UR-UVGI systems are most likely to yield measurable benefits in settings where the risk of disease transmission is substantially greater than in other settings.

UR-UVGI has also been demonstrated to be effective for reduction of transmission of influenza [35] and is commonly used for control of tuberculosis (TB) [1, 25, 36]. At least two important factors have motivated the use of UV for control of TB. First, TB is an important infection among AIDS patients. Second, strains of *M. tuberculosis* (the etiologic agent for TB) have emerged that can cause forms of the disease that are difficult to treat. The outbreak of the COVID-19 pandemic also motivated applications of UR-UVGI, as well as other UV-based methods for inactivating this virus.

UR-UVGI systems complement natural and forced ventilation in indoor settings. Under the assumption of a well-mixed air space, movement of air between the irradiated and nonirradiated spaces in a room leads to (essentially) first-order decay of the concentration of airborne pathogens, and other airborne contaminants. Details of the calculations that are used to quantify this behavior will be presented later in this chapter.

### 11.1.3 Lower-Room UVGI (LR-UVGI)

Lower-room UVGI (LR-UVGI) systems are designed to expose the near-floor area of a room to UVC radiation. The devices themselves may have similar construction to an UR-UVGI system but are located much lower within the air space of a room. LR-UVGI systems can also involve autonomous devices (i.e. robots) with mobile UV sources that are designed to irradiate the floor and near-floor areas of rooms. LR-UVGI systems can be effective for inactivation of pathogens that are associated with surfaces, as well as those that are associated with large droplets or spores, which tend to settle to the lower parts of a room soon after their release to the air space. LR-UVGI systems are typically designed to irradiate (roughly) the lowest 0.5 m of space within a room. In rooms where LR-UVGI systems are installed, patrons should wear clothing that prevents exposure of the skin on their lower bodies. Common pants, socks, and closed-toe shoes will be effective for preventing exposure to UVC radiation from LR-UVGI systems. It should be noted that LR-UVGI systems may not be applicable in settings where infants and young children are likely to be present.

Beyond their heightened potential for human exposure to germicidal UV radiation, LR-UVGI systems also demonstrate other important differences from UR-UVGI systems. Specifically, their performance is limited by the spatial distribution of radiant energy within the system. Pathogens that are associated with floors tend to be sessile. As such, the potential for inactivation of these pathogens is constrained by their location, relative to the UV sources in the room. If the UV sources are relatively distant from the location of a pathogen, the likelihood of inactivation of the pathogen will be diminished. If the floor-associated pathogens are sessile, then mixing conditions in the room will not affect their exposure to UV radiation. The potential for pathogen exposure to UV will be enhanced by resuspension events, but the phenomena that lead to resuspension will also increase the likelihood of human exposure to the pathogens.

### 11.1.4 In-Duct Systems

These systems are installed within the ductwork of heating, ventilation, and air conditioning (HVAC) systems as a means of disinfecting recirculating air. The UV sources, which are usually low-pressure Hg lamps, are typically placed downstream of an air filter with a MERV rating[b] of 6–15, and often immediately upstream of cooling coils to limit condensation on the lamp surface and to allow for disinfection of cooling coil surfaces.

The strategy of applying UV disinfection downstream of a physical separation device is conceptually similar to the "multiple-barrier concept" that is foundational to water treatment. For

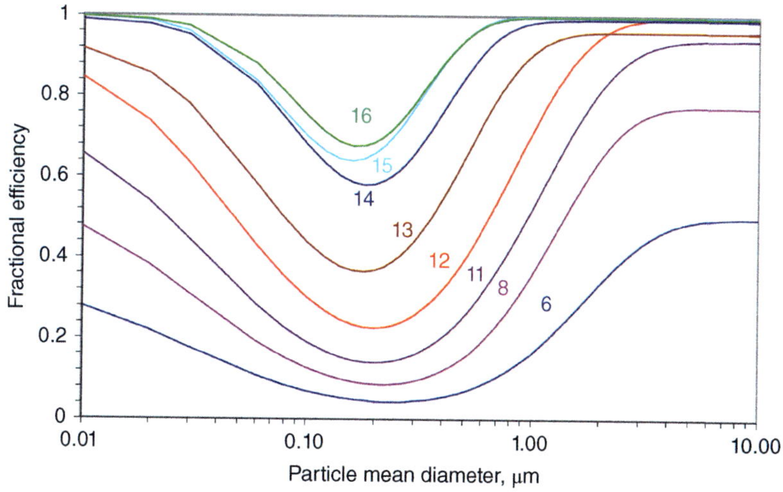


**Figure 11.3** Calculations of particle removal efficiency as a function of particle size for air filters with a range of MERV ratings. *Source:* Kowalski and Bahnfleth 2002/ The Pennsylvania State University.

disinfection of air, such an approach will result in improved performance of the UV disinfection system as a result of removal of large particles, which are likely to provide protection to airborne microbes and viruses.

The performance of air filters depends on several factors, including the materials of construction, the geometry of the collecting media, and the applied air flow rate. For most air filters, there are several mechanisms that contribute to particle separation from air, including diffusion, interception, and inertial impaction. In general terms, the diffusion mechanism becomes increasingly important as particle size decreases, while the opposite is true for interception and inertial impaction. As a result, air filters are typically characterized by a "window" of performance, such that a range of particle sizes can be identified for which particle removal is a minimum, and likely to control filter design.

Figure 11.3 illustrates particle capture efficiency of air filters for a fixed set of operating conditions, over a wide range of MERV ratings. The performance "window" for these filters is evident for particle sizes in the range of roughly 0.1–1.0 μm, which happens to be a size range where UV disinfection works well. In that sense, air filters and UV disinfection display complementary behavior.

In-duct UV disinfection systems are typically applied in settings where an air velocity of 2–3 $\frac{m}{s}$ can be maintained. This condition allows application of a sufficient UV dose for most applications, using the hardware that is typical of these systems.

Relative to the more common UR-UVGI system, in-duct systems offer advantages of (typically) requiring fewer lamps, while at the same time avoiding most opportunities for human exposure; a notable exception to this is for maintenance personnel or others who are exposed to ductwork. Important disadvantages of these systems, relative to UR-UVGI applications are that the performance of the in-duct system is limited by air recirculation and maintenance is more difficult, given the location of the lamps within a duct.

### 11.1.5 In-Room and Mobile Devices

Another strategy for implementation of UVC radiation for disinfection of room air is to install a device that houses one or more UVC lamps. Common configurations of these devices involve stationary cabinets through which air is circulated past enclosed UVC lamps. These devices can be used to improve indoor air quality by recirculating disinfected air. The figure-of-merit for these devices, sometimes referred to as "UV air cleaners," is the *clean air delivery rate* (CADR), which is defined as the product of volumetric flow rate through the device and fractional inactivation of the target pathogen accomplished in a single pass through the system.

There are mobile versions of these devices as well, sometimes guided by a human (behind a protective screen), sometimes driven robotically. These devices can be used to disinfect air and surfaces in indoor spaces. They are used in hospital and clinical settings, commercial venues, and to disinfect passenger or cockpit areas in aircraft, busses, or trains.

### 11.1.6 Isolated Rooms or Chambers

A common problem with application of germicidal UV radiation for disinfection of media is the need to maximize exposure of the object to be disinfected, while limiting or preventing human exposure to the same radiation. For some media, it is possible to accomplish disinfection in a room that houses one or more UVC lamps, which are illuminated only when the room is unoccupied. This strategy emerged as an important approach to facilitate reuse of some PPE during the COVID-19 pandemic. In particular, shortages of N95 masks prompted the use of UV for disinfection between uses. Figure 11.4 illustrates an example of one such facility, wherein used masks were suspended from a rack system. Vertically oriented "towers" of UVC (low-pressure Hg) lamps were mounted on robotic devices to allow movement of the radiation sources past the masks. Masks were irradiated to accomplish inactivation of pathogens that may have become associated with the mask fiber materials during a previous medical shift. Many similar systems have been developed for this purpose.

**Figure 11.4** Mobile UV tower systems used to irradiate used N95 masks at the M Health Fairview Facility. *Source:* Photo from https://www.mhealth.org/childrens/blog/2020/april-2020/m-health-fairview-rolls-out-new-uv-decontamination-process-for-n95-masks.

## 11.2 Safety, UV Exposure Limits

The primary safety concern associated with the use of UVC radiation for disinfection of air and surfaces is exposure of external tissues to germicidal UV radiation, specifically the eyes and skin. The dominant adverse outcomes of human exposure to UVC radiation are photokeratitis (*aka* "snow blindness" or "welder's flash"), irritation of the conjunctiva (photoconjunctivitis), and skin irritation (erythema) [38]. For UVC radiation, these effects tend to be temporary because radiation in the UVC range does not penetrate beyond the external tissues of the eye (corneal epithelium) and skin (stratum corneum or superficial epidermis). Also, the damaged cells are readily replaced as a natural consequence of sloughing. Strong absorption by these external cell layers prevents penetration of UVC radiation to the germinative layers of the skin. As a consequence, the potential for UVC radiation to cause skin cancer is believed to be very small; much smaller, for example, than from exposure to UVB radiation in ambient solar radiation (i.e. "sunlight") [39].

Figure 11.5 provides an illustration of the potential for UV radiation to penetrate human skin. The images that are presented in this figure were developed using a Monte Carlo modeling approach to simulate UV penetration into the cell layers that comprise human skin [40]. These simulations were based on reported values of the optical properties of the cells within these layers. The results indicated that the radiation from conventional low-pressure Hg lamps ($\lambda = 254$ nm), UV LEDs (typically in the range 250 nm $< \lambda <$ 285 nm), and Far UVC sources, such as optically filtered $KrCl^*$ excimer lamps ($\lambda \approx 222$ nm) is almost entirely absorbed within the stratum corneum and the upper reaches of the epidermis. For Far UVC radiation (i.e. radiation with 200 nm $\lesssim \lambda \lesssim$ 230 nm), more than 99.9% of incident radiation is absorbed within the stratum corneum, a layer of dead skin cells that provide protection to the skin layers below. Similar arguments hold for the eye, where the vast majority of incident UVC photons are absorbed by the outer layers of eye tissue. Again, for Far UVC radiation, virtually all incident photons are absorbed by the epithelial cells of the cornea, thereby protecting the cells in the parts of the eye that could be damaged by this radiation.

Figure 11.6 provides an illustration of the potential for UV radiation to cause (nonmelanoma) skin cancer as a function of wavelength. The ability of UV radiation to cause skin cancer is governed by two competing phenomena: photon penetration to the germinative cell layers and the ability of UV photons to initiate damage, especially within the cellular genome.

For UV radiation, the ability to penetrate to the germinative layer of skin or eyes generally decreases with wavelength; this trend becomes particularly evident for radiation with wavelengths less than about 240 nm, where absorbance by proteins starts to become relevant. The potential to cause photochemical damage that could contribute to skin cancer is largely associated with damage to nucleic acids; by extension, the ability of UV radiation to cause damage to nucleic acids within the germinative cells of the skin or eyes is closely related to the absorption spectra of nucleic acids, which generally peak at about 260 nm and fall off sharply at wavelengths above and below this value. This combination of effects leads to a narrow range of wavelengths that are responsible for nonmelanoma skin cancer. As indicated in Figure 11.6, radiation at about 299 nm is the most effective for promoting these types of cancer, with rapid decreases in the efficiency of this process at wavelengths above or below this value. These observations indicate that UVC sources have minimal potential to promote cancer. However, there are other risks of damage to skin and eyes (described above) for UVC radiation that will require that appropriate precautions be taken to prevent exposure.

Guidelines for exposure to UVC radiation were originally developed by the American Conference of Governmental and Industrial Hygienists (ACGIH). Later, the International Commission on Non-Ionizing Radiation Protection (ICNIRP) published a similar set of guidelines for radiation with

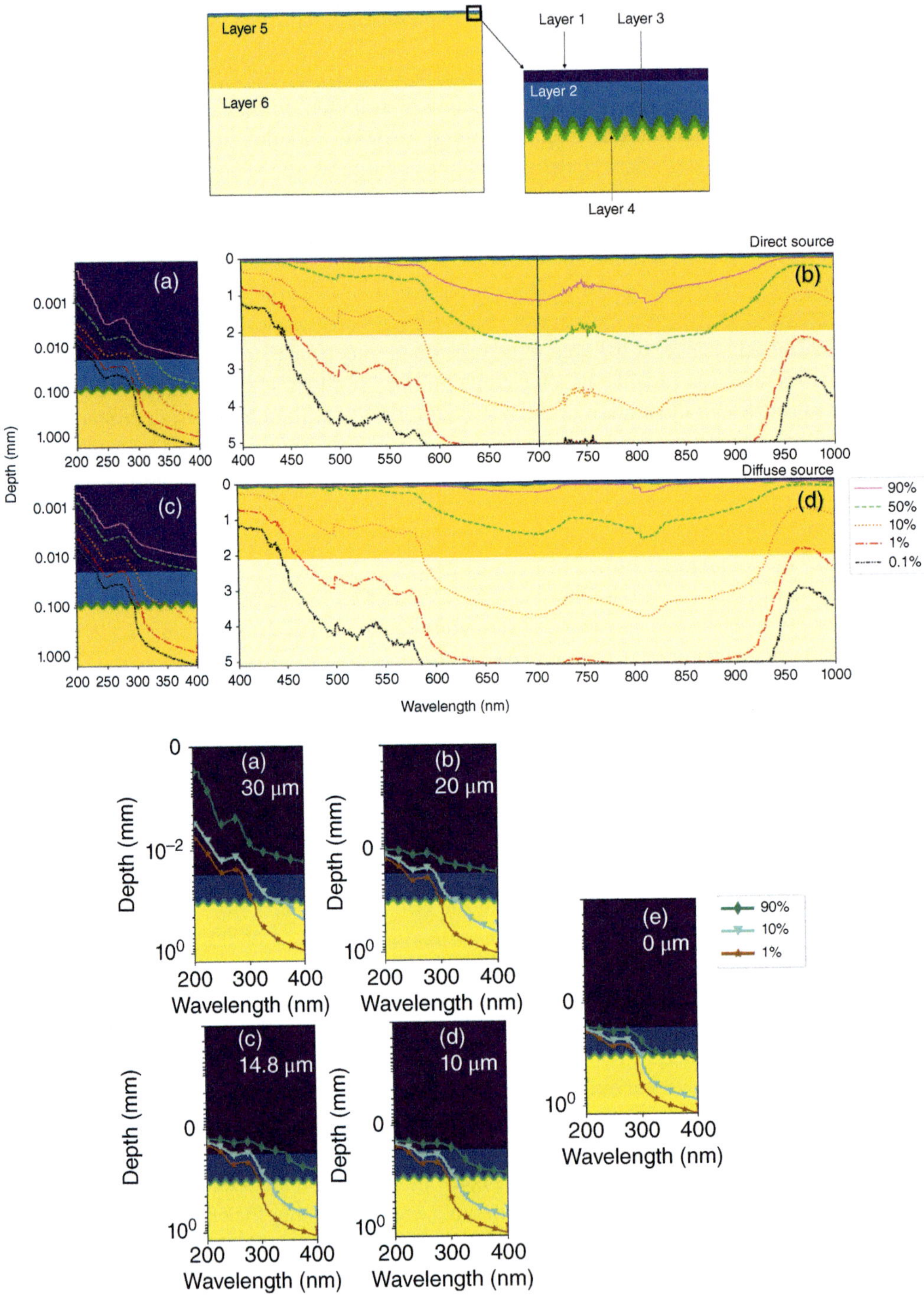


**Figure 11.5** Results of Monte Carlo simulations of UV penetration into human skin. Top panel illustrates six layers of human skin (layer 1: stratum corneum, layer 2: epidermis; layer 3: melanized epidermis; layer 4: basal layer; layer 5: dermis; layer 6: subcutaneous fat). Center panel illustrates fraction of radiation at a given wavelength that penetrates these tissue layers. Bottom panel illustrates penetration into layers for skin with various thicknesses of the stratum corneum layer; stratum corneum thickness values are indicated in each image immediately below the letter. *Source:* Finlayson et al. 2021/John Wiley & Sons/CC BY-4.0.

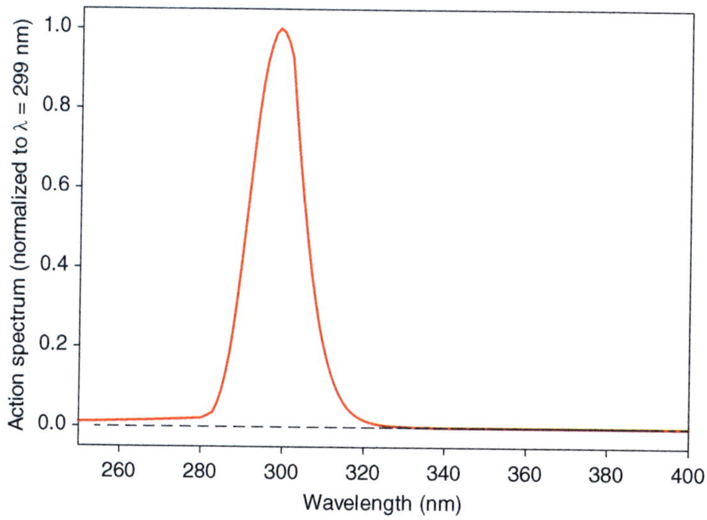


**Figure 11.6** Standard action spectrum for nonmelanoma skin cancer. This action spectrum was developed from laboratory studies of skin cancer and measured spectral transmission of the stratum corneum and upper epidermis. Although it is confined to wavelengths greater than 250 nm, and therefore does not extend into the Far UVC, it implies the significant risk of UVB compared to UVC. The risk for UVC and Far UVC would be further reduced and small. *Source:* Data from Ref. [41].

wavelengths between 180 and 400 nm [42]. Collectively, these guideline documents have been used to establish exposure limits, sometimes referred to as *Threshold Limit Values* (TLVs) for UV radiation. The TLVs for UV exposure have been adopted by many other organizations, including the International Organization for Standardization (ISO), the American National Standards Institute (ANSI), the National Institute for Occupational Safety and Health (NIOSH), and the World Health Organization (WHO).

The UV exposure guidelines were developed by considering the potential for damage to the skin and eyes among the most sensitive people within the general population [42], although eye exposure generally controls [43]. Radiation at approximately 270 nm is the most effective for promoting damage to these tissues. Recognizing this spectral behavior, TLVs are based on "equivalent 270 nm dose" applied. A weighting function, sometimes referred to as "relative spectral effectiveness" ($S_\lambda$) has been developed to account for this wavelength-dependence. Figure 11.7 illustrates $S_\lambda$ as a function of wavelength.

An effective dose of 3 $\frac{\text{mJ}}{\text{cm}^2}$ has been identified as the recommended eight-hour exposure limit for UV radiation. To reiterate, this recommended limit is based on radiation at 270 nm. Exposure to radiation at other wavelengths is normalized to radiation at 270 nm using $S_\lambda$. For example, radiation at 254 nm (i.e. the dominant line from a conventional low-pressure Hg lamp) has a Relative Spectral Effectiveness ($S_\lambda$) of 0.5, meaning that the recommended eight-hours exposure limit for $UV_{254}$ is 6 $\frac{\text{mJ}}{\text{cm}^2}$.

Exposure to polychromatic radiation sources is accounted for by integrating (summing) the effects of exposure over the range of wavelengths applied, as described by Equation 11.1.

$$H_{\text{eff}} = E_{\text{eff}} \cdot t = \sum E_\lambda \cdot S_\lambda \cdot \Delta\lambda \cdot t \tag{11.1}$$

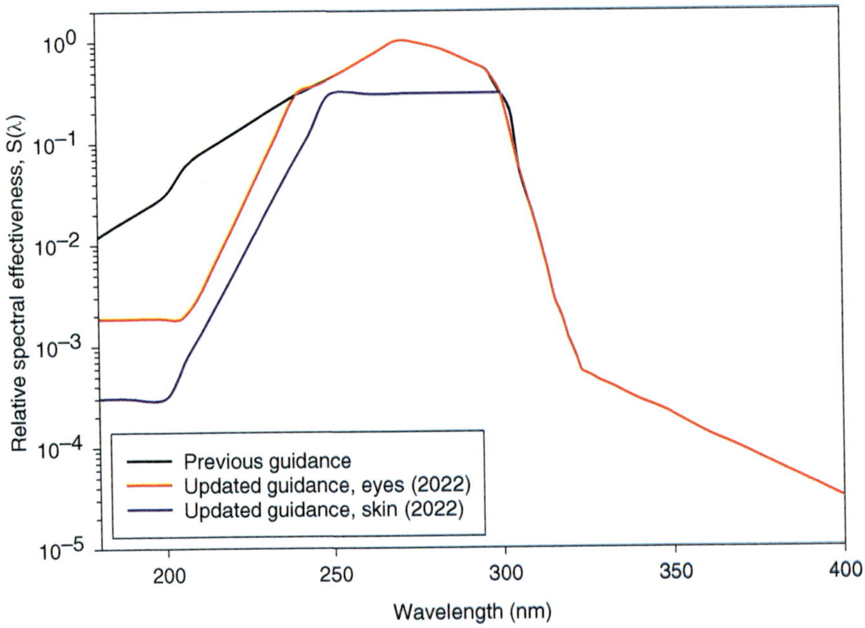


**Figure 11.7** Comparison of previous and current guidance for relative spectral effectiveness as a function of wavelength. Updated guidance values were presented by ACGIH [44]. Updated values for eye exposure are protective of skin; updated values for skin are presented for circumstances where human eyes are protected from exposure.

where

| | | | | |
|---|---|---|---|---|
| $H_{eff}$ | = | effective UV dose (equivalent dose at 270 nm) | [=] | $\frac{\text{mJ}}{\text{cm}^2}$ |
| $E_{eff}$ | = | effective UV fluence rate (equivalent irradiance at 270 nm) | [=] | $\frac{\text{mW}}{\text{cm}^2}$ |
| $t$ | = | exposure time | [=] | seconds |
| $E_\lambda$ | = | spectral irradiance | [=] | $\frac{\text{mW}}{\text{cm}^2 \cdot \text{nm}}$ |
| $\Delta\lambda$ | = | wavelength increment | [=] | nm. |

Equation 11.1 implies that spectral irradiance is constant over the period of exposure. For humans, this is often not the case, as a result of movement. However, it is generally difficult to define time-dependent exposure for humans, especially over long periods of exposure. As such, Equation 11.1 can be applied to define effective dose delivery to specific, worst-case locations within an exposure setting as a means of calculating a conservative estimate of the effective UV dose received.

Exposure limits can be used to define allowable (effective) fluence rate within an exposure setting. A common example is a daily exposure limit for UV, which is typically based on an assumed eight-hour period of exposure, corresponding to a typical workday. The daily (eight-hour) exposure limits for UV radiation are illustrated in Figure 11.8. Note that these curves are mirror images of the Relative Spectral Effectiveness curves, as defined by Equation 11.1; as the value of $S_\lambda$ diminishes, the guideline dose value increases. As a further point of reference, it is worth noting that the TLV for effective dose will be exceeded within a period of 5–10 minutes under exposure to midday sun in the summer months for latitudes between the equator and 40°N [42].

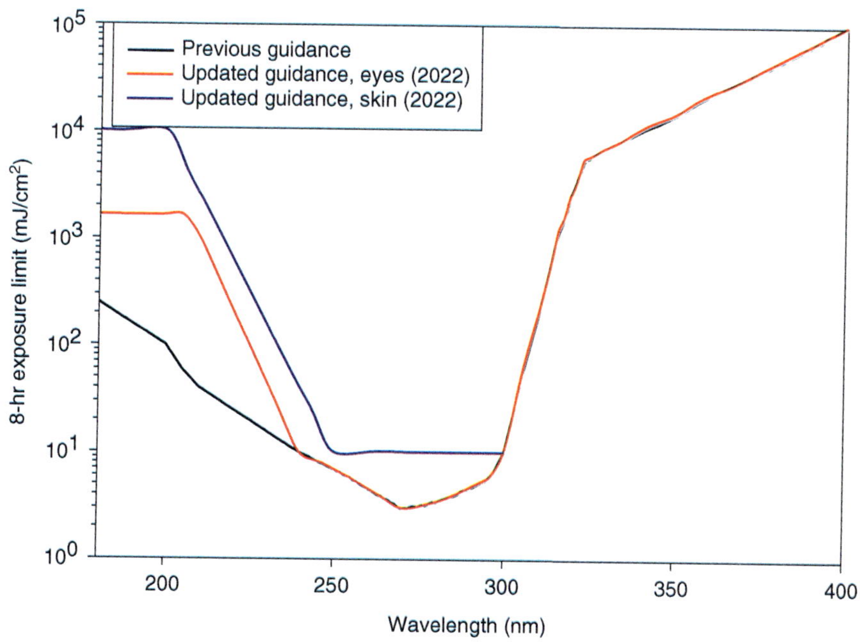


**Figure 11.8** Comparison of previous and updated guidance for eight-hour exposure limit as a function of wavelength. Updated guidance values were presented by ACGIH [44]. Updated values for eye exposure are protective of skin; updated values for skin are presented for circumstances where human eyes are protected from exposure.

Following this logic, a listing of acceptable effective fluence rate can be defined based on the anticipated period of exposure. Table 11.2 provides a list of guideline fluence rate values for various periods of exposure.

The values listed in Table 11.2 are consistent with the third law of photochemistry (dose reciprocity law or Bunsen–Roscoe law) and provide guidance for exposure settings in terms of allowable

**Table 11.2** UV exposure limits based on period of exposure. Exposure limits are presented as equivalent 270 nm fluence rates.

| **Period of exposure** | **Maximum allowable effective fluence rate** $\left(\frac{\mu W}{cm^2}\right)$ |
|---|---|
| 8 hours | 0.1 |
| 4 hours | 0.2 |
| 2 hours | 0.4 |
| 1 hours | 0.8 |
| 30 minutes | 1.7 |
| 15 minutes | 3.3 |
| 10 minutes | 4.0 |
| 5 minutes | 10 |
| 1 minutes | 50 |
| 30 second | 100 |

(effective) fluence rate as a function of anticipated exposure time. Effective fluence rate can be measured using a radiometer (for monochromatic sources) or a spectroradiometer (for polychromatic sources); however, care should be taken to ensure that the sensors used for these measurements are designed for the directional nature of UVC exposure at the point of measurement.

Interest in applications of UV-based devices for disinfection of air and surfaces increased with the onset of the COVID-19 pandemic. The etiologic agent of this disease, SARS-CoV-2, appears to be quite susceptible to inactivation by exposure to UVC radiation. However, the use of UVC-based systems for disinfection and decontamination would be facilitated if they could take place while the target space was occupied and in use.

An emerging group of radiation sources has characteristics that address these concerns. Specifically, Far UVC radiation (informally defined as radiation with wavelengths between 200 and 230 nm) is highly effective for inactivation of most microbial and viral pathogens but appears to be less harmful to human tissues than conventional UVC radiation, such as that emitted by low-pressure Hg lamps or conventional UV LEDs. Radiation from Far UVC sources, such as $KrBr^*$ excimers (peak output at 207 nm) and $KrCl^*$ excimers (peak output at 222 nm) has been shown to be effective for inactivation of microbial and viral pathogens, but to cause minimal damage to artificial human skin or to the skin or eyes of hairless mice [45, 46]. Moreover, Far UVC radiation has been demonstrated to be effective for inactivation of a wide range of microbial and viral pathogens, including methicillin-resistant *Staphylococcus aureus* (MRSA) [46], aerosolized H1N1 influenza virus [12], and coronaviruses [47], including SARS-CoV-2 [48]. Additional research is needed to address some open questions about the viability of Far UVC for applications in spaces occupied by unprotected humans, but available information about these sources is promising.

Based on recent improvements in understanding the effects of UVC radiation on human eyes and skin, the ACGIH has established a new set of exposure guidance values (see Figures 11.7 and 11.8) [44]. The updated values for eye protection are protective of exposed skin, as the eyes are generally more susceptible to damage than the skin. The updated values for skin protection are intended for circumstances where eyes are protected from UV exposure.

### 11.2.1 Ozone Generation

Another safety-related concern connected to UV-based applications is the potential to generate ozone. Photochemical formation of ozone in the gas phase has been extensively investigated, largely because of the roles played by ozone in tropospheric and stratospheric chemistry. In the troposphere, where solar radiation is limited to wavelengths less than about 290 nm, ozone functions as an important oxidant and a component of photochemical smog. In the stratosphere, where atmospheric ozone concentrations are relatively high (i.e. the "ozone layer"), ozone is critical as an absorber of short wavelength solar radiation. In fact, stratospheric ozone is almost entirely responsible for absorption of solar radiation imposed on earth's atmosphere with wavelengths in the range $240\ \text{nm} \lesssim \lambda \lesssim 290\ \text{nm}$ and is responsible for absorption of much of the solar UVB radiation that is incident on our atmosphere [49].

Exposure limits for gas-phase ozone have been established by many governmental and intergovernmental organizations that have responsibilities to ensure public health. Table 11.3 provides a summary of exposure limits for $O_3$ that have been established by several of these agencies. In general, these limits call for gas-phase $O_3$ concentration to be maintained below roughly 0.1 $ppm_V$.

Much of our current understanding of the potential for ozone formation from far UVC sources is based on studies of ozone formation and destruction in the stratosphere, which are initiated by photochemical reactions. The basic reactions that are responsible for photochemically initiated

**Table 11.3** Summary of regulatory and guideline exposure limits for gas-phase $O_3$.

| Agency or organization | $O_3$ limit name or acronym | $O_3$ exposure limit ppm$_v$ $\left(\frac{mg}{m^3}\right)$ | Comments |
|---|---|---|---|
| NIOSH/OSHA | TWA | 0.1 (**0.2**) | Time-weighted average for up to a 10-hours work day during a 40-hours work week |
| | ST | 0.3 (**0.6**) | Short-term TWA |
| EPA | NAAQS | 0.070 (**0.14**) | Primary and secondary NAAQS; annual fourth-highest daily maximum 8-hours concentration, averaged over 3 years |
| WHO | Guideline | 0.1 (**0.2**) | 8-hour mean exposure concentration |

formation and decomposition of ozone were first proposed by Sydney Chapman in 1930. The "Chapman Cycle" comprises the following set of reactions [49, 50]:

$$O_2 + h\nu \rightarrow O + O \tag{11.2}$$

$$O + O_2 + M \rightarrow O_3 + M \tag{11.3}$$

$$O_3 + h\nu \rightarrow O + O_2 \tag{11.4}$$

$$O + O_3 \rightarrow O_2 + O_2 \tag{11.5}$$

$$O + O + M \rightarrow O_2 + M \tag{11.6}$$

Reaction 11.2 is a photochemical reaction that displays strong wavelength-dependence. As with all photochemical reactions, the rate of the reaction depends on the combined effects of absorbance and the quantum yield.

A common unit of concentration for gases in the atmosphere is $\frac{\text{molecules}}{\text{cm}^3}$, often expressed using the symbol $N$. Following this set of units, a form of the Beer–Lambert law applied to gas-phase constituents is:

$$\ln\left(\frac{I_0}{I}\right) = \sigma N \ell \tag{11.7}$$

where

| | | | | |
|---|---|---|---|---|
| $I_0$ | = | irradiance incident on a column of absorbing gas | [=] | $\frac{\text{W}}{\text{cm}^2}$ |
| $I$ | = | irradiance transmitted through a column of absorbing gas | [=] | $\frac{\text{W}}{\text{cm}^2}$ |
| $\sigma$ | = | absorption cross section[c] | [=] | $\text{cm}^2$ |
| $N$ | = | gas-phase concentration of absorbing compound | [=] | $\frac{\text{molecules}}{\text{cm}^3}$ |
| $\ell$ | = | path length through which radiation passes in absorbing gas | [=] | cm. |

Absorption cross sections will vary with wavelength. A collection of absorption cross sections across a range of wavelengths will define an absorption spectrum for an absorbing gas.

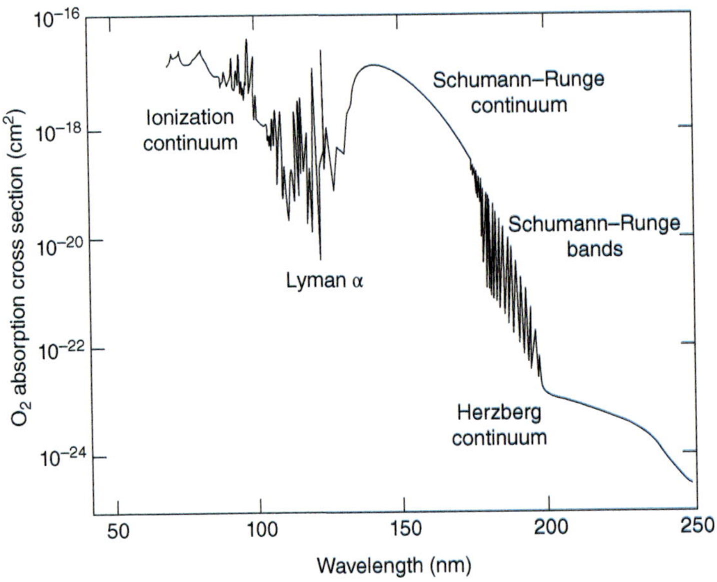


**Figure 11.9** Absorption cross sections for molecular oxygen ($O_2$) for wavelengths that comprise the VUV and UVC ranges. *Source:* Stamnes 2015/with permission of Elsevier.

Absorption cross sections represent measurable physical properties of a chemical. Figure 11.9 illustrates the absorption cross sections (absorption spectrum) for gas-phase molecular oxygen ($O_2$) for wavelengths that include vacuum UV (VUV) and UVC. Absorption starts to become quantifiable at wavelengths less than about 242 nm, which represents the practical upper limit of the wavelength range that initiates reaction 11.2. Within the range of wavelengths that are relevant to applications that involve UVC and VUV radiation, the absorption cross sections of $O_2$ are characterized by the Herzberg continuum, which represents a regular, smooth increase of absorption as wavelength decreases from roughly 240 nm down to 200 nm. For wavelengths from roughly 175 to 200 nm, a banded structure known as the Schumann–Runge bands becomes evident; this structure is attributable to promotion of the $O_2$ molecule from a ground state to excited state transitions [52]. The quantum yield for reaction 11.2 is essentially $1\frac{\text{mole}}{\text{einstein}}$ for wavelengths in the range 175 nm $\lesssim \lambda \lesssim 242$ nm, so the reaction will proceed essentially at the rate of Far UVC or VUV photon absorption by molecular oxygen [53].

The atomic oxygen that is formed from reaction 11.2 will participate in an exothermic reaction with molecular oxygen that results in the formation of $O_3$ (reaction 11.3); the heat given off by reaction 11.3 is transferred to another molecule (M) that participates in this reaction.

Ozone is also photoactive. Wavelengths of radiation less than about 320 nm are sufficiently energetic to initiate reaction 11.4. As with any photochemical reaction, the absorption behavior of the reactant (in this case $O_3$) will play a key role in governing the kinetics of the reaction. Figure 11.10 illustrates recommended values of the absorption cross sections (absorption spectrum) for $O_3$ for temperatures between 293 and 298 K. The spectrum is characterized by several regions, including the Hartley band (200 nm $\leq \lambda \leq$ 300 nm), the Huggins bands (300 nm $\leq \lambda \leq$ 370 nm), and the Chappius band (370 nm $\leq \lambda \leq$ 790 nm). The quantum yield for reaction 11.4 is roughly $0.9\frac{\text{mole}}{\text{einstein}}$ for

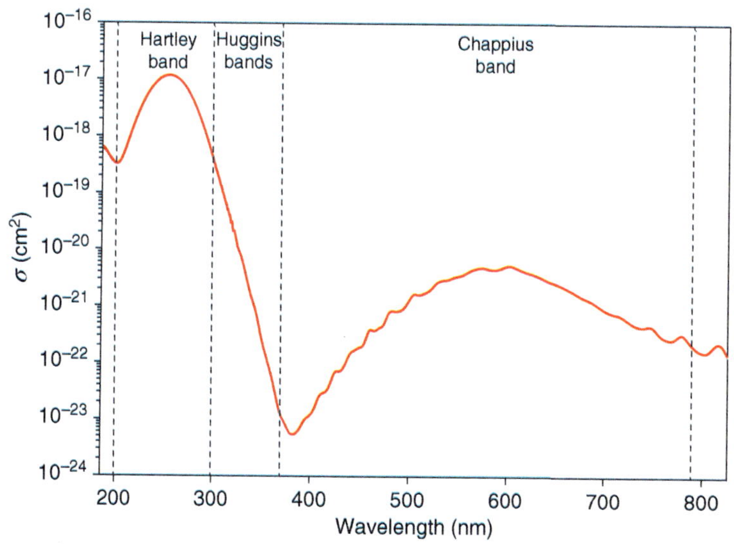


**Figure 11.10** Recommended absorption cross sections for $O_3$ for temperature in the range 293–298 K. *Source:* Data from *Chemical Kinetics and Photochemical Data for Use in Stratospheric Modeling, Evaluation Number 19, JPL* [53].

wavelengths in the range 220 nm $\lesssim \lambda \lesssim$ 305 nm, but decreases steadily from approximately $0.9 \frac{\text{mole}}{\text{einstein}}$ at 225 nm down to $0.48 \frac{\text{mole}}{\text{einstein}}$ at 193 nm [53].

Reaction 11.5 involves the reaction between atomic oxygen and ozone to form two molecules of molecular oxygen. Atomic oxygen (O) and ozone ($O_3$) are sometimes referred to as "odd oxygen"; both forms are highly reactive and generally unstable. Reaction 11.5 results in consumption of two molecules of odd oxygen. Reaction 11.6 has been shown to be a negligible contributor to the Chapman Cycle, largely because it requires collisions between two oxygen atoms, which will be present at very low concentrations in most systems [49].

Accumulation of ozone will depend on a balance between the rates of ozone formation and decay. Since both processes are initiated photochemically and both are characterized by large quantum yields, it is useful to compare the absorption characteristics of the two reactants, $O_2$ and $O_3$. Figure 11.11 provides this comparison for wavelengths of radiation that are relevant to most UVC systems where ozone may be generated, including systems that involve Far UVC sources of radiation. The data presented in Figure 11.11 indicate that ozone is a much stronger absorber of UVC radiation than molecular oxygen; the absorption cross sections for ozone are five to six orders of magnitude larger than those for $O_2$ for wavelengths less than 242 nm. However, the concentration of $O_2$ is likely to be many orders of magnitude larger than the concentration of $O_3$. As described in Chapter 2, the local rate of a photochemical reaction is defined by the product of local fluence rate, absorptivity (here defined by $\sigma$), and the quantum yield. If a gas mixture containing $O_2$ and $O_3$ is exposed to radiation from a Far UVC source that can initiate reactions 11.2 and 11.4, then at a local scale, both $O_2$ and $O_3$ will be exposed to the same local fluence rate, which means that the local ratio of the rates of reactions 11.2 and 11.4 will be governed by the product $\sigma N \Phi$. As an example, if we consider a system operating with an air mixture at a total pressure of 1.0 atm with a temperature of 20 °C and $O_3$ present near the regulatory limit for $O_3$ (0.1 $ppm_v$), with the ambient air being exposed to radiation at 222 nm (the peak wavelength from a $KrCl^*$ excimer lamp),

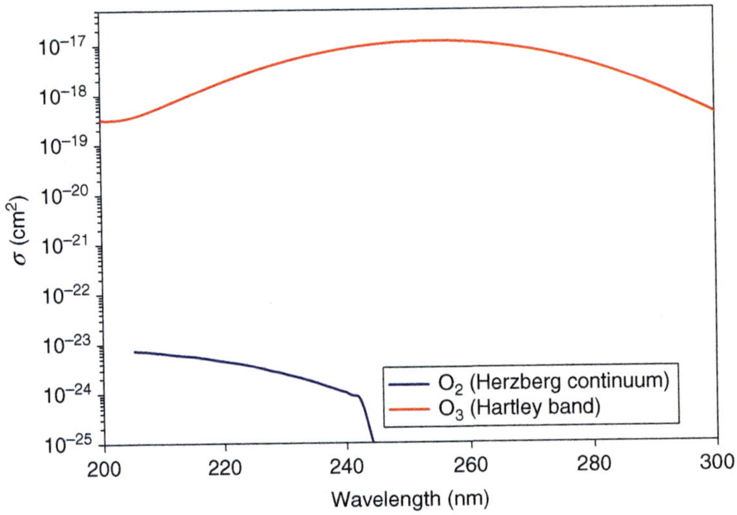


**Figure 11.11** Absorption cross sections for $O_2$ and $O_3$ for the wavelength range 200–300 nm. *Source:* Data from Chemical Kinetics and Photochemical Data for Use in Stratospheric Modeling, Evaluation Number 19, JPL [53].

reaction 11.2 will proceed roughly 4.3 times faster than reaction 11.4. Therefore, the potential exists for generation of $O_3$ from Far UVC sources.

However, it is important to put this information into the proper context, specifically as it relates to Far UVC sources. Continuing with the example of KrCl* excimers as sources, at 222 nm the absorption cross section for $O_2$ is roughly 4 orders of magnitude lower than at 172 nm, the characteristic wavelength of a Xenon excimer $(Xe_2^*)$ lamp. Xenon excimer lamps are commonly used for $O_3$ generation because of the strong absorbance by $O_2$ of radiation at their characteristic emission wavelength, 172 nm. In fact, Xenon excimer lamps can be quite efficient as $O_3$ generators. A KrCl* lamp will also generate $O_3$ photochemically, but the generation rate will be orders of magnitude lower than from a Xenon excimer lamp.

Reaction 11.5 is important because it consumes two molecules of odd oxygen, one of which is a precursor to $O_3$ formation (atomic oxygen, O), the other is $O_3$ itself. However, reaction 11.5 proceeds relatively slowly under most conditions, largely because the concentrations of O and $O_3$ tend to be quite low.

In systems that involve oxygen only, the reaction sequence defined by the Chapman Cycle will govern the rates of ozone formation and decay, which in turn will govern the local steady-state concentration of ozone. In many settings, there are other constituents that may alter the steady-state condition.

An important group of reactions has been identified that will influence $O_3$ dynamics. In particular, there are several compounds that have been identified that will participate in catalytic cycles that mimic reaction 11.5. In general, these catalytic cycles appear as follows, with "*X*" representing the gas-phase catalyst:

$$\begin{aligned} &X + O_3 \rightarrow XO + O_2 \\ &XO + O \rightarrow X + O_2 \\ &\text{net}: O + O_3 \rightarrow O_2 + O_2 \end{aligned} \tag{11.8}$$

Clearly, the net result of the sequence described by 11.8 is the same as reaction 11.5, but the sequence itself follows a different pathway than the elementary reaction described by 11.5. Also, the sequence 11.8 can be orders of magnitude faster than reaction 11.5. As such, sequence 11.8 can profoundly affect the local steady-state $O_3$ concentration.

Several compounds have been identified as being able to serve the function of the catalyst (*X*) described in sequence 11.8. These include: *H*, *HO*, *NO*, *Cl*, and *Br* [49]. The presence of one or more of these catalysts, many of which can themselves be formed photochemically, can dramatically alter (reduce) local gas-phase $O_3$ concentrations. In fact, these catalysts are major contributors to stratospheric ozone depletion and thinning of the ozone layer. In the context of Far UVC applications, the presence of one or more of these catalysts will result in lower local $O_3$ concentrations around Far UVC sources than if the catalysts are not present.

For VUV sources, photolysis of $H_2O$ can be relevant to net $O_3$ production:

$$H_2O + h\nu \rightarrow HO + O \tag{11.9}$$

Reaction 11.9 results in production of hydroxyl radicals (HO), which are known to catalyze $O_3$ destruction (see reaction 11.8). As a result, photochemical $O_3$ generators, which are often based on VUV sources, are generally operated with dry air or dry $O_2$ as a feed gas to suppress this catalytic cycle. As related to Far UVC sources, it is relevant to note that $H_2O$ is a weak absorber of radiation at $\lambda \gtrsim 200$ nm [54] and the quantum yield for reaction 11.9 is essentially zero for radiation at $\lambda \gtrsim$ 193 nm [55].

Reaction 11.10 describes another process by which water vapor will generate HO. The rate constant for reaction 11.10 is large, but the contribution of reaction 11.10 to HO formation, and consequent catalytic destruction of $O_3$, will be constrained by other reactions that oxygen atoms (specifically $O(^1D)$) participate in.

$$H_2O + O \rightarrow HO + HO \tag{11.10}$$

As a result, relative humidity (RH) will play a less significant role in influencing $O_3$ production from Far UVC sources than from VUV sources.

In a broader context, it is important to consider the potential for local photochemical formation of any of the catalysts that can promote $O_3$ decay. The formation of any of these catalysts will reduce effective $O_3$ generation by initiating a catalytic cycle to destroy $O_3$.

$O_3$ and O will also react with other constituents in the gas phase, as well as constituents in condensed phases. These two forms of odd oxygen function as powerful oxidants and will react rapidly with reduced compounds.

Generation of $O_3$ by Far UVC sources will depend on the fundamental chemistry described above, but also local conditions of application. These will include the output power and output spectrum of the Far UVC source, which will govern the fluence rate field that surrounds the Far UVC source, which in turn governs the local rates of the photochemical and thermal reactions described above. In addition, local air circulation and mixing behavior will influence the distribution and accumulation of $O_3$ in any setting. Given these characteristics, it is not possible to define the potential for $O_3$ generation of a Far UVC device as a global statement. However, applications of Far UVC systems should consider the potential for $O_3$ generation, as well as measures that can be included in system design to minimize human exposure. These could include integration of ventilation systems as well as active measures to mitigate $O_3$ production.

Given the fundamental attributes of the Chapman Cycle, essentially any source of radiation that yields UV output below roughly 242 nm has the potential to generate ozone by the photochemical pathway described above. However, $O_3$ generation will generally become more important as the output of relatively short wavelength radiation increases. Examples of common UV sources that have the potential to generate ozone include $KrCl^*$ excimer lamps (peak output at $\lambda = 222$ nm), $KrBr^*$ excimer lamps (peak output at $\lambda = 207$ nm), and low-pressure Hg lamps, with a secondary line at $\lambda = 185$ nm.

Ozone generation can also be a relevant issue for UV sources that involve corona discharge, which is an electrical discharge that results from ionization of a fluid that surrounds a conductor; in the context of UVC-based devices that are used to disinfect air and surfaces, this would involve ionization of air around the electrodes that are used to drive a UVC source. In these systems, the oxygen molecule is split by the electrical current, thereby producing atomic oxygen, which then goes on to produce ozone.

### 11.2.2 UVC-Induced Damage of Materials

Applications of UVC radiation for disinfection of air and surfaces will result in exposure of building materials and other objects to UVC radiation. The potential exists for UVC-induced damage to these materials. However, not all materials can be damaged by UVC exposure.

As described extensively in this text, materials of biological origin can be damaged by exposure to UVC radiation. This damage shows up largely in the form of photochemical damage to nucleic acids and proteins, but other biomolecules can also be damaged.

Among the various categories of materials that could be exposed to UVC radiation, metals and ceramics tend to be largely unaffected [56]. UVC radiation is generally ineffective for promotion of ground-state electrons to higher energy states from pure metals and ceramics, so photochemical reactions tend to be suppressed.

On the other hand, many polymeric materials will undergo UVC-induced degradation. These materials include plastics, paints, and coatings. Within these materials, UVC photon absorption can lead to direct photolysis, especially among C=C bonds in a polymer structure. In addition, photolysis can lead to the formation of organic radicals or peroxy radicals. In turn, these radicals can promote cross-linking among polymers, leading to changes of their physical properties (e.g. strength and brittleness), as well as color [57].

Some materials include additives that may be photoactive. Common examples of photoactive additives include metal oxides, such as zinc oxide and titanium dioxide. ZnO and $TiO_2$ are generally strong absorbers of UV radiation, including radiation in the UVC range. For this reason, they are often included as components of sunscreen agents that are applied to human skin. The benefit of strong absorption is that it can prevent direct photochemical damage to underlying material. However, a potential drawback of this behavior is that these metal oxides can also function as photocatalysts. Therefore, the potential exists for photochemical generation of highly reactive radicals, which may in turn promote localized oxidation or reduction.

Pure glass materials tend to be largely nonresponsive or slow to react to UVC irradiation. However, many glass-like materials contain impurities; sometimes these impurities are intentionally added to alter the properties of glass. These impurities can be photoactive and may allow UVC-induced changes in glass to take place. Additional details of UVC-induced damage to glass can be found in Chapters 4 and 9.

## 11.3 Disinfection Kinetics for Airborne and Surface-Associated Microbes and Viruses

Disinfection of air and surfaces is conducted to reduce the risk of human exposure to microbial and viral pathogens. In both applications, the target pathogen may experience direct contact with air, which can represent a source of stress, largely in the form of desiccation. For some pathogens, nucleic acid conformation can be altered by desiccation, which in turn affects the rates of nucleic acid-based photochemical reactions that result from exposure to UV radiation. Because of this, the intrinsic kinetics of UV disinfection for microbes and viruses in air-exposed samples, whether suspended in air or on surfaces, tend to be faster than the kinetics observed for the same microbe or virus in aqueous suspension. Consequently, UV dose–response behavior for an aqueous suspension of a microbe or virus often represents a conservative estimate of the *intrinsic* kinetics of inactivation of the same microbe or virus in an air-exposed setting.

However, airborne and surface-associated microbes and viruses are often observed within an aerosol or droplet of fluid. For example, many airborne pathogens are found within aerosols that are emitted through breathing, coughing, or sneezing by an infected individual. These aerosols contain proteins, such as mucins, and other compounds that can be strong absorbers of UV radiation. The absorbance characteristics of the droplets will limit the ability of UVC radiation to reach microbes and viruses that are held within them.

Ambient air will contain viruses, bacteria, and possibly other pathogens. The concentrations of bacteria, viruses, fungi, molds, and other airborne microbes are highly variable. Figure 11.12 provides a summary of the concentrations of airborne "virus-like" and "bacterium-like" particles from several indoor settings and ambient outdoor air near a university campus. The particles counted in

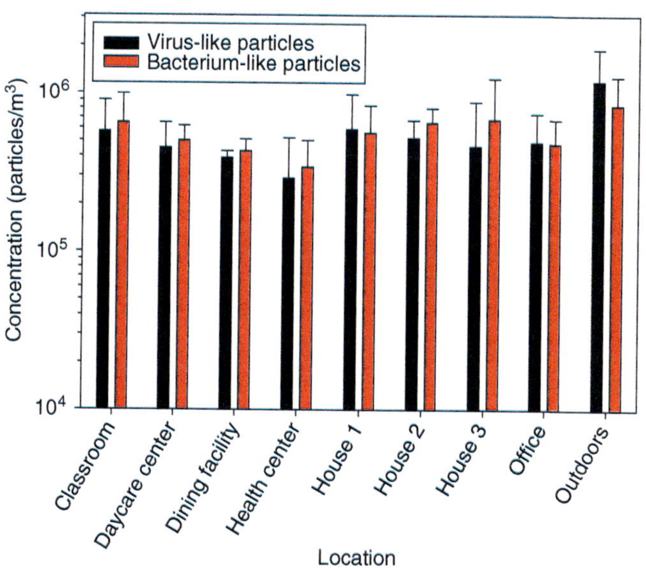


**Figure 11.12** Graphical summary of measured concentrations of airborne virus-like particles and bacterium-like particles from air samples collected from indoor air spaces and ambient outdoor air near a university campus. Bar height and error bars represent the mean and standard deviation, respectively, of three measurements from each location. *Source:* Data from Prussin et al. [58].

this data set responded positively to a nucleic acid (DNA or RNA) stain and had size characteristics that were consistent with viruses or bacteria [58].

The data presented in Figure 11.12 indicate several interesting trends. First, the concentrations of virus-like and bacterium-like particles were consistent across each of the indoor air spaces from which samples were collected. Second, these concentrations were consistently lower than the concentrations of these particles measured in outdoor air samples.

It is worth noting that the measurements presented in Figure 11.12 provide no information about the specific viruses or bacteria that contributed to the measurements or their viability. The majority of airborne viruses and bacteria are nonpathogenic to humans, so these data provide only a rough indication of total viral and microbial burdens to be expected in air samples.

Airborne concentrations of pathogens could be considerably different than the data illustrated in Figure 11.12, under some circumstances. In a crowded indoor space, there could be much higher concentrations of airborne viral or microbial particles. Similarly, in a clinical setting, it is possible for relatively large numbers of infected people to congregate, sometimes carrying and emitting the same pathogen. For example, in flu season or in a TB clinic, the airborne concentrations of influenza viruses or *M. tuberculosis*, respectively, could be quite high.

Since indoor air quality is of concern, in part because of the potential for airborne transmission of diseases, another reference point is the concentration of airborne pathogens in air that is released by an infected individual. As an illustration, Riediker and Tsai developed estimates of the rate of SARS-CoV-2 emission by infected patients associated with regular respiration. They reported wide range of virus emission rates, depending in part on the severity of infection, but indicated values as high as $2.9 \times 10^9 \frac{\text{copies}}{\text{hr}}$ for so-called "high emitting" patients [59]. A similar study by Ma et al. included measured emission rates in the range of $1.03 \times 10^5$–$2.27 \times 10^7 \frac{\text{viruses}}{\text{hr}}$ [60]. Studies have been conducted to quantify emission rates of other pathogens that are transmitted largely or entirely by airborne routes. As a generalization, the rates of pathogen emission are highly variable among infected patients but can be quite high for individuals with extensive infection.

The extent of pathogen inactivation required will vary among applications. "Mathematical sterilization," defined as 6 $\log_{10}$ units of inactivation of a target pathogen, has been suggested as an extreme endpoint of disinfection [17]. This same limit is consistent with recommendations that have been issued by the US Food and Drug Administration (FDA) for decontamination of surgical masks and respirators [61]. The extent of pathogen inactivation required in most applications will be less than limit represented by mathematical sterilization.

A complicating factor in the measurement and reporting of UV dose–response behavior for air-exposed microbes and viruses is the lack of standard methods for conducting these tests. In all such experiments, the goal is to expose subpopulations of the target microbe or virus to single-valued, measurable doses of radiation from the UV source of interest. Unlike experiments with aqueous suspensions of microbes or viruses, for which well-defined apparatus and methods are available [62], there is no standard method that has been adopted for these experiments. However, several devices and experimental methods have been developed to address this need.

Ke et al. developed a device that is a gas-phase analog of a conventional collimated beam used for experiments with liquid suspensions. Figure 11.13 is a schematic illustration of this device.

An aerosolized suspension of the target microbe or virus is introduced via a nebulizer to a vertically oriented cell that includes a mixing fan, which allows the air within the cell to approach a well-mixed condition. A quartz plate is placed across the top of the cell to allow radiation from a collimated beam to be imposed on the gas within the cell. RH within the cell is controlled by inclusion of a separate line that includes humidified or dried air.

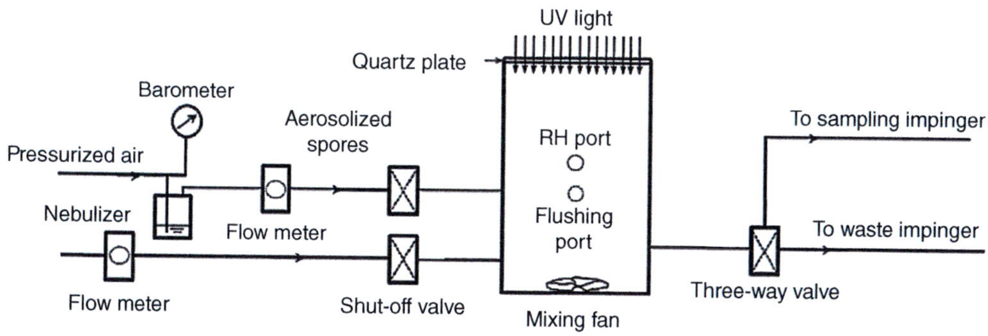


**Figure 11.13** Schematic illustration of collimated beam reactor used for UV exposure of aerosol suspensions of microbes and viruses. *Source:* Ke et al. 2009/Taylor & Francis.

The nebulizer output is applied continuously to allow the target microbe or virus concentration in the cell to reach a steady-state concentration. Air flow to and from the cell is then closed off; the gas-phase suspension is then subjected to UV irradiation from the collimator for a measured exposure period. The average dose delivered to the sample is calculated by accounting for the factors that affect the spatial distribution of radiant energy within the cell, including: lateral beam variability (i.e. roughly the equivalent of the "Petri factor"), vertical beam variability (divergence), reflection, and absorption [62, 63]. The air sample from the cell is then flushed into a sampling impinger to allow collection of airborne microbes or viruses, which are then subjected to an appropriate viability or infectivity assay. The process is then repeated over a range of doses to build the UV dose–response data set.

A strength of this device is that its characteristics are consistent with conventional collimated beam devices that are commonly used for UV dose–response experiments with liquid suspensions. However, the translation of this approach to gas-phase suspensions is limited by the number (concentration) of target microbes or viruses that can be aerosolized and suspended within the cell. In turn, this limits the extent of inactivation that can be measured by this device.

Welch et al. developed a dynamic device that was used to measure UV dose–response behavior of aerosolized targets [12]. The device was similar to devices that had been used by other researchers [64–66]. Figure 11.14 provides a schematic illustration of the device.

The device illustrated in Figure 11.14 operates dynamically, in that a constant flow rate of the gas-phase suspension is introduced to the system. As such, the nominal residence time of the suspension within the irradiated zone is fixed. Nominal dose delivery is varied by inclusion or exclusion of optical filters to attenuate the output from the UV source(s) used with the device.

As a dynamic system, the device illustrated in Figure 11.14 avoids the problem of limited target population size that exists with the batch device described in Figure 11.13. However, the device illustrated in Figure 11.14 is operated based on the assumption that all airborne particles receive the same dose as they pass through the irradiated zone. It is likely that the dose distribution delivered by this system is narrow; however, it is clear that each particle will follow a unique path through the irradiated zone of the device, and the fluence rate delivered to the irradiated zone will demonstrate spatial variability. For small extents of inactivation (e.g. less than 1–2 $\log_{10}$ units), the effect of this dose distribution on disinfection efficacy may be acceptably small, thereby allowing the use of a nominal (average) dose to describe exposure of the target microbe or virus population.

Figure 11.15 illustrates $UV_{222}$ dose–response behavior for three aerosolized viruses and an aerosolized bacterium, as measured by the device illustrated in Figure 11.14. For each of these dose–response data sets, the intrinsic kinetics of inactivation followed essentially first-order behavior.

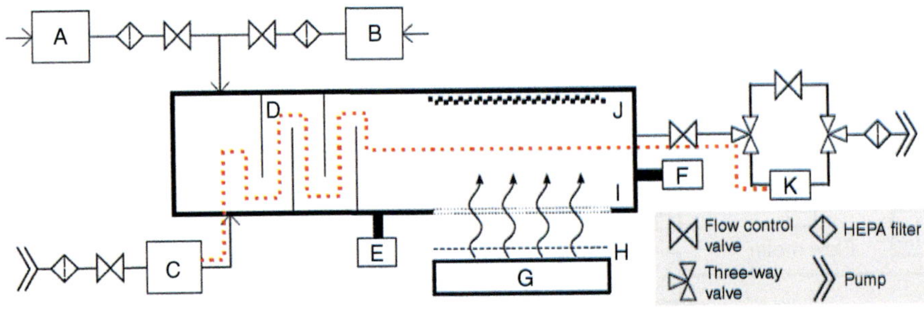


**Figure 11.14** Schematic illustration of continuous-flow reactor used for dynamic UV exposure of aerosol suspensions of microbes and viruses. *Source:* Welch et al. 2018/Springer Nature/CC BY-4.0. System components: water bubbler for humidified air input (A), desiccator for dry air input (B), nebulizer (C), baffles (D), RH and temperature meter (E), particle sizer (F), UV source (G), band pass filters, if needed (H), transmitting window, if needed (I), reflective aluminum surface, if needed (J), BioSampler (K).

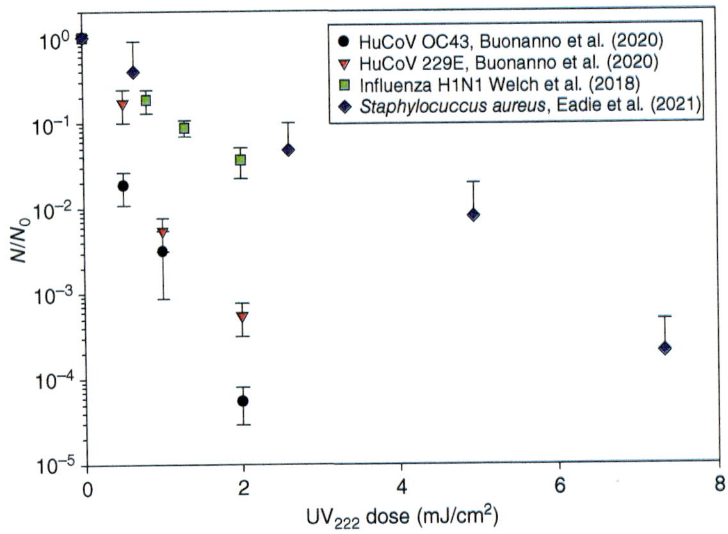


**Figure 11.15** $UV_{222}$ dose–response behavior of aerosolized pathogens, including two coronaviruses (HuCoV OC43 and HuCoV 229E), an influenza virus (H1N1), and a bacterium (*S. aureus*). All measurements were conducted using the apparatus illustrated in Figure 11.14. Error bars represent standard deviation of replicate measurements. *Source:* Data extracted from references indicated in legend [12, 67].

Lim and Blatchley followed an approach that involved application of the target microbe or virus to a membrane filter, followed by drying and equilibration with air at controlled RH, then UV irradiation under a collimated beam [68]. The device used in their experiments is illustrated in Figure 11.16. The membrane filters were assumed to be inert and to not affect disinfection kinetics. The membrane filters were placed in a desiccator that had been modified by the inclusion of a quartz window to allow the output radiation from a collimated UV source to be projected into the desiccator space. RH inside the desiccator was controlled through the use of aqueous salt solutions [69]. Samples of microbes or viruses were exposed to UV radiation from the collimated source for a period of time sufficient to deliver the target dose. The turntable holding the samples was then

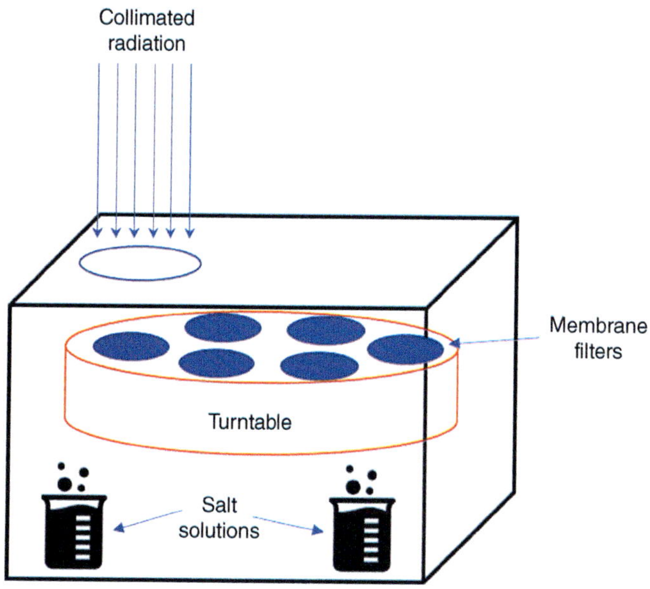


**Figure 11.16** Schematic illustration of device used to conduct UV dose–response experiments with air-exposed microbes and viruses. Image adapted from Lim and Blatchley [68].

rotated to allow the next sample to be placed under the beam to allow a different exposure. Irradiated microbes or viruses were then washed from the UV-exposed membrane filter into an aqueous solution to allow measurement of viability or infectivity, following conventional methods.

The device illustrated in Figure 11.16 allows measurement of UV dose–response behavior over a wide range of RH conditions. It was developed based on the hypothesis that membrane filter materials are inert and play no role in the disinfection process. A limitation of this device is the interplay between the texture of the membrane filter material and the physical dimensions of the microbial or viral targets. In particular, it is possible for some of the target microbes or viruses to be sheltered by the pore structure of the membrane filter or its texture.

Responses of air-exposed microbes and viruses to UV radiation can be influenced by RH. Many studies have indicated that microbes and viruses become more sensitive to UV exposure as RH decreases. For example, Peccia et al. observed increases in $UV_{254}$ inactivation rates of aerosolized *Serratia marcescens*, *Bacillus subtilis*, and *Mycobacterium parafortuitum*, with sharp changes in the responses of these microbes for RH values above roughly 50% [70]. Xu et al. reported a similar sharp decline in disinfection efficacy of a full-room UR-UVGI system against *M. parafortuitum* when RH exceeded 50% [71]. Fletcher presented data to indicate that four gram-negative bacteria (*Burkholderia cepacia*, *S. marcescens*, *Stenotrophomonas maltophilia*, and *Pseudomonas aeruginosa*) were more susceptible to $UV_{254}$ exposure at low RH than high RH [24]. Similar behavior has been reported for airborne viruses, including influenza virus [66] and vaccinia virus [65], the virus that causes smallpox.

Although there are papers that have reported minimal or no effects of RH with respect to the UV sensitivity of airborne microbes or viruses, the majority of such observations have been consistent with the general observations described above; namely, airborne microbes and viruses tend to become more sensitive at low RH. This behavior has been attributed to changes in nucleic acid conformation and the general stress imposed by desiccation.

First-order (single-event) kinetics are assumed for most airborne and air-exposed microbial and viral pathogens. The first-order model is appropriate for most viruses, but deviations from first-order behavior have been observed for bacteria, yeasts, and molds. Table 11.1 includes reported values of the inactivation constant for many microbial and viral pathogens for which an airborne exposure pathway is known to exist. It should be noted that the literature in the area of UVGI applications often reports an inactivation rate constant as a "$Z$ value"; the $Z$ value is a synonym for an inactivation constant ($k$), with the same units and same physical interpretation.

### 11.3.1 Genomic Models

Viability (i.e. culture-based) and infectivity methods represent the gold standard for quantification of microbial and viral UV dose–response behavior. These methods provide a direct assessment of the response of a pathogen or surrogate to UV exposure and the effect of this response on the pathogen's ability to replicate or infect a host. However, under some circumstances, it is difficult or impossible to conduct these experiments. For example, the human norovirus (HuNoV) is believed to be responsible for a large fraction of waterborne and foodborne illness. The virus is highly contagious, but it is difficult to culture. As a result, there are no empirical data to define the UV dose–response behavior of HuNoV, based on infectivity assays. Virulent pathogens also present problems, in that they require access to laboratory facilities and equipment that are designed to limit human exposure during experiments. Examples of viral pathogens that have presented this problem include the Ebola virus and SARS-CoV-2, both of which require Biosafety Level 3 or 4 (BSL-3 or BSL-4) laboratory facilities to conduct experiments.

These and other issues have motivated the development of mathematical models to simulate UV dose–response behavior. The models developed to date have focused on viral pathogens, in part because inactivation is strongly linked to photochemical damage to their genetic material or genome. The models have also focused on responses to $UV_{254}$ exposure, largely because the database of information to describe viral responses at this wavelength is much more extensive than at any other UVC wavelength and because viral damage at this wavelength is largely confined to genomic material.

Lytle and Sagripanti were among the first to develop a genomic model to predict the sensitivity of viruses to $UV_{254}$ exposure [72]. Their model was based on the hypothesis that because viral inactivation at 254 nm is dominated by photochemically induced damage to DNA or RNA, the sensitivity of viruses to $UV_{254}$ exposure is likely to depend on the structure and size of the genome. In general, DNA viruses were expected to be relatively sensitive to UV radiation because they include adenine (A), cytosine (C), guanine (G), and thymine (T), of which T (especially adjacent Ts in a nucleic acid strand) will be most sensitive. In contrast, RNA viruses include A, C, G, as well as uracil (U), which is less UV-sensitive than T. The double-stranded (ds) DNA and RNA viruses were expected to be relatively resistant to UVC exposure because they have access to repair mechanisms that are more efficient than those of the single-stranded (ss) viruses due to the complementary nature of the nucleic acid strands. In addition, the observed UVC dose–response behavior of most viruses is consistent with single-event kinetics, suggesting that a single unit of photochemical damage within the viral genome may be sufficient to prevent replication. Under this model, it was hypothesized that the susceptibility of a virus to UV-based damage will increase with the size of the genome, since larger genomes will present proportionally more targets to damage.

Taking these factors together, the model of Lytle and Sagripanti was based on the concept of "size-normalized sensitivity" (SnS), which was defined formally as:

$$\mathrm{SnS} = D_{37} \cdot \mathrm{GS} \tag{11.11}$$

where

| | | | | |
|---|---|---|---|---|
| SnS | = | size-normalized sensitivity | [=] | $\frac{\text{J}\cdot\text{kb}}{\text{m}^2}$ |
| $D_{37}$ | = | $UV_{254}$ dose that yields a single unit of damage to a virion | [=] | $\frac{\text{J}}{\text{m}^2}$ |
| GS | = | genome size | [=] | kb. |

The parameter $D_{37}$ is a measure of virus sensitivity to $UV_{254}$ radiation. Small values of $D_{37}$ are indicative of viruses that are highly sensitive to $UV_{254}$ exposure. The parameter $D_{37}$ is the inverse of the inactivation constant ($k_{254}$) that is often used in the single-event kinetics model for radiation at 254 nm:

$$k_{254} = \frac{1}{D_{37}} \tag{11.12}$$

The Lytle and Sagripanti model was based further on the assumption that the value of SnS is constant for each genome type. In other words, the value of SnS could be used to predict $D_{37}$ or $k_{254}$ for viruses with a given genome type, based on knowledge of SnS and GS.

Their model was extended to describe the responses of various groups of viruses to solar UVB exposure. This model extension was motivated by interest in being able to predict the rate of inactivation of viruses in the ambient environment. In a general sense, the ability of viruses to retain infectivity in the environment is governed by the combined effects of ambient temperature, RH, and UVB exposure. UVB radiation promotes the same photochemical reactions that lead to genome damage by exposure to radiation within the UVC range; however, UVB radiation is far less efficient than UVC radiation for promoting this damage because of the nature of the absorption spectra of nucleic acids (DNA and RNA) in these ranges of the UV spectrum. On the other hand, the absorption spectra of DNA and RNA are quite similar among virtually all life forms, and because the quantum yield for common photochemical reactions among nucleotides within nucleic acids demonstrates very little variation with wavelength, it is possible to use inactivation constant estimates at 254 nm to predict inactivation responses at longer wavelengths by comparing UV absorbance behavior.

Lytle and Sagripanti assembled a generalized, normalized action spectrum for virus inactivation by compiling measured action spectra for viruses that had been reported in the literature. To a first approximation, their compiled action spectra converged on a single curve. Using this information, they were then able to relate measured or predicted inactivation constants at 254 nm to inactivation constants in the UVB portion of the spectrum. By combining this information with solar spectral irradiance, they were able to predict the rate of viral inactivation resulting from exposure to solar UVB radiation at many locations around the world.

The SnS model was subsequently applied to predict the inactivation responses of coronaviruses and other viruses of concern at 254 nm [21]. Table 11.4 provides a summary of measured inactivation constants ($k_{254}$) for several virus families, along with corresponding estimates of $k_{254}$ based on the SnS model [72]. Table 11.5 provides a summary of measurements and predictions of the SnS model for specific viruses [21].

Pendyala et al. developed a model that accounted for additional details of genome structure among ssRNA viruses, relative to the SnS model. In recognition of the fact that the pyrimidines are substantially more photoreactive than purines, known genomic sequences of viruses were used to define the number of adjacent pyrimidines within the genome (i.e. pyrimidine dinucleotides). They referred to their model as the Pyrimidine Dinucleotide Frequency Value (PyNNFV) Model [20]. Specifically, the PyNNFV model was based on the frequency of dinucleotides (NN) that are present in a viral genome.

**Table 11.4** Summary of virus characteristics, measured inactivation constants ($k_{254}$), along with predictions of $k_{254}$ based on the SnS model.

| Genome type | Virus family | Genome size[a] | Measured $k_{254}$[b] $\left(\frac{m^2}{J}\right)$ | Predicted $k_{254}$[c] $\left(\frac{m^2}{J}\right)$ |
|---|---|---|---|---|
| dsDNA | *Adenoviridae* | 26–45 | 0.0077 (0.0045–0.010) | 0.0056–0.010 |
| | *Asfarviridae* | 170–190 | | 0.067–0.077 |
| | *Herpesviridae* | 125–240 | 0.053 (0.036–0.16) | 0.038–0.077 |
| | *Iridoviridae* | 98–170 | 0.063 | 0.038–0.067 |
| | *Papillomaviridae* | 6.8–8.4 | | 0.0026–0.0033 |
| | *Polyomaviridae* | 4.7–5.3 | 0.0040 | 0.0036–0.0040 |
| | *Poxviridae* | 130–300 | 0.091 | 0.063–0.15 |
| ssDNA | *Circoviridae* | 1.8–2.3 | | 0.039–0.050 |
| | *Hepadnaviridae* | 3.0–3.3 | | 0.067–0.071 |
| | *Parvoviridae* | 4–6 | 0.11 (0.10–0.12) | 0.083–0.13 |
| dsRNA | *Birnaviridae* | 5.7–6.2 | 0.0083 (0.0059–0.0091) | 0.0083–0.0091 |
| | *Reoviridae* | 18.6–26.4 | 0.011 (0.0081–0.022) | 0.010–0.013 |
| (−)ssRNA | *Arenaviridae* | 11 | | 0.077 |
| | *Bornaviridae* | 8.9 | 0.029 | 0.029 |
| | *Bunyaviridae* | 11–19 | | 0.14 |
| | *Deltavirus* | 1.7 | | 0.012 |
| | *Filoviridae* | 19 | | 0.14 |
| | *Orthomyxoviridae* | 10–15 | 0.13 (0.10–0.21) | 0.091–0.14 |
| | *Paramyxoviridae* | 15–16 | 0.091 (0.083–0.10) | 0.091 |
| | *Rhabdoviridae* | 11–15 | 0.23 (0.043–0.91) | 0.22–0.29 |
| (+)ssRNA | *Arterioviridae* | 13–16 | | 0.043–0.056 |
| | *Astroviridae* | 6.8–7.9 | | 0.022–0.027 |
| | *Caliciviridae* | 7.4–8.3 | | 0.025–0.028 |
| | *Coronaviridae* | 20–31 | 0.32 | 0.26–0.40 |
| | *Flaviviridae* | 9.6–12 | | 0.032–0.040 |
| | *'HEV-like' virus* | 7.2 | | 0.024 |
| | *Nodaviridae* | 4.5 | 0.0071 | 0.0071 |
| | *Picornaviridae* | 7–8.5 | 0.020 (0.014–0.040) | 0.019–0.022 |
| | *Togaviridae* | 9–12 | 0.053 (0.043–0.14) | 0.042–0.056 |
| Virus replication by reverse transcriptase | *Retroviridae* | 7–11 | 0.011 (0.0083–0.011) | 0.0091–0.015 |

*Source:* Data from Lytle and Sagripanti [72].

[a] For dsDNA viruses, genome size is presented as the number of kilobase pairs; for ssDNA viruses, the size is the number of kilobases.

[b] For viruses for which more than one value of $k_{254}$ is reported, the values indicated represent the median value, while the parenthetical entries represent the range of reported values.

[c] Predicted values of $k_{254}$ are based on genome size within a virus family. For virus families with a range of reported genome sizes, the predicted values correspond to the upper and lower limits of genome size.

**Table 11.5** Summary of virus characteristics, measured inactivation constants ($k_{254}$), along with predictions of $k_{254}$ for specific viruses based on the SnS model.

| Genome type | Virus family | Virus | Genome size | Measured $k_{254}$ $\left(\frac{m^2}{J}\right)$ | Predicted $k_{254}$ $\left(\frac{m^2}{J}\right)$ |
|---|---|---|---|---|---|
| (+)ssRNA | *Coronaviridae* | SARS-CoV-2 | 29.8 | | 0.33 |
| | | SARS-CoV | 29.7 | | 0.33 |
| | | MERS-CoV | 30.1 | | 0.33 |
| | | MHV | 31.6 | 2.9 | |
| | | EToV | 28.5 | 3.1 | |
| | *Togaviridae* | SINV | 11.7 | 19 | |
| | | VEEV | 11.4 | 23 | |
| | | SFV | 13.0 | 7.2 | |
| (−)ssRNA | *Paramyxiviridae* | NDV | 15.2 | 11–13.5 | |
| | | MeV | 15.9 | 8.8–10.9 | |
| | *Orthomyxoviridae* | FLUAV | 13.6 | | |
| | | Melbourne H1N1 | | 10.2 | |
| | | NIB-4 H3N2-3 | | 11 | |
| | | NIB-6 H1N1 | | 9.6 | |
| | | ISAV | 14.5 | 4.8 | |
| | *Rhabdoviridae* | RABV | 11.9 | 4.3 | |

*Source:* Data from Sagripanti and Lytle [21].

It is important to reiterate that DNA includes the pyrimidine bases cytosine (C) and thymine (T), as well as the purine bases adenine (A) and guanine. In RNA, the complementary base to adenine is uracil (U) rather than thymine (T). Sequencing of RNA viruses is often conducted by applying reverse transcription to the viral RNA to convert the molecule to its complementary DNA (cDNA). Sequencing is then conducted on the cDNA to define the viral genome. Therefore, an RNA viral sequence can be described using T as one of the pyrimidines, although it must be recognized that this pyrimidine actually refers to U in the RNA viral genome.

Pendyala et al. developed a variable to quantify the frequency of NN occurrence within a viral genome. Specifically, they considered the frequency of occurrence of pyrimidine dinucleotides within the cDNA of the viral genome. They developed a parameter, PyNNFV, that was hypothesized to relate directly to the $UV_{254}$ susceptibility of a ssRNA virus:

$$\text{PyNNFV} = \frac{(\text{TT\%})(\text{TC\%})(\text{CT\%})(\text{CC\%})}{\text{genome bp}} \tag{11.13}$$

where the terms in the numerator represent the frequency of occurrence of dinucleotides TT, TC, CT, and CC in the cDNA of a ssRNA virus and the denominator represents the size of the cDNA of the viral genome in units of base pairs (bp). Values of PyNNFV were calculated for 16 ssRNA viruses for which reliable estimates of $D_{90}$ were available (i.e. estimates that were based on standardized methods for measurement of $UV_{254}$ dose–response behavior) [62]. $D_{90}$ was defined as the $UV_{254}$ dose that achieved 1.00 $\log_{10}$ unit of inactivation. Pendyala et al. then demonstrated that the relationship between $D_{90}$ and PyNNFV was nearly linear and had a narrow 95% confidence interval, suggesting that a linear relationship between $D_{90}$ and PyNNFV was appropriate. Based on this analysis, they then applied their model to predict values of $D_{90}$ for a range of ssRNA viruses. Measured and predicted values of $k_{254}$ from their model for many ssRNA viruses are summarized in Table 11.6.

**Table 11.6** Summary of virus characteristics, measured inactivation constants ($k_{254}$), along with predictions of $k_{254}$ for specific viruses based on the PyNNFV model.

| Virus | cDNA genome size (bp) | PyNNFV | Measured $k_{254}$ $\left(\frac{m^2}{J}\right)$ | Predicted $k_{254}$ $\left(\frac{m^2}{J}\right)$ |
|---|---|---|---|---|
| Bacteriophage MS2 | 3,569 | 0.008,04 | 0.013 | 0.013 |
| Murine sarcoma virus | 5,833 | 0.008,07 | 0.012 | 0.013 |
| Human parechovirus | 7,348 | 0.002,10 | 0.031 | 0.044 |
| Murine norovirus | 7,382 | 0.005,70 | 0.023 | 0.019 |
| Coxsackievirus | 7,410 | 0.003,14 | 0.029 | 0.031 |
| Poliovirus | 7,440 | 0.002,63 | 0.032 | 0.037 |
| Hepatitis A virus | 7,476 | 0.002,09 | 0.045 | 0.044 |
| Feline calicivirus | 7,683 | 0.003,63 | 0.038 | 0.028 |
| Moloney murine leukemia virus | 8,332 | 0.005,98 | 0.020 | 0.018 |
| Canine calicivirus | 8,513 | 0.003,45 | 0.034 | 0.029 |
| Semliki forest virus | 11,442 | 0.001,41 | 0.092 | 0.060 |
| Venezuelan equine encephalitis virus | 11,444 | 0.001,53 | 0.042 | 0.056 |
| Western equine encephalomyelitis virus | 11,484 | 0.001,51 | 0.043 | 0.057 |
| Sindbis virus | 11,703 | 0.001,49 | 0.042 | 0.057 |
| Measles virus | 15,894 | 0.001,34 | 0.105 | 0.062 |
| SARS-CoV-1 | 29,751 | 0.000,67 | 0.128 | 0.096 |
| Human norovirus GII | 7,509 | 0.003,934 | N/A | 0.026 |
| Human norovirus GI | 7,654 | 0.002,936 | | 0.033 |
| Human norovirus GIV | 7,839 | 0.003,360 | | 0.030 |
| Human coronavirus 229E | 27,165 | 0.000,489 | | 0.114 |
| Avian infectious bronchitis virus | 27,608 | 0.000,371 | | 0.129 |
| Transmissible gastroenteritis virus | 28,614 | 0.000,391 | | 0.127 |
| Bat coronavirus HKU9-1 | 29,114 | 0.000,465 | | 0.117 |
| Bat coronavirus BM48-31 | 29,276 | 0.000,603 | | 0.102 |
| Canine coronavirus | 29,278 | 0.000,379 | | 0.128 |
| Feline infectious peritonitis virus | 29,357 | 0.000,393 | | 0.126 |
| SARS coronavirus A022 (Civet) | 29,499 | 0.000,640,1 | | 0.099 |
| SARS coronavirus B039 (Civet) | 29,525 | 0.000,640,2 | | 0.099 |

Table 11.6 (Continued)

| Virus | cDNA genome size (bp) | PyNNFV | Measured $k_{254}$ $\left(\frac{m^2}{J}\right)$ | Predicted $k_{254}$ $\left(\frac{m^2}{J}\right)$ |
|---|---|---|---|---|
| SARS-CoV-2 | 29,891 | 0.000,549 | | 0.107 |
| Murine hepatitis virus strain A59 | 29,947 | 0.000,532 | | 0.110 |
| MERS-CoV | 30,033 | 0.000,883 | | 0.082 |
| Bat coronavirus HKU4-1 | 30,286 | 0.000,580 | | 0.105 |
| Murine hepatitis virus strain S | 31,147 | 0.000,515 | | 0.111 |
| Rat coronavirus | 31,274 | 0.000,494 | | 0.113 |
| Murine coronavirus MHV-1 | 31,386 | 0.000,526 | | 0.110 |
| Bat Hp-betacoronavirus | 31,491 | 0.000,691 | | 0.095 |

*Source:* Data from Pendyala et al. [20]. Numerical values above the solid red line represent (mean) measured values from the literature and corresponding values predicted by the PyNNFV model (see references cited by Pendyala et al.). Values listed below the solid red line represent predictions based on the PyNNFV model; literature values for these viruses are generally lacking.

Rockey et al. developed a model that is in some respects similar to the model of Pendyala et al., but more comprehensive in its accounting of genome structure and composition [13]. Specifically, they conducted a review of the literature to identify values of inactivation constants ($k_{254}$) for viruses in aqueous suspension that were developed using experimental methods that conform to accepted standards [62]. They then grouped these estimates based on Baltimore Virus Classification System: ssDNA, dsDNA, (−)ssRNA, (+)ssRNA, and dsRNA. Their motivation for grouping based on genome type and mode of mRNA generation was that viruses in these groups generally share different types of responses to $UV_{254}$ irradiation. For example, the DNA viruses will include pyrimidine bases of cytosine (C) and thymine (T) together with the purine bases adenine (A) and guanine (G). For RNA viruses, the pyrimidine bases will include C and uracil (U), but not T. The distribution of these bases within the genome is important in that the pyrimidines are roughly 10× more sensitive to UVC exposure than the purines [73]. Adjacent pyrimidines are particularly sensitive to exposure to UVC and UVB radiation. Furthermore, dsDNA viruses may undergo repair processes within their host cell, which will reduce inactivation of these viruses relative to those that do not have access to these repair processes.

For two of these virus groups, dsDNA and (+)ssRNA, they then developed models based on multiple linear regression of pyrimidine composition and distribution. For the (+)ssRNA viruses, a regression model based on the numbers of cytosines (C), uracils (U), uracil doublets (UU), and uracil triplets (UUU) resulted in the lowest root mean squared relative prediction error (RMSrPE) for prediction of $k_{254}$. In fact, the value of RMSrPE from their model was lower than the relative inter-experimental error of viruses for which more than one inactivation constant was available from the literature. This suggested that the error from their model predictions for the (+)ssRNA viruses was smaller than the error among independently reported inactivation constants for the same virus.

A conceptually similar multiple linear regression model was developed for dsDNA viruses, with the primary difference being that uracil (U) terms were replaced by equivalent thymine (T) terms. Among the models they evaluated, a multiple linear regression model based on the numbers of thymine doublets (TT), thymine quintuplets (TTTTT), and cytosines (C) performed best.

**Table 11.7** Summary of virus characteristics and predicted inactivation constants ($k_{254}$) from the regression model of Rockey et al. [13].

| Genome type | Virus | $k_{254}\left(\frac{cm^2}{mJ}\right)$ (mean ± 95% margin of error) |
|---|---|---|
| ssRNA | SARS-CoV | $1.9 \pm 0.82$ |
| | SARS-CoV-2 | $2.0 \pm 0.86$ |
| | MERS-CoV | $2.1 \pm 0.91$ |
| | Dengue virus | $0.38 \pm 0.16$ |
| | Zika virus | $0.39 \pm 0.17$ |
| | Human rhinovirus (B14) | $0.34 \pm 0.15$ |
| | Human norovirus (GII.4 Sydney) | $0.28 \pm 0.12$ |
| dsDNA | Herpes simplex virus (Strain 17) | $1.8 \pm 1.1$ |
| | Epstein–Barr virus | $1.9 \pm 1.2$ |
| | Human cytomegalovirus | $3.0 \pm 1.8$ |
| | Variola virus (major) | $2.5 \pm 1.5$ |

The optimized models for (+)ssRNA and dsDNA viruses were then used to predict inactivation constants for viruses in these groups. Reported values of $k_{254}$ based on these multiple linear regression models are summarized in Table 11.7.

Most of the UV dose–response data that are reported in the literature are based on collimated sources of radiation from low-pressure Hg lamps (i.e. $\lambda = 254$ nm) and aqueous suspensions of the target microbe or virus. The models described above generally provide accurate estimates of the $UV_{254}$ dose–response behavior (i.e. intrinsic kinetics) of inactivation of these microbes and viruses. Figure 11.17 provides an illustration of measured and model simulated $UV_{254}$ dose–response behavior for coronaviruses. As described above, the coronaviruses present similar genomes and can be expected to demonstrate similar sensitivity to $UV_{254}$ exposure. The model simulations indicate variability in predictions of coronavirus $UV_{254}$ dose–response behavior that is similar to the reported values from experiments.

## 11.4 Testing and Validation Methods

A wide range of devices for UV-based disinfection of air and surfaces has been developed. Engineers, facility managers, and consumers faced with the responsibility for selecting the performance characteristics of these systems will benefit from a standardized set of protocols to conduct testing and validation of process performance with these systems. Although some progress has been made toward development of these protocols, a need exists to continue this development to address all such devices and their various permutations. The development of standardized protocols is, in some respects, complicated by the wide range of existing and emerging sources of germicidally active UV radiation.

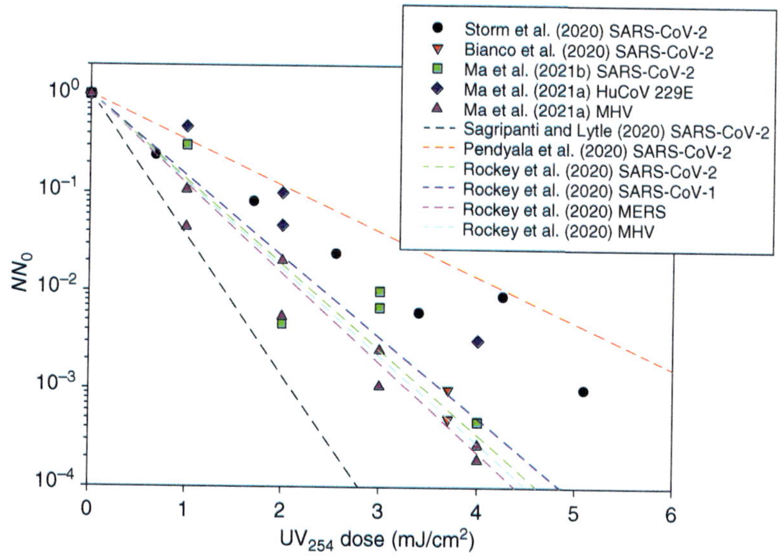


**Figure 11.17** Measured (symbols) and model simulated (dashed lines) $UV_{254}$ dose–response behavior of coronaviruses. Data include only values that were above the limit of detection for the infectivity assay used to quantify viral infectivity.

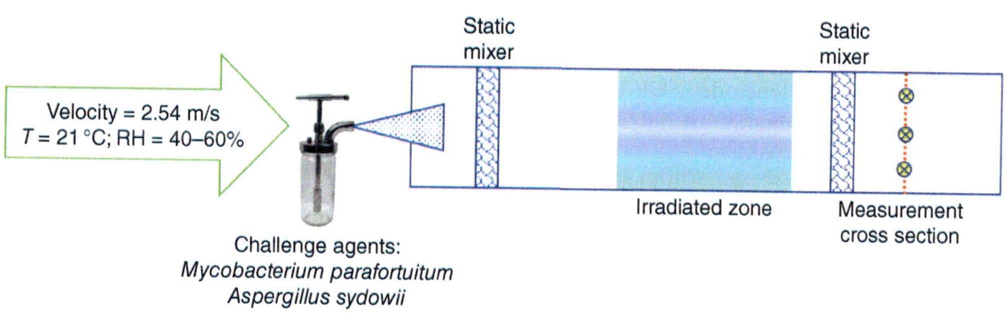


**Figure 11.18** Schematic illustration of validation setup used for testing and validation of an in-duct UV system for air disinfection. The challenge agent is introduced as an aerosol suspension to the device using a nebulizer. The challenge agents defined in ANSI/ASHRAE 185.1-2020 include *Mycobacterium parafortuitum* and *Aspergillus sydowii*; however, other challenge agents are allowed. Static mixers are included upstream and downstream of the test section (irradiated zone) to promote uniform introduction of the challenge agent and to promote uniformity of treated air quality. Multiple sample locations are located across the measurement cross section to allow for quantification of treated air quality, including variability.

ANSI/AHSRAE have developed standards for testing and validation of UV devices that are intended for installations in ducts and air handling systems [74]. A schematic illustration of the test setup is presented in Figure 11.18.

ANSI/AHSRAE 185.1-2020 calls for tests to be conducted with an approach velocity of 2.54 $\frac{\text{m}}{\text{s}}$ and RH of 40–60%. The protocol also calls for *M. parafortuitum* or *Aspergillus sydowii* to be used as challenge agents; however, other challenge agents are allowed. Given that many respiratory diseases involve viral pathogens, it is relevant to consider viral surrogates for these airborne pathogens. Likely candidates as surrogates for use in these tests include several phages: MS2 (non-enveloped (+)ssRNA virus), Φ6 (enveloped dsRNA virus), and T1 (non-enveloped dsDNA virus). MS2 is often chosen because it is among the most-studied of viruses. However, it should be noted that MS2 is more resistant to UVC exposure than most airborne pathogens; therefore, the use of MS2 as a surrogate for airborne viral pathogens can provide misleading results. Φ6 has been suggested as a surrogate for coronaviruses because of the presence of a viral envelope; however, it should be noted that there is no evidence to indicate that the viral envelope plays an important role in UV inactivation of viruses. Despite a lack of structural similarity to the coronaviruses, T1 appears to have $UV_{254}$ dose–response behavior that mimics that of the coronaviruses; as such, it may represent a good surrogate for use in validation of UV-based air disinfection systems. A related protocol (ANSI/ASHRAE 185.2-2020) calls for radiometric measurements to be conducted across measurement cross sections located 1 meter upstream and 1 m downstream of the UV device.

Collectively, the two ANSI/ASHRAE protocols provide a framework for testing and validation of UV-based air disinfection systems; however, many details of these tests are undefined or not addressed within these protocols. For many other UV-based systems that are intended for disinfection of air and surfaces, guidance is generally lacking as to how to validate system performance. On the other hand, some guidelines or "rules-of-thumb" have been developed for a few common UV-based air disinfection devices.

UR-UVGI systems should only be applied in rooms where ceiling height is sufficient to allow UV irradiation of a fraction of the room air volume that lies above the height of most human heads. A minimum ceiling height of 2.4 m (8 ft.) is recommended. In addition, the inclusion of ceiling fans (or other forms of mechanical mixing) is encouraged as a means of promoting well-mixed conditions within a room that includes a UR-UVGI system. Lamp fixtures used in UR-UVGI applications should include a combination of reflecting surfaces and louvers to allow delivery of an average fluence rate of 30–50 $\frac{\mu Wm^2}{cm^2}$ within the upper-room area (based on assumed lamp output at 254 nm), while maintaining a sufficiently low fluence rate in the occupied fraction of the room to limit UV exposures to less than the guideline values provided by ACGIH and other organizations. The benchmark limit for exposure is 30 $\frac{J}{m^2}$ at 270 nm, which is the wavelength that has been reported to be most efficient for promoting damage to human skin and eyes. If an UR-UVGI system is based on low-pressure mercury lamps, which are characterized by essentially monochromatic output at a characteristic wavelength of 254 nm, the spectral sensitivity at this wavelength is 0.50 (relative to 270 nm). As such, the reference $UV_{254}$ dose for an eight-hours exposure period is 60 $\frac{J}{m^2}$ (6 $\frac{mJ}{cm^2}$), which translates to a spatially averaged fluence rate of 0.21 $\frac{\mu W}{cm^2}$ within the occupied fraction of a room. Note that these calculations were based on the previous ACGIH Guidelines for UVC exposure. The updated guidelines (see Figures 11.7 and 11.8) allow for $UV_{254}$ exposures up to 100 $\frac{J}{m^2}$ (10 $\frac{mJ}{cm^2}$), which translates to a spatially averaged fluence rate of 0.35 $\frac{\mu W}{cm^2}$ within the occupied fraction of a room.

The emergence of Far UVC sources has the potential to change how UVC-based devices are implemented for disinfection of air and surfaces. Since the potential for damage to skin and eyes is much lower for Far UVC than for conventional UVC radiation from low-pressure Hg lamps and UV LEDs, the TLVs for Far UVC are substantially higher. This offers the potential to apply Far UVC devices that involve whole-room exposure, rather than limiting exposure to the upper room only.

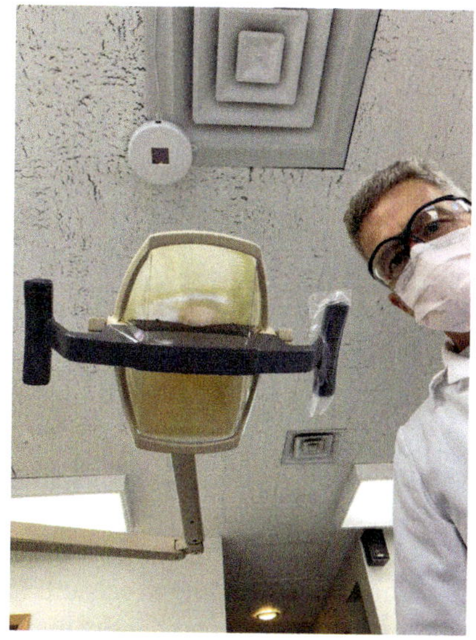

**Figure 11.19** Illustrations of Far UVC applications that involve whole-room irradiation. The Far UVC fixtures are placed on the ceiling to allow irradiation of the entire room. The devices are far enough away from the occupied fraction of the room volume to keep Far UVC exposure to below the TLVs for the radiation that is emitted by these devices. $KrCl^*$ lamps are used as the source of radiation. *Source:* Photos courtesy of Far UV Technologies, Inc.

This can be accomplished by placing Far UVC fixtures on the ceiling of a room. If the ceiling height is sufficient, then the fluence rate delivered to the occupied portion of the room can be low enough to prevent exceedance of the TLVs for Far UVC radiation. Figure 11.19 provides illustrations of Far UVC applications that involve full-room exposure.

## 11.5 Simulations of System Performance

As described in Chapters 3 and 9, as well as in other places in this book, process performance in UV disinfection systems is governed by the combined effects of the dose distribution that is delivered to a microbial or viral population and the intrinsic kinetics of pathogen inactivation for the wavelength(s) of radiation that are involved. Similarly, model simulations of the performance of UV systems that are used for disinfection of air and surfaces must account for the dose distribution delivered to the pathogen population and the intrinsic kinetics of pathogen inactivation.

The most common of UV-based application for disinfection of air is an UR-UVGI system. Figure 11.20 provides a schematic illustration of an application of UR-UVGI system and the process variables that comprise a simple model of IAQ dynamics, as influenced by the combined effects of natural attenuation of airborne pathogens, advective transport into and out of the room, mixing of the air space, and the contributions of UVC exposure to inactivation of airborne pathogens.

Air is forced into a room with fixed volume $\forall$ at a volumetric flow rate $Q$; air leaves the room at the same volumetric flow rate. $C_{in}$ represents the concentration of viable or infective pathogens in the air that is brought into the room $\left(\frac{\text{cfu or pfu}}{\text{m}^3}\right)$; often this concentration will be (essentially) zero,

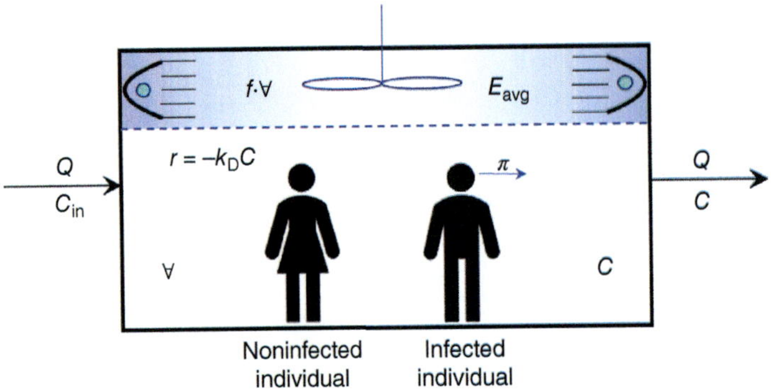


**Figure 11.20** Schematic illustration of model to simulate IAQ dynamics in a room that contains infected individuals and an upper-room UVGI system.

but the model can be used to account for the case where incoming air contains a nonzero concentration of the airborne pathogen of interest. The air space in the room is assumed to be well mixed as a result of convection induced by the HVAC system, localized heating by the occupants and other sources of heat within the system, and mechanical mixing (e.g. ceiling fan, box fan, or other air circulation device). Mechanical mixing is beneficial for destratification of indoor air and generally promotes the approach to a well-mixed condition, which will be assumed for purposes of this model. Operationally, a well-mixed condition is also beneficial as a means of preventing accumulation of pathogens in an air space. An important consequence of the well-mixed condition, from the perspective of modeling, is that the concentration of viable or infective pathogens in the air space, C, is equal to the concentration of viable or infective pathogens leaving the air space $\left(\frac{\text{cfu or pfu}}{\text{m}^3}\right)$.

$I$ represents the number of infected individuals within the room; each infected individual is assumed to emit pathogens at a rate of $\pi$, expressed in units of $\frac{\text{cfu or pfu}}{\text{hr}\cdot\text{person}}$. The per-person pathogen emission rate ($\pi$) can be further decomposed to the product of a pathogen concentration in exhaled breath $\left(\frac{\text{cfu or pfu}}{\text{m}^3}\right)$ and a respiration rate $\left(\frac{\text{m}^3}{\text{hr}}\right)$. Noninfected individuals may also be present in the room; a primary goal for the design of the UVGI system is to minimize the potential for disease transmission from infected to noninfected individuals.

A conventional upper-room UVGI system (i.e. based on low-pressure Hg lamps) will generally comprise multiple lamp housings, each of which may include a reflecting device (e.g. a parabolic reflector) and louvers to direct UVC radiation from the lamps in the housing to the air space above the height of room occupants. A fraction of the total air volume in the room, $f$, will be irradiated; the spatially averaged fluence rate within this irradiated zone is defined as $E_{\text{avg}}\left(\frac{\text{W}}{\text{m}^2}\right)$. For this simulation, it will be assumed that the intrinsic kinetics of UVC-based inactivation of the pathogen of interest follows single-event (i.e. first-order) kinetics with a rate constant of $k\left(\frac{\text{m}^2}{\text{J}}\right)$. For simplicity, the fluence rate within the occupied fraction of the room is assumed to be zero. Although the actual fluence rate in the occupied fraction of a room is unlikely to be zero, the application of an assumption of zero fluence rate within the occupied fraction of the room volume allows a conservative estimate of system performance, in terms of pathogen inactivation.

In addition to the transport behavior and UV-based inactivation reactions, it may also be appropriate to account for natural decay (attenuation) of airborne pathogens. This process, which contributes to loss of airborne pathogens, is attributable to the stress associated with being air-exposed and is likely to follow first-order kinetics, at least for time scales on the order of a few hours, with a first-order rate constant of $k_D$. Equation 11.14 describes the intrinsic kinetics of this process:

$$r = -k_D C \tag{11.14}$$

A mass-balance-based approach is used to simulate the dynamic behavior of this system. The model is developed first for the case of no UV being used in the room. Under these conditions, the overall dynamics of the system are described by:

$$\forall \frac{dC}{dt} = QC_{in} - QC - k_D C\forall + I\pi \tag{11.15}$$

For the normal case of $C_{in} = 0$, Equation 11.15 can be rewritten as:

$$\forall \frac{dC}{dt} + QC + k_D C\forall = I\pi \tag{11.16}$$

Dividing both sides of this equation by $\forall$ and applying the definition for air changes per hour $\left(A = \frac{Q}{\forall}\right)$, we find:

$$\frac{dC}{dt} + (A + k_D) \cdot C = \frac{I\pi}{\forall} \tag{11.17}$$

Equation 11.17 is a linear, first-order ODE that can be solved using the integrating factor method, using the term $\alpha_1 = (A + k_D)$ as the integrating factor. The resulting governing equation is:

$$C = C_0 \cdot e^{-\alpha_1 t} + \frac{I\pi}{\alpha_1 \forall}(1 - e^{-\alpha_1 t}) \tag{11.18}$$

where $C_0$ represents the initial concentration of viable or infective pathogens in the air space. Equation 11.18 describes the dynamic behavior of a system in which the only mechanisms available for reduction in the airborne pathogen concentration are physical air exchange induced by introduction of "fresh air," with exhaust at the same flow rate, and natural decay of the airborne pathogen. Note that in many applications, this is the default option for maintaining IAQ.

When an upper-room UVGI system is turned on, it then becomes possible to actively inactivate airborne pathogens. The basic governing equation can be modified to account for the effects of the upper-room UVGI system:

$$\forall \frac{dC}{dt} = QC_{in} - QC - k_D C\forall + I\pi - fE_{avg}kC\forall \tag{11.19}$$

The last term in Equation 11.19 describes the effects of the intrinsic kinetics of inactivation of airborne pathogens by exposure to UVC radiation. The product $E_{avg}kC$ describes the kinetics of this reaction as an intrinsic reaction rate $\left(\frac{\text{cfu or pfu}}{\text{m}^3 \cdot \text{hr}}\right)$. This term is multiplied by room air volume, $\forall$, to provide a term that accounts for the rate or change in the pathogen number in the room air $\left(\frac{\text{cfu or pfu}}{\text{hr}}\right)$ resulting from exposure to UV radiation, which is dimensionally consistent with the other terms in the equation. This product is multiplied by $f$ to account for the fact that in a well-mixed air space, any airborne particle will spend, on average, $f$ of its total residence time in the room in the irradiated portion of the air space.

The assumption of a well-mixed condition is also used to justify the application of the spatially averaged fluence rate, $E_{avg}$, and in effect the average period of exposure. The well-mixed condition dictates that an airborne particle has an equal chance of passing through any location of the room, including any part of the irradiated zone. This represents a fundamental departure from the more rigorous (and correct) approach of defining a dose distribution applied to the exposure pathogen population. This departure is justified by the relatively slow rates of pathogen exposure and inactivation in these systems, which are controlled largely by mixing in the room and the fact that only a fraction ($f$) of the air space is irradiated. Also, the extent of pathogen inactivation in these systems is typically modest, as compared with UV disinfection systems that are applied to other media, including water and solid surfaces. As such, the error inherent in the use of a volume-averaged model will be reduced. Lastly, this simplified approach is justified by the objective of modeling, which is to simulate the dynamic behavior of an UR-UVGI system based on spatially averaged microbial or viral air quality in the room. This is unlike applications such as water treatment, where a single pass through a UV system is included as part of treatment, and the design objective is to improve microbial or viral quality in the entire fluid leaving the reactor system. In UR-UVGI systems, the objective is to improve air quality in a room where air is recirculated through the irradiated zone to reduce the concentration of viable or infective pathogens in the room.

Following the same logic as in the case of the no UV model, the concentration of pathogens entering the room will be assumed to equal zero. Dividing both sides of this equation by $\forall$ and applying the definition for air changes per hour $\left(A = \frac{Q}{\forall}\right)$, we find:

$$\frac{\mathrm{d}C}{\mathrm{d}t} + \left(A + k_{\mathrm{D}} + fE_{\mathrm{avg}}k\right)C = \frac{I\pi}{\forall} \tag{11.20}$$

The product $fE_{avg}k$ is sometimes referred to as "equivalent air changes per hour" ($A_{eff} = fE_{avg}k$). It is worth noting that the effects of dilution, ambient pathogen decay, and UVC-based inactivation are all first-order with respect to $C$. As such, these effects can be lumped together using a parameter to represent the overall number of air changes per hour:

$$\alpha_2 = A + k_{\mathrm{D}} + A_{\mathrm{eff}} \tag{11.21}$$

Substituting this definition into Equation 11.20, we get:

$$\frac{\mathrm{d}C}{\mathrm{d}t} + \alpha_2 C = \frac{I\pi}{\forall} \tag{11.22}$$

which has exactly the same form as Equation 11.17. As such, the solution is analogous; however, it should be recognized that the physical interpretation of $\alpha_2$ in the closed-form solution, Equation 11.23, includes the effects of UV irradiation, which are not included in the solution presented in Equation 11.18:

$$C = C_0 \cdot \mathrm{e}^{-\alpha_2 t} + \frac{I\pi}{\alpha_2 \forall}\left(1 - \mathrm{e}^{-\alpha_2 t}\right) \tag{11.23}$$

Equation 11.23 can be viewed as a general solution to the model that can be used to simulate IAQ dynamics in a room that is equipped with an UR-UVGI system. When the UV system is not activated, IAQ dynamics can be simulated by assuming a value of $E_{avg} = 0$.

A characteristic of solutions to ODEs that are first-order with respect to time is that the solution generally includes one or more terms to describe transient behavior and a term to describe the steady-state condition. Both behaviors are relevant to the design and operation of UVGI systems.

The steady-state condition will apply after the system has had a long time to operate, under constant conditions. The steady-state term in Equation 11.23 is:

$$C_{SS} = \frac{I\pi}{\alpha_2 \forall} = \frac{I\pi}{Q + k_D \forall + \forall f E_{avg} k} \tag{11.24}$$

The approach to a steady state described by this model is based on the assumption that all operating conditions of the system remain constant for long periods of time. In practical applications, it is unlikely that a true steady-state condition will be met or maintained. However, the steady-state condition provides useful information about the direction air quality will take for a given set of operating conditions.

Equation 11.24 provides an indication of the factors that will affect steady-state IAQ, in terms of concentration of viable or infective pathogens. The numerator of this expression indicates that indoor air quality depends on the number of infected individuals in the room and the rate at which they emit infective particles. Operationally, this implies opportunities for improvement of IAQ by limiting entry of infected individuals to a room and controlling the rate at which they emit infective particles, such as by wearing a face mask or similar device. From a practical and social perspective, limiting the number of infected individuals to a room can be problematic, particularly since some may be asymptomatic. In healthcare facilities, the primary goal is often to provide treatment and comfort to patients who are infected with the pathogen of interest, so their presence in a room is a natural consequence of the operation of the healthcare facility. Historically, the use of face masks and other devices has not been commonly accepted outside of healthcare facilities, although the COVID-19 pandemic appears to have changed that social norm, so opportunities will exist in the future to reduce $\pi$ using simple control measures, like wearing a face mask.

The denominator in Equation 11.24 provides an indication of factors that can be controlled from an engineering perspective. One approach involves an increase in the rate at which outside air is brought into the room, Q. Mechanistically, this allows for reductions in the steady-state pathogen concentration by dilution. However, from a practical perspective, this can be an expensive approach, in that outside air may require heating, cooling, or modification of RH.

The term $k_D$ is a characteristic of a given pathogen and the environmental conditions that exist in the room. For some airborne pathogens, conditions of low RH or high temperature ($T$) can represent sources of stress, which in turn can increase the rate of natural decay. But changes in RH and $T$ from common conditions that exist in indoor spaces can be expensive to maintain and can cause discomfort for room occupants. From a practical perspective, there is little opportunity to improve IAQ by control of the term $k_D$ or the factors that influence it.

The product $\forall f E_{avg} k$ in the denominator of Equation 11.24 indicates the effects of UV irradiation of the air space. Here again, some tradeoffs emerge. Clearly, as the volume of air in the room that is exposed to UVC radiation increases ($f\forall$), the rate of pathogen inactivation increases. However, in occupied rooms, it will be necessary to limit human exposure to UVC radiation (i.e. below TLVs).

The value of the average fluence rate within the irradiated zone, $E_{avg}$, is linearly related to lamp power. Therefore, the contribution of UVC irradiation to airborne pathogen inactivation will be linearly related to UVC power imposed on the upper-room area. However, the capital and operating costs of the UVC sources will also increase (roughly) linearly with the number of lamps and their total output power.

As described previously, Far UVC sources offer the opportunity to increase the fraction of the air space in a room that is irradiated. With proper placement of lamps in rooms, it may be possible to reach a condition where the entire volume of the room is irradiated (i.e. $f = 1$).

Lastly, it is important to recognize that the effects of UVC irradiation in an upper-room UVGI system will not be the same for all pathogens; this effect will be governed by the sensitivity of the pathogen to UVC exposure, which is captured by the inactivation rate constant, $k$ (see Tables 11.1, 11.4, 11.5, 11.6, and 11.7 for examples of reported values of this parameter).

To frame the behavior of UR-UVGI systems in the context of the general governing Equation 11.23, it is useful to consider the range of operating conditions that contribute to the factor $\alpha_2$. We start by recognizing the wide range of outside air exchange rates that are commonly applied in buildings. Recommendations for outside air flow rates to indoor air spaces are provided in ANSI/ASHRAE 62.1-2019 [75]. Under this guidance, the outdoor air flow rate recommended for occupiable spaces ($Q_{bz}$) is defined by the following equation:

$$Q_{bz} = \left(R_p \cdot P_z\right) + \left(R_a \cdot A_z\right) \tag{11.25}$$

where

$R_p$ = outdoor air flow rate required per person $[=] \dfrac{\text{L}}{\text{s} \cdot \text{person}}$

$P_z$ = number of people in the ventilation zone during use $[=]$ people

$R_a$ = outdoor air flow rate per unit floor area $[=] \dfrac{\text{L}}{\text{s} \cdot \text{m}^2}$

$A_z$ = zone floor area $[=]$ $\text{m}^2$.

Table 11.8 provides a summary of outdoor air flow rate guidance for various settings. This information was used together with assumed room dimensions for these applications to develop estimates of the number of air changes per hour that should be applied for facilities that followed the ANSI/ASHRAE 62.1-2019 recommendations. Collectively, the ANSI/ASHRAE guidance suggests operation of an HVAC system in a manner that delivers outdoor air at a flow rate that achieves roughly 0.5–3.5 ach in common residential and commercial settings.

As a class of buildings, medical facilities often call for higher air exchange rates than most other facility types. In recognition of the critical air quality needs of medical facilities, relative to more common applications, the Centers for Disease Control and Prevention have listed a range of air exchange rates for medical and healthcare facilities [76]. A wide range of medical facility types are described, but the recommended range of air exchange is between 2 and 15 ach.

Airborne pathogens, including viruses, are often associated with aerosols, which are small liquid droplets that are emitted by humans as a consequence of breathing, coughing, or sneezing. The ability of airborne pathogens to retain viability or infectivity while inside these droplets depends on the composition of the droplets and local environmental conditions. Specific attributes of aerosol composition that have been demonstrated to affect pathogen viability or infectivity include pH, protein content, and the presence of surfactants [77]. Temperature and RH have also been demonstrated to influence the loss of viability or infectivity of airborne pathogens.

Loss of viral infectivity in aerosols is often observed to follow a biphasic pattern, where initial decay takes place as the aerosol moves toward equilibrium with its surroundings, a process that

**Table 11.8** Examples of air flow rates for recommended for various applications and corresponding number of air changes per hour. Air flow rate definitions and terms are from ANSI/ASHRAE 62.1.

| Facility type | $R_p \left(\frac{L}{s \cdot person}\right)$ | $R_p \left(\frac{L}{s \cdot m^2}\right)$ | Default occupancy $\left(\frac{\#}{100 \cdot m^2}\right)$ | Assumed ceiling height (m) | Assumed floor area ($m^2$) | Air space volume ($m^3$) | Recommended outdoor air flow rate $\left(\frac{m^3}{hr}\right)$ | Air changes per hour ($hr^{-1}$) |
|---|---|---|---|---|---|---|---|---|
| Animal exam room (veterinary office) | 5 | 0.6 | 20 | 4 | 10 | 40 | 58 | 1.44 |
| Correctional facility (cell) | 2.5 | 0.6 | 25 | 3 | 10 | 30 | 44 | 1.47 |
| Educational facility – lecture hall | 3.8 | 0.3 | 150 | 6 | 100 | 600 | 2160 | 3.60 |
| Educational facility – classroom (age 9+) | 5 | 0.6 | 35 | 6 | 100 | 600 | 846 | 1.41 |
| Restaurant dining room | 3.8 | 0.9 | 70 | 6 | 100 | 600 | 1282 | 2.14 |
| Hotel lobby | 3.8 | 0.3 | 30 | 6 | 100 | 600 | 518 | 0.86 |
| Transportation waiting area | 3.8 | 0.3 | 100 | 10 | 200 | 2,000 | 2952 | 1.48 |
| Office space | 2.5 | 0.3 | 5 | 3 | 40 | 120 | 61 | 0.51 |
| Outpatient healthcare – general examination | 3.8 | 0.6 | 20 | 3 | 10 | 30 | 49 | 1.63 |
| Outpatient healthcare – urgent care examination | 3.8 | 0.6 | 20 | 3 | 10 | 30 | 49 | 1.63 |
| Auditorium | 2.5 | 0.3 | 150 | 10 | 200 | 2,000 | 2916 | 1.46 |
| Place of worship | 2.5 | 0.3 | 120 | 10 | 200 | 2,000 | 2376 | 1.19 |
| Retail – mall common area | 3.8 | 0.3 | 40 | 6 | 100 | 600 | 655 | 1.09 |
| Retail – supermarket | 3.8 | 0.3 | 8 | 6 | 200 | 1,200 | 435 | 0.36 |
| Swimming pool/deck | n/a | 2.4 | n/a | 10 | 1000 | 10,000 | 8640 | 0.86 |

**Table 11.9** Reported values of virus recovery at several values of relative humidity (RH).

| Virus | Percent recovery after 15 minutes | | | Inactivation constant, $k_d$ ($hr^{-1}$) | | |
|---|---|---|---|---|---|---|
| | RH = 80% | RH = 50% | RH = 30% | RH = 80% | RH = 50% | RH = 30% |
| Human coronavirus 229 E | 55 | 90.9 | 87 | 2.39 | 0.38 | 0.56 |
| Rotavirus (Wa) | 34.3 | 86.8 | 45.8 | 4.28 | 0.57 | 3.12 |
| Bovine rotavirus (C-486) | 27.3 | 70 | 36.5 | 5.19 | 1.43 | 4.03 |
| Simian rotavirus SA-11 | 24.7 | 78 | 42.1 | 5.59 | 0.99 | 3.46 |
| Bovine rotavirus (UK) | 29.7 | 73.4 | 38.5 | 4.86 | 1.24 | 3.82 |
| Mouse rotavirus | 20.4 | 90.6 | 49 | 6.36 | 0.39 | 2.85 |

*Source:* Data from Ijaz et al. [78]. Inactivation constants ($k_d$) were calculated based on an assumption of first-order decay.

involves desiccation and thermal stabilization. After this initial process, which often takes place on a time scale on the order of an hour or less, viruses are inactivated more slowly. For purposes of this presentation, the focus will be on the initial decay process.

Table 11.9 provides a summary of values of initial loss of several viruses under a range of RH conditions. The initial losses for these experiments were reported after 15 minutes in the aerosolized form. Values of the decay constant, $k_d$, were calculated for each of these viruses and are reported in the table. These data indicate a range of values of $k_d$, which will vary among viruses (in part governed by their structure) and that are influenced by RH. For each of these viruses, the rate of initial loss was influenced by RH, but not in a monotonic fashion. The values of the inactivation constant, $k_d$, ranged from 0.38 to 6.36 $hr^{-1}$. As such, it appears that the contributions of initial loss to changes in airborne virus concentration can be of a similar magnitude to those accomplished by ventilation and UV irradiation; this justifies the inclusion of the natural decay term in the IAQ model.

The effects of airborne pathogen exposure to UVC radiation are governed by the UVC dose distribution delivered to the pathogen population and the intrinsic kinetics of inactivation. For this model, the assumption of a completely mixed air space allows use of a spatially averaged fluence rate to account for UVC exposure. As described previously, current guidance for UR-UVGI systems used to control TB calls for a spatially averaged $UV_{254}$ fluence rate of 30–50 $\frac{\mu W}{cm^2}$ [34]. The values of inactivation constants ($k$) for $UV_{254}$ exposure depend on the structure and composition of the target pathogen, as well as environmental conditions (e.g. RH), and the medium in which the pathogen is suspended (liquid suspension, aerosol, or dried on a surface).

A wide range of inactivation constants have been reported in the literature. Among values listed in Table 11.1, the most resistant pathogens include Adenovirus $\left(k_{254} = 0.023\frac{cm^2}{mJ}\right)$ and the yeast *Cryptococcus neoformans* $\left(k_{254} = 0.047\frac{cm^2}{mJ}\right)$. At the other extreme, the human coronaviruses appear to be among the most sensitive to UVC exposure, with reported values of $k_{254} = 3.3\frac{cm^2}{mJ}$ for SARS-CoV-2 and $k_{222} = 5.9\frac{cm^2}{mJ}$ for Coronavirus OC43, a virus that causes the "common cold."

Using values of the model parameters described above, it is possible to frame the range of relevant values of the product $fE_{avg}k$. Specifically, if we assume that 20% of the airspace in a room is irradiated with UR-UVGI devices, the minimum and maximum values of the parameters $E_{avg}$ and $k$ will yield values that span the range $0.5\ \text{hr}^{-1} < fE_{avg}k < 210\ \text{hr}^{-1}$.

Collectively, this yields a range of practical values of the term $\alpha_2 = (A + k_D + fE_{avg}k)$ that range from $1\ \text{hr}^{-1} \lesssim \alpha_2 \lesssim 230\ \text{hr}^{-1}$. The low end of this range, which would represent a condition in which pathogen concentration is diminished quite slowly, would correspond to a situation in which a pathogen is released to a room that is slow to decay with minimal fresh air circulation and no UV being applied. The conditions described by the upper end of this range would correspond to a pathogen that decays relatively quickly, a room that experiences aggressive introduction of outside air, a pathogen that is sensitive to UV exposure, and that is exposed to UVC radiation in the upper-room space.

Figure 11.21 illustrates the predictions of Equation 11.23 for the range of values of $\alpha_2$ described above. Specifically, this graph illustrates the dynamic behavior of airborne pathogen concentration for an airspace that is initially clean, where a single, highly infected individual is brought into the room at $t = 0$. This simulation is relevant to a clinical setting, where one or more individuals in a room may be identified as "high emitters" of an airborne pathogen.

Figure 11.21 illustrates several aspects of the expected behavior of the system. First, the time required to reach the steady-state condition is a strong function of the conditions of operation, as defined by $\alpha_2$. For small values of $\alpha_2$, the approach to steady state is comparatively slow, and may require a period of an hour or more. On the other hand, for systems that operate with a large value of $\alpha_2$, the approach to steady state is quite rapid, with a characteristic time of 30 seconds or less.

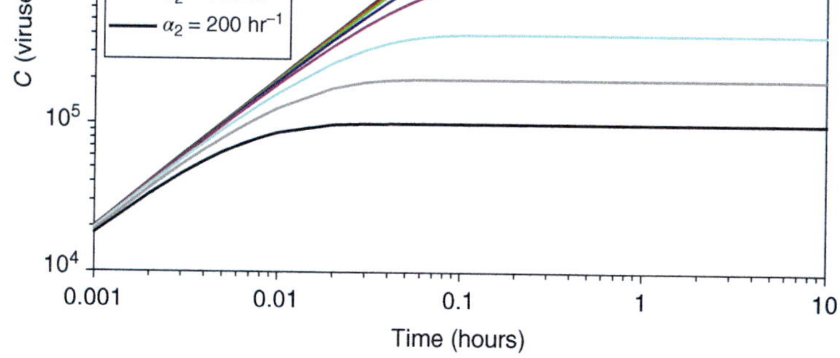


**Figure 11.21** Illustration of dynamic IAQ conditions based on the basic IAQ model (Equation 11.23). The model was applied to simulate IAQ dynamics in a single-patient room in a medical facility. Assumed values of model parameters included: $I = 1$ person, room air volume = 50 m$^3$, initially clean air (i.e. $C_0 = 0$), and $\pi = 10^9\ \frac{\text{pathogens}}{\text{person} \cdot \text{hr}}$. Note that both axes are presented on a $\log_{10}$ scale.

It is important to understand that Equation 11.23 was developed under the assumption of a well-mixed air space. It is unlikely that this condition will ever be met in a real air space, but it is more likely to be approached for small values of $\alpha_2$. For systems that operate with large values of $\alpha_2$, such as systems where a highly UV-sensitive pathogen is exposed to UVC radiation or aggressive introduction of outside air (i.e. high number of air changes per hour), accurate predictions of IAQ dynamics may require the use of a more detailed description of transport behavior and the spatial distribution of radiation energy. Specifically, these simulations would benefit from application of computational fluid dynamics and an accurate fluence rate field model, such as a ray tracing model. It should be noted that the implications of the well-mixed assumption are relevant to all three mechanisms of infective virus concentration reduction in this model (i.e. air replacement by the HVAC system, natural decay, and UV exposure) because all three processes are assumed to demonstrate first-order dependence on infective virus concentration.

Equation 11.23 can also be used to illustrate IAQ dynamics in a situation where a room starts by being heavily contaminated, with measures being introduced to improve IAQ. Figure 11.22 illustrates such a simulation, under an assumption of an initial airborne pathogen concentration of $2 \times 10^7 \frac{\text{viruses}}{\text{m}^3}$, which corresponds to the steady-state condition from the simulation illustrated in Figure 11.21 for the case of $\alpha_2 = 1\ \text{hr}^{-1}$. In physical terms, this is representative of a situation in which a high-emitting infected individual has been in an air space with poor air circulation for a period of an hour or more, such that the steady-state condition has been approached. The time corresponding to $t = 0$ in Figure 11.22 indicates the time when active measures are initiated to reduce the concentration of the airborne pathogen. These measures could include varying degrees of fresh air introduction and application of germicidal UV radiation.

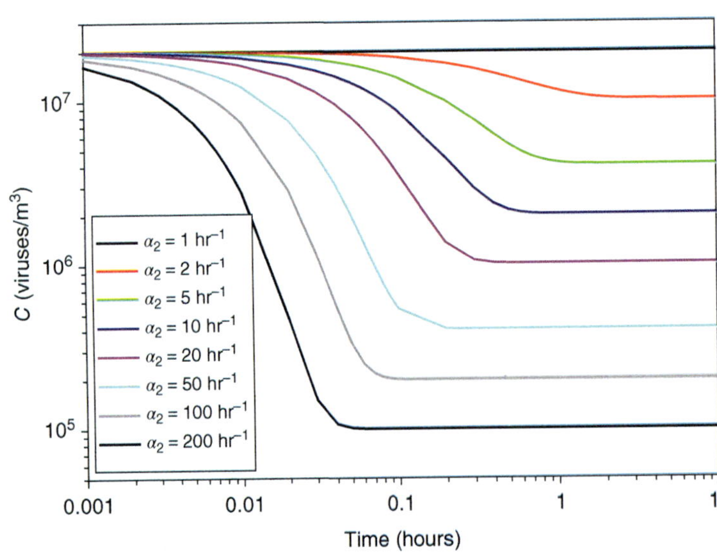


**Figure 11.22** Approach to steady state from an initial condition of heavy contamination. Assumed values of model parameters included: $I = 1$ person, room air volume = 50 m$^3$, initially contaminated air (i.e. $C_0 = 2 \times 10^7 \frac{\text{viruses}}{\text{m}^3}$, the worst-case steady-state condition from Figure 11.20), and $\pi = 10^9 \frac{\text{pathogens}}{\text{person} \cdot \text{hr}}$. Note that both axes are presented on a $\log_{10}$ scale.

Note that the steady-state conditions illustrated in Figures 11.21 and 11.22 are identical, as are the characteristic times required to reach these steady-state conditions. Again, in practical terms, the assumption of a well-mixed air space is a limitation of this model and is almost certainly not realistic for the cases of large values of $\alpha_2$. However, the model simulation results that are indicated in Figures 11.21 and 11.22 provide a first-cut understanding of the effects of fresh air introduction and the use of germicidal UV radiation for control of IAQ.

Figure 11.23 illustrates a third application of the IAQ model for an upper-room configuration. The simulation was based on a room with an air volume of 500 m$^3$, room occupancy of 30, with 10% of the occupants assumed to be infected and emitting an airborne pathogen at a rate of $\pi = 10^5 \frac{\text{viruses}}{\text{person} \cdot \text{hr}}$. It was also assumed that the room initially contained clean air (i.e. $C_0 = 0$) and the air entering the room was clean (i.e. $C_{in} = 0$).

The results of this simulation indicate that steady-state conditions will be reached within one hour or less, depending on the conditions of operation. The steady-state concentration is strongly influenced by the operating conditions as well. Operation at conditions corresponding with $\alpha_a = 200\ \text{hr}^{-1}$ will yield a pathogen concentration of roughly 2/m$^3$.

For perspective, it is relevant to consider the conditions that will prevent or control disease transmission. As an example, an infective dose of SARS-CoV-2 has been reported to be on the order of 300 infective viruses [79]. If we assume that people in the room are respiring at a rate of $14 \frac{\text{breaths}}{\text{min}}$

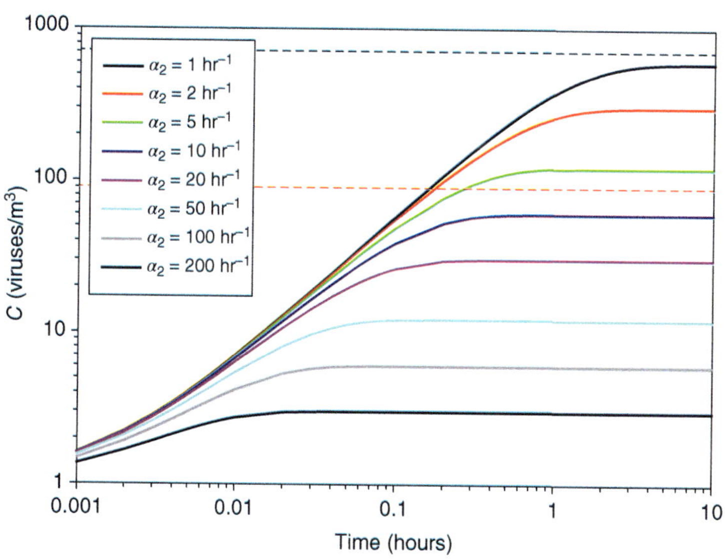


**Figure 11.23** Illustration of dynamic IAQ conditions based on the basic IAQ model (Equation 11.23). The model was applied to simulate IAQ dynamics in a room with 30 occupants, 10% of whom are assumed to be infected and emitting pathogens at a rate of $10^5 \frac{\text{viruses}}{\text{person} \cdot \text{hr}}$. Values of other model parameters included: room air volume = 500 m$^3$, initially clean air (i.e. $C_0 = 0$), and clean air entering the room. Horizontal dashed lines indicate pathogen concentrations that will result in inhalation of 300 infective pathogens in one hour (black) and eight hours of exposure (red) for occupants breathing at $14 \frac{\text{breaths}}{\text{min}}$ with a tidal volume of 0.5 L. Note that both axes are presented on a $\log_{10}$ scale.

with a tidal volume of 0.5 $\frac{\text{L}}{\text{breath}}$, the concentration of airborne viruses that would be required to deliver an infective dose in one hour would be 715 $\frac{\text{viruses}}{\text{m}^3}$. Similarly, the virus concentration required to deliver an infective dose of SARS-CoV-2 in eight hours under the same conditions would be 89 $\frac{\text{viruses}}{\text{m}^3}$. These threshold virus concentrations are illustrated as horizontal, dashed lines in Figure 11.23. It is evident that the conditions required to maintain "safe" air quality using UV-based disinfection of air and forced air exchange are quite modest, requiring roughly eight total air changes per hour. It should be noted, however, that the infective dose for SARS-CoV-2 has not been firmly established and is probably variable among individuals. Moreover, there would almost certainly be a desire to maintain an air phase concentration that is well below the threshold limit described by these calculations, as a measure of conservatism.

The model simulation results presented in Figure 11.23 are representative of a moderately crowded room, such as a classroom or a conference room, where occupants are not displaying severe symptoms of a respiratory disease. These results suggest that UV-based disinfection can be very effective for controlling transmission of airborne pathogens.

Another common mode of application of UV for disinfection of air is a recirculating "air cleaner" device. These devices generally take the form of an enclosed cabinet that contains one or more sources of UVC radiation. The cabinet protects room occupants from exposure to UVC radiation. Room air is circulated through the device using an internal fan. Some of these devices may include a filter located between the fan exhaust and the irradiated zone of the device.

When constant operating conditions, as defined by air flow rate through the device and lamp output power, are delivered by such a device, the dose distribution delivered by the device will remain essentially constant. Under these conditions, the concentration of airborne infective pathogens that retain viability downstream of the irradiated zone will be a constant fraction, $F$, of the concentration of airborne infective pathogens that enter the device. Figure 11.24 provides a schematic illustration of the implementation of such a device and a definition sketch for simulation of IAQ dynamics in a room with air that is treated in this manner.

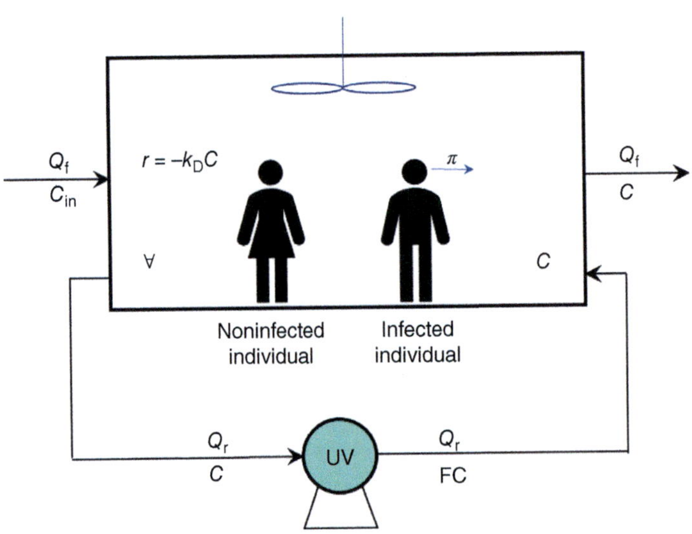


**Figure 11.24** Schematic illustration of model to simulate IAQ dynamics in a room that contains infected individuals and a recirculating UV-based air cleaner.

Using mass-balance principles, it is possible to develop a model to simulate IAQ for a system as illustrated in Figure 11.24. Many of the parameters that will be used to describe IAQ dynamics in this system will be the same as those presented in Figure 11.20 and Equation 11.23. However, the mode of application for UVC radiation will be fundamentally different than that illustrated in Figure 11.20. To account for the effects of the UV-based air cleaner, additional model parameters will be applied:

$Q_f$ = volumetric flow rate of fresh air brought into the room by the HVAC system (equal to the volumetric flow rate of air from the room)
$Q_r$ = volumetric flow rate of air through the air cleaner
$F$ = fraction of airborne pathogens that retain viability or infectivity after passing through the UV air cleaner (assumed to be constant).

Using these terms, it is also possible to define additional factors that describe characteristics of the system illustrated in Figure 11.24. Specifically:

$$A = \frac{Q_f}{\forall} = \text{number of air changes per hour} \quad [=] \quad hr^{-1}$$

$$A_r = \frac{Q_r}{\forall} = \text{number of air changes per hour due to recirculation} \quad [=] \quad hr^{-1}$$

Using these terms and basic mass balance principles, a model to describe IAQ dynamics in a well-mixed room with a UV-based air cleaner can be described as:

$$\forall\frac{dC}{dt} = Q_f C_{in} - Q_f C - Q_r C + Q_r FC - k_D C\forall + I\pi \tag{11.26}$$

As with the previous model, we can generally assume $C_{in} = 0$. Applying this assumption and dividing by $\forall$, we get:

$$\frac{dC}{dt} = -\frac{Q_f}{\forall}C - \frac{Q_r}{\forall}C + \frac{Q_r}{\forall}FC - k_D C + \frac{I\pi}{\forall} \tag{11.27}$$

Substituting the definitions of $A$ and $A_r$ into the governing equation, we find:

$$\frac{dC}{dt} = -AC - (A_r C - A_r FC + k_D C) + \frac{I\pi}{\forall} \tag{11.28}$$

Collecting all terms related to concentration on the left-hand side, we get:

$$\frac{dC}{dt} + C(A + A_r[1-F] + k_D) = \frac{I\pi}{\forall} \tag{11.29}$$

Applying the "integrating factor" method, with an integrating factor of:

$$\alpha_3 = A + A_r[1-F] + k_D \tag{11.30}$$

The term $\alpha_3$ accounts for the combined (additive) effects of physical air exchange, ambient pathogen decay, and the ability of the indoor air cleaning device to return "clean air" to the room, which is closely related to the CADR, a figure-of-merit that is often used to characterize the performance of indoor air cleaning devices.

Using the terminology presented in this model, CADR is calculated as the product of the air flow rate through the device and the fraction of airborne pathogens that are inactivated in a single pass through the device:

$$\text{CADR} = Q_r[1-F] \tag{11.31}$$

Or in other words,

$$\alpha_3 = A + \frac{\text{CADR}}{\forall} + k_\text{D} \tag{11.32}$$

Substituting the definition for $\alpha_3$ into the governing equation, we find:

$$\frac{\text{d}C}{\text{d}t} + \alpha_3 C = \frac{I\pi}{\forall} \tag{11.33}$$

Applying similar logic as presented before in terms of application of the integrating factor method, the closed-form solution to this model to describe IAQ dynamics in a well-mixed room in which air is being treated by a UVC-based air cleaner is described as:

$$C = \frac{I\pi}{\alpha_3 \forall}(1 - \text{e}^{-\alpha_3 t}) + C_0 \text{e}^{-\alpha_3 t} \tag{11.34}$$

Based on the characteristics of common air cleaner devices, realistic values of the parameter $\alpha_3$ will be observed in the range $0.3\ \text{hr}^{-1} \lesssim \alpha_3 \lesssim 25\ \text{hr}^{-1}$. Using this range of values and assumptions of zero viruses in the air entering a room, an initially clean air space ($C_0 = 0$), and a virus emission rate of $10^6 \frac{\text{viruses}}{\text{hr}}$, the model was applied to simulate the time-dependence of infective virus concentration in an indoor room. The results of those calculations are presented in Figure 11.25. These simulation results indicate that UVC-based air cleaners can be effective for controlling airborne pathogens, depending on operating conditions.

A key limitation of the models presented above is the assumption of a well-mixed air space. As with other applications of UV-based systems for treatment of fluids, this limitation can be overcome by application of numerical models that provide more detailed, accurate descriptions of fluid mechanical behavior (i.e. virus transport) and the fluence rate field. Models based on computational fluid dynamics linked to a numerical simulation of the spatial distribution of radiant energy

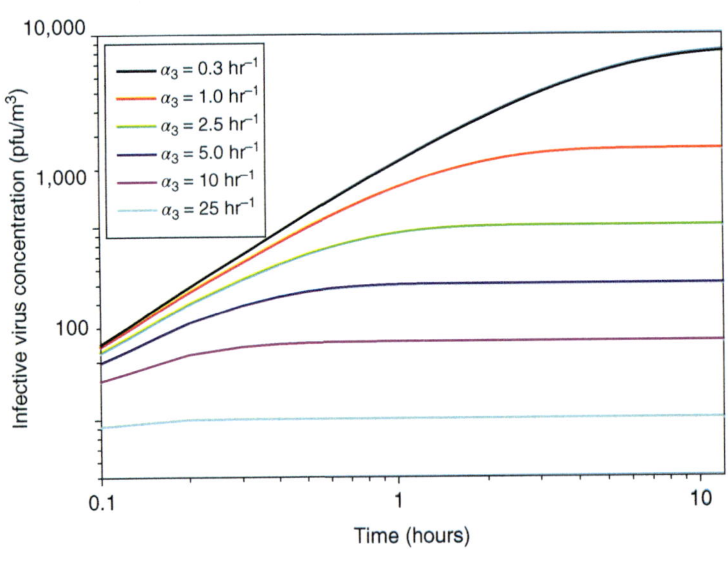


**Figure 11.25** Simulation of IAQ dynamics in a well-mixed room with air treatment using a recirculating air cleaner. Model simulations were based on assumptions of pathogen introduction rate of $10^6$ infective pathogens per hour and room volume of $400\ \text{m}^3$.

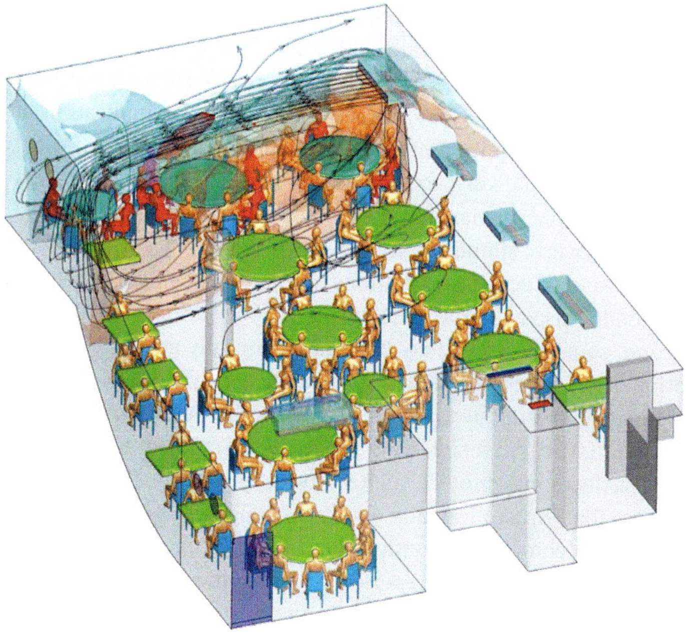

**Figure 11.26** Simulated transport of fine droplets exhaled from an index case (purple) corresponding to a COVID-19 outbreak that is believed to have taken place in a restaurant. Air circulation in the room was maintained by four air handling units, which each promoted circulation that was largely independent of the other air handling devices in this room. *Source:* Li et al. 2021/with permission of Elsevier.

have been developed and applied for this purpose [80]. These models can provide more detailed predictions of the spatial and temporal dynamics of air quality in indoor settings. Of particular relevance are the predictions of airborne pathogen distribution within a room, especially near the source. Also, these models can be used to provide illustrations of the spatial heterogeneity that might be expected in a scenario where one or more infected individuals is present in a room.

As an illustration of true nature of mixing behavior that can exist in indoor spaces, consider the simulation results provided in Figure 11.26. This image, which provides an illustration of the distribution of aerosols and air circulation following a Lagrangian frame of reference, was developed as a forensic exercise to simulate an actual event in which COVID-19 transmission is believed to have occurred in a poorly ventilated space in a restaurant. The room in which transmission is believed to have occurred includes four air handling units that each promote a circulation pattern; these four circulation patterns have minimal overlap and operate more-or-less independently. This leads to a condition that clearly deviates from the well-mixed assumption and probably played an important role in this disease transmission event.

## 11.6 UV Disinfection of Surfaces

UV radiation is commonly used for disinfection of solid objects and surfaces. In principle, UVC radiation can be used to disinfect a wide range of objects and surfaces, including keys, cell phones, countertops, remote control devices, furniture, faucet handles, light switches, doorknobs, medical devices, and other surfaces that are touched frequently by humans.

As with all applications of germicidal UV radiation, disinfection efficacy with these devices will be governed by the ability of the system to deliver germicidally active UV photons to the microbial or viral targets that are associated with or embedded in these surfaces. For surfaces that are flat, on a spatial scale that is consistent with the microbial and viral targets of interest, the application of germicidal UV radiation to achieve disinfection is quite simple, and can be monitored or controlled easily using common tools of radiometry. Alternatively, colorimetric dosimeter labels can be used for quantification of UVC dose delivery to a flat, planar surface. However, this perspective of UV-based disinfection of surfaces is probably overly simplistic in most applications.

At least two issues complicate the delivery of germicidal photons to microbial or viral targets on material surfaces. First, surfaces generally display surface texture that is relevant on the scale of microbial and viral pathogens. Second, some materials may be photochemically active.

Figure 11.27 provides an illustration of the scale of surface texture (roughness) for a wide range of common materials. These measurements illustrate the range of roughness characteristics for common surfaces. These data are relevant to UV-based disinfection processes for surfaces in that they indicate the extent to which material surfaces may provide shelter from exposure to UV radiation. For perspective, it is important to note that viruses range in size from roughly 5 to 300 nm, and display a wide range of shapes. Similarly, common bacteria range in size from roughly 200 to 2000 nm, although some very large bacteria have been reported to be on the order of 750,000 nm. Also, bacteria display many shapes including cells that are roughly spherical, rod-shaped,

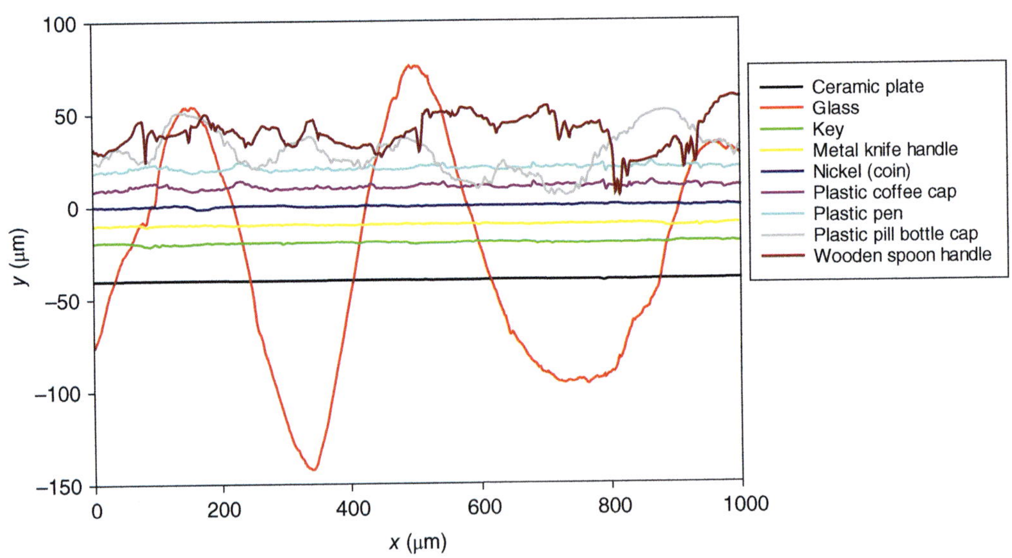


**Figure 11.27** Measured roughness (texture) of common surfaces. Note that the horizontal and vertical scales are different, which leads to apparent distortion of the material surface. *Source:* Data and images for surface profiles courtesy of Zygo Corporation. Appendix 11.A includes three-dimensional and two-dimensional renderings of profiles of each of the surfaces (media) illustrated in Figure 11.27. These images provide more detailed information about the texture/roughness of these surfaces. These data and images were collected using an Optical Profiler (Zygo Corporation). Appendix 11.B includes images to illustrate the sizes and shapes of bacteria and viruses.

comma-shaped, helical, and as filaments. Given these wide ranges in size and shape, it is clear that pathogens can be sheltered from UV radiation (and other forms of external stress) by the texture of material surfaces.

The responses of a surface exposed to UV irradiation can also affect the behavior and efficacy of a surface disinfection device. Specifically, two classes of response could be relevant. First, some materials will reflect UVC radiation. As an illustration, typical wall paints are roughly 3–7% reflective of $UV_{254}$ radiation. However, some commercially available paints have been developed that include metal oxide nanoparticles, which increase $UV_{254}$ reflectivity up to approximately 65% [82]. Experiments conducted with stationary $UV_{254}$ irradiation devices have demonstrated that the application of $UV_{254}$-reflective paint can reduce the time required to inactivate antibiotic-resistant bacteria (or bacterial spores), including MRSA, VRE, and *C. difficile*. These UV-reflective paints improve the spatial distribution of UVC energy within the irradiated room and allow more efficient use of UVC photons for pathogen inactivation than conventional paints or coatings.

Second, some materials will respond photochemically to UVC-irradiation. For example, titanium dioxide ($TiO_2$) and some other metal oxides (e.g. ZnO) function as photocatalysts. $TiO_2$ is commonly used as a white pigment in paints. It is also known to function as a photocatalyst as a result of exposure to UV radiation. The primary process that is responsible for photocatalytic activity by $TiO_2$ results from the photochemical event in which an electron is promoted from the valence band to the conduction band of the photocatalyst, leaving a hole in the valence band (see Figure 11.28) [83].

UVC radiation has also been demonstrated to be effective for disinfection of PPE, which is commonly used in medical and clinical settings. A particularly important class of PPE is filtering facepiece respirators (FFRs); within this product class, the need to disinfect N95 masks (and similar devices) arose as part of the COVID-19 pandemic.

The N95 standard for FFRs requires the ability of the device to physically separate at least 95% of airborne particles of diameter 0.3 μm or larger. A wide range of commercially available devices have been developed to meet the N95 standard or similar standards that are applied in countries other than the United States (e.g. FFP2 standard in the EU; KN95 standard in China; DS/DL2 standard in Japan; KF94 standard in Korea). N95 masks are used in situations where the risk of human

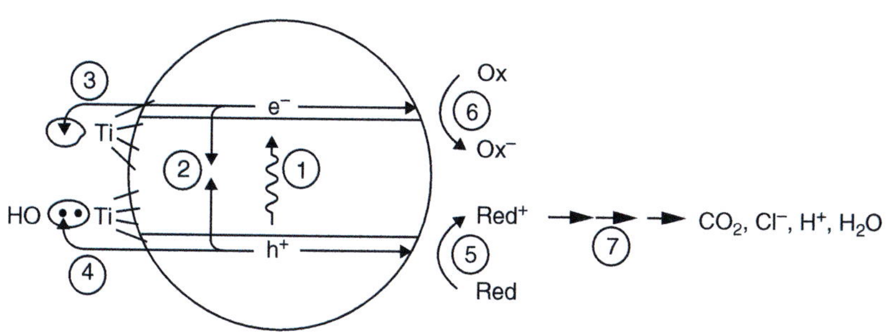


**Figure 11.28** Fundamental steps involved in photocatalysis from metal oxides. (1) Photopromotion of electron from valence band to conduction band, leaving a hole ($h^+$) in the valence band; (2) recombination of the valence electron with the hole; (3) and (4) represent trapping of conduction electron and hole to form reactive intermediates; (5), (6), and (7) reactions involving reactive intermediates to promote oxidation or reduction reactions. *Source:* Schneider et al. 2014/with permission of American Chemical Society.

exposure to airborne (aerosolized) pathogens is high, such as in medical or clinical facilities. There are also situations beyond medical applications where the use of an N95 mask (or similar) is warranted.

During the COVID-19 pandemic, demand for N95 masks exceeded supply, causing shortages of these devices. In addition, N95 masks were originally designed as single-use devices, which creates a medical waste management and disposal problem. For these and other reasons, methods to disinfect/decontaminate N95 masks have been developed. Among the methods that have been approved and applied is UVC irradiation.

Most disinfection strategies, including those based on UVC radiation, involve the application of a reactive agent to induce inactivation of microbial and viral pathogens. However, activity against pathogens is often accompanied by non-disinfection reactions that have the potential to contribute to negative outcomes. In the case of UV disinfection of surfaces, such as N95 masks, these undesirable reactions could lead to degradation of the materials that are used to produce the masks. Therefore, the design of UVC-based systems for disinfection/decontamination of N95 masks needs to demonstrate disinfection efficacy and the absence of detrimental changes in the properties of the materials that comprise the masks or the functional characteristics of the mask.

From the standpoint of disinfection, the key issue is delivery of UVC photons to the microbial or viral targets within the mask structure. Construction of N95 masks varies among manufacturers, but most are composite structures. The outer and inner layers of N95 masks are often made from a nonwoven fabric comprising thin polyester fibers. Inner layers often include electrostatically charged, meltblown polypropylene microfibers; the surface charge expressed by the fibers that comprise the inner (filter) layers of these masks is critical to mask performance [84].

Airborne microbial and viral pathogens are generally associated with aerosol particles. The vast majority of these particles are characterized by aerodynamic diameters less than 1 μm. For example, among influenza patients, 87% of exhaled particles were observed to have a diameter less than 1 μm [85]. A similar study demonstrated that 82% of exhaled particles from patients who were positive for Rhinovirus were in the size range 0.300–0.499 μm [86]. N95 masks and similar devices are effective at capturing aerosols in the size range that is emitted by humans. However, the ability of these masks to physically separate aerosols from air depends on the physical characteristics and structural integrity of the mask materials.

Germicidal UV radiation has the potential to alter the materials that comprise FFRs, including N95 masks. However, experiments involving exposure of N95 and surgical masks to $UV_{254}$ or $UV_{265}$ radiation at doses of 1–10 $\frac{J}{cm^2}$ resulted in no quantifiable loss of structural integrity, ability to act as a filter, or changes in the structure, morphology, or hydrophobicity of mask materials. Moreover, no changes in pressure drop or material tensile strength were observed in these tests [87]. Similar tests involving $UV_{254}$ doses as high as 950 $\frac{J}{cm^2}$ indicated essentially no change in pressure drop across N95 masks, but quantifiable increases in particle penetration into the filter, as well as reductions in material strength [88]. However, it should be noted that a typical dose applied for N95 mask disinfection is on the order of 1 $\frac{J}{cm^2}$; as such, the doses applied in this experiment are well beyond those that are likely to be applied in practice, even for multiple disinfection cycles.

The doses applied in the experiments described above are considerably higher than those that would be required to inactivate virtually any microbial or viral pathogen. These doses describe the amount of radiation received at the outside surface of the mask. These large doses are required

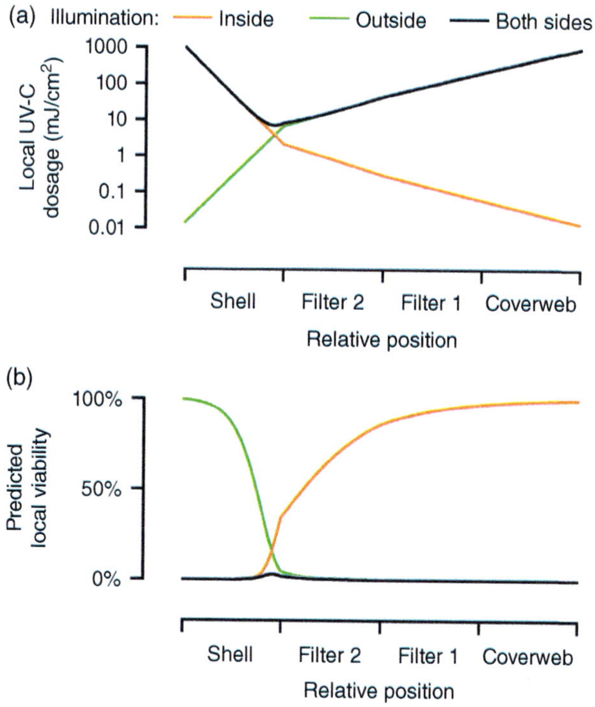


**Figure 11.29** (a) Numerical simulation of $UV_{254}$ penetration into N95 mask layers when a dose of $1\ \frac{J}{cm^2}$ was applied from the inside, outside, and both directions. N95 filter layers from right to left include the coverweb (outside), filter 1 and filter 2 (internal filter layers), and shell (inside). (b) Simulation of SARS-CoV-2 infectivity as a function of position for these three simulated $UV_{254}$ irradiation strategies. *Source:* Image and simulation results from Huber et al. 2021/with permission of Elsevier.

to accomplish effective disinfection of the internal mask layers, which are subjected to much lower exposure due to absorption by the materials that comprise the mask layers.

Numerical models have been developed to simulate penetration of $UV_{254}$ radiation into N95 masks [84, 89]. These models were developed following ray tracing approaches so that the effects of material properties (reflectance, absorption) by individual fiber materials could be accounted for. In a general sense, the material layers that comprise these masks display low, but nonzero transmittance of UVC radiation. As such, delivery of UVC radiation to inner layers of mask materials requires that UVC doses applied to the external layers is on the order of $1\ \frac{J}{cm^2}$, which is at least two orders of magnitude higher than the dose required to accomplish effective inactivation of viruses that are associated with external layers. Figure 11.29 illustrates penetration of $UV_{254}$ radiation into mask materials for cases of $UV_{254}$ doses of $1\ \frac{J}{cm^2}$ being applied to the external layer of the mask, the internal layer of the mask, and both. Also included in this figure is an illustration of SARS-CoV-2 infectivity as function of location within the mask structure. Collectively, these simulation results indicate the $UV_{254}$ irradiation of N95 masks from both sides can be an effective strategy for inactivation of viral pathogens. Similar simulation strategies may be warranted for design of other UV-based disinfection processes for surfaces and solid materials.

## 11.7 Summary

UV disinfection of air and surfaces has been recognized as an effective tool for control of transmission of airborne and surface-associated pathogens for many decades. Systems based on UVC radiation are known to be effective for inactivation of a broad spectrum of microbial and viral pathogens that are spread by airborne or fomite-associated pathways. The design and implementation of these systems are governed largely by a balance between the need to deliver a sufficient UV dose to inactivate pathogens, while at the same time protecting humans (and other potential nonpathogenic targets) from being exposed to unacceptably high doses of UV radiation. Many methods have been developed to meet these contrasting goals.

For UV disinfection of air, common application strategies include upper-room configurations and UV-based air cleaners. Under appropriate conditions of mixing in the air space to be disinfected, these strategies can be effective for inactivation of pathogens, while limiting human exposure. The emergence of Far UVC radiation offers the potential to increase human exposure to these sources, possibly to the point of applying whole-room configurations.

Processes to accomplish UV disinfection of surfaces will be influenced by the interactions of UV radiation with the surface materials. Common examples of interactions that could influence the efficiency of these processes include physical (optical) reactions, such as reflection, refraction, or absorption. UV photons may also initiate photochemical reactions at a material boundary that could influence process outcomes.

## 11.A Images to Illustrate Surface Roughness and Texture

Images and data presented below were generously provided by Zygo Corporation (a Division of AMETEK Ultra Precision Technologies), who performed measurements of roughness/texture on several common material surfaces. For each material, a three-dimensional image is provided together with a surface profile through a transect of the surface. The surface profile is presented for a transect of horizontal length 1000 μm. Note that the vertical axis scales for the three-dimensional and two-dimensional profiles are different than the corresponding horizontal scales, which results in distortion of the images. Note also that the vertical axis scales for each material are different to allow the range of vertical locations to be clearly illustrated.

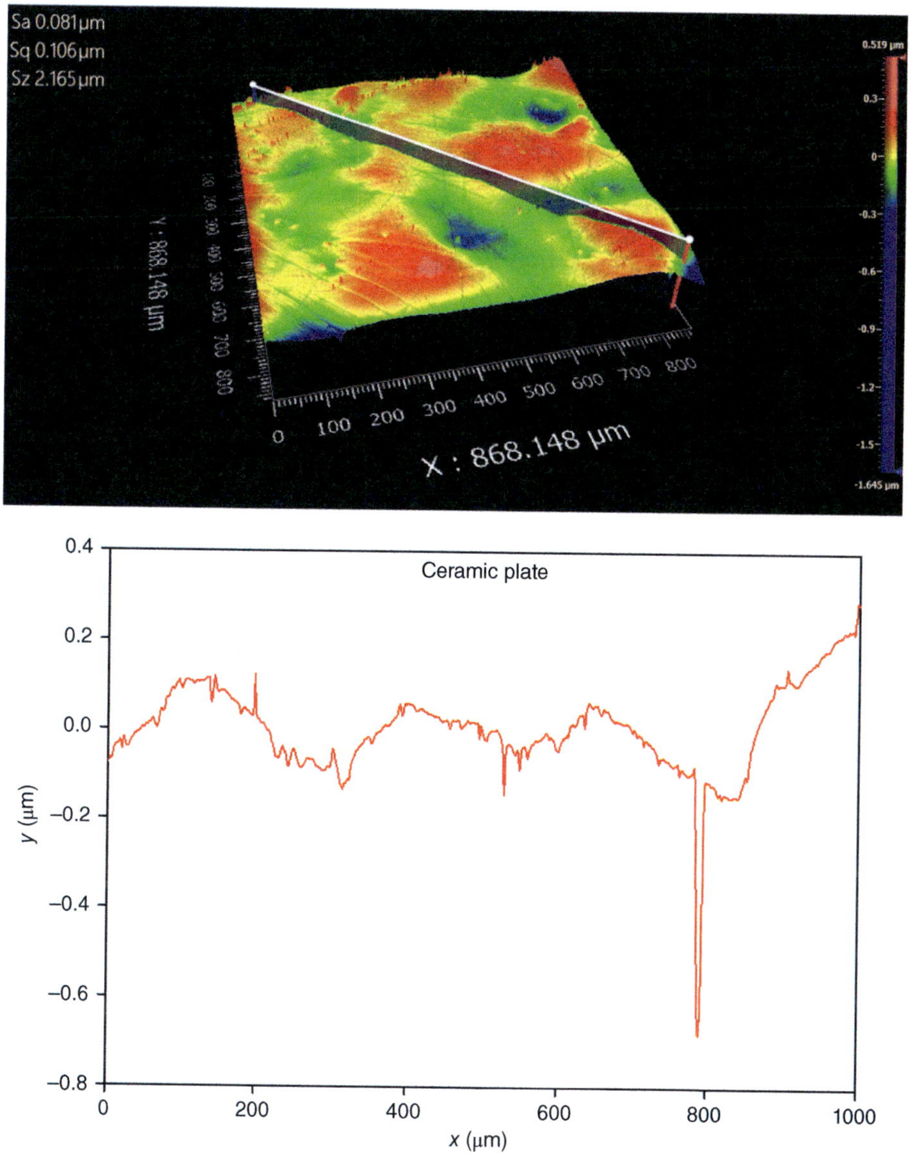


**Figure 11.A.1** (Top) Three-dimensional rendering of the surface roughness/texture of a ceramic plate. (Bottom) Two-dimensional rendering of the surface profile for a 1000 µm segment of the transect illustrated in the top image.

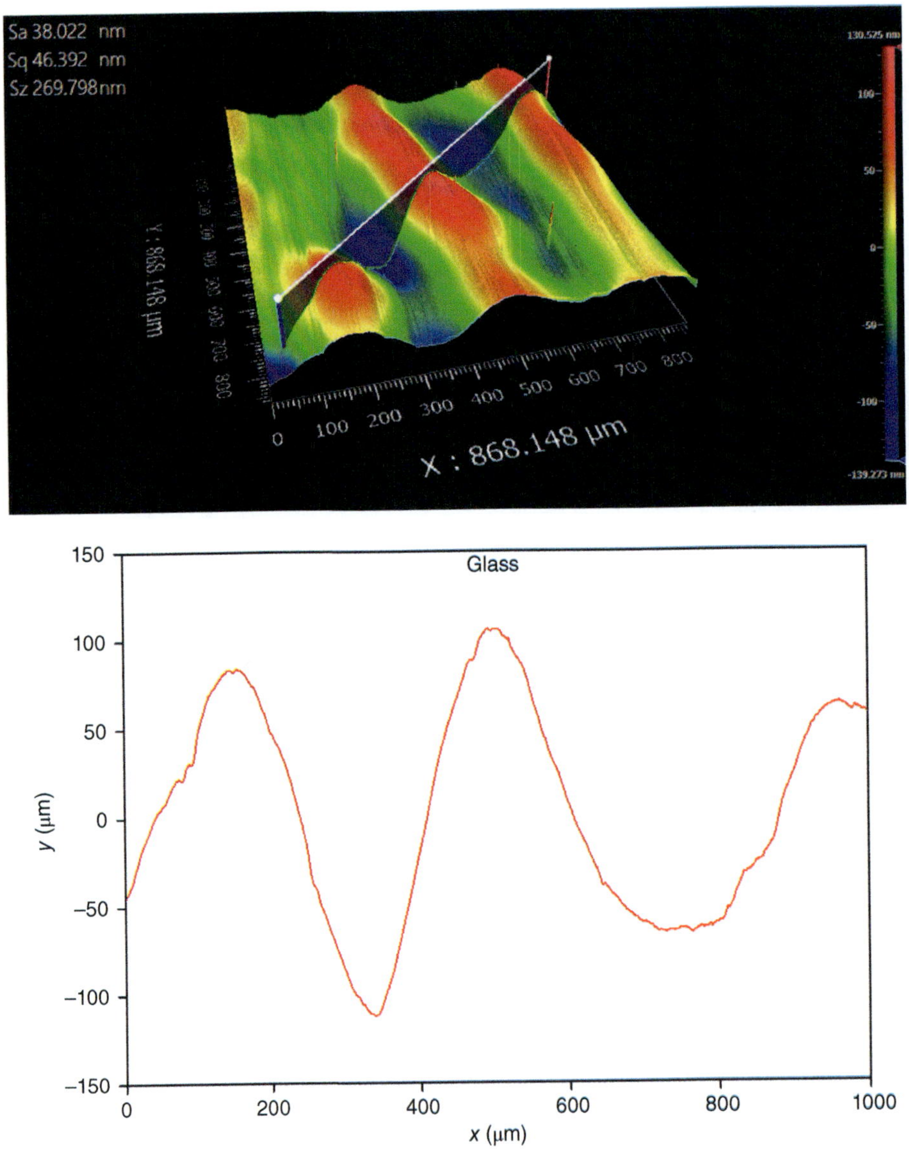


**Figure 11.A.2** (Top) Three-dimensional rendering of the surface roughness/texture of a glass plate. (Bottom) Two-dimensional rendering of the surface profile for a 1000 μm segment of the transect illustrated in the top image.

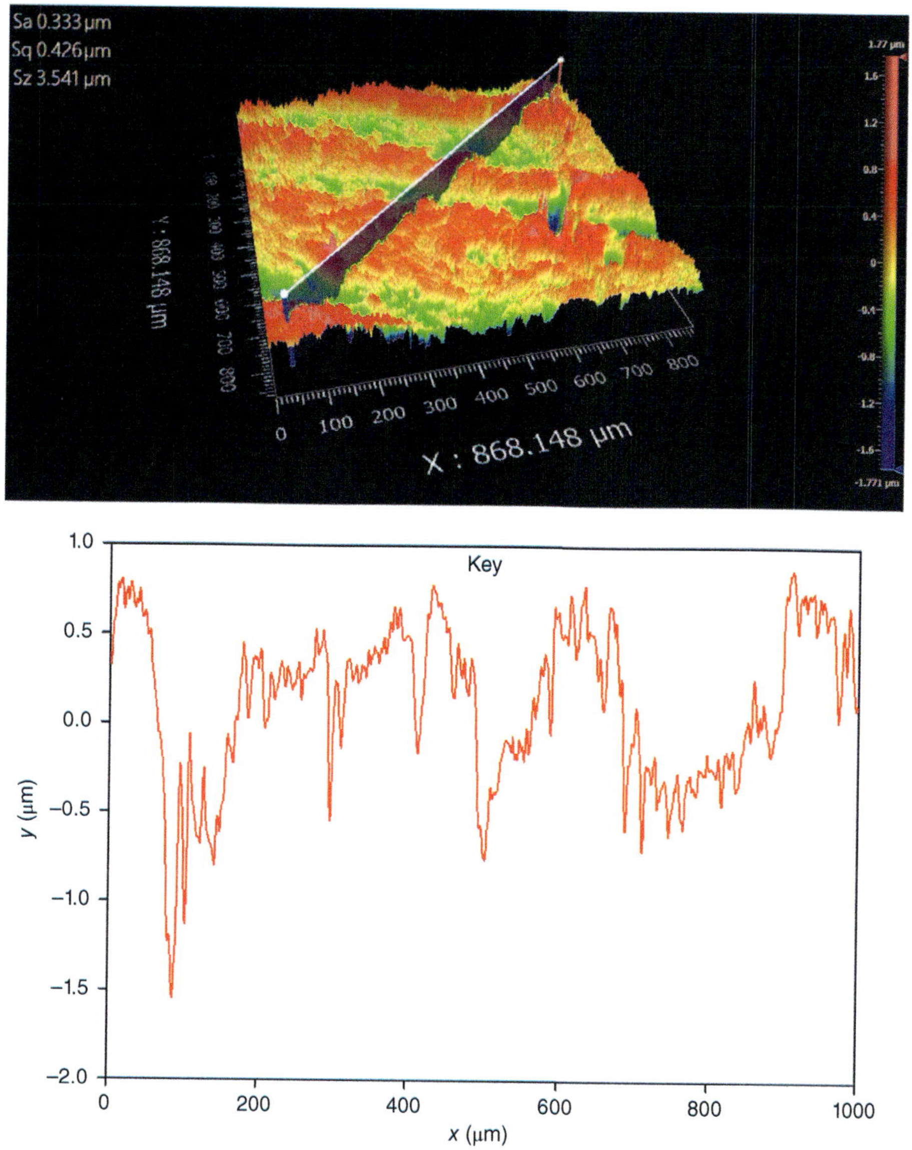


**Figure 11.A.3** (Top) Three-dimensional rendering of the surface roughness/texture of a key. (Bottom) Two-dimensional rendering of the surface profile for a 1000 µm segment of the transect illustrated in the top image.

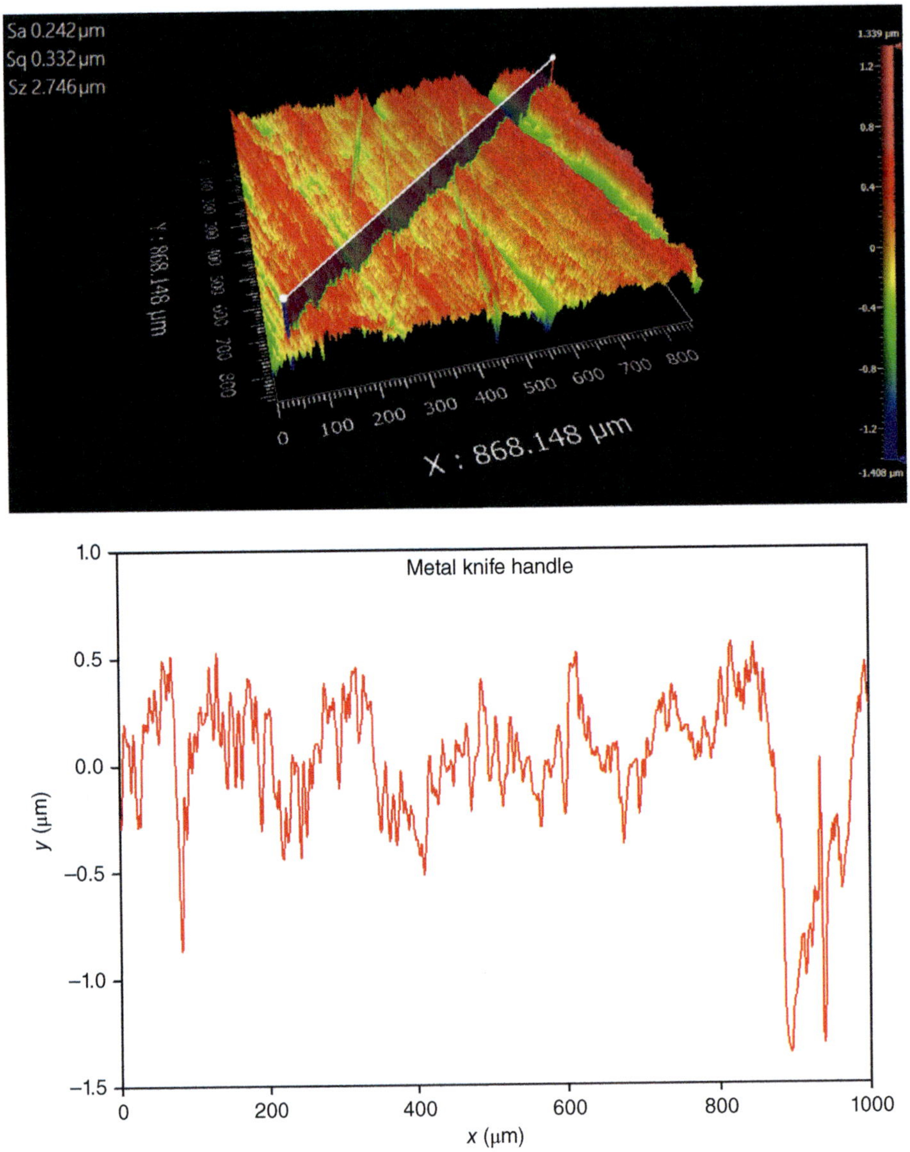


**Figure 11.A.4** (Top) Three-dimensional rendering of the surface roughness/texture of a metal knife handle. (Bottom) Two-dimensional rendering of the surface profile for a 1000 μm segment of the transect illustrated in the top image.

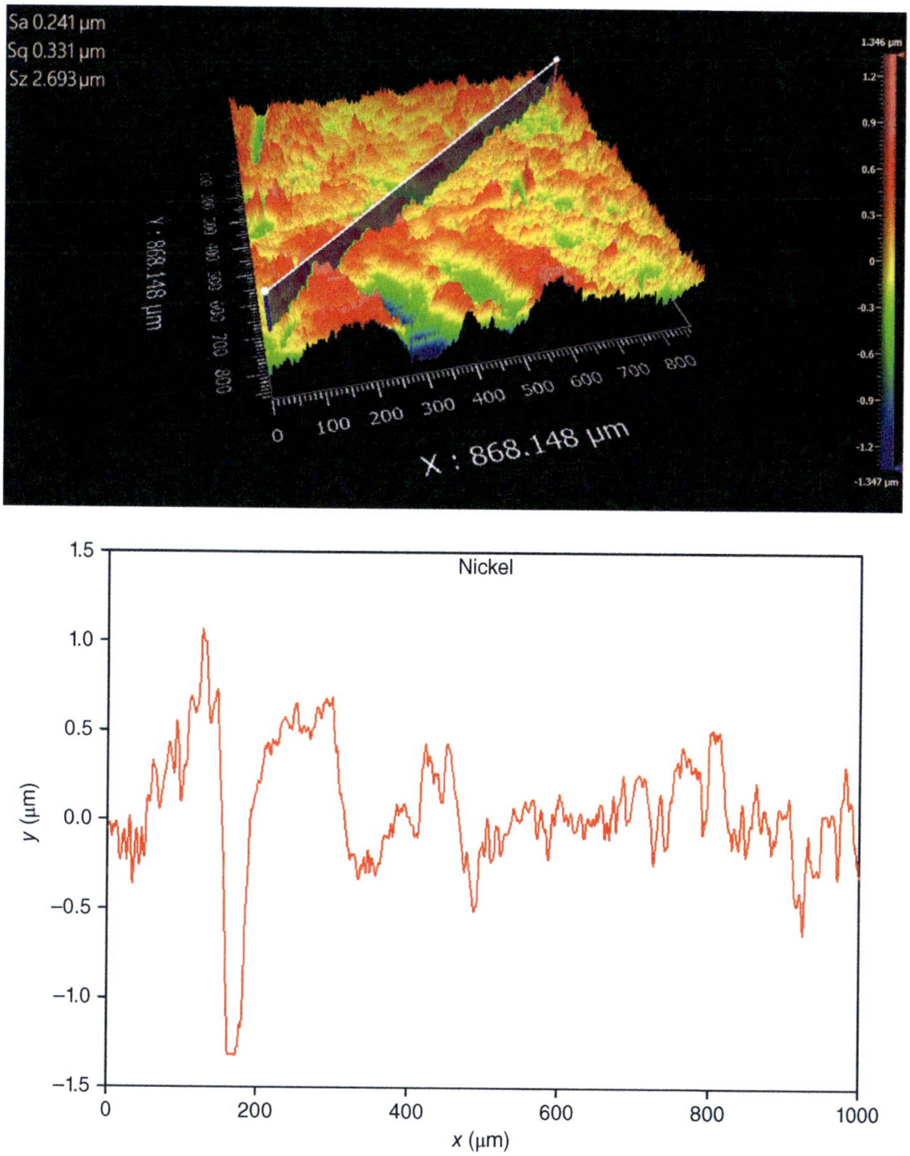


**Figure 11.A.5** (Top) Three-dimensional rendering of the surface roughness/texture of a nickel (coin). (Bottom) Two-dimensional rendering of the surface profile for a 1000 µm segment of the transect illustrated in the top image.

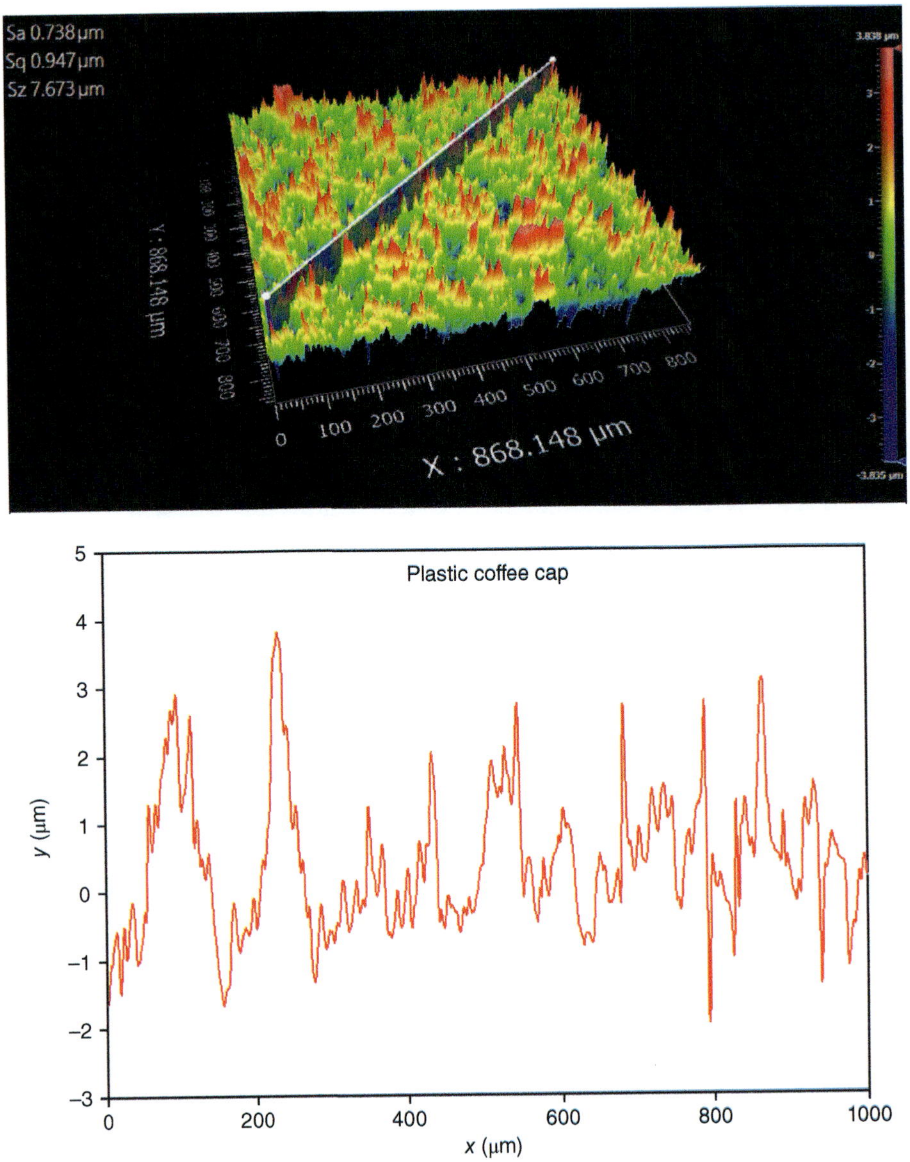


**Figure 11.A.6** (Top) Three-dimensional rendering of the surface roughness/texture of a plastic coffee cup cap. (Bottom) Two-dimensional rendering of the surface profile for a 1000 µm segment of the transect illustrated in the top image.

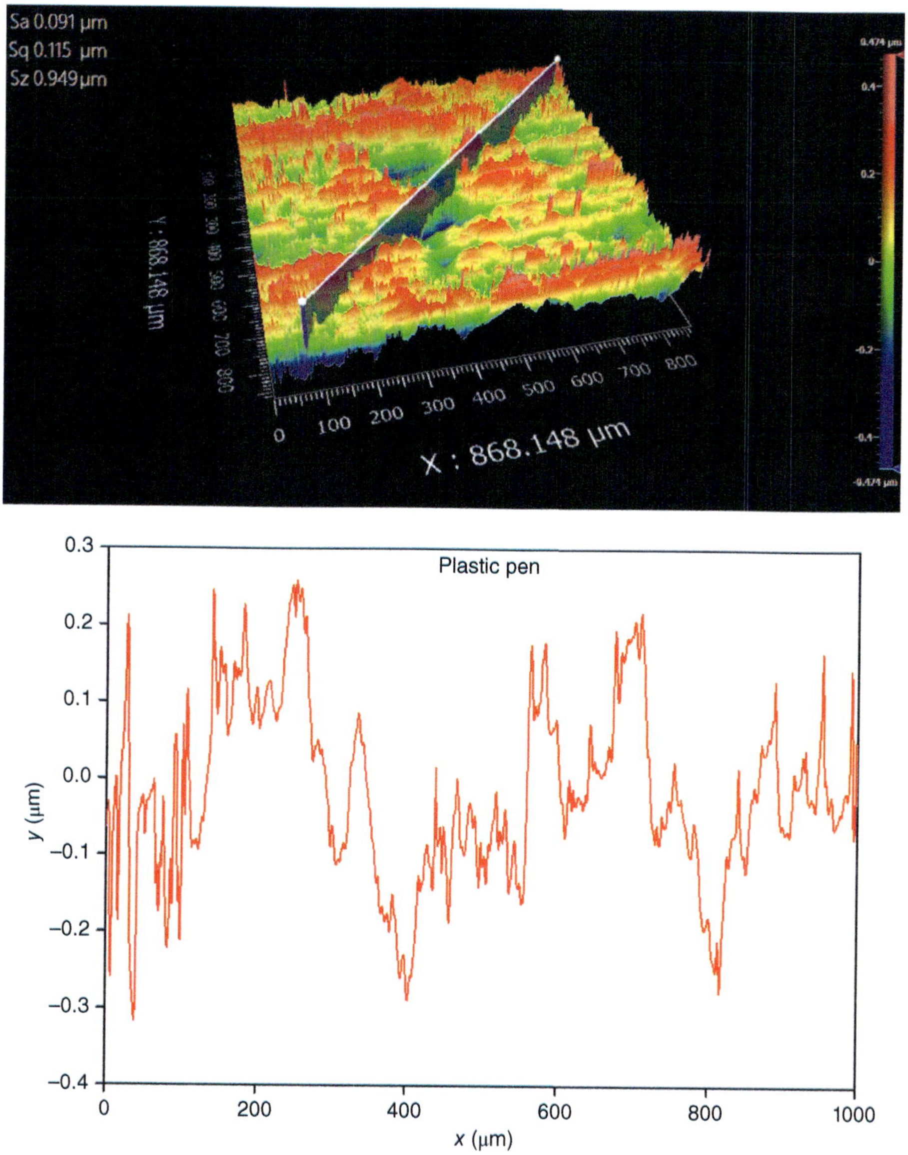


**Figure 11.A.7** (Top) Three-dimensional rendering of the surface roughness/texture of a plastic pen. (Bottom) Two-dimensional rendering of the surface profile for a 1000 μm segment of the transect illustrated in the top image.

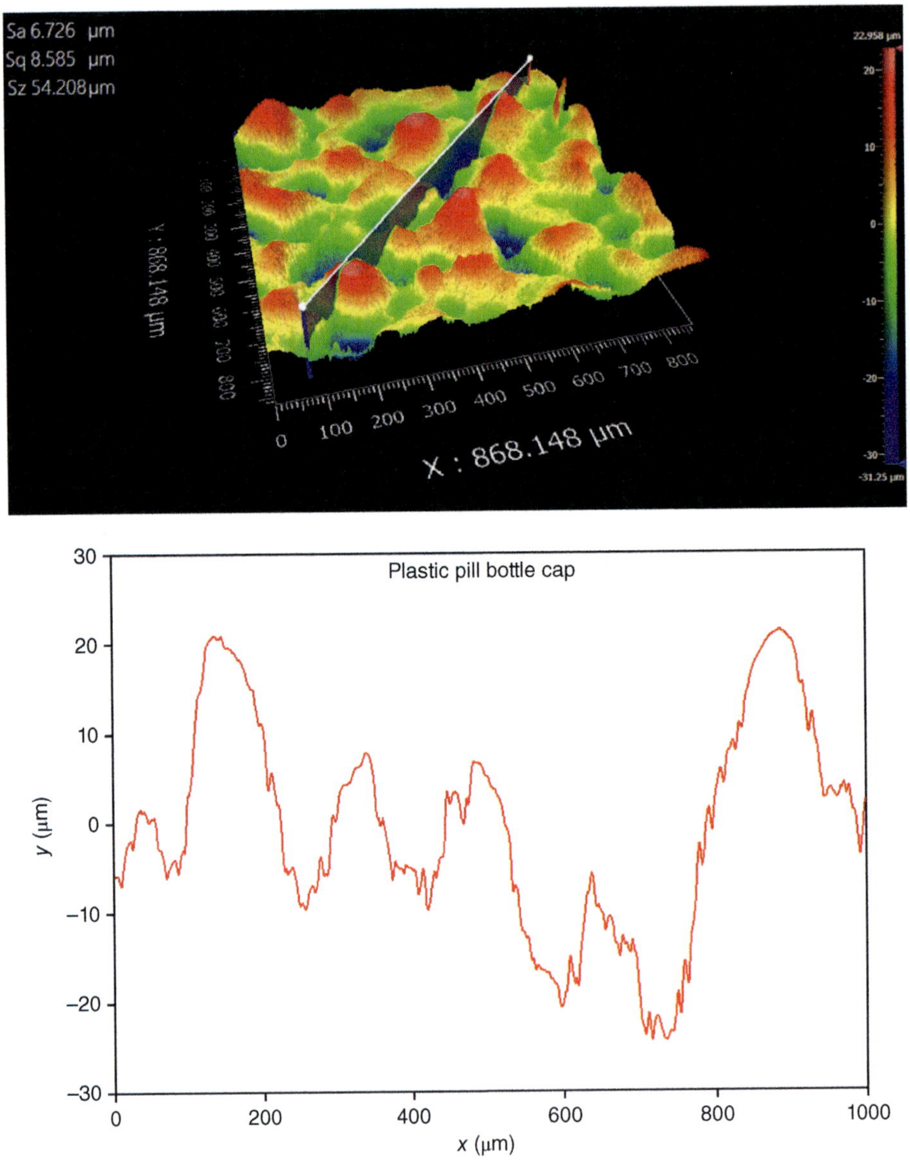


**Figure 11.A.8** (Top) Three-dimensional rendering of the surface roughness/texture of a plastic pill bottle cap. (Bottom) Two-dimensional rendering of the surface profile for a 1000 μm segment of the transect illustrated in the top image.

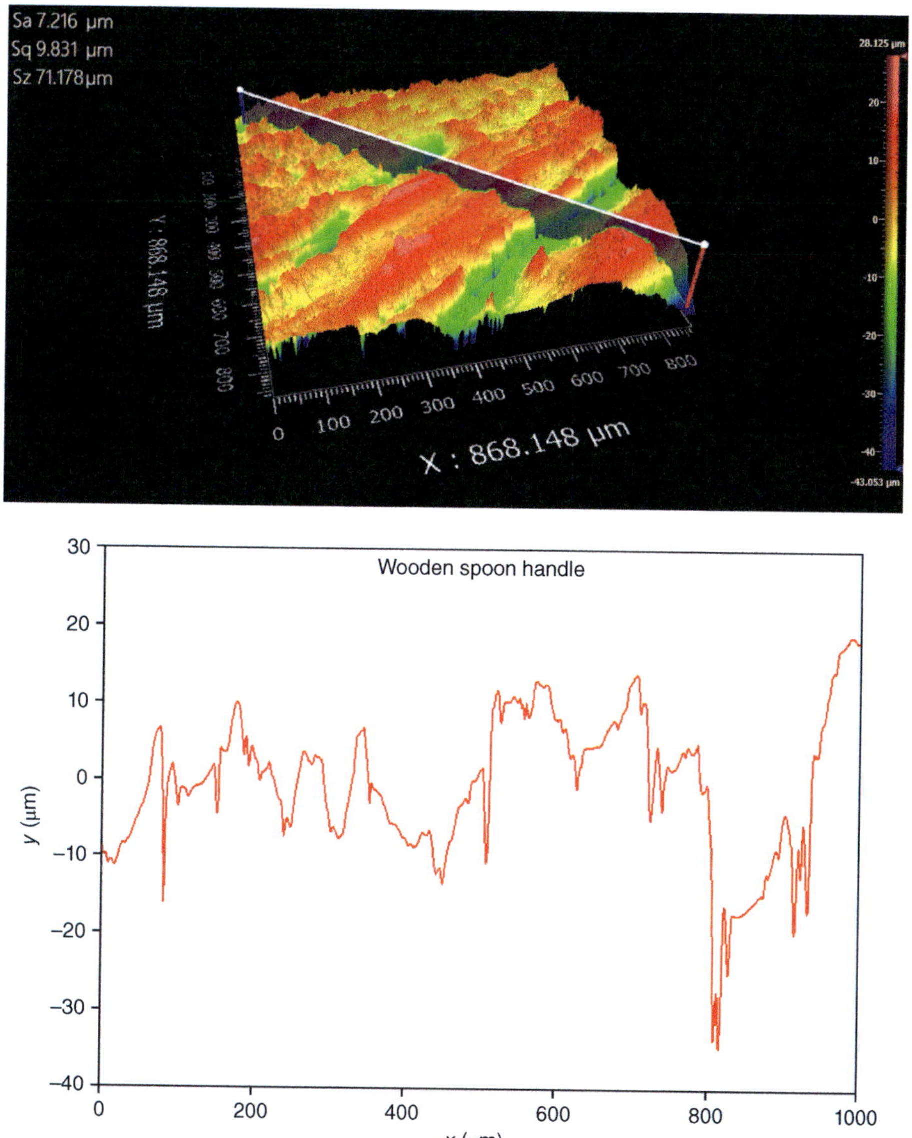


**Figure 11.A.9** (Top) Three-dimensional rendering of the surface roughness/texture of a plastic pill bottle cap. (Bottom) Two-dimensional rendering of the surface profile for a 1000 µm segment of the transect illustrated in the top image.

## 11.B Illustrations of Sizes and Shapes of Bacteria and Viruses

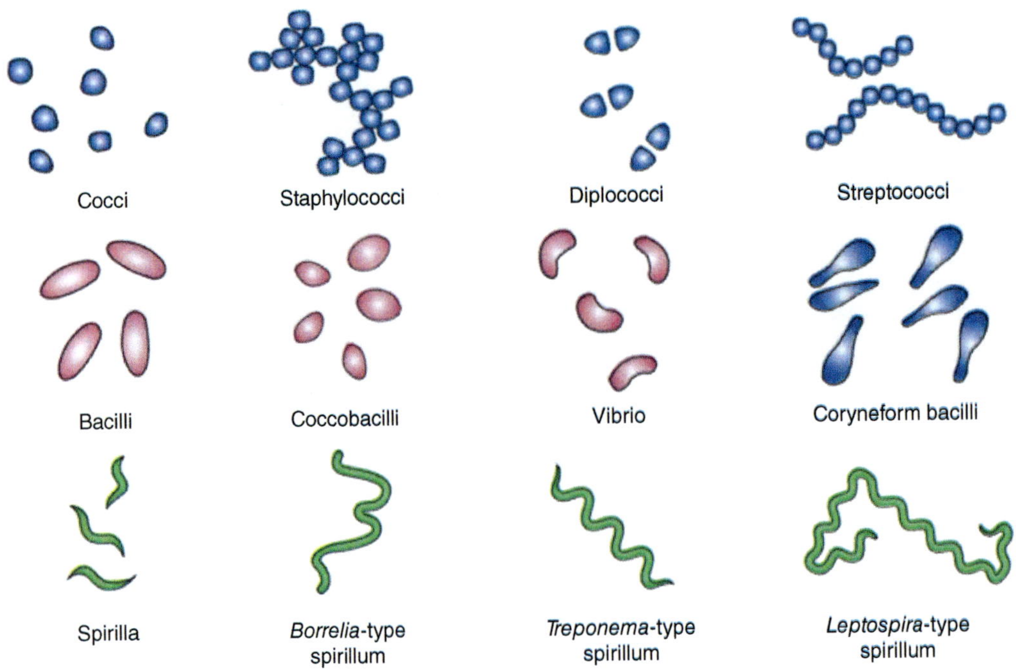


**Figure 11.B.1** Illustrations of typical bacterial shapes. *Source:* Image from Gary et al. 2017/Wolters Kluwer. Link: https://microbiologyinfo.com/different-size-shape-and-arrangement-of-bacterial-cells/.

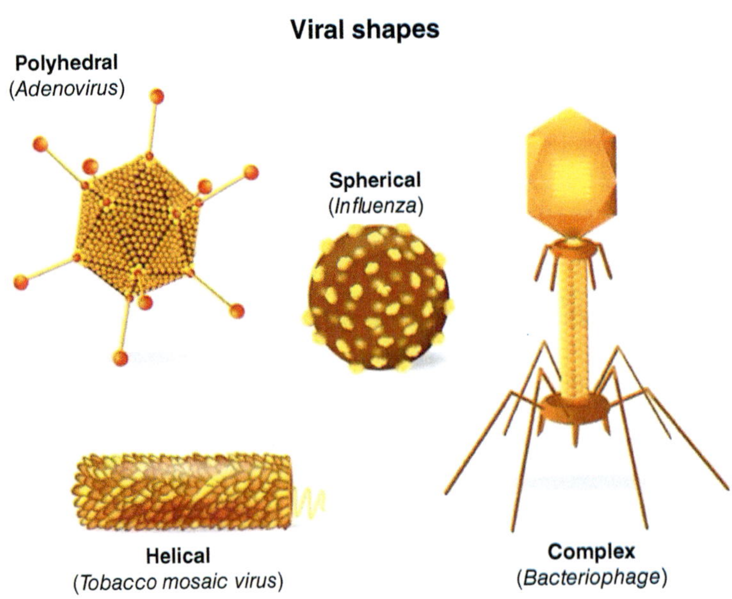


**Figure 11.B.2** Sizes and shapes of viruses within various groups. *Source:* Designua/Adobe Stock.

## Notes

**a** Note that alternative size-based definitions of droplet nuclei have been developed; specifically, some references identify the size cutoff as an aerodynamic diameter of less than 5 μm.

**b** The Minimum Efficiency Reporting Value (MERV) scale is used to rate air filters in terms of their ability to remove particles in the size range of 0.3–10 μm from air. In general, particle removal efficiency increases with MERV rating, as does the pressure drop across the filter.

**c** A more formal and correct designation of the units for absorption cross-sections is $cm^2$/molecule, but often the term "molecule" is omitted.

## References


1 Reed, N.G. (**2010**). The history of ultraviolet germicidal irradiation for air disinfection. *Public Health Reports 125* (1): 15–27. https://doi.org/10.1177/003335491012500105.

2 Tellier, R., Li, Y.G., Cowling, B.J., and Tang, J.W. (**2019**). Recognition of aerosol transmission of infectious agents: a commentary. *BMC Infectious Diseases 19*: https://doi.org/10.1186/s12879-019-3707-y.

3 Nicholson, W.L. and Galeano, B. (**2003**). UV resistance of *Bacillus anthracis* spores revisited: validation of *Bacillus subtilis* spores as UV surrogates for spores of B-anthracis Sterne. *Applied and Environmental Microbiology 69* (2): 1327–1330. https://doi.org/10.1128/aem.69.2.1327-1330.2003.

4 Blatchley III, E.R., Meeusen, A., Aronson, A.I., and Brewster, L. (**2005**). Inactivation of *Bacillus* spores by ultraviolet or gamma radiation. *Journal of Environmental Engineering 131* (9): 1245–1252. https://doi.org/10.1061/(asce)0733-9372(2005)131:9(1245).

5 Oliveira, B.R., Crespo, M.T.B., and Pereira, V.J. (**2020**). Small but powerful: light-emitting diodes for inactivation of *Aspergillus* species in real water matrices. *Water Research 168*: 115108. https://doi.org/10.1016/j.watres.2019.115108.

6 Kim, D.K. and Kanga, D.H. (**2018**). UVC LED irradiation effectively inactivates aerosolized viruses, bacteria, and fungi in a chamber-type air disinfection system. *Applied and Environmental Microbiology 84* (17): e00944-18. https://doi.org/10.1128/aem.00944-18.

7 Nourmoradi, H., Nikaeen, M., Stensvold, C.R., and Mirhendi, H. (**2012**). Ultraviolet irradiation: an effective inactivation method of *Aspergillus* spp. in water for the control of waterborne nosocomial aspergillosis. *Water Research 46* (18): 5935–5940. https://doi.org/10.1016/j.watres.2012.08.015.

8 Beck, S.E., Rodriguez, R.A., Linden, K.G. et al. (**2014**). Wavelength dependent UV inactivation and DNA damage of adenovirus as measured by cell culture infectivity and long range quantitative PCR. *Environmental Science & Technology 48* (1): 591–598. https://doi.org/10.1021/es403850b.

9 Battigelli, D.A., Sobsey, M.D., and Lobe, D.C. (**1993**). The inactivation of hepatitis A virus and other model viruses by UV irradiation. *Water Science and Technology 27* (3–4): 339–342.

10 Powell, W.F. and Setlow, R.B. (**1956**). The effect of monochromatic ultraviolet radiation on the interfering property of influenza virus. *Virology 2* (3): 337–343. https://doi.org/10.1016/0042-6822(56)90028-9.

11 Budowsky, E.I., Bresler, S.E., Friedman, E.A., and Zheleznova, N.V. (**1981**). Principles of selective inactivation of viral genome. 1. UV-induced inactivation of influenza-virus. *Archives of Virology 68* (3–4): 239–247. https://doi.org/10.1007/bf01314577.

12 Welch, D., Buonanno, M., Grilj, V. et al. (**2018**). Far-UVC light: a new tool to control the spread of airborne-mediated microbial diseases. *Scientific Reports 8*: 2752. https://doi.org/10.1038/s41598-018-21058-w.

13 Rockey, N.C., Henderson, J.B., Chin, K. et al. (**2020**). Predictive modeling of virus inactivation by UV. *Environmental Science and Technology 55*: 3322–3332. https://doi.org/10.1101/2020.10.27.355479v1.

14 Gerchman, Y., Mamane, H., Friedman, N., and Mandelboim, M. (**2020**). UV-LED disinfection of coronavirus: wavelength effect. *Journal of Photochemistry and Photobiology B-Biology 212*: 112044. https://doi.org/10.1016/j.jphotobiol.2020.112044.

15 Buonanno, M., Welch, D., Shuryak, I., and Brenner, D.J. (**2020**). Far-UVC light (222 nm) efficiently and safely inactivates airborne human coronaviruses. *Scientific Reports 10*: 10285. https://doi.org/10.1038/s41598-020-67211-2.

16 Oguma, K., Katayama, H., and Ohgaki, S. (**2004**). Photoreactivation of *Legionella pneumophila* after inactivation by low- or medium-pressure ultraviolet lamp. *Water Research 38* (11): 2757–2763. https://doi.org/10.1016/j.watres.2004.03.024.

17 Kowalski, W. (**2009**). *Ultraviolet Germicidal Irradiation Handbook: UVGI for Air and Surface Disinfection*. Springer-Verlag.

18 Sagripanti, J.L. and Lytle, C.D. (**2011**). Sensitivity to ultraviolet radiation of Lassa, vaccinia, and Ebola viruses dried on surfaces. *Archives of Virology 156* (3): 489–494. https://doi.org/10.1007/s00705-010-0847-1.

19 Martinez, L.R. and Casadevall, A. (**2007**). *Cryptococcus neoformans* biofilm formation depends on surface support and carbon source and reduces fungal cell susceptibility to heat, cold, and UV light. *Applied and Environmental Microbiology 73* (14): 4592–4601. https://doi.org/10.1128/aem.02506-06.

20 Pendyala, B., Patras, A., Pokharel, B., and D'Souza, D. (**2020**). Genomic modeling as an approach to identify surrogates for use in experimental validation of SARS-CoV-2 and HuNoV inactivation by UV-C treatment. *Frontiers in Microbiology 11*: 572331. https://doi.org/10.3389/fmicb.2020.572331.

21 Sagripanti, J.L. and Lytle, C.D. (**2020**). Estimated inactivation of coronaviruses by solar radiation with special reference to COVID-19. *Photochemistry and Photobiology 96* (4): 731–737. https://doi.org/10.1111/php.13293.

22 Centers for Disease Control and Prevention (**2017**). Healthcare-Associated Infections (HAIs). https://www.cdc.gov/winnablebattles/report/HAIs.html (accessed 2021).

23 Marra, A.R., Schweizer, M.L., and Edmond, M.B. (**2018**). No-touch disinfection methods to decrease multidrug-resistant organism infections: a systematic review and meta-analysis. *Infection Control and Hospital Epidemiology 39* (1): 20–31. https://doi.org/10.1017/ice.2017.226.

24 Fletcher, L.A. (**2004**). The Influence of Relative Humidity on the UV Susceptibility of Airborne Gram Negative Bacteria. IUVA News, vol. *6*, 12–19. International Ultraviolet Association.

25 Macher, J.M. (**1993**). The use of germicidal lamps to control tuberculosis in healthcare facilities. *Infection Control and Hospital Epidemiology 14* (12): 723–729.

26 Kujundzic, E., Hernandez, M., and Miller, S.L. (**2007**). Ultraviolet germicidal irradiation inactivation of airborne fungal spores and bacteria in upper-room air and HVAC in-duct configurations. *Journal of Environmental Engineering and Science 6* (1): 1–9. https://doi.org/10.1139/s06-039.

27 Wells, W.F., Wells, M.W., and Wilder, T.S. (**1942**). The environmental control of epidemic contagion I an epidemiologic study of radiant disinfection of air in day schools. *American Journal of Hygiene 35* (1): 97–121.

28 Nardell, E. and Nathavitharana, R. (**2019**). Air disinfection in measles transmission hotspots. *Lancet 394* (10203): 1009–1010. https://doi.org/10.1016/s0140-6736(19)31889-6.

29 (a) Hessling, M., Hones, K., Vatter, P., and Lingenfelder, C. (**2020**). Ultraviolet irradiation doses for coronavirus inactivation – review and analysis of coronavirus photoinactivation studies. *GMS Hygiene and Infection Control 15*: Doc08. https://doi.org/10.3205/dgkh000343; (b) Patterson, E.I., Prince, T., Anderson, E.R. et al. (**2020**). Methods of inactivation of SARS-CoV-2 for downstream biological assays. *The Journal of Infectious Diseases 222* (9): 1462–1467. https://doi.org/10.1093/infdis/jiaa507. (c) Storm, N., McKay, L., Downs, S. et al. Rapid and complete inactivation of SARS-CoV-2 by ultraviolet-C irradiation. *Scientific Reports 10*: 22421.

30 Smith, J.A.E., Hopkins, S., Turner, C., et al. (**2021**). Public health impact of mass sporting and cultural events in a rising COVID-19 prevalence in England. https://khub.net/documents/135939561/338928724/Public+health+impact+of+mass+sporting+and+cultural+events+in+a+rising+COVID-19+prevalence+in+England.pdf/05204895-1576-1ee7-b41e-880d5d6b4f17 (accessed 2021).

31 van Doremalen, N., Bushmaker, T., Morris, D.H. et al. (**2020**). Aerosol and surface stability of SARS-CoV-2 as compared with SARS-CoV-1. *New England Journal of Medicine 382* (16): 1564–1567. https://doi.org/10.1056/NEJMc2004973.

32 (a) Gavi (**2020**). 5 reasons why pandemics like COVID-19 are becoming more likely. https://www.gavi.org/vaccineswork/5-reasons-why-pandemics-like-covid-19-are-becoming-more-likely (accessed 2021). (b) Tollefson, J. (**2020**). Why deforestation and extinctions make pandemics more likely. *Nature 584* (7820): 175–176. https://doi.org/10.1038/d41586-020-02341-1.

33 ASHRAE (**2019**). Ultraviolet air and surface treatment. In: *2019 ASHRAE Handbook Heating, Ventilating, and Air-Conditioning Applications*, S-1e, 62.61–62.17. The American Society of Heating, Refrigerating and Air-Conditioning Engineers (ASHRAE).

34 Department of Health and Human Services, Centers for Disease Control, and National Institute for Occupational Safety and Health (**2009**). *Basic Upper-Room Ultraviolet Germicidal Irradiation for Healthcare Settings*. Environmental Control for Tuberculosis.

35 McLean, R.L. (**1956**). Comments on reducing influenza epidemics among hospitalized veterans by UV irradiation of droplets in the air. *The American Review of Respiratory Disease Supplement 83*: 36–38.

36 (a) Mphaphlele, M., Dharmadhikari, A.S., Jensen, P.A. et al. (**2015**). Institutional tuberculosis transmission controlled trial of upper room ultraviolet air disinfection: a basis for new dosing guidelines. *American Journal of Respiratory and Critical Care Medicine 192* (4): 477–484. https://doi.org/10.1164/rccm.201501-0060OC; (b) Escombe, A. R., Moore, D. A. J., Gilman, R. H., Navincopa, M., Ticona, E., Mitchell, B., Noakes, C., Martinez, C., Sheen, P., Ramirez, R. et al. Upper-room ultraviolet light and negative air ionization to prevent tuberculosis transmission. *Plos Medicine* **2009**, *6* (3).e43. https://doi.org/10.1371/journal.pmed.1000043. (c) Miller, S.L. and Macher, J.M. (**2000**). Evaluation of a methodology for quantifying the effect of room air ultraviolet germicidal irradiation on airborne bacteria. *Aerosol Science and Technology 33* (3): 274–295. https://doi.org/10.1080/027868200416259. (d) Nardell, E.A., Bucher, S.J., Brickner, P.W. et al. (**2008**). Safety of upper-room ultraviolet germicidal air disinfection for room occupants: results from the tuberculosis ultraviolet shelter study. *Public Health Reports 123* (1): 52–60. https://doi.org/10.1177/003335490812300108.

37 Kowalski, W.J. and Bahnfleth, W.P. (**2002**). *MERV Filter Models for Aerobiological Applications*, 8. The Pennsylvania State University.

38 Sliney, D. (**2013**). Balancing the risk of eye irritation from UV-C with infection from bioaerosols. *Photochemistry and Photobiology 89* (4): 770–776. https://doi.org/10.1111/php.12093.

39 CIE and International Commission of Illumination (**2010**). *UV-C Photocarcinogenesis Risks from Germicidal Lamps*. CIE.

40 Finlayson, L., Barnard, I.R.M., McMillan, L. et al. (**2021**). Depth penetration of light into skin as a function of wavelength from 200 nm–1000 nm. *Photochemistry and Photobiology* https://doi.org/10.1111/php.13550.

41 International Organization for Standardization (**2016**). *Photocarcinogenesis Action Spectrum (Non-Melanoma Skin Cancers)*. Geneva: International Organization for Standardization (ISO) and the International Commission of Illumination (CIE).

42 ICNIRP (**2004**). Guidelines on limits of exposure to ultraviolet radiation of wavelengths between 180 nm and 400 nm (incoherent optical radiation). *Health Physics 87* (2): 171–186.

**43** NIOSH (**1972**). Criteria for a recommended standard: occupational exposure to ultraviolet radiation https://www.cdc.gov/niosh/docs/73-11009/default.html (accessed 2021).

**44** American Conference of Governmental Industrial Hygienists (**2022**). *2022 Threshold Limit Values (TLVs) and Biological Exposure Indices (BEIs)*. Cincinnati, OH: ACGIH.

**45** (a) Buonanno, M., Randers-Pehrson, G., Bigelow, A.W. et al. (**2013**). 207-nm UV light – a promising tool for safe low-cost reduction of surgical site infections. I: in vitro studies. *Plos One 8* (10): e76968. https://doi.org/10.1371/journal.pone.0076968; (b) Buonanno, M., Stanislauskas, M., Ponnaiya, B. et al. (**2016**). 207-nm UV light-a promising tool for safe low-cost reduction of surgical site infections. II: in-vivo safety studies. *Plos One 11* (6): e76968. https://doi.org/10.1371/journal.pone.0138418.

**46** Buonanno, M., Ponnaiya, B., Welch, D. et al. (**2017**). Germicidal efficacy and mammalian skin safety of 222-nm UV light. *Radiation Research 187* (4): 483–491. https://doi.org/10.1667/rr0010cc.1.

**47** Buonanno, M., Welch, D., Shuryak, I., and Brenner, D.J. (**2020**). Far-UVC light (222 nm) efficiently and safely inactivates airborne human coronaviruses. *Scientific Reports 10* (1): 10285. https://doi.org/10.1038/s41598-020-67211-2.

**48** Kitagawa, H., Nomura, T., Tanuz, N. et al. (**2020**). Effectiveness of 222-nm ultraviolet light on disinfecting SARS-CoV-2 surface contamination. *American Journal of Infection Control 49* (3): 299–301. https://doi.org/10.1016/j.ajic.2020.08.022.

**49** Wayne, R.P. (**1985**). *Chemistry of Atmospheres*. Clarendon Press.

**50** Chapman, S. (**1930**). A theory of upper-atmospheric ozone. *Memoirs of the Royal Meteorological Society 3* (26): 103–125.

**51** Stamnes, K. (**2015**). Radiation transfer in the atmosphere: ultraviolet radiation. In: *Encyclopedia of Atmospheric Sciences*, 2e, vol. *5* (ed. G.R. North, J.A. Pyle and F. Zhang), 37–44. Elsevier.

**52** Finlayson-Pitts, B.J. and Pitts, J.N.J. (**1986**). *Atmospheric Chemistry: Fundamentals and Experimental Technniques*. Wiley.

**53** Burkholder, J.B., Sander, S.P., Abbatt, J.P.D. et al. (**2019**). *Chemical Kinetics and Photochemical Data for Use in Atmospheric Studies*. Pasadena: JPL Publication, Jet Propulsion Laboratory http://jpldataeval.jpl.nasa.gov, Evaluation Number 19.

**54** (a) Barrett, J. and Baxendale, J.H. (**1960**). The photolysis of liquid water. *Transactions of the Faraday Society 56* (1): 37–43. https://doi.org/10.1039/tf9605600037. (b) Weeks, J.L., Maeburn, G.M.A.C., and Gordon, S. (**1963**). Absorption coefficients of liquid water and aqueous solutions in the far ultraviolet. *Radiation Research 19* (3): 559–567.

**55** Gonzalez, M.G., Oliveros, E., Worner, M., and Braun, A.M. (**2004**). Vacuum-ultraviolet photolysis of aqueous reaction systems. *Journal of Photochemistry and Photobiology C-Photochemistry Reviews 5* (3): 225–246. https://doi.org/10.1016/j.jphotochemrev.2004.10.002.

**56** Rockett, C. (**2019**). UV degradation effects in materials: an elementary overview. In: *UV Solutions*. IUVA.

**57** Hare, C.H. (**1992**). The degradation of coatings by ultraviolet light and electromagnetic radiation. *Journal of Protective Coatings & Linings* .

**58** Prussin, A.J., Garcia, E.B., and Marr, L.C. (**2015**). Total concentrations of virus and bacteria in indoor and outdoor air. *Environmental Science & Technology Letters 2* (4): 84–88. https://doi.org/10.1021/acs.estlett.5b00050.

**59** Riediker, M. and Tsai, D.H. (**2020**). Estimation of viral aerosol emissions from simulated individuals with asymptomatic to moderate coronavirus disease 2019. *JAMA Network Open 3* (7): e2013807. https://doi.org/10.1001/jamanetworkopen.2020.13807.

**60** Ma, J., Qi, X., Chen, H. et al. (**2020**). Coronavirus disease 2019 patients in earlier stages exhaled millions of severe acute respiratory syndrome coronavirus 2 Per hour. *Clinical Infectious Diseases 72*: e652–e654. https://doi.org/10.1093/cid/ciaa1283.

**61** Fort Aministration Department (**2020**). Recommendations for Sponsors Requesting EUAs for Decontamination and Bioburden Reduction Systems for Surgical Masks and Respirators During the Coronavirus Disease 2019 (COVID-19) Public Health Emergency: Guidance for Industry and Food and Drug Administration Staff. Washington, DC.

**62** Bolton, J.R. and Linden, K.G. (**2003**). Standardization of methods for fluence (UV dose) determination in bench-scale UV experiments. *Journal of Environmental Engineering ASCE 129* (3): 209–215. https://doi.org/10.1061/(asce)0733-9372(2003)129:3(209).

**63** Ke, Q.S., Craik, S.A., El-Din, M.G., and Bolton, J.R. (**2009**). Development of a protocol for the determination of the ultraviolet sensitivity of microorganisms suspended in air. *Aerosol Science and Technology 43* (4): 284–289. https://doi.org/10.1080/02786820802616893.

**64** (a) Ko, G., First, M.W., and Burge, H.A. (**2000**). Influence of relative humidity on particle size and UV sensitivity of *Serratia marcescens* and *Mycobacterium bovis* BCG aerosols. *Tubercle and Lung Disease 80* (4–5): 217–228. https://doi.org/10.1054/tuld.2000.0249; (b) Lai, K.M., Burge, H.A., and First, M.W. (**2004**). Size and UV germicidal irradiation susceptibility of Serratia marcescens when aerosolized from different suspending media. *Applied and Environmental Microbiology 70* (4): 2021–2027. https://doi.org/10.1128/aem.70.4.2021-2027.2004.

**65** McDevitt, J.J., Lai, K.M., Rudnick, S.N. et al. (**2007**). Characterization of UVC light sensitivity of vaccinia virus. *Applied and Environmental Microbiology 73* (18): 5760–5766. https://doi.org/10.1128/aem.00110-07.

**66** McDevitt, J.J., Rudnick, S.N., and Radonovich, L.J. (**2012**). Aerosol susceptibility of influenza virus to UV-C light. *Applied and Environmental Microbiology 78* (6): 1666–1669. https://doi.org/10.1128/aem.06960-11.

**67** (a) Buonanno, M., Welch, D., Shuryak, I., and Brenner, D.J. (**2020**). Far-UVC light (222 nm) efficiently and safely inactivates airborne human coronaviruses. *Scientific Reports 10*: 10285. https://doi.org/10.1038/s41598-020-67211-2; (b) Eadie, E., Hiwar, W., Fletcher, L., Tidswell, E. O'Mahoney, P. B., Welch, D., Adamson, C. S., Brenner, D. J., Noakes, C., and Wood, K. (**2021**). Far-UVC Efficiently Inactivates an Airborne Pathogen in a Room-Sized Chamber. *Research Square*. https://assets.researchsquare.com/files/rs-908156/v1/a54f46e0-47eb-4424-942f-b162aa4970fe.pdf?c=1631889559. https://doi.org/10.21203/rs.3.rs-908156/v1.

**68** Lim, S. and Blatchley III, E.R. (**2012**). UV dose–response behavior of air-exposed microorganisms. *Journal of Environmental Engineering-Asce 138* (7): 780–785. https://doi.org/10.1061/(asce)ee.1943-7870.0000535.

**69** ASTM (**2007**). *Standard Practice for Maintaining Constant Relative Humidity by Means of Aqueous Solutions*. American Society for Testing and Materials.

**70** Peccia, J., Werth, H.M., Miller, S., and Hernandez, M. (**2001**). Effects of relative humidity on the ultraviolet induced inactivation of airborne bacteria. *Aerosol Science and Technology 35* (3): 728–740. https://doi.org/10.1080/02786820152546770.

**71** Xu, P., Kujundzic, E., Peccia, J. et al. (**2005**). Impact of environmental factors on efficacy of upper-room air ultraviolet germicidal irradiation for inactivating airborne mycobacteria. *Environmental Science & Technology 39* (24): 9656–9664. https://doi.org/10.1021/es0504892.

**72** Lytle, C.D. and Sagripanti, J.L. (**2005**). Predicted inactivation of viruses of relevance to biodefense by solar radiation. *Journal of Virology 79* (22): 14244–14252. https://doi.org/10.1128/jvi.79.22.14244-14252.2005.

**73** Smith, K.C. and Hanawalt, P.C. (**1969**). *Molecular Photobiology: Inactivation and Recovery*. Academic Press.

**74** ANSI/ASHRAE (**2020**). *Method of Testing Ultraviolet Lamps for Use in HVAC&R Units or Air Ducts to Inactivate Microorganisms on Irradiated Surfaces*, Atlanta, GA.

ANSI/ASHRAE (**2020**). *Method of Testing UV-C Lights for Use in Air-Handling Units or Air Ducts to Inactivate Airborne Microorganisms*, Atlanta, GA.

**75** ANSI/ASHRAE (**2019**). *Ventilation for Acceptable Indoor Air Quality*. American National Standards Institute, The American Society of Heating, Refrigerating and Air-Conditioning Engineers https://ashrae.iwrapper.com/ASHRAE_PREVIEW_ONLY_STANDARDS/STD_62.1_2019.

**76** Centers for Disease Control and Prevention (**2019**). Ventilation Specifications for Health-Care Facilities. https://www.cdc.gov/infectioncontrol/guidelines/environmental/appendix/air.html (accessed 2021).

**77** (a) Akers, T.G. and Hatch, M.T. (**1968**). Survival of a picornavirus and its infectious ribonucleic acid after aerosolization. *Applied Microbiology 16* (11): 1811. https://doi.org/10.1128/aem.16.11.1811-1813.1968; (b) Zhao, Y., Aarnink, A.J.A., Dijkman, R. et al. (**2012**). Effects of temperature, relative humidity, absolute humidity, and evaporation potential on survival of airborne Gumboro vaccine virus. *Applied and Environmental Microbiology 78* (4): 1048–1054. https://doi.org/10.1128/aem.06477-11. (c) Lin, K.S., Schulte, C.R., and Marr, L.C. (**2020**). Survival of MS2 and phi 6 viruses in droplets as a function of relative humidity, pH, and salt, protein, and surfactant concentrations. *Plos One 15* (12): e0243505. https://doi.org/10.1371/journal.pone.0243505.

**78** Ijaz, M.K., Karim, Y.G., Sattar, S.A., and Johnsonlussenburg, C.M. (**1987**). Development of methods to study the survival of airborne viruses. *Journal of Virological Methods 18* (2–3): 87–106. https://doi.org/10.1016/0166-0934(87)90114-5.

**79** Nardell, E.A. (**2021**). Air disinfection for airborne infection control with a focus on COVID-19: why germicidal UV is essential. *Photochemistry and Photobiology 97* (3): 493–497. https://doi.org/10.1111/php.13421.

**80** Buchan, A.G., Yang, L., and Atkinson, K.D. (**2020**). Predicting airborne coronavirus inactivation by far-UVC in populated rooms using a high-fidelity coupled radiation-CFD model. *Scientific Reports 10* (1): 19659. https://doi.org/10.1038/s41598-020-76597-y.

**81** Li, Y., Qian, H., Hang, J. et al. (**2021**). Probable airborne transmission of SARS-CoV-2 in a poorly ventilated restaurant. *Building and Environment 196*: 107788.

**82** (a) Jelden, K.C., Gibbs, S.G., Smith, P.W. et al. (**2017**). Ultraviolet (UV)-reflective paint with ultraviolet germicidal irradiation (UVGI) improves decontamination of nosocomial bacteria on hospital room surfaces. *Journal of Occupational and Environmental Hygiene 14* (6): 456–460. https://doi.org/10.1080/15459624.2017.1296231; (b) Rutala, W.A., Gergen, M.F., Tande, B.M., and Weber, D.J. (**2013**). Rapid hospital room decontamination using ultraviolet (UV) light with a nanostructured UV-reflective wall coating. *Infection Control and Hospital Epidemiology 34* (5): 527–529. https://doi.org/10.1086/670211.

**83** Schneider, J., Matsuoka, M., Takeuchi, M. et al. (**2014**). Understanding $TiO_2$ photocatalysis: mechanisms and materials. *Chemical Reviews 114* (19): 9919–9986. https://doi.org/10.1021/cr5001892.

**84** Huber, T., Goldman, O., Epstein, A.E. et al. (**2021**). Principles and practice for SARS-CoV-2 decontamination of N95 masks with UV-C. *Biophysical Journal 120*: 1–16.

**85** Fabian, P., McDevitt, J.J., DeHaan, W.H. et al. (**2008**). Influenza virus in human exhaled breath: an observational study. *Plos One 3* (7): e2691. https://doi.org/10.1371/journal.pone.0002691.

**86** Fabian, P., Brain, J., Houseman, E.A. et al. (**2011**). Origin of exhaled breath particles from healthy and human rhinovirus-infected subjects. *Journal of Aerosol Medicine and Pulmonary Drug Delivery 24* (3): 137–147. https://doi.org/10.1089/jamp.2010.0815.

**87** Zhao, Z., Zhang, Z.B., Lanzarini-Lopes, M. et al. (**2020**). Germicidal ultraviolet light does not damage or impede performance of N95 masks upon multiple uses. *Environmental Science & Technology Letters* *7* (8): 600–605. https://doi.org/10.1021/acs.estlett.0c00416.

**88** Lindsley, W.G., Martin, S.B., Thewlis, R.E. et al. (**2015**). Effects of ultraviolet germicidal irradiation (UVGI) on N95 respirator filtration performance and structural integrity. *Journal of Occupational and Environmental Hygiene 12* (8): 509–517. https://doi.org/10.1080/15459624.2015.1018518.

**89** Lilge, L., Manalac, A., Weersink, M. et al. Light propagation within N95 filtered face respirators: a simulation study for UVC decontamination. *Journal of Biophotonics 13*: e202000232. https://doi.org/10.1002/jbio.202000232.

# Index

Page numbers referring to figures are presented in *italics* and those referring to tables are presented in **bold**.





Companion website: www.wiley.com/go/blatchley/photochemicalreactors

## s



## t

Printed and bound by CPI Group (UK) Ltd, Croydon, CR0 4YY
21/09/2026
14965822-0001